MEDICINAL PLANTS AS ANTI-INFECTIVES

MEDICINAL PLANTS AS ANTI-INFECTIVES

Current Knowledge and New Perspectives

Edited by

FRANÇOIS CHASSAGNE

UMR 152 PharmaDev, IRD, UPS, Université de Toulouse, Toulouse, France

Academic Press is an imprint of Elsevier
125 London Wall, London EC2Y 5AS, United Kingdom
525 B Street, Suite 1650, San Diego, CA 92101, United States
50 Hampshire Street, 5th Floor, Cambridge, MA 02139, United States
The Boulevard, Langford Lane, Kidlington, Oxford OX5 1GB, United Kingdom

Notices

Knowledge and best practice in this field are constantly changing. As new research and experience broaden our understanding, changes in research methods, professional practices, or medical treatment may become necessary.

Practitioners and researchers must always rely on their own experience and knowledge in evaluating and using any information, methods, compounds, or experiments described herein. In using such information or methods they should be mindful of their own safety and the safety of others, including parties for whom they have a professional responsibility.

To the fullest extent of the law, neither the Publisher nor the authors, contributors, or editors, assume any liability for any injury and/or damage to persons or property as a matter of products liability, negligence or otherwise, or from any use or operation of any methods, products, instructions, or ideas contained in the material herein.

ISBN: 978-0-323-90999-0

Publisher: Nikki P. Levy
Acquisitions Editor: Nancy J. Maragioglio
Editorial Project Manager: Andrea R. Dulberger
Production Project Manager: Swapna Srinivasan
Cover Designer: Miles Hitchen

Typeset by MPS Limited, Chennai, India

Contents

List of contributors xi
Preface xv

I

Medicinal plants as anti-infectives: an appraisal of current knowledge worldwide

1. A review of medicinal plants used as antimicrobials in Colombia 3

Yina Pájaro-González, Andrés Felipe Oliveros-Díaz, Julián Cabrera-Barraza, José Cerra-Dominguez and Fredyc Díaz-Castillo

Introduction 3
Materials and methods 6
Plants traditionally used in Colombia as antimicrobials 6
Jacaranda caucana Pittier (Bignoniaceae) 22
Solanum nudum Dunal (Solanaceae) 22
Hymenaea courbaril L. (Leguminosae) 22
Xanthium strumarium L. (Asteraceae) 23
Guazuma ulmifolia Lam. (Malvaceae) 23
Cymbopogon citratus (DC.) Stapf (Poaceae) 23
Austroeupatorium inulaefolium (Kunth) R.M.King & H.Rob. (Asteraceae) 23
Biological evaluation as antimicrobials of plant extracts in Colombia 24
Antibacterial activity 44
Antiviral activity 47
Antibacterial evaluation of 25 native plants of the Colombian Caribbean region against strains of *Escherichia coli*, *Klebsiella pneumonia*, and *Pseudomonas aeruginosa* resistant to carbapenems and *Staphylococcus aureus* resistant to methicillin (Colciencias, Project code: 110777757752) 47
Mammea americana L. (Calophyllaceae) 49
Maclura tinctoria L. D.Don ex Steud. (Moraceae) 50
Conclusions 51
References 51

2. Plants used in Lebanon and the Middle East as Antimicrobials 59

Roula M. Abdel-Massih and Marc El Beyrouthy

Introduction: overview on medicinal plants and their traditional uses as antimicrobials in Lebanon 59
Lebanese plants with antimicrobial activity 62
Amaryllidaceae 62
Anacardiaceae 63
Apiaceae 64
Asteraceae/Compositae 65
Berberidaceae 65
Cannabaceae 66
Cistaceae 67
Conifers 67
Lamiaceae 68
Myrtaceae 73
Portulacaceae 74
Ranunculaceae 75
Rutaceae 76
Rosaceae 77
Conclusion 77
Acknowledgment 84
References 84

3. Medicinal plants in the Balkans with antimicrobial properties 103

Sarah Shabih, Avni Hajdari, Behxhet Mustafa and Cassandra L. Quave

Introduction 103
Medicinal plants with antimicrobial properties 104
Amaryllidaceae 104
Apiaceae 127

Asteraceae 127
Betulaceae 128
Lamiaceae 129
Malvaceae 130
Pinaceae 131
Rosaceae 131
Urticaceae 132
Conclusions 133
References 133

4. Medicinal plants used in South Africa as antibacterial agents for wound healing 139

Samantha Rae Loggenberg, Danielle Twilley, Marco Nuno De Canha and Namrita Lall

Introduction 139
Pathophysiology of wound healing 141
Wound infection 142
Currently available treatments and products 142
Topical creams 142
Transdermal drug delivery systems 143
Bacteria associated with infections of dermal wounds 143
Bacillus subtilis 144
Staphylococcus aureus 144
Staphylococcus epidermidis 145
Pseudomonas aeruginosa 145
South African medicinal plant species with activity against wound-associated bacteria 146
Aloe barberae Dyer 146
Aloe excelsa Berger 147
Aloe ferox Miller 161
Elephantorrhiza elephantina (Burch.) Skeel 162
Erythrina lysistemon Hutch 162
Galenia africana L 163
Grewia occidentalis L 163
Melianthus comosus Vahl. 164
Plectranthus fruticosus L'Hér 164
Polystichum pungens (Kaulf.) C. Presl 165
Sutherlandia frutescens (L.) R.Br. 165
Urtica urens L. 166
Compounds present in plants traditionally used for wound healing in South Africa 166
Aloe species 166
Elephantorrhiza elephantina 167
Erythrina lysistemon 169
Galenia africana 170
Melianthus comosus 170
Plectranthus fruticosus 171
Sutherlandia frutescens 171
Discussion 172
Conclusion 175
Index 175
Glossary 175
References 176

5. The use of South African medicinal plants in the pursuit to treat gonorrhea and other sexually transmitted diseases 183

Tanyaradzwa Tiandra Dembetembe, Namrita Lall and Quenton Kritzinger

Introduction 183
Background on gonorrhea 185
The causal agent: *Neisseria gonorrhoeae* 185
Status of available treatments for gonorrhea 189
Selected South African plants used in traditional medicine for the treatment of sexually transmitted diseases and their bioactivity 192
Aloe ferox 192
Cassia abbreviata 194
Combretum molle 194
Elaeodendron transvaalense 195
Hypoxis hemerocallidea 195
Peltophorum africanum 196
Tabernaemontana elegans 197
Terminalia sericea 198
Conclusion 198
References 199

6. Antibacterial activity of some selected medicinal plants of Pakistan 209

Zia Ur Rehman Mashwani, Rahmat Wali, Muhammad Faraz Khan, Fozia Abasi, Nadia Khalid and Naveed Iqbal Raja

Introduction 209
Antibacterial properties of different medicinal plants from Pakistan 210
Conclusion 232
References 232

7. Medicinal plants used as antidiarrheal agents in the lower Mekong basin 235

François Chassagne

Introduction 235
Traditional medicine for diarrheal diseases in the Mekong Basin 237
The role of traditional medicine in the management of diarrhea 237
The cultural belief system of people living in the Mekong area 238
Pharmacological validation of plants used for diarrhea 239
Models assessing the effect of plants on the signs and symptoms of diarrhea 240
Antidiarrheal effect 240
Spasmolytic activity 240
Models assessing the antimotility and antisecretory activities 241
Antimotility activity 241
Antisecretory activity 241
Models assessing the antiinfective properties 242
Antibacterial activity 242
Antiviral and antiparasitic activity 243
Other models 243
Medicinal plants used for diarrhea in the lower Mekong basin 243
Literature search methodology 243
Overview of the dataset 244
Discussion of some selected plant species 248
Psidium guajava 248
Chromolaena odorata 249
Alstonia scholaris 250
Allium sativum 250
Centella asiatica 251
Punica granatum 251
Caesalpinia sappan 252
Mangifera indica 252
Holarrhena pubescens 253
Oroxylum indicum 254
Conclusion 254
References 255

8. Medicinal plants from West Africa used as antimalarial agents: an overview 267

Agnès Aubouy, Aissata Camara and Mohamed Haddad

Introduction 267
Traditional use of medicinal plants in West Africa 269
Plant extracts and plant compounds validated by in vitro and/or in vivo approach 270
In vitro antimalarial evaluation of plant extracts 272
In vivo antimalarial evaluation of plant extracts 273
In vitro and in vivo evaluation of antimalarial compounds 273
Clinical trials in humans for the evaluation of antimalarial plants and compounds 295
The case of Artemisia in West Africa 298
Conclusion 300
References 300

II

Medicinal plants as anti-infectives: recent innovations and regulations

9. Mycobacterial quorum quenching and biofilm inhibition potential of medicinal plants 309

Jonathan L. Seaman, Carel B. Oosthuizen, Lydia Gibango and Namrita Lall

Introduction 309
Significance of quorum quenching research 311
Current state of quorum quenching research 312
Quorum sensing versus quorum quenching 315
Biofilms 315
Background on biofilms 315
Biofilms and *Mycobacterium tuberculosis* 317
Virulence factors 319
Background on virulence factors 319
Virulence factors and *Mycobacterium tuberculosis* 319
Medicinal plants as quorum quenching agents 321
Medicinal plants and mycobacterial quorum quenching 322
Phytochemicals used in bacterial quorum quenching 326
Conclusion 329
References 329

10. Untargeted metabolomics for the study of antiinfective plants 335

Joshua J. Kellogg

Introduction 335
Plants as sources of antiinfective agents 335
Bioassay-guided fractionation 336
Metabolomics 336
Methods of detection 338
Data analysis 339
Biochemometrics 339
Metabolomics-driven antiinfective discovery from plants 340
Challenges and future directions 345
Metabolome coverage 345
Annotation/identification 349
Synergy 352
Conclusions 352
References 353

11. Value chains and DNA barcoding for the identification of antiinfective medicinal plants 361

Seethapathy G. Saroja, Remya Unnikrishnan, Santhosh Kumar J. Urumarudappa, Xiaoyan Chen and Jiangnan Peng

Introduction 361
Taxonomy and DNA barcoding 361
Infectious diseases and antiinfective plants 362
Herbal products, commercialization, and quality issues of antiinfective plants 363
Advancements in quality control methods 363
Materials and methods 365
Results and discussion 365
Embelia ribes—anthelmintic plant 365
Swertia chirayita—antiviral plant 367
Picrorhiza kurroa—antiviral plant 369
Paris polyphylla—anthelmintic plant 370
Saussurea costus—anthelminthic/antiparasitic plant 372
Syzygium aromaticum—antimicrobial plant 373
Andrographis paniculata—antimicrobial plant 374
Future perspectives 375
References 376

12. Fungal endophytes: a source of antibacterial and antiparasitic compounds 383

Romina Pacheco, Sergio Ortiz, Mohamed Haddad and Marieke Vansteelandt

Introduction 383
Current situation of microbial infections 383
Microbial natural products as sources of new drugs 384
Endophytic fungi 385
Antimicrobial compounds from endophytic fungi 388
Antibacterial compounds 388
Antivirulence compounds 404
Antiparasitic compounds 405
Discussion and conclusion 425
References 428

13. Antiviral potential of medicinal plants: a case study with guava tree against dengue virus using a metabolomic approach 439

Thomas Vial, Chiobouaphong Phakeovilay, Satoru Watanabe, Kitti Wing Ki Chan, Minhua Peng, Eric Deharo, François Chassagne, Subhash G. Vasudevan and Guillaume Marti

Introduction 439
Dengue disease 439
Conventional treatment 441
Medicinal plants 442
Case study: metabolomics reveal antidengue compounds isolated from *Psidium guajava* 442
Introduction 442
Objectives 445
Results 445
Discussion 451
Materials and methods 452
Acknowledgments 455
References 455

14. How history can help present research of new antimicrobial strategies: the case of cutaneous infections' remedies containing metals from the Middle Age Arabic pharmacopeia 459

Véronique Pitchon, Elora Aubert, Catherine Vonthron and Pierre Fechter

Introduction 459

Brief history of Arabic medicine 460
Principles of Arab medicine: theoretical aspects 461
Arab sources of pharmacology: the aqrābādhīn, a constituted literature 462
Cutaneous infections and medications 463
The specificity of skin and eye diseases in the pharmacopeias and the nature of the diseases treated 463
Plants and metals useful for skin diseases 464
Toxicity of metals 468
Renewed interest in metal–organic molecule combinations 470
Elementary metal particle 470
Organometallic molecule 472
Metal nanoparticles 473
Characterization of plant–metal combinations 474
Conclusion 475
References 475

15. Improved traditional medicine for infectious disorders in Mali 479

Rokia Sanogo, Mahamane Haïdara and Adama Dénou

Introduction 479
General information on improved traditional medicines 480
Definition 480
Regulatory framework 480
Categories of improved traditional medicines 481
Marketing authorization files for ITMs in Mali 481
Historical development of ITMs in Mali 484
ITMs in the management of infectious diseases 484
ITMs for the management of malaria 485
ITM for the management of dysentery 489
ITM for the management of viral hepatitis 490
ITM for the management of gastric ulcer associated with *Helicobacter pylori* 491
ITM for the management of dermatosis 492
Conclusion and perspectives 493
Index of ITMs 494
References 494

16. Selecting the most promising local treatments: retrospective treatment-outcome surveys and reverse pharmacology 501

Joëlle Houriet, Jean-Luc Wolfender and Bertrand Graz

Introduction 501
Clinical efficacy 503
Reverse pharmacology approach 504
Step 1: Retrospective treatment outcome study 505
Steps 2 and 3: clinical evaluations 508
Step 4: Laboratory stage 509
New and promising approaches for the laboratory stage in a reverse pharmacology approach 510
Changes of paradigms 511
Pharmacokinetics 513
Models to study absorption and biotransformation of natural products and herbal preparations 514
Deciphering the mode of action of herbal preparations: successes and limitations 516
Examples of application of metabolomic studies in human 517
Case study: perspectives on the *Phaleria nisidai* decoction study 519
Conclusion and perspectives 520
Acknowledgments 522
References 522

17. Nagoya Protocol and access to genetic resources 529

Bruno David

Introduction 529
History and evolution of concepts 529
Development of environmental awareness 529
Concept of biodiversity 530
The Convention on Biodiversity 531
Problems left unsolved by the CBD 533
The Nagoya Protocol 534
The national biodiversity legislations 536
Practical advice 538
Discussion 540
Non stabilized and heterogeneous regulations 540
Ambiguities 541
Some paradoxical effects 543

Legal certainty 543
New trends and evolutions 544
The curious case of pathogens 545
Digital sequence information 545
Biodiversity beyond national jurisdiction 546
Conclusion 547
Acknowledgments 548
References 549

Index 555

List of contributors

Fozia Abasi Department of Botany, PMAS-Arid Agriculture University, Rawalpindi, Pakistan

Roula M. Abdel-Massih Department of Biology, Faculty of Arts and Sciences, University of Balamand, El-Koura, Lebanon

Elora Aubert CNRS, UMR 7200, Laboratory of Therapeutic Innovation, Medalis LabEx, Faculty of Pharmacy, Strasbourg University, Strasbourg, France

Agnès Aubouy UMR 152 PharmaDev, Université de Toulouse, IRD, UPS, Toulouse, France

Julián Cabrera-Barraza Phytochemical and Pharmacological Research Laboratory of the University of Cartagena (LIFFUC), Faculty of Pharmaceutical Sciences, University Cartagena, Cartagena, Colombia

Aissata Camara Institute for Research and Development of Medicinal and Food Plants of Guinea (IRDPMAG), Dubréka, Guinea

José Cerra-Dominguez Phytochemical and Pharmacological Research Laboratory of the University of Cartagena (LIFFUC), Faculty of Pharmaceutical Sciences, University Cartagena, Cartagena, Colombia

Kitti Wing Ki Chan Programme in Emerging Infectious Diseases, Duke-NUS Medical School, Singapore

François Chassagne UMR 152 PharmaDev, IRD, UPS, Université de Toulouse, Toulouse, France

Xiaoyan Chen Department of Chemistry, School of Computer, Mathematical and Natural Sciences, Morgan State University, Baltimore, MD, United States

Bruno David Green Mission Pierre Fabre, Pierre Fabre Research Institute, Toulouse, France

Marco Nuno De Canha Department of Plant and Soil Sciences, Faculty of Natural and Agricultural Sciences, University of Pretoria, Pretoria, South Africa

Eric Deharo UMR 152 PharmaDev, IRD, UPS, Université de Toulouse, Toulouse, France; Institut de Recherche pour le Développement, Vientiane, Lao PDR

Tanyaradzwa Tiandra Dembetembe Department of Plant and Soil Sciences, Faculty of Natural and Agricultural Sciences, University of Pretoria, Pretoria, South Africa

Adama Dénou Faculty of Pharmacy, University of Sciences, Techniques and Technologies of Bamako, Mali

Fredyc Díaz-Castillo Phytochemical and Pharmacological Research Laboratory of the University of Cartagena (LIFFUC), Faculty of Pharmaceutical Sciences, University Cartagena, Cartagena, Colombia

Marc El Beyrouthy Department of Agricultural Sciences, Holy Spirit University of Kaslik, Beirut, Lebanon

Pierre Fechter CNRS, UMR 7242, Biotechnology and Cell Signaling, Strasbourg University, Illkirch-Graffenstaden, France

Lydia Gibango Department of Plant and Soil Sciences, Faculty of Natural and Agricultural Sciences, University of Pretoria, Pretoria, South Africa

Bertrand Graz Antenna Foundation, Geneva, Switzerland

Mohamed Haddad UMR 152 PharmaDev, Université de Toulouse, IRD, UPS, Toulouse, France

Mahamane Haïdara Faculty of Pharmacy, University of Sciences, Techniques and Technologies of Bamako, Mali

Avni Hajdari Department of Biology, University of Prishtina, Prishtinë, Kosovo

Joëlle Houriet School of Pharmaceutical Sciences, University of Geneva, CMU, Geneva, Switzerland; Institute of Pharmaceutical Sciences of Western Switzerland, University of Geneva, CMU, Geneva, Switzerland

Joshua J. Kellogg Department of Veterinary and Biomedical Sciences, Pennsylvania State University, University Park, PA, United States

Nadia Khalid Department of Botany, PMAS-Arid Agriculture University, Rawalpindi, Pakistan

Muhammad Faraz Khan Department of Botany, PMAS-Arid Agriculture University, Rawalpindi, Pakistan

Quenton Kritzinger Department of Plant and Soil Sciences, Faculty of Natural and Agricultural Sciences, University of Pretoria, Pretoria, South Africa

Namrita Lall Department of Plant and Soil Sciences, Faculty of Natural and Agricultural Sciences, University of Pretoria, Pretoria, South Africa; School of Natural Resources, University of Missouri, Columbia, MO, United States; College of Pharmacy, JSS Academy of Higher Education and Research, Mysuru, India

Samantha Rae Loggenberg Department of Plant and Soil Sciences, Faculty of Natural and Agricultural Sciences, University of Pretoria, Pretoria, South Africa

Guillaume Marti UMR 152 PharmaDev, IRD, UPS, Université de Toulouse, Toulouse, France; Laboratoire de Recherche en Sciences Végétales and Metatoul-AgromiX Platform, MetaboHUB, National Infrastructure for Metabolomics and Fluxomics, LRSV, Université de Toulouse, CNRS, UPS, Toulouse, France

Zia Ur Rehman Mashwani Department of Botany, PMAS-Arid Agriculture University, Rawalpindi, Pakistan

Behxhet Mustafa Department of Biology, University of Prishtina, Prishtinë, Kosovo

Andrés Felipe Oliveros-Díaz Phytochemical and Pharmacological Research Laboratory of the University of Cartagena (LIFFUC), Faculty of Pharmaceutical Sciences, University Cartagena, Cartagena, Colombia

Carel B. Oosthuizen Department of Plant and Soil Sciences, Faculty of Natural and Agricultural Sciences, University of Pretoria, Pretoria, South Africa

Sergio Ortiz UMR 152 PharmaDev, Université de Toulouse, IRD, UPS, Toulouse, France; UMR 7200 Laboratoire d'Innovation Thérapeutique, Université de Strasbourg, CNRS, Strasbourg Drug Discovery and Development Institute (IMS), Illkirch-Graffenstaden, France

Romina Pacheco UMR 152 PharmaDev, Université de Toulouse, IRD, UPS, Toulouse, France

Yina Pájaro-González Phytochemical and Pharmacological Research Laboratory of the University of Cartagena (LIFFUC), Faculty of Pharmaceutical Sciences, University Cartagena, Cartagena, Colombia; Research Group in Healthcare Pharmacy and Pharmacology, Faculty of Chemistry and Pharmacy, University of Atlántico, Barranquilla, Colombia

Jiangnan Peng Department of Chemistry, School of Computer, Mathematical and Natural Sciences, Morgan State University, Baltimore, MD, United States

Minhua Peng Programme in Emerging Infectious Diseases, Duke-NUS Medical School, Singapore

Chiobouaphong Phakeovilay UMR 152 PharmaDev, IRD, UPS, Université de Toulouse, Toulouse, France; Laboratoire de Recherche en Sciences Végétales and Metatoul-AgromiX Platform, MetaHUB, National Infrastructure for Metabolomics and Fluxomics, LRSV, Université de Toulouse, CNRS, UPS, Toulouse, France

Véronique Pitchon CNRS, UMR 7044, Archaeology and Ancient History: Mediterranean - Europe, MISHA, Strasbourg University, Strasbourg, France

Cassandra L. Quave Center for the Study of Human Health, Emory University, Atlanta, GA, United States; Department of Dermatology and Center for the Study of Human Health, Emory University, Atlanta, GA, United States

Naveed Iqbal Raja Department of Botany, Faculty of Sciences, PMAS-Arid Agriculture University, Rawalpindi, Pakistan

Rokia Sanogo Department of Traditional Medicine, Bamako, Mali; Faculty of Pharmacy, University of Sciences, Techniques and Technologies of Bamako, Mali

Seethapathy G. Saroja Department of Chemistry, School of Computer, Mathematical and Natural Sciences, Morgan State University, Baltimore, MD, United States

Jonathan L. Seaman Department of Plant and Soil Sciences, Faculty of Natural and Agricultural Sciences, University of Pretoria, Pretoria, South Africa

Sarah Shabih Center for the Study of Human Health, Emory University, Atlanta, GA, United States

Danielle Twilley Department of Plant and Soil Sciences, Faculty of Natural and Agricultural Sciences, University of Pretoria, Pretoria, South Africa

Remya Unnikrishnan Forest Genetics and Biotechnology Division, Kerala Forest Research Institute, Thrissur, India

Santhosh Kumar J. Urumarudappa Research Unit of DNA Barcoding of Thai Medicinal Plants, Department of Pharmacognosy and Pharmaceutical Botany, Faculty of Pharmaceutical Sciences, Chulalongkorn University, Bangkok, Thailand

Marieke Vansteelandt UMR 152 PharmaDev, Université de Toulouse, IRD, UPS, Toulouse, France

Subhash G. Vasudevan Programme in Emerging Infectious Diseases, Duke-NUS Medical School, Singapore

Thomas Vial UMR 152 PharmaDev, IRD, UPS, Université de Toulouse, Toulouse, France; Programme in Emerging Infectious Diseases, Duke-NUS Medical School, Singapore

Catherine Vonthron CNRS, UMR 7200, Laboratory of Therapeutic Innovation, Medalis LabEx, Faculty of Pharmacy, Strasbourg University, Strasbourg, France

Rahmat Wali Department of Botany, PMAS-Arid Agriculture University, Rawalpindi, Pakistan

Satoru Watanabe Programme in Emerging Infectious Diseases, Duke-NUS Medical School, Singapore

Jean-Luc Wolfender School of Pharmaceutical Sciences, University of Geneva, CMU, Geneva, Switzerland; Institute of Pharmaceutical Sciences of Western Switzerland, University of Geneva, CMU, Geneva, Switzerland

Preface

Medicinal plants have been used for centuries to treat various diseases including infectious disorders. Several traditional medical systems (e.g., Traditional Chinese Medicine, Ayurveda, Greek medicine) use plant species to cure disorders known to be caused or worsened by microbial infections such as fever, abscess, wounds, tuberculosis, and urinary tract infections. To validate these traditional uses and discover new drugs, scientific investigations of medicinal plants with anti-infective properties have been carried out since the last century. One of the most famous examples of drug discovery from anti-infectives plants is the compound artemisinin isolated from *Artemisia annua* L.,[1] which was used in the traditional Chinese medicine to treat various type of fevers, and then investigated by Youyou Tu in the 1970s to cure malaria. Forty years later, artemether,[1] an artemisinin derivative, led to the development of an antimalarial drug approved by the FDA, which is now listed on the WHO's List of Essential Medicines. Because other medicinal plants led to the discovery of new anti-infective drugs and some other are currently under investigation (e.g., berberine isolated from *Berberis* species such as *Berberis vulgaris*[1]), an appraisal of the current knowledge of plant-derived compounds investigated for their antimicrobial properties is necessary. This includes plants studied for their antibacterial, antifungal, antimalarial, and antiviral properties.

[1] Pictures of these plant species and representation of the compound are shown on the book front cover.

Historically, the validation of traditional usage and the discovery of drugs from plants is based on interdisciplinary fields such as ethnobotany (recordings of traditional medical knowledge, and collection/identification of plant species), phytochemistry (plant extracts preparation and chemical characterization of bioactive compounds), and pharmacology (biological evaluation of plant extracts). However, recent innovations and regulations in each field are changing the way to study medicinal plants. For example, metabolomics allow for a faster identification of active compounds, DNA barcoding help to avoid misidentification of medicinal plants, new pharmacological tools (e.g., quorum sensing, antibiofilm) lead to a better understanding of the mechanism of action, new targets (e.g., fungal endophytes) uncover new drugs from plants, new approaches (e.g., reverse pharmacology) simplify and speed up the process of safety and efficacy evaluation, and new regulations (i.e., Nagoya Protocol) induce a change in the documentation of traditional knowledge and collection of genetic resources. Finally, the paradigm of the higher therapeutic value of single compounds from plants versus plant extracts is changing, and phytomedicines are becoming more attractive.

The book *Medicinal Plants as Anti-infectives: Current Knowledge and New Perspectives* is organized in two independent sections which aims to:

- describe the medicinal plants and plant-derived compounds investigated for

their anti-infective properties in different geographic area (i.e., the Balkans, Colombia, India, Lebanon, Mali, Pakistan, South Africa, Southeast Asia and West Africa); and
- provide an overview of the main recent innovations and regulations for selecting, accessing, evaluating, identifying, and legalizing anti-infective medicinal plants.

This book is intended to target a large audience including scientists in the field of phytomedicine, pharmacognosy, ethnobotany, ethnopharmacology, and phytochemistry; students in schools of pharmacy following courses on pharmacognosy and undergraduate students following courses on ethnobotany, biology, and chemistry; international and national agencies (e.g., WHO, FDA in the US, ANSES in France) responsible for assessing the benefits and risks associated with the use of health products including plants and plant-derived compounds; and pharmaceutical companies or any private companies interesting in legal rules to access biodiversity and new technologies for drug discovery purposes.

I would like to sincerely thank all the experts for sharing their knowledge and contributing to this book. In spite of the Covid-19 pandemic, they all made great effort to finish their chapters on time and provide a meticulous examination of their field.

François Chassagne

UMR 152 PharmaDev, IRD, UPS, Université de Toulouse, Toulouse, France

PART I

Medicinal plants as anti-infectives: an appraisal of current knowledge worldwide

CHAPTER

1

A review of medicinal plants used as antimicrobials in Colombia

Yina Pájaro-González[1,2], Andrés Felipe Oliveros-Díaz[1], Julián Cabrera-Barraza[1], José Cerra-Dominguez[1] and Fredyc Díaz-Castillo[1]

[1]Phytochemical and Pharmacological Research Laboratory of the University of Cartagena (LIFFUC), Faculty of Pharmaceutical Sciences, University Cartagena, Cartagena, Colombia

[2]Research Group in Healthcare Pharmacy and Pharmacology, Faculty of Chemistry and Pharmacy, University of Atlántico, Barranquilla, Colombia

Introduction

Colombia is located in the northwestern corner of South America, in the connection area of this subcontinent with Central America. It is the only South American country that has coasts on the Atlantic (1600 km) and Pacific (1300 km) oceans. Its continental territory extends from 12 degrees 26′ 46″ N in Punta Gallinas, La Guajira (the northernmost point of South America), to 4 degrees 12′ 30″ S, at the mouth of the San Antonio Creek on the Amazon River, in the city of Leticia, and from 60 degrees 50′ 54″ W, on the island of San José, on the Negro River (point of convergence of the borders of Colombia, Brazil, and Venezuela) to 79 degrees 02′ 33″ or at Cape Manglares, at the mouth of the Mira River, in the Pacific Ocean. The equatorial line crosses the south of the country, in such a way that 87% of the continental territory is in the Northern Hemisphere. The Colombian territory also includes the Gorgona (26 km^2) and Gorgonilla (1.1 km^2) islands, located in the Pacific Ocean at 2 degrees 58′N, about 29 km from the coast, and Malpelo Island (0.35 km^2), an oceanic rock located at 4 degrees N and 465 km west of the continent (Bernal, Gradstein, & Celis, 2016).

In the Caribbean Sea, the Colombian territory also includes the archipelago of San Andrés and Providencia, located off the coast of Nicaragua, between 12 degrees and 16 degrees 30′N and 78 degrees and 82 degrees W. This archipelago includes the islands of San Andrés (26 km^2), Providencia (17 km^2), and Santa Catalina (1 km^2), along with several

Medicinal Plants as Anti-infectives
DOI: https://doi.org/10.1016/B978-0-323-90999-0.00005-7

smaller, uninhabited keys. Additionally, in the Caribbean are Isla Fuerte (3.25 km^2), the San Bernardo archipelagos (10 small islands, 4.5 km^2) and the Rosario Islands (28 small islands, 0.2 km^2), as well as those of Barú and Tierrabomba. All of them are close to the coast and linked to the continent (Bernal et al., 2016).

Colombia is a megadiverse country, ranking first in the world in number of species of birds and orchids, second in plants, amphibians, and freshwater fish, third in palms and reptiles, and sixth in mammals (Borja-Acosta et al., 2019; Vásquez, 2014). The registration and knowledge of the Colombian flora began with the investigative activity of Don José Celestino Mutis, for more than 40 years, who was the promoter of the "Royal Botanical Expedition of the New Kingdom of Granada," a work interrupted in 1816 during the period of the Reconquest and that motivated the visit of Alexander von Humboldt and Bonpland to Santa Fe de Bogotá. Humboldt traveled a good part of the Colombian territory, publishing the results of the botanical part of his trip in two important works, the first one dedicated by him and Bonpland to Mutis and the second, published by Carl Segismund Kunth under the title "*Nova genera et species plantarum*" (Barriga, 1986).

It is also important to highlight the contribution of Francisco José de Caldas, Eloy Valenzuela, Francisco Javier Matis, Francisco Antonio Zea, Jorge Tadeo Lozano, Sinforoso Mutis, Salvador Rizo, and Pablo Antonio García made to Colombian floristic knowledge. On the other hand, Hernando García Barriga, considered as the father of botany in Colombia, dedicated his life to the study of Colombian flora and was an employee of the Colombian National Herbarium. He traveled the whole territory collecting about 21,400 plants that enrich various herbariums in the world. The result of his activity is the publication of numerous works, among which the "Medicinal Flora of Colombia" stands out (Barriga, 1986).

Currently, the Alexander von Humboldt Biological Resources Research Institute, a corporation linked to the Ministry of the Environment and Sustainable Development, carries out the work of permanent registration and monitoring of plant species and biodiversity in general. This institution published in 2016, for the first time in the history of Colombia, the Catalog of Plants and Lichens of Colombia, a floristic inventory prepared by researchers from the National University of Colombia and researchers from other countries (Bernal et al., 2016). The latest report on plants in Colombia is of 29,917 species distributed in more than 400 families and 3000 genera, the most abundant families being Orchidaceae, Fabaceae, Asteraceae, Rubiaceae and Melastomataceae with more than a thousand species each, and with the greatest diversity in the group of orchids with more than 3000 species (Raz and Zamora, 2020). In Colombia, knowledge of the flora is still incomplete and there are vast regions of the country that have been scarcely explored. Every year several dozen species new to science are discovered and the range known to many others is expanded, while researchers discover new applications for our plants day after day. The Humbolt Institute has also characterized medicinal species, which will be discussed later in this chapter. The use of these species in Colombia is regulated by Decrees 2266 of 2004, 3553 of 2004, 4927 of 2009 and 1156 of 2008, issued by the Ministry of Health and Social Protection. Through Resolution 2834 of 2008, the Colombian Vademecum of Medicinal Plants is adopted as a mandatory reference to issue the sanitary registration of traditional phytotherapeutic products. Finally Decree 1156 of 2018, the health registration regime of phytotherapeutic products and a list of medicinal plants is adopted, built from the monographs of the World Health Organization (WHO), the European Health Agency (EMA) and the Colombian Vademecum of Medicinal Plants (Fig. 1.1).

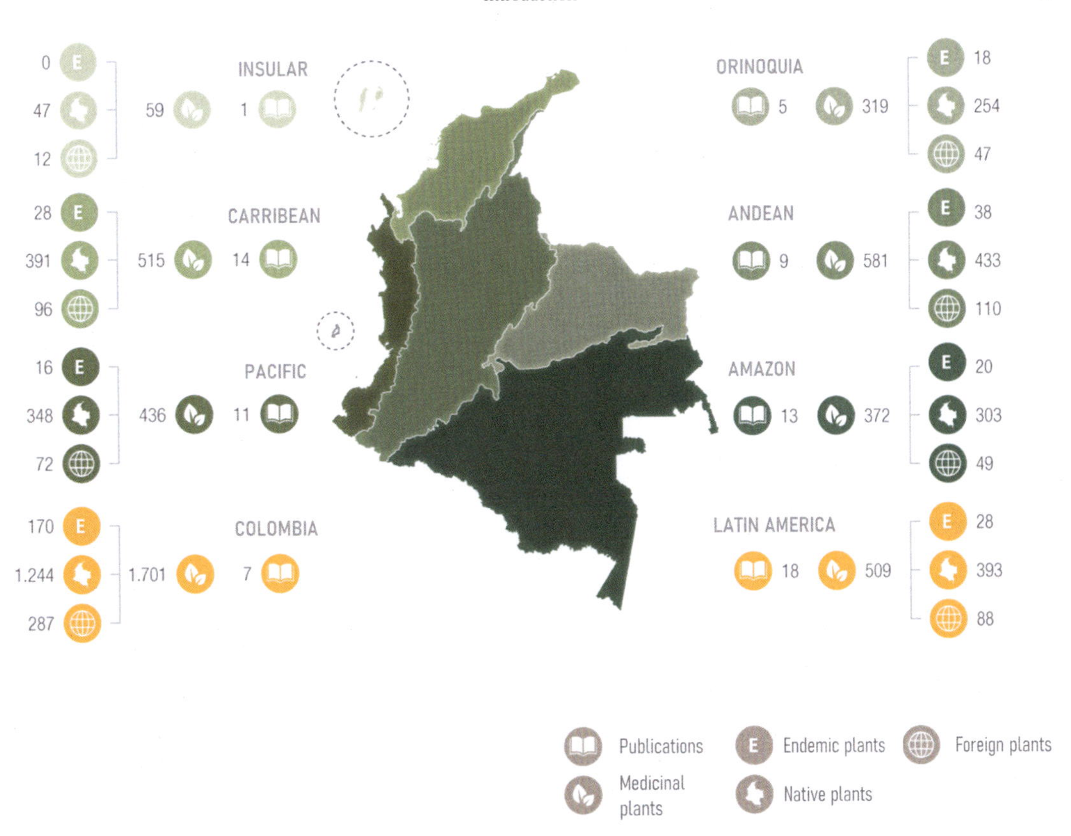

FIGURE 1.1 Natural regions of Colombia and knowledge of medicinal plants. Source: *From Vásquez, C. (2014). Avances en la investigación sobre plantas medicinales. En: Bello et al. (ed). Biodiversidad 2014. Estado y tendencias de la biodiversidad continental en Colombia. Instituto Alexander von Humboldt. Bogotá D.C., Colombia. (2014). http://reporte.humboldt.org.co/biodiversidad/2014/cap1/104/#seccion4. (Accessed 10 February 2021). Extracted from: Alexander von Humboldt Biological Resources Research Institute.*

This chapter deals with the use of medicinal plants as antimicrobials in Colombia. We have tried to review the literature on plants that have been traditionally used in the different regions of Colombia in order to alleviate or treat the symptoms caused by different microorganisms (bacteria, fungi, viruses, or microscopic parasites). Similarly, another review was performed on Colombian research works from the last 30 years (1990–2020) by focusing on antimicrobial activity testing of plants. Based on the results of this review, we arbitrarily selected studies whose results were considered to be the most promising for the search for natural alternatives against the different microorganisms mentioned. Finally, this chapter summarizes some of the advances that our research group has been making on the bioguided fractionation of plant extracts from the Colombian Caribbean region in order to obtain antibacterial molecules against strains of *Escherichia coli*, *Klebsiella pneumonia*, and *Pseudomonas aeruginosa* resistant to carbapenems and *Staphylococcus aureus* resistant to methicillin.

Materials and methods

A bibliographic review focusing on traditional uses and bioassays of antimicrobial activity of Colombian plants was carried out by consulting electronic databases such as PubMed, Science Direct, and Scifinder. Academic specific keywords were used such as "plants of traditional use in Colombia," "antiinfectious plants in Colombia," "antimicrobial plants in Colombia," "antifungal plants in Colombia," "antileishmanial plants in Colombia," "antimalarial plants in Colombia," "antiviral plants in Colombia," and "biological activity of plants in Colombia." We focused our search on the last 30 years (1990–2020). Overall, a total of 2411 documents were obtained. Subsequently, the selection of the information was carried out by reviewing the titles and abstracts of those works including only studies reporting a biological activity of plant extracts less than 600 μg/mL The parameters that were considered for the expression of biological activity in this review were the following: minimum inhibitory concentration (MIC), inhibitory concentration 50% (IC_{50}), lethal concentration 50% (LC_{50}), and effective dose 50% (ED_{50}). Of all the scientific reports reviewed, 43 documents were finally selected, which were classified by years, from 1990 to 2020. In this part of the chapter we present the taxonomic data, collection areas, plant part used, type of extract, biological activity, microorganism evaluated, and the respective bibliographic reference. We hope that the information presented will become an initial support tool for the realization of new studies that allow the documentation on the Colombian flora, its uses and its conservation.

Plants traditionally used in Colombia as antimicrobials

Traditional medicine in Colombia is the product of a triple legacy: indigenous, African, and Spanish (Arias, 2016; Martínez, Chiguasuque, & Casallas, 2006) and are mainly found in rural areas (Pérez & Matiz-Guerra, 2017). The most important contribution of the medicinal use of plants in Colombia, and Latin America in general, comes from the indigenous peoples. There are still 102 indigenous peoples in Colombia, distributed all over the country's departments, who keep their customs in force. For example, the Sibundoy Valley (Alto Putumayo), where the Inga, Kamentzá, and Quillacinga indigenous communities coexist, is classified as one of the places in the world that has a high concentration of cultivated magical and medicinal plants and as an important reserve of ancestral knowledge about medicine and botany (Cardona-Arias, Rivera-Palomino, & Carmona-Fonseca, 2015; Hidalgo, Forero, & Velasco, 2016). The collection, registration, and dissemination of this ancestral knowledge were carried out by some scientists such as Hernando García Barriga and Father Enrique Pérez Arbeláez, founder of the Botanical Garden of Bogotá. On the other hand, Richard Evans Schultes, a North American ethnobotanist, reflected his research in the Colombian Amazonian northwest in various works: *De plantis toxicaris, Las plantas de los dioses* (1982), *The Healing Forest* (1990), and *Vine of the Soul* (1992) (Ramirez, 1994).

From all the bibliographic records made, it has been suggested that some 6000 species are used by our indigenous people and farmers to combat a wide spectrum of diseases

(Fonnegra, 2007; Merchán, 2017). However, an investigation carried out by the Ministry of Environment, Housing and Territorial Development and the Alexander von Humboldt Biological Resources Research Institute has identified to date, 2768 recognized medicinal species, of which 2333 (84.3%) and 435 (15.7%) species are native and foreign to the Neotropics, respectively (http://reporte.humboldt.org.co/biodiversidad/2014/cap1/104/#sección1). There are 227 medicinal species exclusive to Colombia, these medicinal plants belong to 202 botanical families, the most frequently mentioned being the Asteraceae (Compositae) family, followed by the Fabaceae (Leguminosae), Rubiaceae, Solanaceae, Lamiaceae (Labiatae), Euphorbiaceae, Piperaceae, and Rosaceae, among others (Bernal, García, & Quevedo, 2011; Vásquez, 2014). In the scientific literature there is little information on medicinal species from Colombia, as only 178 endemic species (78.4% of the total) are reported as medicinal and only five species (2.2%) are reported in four or more publications. In contrast, foreign plants are referred to as medicinal more repeatedly and exhibit a higher percentage of species as well as a high number of records, compared to medicinal plants native to the Neotropics and endemic to Colombia (Vásquez, 2014). Of the 2768 species, only 4.3% (119 species) have been included in the Colombian Vademecum of Medicinal Plants (2008), of which more than 67% are foreign species and less than 33% are native from Colombia.

The trade in medicinal plants in Colombia remains a very important activity. A total of 156 species of medicinal and aromatic plants are produced and marketed in the country (Restrepo, 2011), of which 41% are native, 50% foreign, and 9% naturalized (Rugeles, Ortiz, Huertas, & Guaitero, 2012). In recent years, the legalization of the use of Colombian Vademecum of Medicinal Plants and the official list issued by National Institute for Drug and Food Surveillance (INVIMA, for its acronym in Spanish) has allowed the commercial boom. The distribution takes place in small health food stores, some laboratories, and street vendors. The fresh plant trade occurs in all market stalls in the country. For example, in Cota (a small municipality in the Bogotá savannah), farmers collect aromatic and medicinal plants such as chamomile, calendula, verbena, and rue, and sell them in the Corabastos marketplace in Bogotá. The Samper Mendoza marketplace in Bogotá, the most representative of the district and the country, is a rural market where more than 500 plant merchants come from different regions of Colombia. This market is considered the main distributor of plants nationwide by the diversity of species that can be found (Quintero, Sanchez, & Estupiñan, 2018). The species with the highest volume of commercialization in the country is calendula (*Calendula officinalis* L.), followed by artichoke (*Cynara scolymus* L.), and valerian (*Valeriana officinalis* L.)

In order to describe the Colombian medicinal plants in general and those that are indicated for infectious diseases, it is necessary to mention that Colombia is a multicultural country besides the combination of biotic and abiotic factors that give rise to a spectacular diversity of climates, soils, and topography. Geographically, the country is divided into five large continental natural regions (Caribbean, Pacific, Amazon, Orinoquia, and Andean) and an insular region. Each of the country's regions is subdivided into subregions, with their own characteristics in terms of economic, social, and cultural structure as well as their variety of climates, ecosystems, and species (Romero, Cabrera, & Ortiz, 2006). Therefore this part of the review was based on the study of plants traditionally used in each region of Colombia for the treatment of infectious or infectious-related diseases

(Table 1.1). Similarly, in this review we wanted to highlight the native species with indications for antimicrobial use in the Colombian Vademecum, as well as the species with the highest number of antimicrobial uses in traditional Colombian medicine, such as *Jacaranda caucana* Pittier, *Solanum nudum* Hassl (Solanaceae), *Hymenaea courbaril* L. (Leguminosae), *Xanthium strumarium* L. (Asteraceae), *Guazuma ulmifolia* Lam. (Malvaceae), *Cymbopogon citratus* (DC.) Stapf (Poaceae) and *Austroeupatorium inulifolium* Kunth.

In our review we found different disorders related to the four most mentioned groups of infectious diseases. Among which the most cited were: venereal diseases (42 citations, 15.7%), respiratory diseases (60 cit., 22.5%), gastrointestinal diseases (89 cit., 33.3%), skin diseases (76, 28.5%).

Some medicinal plants used to treat sexually transmitted diseases that we found in this review were, for example, *Persea americana* (Lauraceae) and *Uncaria tomentosa* (Rubiaceae) used as decoction of the aerial parts for the treatment of AIDS. Additionally, some plant species are used to heal conditions such as gonorrhea, syphilis, ulcers, and vaginal infections, among others. *Jacaranda caucana* (Bignoniaceae) is used in the Andean region to mitigate the symptoms of syphilis, gonorrhea, and chancres. Finally, we can note that these medicinal plants prevail in the Amazon and Pacific regions.

Respiratory diseases are one of the most treated by medicinal plants in Colombia. In Table 1.1, 51 plants are reported to be used for a range of respiratory diseases, from the common flu to serious pulmonary tuberculosis infections. In the Orinoquia region, the *Carapichea ipecacuanha* (Rubiaceae) species is used to treat bronchitis, bronchopneumonia, and lung congestion, among other non-infectious diseases. The plants *Mentha piperita* (Lamiaceae), *Aloysia citriodora* (Verbenaceae), *Cymbopogon citratus* (Poaceae), *Sambucus nigra* (Adoxaceae), *Furcraea cabuya* (Asparagaceae) among other species found in the Andean region are used for the treatment of the common flu, pneumonia and whooping cough.

Among skin diseases treatable by traditional medicine, we were able to find sores, measles, chickenpox, acne, scabies, even leprosy. Species like *Coccoloba obtusifolia* Jacq. (Polygonaceae) in the Orinoquia region is used specifically to treat skin conditions such as pimples and rashes, its

TABLE 1.1 Plants used for the treatment of symptoms associated with infectious diseases caused by different microorganisms in regions of Colombia.

Amazon region				
Scientific name	**Common name**	**Plant part**	**Medicinal use**	**References**
Ficus insipida Willd. (Moraceae)	Higuerón, higuerote, cachinguba, caucho, caucho menudito, chibechi, leche de casingua, ojé, cumacanae, lipanae, damagua, bibosi, cocoba, ficus, gomelero, ira, sacha ooé, copei	Latex	Anthelmintic	Schultes and Raffauf (1990)
Justicia blackii Leonard. (Acanthaceae)	Ke-re'-ma	The Tikuna Indians use the leaves (infusion in the form of tea)	Chronic sinusitis (infectious)	

(Continued)

TABLE 1.1 (Continued)

Amazon region				
Scientific name	**Common name**	**Plant part**	**Medicinal use**	**References**
Asclepias curassavica L. (Apocynaceae)	Bencenuco, jalapa, lombricera, viborana		Purgative, gonorrhea, vermifuge	Laferriere (1994).
Justicia chlorostachya Leonard. (Acanthaceae)	Mee-kee-ña-ta'-ree	The Makuna Indians of the Piraparaná River use the root powder mixed with lard and rubbed on the affected skin	Fungal skin infections	Schultes and Raffauf (1990)
Justicia pectoralis Jacq. (Acanthaceae)	Minoei	The Indians in the Vaupés and Amazonas of Colombia take a whole plant decoction	Lung diseases, especially pneumonia	
Ruellia malacosperma Greenm. (Acanthaceae)	Minerva	The Tikuna use the decoction of the leaves, they cut and boiled in water for 5 min, cooled and drunk	Diarrhea, measles, fever	
Minquartia guianensis Aubl. (Olacaceae)	Acapu, cheé	Bark decoction (orally)	Vermifuge, malaria, diarrhea	Quintana Arias (2012)
Persea americana Mill. (Lauraceae)	Aguacate	Leaves decoction (orally)	AIDS	
Uncaria tomentosa (Willd. ex Schult.) DC. (Rubiaceae)	Uña de gato	Bark and leaves decoction (orally)	AIDS	
Verbena litoralis Kunth (Verbenaceae)	Verbena	Leaves (orally/bath)	Flu, fever	

Andean region				
Scientific name	**Common name**	**Plant part**	**Medicinal use**	**References**
Piper aduncum L. (Piperaceae)	Cordoncillo, pipilongo		Gonorrhea	Perez-Arbelaez (1978)
Jacaranda caucana Pittier (Bignoniaceae)	Gualanday	Leaves, bark	Syphilis, gonorrhea, chancres, other venereal diseases	
Gnaphalium bogotense Kunth. (Compositae)	Donalonso	Whole plant	Diarrhea	Hirschhorn (1981).
Jacaranda caroba (Vell.) DC. (Bignoniaceae)	Curnique, aceituna	Whole plant	Syphilis	
Berberis glauca Kunth. (Berberidaceae)	Espuelo, tachuelo	Whole plant	Febrifuge, purgative	
Alchemilla pectinata (Kunth) Rothm. (Rosaceae)	Orejuela, plegadera, cargarrocio	Whole plant	Enteritis in children, diarrhea, dysentery	

(*Continued*)

TABLE 1.1 (Continued)

Andean region				
Scientific name	**Common name**	**Plant part**	**Medicinal use**	**References**
Senna multiglandulosa (Jacq.) H.S.Irwin & Barneby (Leguminosae)	Alcaparro	Leaves (orally)	Typhoid fever	Jaramillo Gómez (2003)
Sambucus nigra L. (Adoxaceae)	Sauco, tilo		Typhoid fever, colds, tuberculosis, measles, parasites	
Vasconcellea pubescens A.DC. (Caricaceae)	Papayuelo o papayo	Fruits together with *Rubus glaucus* and *Sambucus nigra* flower (orally)	Flu	
Melissa officinalis L. (Lamiaceae)	Toronjil	Leaves infusion prepared in milk (orally)	Measles and chickenpox	
Cymbopogon citratus (DC.) Stapf (Poaceae)	Limonaria	Leaves (orally)	Flu	
Verbena litoralis Kunth (Verbenaceae)	Verbena	Branches with leaves, flowers	Fever, typhoid fever, malaria	
Mentha x piperita L. (Lamiaceae)	Yerbabuena	Leaves (orally)	Antibacterial, flu	Vera Olave (2007)
Aloe vera (L.) Burm.f. (Xanthorrhoeaceae)	Sábila	Inner aloe (topic)	Antibacterial	
Aloysia citriodora Palau (Verbenaceae)	Cidrón	Leaves (orally)	Flu, fever	
Cymbopogon citratus (DC.) Stapf (Poaceae)	Limonaria	Leaves (orally)	Flu, fever	
Calendula officinalis L. (Compositae)	Caléndula	Leaves (orally/topic)	Antibacterial, healing	
Sambucus nigra L. (Adoxaceae)	Sauco	Leaves, stems and flowers (orally)	Flu	
Solanum lycopersicum L. (Solanaceae)	Tomate	Fruit (orally)	Antiviral	
Desmodium adscendens (Sw.) DC. (Leguminosae)	Amor seco	Branches, leaves and roots	The decoction is used to wash wounds, pimples, malaria	Fonnegra Gómez, Villa Londoño, and Monsalve Fonnegra (2012)
Petiveria alliacea L. (Phytolaccaceae)	Anamú	Branches, leaves, and flowers	Flu, infections, fever	
Oxalis scandens Kunth (Oxalidaceae); *Oxalis corniculata* L. (Oxalidaceae)	Acederilla, acedera trébol	Whole plant including the roots (orally)	Intestinal parasites, diarrhea, fever	
Bixa orellana L. (Bixaceae)	Achote	Leaves, seeds	Tonsillitis, sore throat	

(*Continued*)

TABLE 1.1 (Continued)

Andean region				
Scientific name	**Common name**	**Plant part**	**Medicinal use**	**References**
Persea americana Mill. (Lauraceae)	Aguacate	Bark (bath), Fruit	Wash infected wounds. Anthelmintic. Abscesses, boils, skin rash	
Cucurbita maxima Duchesne (Cucurbitaceae)	Ahuyama	Seeds	Anthelmintic (tapeworm)	
Hymenaea courbaril L. (Leguminosae)	Algarrobo	Bark, resin	Sinusitis, diarrhea, vermifuge. The resin is used in lung conditions, as an antibacterial, antifungal, and antiparasitic	
Myrcia popayanensis Hieron. (Myrtaceae)	Arrayan de hoja grande	Finely chopped buds (orally)	They are used together with sour guava against cholerin (infectious diarrhea)	Fonnegra Gómez et al. (2012)
Momordica charantia L. (Cucurbitaceae)	Balsamina	Leaves, branches, stems, fruits.	Plasters and poultices against skin conditions (thrush). Ripe fruits in sirup with honey bee are used against fever and malaria	
Clibadium surinamense L. (Compositae)	Basaica, salvión, margaritona	Leaves (topic)	Fungal infection on feets, hands, and other parts of the body. An alcoholic maceration is rubbed on the affected part	
Phlebodium aureum (L.) J.Sm. (Polypodiaceae)	Calaguala	Rhizomes (orally)	Anthelmintic, syphilis	
Tropaeolum majus L. (Tropaeolaceae)	Capuchina, alcaparra colombiana,	Flowers and leaves	Antibiotic	
Tropaeolum tuberosum Ruiz & Pav. (Tropaeolaceae)	Cubios, abio, nabos	Tuber, flowers, and leaves (orally)	Tuber decoction is used against shingles (herpes zoster). Flowers and leaves are used as an antibiotic and for hepatitis	Fonnegra Gómez et al. (2012)
Artemisia absinthium L. (Compositae)	Curahigado	Branches with leaves (orally)	Liver diseases	
Justicia pectoralis Jacq. (Acantaceae)	Curîbano	Branches with leaves (orally)	Amebas, diarrhea, hepatitis, tonsillitis	

(*Continued*)

TABLE 1.1 (Continued)

Andean region				
Scientific name	**Common name**	**Plant part**	**Medicinal use**	**References**
Croton mutisianus Kunth (Euphorbiaceae)	Drago, sangregao	Latex, bark	Tonsillitis, shingles (herpes zoster), ulcers	
Saurauia ursina Triana & Planch. (Actinidiaceae)	Dulumoco	Branches with leaves (orally)	Flu	
Sisyrinchium micranthum Cav. (Iridaceae)	Espadilla	Whole plant (orally)	Fever, pneumonia, purgative, flu, measles, chickenpox	
Phyllanthus niruri L. (Phyllanthaceae)	Florescondida, cancrapiedra, sacapiedra	Whole plant (orally)	Hepatitis (A, B and C); herpes	
Jacaranda mimosifolia D.Don (Bignoniaceae)	Gualanday	Bark, leaves, and flowers	Antibiotic, venereal diseases, boils, chickenpox	
Annona muricata L. (Annonaceae)	Guanábano	Seeds (orally)	Intestinal parasites	
Psidium guineense Sw. (Myrtaceae)	Guayabo agrio	Buds (orally)	Diarrhea, dysentery	
Hedera helix L. (Araliaceae)	Hiedra	Branches with leaves (orally)	Bronchitis, pertussis	
Gunnera brephogea Linden & André (Gunneraceae)	Hoja de pantano	Leaves (orally)	Dysentery	
Rumex crispus L. (Polygonaceae)	Lengua de vaca, arracachuelo, tomasa, ruibarbo	Whole plant (orally)	Purgative, skin wash, febrifuge, venereal diseases. Hepatitis	Fonnegra Gómez et al., (2012), Vera Marín and Sánchez Sáenz (2015)
Cymbopogon citratus (DC.) Stapf (Poaceae)	Limoncillo, caña de limón, caña pauta, limonaria, limonera	Whole plant (orally)	Febrifuge, diarrhea, malaria, bronchitis, measles, chickenpox	Fonnegra Gómez et al. (2012)
Plantago major L. (Plantaginaceae)	Llantén	Whole plant or leaves (orally)	Respiratory diseases, infected wounds. In sirup with honey, it is recommended for bronchitis	
Malva parviflora L. (Malvaceae)	Malva	Branches or leaves (orally)	Antibiotic, respiratory diseases, fever, bronchitis, flu, vaginitis, conjunctivitis	
Tanacetum parthenium (L.) Sch.Bip. (Compositae)	Manzanillón	Flowers (orally)	Liver disease, vermifuge	
Bidens pilosa L. (Compositae)	Mazequia, Pataperro, cadillo	Whole plant (orally)	Flu, liver disease. In poultices it is used for leprosy and other skin diseases	

(*Continued*)

TABLE 1.1 (Continued)

Andean region				
Scientific name	**Common name**	**Plant part**	**Medicinal use**	**References**
Tragia volubilis L. (Euphorbiaceae)	Ortiga, ortiga blanca	Branches with leaves (orally)	Wound infections	
Cinchona pubescens Vahl (Rubiaceae)	Quina	Bark (orally)	Antibiotic, antiseptic, febrifuge, malaria	
Vasconcellea pubescens A.DC. (Caricaceae)	Papayuela	Fruits	Amebiasis	Fonnegra Gómez et al. (2012)
Phenax rugosus (Poir.) Wedd. (Urticaceae)	Parietaria	Whole plant (orally)	Liver diseases, to wash infected wounds	
Iresine diffusa Humb. and Bonpl. ex Willd. (Amaranthaceae)	Penicilina	Branches with leaves (orally)	Urinary tract infections, diarrhea, infected wounds, antibiotic	
Monnina phytolaccifolia Kunth. (Polygalaceae)	Pito, cargamanta, cargamento	Branches and leaves macerated for 3 days	Sinusitis	
Musa x paradisiaca L. (Musaceae)	Plátano guineo	Exudate dissolved in hot water	Tuberculosis, diarrhea	
Satureja brownei (Sw.) Briq. (Lamiaceae)	Poleo, poleo macho	Branches with leaves (orally)	Pneumonia, bronchitis, fever	
Lippia alba (Mill.) N.E.Br. ex Britton & P.Wilson (Verbenaceae)	Prontoalivio, curalotodo, oregano de cerro	Branches with leaves (orally)	Flu, disinfectant, febrifuge, diarrhea	
Austroeupatorium inulaefolium (Kunth) R.M.King & H.Rob. (Compositae)	Salvia blanca, salvia amarga, venadillo	Branches with leaves (orally)	Dental abscesses, amoebae, gingivitis, washing infected wounds, tuberculosis, liver disease, tonsillitis, malaria	Fonnegra Gómez et al. (2012)
Alternanthera pubiflora (Benth.) Kuntze (Amaranthaceae)	Sanguinaria, guardaparque, menudita	Leaves (orally)	Meningitis	
Salix humboldtiana Willd. (Salicaceae)	Sauce macho, sausa	Bark (orally)	Fever and malaria	
Sambucus nigra L. (Adoxaceae)	Sauco	Flowers (orally)	The flowers decoction is mixed with honey bee. It is employed to treat fever, flu, respiratory diseases, conjunctivitis, and intestinal parasites	
Tibouchina lepidota (Bonpl.) Baill. (Melastomataceae)	Sietecueros	Raw flowers (orally)	Measles, severe diarrhea, vaginal infections	
Pseudelephantopus spicatus (B. Juss. ex Aubl.) Rohr ex C.F. Baker (Compositae)	Sueldaconsuelda, totumillo, contrayerba	Whole plant (orally)	Malaria, dysentery, diarrhea	Fonnegra Gómez et al. (2012)

(Continued)

TABLE 1.1 (Continued)

Andean region				
Scientific name	**Common name**	**Plant part**	**Medicinal use**	**References**
Bocconia frutescens L. (Papaveraceae)	Trompeto, curador, mano de tigre	Leaves and fruits (orally/topic), latex (topic) and roots (orally)	Leaves and fruits: rashes, scabies in skin infections, sores, fungicidal. Roots: liver diseases. Latex is recommended as a purgative and vermifuge. From the fruits an oil is obtained that kills lice and against scabies	
Verbena litoralis Kunth (Verbenaceae)	Verbena, verbena blanca	Inflorescences (orally)	Vaginal infections, wounds and herpes zoster, antibiotic, typhoid fever, febrifuge	
Achyrocline satureioides (Lam.) DC. (Compositae)	Vira-vira macho, botón de oro, venadillo, juan blanco, suso	Aerial parts (orally)	Febrifuge, typhoid fever, flu, lung diseases, bronchitis, skin diseases	
Solanum americanum Mill. (Solanaceae)	Hierbamora, yerbamora	Branches with leaves, flowers, and fruits	Abscesses, infected wounds	Fonnegra Gómez et al. (2012)
Acmella oppositifolia (Lam.) R. K.Jansen (Compositae)	Yuyo quemado, chisacá, botón de oro.	Flowers (orally)	Herpes zoster, gingivitis, liver disease, diarrhea, vermicide	
Allium sativum L. (Amaryllidaceae)	Ajo	Bulbs (topic)	Intestinal parasites	
Pimpinella anisum L. (Apiaceae)	Anís	Leaves or fruits (orally)	Flu	
Borago officinalis L. (Boraginaceae)	Borraja	Flowers mixed with citron or with aloe vera and honey (orally)	Flu	
Furcraea cabuya Trel. (Asparagaceae)	Cabuya	Flowers (orally)	Flu, whooping cough, respiratory disorders	
Impatiens walleriana Hook.f. (Balsaminaceae)	Caracucho blanco	Flowers mixed with milk (orally)	Flu and pneumonia	
Aloysia citriodora Palau (Verbenaceae)	Cidrón	Leaves (orally)	Flu, bronchitis, diarrhea	
Daucus montanus Humb. & Bonpl. exSchult. (Apiaceae)	Fumaria	Branches with leaves (orally)	Tuberculosis	Fonnegra Gómez et al. (2012)

(*Continued*)

TABLE 1.1 (Continued)

Andean region				
Scientific name	**Common name**	**Plant part**	**Medicinal use**	**References**
Salvia scutellarioides Kunth (Lamiaceae)	Mastranto, hierba de sapo	Whole plant (orally)	Herpes	
Gliricidia sepium (Jacq.) Walp. (Leguminosae)	Matarratón	Leaves (orally/bath)	Fever, dengue, liver diseases, flu	
Aloe vera (L.) Burm.f. (Xanthorrhoeaceae)	Sábila	Inner aloe mixed with honey (orally)	Flu, whooping cough, bronchial problems, and pneumonia	
Croton mutisianus Kunth. (Euphorbiaceae)	Sangre de drago	Latex, branches with leaves (topic)	Herpes zoster	
Eryngium foetidum L. (Apiaceae)	Cilantro sabanero	Whole plant (orally)	Hepatitis	Vera Marín and Sánchez Sáenz (2015)
Solanum lycopersicum L. (Solanaceae)	Tomatera	Leaves and stems (orally)	Ulcers and intestinal infections	

Caribbean region				
Scientific name	**Common name**	**Plant part**	**Medicinal use**	**References**
Aegiphila elata Sw. (Lamiaceae)			Ulcers, diarrhea, dysentery	López-Palacios (1986)
Mangifera indica L. (Anacardiaceae)	Mango		Malaria	Jiménez-Escobar and Estupiñán-González (2011)
Spondias mombin L. (Anacardiaceae)	Jobo, cocote		Malaria	
Annona purpurea Moc. & SesséDunal (Annonaceae)	Cabezona, gallina gorda		Fever	
Elaeis oleifera (Kunth) Cortés (Arecaceae)	Corozo, anolí, palmaecorozo		Purgative	
Croton malambo H.Karst. (Euphorbiaceae)	Malambo		Purgative	
Senna reticulata (Willd.) H.S. Irwin & Barneby (Leguminosae)	Bajagua		Purgative	
Gliricidia sepium (Jacq.) Walp. (Leguminosae)	Matarratón	Leaves	Measles, fever	Jiménez-Escobar and Estupiñán-González (2011)
Simaba cedron Planch. (Simaroubaceae)	Cedrón		Malaria	
Guaiacum officinale L. (Zygophyllaceae)	Guayacán		Venereal diseases	

(Continued)

TABLE 1.1 (Continued)

Caribbean region				
Scientific name	**Common name**	**Plant part**	**Medicinal use**	**References**
Buxus citrifolia (Willd.) Spreng. (Buxaceae)	Cafetillo	Leaves (orally/topic)	Antiseptic	Mesa-S, Santamaría, García, and Aguilar-Cano (2016)
Prioria copaifera Griseb. (Leguminosae)	Cativo, canime, trementino	Resin	Infected wounds	
Rhizophora mangle L. (Rhizophoraceae)	Mangle, mangle colorado, mangle piñón, mangle rojo, mangle rosado	Bark (powdered)	Febrifuge, leprosy, and tuberculosis	
Cecropia sciadophylla Mart. (Urticaceae)	Yarumo	Whole plant (bath)	Flu	Quintana Arias (2016)
Myroxylon balsamum (L.) Harms (Leguminosae)	Bálsamo, olor, talú, caraña, bálsamo blanco, bálsamo de tolú, bálsamo del perú, bálsamo rubio, árbol de tolú, árbol de caraña, carano, elemí, guayacán, guayacán tomé, sarrapio, tache, yoya	Fruits	Scabies	Mesa-S et al. (2016)
Bravaisia integerrima (Spreng.) Standl. (Acanthaceae)	Palo de agua	Leaves (orally/bath)	Fever	Quintana Arias (2016)
Eryngium foetidum L. (Apiaceae)	Cilantro	Leaves (orally)	Anthelmintic	
Crescentia cujete L. (Bignoniaceae)	Totumo	Fruit, flower, pulp and branches (orally)	Flu, antifungal	
Bixa orellana L. (Bixaceae)	Achote	Leaves and seeds (orally)	Purgative	
Heliotropium indicum L. (Boraginaceae)	Verbena blanca	Whole plant (orally)	Purgative	Quintana Arias (2016)
Chenopodium ambrosioides L. (Amaranthaceae)	Paico	Leaves	Purgative	
Momordica charantia L. (Cucurbitaceae)	Balsamina	Whole plant (topic)	Fever	
Sterculia apetala (Jacq.) H. Karst. (Malvaceae)	Camajón	Seeds (orally)	Anthelmintic	
Psidium guajava L. (Myrtaceae)	Guayaba	Leaves and fruits (orally)	Febrifuge, diarrhea	
Petiveria alliacea L. (Phytolaccaceae)	Anamú	Leaves (orally)	Flu, antiseptic	
Scoparia dulcis L. (Plantaginaceae)	Escobilla menuda, pimientica	Whole plant (orally)	Purgative, febrifuge	
Quassia amara L. (Simaroubaceae)	Cruceto morado	Stem	Malaria, flu	

TABLE 1.1 (Continued)

Orinoquia region				
Scientific name	**Common name**	**Plant part**	**Medicinal use**	**References**
Cuphea antisyphilitica Kunth. (Lythraceae)	Chiaga, moraditas		Venereal diseases	Hirschhorn (1981)
Selaginella asperula Spring. (Selaginellaceae)	Árbolito de mano	Leaves	Sores	Ortiz Gómez (1989)
Tabebuia ochracea (Cham) Standl. (Bignoniaceae)			Febrifuge	
Protium calanense Cuatrec. (Burseraceae)	Caraño, anime	Bark	Tetanos	
Chelonanthus alatus (Aubl.) Pulle. (Gentianaceae)	Arbol cabeza de rey zarumo		Sores	
Protium llanorum Cuatrec. (Burseraceae)	Caraño	Bark	Flu	
Miconia rufescens (Aubl.) DC. (Melastomataceae)		Whole plant	Lices	
Heteropterys beecheyana A.Juss. (Malpighiaceae)	Bejuco de gavilán	Bark	Gonorrhea	
Davilla nitida (Vahl) Kubitzki (Dilleniaceae)	Chaparrito	Leaves	Febrifuge	
Conyza canadensis (L.) Cronquist (Compositae)	Planta de la manta religiosa		Fever	
Handroanthus ochraceus (Cham.) Mattos (Bignoniaceae)	Ktsubarunae		Febrifuge	
Virola elongata (Benth.) Warb. (Myristicaceae)	Maranae	Roots	Malaria	
Psychotria lupulina Benth. (Rubiaceae)	Sisipiyanae	Roots	Gonorrhea	
Paullinia sessiliflora Radlk. (Sapindaceae)	Tsukuiiniibo		Tuberculosis, flu, anthelmintic	
Siparuna guianensis Aubl. (Siparunaceae)	Romadizo	Bark	Flu, skin diseases	Ortiz Gómez (1989)
Serjania clematidea Triana & Planch. (Sapindaceae)	Bejuco gonorrea		American gonorrhea	
Turnera sp. (Passifloraceae)	Roquito	Roots	Fever, infections	
Guazuma ulmifolia Lam. (Malvaceae)	Guácimo	Bark	Diarrhea, purgative	
Costus spiralis (Jacq.) Roscoe (Costaceae)	Caña agria, caña de sapo		Candidiasis, fever, bronchitis	

(*Continued*)

TABLE 1.1 (Continued)

Orinoquia region				
Scientific name	Common name	Plant part	Medicinal use	References
Heliopsis buphthalmoudes (Jacq.) Dunal. (Compositae)	Pestaña del culo de munuanu	Roots and leaves	Diarrhea, tetanus	
Virola cuspidata (Benth.) Warb. (Myristicaceae)	Maranae	Roots	Malaria	
Eugenia biflora (L.) DC. (Myrtaceae)	Árbol maraca	Roots	Diarrhea	
Myrcia sylvatica (G.Mey.) DC. (Myrtaceae)	Arrayán, espinazo de picure	Bark	Fever	Ortiz Gómez (1989)
Heliconia spp. (Heliconiaceae)	Platanillo, rabo de tigre	Leaves	Whooping cough	
Cephaelis pubescens Hoffm. (Rubiaceae)	Mano de danta	Roots	Gonorrhea	
Cephaelis tomentosa (Aubl.) Vahl. (Rubiaceae)	Pestaña de espiritu		Sores, infected wounds	
Psychotria lupulina Benth. (Rubiaceae)	Sisipiyanae	Roots	Gonorrhea	
Lycoseris mexicana (L.f.) Cass. (Compositae)	Árbolito de pajarito		Antiviral	
Caladium macrotites Schott. (Araceae)		Roots and leaves	Antiviral, infected wounds	
Xylopia amazonica R.E.Fr. (Annonaceae)	Árbol copete, Malagueto	Bark	Diarrhea	
Xanthium strumarium L. (Compositae)	Bardana menor, lapa menor, cachurro o cacharrera, higueruela, lamparones	Whole plant, roots, or fruits	Anthelmintic, liver diseases, hepatitis, sore throats, thrush, dysentery, lung conditions, fever, skin rashes	Munera (2005)
Cedrela odorata L. (Meliaceae)	Cedro	Leaves, bark	Febrifuge	
Copaifera officinalis L. (Leguminosae)	Amacey, cabimbo, camíbar, copayero, currucay, marano, calo de Bálsamo, calo del aceite, tacamaca	Essential oil, resin	Gonorrhea, urinary antiseptic. Dermal infections such as athlete's foot, psoriasis, and infected wounds	
Momordica charantia L. (Cucurbitaceae)	Sibicogen, balsamina, mavilla, bejuco e culebra	Whole plant (orally), fruit cooked in olive oil (topic)	Purgative, vermifuge, sinusitis. It's also used to treat wounds, burns, cholera, and prolonged fevers	
Senna occidentalis (L.) Link (Leguminosae)	Ajotillo grande, bicho de café, café furrusco, chilinchile, yerba de gallinazo	Leaves and seeds (orally)	Febrifuge, chronic dysentery	Munera (2005), Hirschhorn (1981)

(*Continued*)

TABLE 1.1 (Continued)

Orinoquia region				
Scientific name	**Common name**	**Plant part**	**Medicinal use**	**References**
Ceiba pentandra (L.) Gaertn. (Malvaceae)	Bona, ceiba, yunque	Leaves (bath)	Skin diseases and spots	Munera (2005)
Erythrina poeppigiana (Walp.) O.F.Cook (Leguminosae)	Ceibo, ceibo bucare, cachimbo, písamo, cámulo y bucaro	Flowers (orally)	Flu	
Bixa urucurana Willd. (Bixaceae)	Achote	Seeds	Antiviral, antifungal	Ortega David (2015)
Guazuma ulmifolia Lam. (Malvaceae)	Guácimo	Barks, leaves, root, and fruits	Sores, syphilis, malaria, fever, flu, elephantiasis, disinfectant	Munera (2005), Ortega David (2015)
Ricinus communis L. (Euphorbiaceae)	Alcherva, castor, catapucia mayor, cherva, crotón, higuera del diablo, higuerillo, palma de cristo, piojo del diablo, querva, tártago de venezuela	Seeds oil	Eczema, herpes, skin rashes, wounds, burns. It is also used against baldness, both in lotions and in poultices	Munera (2005)
Carapichea ipecacuanha (Brot.) L.Andersson (Rubiaceae)	Anillada menor, bejuquillo, picahonda, poaja, raicilla, radix brasilensis	Roots	Bronchitis, bronchopneumonia, lung congestion, corneal ulcers, diarrhea, dysentery (amebic), intermittent fever, malaria, among other noninfectious diseases	
Myrtus communis L. (Myrtaceae)	Arrayán, murta, murtinhos, murtra, mirta, miltra, murteira, murtera	Leaves essential oil	Antiseptic and antibiotic (comparable to penicillin and other antibiotics with a similar spectrum of action)	
Luffa cylindrica (L.) M.Roem. (Cucurbitaceae)	Estropajo, cabacina, buchina, esponjilla, esponja vegetal	Leaves, seeds oil	Ameba, herpes, chronic sinusitis. Infused leaves are used for dysentery; the fresh juice instilled in the eyes is used for conjunctivitis. The oil extracted by pressure or boiling of the peeled seeds is used in skin diseases	
Eucalyptus globulus Labill. (Myrtaceae)	Eucalipto	Leaves, seeds	Antiseptic for urinary tract infections. Throat and bronchial conditions	
Annona muricata L. (Annonaceae)	Guanábano	Flowers	Pectoral, febrifuge, flu.	

(*Continued*)

TABLE 1.1 (Continued)

Orinoquia region				
Scientific name	**Common name**	**Plant part**	**Medicinal use**	**References**
Coccoloba obtusifolia Jacq. (Polygonaceae)	Uvero		Infected wounds, pimples, and skin rash.	Ortega David (2015)
Jacaranda mimosifolia D.Don (Bignoniaceae)	Gualanday		Malaria, spot healing, acne, skin rash.	
Trichanthera gigantea (Humb. & Bonpl.) Nees (Acanthaceae)	Cajeto		Skin infection	Ortega David (2015)
Gliricidia sepium (Jacq.) Walp. (Leguminosae)	Matarratón	Leaves (orally and bath)	Febrifuge, skin rash, dermatitis	
Petiveria alliacea L. (Phytolaccaceae)	Anamú	Leaves (orally), roots (bath)	Skin conditions, intestinal and respiratory infections	
Anacardium occidentale L. (Anacardiaceae)	Marañón	Sap	Bronchitis, lung diseases	
Xylopia aromatica (Lam.) Mart. (Annonaceae)	Malagueto	Fruits	Malaria	
Caryodendron orinocense H.Krast.(Euphorbiaceae)		Seeds	Germicide, wound infections	
Hyptis brachiata Briq. (Lamiaceae)	Mastranto	Leaves (orally)	Antiparasitic, diarrhea	

Pacific region				
Scientific name	**Common name**	**Plant part**	**Medicinal use**	**Reference**
Mammea americana L. (Calophyllaceae)	Mamejo, mamey	Seeds	Insecticide	Duke (1970)
Momordica charantia L. (Cucurbitaceae)	Balsamina		Febrifuge	
Neurolaena lobata (L.) R. Br. ex Cass. (Compositae)	Contragavilán		Gonorrhea, malaria	
Prioria copaifera Griseb. (Leguminosae)	Cativo	Resin	Venereal ulcers	
Psychotria brachiata Sw. (Rubiaceae)	Puga, baca, pacuru	Leaves	Purgative	
Quassia amara L. (Simaroubaceae)	Hombre grande		Malaria	
Spigelia anthelmia L. (Loganiaceae)	Lombricera		Purgative, vermifuge	
Tussacia friedrichsthaliana Hanst (Gesneriaceae)	Desbaratador		Diarrhea	
Anacardium excelsum (Bertero ex Kunth) Skeels (Anacardiaceae)	Espavé	Bark decoction (orally)	Diarrhea	

(*Continued*)

TABLE 1.1 (Continued)

Pacific region				
Scientific name	**Common name**	**Plant part**	**Medicinal use**	**Reference**
Cassia occidentalis L. (Leguminosae)	Potra	Whole plant decoction (orally)	Vermifuge	
Cassia reticulata Willd. (Leguminosae)	Laureño		Purgative	
Cephaelis ipecacuanha (Brot.) A.Rich. (Rubiaceae).	Raicilla	Roots	Amebicide	
Couma macrocarpa Barb.Rodr. (Apocynaceae)	Popa	Latex	Amebicide	
Cymbopogon nardus L. Rendle (Poaceae)	Hierba de limón		Fever, malaria	
Gliricidia sepium (Jacq.) Walp. (Leguminosae)	Mataratón	Leaves	Febrifuge	
Uragoga emética (L.f.) Baill. (Rubiaceae).	Moncoa, muncua		Febrifuge	Duke (1970)
Malachra alceifolia Jacq. (Malvaceae)	Malva		Febrifuge, venereal diseases	
Hamelia patens Jacq. (Rubiaceae)	Hoja morada, bencenuco	Leaves (tea)/roots	Fever, sanguinolent diarrhea, purgative, syphilis	Duke (1970), Perez-Arbelaez (1978)
Elaeagia utilis (Goudot) Wedd. (Rubiaceae)	Barniz de pasto, árbol de cera, mopa		Pulmonary tuberculosis	Perez-Arbelaez (1978)
Rhizophora mangle L. (Rhizophoraceae)	Mangle	Bark (pulverized)	Febrifuge, leprosy, tuberculosis	
Coutarea hexandra (Jacq.) K.Schum. (Rubiaceae)	Mediagola	Bark	Febrifuge	
Simaba cedron Planch. (Simaroubaceae)	Cedron	Seeds. A preparation called "curarina" is obtained from this organ.	Febrifuge	
Simarouba amara Aubl. (Simaroubaceae)	Marouba	Bark	Febrifuge	
Costus villosissimus Jacq. (Costaceae)	Cañagria	Stem	Febrifuge (including typhoid fever)	

topical use favors the disinfection and cleaning of open wounds. In the Andean region, the tuber of the species *Tropaeolum tuberosum* (Tropaeolaceae) and the branches of *Croton mutisianus* (Euphorbiaceae) are used to treat Herpes zoster. While in the Caribbean and Pacific regions a predominant species such as the mangrove (*Rhizophora mangle* L.) is employed for the treatment of leprosy. Most of the treatments consist of oral decoctions and topical ointments.

On the other hand, one of the main causes of mortality in children under 5 years of age is enteric and diarrheal diseases. The high impact of these diseases is a consequence of the lack of aqueduct and sewage systems in certain areas of the country, leading to a low quality of drinking water and exposure to pathogens (Overgaard et al., 2012; Solano-Aguilar et al., 2013). In the Pacific region, the tea made from the leaves of *Hamelia patens* Jacq. (Rubiaceae) is used to treat acute diarrhea and bloody diarrhea (traces of blood in the feces). The poor diet and hygiene conditions also lead to the acquisition of parasites, especially among children, which explain why it is one of the most common conditions treated by healers or shamans. We found a record of 41 plants that fulfill this purpose, functioning either as an anthelmintic or a purgative. The use of decoctions and infusions for the treatment of intestinal parasites can be found all over the country, however, more reports were found in the Caribbean region.

Among the plants found in this review, we next highlight those known for their healing properties for several types of diseases, meaning a wide spectrum of activities including fever, hepatitis, dysentery, lung conditions, venereal diseases, and skin rashes.

Jacaranda caucana Pittier (Bignoniaceae)

Jacaranda caucana and *Jacaranda mimosifolia* species are recognized for the treatment of venereal infections and skin diseases in Colombia. However, no rigorous pharmacological and clinical studies were found to confirm this activity. The antiprotozoal activity against *Plasmodium falciparum*, *Leishmania* sp., and *Trypanosoma cruzi* has been proven (Weniger et al., 2001). The Colombian Vademecum has approved the use of *Jacaranda caucana* as an antiseptic. Some of the compounds isolated from *Jacaranda* species are triterpenoids and phenylethanoid glycosides and neolignans (Martin, Quinteros, Hay, Gupta, & Hostettmann, 2008; Martin et al., 2009; Ogura, Cordell, & Farnsworth, 1977).

Solanum nudum Dunal (Solanaceae)

Solanum nudum is a plant growing in South America and used in Tumaco (southwest of Colombia) for the treatment of fevers associated with malaria. Some steroidal compounds isolated from this plant such as tumacoside A and tumaquenone displayed antimalarial activity in vitro against *Plasmodium falciparum*, a chloroquine-resistant FCB-1 strain, with an IC_{50} of 27 and 16 μM (17.02 and 9.54 μg/mL), respectively (Saez et al., 1998). These compounds have been equally effective against clinical isolates (Arango, Carmona, & Blair, 2008), without showing mutagenicity, clastogenicity, and cytotoxicity effects (Alvarez, Pabón, Carmona, & Blair, 2004; García-Huertas, Pabón, Arias, & Blair, 2013; Londoño et al., 2006). The extracts of the fruits and leaves of this plant present IC_{50} values of 15.1 and 18.7 μg/mL, respectively (Monzote et al., 2016). The Colombian Vademecum has approved the use of *Solanum nudum* as an antiseptic.

Hymenaea courbaril L. (Leguminosae)

Hymenaea courbaril L., known as algarrobo, guápinol, locust, jutaby, or courbaril, is an imposing tree distributed from the western coast of central Mexico south to Bolivia and the south of central Brazil. The ethanol extract and fractions obtained of *H. courbaril* (barks) were investigated against bacterial clinical isolates, showing MIC values of 125, 250, 500, 750, and 1000 μg/mL against *E. faecalis* 3110, *E. coli* 3004, *S. aureus* 8066, *A. baumannii* 7810 and *K. pneumoniae* 7845, respectively. Among 14 species selected based on their ethnopharmacological use for treating diarrhea, *Hymenaea courbaril* L. showed significant antiviral activity against rotavirus. Fisetin obtained from xylem sap showed activity against the fungi *Microsporum gypseum*, *Trichophyton mentagrophytes*, *T. rubrum*, *T. tonsurans*, *Cryptococcus neoformans*, and *C. gattii* at concentrations of 32–128 μg/mL (da Costa et al., 2014).

Xanthium strumarium L. (Asteraceae)

The whole plant can be used to prepare the medicine, although the roots and fruits are also used separately. It has anthelmintic properties and is used against respiratory, skin, and gastrointestinal diseases. The components of its essential oil significantly inhibited bacteria and fungi such as *Staphylococcus aureus*, *Bacillus subtilis*, *Klebsiella pneumoniae*, *Pseudomonas aeruginosa*, *Candida albicans*, and *Aspergillus niger*. Its phytochemical composition includes the presence of flavones, saponins, and bioactive sesquiterpene lactones (Sharifi-Rad et al., 2015; Yang et al., 2017).

Guazuma ulmifolia Lam. (Malvaceae)

This plant grows in the Orinoquia region and is used in cases of sores, fever, cold, gastrointestinal problems, elephantiasis, and malaria. Its main components are flavonoids and phenolic acids which give it antimicrobial and antioxidant activity (Pereira, 2019; Morais et al., 2017).

Cymbopogon citratus (DC.) Stapf (Poaceae)

This species is well distributed throughout the national territory, albeit it is more used in the Andean region. Its pharmacological and therapeutic properties have been widely reported. Shah et al. (2011) conducted a review in which they show antiamebic, antibacterial, antidiarrheal, antifilarial, antifungal, antimalarial, antimutagenicity, antimycobacterial, antinociceptive, ascaricidal, and antiprotozoal activities. Triterpenoids and phenolic compounds such as quinone flavonoids are part of its phytochemical constituents.

Austroeupatorium inulaefolium (Kunth) R.M.King & H.Rob. (Asteraceae)

This plant is called "salvia amarga" by Andean region natives, and its uses are wide including the treatment of dental access, tuberculosis, tonsillitis, and malaria. It is considered an invasive species that is distributed throughout the world and there are reports of its cytotoxic, antioxidant, and antifungal properties (Madawala, Chandrasiri, Diwakara, Wijesundara, & Karunaratne, 2014). Recently, it was shown that its essential oil, rich in oxygenated sesquiterpenes, is a good inhibitor of Gram-negative bacteria growth such as *Klebsiella pneumoniae*, *Escherichia coli*, and *Pseudomonas aeruginosa* (Lucena et al., 2019).

Biological evaluation as antimicrobials of plant extracts in Colombia

Table 1.2 shows a list of *in vitro* antimicrobial studies of 98 plant species, collected in Colombia, and distributed in the following 41 families: Adoxaceae (1), Alliaceae (1), Amaranthaceae (1), Anacardiaceae (2), Annonaceae (8), Apocynaceae (3), Aquifoliaceae (1), Asteraceae (12), Bignoniaceae (1), Bombacaceae (1), Boraginaceae (1), Burseraceae (1), Caesalpiniaceae (1), Clusiaceae (4), Commelinaceae (1), Cucurbitaceae (2), Euphorbiaceae (7), Fabaceae (2), Labiaceae (1), Lamieaceae (6), Lecythidaceae (1), Magnoliaceae (1), Malpighiaceae (1), Melastomataceae (1), Melastomataceae (2), Meliaceae (1), Menispermaceae (2), Myristicaceae (5), Myrtaceae (6), Onagraceae (1), Papaveraceae (1), Piperaceae (9), Poaceae (1), Polygonaceae (2), Rhamnaceae (1), Rubiaceae (2), Rutaceae (2), Scrophulariaceae (1), Solanaceae (5), Urticaceae (1), Verbenaceae (3), Zingiberaceae (1).

This institution published in 2016, for the first time in the history of Colombia, the Catalog of Plants and Lichens of Colombia, a floristic inventory prepared by researchers from the National University of Colombia and researchers from other countries (Bernal et al., 2016). The latest report on plants in Colombia is of 29,917 species distributed in more than 400 families and 3000 genera, the most abundant families being Orchidaceae, Fabaceae, Asteraceae, Rubiaceae and Melastomataceae with more than a thousand species each, and with the greatest diversity in the group of orchids with more than 3000 species (Raz and Zamora., 2020). In Colombia, knowledge of the flora is still incomplete and there are vast regions of the country that have been scarcely explored. Every year several dozen species new to science are discovered and the range known to many others is expanded, while researchers discover new applications for our plants day after day. The Humbolt Institute has also characterized medicinal species, which will be discussed later in this chapter. The use of these species in Colombia is regulated by Decrees 2266 of 2004, 3553 of 2004, 4927 of 2009 and 1156 of 2008, issued by the Ministry of Health and Social Protection. Through Resolution 2834 of 2008, the Colombian Vademecum of Medicinal Plants is adopted as a mandatory reference to issue the sanitary registration of traditional phytotherapeutic products and finally. Through Decree 1156 of 2018, the health registration regime of phytotherapeutic products and a list of medicinal plants is adopted, built from the monographs of the World Health Organization (WHO), the European Health Agency (EMA) and the Colombian Vademecum of Medicinal Plants

Similarly, in this review we wanted to highlight the native species with indications for antimicrobial use in the Colombian Vademecum, as well as the species with the highest number of antimicrobial uses in traditional Colombian medicine, such as *Jacaranda caucana* Pittier, *Solanum nudum* Hassl (Solanaceae), *Hymenaea courbaril* L. (Leguminosae), *Xanthium strumarium* L. (Asteraceae), *Guazuma ulmifolia* Lam. (Malvaceae), *Cymbopogon citratus* (DC.) Stapf (Poaceae) and *Austroeupatorium inulifolium* Kunth.

Respiratory diseases are one of the most treated by medicinal plants in Colombia. In Table 1.1, 51 plants are reported to be used for a range of respiratory diseases, from the common flu to serious pulmonary tuberculosis infections. In the Orinoquia region, the *Carapichea ipecacuanha* (Rubiaceae) species is used to treat bronchitis, bronchopneumonia, and lung congestion, among other non-infectious diseases. The plants *Mentha piperita* (Lamiaceae), *Aloysia citriodora* (Verbenaceae), *Cymbopogon citratus* (Poaceae), *Sambucus nigra*

TABLE 1.2 Biological evaluation as antimicrobials of plant extracts in Colombia.

Scientific name (family)	Common name	Collection place	Type of extract	Biological activity	Microorganism	Result	Reference
Thymus vulgaris L. (Lamiaceae)	Tomillo	Bucaramanga, Santander	Essential oil (leaves, stem, and flowers)	Antibacterial	*Porphyromonas gingivalis*	MIC = 250 μg/mL	Parra Sepúlveda (2020)
Elettaria cardamomum L. Maton (Zingiberaceae)	Cardamomo		Essential oil (seeds)			MIC = 500 μg/mL	
Ilex paraguariensis. A St-Hil. (Aquifoliaceae)	Yerbamate	Viotá, Cundinamarca	Ethanol extract (leaves)		*Helicobacter pylori*	MIC = 500 μg/mL	Portela Quintero (2020)
Rosmarinus officinalis L. (Lamiaceae)	Romero	Bucaramanga, Santander	Essential oil (whole plant)		*Porphyromonas gingivalis*	MIC = 375–700 μg/mL	Duarte Velandia (2020)
Austroeupatorium inulaefolium Kunth. R.M.King & H. Rob. (Compositae)	Salvia de castilla, salvia de hora	Bogotá, Cundinamarca	Chloroform extract (aerial parts)	Antiparasitic	*Plasmodium falciparum*	CL_{50} = 2.27 μg/mL	Sosa Puerto (2019)
Miconia theizans Bonpl. Cogn. (Melastomataceae)	Tuno blanco		Chloroform extract (aerial parts)			CL_{50} = 0.56 μg/mL	
			Ethanol extract (aerial parts)			CL_{50} = 0.63 μg/mL	
Otholobium mexicanum L.f. J.W. Grimes. (Leguminosae)	Ruchica	Cerinza, Boyacá	Ethanol extract (leaves)	Antibacterial	*Staphylococcus aureus*	MIC = 0.1 μg/mL	Ramírez Romero and Prieto Munévar (2019)
Cymbopogon citratus DC. Stapf (Poaceae)	Limonaria	Sierra nevada de Santa Marta, Magdalena	Essential oil (leaves)		*Streptococcus mutans*	MIC = 1.0 μg/mL	Cuadros, Rivera, Merini, and Pabon (2018)
Solanum nudum Dunal (Solanaceae)	Saúco y zapata	Tumaco, Nariño	hexane extract (leaves)	antiparasitic	*Plasmodium falciparum*	IC_{50} = 17.4 μg/mL	Barrios et al. (2018)
			Corconá, Antioquia			IC_{50} = 100 μg/mL	

(*Continued*)

TABLE 1.2 (Continued)

Scientific name (family)	Common name	Collection place	Type of extract	Biological activity	Microorganism	Result	Reference
Rosmarinus officinalis L. (Lamiaceae)	Romero	Chía, Cundinamarca	Ethanol extract (leaves)	Antibacterial	*Staphylococcus aureus*	MIC = 5.2 μg/mL	Ortíz-Ardila et al. (2017)
					Pseudomonas aeruginosa	MIC = 10.1 μg/mL	
					Salmonella spp.	MIC = 22.4 μg/mL	
					Bacillus subtilis	MIC = 15.6 μg/mL	
Cecropia mutisiana. Mildbr (Urticaceae)	Calentano, guarumo	Viotá. Cundinamarca	Ethanol extract (leaves)		*Staphylococcus aureus*	MIC = 17.6 μg/mL	
			Ethanol extract (leaves)		*Salmonella* spp.	MIC = 11.7 μg/mL	
Cucurbita moschata Duchesne. (Cucurbitaceae)	Ahuyama	Atlántico	Ethanol extract (leaves)		*Staphylococcus aureus* methicillin sensitive	MIC = 19.1 μg/mL	del Castillo Pereira, Molinares Moscarella, Campo Urbina, and Bettin Martínez (2017)
					Staphylococcus aureus methicillin resistant	MIC = 0.16 μg/mL	
Momordica charantia L. (Cucurbitaceae)	Balsamina	Bolívar	Dichlomethane extract (leaves)		*Staphylococcus aureus*	MIC = 62.5 μg/mL	Cervantes Ceballos, Sánchez Hoyos, and Gómez Estrada (2017)
			Methanol extract (leaves)			MIC = 125 μg/mL	
Heliotropium indicum L. (Boraginaceae)	Verbena blanca		Ethanol extract (whole plant)			MIC > 512 μg/mL	
Sambucus nigra L. (Adoxaceae)	Sauco	San Antonio de Tequendama, Cundinamarca	Ethanol extract (flowers)		*Streptococcus pneumoniae*	MIC > 500 μg/mL	Pava, Sanabria, and Leal (2017)
					Enterococcus faecium vancomycin resistant	MIC > 500 μg/mL	

					Klebsiella pneumoniae with KPC (carbapenemase producer)	MIC > 500 μg/mL	
					Providencia rettgeri ESBL (extended spectrum betalactamases)	MIC > 500 μg/mL	
					Pseudomonas aeruginosa	MIC > 500 μg/mL	
					Enterobacter cloacae	MIC > 500 μg/mL	
					Escherichia coli	MIC > 500 μg/mL	
					Staphylococcus aureus	MIC > 500 μg/mL	
					Candida albicans	MIC > 500 μg/mL	
Ulex europaeus L. (Leguminosae)	Retamo espinoso	Monserrate, Cundinamarca	Dichloromethane fraction (leaves)		*Staphylococcus aureus*	MIC = 367.9 μg/mL	Parra Garzón (2017)
			Ethyl acetate fraction (leaves)		*Escherichia coli*	MIC = 219.1 μg/mL	
Mammea americana L.(Calophyllaceae)	Mamey	Turbaco, Bolívar	Coumarin A (ethanol extract)	Antiviral	Dengue virus	EC_{50} = 9.6 μg/mL	Martínez Vega et al. (2017)
			Coumarin B (ethanol extract)			EC_{50} = 2.6 μg/mL	
Tabernaemontana cymosa Jacq. (Apocynaceae)	Bola de puerco		Lupeol acetate (ethanol extract)		Dengue virus	EC_{50} = 37.5 μg/mL	
			Voacangine (ethanol extract)			EC50 = 10.1 μg/mL	

(*Continued*)

TABLE 1.2 (Continued)

Scientific name (family)	Common name	Collection place	Type of extract	Biological activity	Microorganism	Result	Reference
Solanum lycopersicum L. (Solanaceae)	Tomate	Barranquilla, Atlántico	Ethanol extract (fruits and seeds)	Antibacterial	*Streptococcus mutans*	MIC = 500 μg/mL	Puerta Domínguez and Vargas Hernández (2016)
Minthostachys mollis (Benth.) Griseb. (Lamiaceae)	Muña	Pamplona, Norte de Santander	Ethanol extract (leaves)		*Escherichia coli*	MIC = 500 μg/mL	Torrenegra-Alarcón et al. (2016)
					Staphylococcus aureus	MIC = 500 μg/mL	
					Staphylococcus epidermidis	MIC = 600 μg/mL	
Gnaphalium polycephalum Michx. (Compositae)	Siempreviva	Tunja, Boyacá	Ethanol extract (leaves)		*Staphylococcus aureus*	MIC = 500 μg/mL	Ramírez-Rueda and Mojica-Ávila (2014)
					Escherichia coli	MIC = 500 μg/mL	
Croton leptostachyus Kunth (Euphorbiaceae)	Mosquero	Alvarado, Tolima	Ethanol extract (leaves)	Antiparasitic	*Leishmania panamensis*	CI_{50} = 7.39 μg/mL	Neira, Stashenko, and Escobar (2014)
					Leishmania braziliensis	CI_{50} = 22.7 μg/mL	
Croton pedicellatus Kunth (Euphorbiaceae)	Ricino				*Leishmania panamensis*	CI_{50} = 7.14 μg/mL	
					Leishmania braziliensis	CI_{50} = 19.6 μg/mL	
Phyllanthus acuminatus Vahl (Phyllanthaceae)	Jobillo, barsbaco	Puerto López, Metazdsa			*Leishmania panamensis*	CI_{50} = 19.6 μg/mL	
					Leishmania braziliensis	CI_{50} = 91.2 μg/mL	
Piper holtonii C.DC. (Piperaceae)	Cordoncillo	Mesitas, Cundinamarca	Ethanol extract (aerial parts)	Antiparasitic	*Plasmodium falciparum* chloroquine-resistant strain	CI_{50} = 12.0 μg/mL	Rojas Cardozo, Garavito Cárdenas, and Rincón Velandia (2014)

Lippia alba Mill. N. E.Br. ex Britton & P.Wilson (Verbenaceae)	Prontoalivio	Different zones of Colombia	Essential oil	Antibacterial	*Salmonella gallinarum*	MIC = 10–50 μg/mL	Aristizabal and Marín (2012)
					Escherichia coli amoxicillin resistant	MIC = 10–50 μg/mL	
					Escherichia coli susceptible to amoxicillin	MIC = 50 μg/mL	
Lippia origanoides Kunth (Verbenaceae)	Orégano de monte				*Staphylococcus aureus*	MIC = 10–50 μg/mL	
					Salmonella typhimurium	MIC = 10–50 μg/mL	
					Bacillus cereus	MIC = 30–50 μg/mL	
					Salmonella gallinarum	MIC = 10–50 μg/mL	
					Amoxicillin-resistant *Escherichia coli*	MIC = 10–50 μg/mL	
					Escherichia coli susceptible to amoxicillin	MIC = 30–50 μg/mL	
Cananga odorata Lam. Hook.f. & Thomson (Annonaceae)	Flor de cananga				*Staphylococcus aureus*	MIC = 30–50 μg/mL	
					Salmonella typhimurium	MIC = 10–50 μg/mL	
					Bacillus cereus	MIC = 50 μg/mL	
					Salmonella gallinarum,	MIC = 10–50 μg/mL	

(*Continued*)

TABLE 1.2 (Continued)

Scientific name (family)	Common name	Collection place	Type of extract	Biological activity	Microorganism	Result	Reference
					Amoxicillin-resistant *Escherichia coli*	MIC = 10–50 μg/mL	
					Escherichia coli susceptible to amoxicillin	MIC = 30–50 μg/mL	
Tagetes lúcida Cav. (Compositae)	Pericón o yerbaanís				*Staphylococcus aureus*	MIC = 30–50 μg/mL	
					Salmonella typhimurium	MIC = 10–50 μg/mL	
					Bacillus cereus	MIC = 50 μg/mL	
					Salmonella gallinarum	MIC = 10–50 μg/mL	
					Amoxicillin-resistant *Escherichia coli*	MIC = 30–50 μg/mL	
					Escherichia coli susceptible to amoxicillin	MIC = 50 μg/mL	
Eucalyptus citriodora Hook. (Myrtaceae)	Eucalipto aromático	Pereira, Risaralda			*Staphylococcus aureus*	MIC = 10–50 μg/mL	
					Salmonella typhimurium	MIC = 10–50 μg/mL	
					Bacillus cereus	MIC 10–50 μg/mL	
					Salmonella gallinarum	MIC = 30–50 μg/mL	
					Amoxicillin-resistant *Escherichia coli*	MIC = 50 μg/mL	
					Escherichia coli susceptible to amoxicillin	MIC = 50 μg/mL	

Cymbopogon citratus DC. Stapf (Poaceae)	Limonaria				*Staphylococcus aureus*	MIC = 10–50 μg/mL	
					Salmonella typhimurium	MIC = 30–50 μg/mL	
					Bacillus cereus	MIC = 10–50 μg/mL	
					Salmonella gallinarum	MIC = 10–50 μg/mL	
					Amoxicillin-resistant *Escherichia coli*	MIC = 10–50 μg/mL	
					Escherichia coli susceptible to amoxicillin	MIC = 30–50 μg/mL	
Salvia rubriflora Epling (Lamiaceae)		Ethanol extract			*Bacillus cereus*	MIC = 50 μg/mL	
	Sigesbeckia agrestis Poepp (Compositae)	Ton-tzun			*Staphylococcus aureus*	MIC = 10–50 μg/mL	
					Bacillus cereus	MIC = 10–50 μg/mL	
Hyptis perbullata Fern. Alonso (Lamiaceae)					*Bacillus cereus*	MIC = 50 μg/mL	
Conobea scoparioides Cham. & Schltdl. Benth (Plantaginaceae)	Hierba de sapo	Chocó	Essential oil		*Staphylococcus aureus* ATCC 33591	MIC = 6.4 μg/mL	Zaraza Moncayo (2012)
					Bacillus cereus	MIC = 3.2 μg/mL	
					Salmonella typhimurium	MIC = 12.8 μg/mL	
					Pseudomonas aeruginosa	MIC = 16.7 μg/mL	
					Escherichia coli	MIC = 12.8 μg/mL	
Eucalyptus globulus Labill. (Myrtacea)	Eucalipto azul	Pamplona, Norte de Santander	Essential oil (leaves)		*Escherichia coli*	MIC = 34.2 μg/mL	Rueda and Mogollón (2012)

(*Continued*)

TABLE 1.2 (Continued)

Scientific name (family)	Common name	Collection place	Type of extract	Biological activity	Microorganism	Result	Reference
					Salmonella enteritidis	MIC = 57.5 μg/mL	
					Bacillus subtilis	MIC = 16.5 μg/mL	
					Enterococcus faecalis	MIC = 12.4 μg/mL	
					Staphylococcus aureus	MIC = 12.4 μg/mL	
Eucalyptus camaldulensis Dehnh. (Myrtaceae)	Eucalipto rojo				*Escherichia coli*	MIC = 85.6 μg/mL	
					Salmonella enteritidis	MIC = 51.3 μg/mL	
					Bacillus subtilis	MIC = 31.3 μg/mL	
					Enterococcus faecalis	MIC = 30 μg/mL	
					Staphylococcus aureus	MIC = 33.2 μg/mL	
Lippia origanoides Kunth (Verbenaceae)	Orégano de monte	Different zones of Colombia			*Salmonella gallinarum*	MIC = 64–1024 μg/mL	Ramírez, Isaza, Veloza, Stashenko, and Marín (2011)
					Salmonella typhimurium	MIC = 256–1024 μg/mL	
					Escherichia coli amoxicillin-resistant	MIC = 256–1024 μg/mL	
					Escherichia coli	MIC = 256–1024 μg/mL	
					Pseudomonas aeruginosa	MIC > 1024 μg/mL	
					Staphylococcus aureus	MIC = 64–1024 μg/mL	

					Bacillus cereus	MIC = 256–1024 μg/mL	
Piper cumanense Kunth (Piperaceae)	Cordoncillo	Zapatoca, Santander	Ethanol extract	Antiparasitic	*Leishmania panamensis*	CE = 382.9 μg/mL	Sánchez-Suárez et al. (2010)
Piper holtonii C.DC (Piperaceae)	Cordoncillo	La Mesa, Cundinamarca			*Leishmania panamensis*	CE = 280.9 μg/mL	
Rosmarinus officinalis L. (Lamiaceae)	Romero	Medellín, Antioquia	Essential oil (leaves)	Antibacterial	*Shigella sonnei*	MIC = 512 μg/mL	Castaño, Gelmy, Zapata, and Jiménez (2010)
					Salmonella typhimurium	MIC = 512 μg/mL	
Annona muricata L. (Annonaceae)	Guanábana	Anapoima, Cundinamarca	Ethanol extract (leaves)	Antiparasitic	*Trypanosoma cruzi*	IC_{50} = 10 μg/mL	Calderón et al. (2010)
					Plasmodium falciparum	IC_{50} = 20 μg/mL	
Calea peruviana Kunth Benth. ex S. F. Blake (Compositae)	Carrasposa	Sabana de Bogotá, Cundinamarca				IC_{50} = 22 μg/mL	
Calea jamaicensis L. (Compositae)		Cunday, Cundinamarca	Ethanol extract (aerial parts)		*Trypanosoma cruzi*	IC_{50} = 30 μg/mL	
Chromolaena leivensis Hieron. R. M. King & H. Rob. (Compositae)	Sanalo-todo	La Mesa. Cundinamarca	Ethanol extract (aerial parts)			IC_{50} = 8 μg/mL	
Critonia morifolia Mill. R.M. King & H. Rob. (Compositae)	Lengua de vaca	La Mesa, Cundinamarca	Ethanol extract (fruits)			IC_{50} = 29 μg/mL	

(*Continued*)

TABLE 1.2 (Continued)

Scientific name (family)	Common name	Collection place	Type of extract	Biological activity	Microorganism	Result	Reference
Critonia morifolia Mill. R.M. King & H. Rob. (Compositae)	Lengua de vaca	La Mesa, Cundinamarca	Ethanol extract (fruits)		*Plasmodium falciparum*	IC50 = 17 μg/mL	
Eirmocephala brachiata Benth. H. Rob. (Compositae)		Cundinamarca	Ethanol extract (fruits)		*Trypanosoma cruzi*	IC_{50} = 33 μg/mL	
						IC_{50} = 31 μg/mL	
Miconia buxifolia Naudin (Melastomataceae)	Gallinazo	La Mesa, Cundinamarca	Ethanol extract (leaves)			IC_{50} = 24 μg/mL	
Monochaetum myrtoideum Naudin (Melastomataceae)	Saltón	Vereda San Antonio, Cundinamarca	Ethanol extract (leaves)			IC_{50} = 5 μg/mL	
Fuchsia boliviana Carrière (Onagraceae)	Fucsia	Vereda de San Antonio, Cundinamarca	Ethanol extract (leaves)			IC_{50} = 20 μg/mL	
Bocconia integrifolia Bonpl. (Papaveraceae)	Llora sangre, mano de león	La Mesa, Cundinamarca	Ethanol extract (leaves)			IC_{50} = 23 μg/mL	
Cinchona pubescens Vahl (Rubiaceae)	Quina	La Mesa, Cundinamarca	Ethanol extract (leaves)			IC_{50} = 27 μg/mL	
Acnistus arborescens L. Schltdl. (Solanaceae)	Tomatoquina	San Francisco, Cundinamarca	Ethanol extract (leaves)			IC_{50} = 4 μg/mL	
Solanum cornifolium Dunal (Solanaceae)	Tinto	Bogotá, Cundinamarca	Ethanol extract (leaves)		*Plasmodium falciparum*	IC_{50} = 12 μg/mL	
Eugenia caryophyllata Thunb. (Myrtaceae)	Clavo de olor	Manizales, Caldas	Ethanol extract (seeds)	Antibacterial	*Clostridium perfringens*	MIC = 125 μg/mL	Ardila, Vargas, Pérez, and Mejía (2009)
Allium sativum L. (Amaryllidaceae)	Ajo		Hexane extract (fruits)		*Streptococcus mutans*	MIC = 500 μg/mL	

					Porphyromonas gingivalis	MIC = 500 μg/mL	
Lippia origanoides Kunth (Verbenaceae)	Orégano del monte	Pedregal, Nariño	Essential oil (leaves and stem)	Antibacterial	*Mycobacterium tuberculosis*	MIC = 125 μg/mL	Bueno-Sánchez, Martínez-Morales, Stashenko, and Ribón (2009)
		Los Santos, Santander				MIC 400 μg/mL	
		Piedecuesta Santander				MIC = 160 μg/mL	
Cananga odorata Lam. Hook.f. & Thomson. (Annonaceae)	Ylang-ylang	Bucaramanga, Santander	Essential oil (flowers)			MIC = 300 μg/mL	
Swinglea glutinosa Blanco. Mer. (Rutaceae)	Limón africano	Bucaramanga, Santander	Essential oil (fruits)			MIC = 100 μg/mL	
Hyptis mutabilis Rich. Briq. (Lamiaceae)	Mastranto	Villavicencio, Meta	Essential oil (leaves and stem)			MIC = 125 μg/mL	
Piper auritum Kunth. (Piperaceae)	Anisillo, hierba santa	Cali, Valle del Cauca	Essential oil (leaves)			MIC = 400 μg/mL	
Achyrocline alata Kunth. DC. (Compositae)	Vira-vira	Potosí, Nariño	Essential oil (leaves and stem)			MIC = 62.5 μg/mL	
Lippia alba Mill. N. E.Br. ex Britton & P. Wilson (Verbenaceae)	Pronto alivio	Venadillo, Tolima	Essential oil (leaves and stem)			MIC = 200 μg/mL	
		Bucaramanga. Santander				MIC = 130 μg/mL	
Piper bogotense C.D. C. (Piperaceae)	Matico	Ipiales, Nariño	Essential oil (leaves)			MIC = 130 μg/mL	

(*Continued*)

TABLE 1.2 (Continued)

Scientific name (family)	Common name	Collection place	Type of extract	Biological activity	Microorganism	Result	Reference
Virola calophylla Spruce. Warb. (Myristicaceae)	Cumala blanca	Puerto Berrio, Antioquia	Ethanol extract (stem)	Antibacterial	*Mycobacterium tuberculosis*	MIC = 128 μg/mL	Salazar et al. (2007)
Virola flexuosa A.C. Sm. (Myristicaceae)	Soto		Ethanol extract (leaves)			MIC = 128 μg/mL	
Piper sp. (Piperaceae)	Palo soldado		Ethanol extract (bark)			MIC = 128 μg/mL	
Dugandiodendron sp.(Magnoliaceae)	Alma negra		Ethanol extract (bark)			MIC = 128 μg/mL	
Odontocarya paupera Griseb. Diels. (Menispermaceae)	Pega-pega	Ciénaga-Barranquilla, Atlántico	Ethanol extract (leaves and stem)	Antibacterial	*Micrococcus luteus*	MIC = 200–400 μg/mL	Torres, Vizcaino, and Jurado (2007)
					Staphylococcus aureus	MIC = 200–400 μg/mL	
					Bacillus cereus	MIC = 200–400 μg/mL	
					Deinococcus radiophy	MIC = 200–400 μg/mL	
					Listeria monocytogenes	MIC = 200–400 μg/mL	
Piper sancti-felicis Trel.(Piperaceae)		Different regions of Colombia	Essential oil (leaves and stem)	Antifungal	*Candida krusei*	MIC = 125 μg/mL	Benítez et al. (2007)
Piper auduncum L. (Piperaceae)	Rodilla vieja					MIC = 250 μg/mL	
Piper spp. (Piperaceae)	Ricino					MIC = 250 μg/mL	
Piper bogotense C. DC. (Piperaceae)	Matico				*Aspergillus fumigatus*	MIC = 250 μg/mL	

Caryodendron orinocense H.Karst. (Euphorbiaceae)	Inchi	Medellín, Antioquia	Hexane extract (leaves)	Antiviral	Herpes simplex type 2	IC = 125 μg/mL	Arboleda, Cañas, López, and Forero (2007)
			Ethyl acetate extract (leaves)		Herpes simplex type 2	IC = 125 μg/mL	
			Hexane extract (leaves)		Bovine Herpes Virus type 1	IC = 62.5 μg/mL	
			Ethyl acetate extract (leaves)		Bovine Herpes Virus type 1	IC = 62.5 μg/mL	
Phyllanthus niruri L.(Phyllanthaceae)	Chanca piedra	Medellín, Antioquia	Hexane extract (aerial parts)		Herpes simplex type 2	IC = 250 μg/mL	
			Ethyl acetate extract (aerial parts)		Herpes simplex type 2	IC = 125 μg/mL	
			Hexane extract (aerial parts)		Bovine Herpes Virus type 1	IC = 250 μg/mL	
			Ethyl acetate extract (aerial parts)		Bovine Herpes Virus type 1	IC = 125 μg/mL	
Psidium guajava L. (Myrtaceae)	Guayaba	Moniquira and Raquira, Boyaca	Ripe peel	Antibacterial	*E. coli* enterotoxigenic	MIC > 400 μg/mL	Neira-González and Ramírez-González (2005)
					Streptococcus mutans	MIC > 400 μg/mL	
			Leaves		*E. coli* enterotoxigenic	MIC > 400 μg/mL	
Annona purpurea Moc. & Sessé ex Dunal (Annonaceae)	Cabezona, gallina gorda	Sucre and Cordoba	Ethanol extract (leaves)	Antiparasitic	*Leishmania panamensis*	ED_{50} = 0.96 μg/mL	Cárdenas, Lora, Márquez-Vizcaíno, and Blanco (2005)

(*Continued*)

TABLE 1.2 (Continued)

Scientific name (family)	Common name	Collection place	Type of extract	Biological activity	Microorganism	Result	Reference
Alternanthera williamsii Standl. Standl. (Amaranthaceae)	Sanguinaria, abrojo rojo.	Villa maría, Caldas	Leaves	Antibacterial	*Staphylococcus aureus*	MIC = 250 μg/mL	Perez et al. (2004)
Croton leptostachyus Kunth (Euphorbiaceae)	Mosquero, mosquerito	different regions of Colombia	Ethanol extract (aerial parts)	Antiparasitic	*Leishmania donovani*	IC_{50} = 37.5 μg/mL	Ruiz et al. (2004)
					Leishmania braziliensis	IC_{50} = 46.9 μg/mL	
					Leishmania amazonensis	IC_{50} = 37.5 μg/mL	
					Trypanosoma cruzi	IC_{50} = 37.5 μg/mL	
Acnistus arborescens L. Schltdl. (Solanaceae)	Tomatoquina, tabalgue			Antiparasitic	*Leishmania donovanni*	IC_{50} < 12.5 μg/mL	
					Leishmania braziliensis	IC_{50} < 12.5 μg/mL	
					Leishmania amazonensis	IC_{50} < 12.5 μg/mL	
					Trypanosoma cruzi	IC_{50} < 12.5 μg/mL	
Piper cumanense Kunth (Piperaceae)	Cordoncillo		Ethanol extracts (leaves and fruits)	Antiparasitic	*Leishmania donovanni*	IC_{50} < 12.5 μg/mL	
					Leishmania braziliensis	IC_{50} < 12.5 μg/mL	
					Leishmania amazonensis	IC_{50} < 12.5 μg/mL	
					Trypanosoma cruzi	IC_{50} < 12.5 μg/mL	
			Ethanol extract (aerial parts)	Antiparasitic	*Leishmania donovanni*	IC_{50} = 25 μg/mL	

					Leishmania braziliensis	$IC_{50} = 25\ \mu g/mL$
					Leishmania amazonensis	$IC_{50} = 25\ \mu g/mL$
					Trypanosoma cruzi	$IC_{50} = 41.7\ \mu g/mL$
Acacia farnesiana L. Willd. (Leguminosae)	Pela, cují, aromo, cují cimarrón, espinoso		Ethanol Extract (Leaves)	Antiparasitic	*Leishmania donovanni*	$IC_{50} < 12.5\ \mu g/mL$
					Leishmania braziliensis	$IC_{50} < 37.5\ \mu g/mL$
					Leishmania amazonensis	$IC_{50} < 25\ \mu g/mL$
					Trypanosoma cruzi	$IC_{50} < 12.5\ \mu g/mL$
			Ethanol extract (bark)	Antiparasitic	*Leishmania donovanni*	$IC_{50} = 52.8\ \mu g/mL$
					Leishmania braziliensis	$IC_{50} = 52.8\ \mu g/mL$
					Leishmania amazonensis	$IC_{5} = 52.8\ \mu g/mL$
					Trypanosoma cruzi	$IC_{50} = 18.8\ \mu g/mL$
Xylopia aromatica Lam. Mart. (Annonaceae)	Malaguita, anchón, fruta de burro, pepemato		Ethanol extract (aerial parts)	Antiparasitic	*Leishmania donovanni*	$IC_{50} = 41.7\ \mu g/mL$
					Leishmania braziliensis	$IC_{50} = 41.7\ \mu g/mL$
					Leishmania amazonensis	$IC_{50} = 41.7\ \mu g/mL$
					Trypanosoma cruzi	$IC_{50} < 12.5\ \mu g/mL$

(*Continued*)

TABLE 1.2 (Continued)

Scientific name (family)	Common name	Collection place	Type of extract	Biological activity	Microorganism	Result	Reference
Abuta grandifolia Mart. Sandwith. (Menispermaceae)	Vibuajeinia. Taque-taque	Amazonas	Ethanol extract (leaves and bark)	Antiparasitic	*Leishmania donovanni*	$IC_{50} > 100\ \mu g/mL$	
					Leishmania braziliensis	$IC_{50} > 100\ \mu g/mL$	
					Leishmania amazonensis	$IC_{50} > 100\ \mu g/mL$	
					Trypanosoma cruzi	$IC_{50} = 46.8\ \mu g/mL$	
Austroeupatorium inulaefolium Kunth R.M.King & H. Rob. (Compositae)	Salvia amarga	Caldas	Hexane extract (leaves and stem)	Antiparasitic	*Plasmodium falciparum*	$CL_{50} = 0.06\ \mu g/mL$	Blair et al. (2002)
Mangifera indica L. (Anacardiaceae)	Mango	Bogotá	Seeds	Antibacterial	*Staphylococcus aureus*	$MIC = 430\ \mu g/mL$	Galindo, Cárdenas, and Parroquiano (2002)
Vismia macrophylla Kunth. (Hypericaceae)	Carate	Caquetá, Putumayo, Vichada and Buenaventura	Resin	Antiviral	Herpes simplex HSV	$MIC = 5.5\ \mu g/mL$	Lopez, Hudson, and Towers (2001)
Symphonia globulifera L.f. (Clusiaceae)	Varillo, leche amarilla		Root bark			$MIC = 25\ \mu g/mL$	
Eschweilera rufifolia S.A.Mori (Lecythidaceae)		Carguero				$MIC = 8\ \mu g/mL$	
Byrsonima verbascifolia L. DC. (Malpighiaceae)		Guayabito				$MIC = 6.5\ \mu g/mL$	
			Leaves			$MIC = 2.5\ \mu g/mL$	
Iryanthera megistophylla A.C. Sm. (Myristicaceae)	Cabo de indio		Root bark			$MIC = 10\ \mu g/mL$	
Virola multinervia Ducke (Myristicaceae)	Ambil del monte		Resin			$MIC = 11.5\ \mu g/mL$	

			Root			MIC = 17 μg/mL	
Myrteola nummularia Lam. O.Berg (Myrtaceae)	Guayabilla		Aerial parts			MIC = 10.5 μg/mL	
Polygonum punctatum Elliott (Polygonaceae)	Picantillo					MIC = 20 μg/mL	
Adiantum latifolium Lam. (Pteridaceae)	Montañero					MIC = 11.5 μg/mL	
Ampelozizyphus amazonicus Ducke. (Rhamnaceae)	Saracura-mira		Leaves			MIC = 22 μg/mL	
Duroia hirsuta (Poepp.) K.Schum. (Rubiaceae)	Huitillo, turma de mono		Leaves			MIC = 10.5 μg/mL	
Campnosperma panamense Standl. (Anacardiaceae)	Sajo	Buenaventura and Cali	Dichloromethane extract (leaves)	Antiparasitic	*Plasmodium falciparum*	IC_{50} = 3 μg/mL	Weniger et al. (2001)
				Methanol extract (leaves)		IC_{50} = 15 μg/mL	
Guatteria amplifolia Triana & Planch. (Annonaceae)	Cargadero			Methanol extract (aerial parts)		IC_{50} = 1.9 μg/mL	
Aspidosperma megalocarpon Mull. Arg (Apocynaceae)	Costillo acanalado			Methanol extract (roots)		IC_{50} = 25 μg/mL	
Tabernaemontana obliqua Miers Leeuwenb. (Apocynaceae)	Mierda de guagua			Methanol extract (leaves)		IC_{50} = 25 μg/mL	

(*Continued*)

TABLE 1.2 (Continued)

Scientific name (family)	Common name	Collection place	Type of extract	Biological activity	Microorganism	Result	Reference
Jacaranda caucana Pittier (Bignoniaceae)	Gualanday		Methanol extract (leaves)			$IC_{50} = 14\ \mu g/mL$	
Huberodendron patinoi Cuatrec. (Malvaceae)	Carrá		Methanol extract (roots)			$IC_{50} = 3\ \mu g/mL$	
Protium amplum Cuatrec (Burseraceae)	Anime		Dichloromethane extract fruits			$IC_{50} = 32\ \mu g/mL$	
Marila laxiflora Rusby (Calophyllaceae)	Aceitillo		Dichloromethane (leaves)			$IC_{50} = 20\ \mu g/mL$	
Guarea guidonia L. Sleumer (Meliaceae)	Cedro macho		Dichloromethane extract (seeds)			$IC_{50} = 10\ \mu g/mL$	
Otoba novogranatensis Moldenke (Myristicaceae)	Otobo		Methanol extract (leaves)			$IC_{50} = 20\ \mu g/mL$	
Swinglea glutinosa Blanco Merr. (Rutaceae)	Swinglea		Dichloromethane extract (root)			$IC_{50} = 2.5\ \mu g/mL$	
			Methanol extract (root)			$IC_{50} = 10\ \mu g/mL$	
Annona muricata L. (Annonaceae)	Guanabana	Antioquia	Hexane extract (seeds)	Antiparasitic	*Leishmania panamensis*	$DL_{50} = 2.1\ \mu g/mL$	Arango, Dickson, Vélez, and Muñoz (2000)
Callisia grasilis Kunth D.R.Hunt. (Commelinaceae)	Crespinillo	Valle del Cauca	Ethanol extract (stem)	Antiviral	Herpes simplex HSV	$EC_{50} = 10.5\ \mu g/mL$	Betancur-Galvis, Saez, Granados, Salazar, and Ossa (1999)
Annona spp. (Annonaceae)	Guanabanito	Montería, Córdoba	Methanol extract (seeds)			$EC_{50} = 0.049\ \mu g/mL$	
Phyllanthus acuminatus Vahl. (Phyllanthaceae)	Jobillo, barsbaco	Villeta, Cundinamarca	Aerial parts	Antifungal	*Zygorrinchus* sp.	$MIC = 312\ \mu g/mL$	de García, Sanchez, and Gómez (1995)

					Penicillium sp.	MIC = 312 μg/mL	
					Aspergillus fumigatus	MIC = 312 μg/mL	
Ageratina pichinchensis (Kunth) R. M. King & H. Rob. (Compositae)	Hierba de cuy	Cáqueza, Cundinamarca	Aerial parts	Antibacterial	*Staphylococcus epidermidis*	MIC = 289 μg/mL	Galindo, Sarmiento, and Rodríguez (1995)
					Staphylococcus aureus	MIC = 186 μg/mL	
					Mycobacterium sp.	MIC = 190 μg/mL	
Raimondia quinduensis (Kunth) Saff. (Annonaceae)	Anón de monte	Santandercito, Cundinamarca	Fraction rich in alkaloids		*Mycobacterium fortuitum*	MIC = 62.5 μg/mL	
					Brucella spp.	MIC = 125 μg/mL	
					Bacillus anthracis	MIC = 125 μg/mL	
					Bacillus subtilis	MIC = 250 μg/mL	
					Mariniluteicoccus flavus	MIC = 125 μg/mL	
					Micrococcus luteus	MIC = 125 μg/mL	
					Streptococcus pneumoniae	MIC = 500 μg/mL	
					Staphylococcus epidermis	MIC = 125 μg/mL	
					Staphylococcus aureus	MIC = 500 μg/mL	

(Adoxaceae), *Furcraea cabuya* (Asparagaceae) among other species found in the Andean region are used for the treatment of the common flu, pneumonia and whooping cough.

Table 1.2 shows a list of *in vitro* antimicrobial studies of 98 plant species, collected in Colombia, and distributed in the following 41 families: Adoxaceae (1), Alliaceae (1), Amaranthaceae (1), Anacardiaceae (2), Annonaceae (8), Apocynaceae (3), Aquifoliaceae (1), Asteraceae (12), Bignoniaceae (1), Bombacaceae (1), Boraginaceae (1), Burseraceae (1), Caesalpiniaceae (1), Clusiaceae (4), Commelinaceae (1), Cucurbitaceae (2), Euphorbiaceae (7), Fabaceae (2), Labiaceae (1), Lamieaceae (6), Lecythidaceae (1), Magnoliaceae (1), Malpighiaceae (1), Melastomataceae (1), Melastomataceae (2), Meliaceae (1), Menispermaceae (2), Myristicaceae (5), Myrtaceae (6), Onagraceae (1), Papaveraceae (1), Piperaceae (9), Poaceae (1), Polygonaceae (2), Rhamnaceae (1), Rubiaceae (2), Rutaceae (2), Scrophulariaceae (1), Solanaceae (5), Urticaceae (1), Verbenaceae (3), Zingiberaceae (1).

From our literature review, we present here a summary of the anti-infectious activity of the 98 plants selected.

Finally, as it is shown in Table 1.2, six plants were tested for their antifungal activity, but none of them presented an MIC value $\leq$ 100 μg/mL. So, we did not discuss these species as we consider that no promising antifungal extracts were found.

Table 1.3 summarizes a part of the partial results that we have been obtained so far. According to these results, 23% of the extracts evaluated, namely *Ludwigia erecta* (L.) H. Hara (Leaves), *Ludwigia helminthorriza* (whole plant), *Maclura tinctoria* (Leaves), *Mammea americana* (episperm), *Mammea americana* (Leaves), *Mammea americana* (seeds) , *Sagittaria intermedia* (seeds), and *Ludwigia leptocarpa* (Leaves) showed good activity against sensitive and resistant strains of *Staphylococcus aureus*.

On the other hand, by correlating the results of Table 1.1 and Table 1.2, we were able to establish that only twelve of the plants traditionally used in Colombia to treat infections have at least one biological evaluation study that partially supports the scientific validation of their traditional use. These plants are the following: *Annona muricata* (L.) (Annonaceae), *Austroeupatorium inulaefolium* (Kunth) R.M.King & H.Rob., (Compositae), *Cinchona pubescens* Vahl (Rubiaceae), *Cymbopogon citratus* (DC.) Stapf (Poaceae), *Eucalyptus globulus* Labill. (Myrtaceae), *Lippia alba* (Mill.) N.E.Br. ex Britton & P.Wilson (Verbenaceae), *Momordica charantia* (L.) (Cucurbitaceae), *Phyllanthus niruri* (L.) (Phyllanthaceae), *Sambucus nigra* (L.) (Adoxaceae), *Solanum lycopersicum* (L.) (Solanaceae), Xylopia aromatica (Lam.) Mart. (Annonaceae).

The review of the 43 documents allowed us to know that the extracts were prepared by solid/liquid extraction processes, using maceration with solvents of different polarities, hot extraction using Soxhlet, and microwave hydrodistillation to obtain essential oils. Furthermore, the microorganisms evaluated, sensitive or resistant, were identified both from registered strains and from clinical isolates.

From our literature review, we present here a summary of the anti-infectious activity of the 98 plants selected.

Antibacterial activity

Ten plants exhibited good antibacterial activity (defined as an MIC of 100 μg/mL or less), and five showed promising antibacterial activity (defined as an MIC of 5 μg/mL or less). The latter are further discussed:

Otholobium mexicanum (L.f.) J.W. Grimes. (Fabaceae)

O. mexicanum, native to the Neotropics, presented MIC of 0.1 μg/mL against *Staphylococcus aureus* (Romero & Munévar, 2019). Compounds such as nonacosane, octacosane, heptacosane and hexatriacontane, bakuchiol, 3-hydroxybakuchiol, and two isoflavone glycosides, daidzin and genistin have been isolated from this plant (Jáuregui, 2014).

Cucurbita moschata Duchesne (Cucurbitaceae)

C. moschata is a native plant of China and India but cultivated in Colombia in all thermal ranges. It presented a MIC of 0.16 μg/mL against methicillin-resistant *Staphylococcus aureus* (Pereira, M. Moscarella, C. Urbina, & Martínez, 2017). Sterol-type compounds such as spinasterol, stigmasterol and β-sitosterol and oxygenated tetracyclic triterpenes called cucurbitacins have been isolated from this species (Salama, 2006).

Cymbopogon citratus (DC.) Stapf (Poaceae)

This plant is from Indian and Mediterranean origin and is cultivated in Colombia. It presented a MIC of 1 μg/mL against *Streptococcus mutans* (Cuadros et al., 2018). Terpenic compounds such as geranial, neral, and β-myrcene have been isolated from this species (Quintanilla et al., 2012).

Conobea scoparioides (Cham. & Schltdl.) Benth (Scrophulariaceae)

C. scoparioides is native to the Neotropics; this species presented a MIC of 3.2 μg/mL against *Bacillus cereus* (Moncayo, 2012). Thymol and methyl thymol have been isolated from this plant, as main compounds of the essential oil and oxygenated tetracyclic triterpenes such as cucurbitacin E (Mina & Montaño, 2010).

Rosmarinus officinalis Govaerts. (Lamiaceae)

Rosmarinus officinalis is cultivated in Colombia (Mediterranean origin), and it presented a MIC of 5.2 μg/mL against *Staphylococcus aureus* (Ortíz-Ardila et al., 2017). Terpenoidal compounds such as α-pinene, 1,8-cineole, camphor, verbenone, borneol, eucalyptol, β-caryophyllene, bornyl acetate, camphene, β-pinene, limonene, phelandrene, and γ-terpinene have been isolated from this plant (María, Miguel, & Leonardo, 2018; Bonilla et al., 2016).

Antiparasitic activity

Thirty-one species of plants exhibited good antiparasitic activity ($MIC \leq 20$ μg/mL), of which nine showed promising biological activity ($MIC \leq 5$ μg/mL):

Miconia theaezans (Bonpl.) Cogn. (Melastomataceae)

M. theaezans is a Colombian endemic species, whose chloroform and ethanol extracts presented LC_{50} values equal to 0.56 and 0.63 μg/mL against *Plasmodium falciparum*, respectively (Sosa Puerto, 2019).

Annona purpurea Dunal (Annonaceae)

A. purpurea is native to the Neotropics, presented an ED_{50} value equal to 0.96 μg/mL against *Leishmania panamensis* (Cárdenas et al., 2005). Different types of compounds have been isolated from this plant, among which some acetogenins, terpenoids, alkaloids, megastigmanans, lectins, flavonoids, essential oils, cyclopeptides, and fatty acids stand out (Luna-Cazáres & González-Esquinca, 2015; Muñoz-Acevedo et al., 2016).

Guatteria amplifolia Triana & Planch. (Annonaceae)

This species has an altitudinal distribution range of 45–1350 m in the Andean and Pacific regions and is distributed from Mexico to Colombia. It presented an IC_{50} value of 1.9 μg/mL against *Plasmodium falciparum* (Weniger et al., 2001).

Annona muricata Linn. (Annonaceae)

A. muricata, native to the Neotropics, presented an LD_{50} of 2.1 μg/mL against *Leishmania panamensis* (Arango et al., 2000).

Austroeupatorium inulifolium (Kunth) R.M. King & H. Rob. (Compositae)

A. inulifolium, native to South America, presented an LC_{50} of 2.27 μg/mL against *Plasmodium falciparum* (Sosa Puerto, 2019). Compounds of the flavonoid type such as 5,6,3′-trihydroxy-7,4′-dimethoxyflavone, pedalitin, and terpenoids such as β-caryophyllene, α-caryophyllene, germacrene D, δ-elemene, limonene, patchoulene, and viridiflorol have been isolated from this plant (Cárdenas, Mejía, & Cárdenas, 2013).

Campnosperma panamense Standl. (Anacardiaceae)

C. panamense can be found in South America, and it presented an IC_{50} of 3 μg/mL against *Plasmodium falciparum* (Weniger et al., 2001). Phytochemical studies show abundance of triterpenes, sterols, and polar flavonoids in this plant species (Weniger et al., 2010).

Huberodendron patinoi Cuatrec. (Bombacaceae)

H. patinoi, a native plant of Colombia and Ecuador, presented an IC_{50} value of 3 μg/mL against *Plasmodium falciparum* (Weniger et al., 2001). Xylopine was isolated from this species, a compound that has a high activity against parasites of the *Leishmania* genus (Fernández-Alonso, 2002).

Monochaetum myrtoideum (Bonpl.) (Melastomataceae) and *Acnistus arborescens* Linn. Schltdl. (Solanaceae)

Two of the plant species native to Latin America found in Colombia are *Monochaetum myrtoideum* and *Acnistus arborescens*, which showed promising activities against *Trypanosoma cruzi* with IC_{50} values of 5 μg/mL and IC_{50} of 4 μg/mL, respectively (Calderón et al., 2010).

Swinglea glutinosa Merr (Rutaceae)

S. glutinosa is cultivated in Colombia (Asian origin) presented an IC_{50} value of 2.5 μg/mL against *Plasmodium falciparum* (Weniger et al., 2010). From this species, β-pinene and nerolidol have been identified as the most abundant compounds in the essential oil of this plant (Cam & Le, 2012).

Antiviral activity

Thirteen plants exhibited good antiviral activity (MIC ≤ 25 μg/mL), of which three stood out with a promising antiviral activity value (MIC ≤ 5.5 μg/mL):

Annona sp. (Annonaceae)

This plant species presented an EC_{50} value of 0.0059 μg/mL against Herpes simplex virus HSV (Betancur-Galvis et al., 1999).

Byrsonima verbascifolia L. DC (Malpighiaceae)

B. verbascifolia is a native plant of the Neotropics and its leaves presented a MIC of 2.5 μg/mL (Lopez et al., 2001). The isolation of flavonoids (flavanones, biflavonoids, catechins and epicatechins and proanthocyanidins), derivatives of gallic acid and quinic acid and triterpenes have been reported from the genus *Byrsonima* (Guilhon-Simplicio & De Meneses Pereira, 2011).

Vismia macrophylla Kunth. (Clusiaceae)

V. macrophylla is a native plant of the Neotropics and its resin presented a MIC of 5.5 μg/mL against *Herpes simplex* HSV (Lopez et al., 2001). Some secondary metabolites have been isolated from this plant species, such as terpenoids, lignans, flavonoids, and anthracenic compounds such as anthrones and xanthones (Rojas, Buitrago, Rojas, & Morales, 2011).

Finally, as it is shown in Table 1.2, six plants were tested for their antifungal activity, but none of them presented an MIC value ≤ 100 μg/mL. So, we did not discuss these species as we consider that no promising antifungal extracts were found.

Antibacterial evaluation of 25 native plants of the Colombian Caribbean region against strains of *Escherichia coli*, *Klebsiella pneumonia*, and *Pseudomonas aeruginosa* resistant to carbapenems and *Staphylococcus aureus* resistant to methicillin (Colciencias, Project code: 110777757752)

The objective of this research is to evaluate the antibacterial activity of 25 native plants of the Colombian Caribbean region against *Escherichia coli*, *Klebsiella pneumonia*, and *Pseudomonas aeruginosa* resistant to carbapenems and methicillin-resistant *Staphylococcus aureus*. Thirty of the 75 ethanol extracts proposed in the project have been evaluated against ATCC reference strains for bacteria (Table 1.3). Table 1.3 summarizes a part of the partial results that we have been obtained so far. According to these results, 23% of the extracts evaluated, namely

TABLE 1.3 Growth inhibition (%) of *S. aureus*, *K. pneumoniae*, *P. aeruginosa*, *E. coli* (antibiotic-resistant and sensitive strains), produced by 30 plant extracts found in the Colombian Caribbean region.

	Staphylococcus aureus		*Klebsiella pneumoniae*		*Pseudomonas aeruginosa*		*Escherichia coli*	
	ATCC 29213	**ATCC 43300**	**ATCC 700603**	**ATCC BAA 1705**	**ATCC 27853**	**ATCC BAA-2108**	**ATCC 25922**	**ATCC BAA-2452**
Plant species (Plant organ)	**Mean ± SEM**[a]		**Mean ± SEM**		**Mean ± SEM**		**Mean ± SEM**	
Cochlospermum vitifolium (fruits)	73.8 ± 9.2	64.3 ± 3.0	68.9 ± 11.8	14.5 ± 2.4	81.2 ± 2.1	34.7 ± 3.8	61.3 ± 1.7	43.9 ± 1.6
Croton malambo (stem bark)	52.7 ± 5.5	18.9 ± 5.2	0	0	17.1 ± 3.0	0	3.2 ± 2.9	0
Dicliptera sexangularis (leaves)	0	0	18.7 ± 2.4	7.9 ± 4.3	0	28.0 ± 4.5	0	36.8 ± 1.9
Dicliptera sexangularis (stem)	60.2 ± 6.9	40.8 ± 1.3	13.9 ± 3.2	0.9 ± 2.6	17.2 ± 1.9	13.6 ± 7.2	4.6 ± 4.9	45.0 ± 2.0
Echinodorus paniculatus (leaves)	37.6 ± 9.9	32.3 ± 0.6	0	0	71.4 ± 0.7	41.7 ± 1.6	33.6 ± 1.6	31.0 ± 2.0
Echinodorus paniculatus (stem)	0	2.1 ± 5.9	0	2.3 ± 2.7	39.2 ± 1.0	0	0	0
Echinodorus tunicatus (leaves)	44.4 ± 1.7	19.4 ± 11.3	0	0	13.2 ± 8.1	0	0	0
Echinodorus tunicatus (seeds)	60.1 ± 8.2	22.8 ± 1.1	0	0	30.4 ± 1.3	33.4 ± 11.0	0	6.0 ± 0.1
Echinodorus tunicatus (stem)	23.0 ± 3.9	17.5 ± 0.9	0	0	0	5.6 ± 25.6	0	28.6 ± 25.6
Eleocharis elegans (leaves)	21.9 ± 9.9	17.3 ± 1.6	0	0	30.0 ± 1.0	0	0	0
Eleocharis elegans (flowers)	61.9 ± 0.9	63.7 ± 5.7	51.8 ± 6.9	26.3 ± 5.1	45.2 ± 2.1	42.2 ± 1.5	0	39.7 ± 1.7
Ludwigia erecta (stem)	23.1 ± 1.8	12.2 ± 0.7	26.0 ± 4.8	0.4 ± 1.8	22.4 ± 2.5	9.0 ± 34.3	39.2 ± 4.7	21.7 ± 34.3
Ludwigia erecta (leaves)	84.9 ± 2.1	67.8 ± 7.2	19.7 ± 1.4	20.1 ± 1.0	58.5 ± 1.4	55.9 ± 5.1	28.1 ± 2.0	52.3 ± 0.5
Ludwigia erecta (roots)	61.6 ± 4.7	46.1 ± 7.5	42.2 ± 0.5	32.2 ± 3.8	54.5 ± 1.1	43.5 ± 5.8	40.2 ± 2.5	54.1 ± 0.4
Ludwigia helminthorriza (whole plant)	96.4 ± 1.6	79.0 ± 10.7	41.3 ± 6.6	25.6 ± 1.2	48.1 ± 1.9	9.7 ± 2.0	0	11.8 ± 1.6
Ludwigia leptocarpa (leaves)	87.0 ± 3.6	46.0 ± 1.4	73.2 ± 7.7	23.6 ± 3.8	89.2 ± 2.6	47.2 ± 3.5	67.6 ± 1.9	50.9 ± 5.7
Ludwigia leptocarpa (roots)	35.5 ± 2.7	29.3 ± 2.1	29.5 ± 2.2	12.9 ± 2.4	33.5 ± 2.7	22.4 ± 17.3	42.8 ± 2.0	33.8 ± 17.3
Ludwigia leptocarpa (stem)	40.6 ± 1.5	34.5 ± 1.4	48.8 ± 4.1	22.2 ± 1.9	64.2 ± 1.8	37.8 ± 16.7	54.8 ± 2.5	38.3 ± 16.7
Maclura tinctoria (stem bark)	52.6 ± 0.6	31.8 ± 0.5	22.8 ± 6.2	0	21.5 ± 1.1	2.9 ± 6.0	14.4 ± 5.2	17.7 ± 2.0
Maclura tinctoria (leaves)	97.4 ± 11.5	100.0 ± 10.9	41.2 ± 10.2	16.0 ± 3.3	100.0 ± 0.9	61.7 ± 5.8	60.7 ± 1.9	44.0 ± 3.7
Mammea americana (episperm)	71.4 ± 4.3	75.0 ± 1.0	46.7 ± 5.5	26.9 ± 3.6	38.9 ± 1.4	0	37.2 ± 1.6	18.8 ± 5.2

(*Continued*)

TABLE 1.3 (Continued)

Plant species (Plant organ)	*Staphylococcus aureus* ATCC 29213	*Staphylococcus aureus* ATCC 43300	*Klebsiella pneumoniae* ATCC 700603	*Klebsiella pneumoniae* ATCC BAA 1705	*Pseudomonas aeruginosa* ATCC 27853	*Pseudomonas aeruginosa* ATCC BAA-2108	*Escherichia coli* ATCC 25922	*Escherichia coli* ATCC BAA-2452
	Mean ± SEM[a]		Mean ± SEM		Mean ± SEM		Mean ± SEM	
Mammea americana (leaves)	76.4 ± 2.1	92.9 ± 2.2	23.5 ± 7.0	21.0 ± 2.7	38.0 ± 1.7	24.4 ± 3.5	25.0 ± 2.4	33.7 ± 0.7
Mammea americana (seeds)	100.0 ± 1.7	100.0 ± 0.8	28.8 ± 2.2	0	75.8 ± 0.7	29.5 ± 3.0	41.6 ± 0.8	17.8 ± 2.0
Nymphaea novogranatensis (whole plant)	50.7 ± 5.8	21.5 ± 7.0	28.2 ± 6.0	0	47.6 ± 3.0	30.5 ± 3.8	37.3 ± 0.8	0
Sagittaria intermedia (leaves)	37.0 ± 4.2	36.1 ± 2.1	0	0	0	0	17.9 ± 1.4	0
Sagittaria intermedia (seeds)	92.8 ± 2.5	83.7 ± 5.1	22.7 ± 8.5	24.3 ± 0.5	3.6 ± 7.3	29.9 ± 2.1	43.9 ± 1.4	3.5 ± 2.1
Sagittaria lancifolia (leaves)	16.3 ± 3.3	8.1 ± 3.1	1.3 ± 2.5	3.2 ± 2.7	21.4 ± 3.7	0	12.9 ± 2.9	0
Sagittaria lancifolia (roots)	47.8 ± 2.1	35.9 ± 3.6	5.9 ± 3.0	0	0	0	0	30.7 ± 0.8
Salvinia spp. (whole plant)	14.3 ± 4.7	7.9 ± 1.9	0	0	0	0	0	48.5 ± 1.1
Utricularia foliosa (whole plant)	66.8 ± 4.9	28.9 ± 1.4	0	0	39.5 ± 0.7	0	1.2 ± 4.2	0

[a]*SEM: Standard Error of the Mean. The results are expressed according to criteria established in our research group (LIFFUC) based on the reviewed literature; the extracts were evaluated at 512 μg/mL (50 μg/mL for M. americana-seeds against Staphylococcus aureus) and were classified according to their activity as follows: (a) between 0% and 24% inhibition of bacterial growth is considered as null antibacterial activity; (b) between 25% and 49% is considered as low antibacterial activity; (c) between 50% and 74% is considered as moderate antibacterial activity; (d) between 75% and 100% is considered as good antibacterial activity.*

Ludwigia erecta (L.) H. Hara (Leaves), *Ludwigia helminthorriza* (whole plant), *Maclura tinctoria* (Leaves), *Mammea americana* (episperm), *Mammea americana* (Leaves), *Mammea americana* (seeds), *Sagittaria intermedia* (seeds), and *Ludwigia leptocarpa* (Leaves) showed good activity against sensitive and resistant strains of *Staphylococcus aureus.*

Thirteen percent of the extracts evaluated, namely, *Cochlospermum vitifolium* (bark and fruits), *Ludwigia leptocarpa* (leaves), *Maclura tinctoria* (leaves), and *Mammea americana* (seeds) showed good activity against the sensitive strain of *P. aeruginosa* bacteria. *Cochlospermum vitifolium* fruit extracts, *Maclura tinctoria* leaves, *Ludwigia erecta* leaves, and *Ludwigia leptocarpa* leaves can be considered as broad-spectrum extracts due to their moderate (to good) antibacterial activity against sensitive and resistant strains. This makes them promising extracts for bioguided fractionation and the isolation of active ingredients against Gram-positive and Gram-negative bacteria.

The extracts of the plants *Mammea americana* and *Sagittaria intermedia* showed a certain degree of selectivity against Gram-positive bacteria, since it can be observed that they do not have activity against Gram-negative strains.

Mammea americana L. (Calophyllaceae)

Mammea americana is a tree probably native to the Antillas Menores and naturalized in the rest of South America in prehistoric times. It is known as mamey tree, albaricoque de santo domingo, albaricoque, mamao, mamey amarillo, mamey de cartagena, mamey de santo domingo, mamey dominicano, mameyo, martín sapote, mata serrano, zapote de santo domingo, zapote de niño, zapote domingo, zapote mame, zapote mamey, in the different countries where it is located. In traditional medicine, this plant is best known for its insecticidal properties. Yasunaka et al., in 2005, evaluated the antibacterial activity of a methanolic extract of seeds against strains of *Staphylococcus aureus* FDA 209P sensitive to methicillin (MIC = 2 μg/mL) and strains no. 3208 and no. 80401 resistant to methicillin (MIC = 4–8 μg/mL). In addition, the activity against *Escherichia coli* K12 (MIC = 256 μg/mL) was evaluated in the same study, and the authors isolated mammea A/AA, the active compound with an MIC of 8 μg/mL against the three strains of *S. aureus*. Canning, Sun, Ji, Gupta, and Zhou (2013) also reported the antibacterial activity of this compound against *Campylobacter jejuni* (MIC = 0.25 μg/mL), *Clostridium difficile* (MIC = 0.25 μg/mL), and *Streptococcus pneumoniae* (MIC = 0.25 μg/mL). Herrera Herrera, Franco Ospina, Fang, and Díaz Caballero (2014) evaluated the antibacterial activity of the oily and ethanolic phase of an extract of *Mammea americana* seeds against *Streptococcus mutans* (ATCC 25175) and *Porphyromonas gingivalis* (ATCC 33277), bacteria of the oral cavity, and they found that both extracts had higher activity (MIC = 15.6 μg/mL for the oil phase and 62.5 μg/mL for the ethanol phase), against *S. mutans*, than against *P. gingivalis* (MIC = 250 and 500 μg/mL, respectively) (Canning et al., 2013; Herrera Herrera et al., 2014; Yasunaka et al., 2005).

In a screening of antibacterial activity against *S. aureus* carried out with ethanol extracts of plants from the Colombian Caribbean region in our laboratories, the extract from mamey seeds was found to be the most active with a MIC of 2 μg/mL.

Maclura tinctoria L. D.Don ex Steud. (Moraceae)

The organic extract of the plant *Maclura tinctoria* exhibited moderate anti-HIV activity. Macluraxanthones B and C showed the best potential anti-HIV, with EC_{50} values of 1–2.2 μg/mL; however, they exhibited very high toxicity toward the CEM-SS host cells, with IC_{50} values of 2.2–17 μg/mL (Groweiss, Cardellina, & Boyd, 2000). Chalcones such as 2′,4′,4,2″-tetrahydroxy-[3′-methylbut-3″-enyl]-chalcone and 2′,4′,4-trihydroxy-3′-[3″-methylbut-3″-enyl]chalcone (isobavachalcone) showed inhibitory activity against *C. albicans* (IC_{50} of 15 and 3 mg/mL, respectively) and *C. neoformans* (IC_{50} of 7 mg/mL) (ElSohly, Joshi, Nimrod, Walker, & Clark, 2001). The stem and bark extracts of *M. tinctoria* were active against *Streptococcus sanguinis* (ATCC 10556), *Streptococcus mitis* (ATCC 49456), and *Streptococcus mutans* (ATCC 25175) at concentrations in the range of 80–400 μg/mL (Lamounier et al., 2012). Morin hydrate (also called 2′,3,4′,5,7-pentahydroxyflavone), a flavonoid found in *Maclura pomifera* (osage orange), *Maclura tinctoria* (old fustic), and in the leaves of *Psidium guajava* (common guava), was found to inhibit the self-assembly of the heptameric transmembrane pore of hemolysin-α (a staphylococcal toxin) in mouse model of pneumonia, and thus to inhibit hemolytic activity (Wang et al., 2015). A fraction enriched with flavonoids (prenylated flavonoids) from the extract of *M. tinctoria* leaves showed a MIC between 10 and 40 μg/mL against *S. aureus* (Das Chagas Almeida et al., 2019). Compound 5,7,3′,4′-tetrahydroxy-6,8-diprenylisoflavone obtained from *Maclura tinctoria* leaves extracts have antileishmanial action. It showed IC_{50} values against *L. infantum* and *L. amazonensis* of 6.36 and 2.45 μg/mL (for promastigotes), and 2.70 and 1.12 μg/mL (for amastigotes), respectively. Also, this compound was effective in the treatment of infected macrophages and caused alterations in the parasite mitochondria. In *L. infantum*-infected mice, it also showed a significant reduction in the parasite load in distinct organs, when compared to the control groups (Pereira et al., 2020). In our screening for antibacterial activity against *S. aureus*, the extract from the leaves of *Maclura tinctoria* was one of the most active extract with a MIC of 64 μg/mL.

Conclusions

Of the 317 plants selected when making the bibliographic review, it was found that the vast majority of them, despite having at least one report of traditional use in Colombia, do not have biological evaluation studies against pathogenic microorganisms that contribute to validate the traditional use that they are credited.

On the other hand, by correlating the results of Table 1.1 and Table 1.2, we were able to establish that only twelve of the plants traditionally used in Colombia to treat infections have at least one biological evaluation study that partially supports the scientific validation of their traditional use. These plants are the following: *Annona muricata* (L.) (Annonaceae), *Austroeupatorium inulaefolium* (Kunth) R.M.King & H.Rob., (Compositae), *Cinchona pubescens* Vahl (Rubiaceae), *Cymbopogon citratus* (DC.) Stapf (Poaceae), *Eucalyptus globulus* Labill. (Myrtaceae), *Lippia alba* (Mill.) N.E.Br. ex Britton & P.Wilson (Verbenaceae), *Momordica charantia* (L.) (Cucurbitaceae), *Phyllanthus niruri* (L.) (Phyllanthaceae), *Sambucus nigra* (L.) (Adoxaceae), *Solanum lycopersicum* (L.) (Solanaceae), Xylopia aromatica (Lam.) Mart. (Annonaceae).

In the same way, it could be established that of the plant species *Allium sativum* (L.) (Amaryllidaceae), *Annona purpurea* Moc. & Sessé ex Dunal (Annonaceae), *Caryodendron orinocense* H.Karst. (Euphorbiaceae), *Jacaranda caucana* Pittier (Bignoniaceae), *Heliotropium indicum* (L.) (Boraginaceae), *Lippia alba* (Mill.) N.E.Br. ex Britton & P.Wilson (Verbenaceae), *Mammea americana* (L.) (Calophyllaceae), *Mangifera indica* (L.) (Anacardiaceae), *Piper aduncum* (L.) (Piperaceae), evaluated by Colombian research groups using bioassays against pathogenic microorganisms, none presented reports in the reviewed literature on some traditional use in Colombia.

References

Alvarez, G., Pabón, A., Carmona, J., & Blair, S. (2004). Evaluation of clastogenic potential of the antimalarial plant *Solanum nudum*. *Phytotherapy Research, 18*, 845–848. Available from https://doi.org/10.1002/ptr.1534.

Arango, E. M., Carmona, J., & Blair, S. (2008). In vitro susceptibility of Colombian *Plasmodium falciparum* isolates to three *Solanum nudum* Dunal (Solanaceae) steroids. *Vitae, 15*, 150–156.

Arango, G. J., Dickson, J. S., Vélez, I. D., & Muñoz, D. L. (2000). Actividad Leishmanicida de Annonacina, Aislada de Annona muricata contra Leishmania panamensis. *Vitae, 7*(1).

Arboleda, D., Cañas, A. L., López, A., & Forero, J. E. (2007). Evaluación de la actividad antiviral in vitro de cuatro extractos de las especies Caryodendron orinocense y Phyllanthus niruri de la familia Euphorbiaceae contra los virus Herpes Bovino tipo 1 y Herpes Simplex tipo 2. *Vitae, 14*(1), 55–60.

Ardila, M., Vargas, A., Pérez, J., & Mejía, L. (2009). Ensayo preliminar de la actividad antibacteriana de extractos de *Allium sativum, Coriandrum sativum, Eugenia Caryophyllata, Origanum vulgare, Rosmarinus officinalis* y *Thymus vulgaris* frente a *Clostridium perfringens*. *Biosalud, 8*(3), 47–57.

Arias, R. F. Q. (2016). *Medicina tradicional en la comunidad de San Basilio de Palenque.*

Aristizabal, L. S. R., & Marín, D. (2012). Evaluación de la actividad antibacteriana de aceites esenciales y extractos etanólicos utilizando métodos de difusión en agar y dilución en pozo. *Scientia et Technica, 17*(50), 152–157.

Barriga, G. (1986). Un movimiento en pro de las Ciencias Botánicas. *Caldasia, 15*, 41–46.

Barrios, E. P. L., Vidal, A. L. P., Morales, P. A. M., Trujillo, S. B., Jaramillo, C. A. P., & Marín, P. A. (2018). Actividad antiplasmodial in vitro de metabolitos secundarios de Solanum nudum provenientes de dos regiones de Colombia. *Revista EIA, 15*(30), 25–39.

Benítez, N. P., Bueno, J. G., Zapata, B., Stashenko, E. E., Arango, A. M., Montiel, J., & Martínez, C. (2007). Actividad in vitro anti-candida y anti-aspergillus de aceites esenciales de plantas de la familia piperaceae. *Scientia et technica, 1*(33), 247–249.

Bernal, H.Y., García, H., Quevedo, G. (2011). Pautas para el conocimiento, conservación y uso sostenible de las plantas medicinales nativas en Colombia. Estrategia Nacional para la conservación de plantas.

Bernal, R., Gradstein, R., & Celis, M. (2016). *Catalogo de plantas y Líquenes de Colombia* (Vol. 1). Universidad Nacional de Colombia.

Betancur-Galvis, L. A., Saez, J., Granados, H., Salazar, A., & Ossa, J. E. (1999). Antitumor and antiviral activity of Colombian medicinal plant extracts. *Memórias do Instituto Oswaldo Cruz, 94*(4), 531–535. Available from https://doi.org/10.1590/S0074-02761999000400019.

Blair, S., Mesa, J., Correa, A., Carmona-Fonseca, J., Granados, H., & Sáez, J. (2002). Antimalarial activity of neurolenin B and derivates of Eupatorium inulaefolium (Asteraceae). *Die Pharmazie, 57*(6), 413.

Borja-Acosta, K.G., Diaz, A., Murillo-Bedoya, D., Acevedo-Charry, O., DoNascimiento, C., Lozano-Flórez, J., ... Gómez-Posada, C. (2019), Conocimiento e innovación en las colecciones biológicas del Instituto Humboldt | Biodiversidad.

Bueno-Sánchez, J. G., Martínez-Morales, J. R., Stashenko, E. E., & Ribón, W. (2009). Anti-tubercular activity of eleven aromatic and medicinal plants occurring in Colombia. *Biomédica, 29*(1), 51–60.

Calderón, A. I., Romero, L. I., Ortega-Barría, E., Solís, P. N., Zacchino, S., Gimenez, A., ... Gupta, M. P. (2010). Screening of Latin American plants for antiparasitic activities against malaria, Chagas disease, and leishmaniasis. *Pharmaceutical Biology, 48*(5), 545–553. Available from https://doi.org/10.3109/13880200903193344.

Cam, T. & Le, T. (2012). Evaluación de la composición química y actividad antioxidante de la hoja y del fruto de Swinglea Glutinosa (Ph.D. dissertation).

Canning, C., Sun, S., Ji, X., Gupta, S., & Zhou, K. (2013). Antibacterial and cytotoxic activity of isoprenylated coumarin mammea A/AA isolated from *Mammea africana*. *Journal of Ethnopharmacology, 147*, 259–262. Available from https://doi.org/10.1016/j.jep.2013.02.026.

Cárdenas, A., Mejía, G. I., & Cárdenas, J. E. P. (2013). Especies vegetales investigadas por sus propiedades antimicrobianas, inmunomoduladoras e hipoglucemiantes en el departamento de Caldas (Colombia-Sudamérica). *Revista Biosalud Manizales, 12*, 59–82.

Cárdenas, D. L., Lora, J. A., Márquez-Vizcaíno, R. L., & Blanco, P. J. (2005). Actividad leishmanicida de Annona purpurea. *Actualidades biológicas, 27*(1), 35–37.

Cardona-Arias, J. A., Rivera-Palomino, Y., & Carmona-Fonseca, J. (2015). Expresión de la interculturalidad en salud en un pueblo emberá-chamí de Colombia. *Revista Cubana de Salud Pública, 41*.

Castaño, H. I., Gelmy, C., Zapata, J. E., & Jiménez, S. L. (2010). Actividad bactericida del extracto etanólico y del aceite esencial de hojas de *Rosmarinus officinalis* L. sobre algunas bacterias de interés alimentario. *Vitae, 17*(2), 149–154.

Cervantes Ceballos, L., Sánchez Hoyos, F., & Gómez Estrada, H. (2017). Antibacterial activity of *Cordia dentata* Poir, *Heliotropium indicum* Linn and *Momordica charantia* Linn from the Northern Colombian Coast. *Revista Colombiana de Ciencias Químico-Farmacéuticas, 46*(2), 143–159.

Cuadros, M. O., Rivera, A. P. T., Merini, L. J., & Pabon, M. C. M. (2018). Antimicrobial activity of *Cymbopogon citratus* (Poaceae) essential oil on *Streptococcus mutans* biofilm and cytotoxic effect on keratinocytes and fibroblasts. *Revista de Biología Tropical, 66*(4), 1519–1529.

da Costa, M. P., Bozinis, M. C. V., Andrade, W. M., Costa, C. R., da Silva, A. L., Alves de Oliveira, C. M., ... Silva, M. d. R. R. (2014). Antifungal and cytotoxicity activities of the fresh xylem sap of *Hymenaea courbaril* L. and its major constituent fisetin. *BMC Complementary and Alternative Medicine, 14*. Available from https://doi.org/10.1186/1472-6882-14-245.

Das Chagas Almeida, A., Azevedo Rodrigues, L., Dos Santos Paulino, G., Pereira Aguilar, A., Andrade Almeida, A., Olavo Ferreira, S., ... De Oliveira Barros Ribon, A. (2019). Prenylated flavonoid-enriched fraction from *Maclura tinctoria* shows biological activity against *Staphylococcus aureus* and protects *Galleria mellonella* larvae from bacterial infection. *BMC Complementary and Alternative Medicine, 19*. Available from https://doi.org/10.1186/s12906-019-2600-y.

de García, C. L. G., Sanchez, Y., & Gómez, E. (1995). Actividad antimicrobiana de Phyllanthus acuminatus Vahl. *Revista Colombiana de Ciencias Químico-Farmaceúticas, 24*(1), 35–39.

del Castillo Pereira, A., Molinares Moscarella, P., Campo Urbina, M., & Bettin Martínez, A. (2017). Actividad antibacteriana del extracto total de hojas de *Cucurbita moschata* Duchesne (Ahuyama). *Revista Cubana de Plantas Medicinales, 22*(1).

Duarte Velandia, L. V. (2020) (s. f.). Determinación de la actividad antibacteriana y fitotoxicidad de los aceites esenciales de Anís (*Pimpinella anisum*) y Romero (*Rosmarinus officinalis*).

Duke, J. A. (1970). Ethnobotanical observations on the Chocó Indians. *Economic Botany, 24*(3), 344–366.

ElSohly, H. N., Joshi, A. S., Nimrod, A. C., Walker, L. A., & Clark, A. M. (2001). Antifungal chalcones from *Maclura tinctoria. Planta Medica, 67*, 87–89. Available from https://doi.org/10.1055/s-2001-10621.

Fernández-Alonso, J. L. (2002). Bombacaceae neotropicae novae vel minus cognitae III. Nuevas especies de Matisia y Quararibea de Colombia. *Novon, 12*, 343–351. Available from https://doi.org/10.2307/3393077.

Fonnegra Gómez, R., Villa Londoño, J., & Monsalve Fonnegra, Z. I. (2012). Plantas usadas como medicinales en el Altiplano del Oriente antioqueño-Colombia. Universidad de Antioquia. <https://issuu.com/herbariohua/docs/plantas_usadas_como_medicinales_en>.

Fonnegra, F.G. (2007). *Plantas medicinales aprobadas en Colombia*.

Galindo, A. S., Cárdenas, L. A., & Parroquiano, M. L. (2002). Actividad antimicrobiana y examen fitoquímico preliminar de siete angiospermas y una muestra de propóleo. *Revista Colombiana de Ciencias Químico-Farmacéuticas, 31*(1).

Galindo, A. S., Sarmiento, G. R., & Rodríguez, M. A. (1995). Actividad antifúngica y antibacteriana de Ageratina ibaguensis. *Revista Colombiana de Ciencias Químico-Farmacéuticas, 23*(1), 58–63.

García-Huertas, P., Pabón, A., Arias, C., & Blair, S. (2013). Evaluación del efecto citotóxico y del daño genético de extractos estandarizados de Solanum nudum con actividad anti-Plasmodium. *Biomedica: Revista del Instituto Nacional de Salud, 33*, 78–87. Available from https://doi.org/10.7705/biomedica.v33i1.838.

Groweiss, A., Cardellina, J. H., & Boyd, M. R. (2000). HIV-Inhibitory prenylated xanthones and flavones from *Maclura tinctoria. Journal of Natural Products, 63*, 1537–1539. Available from https://doi.org/10.1021/np000175m.

Guilhon-Simplicio, F., & De Meneses Pereira, M. (2011). Aspectos químicos e farmacológicos de Byrsonima (Malpighiaceae). *Quimica Nova, 34*, 1032–1041. Available from https://doi.org/10.1590/S0100-40422011000600021.

Herrera Herrera, A., Franco Ospina, L., Fang, L., & Díaz Caballero, A. (2014). Susceptibility of *Porphyromonas gingivalis* and *Streptococcus mutans* to antibacterial effect from *Mammea americana. Advances in Pharmacological Sciences, 2014*. Available from https://doi.org/10.1155/2014/384815.

Hidalgo, C., Forero &. Velasco, M. (2016). La etnobotánica y su importancia como herramienta para la articulación entre conocimientos ancestrales y científicos.

Hirschhorn, H. H. (1981). Botanical remedies of south and central America, and the Caribbean: An archival analysis. Part I. *Journal of Ethnopharmacology, 4*(2), 129–158.

Jaramillo Gómez, A. (2003). Plantas medicinales en los jardines de las veredas Mancilla, La Tribuna, Pueblo viejo y Tierra morada (Facatativá Cundinamarca). Pontificia Universidad Javeriana.

Jáuregui. (2014). Aislamiento y caracterización de compuestos activos anti-helicobacter pylori a partir de la planta Otholobium Mexicanum (LF) JW Grimes., Aislamiento y Caracterización de Compuestos Activos Anti-Helicobacter Pylori a Partir de La Planta Otholobium Mexicanum (L. F.) J. W. Grimes. 16. <http://repositorio.upch.edu.pe/handle/upch/1323?locale-attribute = en>. (Accessed 16 January 2021).

Jiménez-Escobar, N. D., & Estupiñán-González, A. C. (2011). Useful trees of the Caribbean region of Colombia. *Bioremediation, Biodiversity and Bioavailability, 5*(1), 65–79.

Laferriere, J. E. (1994). Medicinal plants of the lowland Inga people of Colombia. *International Journal of Pharmacognosy, 32*(1), 90–94.

Lamounier, K. C., Cunha, L. C. S., De Morais, S. A. L., De Aquino, F. J. T., Chang, R., Do Nascimento, E. A., ... Cunha, W. R. (2012). Chemical analysis and study of phenolics, antioxidant activity, and antibacterial effect of the wood and bark of *Maclura tinctoria* (L.) D. Don ex Steud. *Evidence-Based Complementary and Alternative Medicine, 2012*. Available from https://doi.org/10.1155/2012/451039.

Londoño, B., Arango, E., Zapata, C., Herrera, S., Saez, J., Blair, S., & Carmona-Fonseca, J. (2006). Effect of *Solanum nudum* Dunal (Solanaceae) steroids on hepatic trophozoites of *Plasmodium vivax*. *Phytotherapy Research, 20*, 267–273. Available from https://doi.org/10.1002/ptr.1849.

Lopez, A., Hudson, J. B., & Towers, G. H. N. (2001). Antiviral and antimicrobial activities of Colombian medicinal plants. *Journal of Ethnopharmacology, 77*(2–3), 189–196. Available from https://doi.org/10.1016/S0378-8741(01)00292-6.

López-Palacios, S. (1986). Lista preliminar de las Verbenaceae existentes en Colombia con algunos de sus usos y nombres vulgares. *Caldasia, 15*(71/75), 155–176.

Lucena, M. E., Contreras, M. E., Moreno, V. G., Rojas-Fermín, L., de Rojas, Y. C., Fajardo, F. J. U., ... Torres, S. (2019). Composición y actividad antibacteriana del aceite esencial de austroeupatorium inulifolium (Kunth) king & robinson (asteraceae). *Revista Cubana de Farmacia, 52*, 1–16. Available from http://revfarmacia.sld.cu/index.php/far/article/view/369/272.

Luna-Cazáres, L. M., González-Esquinca, A. R. (2015). La chincuya (*Annona purpurea* Moc. & Sessé ex Dunal): Una planta mesoamericana.

Pereira, Gustavo Araujo, Araujo, Nayara Macêdo Peixoto, Arruda, Henrique Silvano, de Paulo Farias, David, Molina, Gustavo, & Pastore, Glaucia Maria (2019). Phytochemicals and biological activities of mutamba (*Guazuma ulmifolia* Lam.): A review. *Food Research International, 126*, 108713.

Bonilla, D. M., Yulitza, M., E., M. C., Ozkarina, M., Ángela, P. M., Jairo, C., ... Lina, N. (2016). Efecto del aceite esencial de Rosmarinus officinalis sobre Porphyromonas gingivalis cultivada in vitro. *Revista Colombiana de Ciencias Químico-Farmacéuticas, 45*(2), 275. Available from https://doi.org/10.15446/rcciquifa.v45n2.59942.

Madawala, S., Chandrasiri, I., Diwakara, S., Wijesundara, S., & Karunaratne, V. (2014). Bioactivities of invasive plant *Austroeupatorium inulifolium*.

María, P. A. A., Miguel, R. L. L., & Leonardo, R. C. J. (2018). Identificación de componentes químicos del aceite esencial de romero (*Rosmarinus officinalis* L.) proveniente de cultivos orgánicos en la zona alta andina. *Revista Colombiana de Investigaciones Agroindustriales*, 6–15. Available from https://doi.org/10.23850/24220582.658.

Martin, F., Hay, A. E., Condoretty, V. R. Q., Cressend, D., Reist, M., Gupta, M. P., ... Hostettmann, K. (2009). Antioxidant phenylethanoid glycosides and a neolignan from *Jacaranda caucana*. *Journal of Natural Products, 72*, 852–856. Available from https://doi.org/10.1021/np900038j.

Martin, F., Quinteros, V., Hay, A. E., Gupta, M. P., & Hostettmann, K. (2008). New phenylethanoid glycosides from *Jacaranda caucana* (Bignoniaceae). *Planta Medica, 74*.

Martínez Vega, R. A., Barreto Dos Santos, F., Galvão de Araujo, J. M., Joint, G., Sarti, E., & Ramos Castañeda, J. (2017). Antiviral effect of compounds derived from the seeds of *Mammea americana* and *Tabernaemontana cymosa* on Dengue and Chikungunya virus infections. *PLoS Neglected Tropical Diseases*.

Martínez, S., Chiguasuque, M. N., & Casallas, R. (2006). Ziscagoscua. Manual médico para la comunidad indígena Muisca de Bosa.

Merchán. (2017). Caracterización del mercado colombiano de plantas medicinales y aromáticas.

Mesa-S, L. M., Santamaría, M., García, H., & Aguilar-Cano, J. (2016). Catálogo de biodiversidad de la región caribe. Volumen 3. Serie Planeación ambiental para la conservación de la biodiversidad en áreas operativas de Ecopetrol. Proyecto Planeación ambiental para la conservación de la biodiversidad en las áreas operativas de Ecopetrol (Primera). Instituto de Investigación de Recursos Biológicos Alexander von Humboldt. <http://observatorio.epacartagena.gov.co/ftp-uploads/pub-Vol3-Caribe_baja.pdf>.

Mina, R. T. G., & Montaño, A. M. H. (2010). Primeros ensayos para el cultivo y caracterización del aceite esencial de Conobea scoparioides, (Cham. & Schltdl.) Benth. Para El Pacífico Colombiano. Entramado (pp. 24–35).

Moncayo. (2012). Actividad antibacteriana del aceite esencial de la conobea scopariodes frente a cinco cepas bacterianas de interés clínico en Colombia.

Monzote, L., Jiménez, J., Cuesta-Rubio, O., Márquez, I., Gutiérrez, Y., da Rocha, C. Q., ... Vilegas, W. (2016). In vitro assessment of plants growing in Cuba belonging to Solanaceae family against *Leishmania amazonensis*. *Phytotherapy Research, 30*, 1785–1793. Available from https://doi.org/10.1002/ptr.5681.

Morais, S. M., Calixto-Júnior, J. T., Ribeiro, L. M., Sousa, H. A., Silva, A. A. S., Figueiredo, F. G., ... Coutinho, H. D. M. (2017). Phenolic composition and antioxidant, anticholinesterase and antibiotic-modulating antifungal activities of *Guazuma ulmifolia* Lam. (Malvaceae) ethanol extract. *South African Journal of Botany, 110*, 251–257. Available from https://doi.org/10.1016/j.sajb.2016.08.003.

Munera, J. D. (2005). Inventario de plantas medicinales del municipio de Arauca. Retrieved from <https://www.academia.edu/26165022/INVENTARIO_DE_PLANTAS_MEDICINALES_DEL_MUNICIPIO_DE_ARAUCA?auto = download>.

Muñoz-Acevedo, A., Aristizábal-Córdoba, S., Rodríguez, J. D., Torres, E. A., Molina, A. M., Guitiérrez, R. G., & Kouznetsov, V. V. (2016). Citotoxicidad/capacidad antiradicalaria in-vitro y caracterización estructural por GC-MS/1H-13C-NMR de los aceites esenciales de hojas de árboles joven/adulto de *Annona purpurea* Moc. & Sessé ex Dunal de Repelón (Atlántico, Colombia). *Boletin Latinoamericano y del Caribe de Plantas Medicinales y Aromaticas, 15*, 99–111. Available from http://www.blacpma.usach.cl/images/docs/015-002/005_articulo_4.pdf.

Neira, L. F., Stashenko, E., & Escobar, P. (2014). Actividad antiparasitaria de extractos de plantas colombianas de la familia Euphorbiaceae. Revista de la Universidad Industrial de Santander. *Salud, 46*(1), 15–22.

Neira-González, A. M., & Ramírez-González, M. B. (2005). Actividad antibacteriana de extractos de dos especies de Guayaba contra *Streptococcus matans* y *Escherichia coli*. *Actualidades biológicas, 27*(1), 27–30.

Ogura, M., Cordell, G. A., & Farnsworth, R. (1977). Potential anticancer agents. IV. Constituents of *Jacaranda caucana* Pittier (Bignoniaceae). *Lloydia, 40*, 157–168.

Ortega David, E. H. (2015). Usos tradicionales de las plantas de la Orinoquia colombiana. *UG-Ciencia, 21*, 16–28.

Ortiz Gómez, F. (1989). Botánica médica Guahibo. Plantas medicinales, mágicas y psicotrópicas utilizadas por los Sikuani y Cuiba (Llanos orientales de Colombia). *Caldasia, 16*(76). Available from https://revistas.unal.edu.co/index.php/cal/article/view/35456/35833.

Ortíz-Ardila, A. E., Correa-Cuadros, J. P., Celis-Zambrano, C. A., Rodríguez-Bocanegra, M. X., Robles-Camargo, J., & Sequeda-Castañeda, L. G. (2017). Antioxidant and antimicrobial capacity *Cecropia mutisiana* Mildbr. (Cecropiaceae) leave extracts. *Emirates Journal of Food and Agriculture, 29*, 25–35. Available from https://doi.org/10.9755/ejfa.2016-07-915.

Overgaard, H. J., Alexander, N., Mátiz, M. I., Jaramillo, J. F., Olano, V. A., Vargas, S., ... Stenström, T. A. (2012). Diarrhea and dengue control in rural primary schools in Colombia: Study protocol for a randomized controlled trial. *Trials, 13*. Available from https://doi.org/10.1186/1745-6215-13-182.

Parra Garzón, C.A. (2017). (s. f.). Actividad Antimicrobiana y Caracterización Química del Aceite Esencial de Ulex Europaeus L. (Fabaceae).

Parra Sepúlveda, S.F. (2020). (s. f.). Estudio de la composición química y actividad biológica de aceites esenciales de Cardamomo (*Elettaria cardamomum*) y Tomillo (*Thymus vulgaris*).

Pava, C. N. R., Sanabria, A. G. Z., & Leal, L. C. S. (2017). Actividad antimicrobiana de cuatro variedades de plantas frente a patógenos de importancia clínica en Colombia. *Nova, 15*(27), 119–129.

Pereira, I. A. G., Mendonça, D. V. C., Tavares, G. S. V., Lage, D. P., Ramos, F. F., Oliveira-da-Silva, J. A., ... Gonçalves, D. U. (2020). Parasitological and immunological evaluation of a novel chemotherapeutic agent against visceral leishmaniasis. *Parasite Immunology, 42*. Available from https://doi.org/10.1111/pim.12784.

Pereira, M., Moscarella, P., Urbina, C., & Martínez, M. (2017). Antibacterial activity of total extract from leaves of *Cucurbita moschata* Duchesne (ahuyama). In: Revista Cubana de Plantas Medicinales.

Pérez, D., & Matiz-Guerra, L. C. (2017). Uso de las plantas por comunidades campesinas en la ruralidad de Bogotá D.C., Colombia. *Caldasia, 39*, 68–78. Available from https://doi.org/10.15446/caldasia.v39n1.59932.

Perez, J. E., Mejía, G. I., Bueno, J. G., Arango, M. C., Hincapié, B. L., Nieto, A. M., & Londoño, D. P. (2004). Efecto de los extractos de *Phenax rugosus*, *Tabebuia chrysantha*, *Althernantera illiamsii* y *Solanum dolichosepalum* sobre el leucograma y la producción de anticuerpos en ratas. *Revista médica de Risaralda, 10*(2), 3.

Perez-Arbelaez, E. (1978). Plantas útiles de Colombia (Cuarta). Litografía Arco.

Portela Quintero, L.C. (2020) (s. f.). Evaluación de la actividad antimicrobiana in vitro de extractos vegetales de Ilex guayusa e Ilex paraguariensis frente a *Helicobacter pylori*.

Puerta Domínguez, M. A., & Vargas Hernández, M. A. (2016). Efecto antibacteriano de extracto acuoso de solanum lycopersicum sobre streptococcus mutans y porphyromonas gingivalis (Ph.D. Thesis). Universidad de Cartagena.

Quintana Arias, R. F. (2012). Estudio de plantas medicinales usadas en la comunidad indígena Tikuna del alto Amazonas, Macedonia. *Nova, 10*(18), 181–193.

Quintana Arias, R. F. (2016). Medicina tradicional en la comunidad de San Basilio de Palenque. *Nova, 14*(25), 67–93.

Quintanilla, R. R., Carlos, R., Moyano, G. A., Salazar, H. C., Martínez, J., & Stashenko, E. (2012). Estudio comparativo de la composición de los aceites esenciales de cuatro especies del género Cymbopogon (Poaceae) cultivadas en Colombia. In: Boletín Latinoamericano y Del Caribe de Plantas Medicinales y Aromáticas (pp. 77–85).

Quintero, T., Sanchez, C. F. G., & Estupiñan, L. E. (2018). Análisis del uso tradicional de plantas medicinales que se comercializan en Bogotá, Colombia; un abordaje desde las ciencias ambientales.
Ramírez Romero, V., & Prieto Munévar, A. (2019). Evaluación del halo de inhibición de la planta *Otholobium mexicanum* frente a Staphylococcuss aureaus ATCC 25923.
Ramírez, L. S., Isaza, J. H., Veloza, L. Á., Stashenko, E., & Marín, D. (2011). Actividad antibacteriana de aceites esenciales de Lippia origanoides de diferentes orígenes de Colombia. *Ciencia*, *17*(4).
Ramirez, Z. (1994). El aprendizaje de las plantas en la senda de un conocimiento olvidado.
Ramírez-Rueda, R. Y., & Mojica-Ávila, D. N. (2014). Actividad antibacteriana de extractos de *Gnaphalium polycephalum* Michxcontra *Staphilococcus aureus*, *Escherichia coli* y *Pseudomonas aeruginosa*. Revista Investigación en Salud Universidad de Boyacá (pp. 63–71).
Raz, L., & Zamora, H. A. (2020). Catálogo de Plantas y Líquenes de Colombia. <https://doi.org/10.15472/7avdhn>.
Restrepo, J. J. A. (2011). Plantas aromáticas y medicinales: enfermedades de importancia y sus usos terapéuticos: medidas para la temporada invernal. ICA, Instituto Colombiano Agropecuario.
Rojas Cardozo, M. A., Garavito Cárdenas, G., & Rincón Velandia, J. (2014). Actividad antiplasmódica in vitro de *Piper holtonii* (cordoncillo). *Revista Cubana de Plantas Medicinales*, *19*(1), 69–75.
Rojas, J., Buitrago, A., Rojas, L., & Morales, A. (2011). Essential oil composition of *Vismia macrophylla* leaves (Guttiferae). *Natural Product Communications*, *6*(1). Available from https://doi.org/10.1177/1934578x1100600120.
Romero, M., Cabrera, E., & Ortiz, N. (2006). Informe sobre el estado de la biodiversidad en Colombia.
Romero, R., & P. Munévar (2019). Repositorio—Universidad de Ciencias Aplicadas y Ambientales UDCA: Evaluación del halo de inhibición de la planta *Otholobium mexicanum* frente a *Staphylococcuss aureaus* ATCC 25923.
Rueda, X. Y., & Mogollón, O. F. C. (2012). Composición química y actividad antibacteriana del aceite esencial de las especies *Eucalyptus globulus* y *E. camaldulensis* de tres zonas de Pamplona (Colombia). *Bistua: Revista de la Facultad de Ciencias Básicas*, *10*(1), 52–61.
Rugeles, L., Ortiz, J., Huertas, A., & Guaitero, B. (2012). La cadena de valor de los ingredientes naturales del Biocomercio en las industrias farmacéutica, alimentaria y cosmética-FAC.
Ruiz, P. G., Garavito, G., Acebey, C. L., Arteaga, L., Pinzon, R., & Gimenez, T. A. (2004). Actividad leishmanicida y tripanocida de algunas plantas reportadas como medicinales en Colombia. *Biofarbo*, *12*(12), 27–30.
Saez, J., Cardona, W., Espinal, D., Blair, S., Mesa, J., Bocar, M., & Jossang, A. (1998). Five new steroids from Solanum nudum. *Tetrahedron*, *54*, 10771–10778. Available from https://doi.org/10.1016/S0040-4020(98)00632-2.
Salama. (2006). Las Cucurbitáceas. Importancia económica, bioquímica y medicinal.
Salazar, E. B., Benavides, J., Sepulveda, L., Quiñones, W., Torres, F., Cardona, D., ... Franzblau, S. (2007). Actividad antimicobacteriana de algunas plantas de la flora colombiana. *Scientia et Technica*, *13*(33), 133–136.
Sánchez-Suárez, J., Albarracín, D., Rojas, M., Rincón, J., Robledo, S., Muñoz, D. L., ... Delgado, G. (2010). Evaluation of the cytotoxic and leishmanicidal activity of extracts and fractions of *Piper cumanense* and *Piper holtonii*. *Revista Colombiana de Ciencias Químico-Farmacéuticas*, *39*(1), 21–29.
Schultes, R. E., & Raffauf, R. F. (1990). The healing forest medicinal and toxic plants of the northwest Amazonia (T. R. Dudley, Ed.). Dioscorides Press.
Shah, G., Shri, R., Panchal, V., Sharma, N., Singh, B., & Mann, A. S. (2011). Scientific basis for the therapeutic use of *Cymbopogon citratus*, stapf (Lemon grass). *Journal of Advanced Pharmaceutical Technology and Research*, *2*, 3–8. Available from https://doi.org/10.4103/2231-4040.79796.
Sharifi-Rad, J., Hoseini-Alfatemi, S. M., Sharifi-Rad, M., Sharifi-Rad, M., Iriti, M., Sharifi-Rad, M., ... Raeisi, S. (2015). Phytochemical compositions and biological activities of essential oil from *Xanthium strumarium* L. *Molecules (Basel, Switzerland)*, *20*, 7034–7047. Available from https://doi.org/10.3390/molecules20047034.
Solano-Aguilar, G., Fernandez, K. P., Ets, H., Molokin, A., Vinyard, B., Urban, J. F., & Gutierrez, M. F. (2013). Evaluación de la actividad antiplasmodial de las partes aéreas de plantas colectadas en Colombia. (s. f.) *Journal of Pediatric Gastroenterology and Nutrition*, *56*.
Sosa Puerto, W. E. (2019) (s. f.). *Evaluación de la actividad antiplasmodial de las partes aéreas de plantas colectadas en Colombia* (Ph.D. thesis). Universidad Nacional de Colombia-Sede Bogotá. <https://repositorio.unal.edu.co/handle/unal/77076>.
Torrenegra-Alarcón, M., Granados-Conde, C., Durán-Lengua, M., León-Méndez, G., Yáñez-Rueda, X., Martínez, C., & Pájaro-Castro, N. (2016). Composición química y actividad antibacteriana del aceite esencial de Minthostachys mollis. *Orinoquia*, *20*(1), 69–74.

Torres, C. D. L. R., Vizcaino, R. L. M., & Jurado, C. L. (2007). Determinacion química actividad antimicotica y antimicrobiana de odontocarya paupera. *Scientia et Technica*, *13*(33), 427–429.

Vásquez, C. (2014). Avances en la investigación sobre plantas medicinales. En: Bello et al. (ed). *Biodiversidad 2014. Estado y tendencias de la biodiversidad continental en Colombia*. Instituto Alexander von Humboldt. Bogotá D.C., Colombia. (2014).

Vera Marín, B., & Sánchez Sáenz, M. (2015). Registro de algunas plantas medicinales cultivadas en San Cristóbal, municipio de Medellín (Antioquia-Colombia). *Revista Facultad Nacional de Agronomía Medellín*, *68*(2), 7647–7658.

Vera Olave, K. T. (2007). Estado de conocimiento científico de las plantas medicinales usadas en dos veredas del mucipio de Vélez, Santander [Pontificia Universidad Javeriana]. <https://repository.javeriana.edu.co/bitstream/handle/10554/33756/VeraOlaveKarolTatiana2017.pdf?sequence = 4&isAllowed = y>.

Wang, J., Zhou, X., Liu, S., Li, G., Shi, L., Dong, J., ... Niu, X. (2015). Morin hydrate attenuates *Staphylococcus aureus* virulence by inhibiting the self-assembly of α-hemolysin. *Journal of Applied Microbiology*, *118*, 753–763. Available from https://doi.org/10.1111/jam.12743.

Weniger, B., Seri, G., Stiebing, S., Kerhuel, A., Collot, V., Schmitt, M., ... Vonthron-Sénécheau, C. (2010). Antiplasmodial evaluation and pharmacomodulation of lanaroflavone, a biflavonoid isolated from *Campnosperma panamense* Standl. (Anacardiaceae). *Planta Medica*. Available from https://doi.org/10.1055/s-0030-1264718.

Weniger, B., Robledo, S., Arango, G. J., Deharo, E., Aragón, R., Muñoz, V., ... Anton, R. (2001). Antiprotozoal activities of Colombian plants. *Journal of Ethnopharmacology*, *78*(2–3), 193–200. Available from https://doi.org/10.1016/S0378-8741(01)00346-4.

Yang, B., Wang, F., Cao, H., Liu, G., Zhang, Y., Yan, P., & Li, B. (2017). Caffeoylxanthiazonoside exerts cardioprotective effects during chronic heart failure via inhibition of inflammatory responses in cardiac cells. *Experimental and Therapeutic Medicine.*, *14*, 4224–4230. Available from https://doi.org/10.3892/etm.2017.5080.

Yasunaka, K., Abe, F., Nagayama, A., Okabe, H., Lozada-Pérez, L., López-Villafranco, E., ... Reyes-Chilpa, R. (2005). Antibacterial activity of crude extracts from Mexican medicinal plants and purified coumarins and xanthones. *Journal of Ethnopharmacology*, *97*, 293–299. Available from https://doi.org/10.1016/j.jep.2004.11.014.

Zaraza Moncayo, Z. (2012). *Actividad antibacteriana del aceite esencial de la conobea scopariodes frente a cinco cepas bacterianas de interés clínico en Colombia* (Ph.D. thesis). Universidad del Rosario.

CHAPTER

2

Plants used in Lebanon and the Middle East as Antimicrobials

Roula M. Abdel-Massih[1] *and Marc El Beyrouthy*[2]

[1]Department of Biology, Faculty of Arts and Sciences, University of Balamand, El-Koura, Lebanon [2]Department of Agricultural Sciences, Holy Spirit University of Kaslik, Beirut, Lebanon

Introduction: overview on medicinal plants and their traditional uses as antimicrobials in Lebanon

The overuse or misuse of antimicrobial agents or food preservatives and the lack of development of new drugs in the pharmaceutical industry led to a growing incidence of microbial resistance worldwide. Around 80% of antibiotics in the US are used as growth supplements or to prevent infection in livestock and hence the spread of resistance to humans (Bartlett, Gilbert, & Spellberg, 2013). Annual deaths due to antimicrobial resistance continue to increase worldwide and are projected to be over 10 million by 2050 (O'Neill, 2016). There is a current clear trend to turn to nature and plants as a safe alternative for new antimicrobial agents.

Medicinal plants have been used for centuries as remedies for different human diseases and this interest has been revived due to problems associated with resistance to antibiotics. Plants are being screened worldwide in search for new effective antimicrobial compounds. It is estimated that there are at least 250,000 species of plants in the world today (Cowan, 1999) and it has long been known that the highly heterogeneous soil and climatic conditions of the Mediterranean area have resulted in a great diversity of medicinal plants in the region (Daglia, 2012; Kutlu, 2003; Pan et al., 2014; Yasar, Sagdic, & Kisioglu, 2005). Previous reports suggest that some plant extracts possess beneficial biological effects, including antimicrobial properties (Abdel-Massih, Abdou, Baydoun, & Daoud, 2010; Correia et al., 2004; Daoud, Abdo, & Abdel-Massih, 2011; Ohsaki, Takashima, Chiba, & Kawamura, 1999; Trouillas et al., 2003). Some of these activities have been attributed to the action of phenolics (Abi-Khattar et al., 2019; Korukluoglu, Sahan, Yigit, & Ozer, 2008),

Medicinal Plants as Anti-infectives
DOI: https://doi.org/10.1016/B978-0-323-90999-0.00012-4

terpenoids, alkaloids, flavonoids (Puupponen-Pimiä et al., 2001), pectins (Daoud, Sura, & Abdel-Massih, 2013; Menshikov et al., 1997), and other plant polymers that help plants to respond against biotic or abiotic stresses (Simões, Bennett, & Rosa, 2009).

The Mediterranean flora is highly diverse with around 25,000 plant species (10% of higher plants) many of which are endemic to the region (Baydoun, Chalak, Dalleh, & Arnold, 2015; Medail & Quezel, 1999). Ten plant biodiversity hotspots have been identified in the Mediterranean with Lebanon as one of them (Medail & Quezel, 1999). Lebanon is a small country of 10,452 km^2 in the Middle East situated at the hinge of three continents: Asia, Europe, and Africa. It stretches along the eastern shore of the Mediterranean Sea and has a rectangular shape with its length greater than its width. Two mountain ranges parallel to the coast, Mount Lebanon and Anti-Lebanon, are spread over the length of the country and separated by a plain, the Bekaa Valley. Although Lebanon is a relatively small Mediterranean country, it consists of different geographic territories that allow a variety of natural environments and habitats. The combination of climate and geological variation allows plant diversity in the region (Mouterde, 1966, 1970, 1983). Relative to its size, Lebanon hosts one of the highest densities of plant diversity in the Mediterranean basin with around 3000 species, 92 of which are endemic to Lebanon (Ministry of Agriculture-Lebanon, 1996). Many indigenous and endemic species are threatened due to climate effects, rapid urban expansion, and overgrazing (Bou Dagher-Kharrat, El Zein, & Rouhan, 2018) (Fig. 2.1).

The narrow coastal plain and lower part of the Lebanese mountains correspond to the hot Mediterranean area and are rich in carob, pistachio, and myrtle. Higher parts of Mount Lebanon are rich in pines, oaks, shrubs, and herbaceous plants. In some specific areas at an altitude of 1400 m, cedar trees grow in addition to some trees and shrubs such as *Berberis libanotica* Ehrenb. ex C.K.Schneid., *Lonicera nummulariifolia* Juab. & Spach., and *Quercus brantii* Lindl. (Hajar, Haïdar-Boustani, Khater, & Cheddadi, 2010). However, plants adapted to extreme conditions, icy in winter and terribly arid in summer, grow in Northeast of Bekaa (Fahed, 2016; Medail & Quezel, 1999).

Lebanon is rich in many aromatic or medicinal plants that are widely used in cooking, as herbal tea mixtures, or for treatment of various illnesses (especially in rural areas) (El Beyrouthy, 2009). Around 529 species (102 families) are traditionally used as medicinal plants many of which have antimicrobial properties (El Beyrouthy, Arnold, Delelis-Dusollier, & Dupont, 2008). Many Lebanese endemic species were used traditionally in folk medicine for their antimicrobial properties including *B. libanotica*, *Origanum libanoticum* Boiss., *Lactuca triquetra* (Labill.) Boiss., *Ferula elaeochytris* Korovin, *Prangos asperula* Boiss., *Origanum ehrenbergii* Boiss., *Marrubium globosum* Montbr. & Auch. ex Benth. ssp. libanoticum Boiss. Other traditional uses involved extracting the essential oils from plants for different biological applications. Common essential oils are recognized for their broad-spectrum antimicrobial activity at concentrations of 1–128 μg/mL such as oils of basil (Moghaddam, Shayegh, Mikaili, & Sharaf, 2011), cinnamon (Ooi et al., 2006), coriander (Galvão et al., 2012), thyme (Çetin, Çakmakçi, & Çakmakçi, 2011), oregano (Çetin et al., 2011), and lemongrass (Singh, Singh, Singh, & Ebibeni, 2011). The oxygenated form of terpenes (Dorman & Deans, 2000), phenolic compounds, or alcohols isolated from plants are also known for their strong antimicrobial activities (Giweli, Džamic, Sokovic, Ristic, & Marin, 2012; Kordali et al., 2008; Nostro et al., 2004).

FIGURE 2.1 Map of Lebanon. *Source: Wikimedia Commons. https://commons.wikimedia.org/wiki/File:Lebanon_Base_Map.png.*

Lebanese plants with antimicrobial activity

Different plant families contribute to the rich flora in Lebanon in particular or to the Mediterranean region in general. They are also well known through folk medicine or tested experimentally for their antimicrobial activity (Fig. 2.2).

Amaryllidaceae

Allium cepa/Allium sativum

Allium cepa L. (onion) and *Allium sativum* L. (garlic) are two plants from the Amaryllidaceae family that are known for their traditional medicinal effects. The bulbs of *A. cepa* are usually used raw, as decoction, or cooked for the treatment of wounds, infections, bronchitis, inflammation, constipation, hair loss, and rheumatism (Eddouks, Ajebli, & Hebi, 2017). *A. cepa* is traditionally used in many countries such as Algeria, Albania, Cyprus, Italy, and Spain for the treatment of skin, respiratory, and kidney disorders (González-Tejero et al., 2008). It is also used in Palestine against diabetes (Ali-Shtayeh, Jamous, & Jamous, 2012; Miara et al., 2019). Onion skin extracts, which are considered as waste, exhibit high antimicrobial activity against different bacteria (*Bacillus cereus, Pseudomonas fluorescens*, and *Escherichia coli*) and fungi (*Penicillium cyclopium, Trichoderma viride*, and *Aspergillus niger*) (Ahiabor, Gordon, Ayittey, & Agyare, 2016). Polyphenols extracted from the skin or the edible part of *A. cepa* exhibit both antimicrobial and antibiofilm activities (Lee et al., 2013; Škerget, Majhenič, Bezjak, & Knez, 2009); however, they can be toxic to animals if the amount ingested is more than 0.5% of their bodyweight (Cope, 2005). Sharma, Mahato, and Lee (2018) found that the amount of total phenolics and

FIGURE 2.2 Selected endemic Lebanese medicinal plants. (A) *Berberis libanotica*, (B) *Cedrus libani*, (C) *Cyclotrichium origanifolium*, (D) *Origanum ehrenbergii*, (E) *Origanum libanoticum*, (F) *Prangos asperula*.

flavonoids (especially quercetin) in onion increases after 3 months of storage that might contribute to higher antioxidant and antimicrobial activity. Red onions show stronger anti-biofilm and antimicrobial activity compared to yellow onions. White onions exhibit negligible antimicrobial activity (Sharma et al., 2018).

Allium sativum L. is a perennial flowering plant with a pungent smell or aroma and strong diverse flavors. Garlic bulbs, raw or prepared through decoction or maceration have antiseptic, antiinflammatory, antispasmodic, anticoagulant, or hypotensive activity. They are traditionally used against influenza, intestinal parasites, respiratory diseases, ear and eye diseases, hemorrhoids, diabetes, hypertension, kidney disorders, and dandruff (Miara, Hammou, & Aoul, 2013). Different functions have been attributed to its organosulfur compounds that are formed during garlic processing. The various garlic preparations include raw or cooked, fresh or aged, oil or juice, and each with different pharmacological effects (Habtemariam, 2019). Many natural spices have strong antibacterial activity against multiple-drug-resistant clinical pathogens. Karuppiah and Rajaram (2012) tested the activity of the ethanolic extract of *A. sativum* and *Zingiber officinale* Roscoe (ginger) against two Gram-positive and five Gram-negative multidrug resistant bacteria. Garlic exhibited the highest inhibition zone (19.45 mm) and the lowest minimum inhibitory concentration (MIC) 67.00 μg/mL against *Pseudomonas aeruginosa*. This antibacterial activity is attributed to different compounds found in garlic cloves such as alliin (a non-protein amino acid), ajoene, and diallyl sulfides (Oosthuizen, Reid, & Lall, 2017). When garlic is chopped or chewed, alliin mixes with the enzyme alliinase present in the bulb to form allicin. Allicin is a widely studied bioactive component that is responsible for the pungent smell. Aqueous garlic extract had an MIC of 80−120 mg/mL and a minimum bactericidal concentration (MBC) between 120 and 160 mg/mL using the zone inhibition method. A toxicological study revealed that the extract was safe at a dose lower than 300 mg/kg (Lawal et al., 2016).

Anacardiaceae

Pistacia species

Pistacia lentiscus L. (mastic tree) belongs to the Anacardiaceae family, commonly known as sumac or cashew family. Anacardiaceae family consists of different genera that are important economically such as sumac, mango, poison ivy, smoke tree, cashew, pistachio, and others. Genus *Pistacia* includes about 20 species (Al-Saghir & Porter, 2012; Rauf et al., 2017) including *P. lentiscus* (from which plant mastic resin is derived), *Pistacia vera* L. (known for its edible seeds, pistachios), *Pistacia terebinthus* L. (source of terebinth resin), and *Pistacia chinensis* Bunge (grown as decorative tree, Chinese pistache). *P. lentiscus* are mostly found in the Mediterranean areas and its aromatic resin (Mastic Gum) is mostly cultivated on the Greek island of Chios (Pachi et al., 2020). *P. lentiscus* was reported to have antiinflammatory, antioxidative, neuroprotective, wound-healing, diuretic, antimicrobial, and anticancer effects (Benhammou, Bekkara, & Panovska, 2008; Dimas, Pantazis, & Ramanujam, 2012; Paraschos, Mitakou, & Skaltsounis, 2012; Rauf et al., 2017; Yemmen, Landolsi, Ben Hamida, Mégraud, & Trabelsi Ayadi, 2017; Zorzan et al., 2019). Mastic gum is used traditionally in cosmetics, in oral hygiene, as spices, or as treatment for diabetes, gastrointestinal problems, dental decay, or for regulating blood cholesterol levels (Ali-Shtayeh, Yaniv, & Mahajna, 2000; Hanlidou, Karousou, Kleftoyanni, & Kokkini, 2004; Kartalis et al., 2016; Triantafyllou, Chaviaras, Sergentanis, Protopapa, & Tsaknis, 2007). Mastic gum or oil is also used in dermatological products, toothpastes, drinks, or food products (Pachi et al., 2020). Even low concentrations of ethanol extracts of mastic gum inhibited the growth of *Helicobacter pylori* (NCTC 11637) (Huwez, Thirlwell, Cockayne, & Ala'Aldeen, 1998) by altering bacterial cell wall as seen using a transmission electron microscope (Marone et al., 2001). Other studies contradict this finding where mice given 2 g of mastic gum twice daily for a week did not exhibit strong antibacterial activity with MIC (7.8 mg/L) and MBC (31.25 mg/L) against *H. pylori* strains tested (Loughlin, Ala'Aldeen, & Jenks, 2003). Other studies show antibacterial activity of mastic gum and mastic oil against *Staphylococcus aureus*, *Salmonella enteritidis*, and other food-borne microorganisms at 0.1–1.5 vol./vol. oil concentrations (Tassou & Nychas, 1995). Aqueous mastic extracts also exhibit antifungal activity (Ali-Shtayeh & Abu Ghdeib, 1999). Resin essential oil was more effective against different Gram-positive bacteria, Gram-negative bacteria, and fungi (*Candida albicans*, *Candida tropicalis*, and *Torulopsis glabrata*) with MIC ranging from 1.25 to 9 mg/mL compared to essential oil from twigs and leaves (Magiatis, Melliou, Skaltsounis, Chinou, & Mitaku, 1999). Mastic gum has an MBC of 0.07–10 mg/mL activity against Gram-negative anaerobic bacteria or other microorganisms that cause oral cavities (Karygianni et al., 2014). They are proposed to have potential antiplaque activity (Takahashi et al., 2003). Different compounds are responsible for mastic gum antimicrobial activity such as linalool and alpha-terpineol with MBC 3.05 and 2.43 mg/mL, respectively, against *E. coli* (Paraschos et al., 2011).

Apiaceae

Prangos asperula

Prangos asperula Boiss. (Farsh Al Dabe'e, Rough Prangos) belongs to the Apiaceae family and its aerial parts are traditionally used in the Mediterranean and the Middle East countries to treat digestive disorders, dermatological problems, hemorrhoids, and blood pressure problems (Loizzo et al., 2008). It also has wound healing, antidiabetic, antimicrobial, antiviral, and antioxidant activities (Mneimne, Baydoun, Nemer, & Arnold, 2016). *P. asperula* grows widely in Bakish (Loizzo et al., 2008), Ehden, Shouf, and other Lebanese high mountains at altitudes of 1200–1500 m (Bouaoun, Hilan, Garabeth, & Sfeir, 2007; Mneimne et al., 2016). Ethyl acetate and hexane extracts from *Prangos pabularia* Lindl. consist of coumarins, terpenoids, glycosides, pyrone derivatives, furanocoumarin derivatives, and other compounds (Tada et al., 2002). Alkaloids, coumarins, and terpenoids were isolated from oils from different plant parts (roots, flowers, or fruits) of *Prangos* species with promising biological activity. The essential oils from *P. asperula* exhibit antimicrobial activity with the strongest activity against *S. aureus* (15.06 mm), followed by *E. coli* (11.80 mm), and *Aspergillus fumigatus* (9.16 mm) using the disc diffusion method (Mneimne et al., 2016). Bouaoun et al. (2007) also found strong activity of the essential oil against Gram-positive bacteria (*S. aureus, S. faecalis*), Gram-negative bacteria (*E. coli* and *Salmonella typhi*), and fungi (*C. albicans*).

Asteraceae/Compositae

Matricaria species

Matricaria species are flowering plants of the family Asteraceae that includes *Lactuca sativa* L. (lettuce), *Helianthus annuus* L. (sunflowers), *Cynara scolymus* L. (artichoke), and *Matricaria recutita* (L.) Rauschert (chamomiles). *Matricaria* are well known for their use as herbal teas, in cosmetics, in folk medicine, or for aromatic purposes (Mohammad, 2011; Sharifi-Rad, Nazaruk et al., 2018). *M. recutita* or *Matricaria chamomilla* (German chamomile) are rich in secondary metabolites such as lactones, flavonoids, coumarins, and phenolic compounds (Kazemi, 2014). Different applications include using chamomile for the treatment of headaches, conjunctivitis, chronic fever, inflammation, and dermatitis (Mahdizadeh, Ghadiri, & Gorji, 2015). Chamomile and different plants of the *Matricaria* genus are also known for their antimicrobial, analgesic, and antiinflammatory effects (Miraj & Alesaeidi, 2016; Shoara et al., 2015; Singh, Khanam, Misra, & Srivastava, 2011). The essential oils of these plants may help in natural preservation of food (Soković, Glamočlija, Marin, Brkić, & van Griensven, 2010) and have an antibacterial effect against Gram-positive and Gram-negative bacteria (Kazemi, 2014). These oils mainly exhibit a bacteriostatic effect against Gram-positive bacteria (Marino, Bersani, & Comi, 2001).

Fabri, Nogueira, Dutra, Bouzada, and Scio (2011) studied the antimicrobial activity of the methanolic extract of *M. recutita* leaves using the broth microdilution method and attributed this activity to the presence of phenols, tannins, triterpenoids, and flavonoids. However, in another study, both the cyclohexane and ethanolic extracts of *M. chamomilla* flowers did not exhibit much antibacterial activity (Carvalho, Silva, Silva, Scarcelli, & Manhani, 2014). Subcritical water extracts of *Matricaria* species exhibit strong antimicrobial activity (MIC 39 μg/mL) against *E. coli* and *A. niger* (Metrouh-Amir, Duarte, & Maiza, 2015). Both flowers and leaves of a 50% hydroalcoholic extract of *M. recutita* extracts showed antimicrobial activity of their fractionated extracts against *Pseudomonas syringae* (Móricz et al., 2012), *S. aureus*, and *Staphylococcus epidermidis* strains (Jesionek et al., 2015). Stronger antibacterial activity was observed from essential oils or extracts of *Matricaria* species against Gram-positive bacteria (Munir et al., 2014). Many factors such as environmental conditions, extraction methods, and extraction solvents affect the composition and biological activity of plant extracts (Sharifi-Rad, Nazaruk et al., 2018). The amount of α-bisabolol in essential oils of *Matricaria* species contributes to its antifungal (Pauli, 2006) and antibacterial activity (Mekonnen, Yitayew, Tesema, & Taddese, 2016).

Berberidaceae

Berberis libanotica

Berberis species (barberry) belong to the Berberidaceae family. They are spiny shrubs present in altitudes of around 1200 m. More than 450 species exist (Irshad, Pervaiz, Abrar, Fahelboum, & Awen, 2013; Malik et al., 2017) and are attributed different biological activities such as antirheumatic and antineuralgic (El Beyrouthy et al., 2008), anticancer, antiinflammatory, antimicrobial, and antiseptic (Ali et al., 2013). They are known for the treatment of heart disease, gastrointestinal disorders, and hemorrhoids (Tetik, Civelek, & Cakilcioglu, 2013). *Berberis libanotica* Ehrenb. ex C.K.Schneid. is traditionally prepared as a decoction mixed with flour, as a maceration in alcohol, or as a decoction in association with *Geranium robertianum* L., *Thymus syriacus* Boiss., and *Melissa officinalis* L. (El Beyrouthy et al., 2008).

Berberis species are rich in alkaloids, phenols, and flavonoids. The major alkaloid is berberine; other alkaloids are found in lower percentages such as berbamine, jatrorrhizine, palmatine (Ali et al., 2013). Malik et al. (2017) found a higher phenol and flavonoid content and a lower berberine content in *B. libanotica* compared to *Berberis aetnensis* C. Presl. Berberine isolated from *Berberis heterophylla* Juss. ex Poir. showed antimicrobial activity against clinical isolates of *S. aureus* and different *Candida* species; whereas the aqueous extract did not exhibit any activity (Freile et al., 2003). *B. libanotica* methanol fractions showed greater antioxidant activity than *B. aetnensis* fractions; both root extracts inhibit acetylcholinesterase and butyrylcholinesterase activity (Wojtunik-Kulesza, Oniszczuk, Oniszczuk, & Waksmundzka-Hajnos, 2016).

Cannabaceae

Humulus lupulus

Humulus lupulus L. is a flowering plant in the Cannabinaceae (hemp) family known for its important role in the beer brewing industry (Zanoli & Zavatti, 2008). The female flower cones (hops) are fragrant and have a bitter flavor. They provide aromatic, medical, antimicrobial, and preservative qualities due to their polyphenolic and acyl phloroglucides content (Moir, 2000; Morcol, Negrin, Matthews, & Kennelly, 2020). The flowers are used as a dye source, food flavoring, perfume, spice, and to make beer or other beverages. Traditionally, *H. lupulus* flowers are known for their effect in treating sleep disturbances; however, this sedative activity was not clinically proven. Native American tribes used hops as a sedative, analgesic, antirheumatic, and as an antiinflammatory agent (Zanoli & Zavatti, 2008). Many studies report the antibacterial and antifungal activity of essential oils or extracts from *H. lupulus* flowers, leaves, and seeds (Abram et al., 2015; Nionelli, Pontonio, Gobbetti, & Rizzello, 2018). A stronger antifungal activity was observed from hop cone extract against *Penicillium* and *Aspergillus* strains (Alonso-Esteban et al., 2019; Nionelli et al., 2018). Hop seed extracts gave strong antibacterial activity with MICs ranging from 0.01 mg/mL against *B. cereus* and 0.15 mg/mL against *E. coli* and *Salmonella typhimurium* (Alonso-Esteban et al., 2019). Different active compounds from hops such as humulone and lupulone (Oshugi et al., 1997; Simpson & Smith, 1992; Teuber & Schmalreck, 1973; Zanoli & Zavatti, 2008) exhibit antibacterial (mostly against Gram-positive bacteria) and antifungal activity (Mizobuchi & Sato, 1984). Xanthohumol was also found to have a wide antimicrobial and antiviral activity (Gerhauser, 2005). Hop cone extracts, mainly lupulone and xanthohumol, exhibit antibacterial (Yamaguchi, Satoh-Yamaguchi, & Ono, 2009) and antibiofilm activity against different *Staphylococcus* species with different profiles of resistance (Bogdanova et al., 2018). Abram et al. (2015) suggest that other compounds such as catechins present in hop extracts may be responsible for antimicrobial activity against *S. aureus* (MIC < 0.003 mg/mL).

Cistaceae

Cistus species

Cistus species are perennial shrubs, native to the Mediterranean area that are known in folk medicine for their role against different disorders and as antiinfective agents (Bouamama, Noël, Villard, Benharref, & Jana, 2006; Ustün, Ozçelik, Akyön, Abbasoglu, & Yesilada, 2006). Decoctions from plant leaves, flowers, or aerial parts are an effective remedy against microbial infections (Bassolé & Juliani, 2012; Salin et al., 2011; Viapiana, Konopacka, Waleron, & Wesolowski, 2017; Zidane et al., 2013), skin inflammations (Leto, Tuttolomondo, La Bella, & Licata, 2013), pain, constipation, rheumatism, snakebites, wounds, urinary diseases, and disorders (Gürdal & Kültür, 2013). A survey in Kos, a Greek Mediterranean island where most of Hippocrates medical practices were derived shows that at higher altitudes *Cistus salviifolius* L. and *Teucrium capitatum* L. were the most commonly encountered species (Leto et al., 2013). Both plants are rich in a volatile sesquiterpene, germacrene D that has both antibacterial and cytotoxic activities (Barrajón-Catalán et al., 2011; Leto et al., 2013). However, even within the same plant species, the difference in the fruiting stage, climatic and soil conditions lead to differences in the main sesquiterpene hydrocarbons. Germacrene D was the dominant essential oil found in *C. salviifolius* in Sicily, camphor in Spain and Greece (Demetzos, Angelopoulou, & Perdetzoglou, 2002; Morales-Soto et al., 2015), and β-damascenone in Sardinia (Mastino, Marchetti, Costa, & Usai, 2017).

C. salviifolius is traditionally used as a tea substitute, ointment, or as a cicatrizing or astringent agent. The antibacterial activity of *C. salviifolius* is not only related to its essential oils or terpenes (Gertsch, 2011; Güvenç et al., 2005; Morales-Soto et al., 2015) but also to its polyphenolic compounds. Antimicrobial activity against Gram-negative and Gram-positive bacteria are reported from organic solvent extracts derived from *Cistus creticus* L. (Güvenç et al., 2005) and *Cistus ladanifer* L. (Ferreira et al., 2012) or aqueous extracts derived from *Cistus populifolius* L. and *C. ladanifer* (Barrajón-Catalán et al., 2011). Aqueous extracts from *C. salviifolius* exhibit strong antibacterial or bacteriostatic activity against *S. aureus* that may be attributed primarily to polar compounds and to other flavonols, whereas inhibitory activity against *E. coli* may be linked to flavonols and galloylated flavanols (Tomás-Menor et al., 2013). *C. salviifolius*, *C. populifolius*, *Cistus laurifolius* L., *C. ladanifer*, and *Cistus monspeliensis* L. have strong antifungal activity with a MIC of 0.625 mg/mL (Karim et al., 2017). In addition to their antibacterial and antifungal properties, *Cistus* species also have antiviral, antiparasitic, and insecticidal activities (Hutschenreuther, Birkemeyer, Grötzinger, Straubinger, & Rauwald, 2010; Morales-Soto et al., 2015; Verdeguer, Blázquez, & Boira, 2012).

Conifers

Conifers are mostly evergreen woody plants with around 630 species worldwide. In Lebanon, there are 10 indigenous conifer species (from five genera) (Talhouk, Zurayk, & Khuri, 2001) and others were imported to Lebanon for ornamental purposes such as *Cupressus macrocarpa* (Hartw.) D.P.Little (goldcrest). *Cedrus libani* A. Rich. is the national emblem of Lebanon and its wood was used since ancient times in building of temples and boats (Chaney & Basbous, 1978). Egyptians used *Cedrus libani* essential oils for mummification. Conifers play an important part in Lebanese folk medicine. *Abies cilicica* (Antoine & Kotschy) Carrière was used in folk medicine in Lebanon for the treatment of rheumatism and in Turkey for the treatment of bronchitis, gastrointestinal diseases, and asthma (Sargin, 2015). *Cupressus sempervirens* L. f. horizontalis (Mill.) Voss is known as antiasthmatic, antitussive, and antirheumatic; while *Juniperus excelsa* M. Bieb. and *J. oxycedrus* L. are known as antirheumatic and antineuralgic (El Beyrouthy et al., 2008). Different Lebanese conifer essential oils, harvested in Lebanon, were tested for their antimicrobial activity against different fungi and bacteria. The MICs of conifer essentials oils were determined against a range of bacteria and fungi responsible for skin infections using the broth microdilution technique. The essential oils from *J. oxycedrus*, *J. excelsa*, and *C. libani* show strong activity against *S. aureus* (MIC of 64 μg/mL). Dermatophyte species were sensitive (MIC values 32–64 μg/mL) to essential oils from *C. sempervirens*, *A. cilicica*, *J. excelsa*, *C. libani*, *J. oxycedrus*, and *C. macrocarpa* (Fahed et al., 2017).

Lamiaceae

An ethnopharmacological study to compare traditional medicinal plants in Turkey (Marmaris district) and the Mediterranean countries reveals the presence of 64 medicinal plant species (from 35 families) and the use of nine essential oils. Many of the commonly used traditional plants belong to the Lamiaceae (13 species) and Asteraceae (four species) families. Some of the reported commonly used medicinal plants are *Salvia fruticosa* Mill. (mostly its essential oils), *Origanum onites* L., *Mentha pulegium* L., *Lavandula stoechas* L., and *Satureja thymbra* L. Other medicinal plants that are endemic to Lebanon, Syria, and Turkey include *Sideritis libanotica* (Benth.) Bornm and *Thymus cilicicus* Boiss. & Bal. (Gürdal & Kültür, 2013).

The Lamiaceae family is a large family with around 7886 species (245 genera) with valuable secondary metabolites that have attracted different industries (Bekut et al., 2018; Khoury, Stien, Eparvier, Ouaini, & El Beyrouthy, 2016). They are utilized in the food, pharmaceutical, and cosmetic industries. Although the Lamiaceae family is distributed globally, a large concentration of these plants is found in the Mediterranean region. Some are used as ornamentals (*Ajuga*, *Salvia*, and *Coleus*) and others as culinary herbs such as thyme (*Thymus*), sage (*Salvia*), mint (*Mentha*), rosemary (*Rosmarinus*), oregano or marjoram (*Origanum*), lavender (*Lavandula*), and basil (*Ocimum*). Around 136 species (29 genera) of Lamiaceae species are found in Lebanon (El Beyrouthy, Dhifi, & Arnold, 2013). Many are used in the Lebanese cuisine in salads such as *S. fruticosa*, *Rosmarinus officinalis* L., *Thymbra spicata* L. (Bozkurt, 2006; Eruygur, Çetin, Ataş, & Çevik, 2017), and *Coridothymus capitatus* (L.) Reichenb. fil. Others are mixed with cheese (*Thymus* and *Origanum*) to form "Shanklish" or used to form "manakeesh" a baked dough with a mixture of thyme herbs on the top (*Origanum syriacum* L., *S. thymbra*, and *T. spicata*) in addition to *Rhus coriaria* Linn. (sumac) and sesame seeds (Khalil et al., 2019; Khoury et al., 2016). Other indigenous Lamiaceae species widely used in Lebanese folk medicine include the genera *Salvia*, *Rosmarinus*, *Satureja*, *Origanum*, *Thymus*, *Lavandula*, *Melissa*, and *Mentha* (Khoury et al., 2016). The essential oils of these plants and different extracts (methanol, ethanol, and other solvent extracts) are widely used in Lebanese folk medicine (El Beyrouthy et al., 2008, 2013; El Beyrouthy, 2009). Khoury et al. (2016) tested the antimicrobial activity of the essential oils of 11 Lamiaceae species against standard strains of *S. aureus* (Gram-positive), *E. coli* (Gram-negative), *C. albicans* (yeast), and *Trichophyton rubrum* (dermatophyte clinical isolate). *T. spicata, S. thymbra, C. capitatus, L. stoechas,* and *O. syriacum* exhibited antimicrobial activity. The amount of carvacrol and thymol (Marchese et al., 2016) in the essential oils positively correlated with the antimicrobial activity (Khoury et al., 2016).

Phlomis species

Another member of the Lamiaceae family, the *Phlomis* genus is also known for its culinary usages and biological activity. Its essential oils, flavonoids, phenylethyl alcohol, and iridoids contribute to different pharmacological activities. Extracts are used in the treatment of gastric ulcer, diabetes, hemorrhoids, wounds, and inflammation (Amor et al., 2009). *Phlomis* is part of a very popular traditional herbal tea mixture in Lebanon and Syria. It is commonly known as "Zhourat" or "Zahraa" and it consists of dried flowers or leaves of different plants including *Phlomis syriaca* Boiss. (leaves and flowers), *Rosa damascena* Mill. (flowers), *Aloysia citrodora* Paláu (leaves), *Zea mays* L. (flowers), *M. chamomilla* (flowers), *Elaeagnus angustifolia* L. (flowers and leaves), *Micromeria myrtifolia* Boiss. & Hohen (aerial parts), and other herbs (Carmona, Llorach, Obon, & Rivera, 2005; Obón, Rivera, Alcaraz, & Attieh, 2014). It is believed that the mixture has stronger antioxidant activity (Guimarães, Barros, Carvalho, & Ferreira, 2011) and acts as an effective treatment for colds and bronchitis (Baydoun et al., 2015). Essential oils and methanol extracts from different *Phlomis* species exhibit antibacterial effects against *S. aureus, E. coli, Klebsiella pneumonia,* and other pathogenic bacteria (Morteza-Semnani, Saeedi, Mahdavi, & Rahimi, 2006).

Cyclotrichium species

Cyclotrichium genus is a member of the Lamiaceae family used as spices in different food recipes and well known in the traditional folk medicine (Formisano, Oliviero, Rigano, Saab, & Senatore, 2014). It is an aromatic plant, with a mint-like smell, endemic to Lebanon, Syria, Turkey (Özdemir et al., 2017), Iraq, and Iran (Tepe, Sokmen, Sokmen, Daferera, & Polissiou, 2005). *Cyclotrichium origanifolium* (Labill.) Manden. & Scheng (menthol) and *M. myrtifolia* essential oils have different activities such as antipyretic, antifungal, and antibacterial activities (Formisano, Mignola, et al., 2007; Formisano, Rigano, et al., 2007). *C. origanifolium* essential oils exhibit the strongest antibacterial activity (MIC 0.06–8 mg/mL) followed by nonpolar solvent extracts; whereas the polar extracts (aqueous and ethanol extracts) did not exhibit antimicrobial activity (Tepe et al., 2005). *C. origanifolium* essential oils are rich in pulegone, menthone, and limonene. The methanol extract had the strongest free radical scavenging activity (Tepe et al., 2005).

Salvia species

Salvia species constitute around one-quarter of the Lamiaceae family and are highly diverse with over 900 species worldwide (Kamatou, Makunga, Ramogola, & Viljoen, 2008). Some *Salvia* species are popular as aromatic herbs, essential oils, medicinal plants, ornamental plants, or as food due to their good flavors. Commonly used *Salvia* species in the food market include *Salvia officinalis* L. (common sage), *Salvia hispanica* L. (chia),

Salvia sclarea L. (clary sage), *Salvia miltiorrhiza* Bunge (danshen), and *Salvia lavandulifolia* Vahl (Spanish sage) (Sharifi-Rad, Ozcelik, et al., 2018). Other economically important species such as *Salvia fruticosa* Mill., *Salvia tomentosa* Mill., and *Salvia verbenaca* L. are known for their biological activities such as antimicrobial, antioxidant, antiplasmodial, anticancer, antipyretic, antinociceptive, antihyperglycemia, skin curative agents, and help against the loss of memory. They are also used in the treatment of the common cold, tonsillitis, menstrual pain, abdominal pain, and nausea (Bulut, Haznedaroğlu, Doğan, Koyu, & Tuzlacı, 2017; Gürdal & Kültür, 2013; Kamatou et al., 2008; Sharifi-Rad, Ozcelik, et al., 2018; Zengin et al., 2019).

In South Africa, sages form a fundamental part of traditional healing especially in the treatment of malaria, microbial infections, inflammation, and disinfection (Kamatou et al., 2008). *Salvia africana-lutea* L. leaves, fruits, or flowers are prepared fresh or dried as herbal tea mixtures for the traditional treatment of colds, bronchitis, or tuberculosis (Kamatou et al., 2008; Ulubelen, Topcu, & Johansson, 1997). *Salvia africana-caerulea* L. and other *Salvia* species are also used in the treatment against the flu (Kamatou et al., 2008). The leaves of *Salvia repens* Burch. ex Benth. can be used to treat sores (by adding to a bath), root decoctions can help in the treatment of diarrhea, and the smoke (due to burning of plants or leaves) can help in the disinfection of a room or to repel insects. In Europe, *Salvia runcinata* Benth. decoctions are used for the treatment of allergies (Kamatou et al., 2008). The Chinese *S. miltiorrhiza*, used for treating heart disease, achieved around 205 million USD in the global market in 2008 (Jia, Huang, Zhang, & Leung, 2012). Though many *Salvia* species are cultivated in different locations, around 250 species are endemic to Central Asia and the Mediterranean (Walker, Sytsma, Treutlein, & Wink, 2004). *S. fruticosa* is endemic to the Eastern Mediterranean basin (Sharifi-Rad, Ozcelik, et al., 2018).

Salvia species are rich in different bioactive compounds such as terpene derivatives, essential oils, phenolic compounds, flavonoids, and tannins (Bahadori, Eskandani, De Mieri, Hamburger, & Nazemiyeh, 2018; Sharifi-Rad, Ozcelik, et al., 2018). Different reviews and papers discuss terpenoids isolated from *Salvia* and relate them to the plant's bioactive properties (Jash, Gorai, & Roy, 2016; Sharifi-Rad, Ozcelik, et al., 2018). Many of these compounds from *Salvia* are attributed antimicrobial activities. An abietane diterpene, galdosol (from *Salvia canariensis* L. aerial parts) exhibits antibacterial activity against *S. aureus*, *Bacillus subtilis*, and *Micrococcus luteus* (González et al., 1989). Ethanolic extracts from aerial parts and roots of *S. fruticosa*, *Salvia cilicica* Boiss., *S. officinalis*, and *S. tomentosa* exhibit antifungal activity. Essential oils from *S. fruticosa* had a MIC of 256 and 512 μg/mL against *T. rubrum* and *C. albicans*, respectively (Khoury et al., 2016; Sharifi-Rad, Ozcelik, et al., 2018). Essential oils from *S. officinalis* exhibit antimicrobial activity against *C. albicans*, *E. coli*, and *S. aureus* (Cutillas, Carrasco, Martinez-Gutierrez, Tomas, & Tudela, 2017). Three terpenoids from *Salvia multicaulis* Vahl exhibit activity against tuberculosis (Ulubelen et al., 1997). Different antibacterial and antiplasmodial compounds have been isolated from *Salvia chamelaegnea* Berg., *Salvia radula* Benth., and *S. verbenaca*. Carnosol, oleanolic acid, 7-O-methylepirosmanol, and ursolic acid isolated from *Salvia* species or other members of the Lamiaceae family exhibit antimicrobial activities (Kamatou, Van Vuuren, Van Heerden, Seaman, & Viljoen, 2007). Ursolic acid and carnosol were also isolated from *Rosmarinus officinalis* (Kamatou et al., 2008).

Rosmarinus officinalis

Rosmarinus officinalis L. commonly known as Rosemary is a perennial herb that belongs to the Lamiaceae family. Its secondary metabolites are attributed to different biological activities such as antibacterial (Bozin, Mimica-Dukic, Samojlik, & Jovin, 2007), antiinflammatory, antinociceptive (De Melo et al., 2011; Estévez, Ramírez, Ventanas, & Cava, 2007; González-Trujano et al., 2007), antimutagenic (Tajehmiri, Ghasemi, Sabet, Chakoosari, & Abdolahzadegan, 2014), antioxidant (Rašković et al., 2014), hepatoprotective (Abdel-Wahhab, El-Shamy, El-Beih, Morcy, & Mannaa, 2011), antidiabetic (Bakırel, Bakırel, Keleş, Ülgen, & Yardibi, 2008), and anticancer (Moore, Yousef, & Tsiani, 2016; Ngo, Williams, & Head, 2011). Some of the main bioactive compounds in rosemary are rosmarinic acid, caffeic acid, carnosic acid, carnosol, and triterpenoid acids (Ngo et al., 2011).

Thymol/carvacrol rich species

Essential oils from *Thymus vulgaris* L. and other thyme species exhibit bacteriostatic effect against both Gram-positive and Gram-negative bacteria (Alibi et al., 2020; Marino, Bersani, & Comi, 1999; Rota, Herrera, Martínez, Sotomayor, & Jordán, 2008). The essential oils of *T. vulgaris* exhibit strong antibiofilm, antimicrobial, antiquorum sensing, and antioxidant activities. The inhibition diameter zone was greater than 20 mm for 85.71% of 105 multidrug-resistant strains tested (Alibi et al., 2020). Thymol (2-isopropyl-5-methylphenol) showed the strongest inhibitory activity compared with carvacrol, eugenol, trans-cinnamic acid, and others with MIC 1.0 and 1.2 mmol/L against *S. typhimurium* and *E. coli*, respectively (Olasupo, Fitzgerald, Gasson, & Narbad, 2003). Similarly, thymol from essential oil of *Thymus syriacus* Boiss. also exhibited antibacterial activity against different Gram-negative bacteria (Al-Mariri, Swied, Oda, & Al Hallab, 2013). Thymol showed antibacterial activity against *S. aureus* and *S. epidermidis* clinical isolates with different profiles of resistance with MIC 0.03%–0.06% vol./vol. using the agar dilution method (Nostro et al., 2004). Gram-positive strains (MIC 0.31 mg/mL against *S. aureus*) were more sensitive than Gram-negative strains (MIC 5.00 mg/mL against *E. coli*) as shown using the microdilution method (Trombetta et al., 2005). The activity of thymol may be through affecting the integrity of the lipid membrane. Different reviews on the antibacterial and antifungal effects of thymol are found in the literature (Marchese et al., 2016). Chloroform and methanol extracts of *T. syriacus* exhibit antimycobacterial activity with MIC 6.3 and 50 μg/mL, respectively (Askun, Tumen, Satil, Modanlioglu, & Yalcin, 2012; Jurno, Netto, Duarte, & Machado, 2019). *Thymus leucotrichus* Halácsy methanol and acetone extracts and *T. syriacus* essential oil showed activity against verocytotoxigenic *E. coli* strains (Al-Mariri & Safi, 2014; Nabavi et al., 2015; Ulukanli, Cigremis, & Ilcim, 2011).

Za'atar plants: Satureja thymbra; Origanum syriacum

Different plant species from the Lamiaceae family are referred to in the Arab world by the name "za'atar." These plants are known for their aromatic flavor, culinary uses, and folk medicine preparations (Ali-Shtayeh, Yaghmour, Faidi, Salem, & Al-Nuri, 1998). "Za'atar" includes *Satureja thymbra* L., *Origanum syriacum* L. (white oregano), *C. capitatus*, and *T. spicata* (Shehadeh et al., 2019). *S. thymbra* Lamiaceae grows wild in the Mediterranean area and is also cultivated for commercial use. It has different antibacterial activity due to its essential oils (Azaz, Kürkcüoglu, Satil, Baser, & Tümen, 2005; Chorianopoulos et al., 2006; Marković et al., 2011) and its phenolic compounds (Skoula, Grayer, & Kite, 2005). The combination of essential oils and phenolic extracts was shown to have both antibacterial and antioxidant activity and thus decrease lipid auto-oxidation in fish coating material (Choulitoudi et al., 2016).

Origanum syriacum is a perennial herbaceous wild plant that is native to the Mediterranean region. It has white aromatic flowers and hairy leaves with a strong taste. It is widely used in folk medicine in Lebanon, Syria, Palestine, and Jordan for the treatment against microbial infections, throat pain, abdominal pain, or other disorders (Aburjai, Hudaib, Tayyem, Yousef, & Qishawi, 2007; Ali-Shtayeh et al., 2000; Husein et al., 2014; Shehadeh et al., 2014; Shehadeh, Suaifan, & Darwish, 2017). *Origanum* species essential oils and different solvent extracts exhibit antimicrobial, anticancer, and antioxidant activity (Abdel-Massih, Fares, Bazzi, El-Chami, & Baydoun, 2010; Al Hafi et al., 2016; Benelli et al., 2019; El Gendy et al., 2015; Khoury et al., 2016; Loizzo et al., 2009; Viuda-Martos et al., 2010). *O. syriacum* from different regions in Palestine show MIC with moderate activity (MIC 97–25000 μg/mL) against five standard bacterial strains. Higher lipophilic content or hydrophilic (thymol-rich) essential oils led to increased antibacterial activity against Gram-positive bacteria. However, increase in carvacrol led to stronger inhibitory activity against both Gram-positive and Gram-negative bacteria (Saidi, Ghafourian, Zarin-Abaadi, Movahedi, & Sadeghifard, 2012; Shehadeh et al., 2019). The antimicrobial activity of essential oils of *O. ehrenbergii* Boiss. and *O. libanoticum* Boiss. (endemic to Lebanon) and *O. syriacum* (endemic to the Levantine) is affected by carvacrol content. *O. ehrenbergii* (60.8% carvacrol) and *O. syriacum* (79% carvacrol) showed moderate antimicrobial activity (MIC 400–1200 μg/mL), whereas *O. libanoticum* (0% carvacrol) was inactive against yeast and pathogenic bacteria tested (Al Hafi et al., 2016).

Different Lamiaceae genera

The concentration of different compounds in extracts or essential oils derived from members of the Lamiaceae family (*Origanum, Satureja, Thymbra, Phlomis*) vary with environmental conditions (Baydar, Sağdiç, Özkan, & Karadoğan, 2004; Kizil, 2010; Miguel et al., 2004), soil characteristics (Economou et al., 2014), extraction procedure, and developmental stage at the time of collection (Zhang & Wang, 2008). Among 27 indigenous Lebanese plants, methanol extracts of nine herbs show antimicrobial activity ≥88.8% using the disk diffusion method. *O. libanoticum* (whole plant) and *Verbascum leptostychum* DC. (flower extracts) show 99.9% inhibition. Stronger antimicrobial activity was detected in the methanol extract compared to the water extracts of the tested traditional medicinal plants (Barbour et al., 2004). The methanolic extract from *O. libanoticum* had a MIC of 1:2.5 against *E. coli* and a MIC of 1:3.5 against *Shigella dysenteriae* and *Proteus* species (Barbour et al., 2004). *O. libanoticum, Origanum majorana* L., and *Origanum vulgare* L. also exhibit antifungal activity (El Gendy et al., 2015; Waller et al., 2017).

Thymbra spicata

To prevent multiplication of foodborne bacteria, leafy parts of plants such as thyme, savory, and oregano are traditionally added to meat and food products to extend their shelf life (Sağğdıç & Özcan, 2003). Hydrosols from 15 spices were tested for antibacterial activity to study their potential as food preservatives, and only anise (*Pimpinella anisum* L.), cumin (*Cuminum cyminum* L.), oregano (*O. vulgare*), summer savory (*Satureja hortensis* L.), and black thyme (*T. spicata*) exhibited activity (Sağdıç & Özcan, 2003). *T. spicata* essential oils (75.5% carvacrol) exhibited stronger antibacterial activity compared to other Lamiaceae species such as *Origanum minutiflorum* Schwarz & Davis (wild oregano; endemic to Turkey), *Satureja cuneifolia* Ten. (wild savory), and *O. onites* (oregano). The antibacterial activity may be primarily due to carvacrol and to the hydrocarbons (c-terpinene and p-cymene) present in its essential oils. The essential oils from *T. spicata* were the most active (among those tested) as seen using paper disc diffusion method where the 1/50 solution (in ethanol) inhibited most strains tested (Baydar et al., 2004). Other studies show antimicrobial activity of *T. spicata* decoction and hydrosols against different bacteria and fungi (Kivanc & Akgül, 1988; Özcan & Boyraz, 2000; Sağdıç & Özcan, 2003; Sağdiç, Kuşçu, Özcan, & Özçelik, 2002; Saidi et al., 2012). *T. spicata* is used as a "healthy plant" in Lebanon due to its antimicrobial potential (Bozkurt, 2006; Eruygur et al., 2017). It is rich in phenolic compounds such as rosmarinic acid, carvacrol (Stefanaki, Cook, Lanaras, & Kokkini, 2018), thymol, and flavonoids (Dorman, Bachmayer, Kosar, & Hiltunen, 2004; Hanci, Sahin, & Yilmaz, 2003). Eruygur et al. (2017) compared water extracts from leaves and flowers of *T. spicata* (prepared as the traditional infusion method) with ethanol extracts. Although ethanol extracted more compounds and had stronger antioxidant activity (mainly due to higher carvacrol content), the aqueous extract had more rosmarinic acid content and wound-repair ability (Eruygur et al., 2017). Both phenol and flavonoid compounds are recognized for their antimicrobial activity; however, phenols seem to contribute to stronger activity. A negative correlation was obtained using Spearman's rho test between minimal inhibitory concentration and the presence of these active compounds after screening methanolic extracts of 21 medicinal plants in Spain (Stanković, Radić, Blanco-Salas, Vázquez-Pardo, & Ruiz-Téllez, 2017). The methanol extracts of *T. spicata* exhibited strong antimycobacterial activity against *Mycobacterium tuberculosis* (MIC 196 μg/mL) and moderate activity against *S. typhimurium*, *E. coli*, *S. epidermidis*, and *Enterobacter aerogenes* (Askun, Tumen, Satil, & Ates, 2009). Crude methanol extracts of different plants have higher antimicrobial activity than aqueous extracts as reported in the literature (Askun et al., 2009; Parekh, Jadeja, & Chanda, 2005). These extracts are rich in flavonoids, alkaloids, glycosides, amino acids, phytosterols, tannins, triterpenoids, saponins, and steroids (Abdel-Wahhab et al., 2011; Kumar et al., 2009).

Myrtaceae

Eucalyptus species

The genus *Eucalyptus* belongs to the Myrtaceae family and consists of around 900 species (Gilles, Zhao, An, & Agboola, 2010; Tyagi & Malik, 2011). *Eucalyptus* is well known worldwide as a source of fiber (for paper production) and for its medicinal properties (Luís et al., 2016). Its essential oils are known in folk medicine for their antifungal, antibacterial, analgesic, and antiinflammatory properties (Bussmann et al., 2010; Elaissi et al., 2011; Mulyaningsih, Sporer, Zimmermann, Reichling, & Wink, 2010; Navarro, Villarreal, Rojas, & Lozoya, 1996; Salari, Amine, Shirazi, Hafezi, & Mohammadypour, 2006; Silva et al., 2003). They are also known to be effective against respiratory infections, colds, sinusitis, and influenza (Sadlon & Lamson, 2010). *Eucalyptus globulus* Labill. (blue gum) oil has strong antimicrobial activity (Tohidpour, Sattari, Omidbaigi, Yadegar, & Nazemi, 2010; Tyagi and Malik, 2011). Both essential oils and solvent extracts exhibit additive antibacterial effects with antibiotics (Pereira, Dias, Vasconcelos, Rosa, & Saavedra, 2014). *E. globulus* (70%) methanol extracts exhibit a stronger antimicrobial activity than pure methanol, acetone, dichloromethane, or water extracts due to larger yield of phenolics in the methanol solvent (Pereira et al., 2014). The methanol extracts from *E. globulus* leaves (extracted at 80°C) showed antibacterial activity against *Streptococcus pneumoniae* (MIC 90 of 32 mg/L), *Streptococcus pyogenes* (MIC 90 of 64 mg/L), *S. aureus* (MIC 90 of 128 mg/L), and *Haemophilus influenzae* (MIC 90 of 32 mg/L). MIC 90 is the MIC required to inhibit the growth of 90% of organisms tested. In another study, the chloroform:methanol extract of *E. globulus* showed the strongest antimicrobial activity with MICs ranging from 1 to 500 μg/mL (Zonyane, Van Vuuren, & Makunga, 2013). The methanolic extract of a traditional South African herbal combination made of *E. globulus*, *Agathosma crenulata* (L.) Pillans, and *Dodonaea viscosa* Jacq. showed strong antibacterial activity with MIC 49 μg/mL against *S. aureus* and MIC 98 μg/mL against *K. pneumoniae* and *E. coli* (Zonyane et al., 2013). GC-MS analysis of the methanol extract after subfractionation revealed that the main components with antibacterial activity are citric acid, gallic acid, citramalic acid, and ellagic acid (Luís et al., 2016). The essential oils from *E. globulus* and 1,8-cineole, one of the main bioactive components, exhibit antibiofilm effects, antibacterial activity, and antiquorum sensing against methicillin-resistant *S. aureus* (MRSA). The essential oils exhibit stronger antimicrobial activity than its main isolated compounds (Merghni et al., 2018). Essential oils from *E. globulus* are rich in oxygenated monoterpenes (78.58%) and 1,8-cineole (55.29%). Spathulenol (7.44%) and alpha-terpineol (5.46%) are also found in lower amounts. They exhibit a marked antibacterial activity against periodontopathogenic bacterial species with stronger activity on Gram-negative bacteria (Harkat-Madouri et al., 2015). Conversely, other studies show stronger antibacterial activity of these oils against Gram-positive bacteria due to differences in cell wall architecture (Mulyaningsih, Sporer, Reichling, & Wink, 2011; Salehi et al., 2019). Essential oils from *E. globulus* and other plants in Lebanon (*J. excelsa* and *Micromeria barbata* Boiss. & Kotschy) showed strong antimycobacterial activity (El Omari et al., 2019).

Portulacaceae

Portulaca oleracea

Portulaca oleracea L., common purslane, is an annual weed from the family Portulacaceae. It is native to the Mediterranean region and spread worldwide. The leaves and stems are used in salad and are rich in omega-3 fatty acids (Petropoulos et al., 2015), α-linolenic acid, palmitoleic, palmitic, and other fatty acids (Uddin et al., 2014), carotenoids, lutein, zeaxanthin, vitamins (Dias, Camões, & Oliveira, 2009), phenolic acids, terpenoids, tannins, and other bioactive compounds (Erkan, 2012). However, it is also rich in antinutrients such as oxalates that may lead to kidney stone or other deleterious effects on human health (Petropoulos et al., 2015). In addition to being part of edible leaves in the

"Mediterranean diet" (Morales et al., 2014); it also has an important role in folk medicine (Lev & Amar, 2002; Menale, De Castro, Cascone, & Muoio, 2016; Petropoulos, Karkanis, Martins, & Ferreira, 2018; Ramadan, Schaalan, & Tolba, 2017). *P. oleracea* is known for its antioxidant, antimicrobial, cardioprotective, neuroprotective (inhibitors of acetylcholinesterase), antimutagenic, antiinflammatory, antidiabetic, antinociceptive, and immunomodulatory activity (Alam et al., 2014; Chen, Li, Zhang, Xia, & Zhang, 2016; Chowdhary, Meruva, Naresh, & Elumalai, 2013; Ramadan et al., 2017; YouGuo, ZongJi, & XiaoPing, 2009). Topical use of leaf juice is traditionally utilized in Italy and other areas for abdominal pains, flu, bronchitis, intestinal disorders, inflammation, skin irritation, infections, and for wound treatment (De Feo, Ambrosio, & Senatore, 1992; De Feo, Aquino, Menghini, Ramundo, & Senatore, 1992; Di Novella, Di Novella, De Martino, Mancini, & De Feo, 2013; Di Sanzo, De Martino, Mancini, & De Feo, 2013; Guarino, De Simone, & Santoro, 2008; Guarrera & Leporatti, 2007; Passalacqua, Guarrera, & De Fine, 2007; Savo, Caneva, Maria, & David, 2011; Tuttolomondo et al., 2014).

Ranunculaceae

Clematis vitalba

Different species of the genus *Clematis* or from the Ranunculaceae family (Buttercup family), rich in the glycoside ranunculin, are used in the traditional Lebanese medicine such as *Clematis flammula* L. (virgin's bower), *Clematis cirrhosa* L. (fern-leaved clematis), and *Clematis vitalba* L. (El Beyrouthy et al., 2008). *Clematis* genera are rich in active components such as pentacyclic triterpenoid saponins (used traditionally in Chinese medicine), coumarins, flavonoids, and alkaloids (Chawla, Kumar, & Sharma, 2012). The genus *Clementis* is widespread around the world and used in folk medicine for the treatment of asthma, rheumatism, nervous disorders, malaria, and syphilis. It also has antibacterial, antiinflammatory (Yesilada & Küpeli, 2007), analgesic, and anticancer effects (Chawla et al., 2012). *C. vitalba* L., consumed as food in Italy (Pieroni, Nebel, Quave, & Heinrich, 2002; Pieroni, 1999), is known to have strong antimicrobial activity (Ali-Shtayeh & Abu Ghdeib, 1999; Khan, Kihara, & Omoloso, 2001). Methanol extracts of *C. vitalba* shoots exhibit a broad antifungal and antibacterial activity (MIC 1.6–12.3 μg/mL) as shown using the agar diffusion assay (Buzzini & Pieroni, 2003).

Nigella sativa

Nigella sativa L. from the Ranunculaceae family is commonly known as Habbat-al-baraka, Alhabba-al-sawda, Chouniz or Black cumin (El Beyrouthy et al., 2008). It is widely used as a medicinal plant throughout India and Arabian countries. It is also cultivated in Europe, Lebanon, and Syria. It was considered as a miraculous herb in ancient traditional medicine and is mentioned in the Holy Bible (Old Testament), by Prophet Mohammed, by Avicenna in his book "The Canon of Medicine," by Hippocrates, and by Dioscorides (Yimer, Tuem, Karim, Ur-Rehman, & Anwar, 2019). *N. sativa* seeds are bitter and are traditionally used against rheumatism, jaundice, fever, and skin diseases (Paarakh, 2010). The seed oils were investigated for antiinflammatory (Pise & Padwal, 2017; Zakaria, Jais, & Ishak, 2018), antioxidant, antidiabetic (Daryabeygi-Khotbehsara, Golzarand, Ghaffari, &

Djafarian, 2017; El Rabey, Al-Seeni, & Bakhashwain, 2017), gastroprotective, antimicrobial, antiviral, and antiparasitic activity (Paarakh, 2010; Yimer et al., 2019). Thymoquinone or thymoquinone derivatives from *N. sativa* seed oil extracts show anticancer activities (60%–80% growth inhibition of pancreatic cancer cell lines) when administered for 48 h prior to gemcitabine or oxaliplatin treatment (Banerjee et al., 2009; Breyer, Effenberger, & Schobert, 2009). Crude extracts of *N. sativa* seeds using different solvents exhibit strong antibacterial activity. The ethanol extract had an MIC of 0.2–0.5 mg/mL against MRSA (Hannan, Saleem, Chaudhary, Barkaat, & Arshad, 2008). Seed diethyl ether extracts on filter paper discs showed both antibacterial and antifungal activity (Hanafy & Hatem, 1991; Sokmen, Jones, & Erturk, 1999). Seed essential oils from *N. sativa* also exhibit strong antibacterial (Kokoska et al., 2008) and antifungal activity (Islam, Ahsan, Hassan, & Malek, 1989; Mahmoudvand, Sepahvand, Jahanbakhsh, Ezatpour, & Ayatollahi Mousavi, 2014); this activity is mostly attributed to the volatile oil thymoquinone (Abdallah, 2017; Paarakh, 2010; Toama, El-Alfy, & El-Fatatry, 1974). Thymoquinone had an MIC of 8–32 g/mL against Gram-positive bacteria and inhibited biofilm formation of *S. aureus* and *S. epidermidis* (Chaieb, Kouidhi, Jrah, Mahdouani, & Bakhrouf, 2011). *N. sativa* oils exhibit strong antibacterial activity against MRSA (Al-Jaafary, Al-Atiyah, Al-Khamis, Al-Sultan, & Badger-Emeka, 2016; Gawron et al., 2019) at different oil concentrations (20%–100% oil) but not against *E. coli* and *Acinetobacter baumannii* (Al-Jaafary et al., 2016). Crude *N. sativa* alkaloid and water extracts showed stronger antibacterial activity against Gram-negative bacteria (Morsi, 2000); whereas essential oils (Kokoska et al., 2008) and crude methanol seed extracts (Hasan, Nawahwi, & Malek, 2013) showed stronger antibacterial activity against Gram-positive bacteria.

Rutaceae

Ruta species

Different species of the genus Ruta (*Ruta chalepensis* L., *Ruta graveolens* L., and *Ruta montana* L.) are known in the Mediterranean region for their importance in folk medicine. Their importance is discussed in Hippocrates' work in treating pulmonary and gynecological conditions (Pollio, De Natale, Appetiti, Aliotta, & Touwaide, 2008). The Rutaceae family is commonly known as the rue or citrus family. *R. graveolens* essential oils and leaf extracts are used as a tonic or for the traditional treatment against cold headaches, dementia, and ear pain (Jarić et al., 2011). *R. chalepensis* (fringed rue) leaf infusions (or chewed) are used for treating eye diseases (antiinflammatory), stomach ache, intestinal worms, respiratory problems, menstrual problems, and other conditions (Leto et al., 2013; Menale et al., 2016; Miara et al., 2019). High amounts may lead to some toxic effects such as vomiting, abortion, gastroenteritis, and confusion (Gedif & Hahn, 2003; Haile et al., 2017). The "rue juice" is considered as a repellent to keep scorpions away (González & Vallejo, 2013). *R. chalepensis* has antimicrobial (Kacem et al., 2015), anticancer, and antiinflammatory activity (Khlifi et al., 2013; Miara et al., 2019). It is rich in flavonoids, alkaloids, phenols, coumarins, tannins, and saponins (Kacem et al., 2015). Ethanol extracts reveal stronger antioxidant and antimicrobial activities (195–1562 mg/mL) compared to methanol and methanol/water extracts (Kacem et al., 2015).

Rosaceae

Rosa damascena

Rosa damascena Mill. or "Ward jouri" is a shrub from the Rosaceae family. Their roses have important economic benefits as garden plants, fragrances (from rose petals), and flavorings. *R. damascena* is commonly used in Lebanon in different traditional culinary preparations or beverages such as rose water or "Maward," rose syrup, rose jelly, and jam. Rose dried buds are used as herbal teas and dried rose petals are added on Arabic sweets or in combination with other local herbs in cooking (Najem, El Beyrouthy, Wakim, Neema, & Ouaini, 2011). *R. damascena* has different medicinal benefits according to traditional Lebanese folk medicine for its antimicrobial, carminative, and antioxidant activities (Aridoğan et al., 2002; Basim & Basim, 2003; Ozkan, Sagdic, Baydar, & Baydar, 2004). When applied topically it is known for its antirheumatic activity and its effect against conjunctivitis and dermatitis (El Beyrouthy et al., 2008; El Beyrouthy, 2008). Its essential oils are also known for their applications in perfumery, aromatherapy, and cosmetics (Özel, Göğüş, & Lewis, 2006). It is one of the most expensive oils due to the low oil content in the plant. Although *R. damascena* has been known for a long time in Lebanon, its geographical origin is believed to be from Damscas and hence the name "Damask rose" (Gault & Synge, 1971). The oil from Damask rose in Lebanon has a different composition than that from different countries and this increases its value. It has a higher percentage of monoterpene alcohols that contribute to its aroma (such as β-phenylethanol, citronellol, geraniol, nerol, and linalool) and lower quantities of hydrocarbons as seen in different Lebanese cultivars (Bayrak & Akgül, 1994; Najem et al., 2011).

Essential oils from *R. damascena* exhibit strong antimicrobial activities (Kumar, Bhandari, Singh, & Bari, 2009; Mileva, Kusovski, Krastev, Dobreva, & Galabov, 2014; Moein, Zomorodian, Almasi, Pakshir, & Zarshenas, 2017; Shohayeb, Abdel-Hameed, Bazaid, & Maghrabi, 2014; Yi, Sun, Bao, Ma, & Sun, 2019). Further fractionation of these oils through molecular distillation also yields fractions with strong antibacterial (0.625–1.25 mg/mL) activity against *S. aureus*, *B. subtilis*, *E. coli*, and *P. aeruginosa* and antifungal (6.25–12.5 mg/mL) activities against *A. niger*, *Rhizopus nigricans*, and *Blastocladia pringsheimii* (Yi et al., 2019). The potential antimicrobial compounds are mainly β-phenylethanol, linalool, citronellol, geraniol, eugenol, and methyl eugenol (Yi et al., 2019). Methanol *R. damascena* extracts exhibit antimicrobial activity (MIC 2 mg/mL) against *Propionibacterium acnes* (Azimi, Fallah-Tafti, Khakshur, & Abdollahi, 2012; Tsai, Tsai, Wu, Tseng, & Tsai, 2010).

Conclusion

The main plants used in traditional medicine in Lebanon as antimicrobials are summarized in Table 2.1.

Methods used for the table: Information in the table is obtained from the ethnopharmacological work of Marc Beyrouthy (2004–20). All voucher specimens of the species are deposited at the Herbarium of the Faculty of Agronomy of the Holy Spirit University of Kaslik (Lebanon). The results are recorded in a synoptic table. In the inventory, the families and the species within these families are listed in alphabetical order. The information

TABLE 2.1 List of traditional Lebanese medicinal plants.

Voucher	Scientific name	Vernacular name	Origin	Part used	Traditional medicinal indication	Preparation (I = internal use; E = external use)
	Amaryllidaceae					
MNC052	*Allium cepa* Linn.	Basal	Cultivated	Bulb	Antiseptic	E: local application
				Fresh leaves	Antiseptic	I: eaten alone or mixed with other salads
MNC075	*Allium sativum* Linn.	Toum	Cultivated	Bulb, Leaves	Antimicrobial, antiseptic, antifungal	I: consumed fresh or as an infusion
	Anacardiaceae					
MNU472	*Pistacia lentiscus* Linn.	Sareys, Mastiq, Botom	Indigenous	Aerial parts	Antimicrobial	E: decoction and local application in the form of a compress
	Apiaceae					
MNU621	*Apium graveolens* Linn.	Al-karfass, Krafs, Krafs barri, Chabatbat	Indigenous, Cultivated and imported	Fruit	Antiseptic	I: Infusion or decoction
MNU618	*Prangos asperula* Boiss.	Farsh-al-dabbeh, Farsh-el-dabo'	Endemic of Lebanon and Syria	Aerial parts	Antimicrobial, antifungal	I: decoction
	Asteraceae					
MNV401	*Achillea biebersteinii* Afan.	Habbouk, Kaff mariam Ghadda	Indigenous	Flowering parts	Antimicrobial, antifungal	I
MNV400	*Achillea falcata* Linn.	'Ebaitaran, 'Ebaytaroun, Kaisoun	Indigenous	Flowering parts	Antimicrobial, antifungal	I: decoction: 60 g/l of water
MNP103	*Achillea millefolium* Linn.	Em Alef warka, Akhilia	Imported	Flowering parts	Urinary antiseptic	I: decoction: 60 g/L of water
MNV375	*Dittrichia viscosa* Linn. (Greuter) (Syn: *Inula viscosa* (Linn.) Aiton)	Tayyoun	Indigenous	Roots, Flowers, Leaves, Flowering parts	Antiseptic	I: decoction
				Fresh leaves	Antimicrobial	E: direct application of freshly crushed leaves
MNV418	*Matricaria aurea* (Linn.) Schultz Bip.	Babounej, Babounej ma'rūf	Indigenous	Leaves, Flowers	Antimicrobial	I: infusion, one spoon in a cup of water three times a day
MNV417	*Matricaria recutita* (Linn.) Rauschert (Syn: *M. chamomilla* L.)	Babounej, Babounej 'almani	Indigenous, cultivated and imported	Flowering heads	Antimicrobial	E: poultice and steam inhalation
					Antifungal	E: steam inhalation
	Berberidaceae					
MNU002	*Berberis libanotica* C. K. Schneider	Barbaris, Shelsh-al-barbaris (Roots)	Endemic of Lebanon and Syria	Roots	Antifungal	I: decoction

(*Continued*)

TABLE 2.1 (Continued)

Voucher	Scientific name	Vernacular name	Origin	Part used	Traditional medicinal indication	Preparation (I = internal use; E = external use)
	Brassicaceae					
MNP002	*Anastatica hierochuntica* Linn.	Kaff mariam	Imported	Entire plant	Antiseptic	E: decoction in the form of a poultice
MNU079	*Capparis spinosa* Linn.	Al koubbar, Kabbar, Halak-es-sit	Indigenous	Flowering buds	Antiseptic	I
				Flowering buds, Roots	Renal antiseptic	I
	Cannabaceae					
MNP004	*Humulus lupulus* Linn.	Hachichat-al-dinar, Aljounjel	Cultivated and imported	Flowering buds and entire plant	Antiseptic, antimicrobial	I: decoction
	Caryophyllaceae					
MNP051	*Saponaria officinalis* Linn.	Al sabouniyya, 'Ork-al-halawa, 'Osloj	Cultivated and imported	Leaves	Antiseptic	E: juice used as a compress
	Cistaceae					
MNU550a	*Cistus creticus* Linn.	Ghabra, Ghébré	Indigenous	Leaves	Antimicrobial	E: local application of crushed leaves
MNU550b	*Cistus salviifolius* Linn.	Ghabra, Ghébré	Indigenous	Leaves	Antimicrobial	E: local application of crushed leaves
	Cucurbitaceae					
MNV348	*Citrullus colocynthis* Linn. (Schrad.)	Hanzal, Al 'alkam, Handal	Indigenous and imported	Seeds and fruits	Antiseptic, antifungal	E: feet in the decoction
MNV347	*Ecballium elaterium* (Linn.) A. Rich.	Kit-el-hmar, Khiar-al-hmar, Me'té-el-baghel, Kisa'-el-hmar	Indigenous	Mature fruit	Antiseptic	E: nasal instillation of a few drops of fresh juice sprayed from ripe fruit
	Cupressaceae					
MNI018	*Juniperus excelsa* M.B.	Lezzab	Indigenous	Cones	Antiseptic, antifungal	E: bath with macerate
	Ericaceae					
MNP008	*Arctostaphylos uva-ursi* (Linn.) Spreng.	Habb 'enab-al-debb	Imported	Seeds	Urinary antiseptic	I: decoction and drink after meals for 2 weeks a cup in the morning and in the evening
	Fabaceae					
MNU227a	*Spartium junceum* Linn.	Sitt khadijé, Lisan-el-'asfour, Wazlan	Indigenous	Flowers	Antiseptic	I: infusion
	Fagaceae					
MNI367	*Quercus coccifera* Linn. (Syn: *Q. calliprinos* Webb)	Balllout, Sendayan, Shelsh-el-seendiane (Roots)	Indigenous	Entire plant	Antiseptic	E: decoction in gargle, baths

(Continued)

TABLE 2.1 (Continued)

Voucher	Scientific name	Vernacular name	Origin	Part used	Traditional medicinal indication	Preparation (I = internal use; E = external use)
MNI360	*Quercus infectoria* Olivier	'Afos	Indigenous	Gale	Antiseptic	E: decoction in gargle, baths
	Hypericaceae					
MNU527	*Hypericum lanuginosum* Lam.	Dāzi soufi, Hashishat-al-qalb	Endemic (Lebanon, Syria, Sinai, Jordan, Egypt, Cyprus, Palestine)	Leaves, Flowers	Antifungal	I: maceration
MNU528	*Hypericum perforatum* Linn.	Hioufa raykoune, Hachichat-al-qalb, 'Eshbat al taloul	Indigenous	Flowering parts	Antiseptic	E: maceration in olive oil for 15 days and use the maceration in the form of a compress
	Lamiaceae					
MNV191a	*Coridothymus capitatus* (Linn.) Reichenb. fil	Za'atar farisi, Za'atar 'assal, Za'atar	Indigenous	Flowering tops	Urinary antiseptic	I: Infusion or decoction on an empty stomach: one cup in the morning
MNV180	*Cyclotrichium origanifolium* (Labill.) Manden. & Scheng (Syn: *Calamintha origanifolia* (Labill.) Boiss.)	Na'ana'eyn, Hashishat-al-bahsa, Hashishat-al-bouhaiss, Hashishat-al-jabal, Hashishat-al-daght, Hashishat-al-bhis	Endemic (Asiatic Turkey, Lebanon)	Flowering tops	Antimicrobial, antifungal	I
MNC121	*Lavandula angustifolia* Mill.	Khuzama, Lawanda	Cultivated and imported	Flowering tops	Antiseptic	I: infusion
MNV114	*Lavandula stoechas* Linn.	Astakhoudos, Khuzama Shih	Indigenous	Flowering tops	Antiseptic	I: infusion
MNV116	*Marrubium vulgare* Linn.	Gbayri, 'Eshbat-al-kalb, Frasioun	Indigenous	Flowering tops	Antiseptic	I: infusion
MNV187	*Origanum ehrenbergii* Boiss.	Al zouwayba'a, Za'atar-al-snawbar, Za'atar jordi, Za'aitri	Endemic of Lebanon	Flowering tops	Antimicrobial	E: infusion
MNV185	*Origanum libanoticum* Boiss.	Za'atar lebnan	Endemic of Lebanon	Flowering parts	Antimicrobial	E: infusion
MNP118	*Origanum majorana* Linn.	Mardakouch	Cultivated	Leaves	Antiseptic	I: edible
				Flowering parts	Antimicrobial	I: infusion
				Flowering parts	Antiseptic	I: decoction
MNV188	*Origanum syriacum* Linn.	Zouba', Za'atar	Indigenous	Flowering parts	Antiseptic	I: decoction mixed with honey: two cups/day
				Leaves	Antiseptic	I: infusion: 20 g/L

(*Continued*)

TABLE 2.1 (Continued)

Voucher	Scientific name	Vernacular name	Origin	Part used	Traditional medicinal indication	Preparation (I = internal use; E = external use)
MNV132	*Phlomis syriaca* Boiss.	Mossays, 'Ozzayra souriyé	Endemic (Asiatic Turkey, Palestine, Jordan, Lebanon, Syria)	Flowering parts	Antifungal, antimicrobial	I: decoction
MNV154	*Rosmarinus officinalis* Linn.	Eklil-al-jabal, Hasa-al-bān, Nada-al-bahr	Indigenous, cultivated and imported	Flowering parts, Leaves, Stems	Antifungal, antiseptic	I: decoction or drink a cup of coffee from the distilled water of the flowering plant once a day
MNV159	*Salvia fruticosa* Mill.	Aiza'an, Kassiin, 'Ouaissé, Maryamiyyé	Indigenous, Cultivated and imported	Flowering parts, Leaves	Antimicrobial, antiseptic	I: distilled sage water: three drops + one spoon of honey
MNV173	*Satureja cuneifolia* Ten.	Eshabat-el-wasab, Za'atar farisi	Indigenous	Aerial parts, Stems	Antifungal, antimicrobial	I: decoction
MNV175a	*Satureja myrtifolia* (Boiss. & Hohen.) Greuter & Burdet (Syn: *Micromeria myrtifolia* Boiss. & Hohen.)	Zoufa	Indigenous	Flowering parts, Leaves	Antiseptic	E: infusion
MNV173a	*Satureja thymbra* Linn.	Za'atar khlat, Za'atar bou khlayt, Za'atar rumi, Za'atar franji, Za'atar-al-hamir, Za'atar	Indigenous	Flowering parts	Antiseptic, antimicrobial, antifungal	I: infusion and essential oil
MNV191	*Thymbra spicata* Linn.	Za'atar khlat, Za'atar bou khlayt, Za'atar	Indigenous	Flowering parts	Antimicrobial, antifungal	I: infusion and essential oil
MNV190a	*Thymus syriacus* Boiss.	Za'atar	Indigenous	Aerial parts	Antiseptic, antimicrobial	I: infusion
MNV181a	*Ziziphora clinopodioides* Lam. (*Syn: Z. canescens* Benth.)	Na'na'a jordi	Indigenous	Flowering parts	Antiseptic, antimicrobial	I: infusion
MNV181	*Ziziphora capitata* Linn. subsp. orientalis Sam. ex Rech	Na'na' jordi, Hashishat al'alk	Indigenous	Flowering parts	Antiseptic, antimicrobial	I: infusion
	Lauraceae					
MNU001	*Laurus nobilis* Linn.	Al ghar, Chajarat-al-rand	Indigenous cultivated and imported	Leaves	Antimicrobial	I: infusion of two leaves with a little orange pulp in 200 mL of water
	Lythraceae					
MNP053	*Lawsonia inermis* Roxb.	Hénné, Henna	Imported	Crushed leaves	Antiseptic	E: maceration in olive oil and local application
MNP018	*Punica granatum* Linn.	Roumman	Cultivated and imported	Fruit rind	Antimicrobial	I: sweetened infusion with honey

(Continued)

TABLE 2.1 (Continued)

Voucher	Scientific name	Vernacular name	Origin	Part used	Traditional medicinal indication	Preparation (I = internal use; E = external use)
	Meliaceae					
MNC469	*Melia azedarach* Linn.	Zanzalakht	Cultivated	Seeds	Antiseptic	E: oil in local application
	Myrtaceae					
MNC028	*Eucalyptus globulus* Labill.	Kina, Al Eucalyptous	Cultivated	Leaves	Antiseptic, antimicrobial	E: decoction and inhalation of vapors
MNU563	*Myrtus communis* Linn.	Houmblass, Heblass, Ass, Hounblass, Rihan	Indigenous	Leaves, Fruit	Antiseptic	I: Infusion
MNP021	*Syzygium aromaticum* (Linn.) Merr. & L.M. Perry	Kebech kronfol	Imported	Fruit	Antiseptic, antifungal	E: Mouthwashes: three drops of essential oil diluted in a little alcohol and then added a glass of water
	Myristicaceae					
MNP042	*Myristica fragrans* Houtt.	Jawazat-al-tib, Besbas	Imported	Fruit	Antiseptic	E and I: infusion
	Oleaceae					
MNV019	*Olea europaea* Linn.	Zaytoun	Cultivated	Leaves	Antiseptic	I: decoction or infusion, E: cataplasme
	Oxalidaceae					
MNU433	*Oxalis corniculata* Linn.	Hommayda	Indigenous	Flowering parts	Antimicrobial	I
	Poaceae					
MNC191	*Cymbopogon citratus* (DC) Stapf.	Shay akhdar	Cultivated and imported	Leaves	Antiseptic	E: decoction and inhalation of vapors, I: infusion
	Papaveraceae					
MNV064	*Papaver rhoeas* Linn.	Chakkik ahmar, Chakaik-al-no'man, Fuaisseh, Kechkhach	Indigenous	Flowers, Seeds, Latex	Antiseptic	I: aqueous decoction of the petals for 1 h, infusion (5%), E: local application of the latex
MNV065	*Papaver umbonatum* Boiss. (*Syn: P. syriacum* Boiss. et Bl.)	Kechkhach	Endemic (Palestine, Jordan, Lebanon and Syria)	Flowers	Antiseptic	I: decoction
	Pinaceae					
MNI015a	*Cedrus libani* A. Richard	Arez, Arez lebnan	Endemic (Asiatic Turkey, Syria, Lebanon, Cyprus, Morocco, Algeria)	Wood	Antimicrobial	E: decoction and local application

(*Continued*)

TABLE 2.1 (Continued)

Voucher	Scientific name	Vernacular name	Origin	Part used	Traditional medicinal indication	Preparation (I = internal use; E = external use)
MNI016	*Pinus halepensis* Mill. subsp. brutia (Ten.) E. Murray (*Syn: P. brutia* Ten.)	Snoubar barri	Indigenous	Leaves, Stems	Antiseptic	E: gargle with decoction in diluted vinegar
	Polygonaceae					
MNP028	*Rheum officinale* Baill.	Rawand	Cultivated and imported	Rhizome and Roots	Antiseptic	I: decoction or maceration in cold water for one day, drink three times a day until improvement
MNI406	*Rheum ribes* Linn.	Cherch-al-ribes, Roubass, Chelch-al-ribes	Indigenous	Rhizome	Antimicrobial	I: decoction and drink the decoction on an empty stomach. Soak overnight, filter, and drink
	Portulacaceae					
MNI450	*Portulaca oleracea* Linn.	Bakklé, Farfhine	Indigenous and cultivated	Seeds	Antimicrobial, antiseptic	I: decoction for 10 min
	Ranunculaceae					
MNU054	*Clematis cirrhosa* Linn.	Mar'ān, Habl miskī, Mal'ā 'anamiyyah	Indigenous	Leaves, Roots	Antiseptic	E: decoction, 40 g/L of water for 20 min and poultice
MNP094	*Nigella sativa* Linn.	Habbat-al-baraka, Habba-al-sawda, Chouniz	Imported	Seeds	Antiseptic	I: decoction or infusion, mixed with honey and absorbed in the morning on an empty stomach
	Rosaceae					
MNC071	*Rosa damascena* Mill.	Ward jouri	Cultivated and naturalized	Essential oils	Antimicrobial	I
	Rutaceae					
MNU461	*Ruta chalepensis* Linn. (*Syn: R. chalepensis* Linn. subsp. bracteosa (DC.) Batt.)	Khouft, Sahdab	Indigenous	Flowering parts, Leaves	Antimicrobial	I and E: infusion
	Salicaceae					
MNI350	*Salix alba* Linn.	Safsaf, Safsaf abiad, 'Oud-el-ma', 'Oud libnani	Indigenous and cultivated	Bark	Antiseptic	I and E: infusion
MNI350a	*Salix libani* Bornm.	Safsaf libnani	Endemic (Amanus, Lebanon, Syria)	Bark	Antiseptic	I and E: decoction
	Santalaceae					
MNP005	*Santalum album* Linn.	Sandal abyad, Sandal, Khamia-el-bandal	Imported	Essential oils	Antiseptic	E: fumigation and inhalation

on the medicinal use of plants has been recorded directly on the basis of a detailed survey on herbalists, folk healers, older experienced people, and midwives. All those that might have been obtained information from secondary sources were excluded. Only remedies that were said to be handed down by oral tradition were considered. Furthermore, only species that could be directly indicated and collected by the persons interviewed are cited. All the wild plants used have local vernacular names that point to their rather ancient use. The herbalists were interviewed using a questionnaire similar to the form for botanical and ethnopharmacognosic investigation of plants used in traditional medicine produced by WHO (Penso, 1980). Medicinal plants have been identified based on the "Nouvelle flore du Liban et de la Syrie" Mouterde (1966, 1970, and 1983). We note that the new phylogenetic classification APG II 2003 (Angiosperm Phylogeny Group, 2003) was adopted in order to update the families cited in Mouterde.

Acknowledgment

The authors are grateful to The TWAS Young Affiliates Network (TYAN) that supported the cooperation with the book editor through the TYAN Collaborative Grant Award F.R. 3240307334 that allowed the first author's travel to Emory University, USA.

References

Abdallah, E. M. (2017). Black seed (*Nigella sativa*) as antimicrobial drug: A mini-review. *Novel Approaches in Drug Designing and Development*, *3*(2), 1–5, 555603.

Abdel-Massih, R., Abdou, E., Baydoun, E., & Daoud, Z. (2010). Antibacterial activity of the extracts obtained from *Rosmarinus officinalis*, *Origanum majorana*, and *Trigonella foenum-graecum* on highly drug resistant gram negative bacilli. *Journal of Botany*, *2010*, 464087.

Abdel-Massih, R. M., Fares, R., Bazzi, S., El-Chami, N., & Baydoun, E. (2010). The apoptotic and anti-proliferative activity of *Origanum majorana* extracts on human leukemic cell line. *Leukemia Research*, *34*, 1052–1056. Available from https://doi.org/10.1016/j.leukres.2009.09.018.

Abdel-Wahhab, K. G. E. D., El-Shamy, K. A., El-Beih, N. A. E. Z., Morcy, F. A., & Mannaa, F. A. E. (2011). Protective effect of a natural herb (*Rosmarinus officinalis*) against hepatotoxicity in male albino rats. *Comunicata Scientiae*, *2*, 9–17. Available from http://www.comunicata.ufpi.br/index.php/comunicata/article/view/82/59.

Abi-Khattar, A. M., Rajha, H. N., Abdel-Massih, R. M., Maroun, R. G., Louka, N., & Debs, E. (2019). Intensification of polyphenols extraction from olive leaves using IRed-Irrad®; an environmentally friendly innovative technology. *Antioxidants.*, *8*(7), 227.

Abram, V., Čeh, B., Vidmar, M., Hercezi, M., Lazić, N., Bucik, V., Možina, S., Košir, I. J., Kač, M., Demšar, L., & Poklar Ulrih, N. (2015). A comparison of antioxidant and antimicrobial activity between hop leaves and hop cones. *Industrial Crops and Products*, *64*, 124–134. Available from https://doi.org/10.1016/j.indcrop.2014.11.008.

Aburjai, T., Hudaib, M., Tayyem, R., Yousef, M., & Qishawi, M. (2007). Ethnopharmacological survey of medicinal herbs in Jordan, the Ajloun Heights region. *Journal of Ethnopharmacology*, *110*, 294–304. Available from https://doi.org/10.1016/j.jep.2006.09.031.

Ahiabor, C., Gordon, A., Ayittey, K., & Agyare, R. (2016). In vitro assessment of antibacterial activity of crude extracts of onion (*Allium cepa* L.) and shallot (*Allium aescalonicum* L.) on isolates of *Escherichia coli* (ATCC 25922), *Staphylococcus aureus* (ATCC 25923), and *Salmonella typhi* (ATCC 19430). *International Journal of Advanced Research*, *2*(5), 1029e32.

Al Hafi, M., El Beyrouthy, M., Ouaini, N., Stien, D., Rutledge, D., & Chaillou, S. (2016). Chemical composition and antimicrobial activity of *Origanum libanoticum*, *Origanum ehrenbergii*, and *Origanum syriacum* growing wild in Lebanon. *Chemistry and Biodiversity*, *13*, 555–560. Available from https://doi.org/10.1002/cbdv.201500178.

Alam, M. A., Juraimi, A. S., Rafii, M. Y., Abdul Hamid, A., Aslani, F., Hasan, M. M., Zainudin, M. A. M., & Uddin, M. K. (2014). Evaluation of antioxidant compounds, antioxidant activities, and mineral composition of

13 collected purslane (*Portulaca oleracea* L.) accessions. *BioMed Research International*, 6–10. Available from https://doi.org/10.1155/2014/296063.

Ali, S., Igoli, J., Clements, C., Semaan, D., Almazeb, M., Rashid, M. U., Shah, S. Q., Ferro, V., Gray, A., & Khan, M. R. (2013). Antidiabetic and antimicrobial activities of fractions and compounds isolated from *Berberis brevissima* Jafri and *Berberis parkeriana* Schneid. *Bangladesh Journal of Pharmacology*, *8*(3), 336–342. Available from https://doi.org/10.3329/bjp.v8i3.13888.

Alibi, S., Selma, W. B., Ramos-Vivas, J., Smach, M. A., & Mansour, H. B. (2020). Anti-oxidant, antibacterial, anti-biofilm, and anti-quorum sensing activities of four essential oils against multidrug-resistant bacterial clinical isolates. *Current Research in Translational Medicine*, *68*(2), 59–66. Available from https://doi.org/10.1016/j.retram.2020.01.001.

Ali-Shtayeh, M. S., & Abu Ghdeib, S. I. (1999). Antifungal activity of plant extracts against dermatophytes. *Mycoses*, *42*, 665–672. Available from https://doi.org/10.1046/j.1439-0507.1999.00499.x.

Ali-Shtayeh, M. S., Jamous, R. M., & Jamous, R. M. (2012). Complementary and alternative medicine use amongst Palestinian diabetic patients. *Complementary Therapies in Clinical Practice*, *18*, 16–21. Available from https://doi.org/10.1016/j.ctcp.2011.09.001.

Ali-Shtayeh, M. S., Yaghmour, R. M. R., Faidi, Y. R., Salem, K., & Al-Nuri, M. A. (1998). Antimicrobial activity of 20 plants used in folkloric medicine in the Palestinian area. *Journal of Ethnopharmacology*, *60*, 265–271. Available from https://doi.org/10.1016/S0378-8741(97)00153-0.

Ali-Shtayeh, M. S., Yaniv, Z., & Mahajna, J. (2000). Ethnobotanical survey in the Palestinian area: A classification of the healing potential of medicinal plants. *Journal of Ethnopharmacology*, *73*, 221–232. Available from https://doi.org/10.1016/S0378-8741(00)00316-0.

Al-Jaafary, M., Al-Atiyah, F., Al-Khamis, E., Al-Sultan, A., & Badger-Emeka, L. I. (2016). In-vitro studies on the effect of Nigella sativa Linn. seed oil extract on multidrug resistant Gram positive and Gram negative bacteria. *Journal of Medicinal Plants*, *4*, 195–199.

Al-Mariri, A., & Safi, M. (2014). In vitro antibacterial activity of several plant extracts and oils against some Gram-negative bacteria. *Iranian Journal of Medical Sciences*, *39*, 36–43. Available from http://ijms.sums.ac.ir/index.php/IJMS/article/download/419/117.

Al-Mariri, A., Swied, G., Oda, A., & Al Hallab, L. (2013). Antibacterial activity of *Thymus syriacus* boiss essential oil and its components against some Syrian Gram-negative bacteria isolates. *Iranian Journal of Medical Sciences*, *38*, 180–186. Available from http://ijms.sums.ac.ir/index.php/ijms/article/view/1108/432.

Alonso-Esteban, J. I., Pinela, J., Barros, L., Ćirić, A., & Ferreira, I. C. F. R. (2019). Phenolic composition and antioxidant, antimicrobial and cytotoxic properties of hop (*Humulus lupulus* L.) Seeds. *Industrial Crops and Products*, *134*, 154–159. Available from https://doi.org/10.1016/j.indcrop.2019.04.001.

Al-Saghir, M., & Porter, D. (2012). Taxonomic revision of the Genus Pistacia L. (Anacardiaceae). *American Journal of Plant Sciences*, *3*, 12–32. Available from https://doi.org/10.4236/ajps.2012.31002.

Amor, I. L.-B., Boubaker, J., Sgaier, M. B., Skandrani, I., & Chekir-Ghedira, L. (2009). Phytochemistry and biological activities of *Phlomis* species. *Journal of Ethnopharmacology*, *125*(27), 183–202. Available from https://doi.org/10.1016/j.jep.2009.06.022.

Aridoğan, B. C., Baydar, H., Kaya, S., Demirci, M., Özbaşar, D., & Mumcu, E. (2002). Antimicrobial activity and chemical composition of some essential oils. *Archives of Pharmacal Research*, *25*, 860–864. Available from https://doi.org/10.1007/bf02977005.

Askun, T., Tumen, G., Satil, F., & Ates, M. (2009). In vitro activity of methanol extracts of plants used as spices against *Mycobacterium tuberculosis* and other bacteria. *Food Chemistry*, *116*, 289–294. Available from https://doi.org/10.1016/j.foodchem.2009.02.048.

Askun, T., Tumen, G., Satil, F., Modanlioglu, S., & Yalcin, O. (2012). Antimycobacterial activity some different lamiaceae plant extracts containing flavonoids and other phenolic compounds. In P. J. Cardona (Ed.), *Understanding tuberculosis – New approaches to fighting against drug resistance* (pp. 310–336). Rijeka: InTech.

Azaz, A. D., Kürkcüoglu, M., Satil, F., Baser, K. H. C., & Tümen, G. (2005). In vitro antimicrobial activity and chemical composition of some Satureja essential oils. *Flavour and Fragrance Journal*, *20*, 587–591. Available from https://doi.org/10.1002/ffj.1492.

Azimi, H., Fallah-Tafti, M., Khakshur, A. A., & Abdollahi, M. (2012). A review of phytotherapy of acne vulgaris: Perspective of new pharmacological treatments. *Fitoterapia*, *83*, 1306–1317. Available from https://doi.org/10.1016/j.fitote.2012.03.026.

Bahadori, M. B., Eskandani, M., De Mieri, M., Hamburger, M., & Nazemiyeh, H. (2018). Anti-proliferative activity-guided isolation of clerodermic acid from *Salvia nemorosa* L.: Geno/cytotoxicity and hypoxia-mediated mechanism of action. *Food and Chemical Toxicology, 120*, 155–163. Available from https://doi.org/10.1016/j.fct.2018.06.060.

Bakırel, T., Bakırel, U., Keleş, O. Ü., Ülgen, S. G., & Yardibi, H. (2008). In vivo assessment of antidiabetic and antioxidant activities of rosemary (*Rosmarinus officinalis*) in alloxan-diabetic rabbits. *Journal of Ethnopharmacology, 116*, 64–73. Available from https://doi.org/10.1016/j.jep.2007.10.039.

Banerjee, S., Kaseb, A. O., Wang, Z., Kong, D., Mohammad, M., Padhye, S., Sarkar, F. H., & Mohammad, R. M. (2009). Antitumor activity of gemcitabine and oxaliplatin is augmented by thymoquinone in pancreatic cancer. *Cancer Research, 69*(13), 5575–5583. Available from https://doi.org/10.1158/0008-5472.CAN-08-4235.

Barbour, E. K., Al Sharif, M., Sagherian, V. K., Habre, A. N., Talhouk, R. S., & Talhouk, S. N. (2004). Screening of selected indigenous plants of Lebanon for antimicrobial activity. *Journal of Ethnopharmacology, 93*, 1–7. Available from https://doi.org/10.1016/j.jep.2004.02.027.

Barrajón-Catalán, E., Fernández-Arroyo, S., Roldán, C., Guillén, E., Saura, D., Segura-Carretero, A., & Micol, V. (2011). A systematic study of the polyphenolic composition of aqueous extracts deriving from several Cistus genus species: Evolutionary relationship. *Phytochemical Analysis, 22*, 303–312. Available from https://doi.org/10.1002/pca.1281.

Bartlett, J. G., Gilbert, D. N., & Spellberg, B. (2013). Seven ways to preserve the miracle of antibiotics. *Clinical Infectious Diseases, 56*, 1445–1450. Available from https://doi.org/10.1093/cid/cit070.

Basim, E., & Basim, H. (2003). Antibacterial activity of *Rosa damascena* essential oil. *Fitoterapia, 74*, 394–396. Available from https://doi.org/10.1016/S0367-326X(03)00044-3.

Bassolé, I. H. N., & Juliani, H. R. (2012). Essential oils in combination and their antimicrobial properties. *Molecules (Basel, Switzerland), 17*, 3989–4006. Available from https://doi.org/10.3390/molecules17043989.

Baydar, H., Sağdiç, O., Özkan, G., & Karadoğan, T. (2004). Antibacterial activity and composition of essential oils from Origanum, Thymbra and Satureja species with commercial importance in Turkey. *Food Control, 15*, 169–172. Available from https://doi.org/10.1016/S0956-7135(03)00028-8.

Baydoun, S., Chalak, L., Dalleh, H., & Arnold, N. (2015). Ethnopharmacological survey of medicinal plants used in traditional medicine by the communities of Mount Hermon, Lebanon. *Journal of Ethnopharmacology, 173*, 139–156. Available from https://doi.org/10.1016/j.jep.2015.06.052.

Bayrak, A., & Akgül, A. (1994). Volatile oil composition of Turkish rose (*Rosa damascena*). *Journal of the Science of Food and Agriculture, 64*, 441–448. Available from https://doi.org/10.1002/jsfa.2740640408.

Bekut, M., Brkić, S., Kladar, N., Dragović, G., Gavarić, N., & Božin, B. (2018). Potential of selected Lamiaceae plants in anti(retro)viral therapy. *Pharmacological Research, 133*, 301–314. Available from https://doi.org/10.1016/j.phrs.2017.12.016.

Benelli, G., Pavela, R., Petrelli, R., Cappellacci, L., Bartolucci, F., Canale, A., & Maggi, F. (2019). *Origanum syriacum* subsp. syriacum: From an ingredient of Lebanese 'manoushe' to a source of effective and eco-friendly botanical insecticides. *Industrial Crops and Products, 134*, 26–32. Available from https://doi.org/10.1016/j.indcrop.2019.03.055.

Benhammou, N., Bekkara, F. A., & Panovska, T. K. (2008). Antioxidant and antimicrobial activities of the *Pistacia lentiscus* and *Pistacia atlantica* extracts. *African Journal of Pharmacy and Pharmacology, 2*, 22–28. Available from https://doi.org/10.5897/AJPP.9000056.

Bogdanova, K., Röderova, M., Kolar, M., Langova, K., Dusek, M., Jost, P., Kubelkova, K., Bostik, P., & Olsovska, J. (2018). Antibiofilm activity of bioactive hop compounds humulone, lupulone and xanthohumol toward susceptible and resistant *Staphylococci*. *Research in Microbiology, 169*, 127–134. Available from https://doi.org/10.1016/j.resmic.2017.12.005.

Bou Dagher-Kharrat, M., El Zein, H., & Rouhan, G. (2018). Setting conservation priorities for Lebanese flora—Identification of important plant areas. *Journal for Nature Conservation, 43*, 85–94. Available from https://doi.org/10.1016/j.jnc.2017.11.004.

Bouamama, H., Noël, T., Villard, J., Benharref, A., & Jana, M. (2006). Antimicrobial activities of the leaf extracts of two Moroccan Cistus L. species. *Journal of Ethnopharmacology, 104*, 104–107. Available from https://doi.org/10.1016/j.jep.2005.08.062.

Bouaoun, D., Hilan, C., Garabeth, F., & Sfeir, R. (2007). Étude de l'activité antimicrobienne de l'huile essentielle d'une plante sauvage Prangos asperula Boiss. *Phytothérapie, 5*, 129–134. Available from https://doi.org/10.1007/s10298-007-0238-2.

Bozin, B., Mimica-Dukic, N., Samojlik, I., & Jovin, E. (2007). Antimicrobial and antioxidant properties of Rosemary and Sage (*Rosmarinus officinalis* L. and *Salvia officinalis* L., Lamiaceae) essential oils. *Journal of Agricultural and Food Chemistry, 55*, 7879–7885. Available from https://doi.org/10.1021/jf0715323.

Bozkurt, H. (2006). Utilization of natural antioxidants: Green tea extract and Thymbra spicata oil in Turkish dry-fermented sausage. *Meat Science, 73*, 442–450. Available from https://doi.org/10.1016/j.meatsci.2006.01.005.

Breyer, S., Effenberger, K., & Schobert, R. (2009). Effects of thymoquinone - Fatty acid conjugates on cancer cells. *ChemMedChem, 4*, 761–768. Available from https://doi.org/10.1002/cmdc.200800430.

Bulut, G., Haznedaroğlu, M. Z., Doğan, A., Koyu, H., & Tuzlacı, E. (2017). An ethnobotanical study of medicinal plants in Acipayam (Denizli-Turkey). *Journal of Herbal Medicine, 10*, 64–81. Available from https://doi.org/10.1016/j.hermed.2017.08.001.

Bussmann, R. W., Malca-García, G., Glenn, A., Sharon, D., Chait, G., Díaz, D., Pourmand, K., Jonat, B., Somogy, S., Guardado, G., Aguirre, C., Chan, R., Meyer, K., Kuhlman, A., Ownesmith, A., Effio-Carbajal, J., Frías-Fernandez, F., & Benito, M. (2010). Minimum inhibitory concentrations of medicinal plants used in Northern Peru as antibacterial remedies. *Journal of Ethnopharmacology, 132*, 101–108. Available from https://doi.org/10.1016/j.jep.2010.07.048.

Buzzini, P., & Pieroni, A. (2003). Antimicrobial activity of extracts of *Clematis vitalba* towards pathogenic yeast and yeast-like microorganisms. *Fitoterapia, 74*, 397–400. Available from https://doi.org/10.1016/S0367-326X(03)00047-9.

Carmona, M. D., Llorach, R., Obon, C., & Rivera, D. (2005). "Zahraa," a Unani multicomponent herbal tea widely consumed in Syria: Components of drug mixtures and alleged medicinal properties. *Journal of Ethnopharmacology, 102*, 344–350. Available from https://doi.org/10.1016/j.jep.2005.06.030.

Carvalho, A. F., Silva, D. M., Silva, T. R. C., Scarcelli, E., & Manhani, M. R. (2014). Avaliação da atividade antibacteriana de extratos etanólico e de ciclohexano a partir das flores de camomila (*Matricaria chamomilla* L.). *Revista Brasileira de Plantas Medicinais, 16*, 521–526. Available from https://doi.org/10.1590/1983-084X/12_159.

Çetin, B., Çakmakçi, S., & Çakmakçi, R. (2011). Thyme ve oregano uçucu yağlarının antimikrobiyal etkilerinin araştırılması. *Turkish Journal of Agriculture and Forestry, 35*, 145–154. Available from https://doi.org/10.3906/tar-0906-162.

Chaieb, K., Kouidhi, B., Jrah, H., Mahdouani, K., & Bakhrouf, A. (2011). Antibacterial activity of Thymoquinone, an active principle of *Nigella sativa* and its potency to prevent bacterial biofilm formation. *BMC Complementary and Alternative Medicine, 11*, 29. Available from https://doi.org/10.1186/1472-6882-11-29.

Chaney, W. R., & Basbous, M. (1978). The cedars of lebanon witnesses of history. *Economic Botany, 32*, 118–123. Available from https://doi.org/10.1007/BF02866865.

Chawla, R., Kumar, S., & Sharma, A. (2012). The genus Clematis (Ranunculaceae): Chemical and pharmacological perspectives. *Journal of Ethnopharmacology, 143*, 116–150. Available from https://doi.org/10.1016/j.jep.2012.06.014.

Chen, Y. X., Li, G. Z., Zhang, B., Xia, Z. Y., & Zhang, M. (2016). Molecular evaluation of herbal compounds as potent inhibitors of acetylcholinesterase for the treatment of Alzheimer's disease. *Molecular Medicine Reports, 14*, 446–452. Available from https://doi.org/10.3892/mmr.2016.5244.

Chorianopoulos, N., Evergetis, E., Mallouchos, A., Kalpoutzakis, E., Nychas, G. J., & Haroutounian, S. A. (2006). Characterization of the essential oil volatiles of *Satureja thymbra* and *Satureja parnassica*: Influence of harvesting time and antimicrobial activity. *Journal of Agricultural and Food Chemistry, 54*, 3139–3145. Available from https://doi.org/10.1021/jf053183n.

Choulitoudi, E., Bravou, K., Bimpilas, A., Tsironi, T., Tsimogiannis, D., Taoukis, P., & Oreopoulou, V. (2016). Antimicrobial and antioxidant activity of *Satureja thymbra* in gilthead seabream fillets edible coating. *Food and Bioproducts Processing, 100*, 570–577. Available from https://doi.org/10.1016/j.fbp.2016.06.013.

Chowdhary, C. V., Meruva, A., Naresh, K., & Elumalai, R. K. A. (2013). A review on phytochemical and pharmacological profile of portulaca oleracea linn. (Purslane). *International Journal of Research in Ayurveda and Pharmacy, 4*, 34–37. Available from https://doi.org/10.7897/2277-4343.04119.

Cope, R. B. (2005). Allium species poisoning in dogs and cats. *Veterinary Medicine, 100*, 562–566.

Correia, R. T. P., Mccue, P., Vattem, D. A., Magalhães, M. M. A., MacÊdo, G. R., & Shetty, K. (2004). Amylase and helicobacter pylori inhibition by phenolic extracts of pineapple wastes bioprocessed by rhizopus oligosporus. *Journal of Food Biochemistry, 28*, 419–434. Available from https://doi.org/10.1111/j.1745-4514.2004.06003.x.

Cowan, M. M. (1999). Plant products as antimicrobial agents. *Clinical Microbiology Reviews*, *12*, 564–582. Available from https://doi.org/10.1128/cmr.12.4.564.

Cutillas, A.-B., Carrasco, A., Martinez-Gutierrez, R., Tomas, V., & Tudela, J. (2017). *Salvia officinalis* L. essential oil from Spain: Determination of composition, antioxidant capacity, antienzymatic and antimicrobial bioactivities. *Chemistry and Biodiversity*, *14*(8), e1700102.

Daglia, M. (2012). Polyphenols as antimicrobial agents. *Current Opinion in Biotechnology*, *23*, 174–181. Available from https://doi.org/10.1016/j.copbio.2011.08.007.

Daoud, Z., Abdo, E., & Abdel-Massih, R. M. (2011). Antibacterial activity of *Rheum rhaponticum*, *Olea europaea*, and *Viola odorata* on ESBL producing clinical isolates of *Escherichia coli* and *Klebsiella pneumoniae*. *International Journal of Pharmaceutical Sciences*, *2*, 1669–1678.

Daoud, Z., Sura, M., & Abdel-Massih, R. (2013). Pectin shows antibacterial activity against *Helicobacter pylori*. *Advances in Bioscience and Biotechnology*, *4*, 273–277.

Daryabeygi-Khotbehsara, R., Golzarand, M., Ghaffari, M. P., & Djafarian, K. (2017). *Nigella sativa* improves glucose homeostasis and serum lipids in type 2 diabetes: A systematic review and *meta*-analysis. *Complementary Therapies in Medicine*, *35*, 6–13. Available from https://doi.org/10.1016/j.ctim.2017.08.016.

De Feo, V., Ambrosio, C., & Senatore, F. (1992). Traditional phytotherapy in Caserta province, Campania, Southern Italy. *Fitoterapia*, *63*, 337–349.

De Feo, V., Aquino, R., Menghini, A., Ramundo, E., & Senatore, F. (1992). Traditional phytotherapy in the Peninsula Sorrentina, Campania, Southern Italy. *Journal of Ethnopharmacology*, *36*, 113–125. Available from https://doi.org/10.1016/0378-8741(92)90010-O.

deMelo, G. A. N., Grespan, R., Fonseca, J. P., Farinha, T. O., Silva, E. L., Romero, A. L., Bersani-Amado, C. A., & Cuman, R. K. N. (2011). *Rosmarinus officinalis* L. essential oil inhibits in vivo and in vitro leukocyte migration. *Journal of Medicinal Food*, *14*, 944–949. Available from https://doi.org/10.1089/jmf.2010.0159.

Demetzos, C., Angelopoulou, D., & Perdetzoglou, D. (2002). A comparative study of the essential oils of *Cistus salviifolius* in several populations of Crete (Greece). *Biochemical Systematics and Ecology*, *30*, 651–665. Available from https://doi.org/10.1016/S0305-1978(01)00145-4.

Di Novella, R., Di Novella, N., De Martino, L., Mancini, E., & De Feo, V. (2013). Traditional plant use in the National Park of Cilento and Vallo di Diano, Campania, Southern, Italy. *Journal of Ethnopharmacology*, *145*, 328–342. Available from https://doi.org/10.1016/j.jep.2012.10.065.

Di Sanzo, P., De Martino, L., Mancini, E., & De Feo, V. (2013). Medicinal and useful plants in the tradition of Rotonda, Pollino National Park, Southern Italy. *Journal of Ethnobiology and Ethnomedicine*, *9*(19). Available from https://doi.org/10.1186/1746-4269-9-19.

Dias, M. G., Camões, M. F. G. F. C., & Oliveira, L. (2009). Carotenoids in traditional Portuguese fruits and vegetables. *Food Chemistry*, *113*, 808–815. Available from https://doi.org/10.1016/j.foodchem.2008.08.002.

Dimas, K. S., Pantazis, P., & Ramanujam, R. (2012). Chios mastic gum: A plant-produced resin exhibiting numerous diverse pharmaceutical and biomedical properties. *In Vivo (Athens, Greece)*, *26*, 777–785. Available from http://iv.iiarjournals.org/content/26/5/777.full.pdf + html.

Dorman, H. J. D., Bachmayer, O., Kosar, M., & Hiltunen, R. (2004). Antioxidant properties of aqueous extracts from selected lamiaceae species grown in Turkey. *Journal of Agricultural and Food Chemistry*, *52*, 762–770. Available from https://doi.org/10.1021/jf034908v.

Dorman, H. J. D., & Deans, S. G. (2000). Antimicrobial agents from plants: Antibacterial activity of plant volatile oils. *Journal of Applied Microbiology*, *88*, 308–316. Available from https://doi.org/10.1046/j.1365-2672.2000.00969.x.

Economou, G., Panagopoulos, G., Karamanos, A., Tarantilis, P., Kalivas, D., & Kotoulas, V. (2014). An assessment of the behavior of carvacrol - Rich wild Lamiaceae species from the eastern Aegean under cultivation in two different environments. *Industrial Crops and Products*, *54*, 62–69. Available from https://doi.org/10.1016/j.indcrop.2013.12.044.

Eddouks, M., Ajebli, M., & Hebi, M. (2017). Ethnopharmacological survey of medicinal plants used in Daraa-Tafilalet region (Province of Errachidia), Morocco. *Journal of Ethnopharmacology*, *198*, 516–530. Available from https://doi.org/10.1016/j.jep.2016.12.017.

El Beyrouthy, M. (2008). Contribution à l'ethnopharmacologie libanaise et aux Lamiaceae du Liban, 1 vol. (607 f.).

El Beyrouthy, M. (2009). Contribution a l'ethnopharmacologie libanaise et aux Lamiaceae du Liban. *Acta Botanica Gallica*, *156*, 515–521.

El Beyrouthy, M., Arnold, N., Delelis-Dusollier, A., & Dupont, F. (2008). Plants used as remedies antirheumatic and antineuralgic in the traditional medicine of Lebanon. *Journal of Ethnopharmacology*, 315–334. Available from https://doi.org/10.1016/j.jep.2008.08.024.

El Beyrouthy, M., Dhifi, W., & Arnold, N. (2013). Ethnopharmacological survey of the indigenous Lamiaceae from Lebanon. *Acta Horticulturae, 997*, 257–275.

El Gendy, A. N., Leonardi, M., Mugnaini, L., Bertelloni, F., Ebani, V. V., Nardoni, S., Mancianti, F., Hendawy, S., Omer, E., & Pistelli, L. (2015). Chemical composition and antimicrobial activity of essential oil of wild and cultivated *Origanum syriacum* plants grown in Sinai, Egypt. *Industrial Crops and Products, 67*, 201–207. Available from https://doi.org/10.1016/j.indcrop.2015.01.038.

El Omari, K., Hamze, M., Alwan, S., Osman, M., Jama, C., & Chihib, N. E. (2019). In-vitro evaluation of the antibacterial activity of the essential oils of *Micromeria barbata, Eucalyptus globulus* and *Juniperus excelsa* against strains of *Mycobacterium tuberculosis* (including MDR), *Mycobacterium kansasii* and *Mycobacterium gordonae. Journal of Infection and Public Health, 12*, 615–618. Available from https://doi.org/10.1016/j.jiph.2019.01.058.

El Rabey, H. A., Al-Seeni, M. N., & Bakhashwain, A. S. (2017). The antidiabetic activity of *Nigella sativa* and propolis on streptozotocin-induced diabetes and diabetic nephropathy in male rats. *Evidence-Based Complementary and Alternative Medicine*, 5439645.

Elaissi, A., Salah, K. H., Mabrouk, S., Larbi, K. M., Chemli, R., & Harzallah-Skhiri, F. (2011). Antibacterial activity and chemical composition of 20 Eucalyptus species' essential oils. *Food Chemistry, 129*, 1427–1434. Available from https://doi.org/10.1016/j.foodchem.2011.05.100.

Erkan, N. (2012). Antioxidant activity and phenolic compounds of fractions from *Portulaca oleracea* L. *Food Chemistry, 133*, 775–781. Available from https://doi.org/10.1016/j.foodchem.2012.01.091.

Eruygur, N., Çetin, S., Ataş, M., & Çevik, O. (2017). A study on the antioxidant, antimicrobial and cytotoxic activity of *Thymbra spicata* L. var. spicata ethanol extract. *Cumhuriyet Medical Journal, 39*(531–538). Available from https://doi.org/10.7197/223.v39i31705.347450.

Estévez, M., Ramírez, R., Ventanas, S., & Cava, R. (2007). Sage and rosemary essential oils vs BHT for the inhibition of lipid oxidative reactions in liver pâté. *LWT - Food Science and Technology, 40*, 58–65. Available from https://doi.org/10.1016/j.lwt.2005.07.010.

Fabri, R. L., Nogueira, M. S., Dutra, L. B., Bouzada, M. L. M., & Scio, E. (2011). Potencial antioxidante e antimicrobiano de espécies da família asteraceae. *Revista Brasileira de Plantas Medicinais, 13*, 183–189. Available from https://doi.org/10.1590/S1516-05722011000200009.

Fahed, L. (2016). Diversité chimique et potentiel antimicrobien d'huiles essentielles de plantes libanaises. In *Museum National d'histoire Naturelle - MNHN PARIS*. Français, NNT.

Fahed, L., Khoury, M., Stien, D., Ouaini, N., Eparvier, V., & El Beyrouthy, M. (2017). Essential oils composition and antimicrobial activity of six conifers harvested in Lebanon. *Chemistry and Biodiversity, 14*, e1600235. Available from https://doi.org/10.1002/cbdv.201600235.

Ferreira, S., Santos, J., Duarte, A., Duarte, A. P., Queiroz, J. A., & Domingues, F. C. (2012). Screening of antimicrobial activity of *Cistus ladanifer* and *Arbutus unedo* extracts. *Natural Product Research, 26*, 1558–1560. Available from https://doi.org/10.1080/14786419.2011.569504.

Formisano, C., Rigano, D., Napolitano, F., Senatore, F., Arnold, N. A., Piozzi, F., & Rosselli, S. (2007). Volatile constituents of *Calamintha origanifolia* boiss. Growing wild in Lebanon. *Natural Product Communications, 2*, 1253–1256. Available from https://doi.org/10.1177/1934578x0700201213.

Formisano, C., Mignola, E., Rigano, D., Senatore, F., Bellone, G., Bruno, M., & Rosselli, S. (2007). Chemical composition and antimicrobial activity of the essential oil from aerial parts of *Micromeria fruticulosa* (Bertol.) Grande (Lamiaceae) growing wild in Southern Italy. *Flavour and Fragrance Journal, 22*, 289–292. Available from https://doi.org/10.1002/ffj.1795.

Formisano, C., Oliviero, F., Rigano, D., Saab, A. M., & Senatore, F. (2014). Chemical composition of essential oils and in vitro antioxidant properties of extracts and essential oils of *Calamintha origanifolia* and *Micromeria myrtifolia*, two Lamiaceae from the Lebanon flora. *Industrial Crops and Products, 62*, 405–411. Available from https://doi.org/10.1016/j.indcrop.2014.08.043.

Freile, M. L., Giannini, F., Pucci, G., Sturniolo, A., Rodero, L., Pucci, O., Balzareti, V., & Enriz, R. D. (2003). Antimicrobial activity of aqueous extracts and of berberine isolated from *Berberis heterophylla*. *Fitoterapia, 74*(7), 702–705. Available from https://doi.org/10.1016/S0367-326X(03)00156-4.

Galvão, L. C. D. C., Furletti, V. F., Bersan, S. M. F., da Cunha, M. G., Ruiz, A. L. T. G., de Carvalho, J. E., Sartoratto, A., Rehder, V. L. G., Figueira, G. M., Teixeira Duarte, M. C., Ikegaki, M., de Alencar, S. M., & Rosalen, P. L. (2012). Antimicrobial activity of essential oils against *Streptococcus mutans* and their antiproliferative effects. *Evidence-Based Complementary and Alternative Medicine, Article ID, 751435*, 12. Available from https://doi.org/10.1155/2012/751435.

Gault, S. M., & Synge, P. M. (1971). *The dictionary of roses in colour* (p. 191) London: Rainbird. References Books.

Gawron, G., Krzyczkowski, W., Lemke, K., Ołdak, A., Kadziński, L., & Banecki, B. (2019). *Nigella sativa* seed extract applicability in preparations against methicillin-resistant *Staphylococcus aureus* and effects on human dermal fibroblasts viability. *Journal of Ethnopharmacology, 244*, 112135. Available from https://doi.org/10.1016/j.jep.2019.112135.

Gedif, T., & Hahn, H. J. (2003). The use of medicinal plants in self-care in rural central Ethiopia. *Journal of Ethnopharmacology, 87*, 155–161. Available from https://doi.org/10.1016/S0378-8741(03)00109-0.

Gerhauser, C. (2005). Broad spectrum anti-infective potential of xanthohumol from hop (*Humulus lupulus* L.) in comparison with activities of other hop constituents and xanthohumol metabolites. *Molecular Nutrition Food Research, 49*, 827–831.

Gertsch, J. (2011). Botanical drugs, synergy, and network pharmacology: Forth and back to intelligent mixtures. *Planta Medica, 77*, 1086–1098. Available from https://doi.org/10.1055/s-0030-1270904.

Gilles, M., Zhao, J., An, M., & Agboola, S. (2010). Chemical composition and antimicrobial properties of essential oils of three Australian Eucalyptus species. *Food Chemistry, 119*, 731–737. Available from https://doi.org/10.1016/j.foodchem.2009.07.021.

Giweli, A., Džamic, A. M., Sokovic, M., Ristic, M. S., & Marin, P. D. (2012). Antimicrobial and antioxidant activities of essential oils of satureja thymbra growing wild in libya. *Molecules (Basel, Switzerland), 17*, 4836–4850. Available from https://doi.org/10.3390/molecules17054836.

Gonzalez, A. G., Abad, T., Jimenez, I. A., Ravelo, A. G., Aguiar, J. G. L. Z., Andrés, L., Plasencia, M., Herrera, J. R., & Moujir, L. (1989). A first study of antibacterial activity of diterpenes isolated from some *Salvia* species (Lamiaceae). *Biochemical Systematics and Ecology, 17*, 293–296. Available from https://doi.org/10.1016/0305-1978(89)90005-7.

González, J. A., & Vallejo, J. R. (2013). The scorpion in Spanish folk medicine: A review of traditional remedies for stings and its use as a therapeutic resource. *Journal of Ethnopharmacology, 146*, 62–74. Available from https://doi.org/10.1016/j.jep.2012.12.033.

González-Tejero, M. R., Casares-Porcel, M., Sánchez-Rojas, C. P., Ramiro-Gutierrez, J. M., Molero-Mesa, J., Pieroni, A., Giusti, M. E., Censorii, E., De Pasquale, C., Della, A., Paraskeva-Hadijchambi, D., Hadjichambis, A., Houmani, Z., El-Demerdash, M., ElZayat, M., Hmamouchi, M., & El-Johrig, S. (2008). Medicinal plants in the Mediterranean area: Synthesis of the results of the project Rubia. *Journal of Ethnopharmacology, 116*, 341–357. Available from https://doi.org/10.1016/j.jep.2007.11.045.

González-Trujano, M. E., Peña, E. I., Martínez, A. L., Moreno, J., Guevara-Fefer, P., Déciga-Campos, M., & López-Muñoz, F. J. (2007). Evaluation of the antinociceptive effect of *Rosmarinus officinalis* L. using three different experimental models in rodents. *Journal of Ethnopharmacology, 111*, 476–482. Available from https://doi.org/10.1016/j.jep.2006.12.011.

Guarino, C., De Simone, L., & Santoro, S. (2008). Ethnobotanical study of the Sannio area, Campania, Southern Italy. *Ethnobotany Research and Applications, 6*, 255–317. Available from https://doi.org/10.17348/era.6.0.255-317.

Guarrera, P. M., & Leporatti, M. L. (2007). Ethnobotanical remarks in Capitanata and Salento areas (Puglia, Southern Italy). *Etnobiologìa, 5*, 51–64.

Guimarães, R., Barros, L., Carvalho, A. M., & Ferreira, I. C. F. R. (2011). Infusions and decoctions of mixed herbs used in folk medicine: Synergism in antioxidant potential. *Phytotherapy Research, 25*, 1209–1214. Available from https://doi.org/10.1002/ptr.3366.

Gürdal, B., & Kültür, S. (2013). An ethnobotanical study of medicinal plants in Marmaris (Muğla, Turkey). *Journal of Ethnopharmacology, 146*, 113–126. Available from https://doi.org/10.1016/j.jep.2012.12.012.

Güvenç, A., Yıldız, S., Özkan, A. M., Erdurak, C. S., Coşkun, M., Yılmaz, G., Okuyama, T., & Okada, Y. (2005). Antimicrobiological studies on Turkish *Cistus* species. *Pharmaceutical Biology, 43*(2), 178–183. Available from https://doi.org/10.1080/13880200590919537.

Habtemariam, S. (2019). Chapter 19: The chemical and pharmacological basis of garlic (*Allium sativum* L.) as potential therapy for type 2 diabetes and metabolic syndrome. In: Medicinal foods as potential therapy for type-2 diabetes and associated diseases, 689–749.

Haile, K. T., Ayele, A. A., Mekuria, A. B., Demeke, C. A., Gebresillassie, B. M., & Erku, D. A. (2017). Traditional herbal medicine use among people living with HIV/AIDS in Gondar, Ethiopia: Do their health care providers know? *Complementary Therapies in Medicine, 35*, 14–19. Available from https://doi.org/10.1016/j.ctim.2017.08.019.

Hajar, L., Haïdar-Boustani, M., Khater, C., & Cheddadi, R. (2010). Environmental changes in Lebanon during the Holocene: Man vs. climate impacts. *Journal of Arid Environments, 74*, 746–755. Available from https://doi.org/10.1016/j.jaridenv.2008.11.002.

Hanafy, M. S. M., & Hatem, M. E. (1991). Studies on the antimicrobial activity of Nigella sativa seed (black cumin). *Journal of Ethnopharmacology, 34*, 275–278. Available from https://doi.org/10.1016/0378-8741(91)90047-H.

Hanci, S., Sahin, S., & Yilmaz, L. (2003). Isolation of volatile oil from thyme (*Thymbra spicata*) by steam distillation. *Nahrung - Food, 47*, 252–255. Available from https://doi.org/10.1002/food.200390059.

Hanlidou, E., Karousou, R., Kleftoyanni, V., & Kokkini, S. (2004). The herbal market of Thessaloniki (N Greece) and its relation to the ethnobotanical tradition. *Journal of Ethnopharmacology, 91*, 281–299. Available from https://doi.org/10.1016/j.jep.2004.01.007.

Hannan, A., Saleem, S., Chaudhary, S., Barkaat, M., & Arshad, M. U. (2008). Anti bacterial activity of Nigella sativa against clinical isolates of methicillin resistant *Staphylococcus aureus*. *Journal of Ayub Medical College, Abbottabad, 20*, 72–74.

Harkat-Madouri, L., Asma, B., Madani, K., Si Said, Z. B.-O., & Boulekbache-Makhlouf, L. (2015). Chemical composition, antibacterial and antioxidant activities of essential oil of *Eucalyptus globulus* from Algeria. *Industrial Crops and Products, 78*, 148–153. Available from https://doi.org/10.1016/j.indcrop.2015.10.015.

Hasan, N. A., Nawahwi, M. Z., & Malek, H. A. (2013). Antimicrobial activity of nigella sativa seed extract. *Sains Malaysiana, 42*, 143–147. Available from http://www.ukm.my/jsm/pdf_files/SM-PDF-42-2-2013/04%20NorAishah.pdf.

Husein, A. I., Ali-Shtayeh, M. S., Jamous, R. M., Zaitoun, S. Y. A., Jondi, W. J., & Zatar, N. A.-A. (2014). Antimicrobial activities of six plants used in traditional Arabic Palestinian herbal medicine. *African Journal of Microbiology Research*, 3501–3507. Available from https://doi.org/10.5897/AJMR2014.6921.

Hutschenreuther, A., Birkemeyer, C., Grötzinger, K., Straubinger, R. K., & Rauwald, H. W. (2010). Growth inhibiting activity of volatile oil from *Cistus creticus* L. against *Borrelia burgdorferi* s.s. in vitro. *Die Pharmazie, 65*, 290–295.

Huwez, F. U., Thirlwell, D., Cockayne, A., & Ala'Aldeen, D. A. A. (1998). Mastic gum kills *Helicobacter pylori*. *New England Journal of Medicine, 339*, 1946. Available from https://doi.org/10.1056/NEJM199812243392618.

Irshad, A. H., Pervaiz, A. H., Abrar, Y. B., Fahelboum, I., & Awen, B. Z. S. (2013). Antibacterial activity of *Berberis lycium* root extract. *Trakia Journal of Sciences, 11*, 88–90.

Islam, S. K., Ahsan, M., Hassan, C. M., & Malek, M. A. (1989). Antifungal activities of the oils of *Nigella sativa* seeds. *Pakistan Journal of Pharmaceutical Sciences, 2*, 25–28.

Jarić, S., Mitrović, M., Djurdjević, L., Kostić, O., Gajić, G., Pavlović, D., & Pavlović, P. (2011). Phytotherapy in medieval Serbian medicine according to the pharmacological manuscripts of the Chilandar Medical Codex (15-16th centuries). *Journal of Ethnopharmacology, 137*, 601–619. Available from https://doi.org/10.1016/j.jep.2011.06.016.

Jash, S. K., Gorai, D., & Roy, R. (2016). Salvia genus and triterpenoids. *International Journal of Pharmaceutical Sciences and Research, 7*, 4710–4732.

Jesionek, W., Móricz, Á. M., Ott, P. G., Kocsis, B., Horváth, G., & Choma, I. M. (2015). TLC-direct bioautography and LC/MS as complementary methods in identification of antibacterial agents in plant tinctures from the asteraceae family. *Journal of AOAC International, 98*, 857–861. Available from https://doi.org/10.5740/jaoacint.SGE2-Choma.

Jia, Y., Huang, F., Zhang, S., & Leung, S. W. (2012). Is danshen (*Salvia miltiorrhiza*) dripping pill more effective than isosorbide dinitrate in treating angina pectoris? A systematic review of randomized controlled trials. *International Journal of Cardiology, 157*, 330–340. Available from https://doi.org/10.1016/j.ijcard.2010.12.073.

Jurno, A. C., Netto, L. O. C., Duarte, R. S., & Machado, R. R. P. (2019). The search for plant activity against tuberculosis using breakpoints: A review. *Tuberculosis, 117*, 65–78.

Kacem, M., Kacem, I., Simon, G., Ben Mansour, A., Chaabouni, S., Elfeki, A., & Bouaziz, M. (2015). Phytochemicals and biological activities of *Ruta chalepensis* L. growing in Tunisia. *Food Bioscience, 12*, 73–83. Available from https://doi.org/10.1016/j.fbio.2015.08.001.

Kamatou, G. P. P., Makunga, N. P., Ramogola, W. P. N., & Viljoen, A. M. (2008). South African *Salvia* species: A review of biological activities and phytochemistry. *Journal of Ethnopharmacology, 119*, 664–672. Available from https://doi.org/10.1016/j.jep.2008.06.030.

Kamatou, G. P. P., Van Vuuren, S. F., Van Heerden, F. R., Seaman, T., & Viljoen, A. M. (2007). Antibacterial and antimycobacterial activities of South African *Salvia* species and isolated compounds from *S. chamelaeagnea*. *South African Journal of Botany, 73*, 552–557. Available from https://doi.org/10.1016/j.sajb.2007.05.001.

Karim, H., Boubaker, H., Askarne, L., Cherifi, K., Lakhtar, H., Msanda, F., Boudyach, E. H., & Ait Ben Aoumar, A. (2017). Use of *Cistus* aqueous extracts as botanical fungicides in the control of Citrus sour rot. *Microbial Pathogenesis, 104*, 263–267. Available from https://doi.org/10.1016/j.micpath.2017.01.041.

Kartalis, A., Didagelos, M., Georgiadis, I., Benetos, G., Smyrnioudis, N., Marmaras, H., . . . Andrikopoulos, G. (2016). Effects of Chios mastic gum on cholesterol and glucose levels of healthy volunteers: A prospective, randomized, placebo-controlled, pilot study (CHIOS-MASTIHA). *European Journal of Preventive Cardiology, 23*, 722–729.

Karuppiah, P., & Rajaram, S. (2012). Antibacterial effect of *Allium sativum* cloves and *Zingiber officinale* rhizomes against multiple-drug resistant clinical pathogens. *Asian Pacific Journal of Tropical Biomedicine, 2*, 597–601. Available from https://doi.org/10.1016/S2221-1691(12)60104-X.

Karygianni, L., Cecere, M., Skaltsounis, A. L., Argyropoulou, A., Hellwig, E., Aligiannis, N., Wittmer, A., & Al-Ahmad, A. (2014). High-level antimicrobial efficacy of representative Mediterranean natural plant extracts against oral microorganisms. *BioMed Research International 2014, Article ID, 839019*, 8. Available from https://doi.org/10.1155/2014/839019.

Kazemi, M. (2014). Chemical composition and antimicrobial activity of essential oil of *Matricaria chamomilla*. *Bulletin of Environment, Pharmacology and Life Sciences, 3*, 148–153.

Khalil, M., Khalifeh, H., Baldini, F., Salis, A., & Vergani, L. (2019). Antisteatotic and antioxidant activities of *Thymbra spicata* L. extracts in hepatic and endothelial cells as *in vitro* models of non-alcoholic fatty liver disease. *Journal of Ethnopharmacology, 239*, 111919. Available from https://doi.org/10.1016/j.jep.2019.111919.

Khan, M. R., Kihara, M., & Omoloso, A. D. (2001). Antimicrobial activity of *Clematis papuasica* and *Nauclea obversifolia*. *Fitoterapia, 72*, 575–578. Available from https://doi.org/10.1016/S0367-326X(01)00258-1.

Khlifi, D., Sghaier, R. M., Amouri, S., Laouini, D., Hamdi, M., & Bouajila, J. (2013). Composition and anti-oxidant, anti-cancer and anti-inflammatory activities of *Artemisia herba*-alba, *Ruta chalpensis* L. and *Peganum harmala* L. *Food and Chemical Toxicology, 55*, 202–208. Available from https://doi.org/10.1016/j.fct.2013.01.004.

Khoury, M., Stien, D., Eparvier, V., Ouaini, N., & El Beyrouthy, M. (2016). Report on the medicinal use of eleven lamiaceae species in lebanon and rationalization of their antimicrobial potential by examination of the chemical composition and antimicrobial activity of their essential oils. *Evidence-Based Complementary and Alternative Medicine, 2016*, 2547169. Available from https://doi.org/10.1155/2016/2547169.

Kivanc, M., & Akgül, A. (1988). Inhibitory effect of *Thymbra spicata* L. at various concentrations on growth of *Escherichia coli* under different temperatures. *Turkish Journal of Botany, 12*, 248–253.

Kizil, S. (2010). Determination of essential oil variations of *Thymbra spicata* var. spicata L. naturally growing in the wild flora of East Mediterranean and Southeastern Anatolia regions of Turkey. *Industrial Crops and Products, 32*, 593–600. Available from https://doi.org/10.1016/j.indcrop.2010.07.008.

Kokoska, L., Havlik, J., Valterova, I., Sovova, H., Sajfrtova, M., & Jankovska, I. (2008). Comparison of chemical composition and antibacterial activity of *Nigella sativa* seed essential oils obtained by different extraction methods. *Journal of Food Protection, 71*, 2475–2480. Available from https://doi.org/10.4315/0362-028X-71.12.2475.

Kordali, S., Cakir, A., Ozer, H., Cakmakci, R., Kesdek, M., & Mete, E. (2008). Antifungal, phytotoxic and insecticidal properties of essential oil isolated from Turkish Origanum acutidens and its three components, carvacrol, thymol and p-cymene. *Bioresource Technology, 99*, 8788–8795. Available from https://doi.org/10.1016/j.biortech.2008.04.048.

Korukluoglu, M., Sahan, Y., Yigit, A., & Ozer, E. T. (2008). Antibacterial activity and chemical constitutions. *Food Control, 34*, 383–396.

Kumar, A., Ilavarasan, R., Jayachandran, T., Decaraman, M., Aravindhan, P., Padmanabhan, N., & Krishnan, M. R. V. (2009). Phytochemicals investigation on a tropical plant, *Syzygium cumini* from Kattuppalayam, Erode District, Tamil Nadu, South India. *Pakistan Journal of Nutrition, 8*, 83–85. Available from https://doi.org/10.3923/pjn.2009.83.85.

Kumar, N., Bhandari, P., Singh, B., & Bari, S. S. (2009). Antioxidant activity and ultra-performance LC-electrospray ionization-quadrupole time-of-flight mass spectrometry for phenolics-based fingerprinting of Rose species: *Rosa damascena, Rosa bourboniana* and *Rosa brunonii. Food and Chemical Toxicology, 47*, 361–367. Available from https://doi.org/10.1016/j.fct.2008.11.036.

Kutlu, H. R. (2003). Screening medicinal and aromatic plant extracts from the Mediterranean region for antimicrobial, antioxidant and growth promoter effects to develop safe and sustainable feed additives. FP6-2002-FOOD. STREP (Small or medium-scale focused research projects) project submitted by EPSS.

Lawal, B., Shittu, O. K., Oibiokpa, F. I., Mohammed, H., Umar, S. I., & Haruna, G. M. (2016). Antimicrobial evaluation, acute and sub-acute toxicity studies of *Allium sativum. Journal of Acute Disease, 5*, 296–301. Available from https://doi.org/10.1016/j.joad.2016.05.002.

Lee, J. H., Park, J. H., Cho, H. S., Joo, S. W., Cho, M. H., & Lee, J. (2013). Anti-biofilm activities of quercetin and tannic acid against *Staphylococcus aureus. Biofouling, 29*, 491–499. Available from https://doi.org/10.1080/08927014.2013.788692.

Leto, C., Tuttolomondo, T., La Bella, S., & Licata, M. (2013). Ethnobotanical study in the Madonie Regional Park (Central Sicily, Italy) - Medicinal use of wild shrub and herbaceous plant species. *Journal of Ethnopharmacology, 146*, 90–112. Available from https://doi.org/10.1016/j.jep.2012.11.042.

Lev, E., & Amar, Z. (2002). Ethnopharmacological survey of traditional drugs sold in the Kingdom of Jordan. *Journal of Ethnopharmacology, 82*, 131–145. Available from https://doi.org/10.1016/S0378-8741(02)00182-4.

Loizzo, M. R., Menichini, F., Conforti, F., Tundis, R., Bonesi, M., Saab, A. M., Statti, G. A., de Cindio, B., Houghton, P. J., Menichini, F., & Frega, N. G. (2009). Chemical analysis, antioxidant, antiinflammatory and anticholinesterase activities of *Origanum ehrenbergii* Boiss and *Origanum syriacum* L. essential oils. *Food Chemistry, 117*, 174–180. Available from https://doi.org/10.1016/j.foodchem.2009.03.095.

Loizzo, M. R., Saab, A. M., Tundis, R., Menichini, F., Bonesi, M., Piccolo, V., Statti, G. A., de Cindio, B., Houghton, P. J., & Menichini, F. (2008). In vitro inhibitory activities of plants used in Lebanon traditional medicine against angiotensin converting enzyme (ACE) and digestive enzymes related to diabetes. *Journal of Ethnopharmacology, 119*(1), 109–116. Available from https://doi.org/10.1016/j.jep.2008.06.003.

Loughlin, M. F., Ala'Aldeen, D. A., & Jenks, P. J. (2003). Monotherapy with mastic does not eredicate *Helicobacter pylori* infection from mice. *Journal of Antimicrobial Chemotherapy, 51*, 367–371. Available from https://doi.org/10.1093/jac/dkg057.

Luís, Â., Neiva, D. M., Pereira, H., Gominho, J., Domingues, F., & Duarte, A. P. (2016). Bioassay-guided fractionation, GC–MS identification and in vitro evaluation of antioxidant and antimicrobial activities of bioactive compounds from *Eucalyptus globulus* stump wood methanolic extract. *Industrial Crops and Products, 91*, 97–103. Available from https://doi.org/10.1016/j.indcrop.2016.06.022.

Magiatis, P., Melliou, E., Skaltsounis, A. L., Chinou, I. B., & Mitaku, S. (1999). Chemical composition and antimicrobial activity of the essential oils of *Pistacia lentiscus* var. chia. *Planta Medica, 65*, 749–752. Available from https://doi.org/10.1055/s-2006-960856.

Mahdizadeh, S., Ghadiri, M. K., & Gorji, A. (2015). Avicenna's canon of medicine: A review of analgesics and anti-inflammatory substances. *Avicenna Journal of Phytomedicine, 5*, 182–202.

Mahmoudvand, H., Sepahvand, A., Jahanbakhsh, S., Ezatpour, B., & Ayatollahi Mousavi, S. A. (2014). Evaluation of antifungal activities of the essential oil and various extracts of *Nigella sativa* and its main component, thymoquinone against pathogenic dermatophyte strains. *Journal de Mycologie Medicale, 24*, e155–e161. Available from https://doi.org/10.1016/j.mycmed.2014.06.048.

Malik, T. A., Kamili, A. N., Chishti, M. Z., Ahad, S., Tantry, M. A., Hussain, P. R., & Johri, R. K. (2017). Breaking the resistance of *Escherichia coli*: Antimicrobial activity of *Berberis lycium* Royle. *Microbial Pathogenesis, 102*, 12–20. Available from https://doi.org/10.1016/j.micpath.2016.11.011.

Marchese, A., Orhan, I. E., Daglia, M., Barbieri, R., & Nabavi, S. M. (2016). Antibacterial and antifungal activities of thymol: A brief review of the literature. *Food Chemistry, 21*, 402–414. Available from https://doi.org/10.1016/j.foodchem.2016.04.111.

Marino, M., Bersani, C., & Comi, G. (1999). Antimicrobial activity of the essential oils of *Thymus vulgaris* L. measured using a bioimpedometric method. *Journal of Food Protection, 62*, 1017–1023. Available from https://doi.org/10.4315/0362-028X-62.9.1017.

Marino, M., Bersani, C., & Comi, G. (2001). Impedance measurements to study the antimicrobial activity of essential oils from Lamiaceae and Compositae. *International Journal of Food Microbiology, 67*, 187–195. Available from https://doi.org/10.1016/S0168-1605(01)00447-0.

Marković, T., Chatzopoulou, P., Siljegović, J., Nikolić, M., Glamočlija, J., Ćirić, A., & Soković, M. (2011). Chemical analysis and antimicrobial activities of the essential oils of Satureja Thymbra L. and Thymbra Spicata L. and their main components. *Archives of Biological Sciences*, *63*, 457–464. Available from https://doi.org/10.2298/ABS1102457M.

Marone, P., Bono, L., Leone, E., Bona, S., Carretto, E., & Perversi, L. (2001). Bactericidal activity of *Pistacia lentiscus* mastic gum against *Helicobacter pylori*. *Journal of Chemotherapy*, *13*, 611–614. Available from https://doi.org/10.1179/joc.2001.13.6.611.

Mastino, P. M., Marchetti, M., Costa, J., & Usai, M. (2017). Comparison of essential oils from Cistus species growing in Sardinia. *Natural Product Research*, *31*, 299–307. Available from https://doi.org/10.1080/14786419.2016.1236095.

Medail, F., & Quezel, P. (1999). Biodiversity hotspots in the Mediterranean Basin: Setting global conservation priorities. *Conservation Biology*, *13*, 1510–1513. Available from https://doi.org/10.1046/j.1523-1739.1999.98467.x.

Mekonnen, A., Yitayew, B., Tesema, A., & Taddese, S. (2016). In vitro antimicrobial activity of essential oil of *Thymus schimperi, Matricaria chamomilla, Eucalyptus globulus,* and *Rosmarinus officinalis*. *International Journal of Microbiology*, *2016*, 9545693.

Merghni, A., Noumi, E., Hadded, O., Dridi, N., & Snoussi, M. (2018). Assessment of the antibiofilm and antiquorum sensing activities of *Eucalyptus globulus* essential oil and its main component 1,8-cineole against methicillin-resistant *Staphylococcus aureus* strains. *Microbial Pathogenesis*, *118*, 74–80.

Menale, B., De Castro, O., Cascone, C., & Muoio, R. (2016). Ethnobotanical investigation on medicinal plants in the Vesuvio National Park (Campania, Southern Italy). *Journal of Ethnopharmacology*, *192*, 320–349. Available from https://doi.org/10.1016/j.jep.2016.07.049.

Menshikov, D. D., Lazareva, E. B., Popova, T. S., Shramko, L. U., Tokaev, I. S., Zalogueva, G. V., & Gaponova, I. N. (1997). Antimicrobial properties of pectins and their influence on antibiotic activity. *Antibiotiki i Khimioterapiya*, *42*, 10–15.

Metrouh-Amir, H., Duarte, C. M. M., & Maiza, F. (2015). Solvent effect on total phenolic contents, antioxidant, and antibacterial activities of *Matricaria pubescens*. *Industrial Crops and Products*, *67*, 249–256. Available from https://doi.org/10.1016/j.indcrop.2015.01.049.

Miara, M. D., Bendif, H., Rebbas, K., Rabah, B., Hammou, M. A., & Maggi, F. (2019). Medicinal plants and their traditional uses in the highland region of Bordj Bou Arreridj (Northeast Algeria). *Journal of Herbal Medicine*, *16*, 100262. Available from https://doi.org/10.1016/j.hermed.2019.100262.

Miara, M. D., Hammou, M. A., & Aoul, S. H. (2013). Phytothérapie et taxonomie des plantes médicinales spontanées dans la région de Tiaret (Algérie). *Phytothérapie*, *11*, 206–218. Available from https://doi.org/10.1007/s10298-013-0789-3.

Miguel, G., Simões, M., Figueiredo, A. C., Barroso, J. G., Pedro, L. G., & Carvalho, L. (2004). Composition and antioxidant activities of the essential oils of *Thymus caespititius, Thymus camphoratus* and *Thymus mastichina*. *Food Chemistry*, *86*, 183–188. Available from https://doi.org/10.1016/j.foodchem.2003.08.031.

Mileva, M. M., Kusovski, V. K., Krastev, D. S., Dobreva, A. M., & Galabov, A. S. (2014). Chemical composition, in vitro antiradical and antimicrobial activities of Bulgarian *Rosa alba* L. essential oil against some oral pathogens. *International Journal of Current Microbiology and Applied Sciences*, *3*, 11–20.

Ministry of Agriculture-Lebanon. (1996). Biological diversity of Lebanon - Country Study Report, UNEP.

Miraj, S., & Alesaeidi, S. (2016). A systematic review study of therapeutic effects of *Matricaria recuitta* chamomile (chamomile). *Electronic Physician*, *8*, 3024–3031. Available from https://doi.org/10.19082/3024.

Mizobuchi, S., & Sato, Y. (1984). A new flavanone with antifungal activity isolated from flops. *Agricultural and Biological Chemistry*, *48*, 2771–2775. Available from https://doi.org/10.1080/00021369.1984.10866564.

Mneimne, M., Baydoun, S., Nemer, N., & Arnold, A. N. (2016). Chemical composition and antimicrobial activity of essential oils isolated from aerial parts of *Prangos asperula* Boiss. (Apiaceae) growing wild in Lebanon. *Medicinal Plant Research*, *6*, 1–9. Available from https://doi.org/10.5376/mpr.2016.06.0003.

Moein, M., Zomorodian, K., Almasi, M., Pakshir, K., & Zarshenas, M. M. (2017). Preparation and analysis of *Rosa damascena* essential oil composition and antimicrobial activity assessment of related fractions. *Iranian Journal of Science and Technology, Transaction A: Science*, *41*, 87–94. Available from https://doi.org/10.1007/s40995-017-0220-2.

Moghaddam, A. M. D., Shayegh, J., Mikaili, P., & Sharaf, J. D. (2011). Antimicrobial activity of essential oil extract of *Ocimum basilicum* L. leaves on a variety of pathogenic bacteria. *Journal of Medicinal Plants Research*, *5*, 3453–3456. Available from http://www.academicjournals.org/JMPR/PDF/pdf2011/4Aug/Moghaddam%20et%20al.pdf.

Mohammad, S. M. (2011). Study on cammomile (*Matricaria chamomilla* L.) usage and farming. *Advances in Environmental Biology*, *5*, 1446–1453. Available from http://www.aensionline.com/aeb/2011/1446-1453.pdf.

Moir, M. (2000). Hops - A millennium review. *Journal of the American Society of Brewing Chemists*, *58*, 131–146. Available from https://doi.org/10.1094/asbcj-58-0131.

Moore, J., Yousef, M., & Tsiani, E. (2016). Anticancer effects of rosemary (*Rosmarinus officinalis* L.) extract and rosemary extract polyphenols. *Nutrients*, *8*, 731. Available from https://doi.org/10.3390/nu8110731.

Morales, P., Ferreira, I. C. F. R., Carvalho, A. M., Sánchez-Mata, M. C., Cámara, M., Fernández-Ruiz, V., de Santayana, M. P., & Tardío, J. (2014). Mediterranean non-cultivated vegetables as dietary sources of compounds with antioxidant and biological activity. *LWT - Food Science and Technology*, *55*(1), 389–396. Available from https://doi.org/10.1016/j.lwt.2013.08.017.

Morales-Soto, A., Oruna-Concha, M. J., Elmore, J. S., Barrajón-Catalán, E., Micol, V., Roldán, C., & Segura-Carretero, A. (2015). Volatile profile of Spanish Cistus plants as sources of antimicrobials for industrial applications. *Industrial Crops and Products*, *74*, 425–433. Available from https://doi.org/10.1016/j.indcrop.2015.04.034.

Morcol, T. B., Negrin, A., Matthews, P. D., & Kennelly, E. J. (2020). Hop (*Humulus lupulus* L.) terroir has large effect on a glycosylated green leaf volatile but not on other aroma glycosides. *Food Chemistry*, *321*, 126644. Available from https://doi.org/10.1016/j.foodchem.2020.126644.

Móricz, A. M., Szarka, S., Ott, P. G., Héthelyi, E. B., Szoke, E., & Tyihák, E. (2012). Separation and identification of antibacterial chamomile components using OPLC, bioautography and GC-MS. *Medicinal Chemistry*, *8*, 85–94. Available from https://doi.org/10.2174/157340612799278487.

Morsi, N. M. (2000). Antimicrobial effect of crude extracts of *Nigella sativa* on multiple antibiotics-resistant bacteria. *Acta Microbiologica Polonica*, *49*, 63–74.

Morteza-Semnani, K., Saeedi, M., Mahdavi, M. R., & Rahimi, F. (2006). Antimicrobial studies on extracts of three species of Phlomis. *Pharmaceutical Biology*, *44*, 426–429. Available from https://doi.org/10.1080/13880200600798445.

Mouterde, P. (1966). Nouvelle Flore du Liban et de la Syrie, Tome I, Editions de l'Imprimerie catholique, Beyrouth.

Mulyaningsih, S., Sporer, F., Reichling, J., & Wink, M. (2011). Antibacterial activity of essential oils from Eucalyptus and of selected components against multidrug-resistant bacterial pathogens. *Pharmaceutical Biology*, *49*, 893–899. Available from https://doi.org/10.3109/13880209.2011.553625.

Mulyaningsih, S., Sporer, F., Zimmermann, S., Reichling, J., & Wink, M. (2010). Synergistic properties of the terpenoids aromadendrene and 1,8-cineole from the essential oil of eucalyptus globulus against antibiotic-susceptible and antibiotic-resistant pathogens. *Phytomedicine: International Journal of Phytotherapy and Phytopharmacology*, *17*, 1061–1066. Available from https://doi.org/10.1016/j.phymed.2010.06.018.

Munir, N., Iqbal, A. S., Altaf, I., Bashir, R., Sharif, N., Saleem, F., & Naz, S. (2014). Evaluation of antioxidant and antimicrobial potential of two endangered plant species *Atropa belladonna* and *Matricaria chamomilla*. *African Journal of Traditional, Complementary and Alternative Medicines*, *11*, 111–117. Available from https://doi.org/10.4314/ajtcam.v11i5.18.

Nabavi, S. M., Marchese, A., Izadi, M., Curti, V., Daglia, M., & Nabavi, S. F. (2015). Plants belonging to the genus Thymus as antibacterial agents: From farm to pharmacy. *Food Chemistry*, *173*, 339–347. Available from https://doi.org/10.1016/j.foodchem.2014.10.042.

Najem, W., El Beyrouthy, M., Wakim, L. H., Neema, C., & Ouaini, N. (2011). Essential oil composition of *Rosa damascena* Mill. from different localities in Lebanon. *Acta Botanica Gallica*, *158*, 365–373. Available from https://doi.org/10.1080/12538078.2011.10516279.

Navarro, V., Villarreal, M. L., Rojas, G., & Lozoya, X. (1996). Antimicrobial evaluation of some plants used in Mexican traditional medicine for the treatment of infectious diseases. *Journal of Ethnopharmacology*, *53*, 143–147. Available from https://doi.org/10.1016/0378-8741(96)01429-8.

Ngo, S. N. T., Williams, D. B., & Head, R. J. (2011). Rosemary and cancer prevention: Preclinical perspectives. *Critical Reviews in Food Science and Nutrition*, *51*, 946–954. Available from https://doi.org/10.1080/10408398.2010.490883.

Nionelli, L., Pontonio, E., Gobbetti, M., & Rizzello, C. G. (2018). Use of hop extract as antifungal ingredient for bread making and selection of autochthonous resistant starters for sourdough fermentation. *International Journal of Food Microbiology*, *266*, 173–182. Available from https://doi.org/10.1016/j.ijfoodmicro.2017.12.002.

Nostro, A., Blanco, A., Cannatelli, M., Enea, V., Flamini, G., Morelli, I., Roccaro, A., & Alonzo, V. (2004). Susceptibility of methicillin-resistant staphylococci to oregano essential oil, carvacrol and thymol. *FEMS Microbiology Letters*, *230*, 191–195. Available from https://doi.org/10.1016/S0378-1097(03)00890-5.

O'Neill, J. (2016). Tackling drug-resistant infections globally: Final report and recommendations, review on antimicrobial resistance. https://amr-review.org/sites/default/files/160518_Final%20paper_with%20cover.pdf.

Obón, C., Rivera, D., Alcaraz, F., & Attieh, L. (2014). Beverage and culture. "Zhourat," a multivariate analysis of the globalization of a herbal tea from the Middle East. *Appetite*, *79*, 1–10. Available from https://doi.org/10.1016/j.appet.2014.03.024.

Ohsaki, A., Takashima, J., Chiba, N., & Kawamura, M. (1999). Microanalysis of a selective potent anti-*Helicobacter pylori* compound in a Brazilian medicinal plant, *Myroxylon peruiferum* and the activity of analogues. *Bioorganic and Medicinal Chemistry Letters*, *9*, 1109–1112. Available from https://doi.org/10.1016/S0960-894X(99)00141-9.

Olasupo, N. A., Fitzgerald, D. J., Gasson, M. J., & Narbad, A. (2003). Activity of natural antimicrobial compounds against *Escherichia coli* and *Salmonella enterica* serovar Typhimurium. *Letters in Applied Microbiology*, *37*, 448–451. Available from https://doi.org/10.1046/j.1472-765X.2003.01427.x.

Ooi, L. S. M., Li, Y., Kam, S. L., Wang, H., Wong, E. Y. L., & Ooi, V. E. C. (2006). Antimicrobial activities of Cinnamon oil and cinnamaldehyde from the Chinese medicinal herb Cinnamomum cassia Blume. *American Journal of Chinese Medicine*, *34*, 511–522. Available from https://doi.org/10.1142/S0192415X06004041.

Oosthuizen, C. B., Reid, A. M., & Lall, N. (2017). *Garlic (*Allium sativum*) and its associated molecules, as medicine. Medicinal plants for holistic health and well-being* (pp. 277–295). South Africa: Elsevier. Available from https://doi.org/10.1016/B978-0-12-812475-8.00009-3.

Oshugi, M., Basnet, P., Kadota, S., Isbii, E., Tamora, T., Okumura, Y., & Namba, T. (1997). Antibacterial activity of traditional medicines and an active constituent lupulone from *Humulus lupulus* against *Helicobacter pylori*. *Journal Traditional Medicine*, *14*, 186–191.

Özcan, M., & Boyraz, N. (2000). Antifungal properties of some herb decoctions. *European Food Research and Technology*, *212*, 86–88. Available from https://doi.org/10.1007/s002170000249.

Özdemir, A., Yildiz, M., Senol, F. S., Şimay, Y. D., & Ark, M. (2017). Promising anticancer activity of *Cyclotrichium niveum* L. extracts through induction of both apoptosis and necrosis. *Food and Chemical Toxicology*, *109*, 898–909. Available from https://doi.org/10.1016/j.fct.2017.03.062.

Özel, M. Z., Göğüş, F., & Lewis, A. C. (2006). Comparison of direct thermal desorption with water distillation and superheated water extraction for the analysis of volatile components of *Rosa damascena* Mill. using GCxGC-TOF/MS. *Analytica Chimica Acta*, *566*, 172–177. Available from https://doi.org/10.1016/j.aca.2006.03.014.

Ozkan, G., Sagdic, O., Baydar, N. G., & Baydar, H. (2004). Antioxidant and antibacterial activities of *Rosa damascena* flower extracts. *Food Science and Technology International*, *10*, 277–281.

Paarakh, P. M. (2010). *Nigella sativa* Linn.- A comprehensive review. *Indian Journal of Natural Products and Resources*, *1*, 409–429. Available from http://nopr.niscair.res.in/bitstream/123456789/10825/1/IJNPR%201(4)%20409-429.pdf.

Pachi, V. K., Mikropoulou, E. V., Gkiouvetidis, P., Siafakas, K., Argyropoulou, A., Angelis, A., Mitakou, S., & Halabalaki, M. (2020). Traditional uses, phytochemistry and pharmacology of Chios mastic gum (*Pistacia lentiscus* var. Chia, Anacardiaceae): A review. *Journal of Ethnopharmacology, Article*, 112485. Available from https://doi.org/10.1016/j.jep.2019.112485.

Pan, S. Y., Litscher, G., Gao, S. H., Zhou, S. F., Yu, Z. L., Chen, H. Q., Zhang, S. F., Tang, M. K., Sun, J. N., & Ko, K. M. (2014). Historical perspective of traditional indigenous medical practices: The current renaissance and conservation of herbal resources. *Evidence-Based Complementary and Alternative Medicine, Article ID 525340*, 20. Available from https://doi.org/10.1155/2014/525340.

Paraschos, S., Magiatis, P., Gousia, P., Economou, V., Sakkas, H., Papadopoulou, C., & Skaltsounis, A. L. (2011). Chemical investigation and antimicrobial properties of mastic water and its major constituents. *Food Chemistry*, *129*, 907–911. Available from https://doi.org/10.1016/j.foodchem.2011.05.043.

Paraschos, S., Mitakou, S., & Skaltsounis, A. L. (2012). Chios gum mastic: A review of its biological activities. *Current Medicinal Chemistry*, *19*, 2292–2302. Available from https://doi.org/10.2174/092986712800229014.

Parekh, J., Jadeja, D., & Chanda, S. (2005). Efficacy of aqueous and methanol extracts of some medicinal plants for potential antibacterial activity. *Turkish Journal of Biology*, *29*, 203–210.

Passalacqua, N. G., Guarrera, P. M., & De Fine, G. (2007). Contribution to the knowledge of the folk plant medicine in Calabria region (Southern Italy). *Fitoterapia*, *78*, 52–68. Available from https://doi.org/10.1016/j.fitote.2006.07.005.

Pauli, A. (2006). α-bisabolol from Chamomile - A specific ergosterol biosynthesis inhibitor? *International Journal of Aromatherapy*, *16*, 21–25. Available from https://doi.org/10.1016/j.ijat.2006.01.002.

Penso, G. (1980). The role of WHO in the selection and characterization of medicinal plants (vegetable drugs). *Journal of Ethnopharmacology*, *2*(2), 183–188. Available from https://doi.org/10.1016/0378-8741(80)90013-6.

Pereira, V., Dias, C., Vasconcelos, M. C., Rosa, E., & Saavedra, M. J. (2014). Antibacterial activity and synergistic effects between *Eucalyptus globulus* leaf residues (essential oils and extracts) and antibiotics against several isolates of respiratory tract infections (*Pseudomonas aeruginosa*). *Industrial Crops and Products*, *52*, 1–7. Available from https://doi.org/10.1016/j.indcrop.2013.09.032.

Petropoulos, S., Karkanis, A., Fernandes, Â., Barros, L., Ferreira, I. C. F. R., Ntatsi, G., Petrotos, K., Lykas, C., & Khah, E. (2015). Chemical composition and yield of six genotypes of common purslane (*Portulaca oleracea* L.): An alternative source of omega-3 fatty acids. *Plant Foods for Human Nutrition*, *70*, 420–426. Available from https://doi.org/10.1007/s11130-015-0511-8.

Petropoulos, S. A., Karkanis, A., Martins, N., & Ferreira, I. C. F. R. (2018). Edible halophytes of the Mediterranean basin: Potential candidates for novel food products. *Trends in Food Science and Technology*, *74*, 69–84. Available from https://doi.org/10.1016/j.tifs.2018.02.006.

Pieroni, A. (1999). Gathered wild food plants in the upper valley of the Serchio river (Garfagnana), central Italy. *Economic Botany*, *53*, 327–341. Available from https://doi.org/10.1007/BF02866645.

Pieroni, A., Nebel, S., Quave, C., & Heinrich, M. (2002). Ethnopharmacology of liakra: Traditionally weedy vegatables of the Arbereshe of the vulture area in Southern Italy. *Journal of Ethnopharmacology*, *81*, 165–186.

Pise, H. N., & Padwal, S. L. (2017). Evaluation of anti-inflammatory activity of nigella sativa: An experimental study. *National Journal of Physiology, Pharmacy and Pharmacology*, *7*, 707–711. Available from https://doi.org/10.5455/njppp.2017.7.0204705032017.

Pollio, A., De Natale, A., Appetiti, E., Aliotta, G., & Touwaide, A. (2008). Continuity and change in the Mediterranean medical tradition: *Ruta* spp. (rutaceae) in Hippocratic medicine and present practices. *Journal of Ethnopharmacology*, *116*, 469–482. Available from https://doi.org/10.1016/j.jep.2007.12.013.

Puupponen-Pimiä, R., Nohynek, L., Meier, C., Kähkönen, M., Heinonen, M., Hopia, A., & Oksman-Caldentey, K.-M. (2001). Antimicrobial properties of phenolic compounds from berries. *Journal of Applied Microbiology*, *90*, 494–507. Available from https://doi.org/10.1046/j.1365-2672.2001.01271.x.

Ramadan, B. K., Schaalan, M. F., & Tolba, A. M. (2017). Hypoglycemic and pancreatic protective effects of *Portulaca oleracea* extract in alloxan induced diabetic rats. *BMC Complementary and Alternative Medicine*, *17*, 37. Available from https://doi.org/10.1186/s12906-016-1530-1.

Rašković, A., Milanović, I., Pavlović, N., Ćebović, T., Vukmirović, S., & Mikov, M. (2014). Antioxidant activity of rosemary (*Rosmarinus officinalis* L.) essential oil and its hepatoprotective potential. *BMC Complementary and Alternative Medicine*, *14*, 225. Available from https://doi.org/10.1186/1472-6882-14-225.

Rauf, A., Patel, S., Uddin, G., Siddiqui, B. S., Ahmad, B., Muhammad, N., Mabkhot, Y. N., & Hadda, T. B. (2017). Phytochemical, ethnomedicinal uses and pharmacological profile of genus *Pistacia*. *Biomedicine and Pharmacotherapy*, *86*, 393–404. Available from https://doi.org/10.1016/j.biopha.2016.12.017.

Rota, M. C., Herrera, A., Martínez, R. M., Sotomayor, J. A., & Jordán, M. J. (2008). Antimicrobial activity and chemical composition of *Thymus vulgaris*, *Thymus zygis* and *Thymus hyemalis* essential oils. *Food Control*, *19*, 681–687. Available from https://doi.org/10.1016/j.foodcont.2007.07.007.

Sadlon, A. E., & Lamson, D. W. (2010). Immune-modifying and antimicrobial effects of eucalyptus oil and simple inhalation devices. *Alternative Medicine Review*, *15*, 33–47.

Sağdiç, O., Kuşçu, A., Özcan, M., & Özçelik, S. (2002). Effects of Turkish spice extracts at various concentrations on the growth of *Escherichia coli* O157:H7. *Food Microbiology*, *19*, 473–480. Available from https://doi.org/10.1006/fmic.2002.0494.

Sağdıç, O., & Özcan, M. (2003). Antibacterial activity of Turkish spice hydrosols. *Food Control*, *14*, 141–143. Available from https://doi.org/10.1016/S0956-7135(02)00057-9.

Saidi, M., Ghafourian, S., Zarin-Abaadi, M., Movahedi, K., & Sadeghifard, N. (2012). In vitro antimicrobial and antioxidant activity of black thyme (*Thymbra spicata* L.) essential oils. *Roumanian Archives of Microbiology and Immunology*, *71*, 61–69.

Salari, M. H., Amine, G., Shirazi, M. H., Hafezi, R., & Mohammadypour, M. (2006). Antibacterial effects of *Eucalyptus globulus* leaf extract on pathogenic bacteria isolated from specimens of patients with respiratory tract disorders. *Clinical Microbiology and Infection*, *12*, 194–196. Available from https://doi.org/10.1111/j.1469-0691.2005.01284.x.

Salehi, B., Sharifi-Rad, J., Quispe, C., Llaique, H., & Martins, N. (2019). Insights into *Eucalyptus* genus chemical constituents, biological activities and health-promoting effects. *Trends in Food Science & Technology, 91*, 609–624. Available from https://doi.org/10.1016/j.tifs.2019.08.003.

Salin, O. P., Pohjala, L. L., Saikku, P., Vuorela, H. J., Leinonen, M., & Vuorela, P. M. (2011). Effects of coadministration of natural polyphenols with doxycycline or calcium modulators on acute *Chlamydia pneumoniae* infection in vitro. *Journal of Antibiotics, 64*, 747–752. Available from https://doi.org/10.1038/ja.2011.79.

Sargin, S. A. (2015). Ethnobotanical survey of medicinal plants in Bozyazı district of Mersin, Turkey. *Journal of Ethnopharmacology, 173*, 105–126. Available from https://doi.org/10.1016/j.jep.2015.07.009.

Savo, V., Caneva, G., Maria, G. P., & David, R. (2011). Folk phytotherapy of the Amalfi Coast (Campania, Southern Italy). *Journal of Ethnopharmacology, 135*, 376–392. Available from https://doi.org/10.1016/j.jep.2011.03.027.

Sharifi-Rad, M., Nazaruk, J., Polito, L., Morais-Braga, M. F. B., & Sharifi-Rad, J. (2018a). Matricaria genus as a source of antimicrobial agents: From farm to pharmacy and food applications. *Microbiological Research, 215*, 76–88. Available from https://doi.org/10.1016/j.micres.2018.06.010.

Sharifi-Rad, M., Ozcelik, B., Altın, G., Daşkaya-Dikmen, C., Martorell, M., Ramírez-Alarcón, K., Alarcón-Zapatae, P., Morais-Braga, M. F. B., Carneiro, J. N. P., Leal, A. L. A. B., Coutinho, H. D. M., Gyawali, R., Tahergorabi, R., Ibrahim, S. A., Sahrifi-Rad, R., Sharopov, F., Salehi, B., Contreras, M. D. M., Segura-Carretero, A., Sen, S., Acharyao, K., & Sharifi-Rad, J. (2018b). *Salvia* spp. plants-from farm to food applications and phytopharmacotherapy. *Trends in Food Science & Technology, 80*, 242–263. Available from https://doi.org/10.1016/j.tifs.2018.08.008.

Sharma, K., Mahato, N., & Lee, Y. R. (2018). Systematic study on active compounds as antibacterial and antibiofilm agent in aging onions. *Journal of Food and Drug Analysis, 26*, 518–528. Available from https://doi.org/10.1016/j.jfda.2017.06.009.

Shehadeh, M., Jaradat, N., Al-Masri, M., Zaid, A. N., & Darwish, R. (2019). Rapid, cost-effective and organic solvent-free production of biologically active essential oil from Mediterranean wild *Origanum syriacum*. *Saudi Pharmaceutical Journal, 27*, 612–618. Available from https://doi.org/10.1016/j.jsps.2019.03.001.

Shehadeh, M., Suaifan, G., & Darwish, R. (2017). Complementary and alternative modalities; a new vein in weight control and reduction interventions. A pilot study in Jordan. *International Biological and Biomedical Journal., 2*, 1–5.

Shehadeh, M., Silvio, S., Ghadeer, A., Darwish, R. M., Giangaspero, A., Vassallo, A., Lepore, L., Oran, S. A., Hammad, H., & Tubaro, A. (2014). Topical anti-inflammatory potential of six *Salvia* species grown in Jordan. *Jordan Journal of Pharmaceutical Sciences, 7*, 153–161.

Shoara, R., Hashempur, M. H., Ashraf, A., Salehi, A., Dehshahri, S., & Habibagahi, Z. (2015). Efficacy and safety of topical *Matricaria chamomilla* L. (chamomile) oil for knee osteoarthritis: A randomizedcontrolled clinical trial. *Complementary Therapies in Clinical Practice, 21*, 181–187. Available from https://doi.org/10.1016/j.ctcp.2015.06.003.

Shohayeb, M., Abdel-Hameed, E. S. S., Bazaid, S. A., & Maghrabi, I. (2014). Antibacterial and antifungal activity of *Rosa damascena* MILL. essential oil, different extracts of rose petals. *Global Journal of Pharmacology, 8*, 1–7. Available from https://doi.org/10.5829/idosi.gjp.2014.8.1.81275.

Silva, J., Abebe, W., Sousa, S. M., Duarte, V. G., Machado, M. I. L., & Matos, F. J. A. (2003). Analgesic and anti-inflammatory effects of essential oils of Eucalyptus. *Journal of Ethnopharmacology, 89*, 277–283. Available from https://doi.org/10.1016/j.jep.2003.09.007.

Simões, M., Bennett, R. N., & Rosa, E. A. S. (2009). Understanding antimicrobial activities of phytochemicals against multidrug resistant bacteria and biofilms. *Natural Product Reports, 26*, 746–757. Available from https://doi.org/10.1039/b821648g.

Simpson, W. J., & Smith, A. R. W. (1992). Factors affecting antibacterial activity of hop compounds and their derivatives. *Journal of Applied Bacteriology, 72*, 327–334. Available from https://doi.org/10.1111/j.1365-2672.1992.tb01843.x.

Singh, B. R., Singh, V., Singh, R. K., & Ebibeni, N. (2011). Antimicrobial activity of lemongrass (*Cymbopogon citratus*) oil against microbes of environmental, clinical and food origin. *International Research Journal of Pharmacy and Pharmacology, 1*, 228–236.

Singh, O., Khanam, Z., Misra, N., & Srivastava, M. K. (2011). Chamomile (*Matricaria chamomilla* L.): An overview. *Pharmacognosy Reviews, 5*, 82–95. Available from https://doi.org/10.4103/0973-7847.79103.

Škerget, M., Majhenič, L., Bezjak, M., & Knez, Z. (2009). Antioxidant, radical scavenging and antimicrobial activities of red onion (*Allium cepa* L) skin and edible part extracts. *Chemical and Biochemical Engineering Quarterly, 23*, 435–444.

Skoula, M., Grayer, R. J., & Kite, G. C. (2005). Surface flavonoids in *Satureja thymbra* and *Satureja spinosa* (Lamiaceae). *Biochemical Systematics and Ecology, 33*, 541–544. Available from https://doi.org/10.1016/j.bse.2004.10.003.

Sokmen, A., Jones, B. M., & Erturk, M. (1999). The in vitro antibacterial activity of Turkish medicinal plants. *Journal of Ethnopharmacology, 67*, 79–86. Available from https://doi.org/10.1016/S0378-8741(98)00189-5.

Soković, M., Glamočlija, J., Marin, P. D., Brkić, D., & van Griensven, L. J. L. D. (2010). Antibacterial effects of the essential oils of commonly consumed medicinal herbs using an in vitro model. *Molecules (Basel, Switzerland), 15*, 7532–7546. Available from https://doi.org/10.3390/molecules15117532.

Stanković, M. S., Radić, Z. S., Blanco-Salas, J., Vázquez-Pardo, F. M., & Ruiz-Téllez, T. (2017). Screening of selected species from Spanish flora as a source of bioactive substances. *Industrial Crops and Products, 95*, 493–501. Available from https://doi.org/10.1016/j.indcrop.2016.09.070.

Stefanaki, A., Cook, C. M., Lanaras, T., & Kokkini, S. (2018). Essential oil variation of *Thymbra spicata* L. (Lamiaceae), an East Mediterranean "oregano" herb. *Biochemical Systematics and Ecology, 80*, 63–69. Available from https://doi.org/10.1016/j.bse.2018.06.006.

Tada, Y., Shikishima, Y., Takaishi, Y., Shibata, H., Higuti, T., Honda, G., Ito, M., Takeda, Y., Kodzhimatov, O. K., Ashurmetov, O., & Ohmoto, Y. (2002). Coumarins and γ-pyrone derivatives from *Prangos pabularia*: Antibacterial activity and inhibition of cytokine release. *Phytochemistry, 59*, 649–654. Available from https://doi.org/10.1016/S0031-9422(02)00023-7.

Tajehmiri, A., Ghasemi, M., Sabet, F., Chakoosari, N., & Abdolahzadegan. (2014). Antimutagenic activity of *Rosmarinus officinalis* L. by Ames test. *Scientific Journal of Microbiology, 3*(7), JR_SJM-3-7_002.

Takahashi, K., Fukazawa, M., Motohira, H., Ochiai, K., Nishikawa, H., & Miyata, T. (2003). A pilot study on anti-plaque effects of mastic chewing gum in the oral cavity. *Journal of Periodontology, 74*, 501–505. Available from https://doi.org/10.1902/jop.2003.74.4.501.

Talhouk, S. N., Zurayk, R., & Khuri, S. (2001). Conservation of the coniferous forests of Lebanon: Past, present and future prospects. *Oryx, 35*, 206–215. Available from https://doi.org/10.1046/j.1365-3008.2001.00180.x.

Tassou, C. C., & Nychas, G. J. E. (1995). Antimicrobial activity of the essential oil of mastic gum (*Pistacia lentiscus* var. chia) on Gram positive and Gram negative bacteria in broth and in Model Food System. *International Biodeterioration and Biodegradation, 36*, 411–420. Available from https://doi.org/10.1016/0964-8305(95)00103-4.

Tepe, B., Sokmen, M., Sokmen, A., Daferera, D., & Polissiou, M. (2005). Antimicrobial and antioxidative activity of the essential oil and various extracts of *Cyclotrichium origanifolium* (Labill.) Manden. & Scheng. *Journal of Food Engineering, 69*, 335–342. Available from https://doi.org/10.1016/j.jfoodeng.2004.08.024.

Tetik, F., Civelek, S., & Cakilcioglu, U. (2013). Traditional uses of some medicinal plants in Malatya (Turkey). *Journal of Ethnopharmacology, 146*, 331–346. Available from https://doi.org/10.1016/j.jep.2012.12.054.

Teuber, M., & Schmalreck, A. F. (1973). Membrane leakage in *Bacillus subtilis* 168 induced by the hop constituents lupulone, humulone, isohumulone and humulinic acid. *Archiv Für Mikrobiologie, 94*, 159–171. Available from https://doi.org/10.1007/BF00416690.

Toama, M. A., El-Alfy, T. S., & El-Fatatry, H. M. (1974). Antimicrobial activity of the volatile oil of *Nigella sativa* Linneaus seeds. *Antimicrobial Agents and Chemotherapy, 6*, 225–226. Available from https://doi.org/10.1128/AAC.6.2.225.

Tohidpour, A., Sattari, M., Omidbaigi, R., Yadegar, A., & Nazemi, J. (2010). Antibacterial effect of essential oils from two medicinal plants against Methicillin-resistant *Staphylococcus aureus* (MRSA). *Phytomedicine: International Journal of Phytotherapy and Phytopharmacology, 17*, 142–145. Available from https://doi.org/10.1016/j.phymed.2009.05.007.

Tomás-Menor, L., Morales-Soto, A., Barrajón-Catalán, E., Roldán-Segura, C., Segura-Carretero, A., & Micol, V. (2013). Correlation between the antibacterial activity and the composition of extracts derived from various Spanish Cistus species. *Food and Chemical Toxicology, 55*, 313–322. Available from https://doi.org/10.1016/j.fct.2013.01.006.

Triantafyllou, A., Chaviaras, N., Sergentanis, T. N., Protopapa, E., & Tsaknis, J. (2007). Chios mastic gum modulates serum biochemical parameters in a human population. *Journal of Ethnopharmacology, 111*, 43–49. Available from https://doi.org/10.1016/j.jep.2006.10.031.

Trombetta, D., Castelli, F., Sarpietro, M. G., Venuti, V., Cristani, M., Daniele, C., Saija, A., Mazzanti, G., & Bisignano, G. (2005). Mechanisms of antibacterial action of three monoterpenes. *Antimicrobial Agents and Chemotherapy, 49*, 2474–2478. Available from https://doi.org/10.1128/AAC.49.6.2474-2478.2005.

Trouillas, P., Calliste, C. A., Allais, D. P., Simon, A., Marfak, A., Delage, C., & Duroux, J. L. (2003). Antioxidant, anti-inflammatory and antiproliferative properties of sixteen water plant extracts used in the Limousin countryside as herbal teas. *Food Chemistry*, *80*, 399–407. Available from https://doi.org/10.1016/S0308-8146(02)00282-0.

Tsai, T. H., Tsai, T. H., Wu, W. H., Tseng, J. T. P., & Tsai, P. J. (2010). In vitro antimicrobial and anti-inflammatory effects of herbs against *Propionibacterium acnes*. *Food Chemistry*, *119*, 964–968. Available from https://doi.org/10.1016/j.foodchem.2009.07.062.

Tuttolomondo, T., Licata, M., Leto, C., Letizia Gargano, M., Venturella, G., & La Bella, S. (2014). Plant genetic resources and traditional knowledge on medicinal use of wild shrub and herbaceous plant species in the Etna Regional Park (Eastern Sicily, Italy). *Journal of Ethnopharmacology*, *155*, 1362–1381. Available from https://doi.org/10.1016/j.jep.2014.07.043.

Tyagi, A. K., & Malik, A. (2011). Antimicrobial potential and chemical composition of *Eucalyptus globulus* oil in liquid and vapour phase against food spoilage microorganisms. *Food Chemistry*, *126*, 228–235. Available from https://doi.org/10.1016/j.foodchem.2010.11.002.

Uddin, M. K., Juraimi, A. S., Hossain, M. S., Nahar, M. A. U., Ali, M. E., & Rahman, M. M. (2014). Purslane weed (*Portulaca oleracea*): A prospective plant source of nutrition, omega-3 fatty acid, and antioxidant attributes. *The Scientific World Journal*, *2014*, 951019. Available from https://doi.org/10.1155/2014/951019.

Ulubelen, A., Topcu, G., & Johansson, C. B. (1997). Norditerpenoids and diterpenoids from *Salvia multicaulis* with antituberculous activity. *Journal of Natural Products*, *60*, 1275–1280. Available from https://doi.org/10.1021/np9700681.

Ulukanli, Z., Cigremis, Y., & Ilcim, A. (2011). In vitro antimicrobial and antioxidant activity of acetone and methanol extracts from Thymus leucotrichius (Lamiaceae). *European Review for Medical and Pharmacological Sciences*, *15*, 649–657.

Ustün, O., Ozçelik, B., Akyön, Y., Abbasoglu, U., & Yesilada, E. (2006). Flavonoids with anti-*Helicobacter pylori* activity from *Cistus laurifolius* leaves. *Journal of Ethnopharmacology*, *108*, 457–461. Available from https://doi.org/10.1016/j.jep.2006.06.001.

Verdeguer, M., Blázquez, M. A., & Boira, H. (2012). Chemical composition and herbicidal activity of the essential oil from a *Cistus ladanifer* L. population from Spain. *Natural Product Research*, *26*, 1602–1609. Available from https://doi.org/10.1080/14786419.2011.592835.

Viapiana, A., Konopacka, A., Waleron, K., & Wesolowski, M. (2017). *Cistus incanus* L. commercial products as a good source of polyphenols in human diet. *Industrial Crops and Products*, *107*, 297–304. Available from https://doi.org/10.1016/j.indcrop.2017.05.066.

Viuda-Martos, M., El Gendy, A. E. N. G. S., Sendra, E., Fernández-López, J., El Razik, K. A. A., Omer, E. A., & Pérez-Alvarezj, J. A. (2010). Chemical composition and antioxidant and anti-Listeria activities of essential oils obtained from some Egyptian plants. *Journal of Agricultural and Food Chemistry*, *58*, 9063–9070. Available from https://doi.org/10.1021/jf101620c.

Walker, J. B., Sytsma, K. J., Treutlein, J., & Wink, M. (2004). Salvia (Lamiaceae) is not monophyletic: Implications for the systematics, radiation, and ecological specializations of Salvia and tribe Mentheae. *American Journal of Botany*, *91*, 1115–1125. Available from https://doi.org/10.3732/ajb.91.7.1115.

Waller, S. B., Cleff, M. B., Serra, E. F., Silva, A. L., & Meireles, M. C. A. (2017). Plants from Lamiaceae family as source of antifungal molecules in humane and veterinary medicine. *Microbial Pathogenesis*, *104*, 232–237. Available from https://doi.org/10.1016/j.micpath.2017.01.050.

Wojtunik-Kulesza, K. A., Oniszczuk, A., Oniszczuk, T., & Waksmundzka-Hajnos, M. (2016). The influence of common free radicals and antioxidants on development of Alzheimer's Disease. *Biomedicine and Pharmacotherapy*, *78*, 39–49. Available from https://doi.org/10.1016/j.biopha.2015.12.024.

Yamaguchi, N., Satoh-Yamaguchi, K., & Ono, M. (2009). In vitro evaluation of antibacterial, anticollagenase, and antioxidant activities of hop components (*Humulus lupulus*) addressing acne vulgaris. *Phytomedicine: International Journal of Phytotherapy and Phytopharmacology*, *16*, 369–376. Available from https://doi.org/10.1016/j.phymed.2008.12.021.

Yasar, S., Sagdic, O., & Kisioglu, A. N. (2005). In vitro antibacterial effects of single or combined plant extracts. *Journal of Food, Agriculture and Environment*, *3*, 39–43.

Yemmen, M., Landolsi, A., Ben Hamida, J., Mégraud, F., & Trabelsi Ayadi, M. (2017). Antioxidant activities, anticancer activity and polyphenolics profile, of leaf, fruit and stem extracts of *Pistacia lentiscus* from Tunisia. *Cellular and Molecular Biology*, *63*, 87–95. Available from https://doi.org/10.14715/cmb/2017.63.9.16.

Yesilada, E., & Küpeli, E. (2007). *Clematis vitalba* L. aerial part exhibits potent anti-inflammatory, antinociceptive and antipyretic effects. *Journal of Ethnopharmacology, 110*, 504–515. Available from https://doi.org/10.1016/j.jep.2006.10.016.

Yi, F., Sun, J., Bao, X., Ma, B., & Sun, M. (2019). Influence of molecular distillation on antioxidant and antimicrobial activities of rose essential oils. *LWT, 102*, 310–316. Available from https://doi.org/10.1016/j.lwt.2018.12.051.

Yimer, E. M., Tuem, K. B., Karim, A., Ur-Rehman, N., & Anwar, F. (2019). *Nigella sativa* L. (Black Cumin): A promising natural remedy for wide range of illnesses. *Evidence-Based Complementary and Alternative Medicine, 2019*, 1528635. Available from https://doi.org/10.1155/2019/1528635.

YouGuo, C., ZongJi, S., & XiaoPing, C. (2009). Evaluation of free radicals scavenging and immunity-modulatory activities of Purslane polysaccharides. *International Journal of Biological Macromolecules, 45*, 448–452. Available from https://doi.org/10.1016/j.ijbiomac.2009.07.009.

Zakaria, A., Jais, M. R., & Ishak, R. (2018). Analgesic properties of *Nigella Sativa* and *Eucheuma Cottonii* extracts. *Journal of Natural Science, Biology and Medicine, 9*, 23–26. Available from https://doi.org/10.4103/jnsbm.JNSBM_131_17.

Zanoli, P., & Zavatti, M. (2008). Pharmacognostic and pharmacological profile of *Humulus lupulus* L. *Journal of Ethnopharmacology, 116*, 383–396. Available from https://doi.org/10.1016/j.jep.2008.01.011.

Zengin, G., Atasagun, B., Zakariyyah Aumeeruddy, M., Saleem, H., Mollica, A., Babak Bahadori, M., & Mahomoodally, M. F. (2019). Phenolic profiling and in vitro biological properties of two Lamiaceae species (*Salvia modesta* and *Thymus argaeus*): A comprehensive evaluation. *Industrial Crops and Products, 128*, 308–314. Available from https://doi.org/10.1016/j.indcrop.2018.11.027.

Zhang, Y., & Wang, Z. Z. (2008). Comparative analysis of essential oil components of three Phlomis species in Qinling Mountains of China. *Journal of Pharmaceutical and Biomedical Analysis, 47*, 213–217. Available from https://doi.org/10.1016/j.jpba.2007.12.027.

Zidane, H., Elmiz, M., Aouinti, F., Tahani, A., Wathelet, J., Sindic, M., & Elbachiri, A. (2013). Chemical composition and antioxidant activity of essential oil, various organic extracts of *Cistus ladanifer* and *Cistus libanotis* growing in Eastern Morocco. *African Journal of Biotechnology, 12*, 5314–5320. Available from https://doi.org/10.5897/AJB2013.12868.

Zonyane, S., Van Vuuren, S. F., & Makunga, N. P. (2013). Antimicrobial interactions of Khoi-San poly-herbal remedies with emphasis on the combination; *Agathosma crenulata, Dodonaea viscosa* and *Eucalyptus globulus*. *Journal of Ethnopharmacology, 148*, 144–151. Available from https://doi.org/10.1016/j.jep.2013.04.003.

Zorzan, M., Collazuol, D., Ribaudo, G., Ongaro, A., & Pezzani, R. (2019). Biological effects and potential mechanisms of action of *Pistacia lentiscus* Chios mastic extract in CaCo-2 cell model. *Journal of Functional Foods, 54*, 92–97. Available from https://doi.org/10.1016/j.jff.2019.01.007.

CHAPTER

3

Medicinal plants in the Balkans with antimicrobial properties

Sarah Shabih[1], *Avni Hajdari*[2], *Behxhet Mustafa*[2] *and Cassandra L. Quave*[1,3]

[1]Center for the Study of Human Health, Emory University, Atlanta, GA, United States [2]Department of Biology, University of Prishtina, Prishtinë, Kosovo [3]Department of Dermatology and Center for the Study of Human Health, Emory University, Atlanta, GA, United States

Introduction

The Balkans are a European biodiversity hotspot characterized by high biological diversity, resulting from unique geomorphological, climatic, hydrological, and soil conditions. In addition to being a rich region for biodiversity, the Balkans also represents an attractive center in terms of cultural, linguistic, and religious diversity, making the region a tremendous reservoir of traditional ecological knowledge (TEK) related to wild plants use (Mustafa et al., 2012). According to a recent study (Mustafa, Hajdari, Pulaj, Quave, & Pieroni, 2020), the richness in TEK in this region is due to the complex biocultural diversity, economic development based mainly in small-scale agropastoral activities in rural and mountainous areas, managed mainly by elderly peoples, and the long-held tradition in the collection of wild plants. Although the region is rich in TEK, a limited number of ethnobotanical investigations have been conducted last century. In the last few decades, however, the Western Balkan countries have been the focus of several ethnobotanical surveys. For example, these have been conducted in Albania (Pieroni, Dibra, Grishaj, Grishaj, & Gjon, 2005; Pieroni et al., 2014; Pieroni et al., 2014; Pieroni, Ibraliu, Abbasi, & Papajani-Toska, 2015; Pieroni, 2008), Bosnia and Herzegovina (Šarić-Kundalić, Dobeš, Klatte-Asselmeyer, & Saukel, 2010; Šarić-Kundalić, Dobeš, Klatte-Asselmeyer, & Saukel, 2011), Kosovo (Hajdari, Pieroni, Jhaveri, Mustafa, & Quave, 2018; Mustafa et al., 2012; Mustafa et al., 2012; Mustafa et al., 2015; Mustafa et al., 2020; Pieroni et al., 2017), Montenegro (Menković et al., 2011), North Macedonia (Pieroni et al., 2013; Pieroni et al., 2017; Rexhepi

Medicinal Plants as Anti-infectives
DOI: https://doi.org/10.1016/B978-0-323-90999-0.00013-6

et al., 2013), and Serbia (Jarić et al., 2007; Jarić et al., 2015; Pieroni et al., 2011; Šavikin et al., 2013; Zlatković et al., 2014).

This trend has been fostered by the growing interest of the Western herbal market in medicinal plants traded from this area (Kathe et al., 2003), the need for documenting the last remaining traces of TEK in areas, the increasing economic trends to develop eco-tourism, and other sustainable rural activities based upon local bio-cultural heritage; and the fact that ethnobiologists have deemed this region a unique case study for its tremendous biological, cultural, and ethnic diversity (Mustafa et al., 2012). This intensive TEK research has shed light on the traditional uses of plant species used in different household domains, including for home remedies to treat a variety of infectious diseases. However, despite their common use, little is known concerning the safety and efficacy of many homemade antimicrobial products.

Thus, the main objective of this work is to review the existing literature to identify ethnobotanical knowledge in the Balkans related to antimicrobial uses, which could be of potential interest for further phytochemical, pharmacological, and/or toxicological studies, as well as for local development. Minimum inhibitory concentration (MIC) values, which is defined as the lowest concentration of an antibacterial agent that inhibits the growth of a microorganism in vitro, are used in this review to describe the antimicrobial potency of a selection of medicinal plants (CLSI, 2012). For plant extracts, only MICs $\leq 500\ \mu g/mL$ are considered parameters for good antibacterial activity (Chassagne et al., 2021); however, antimicrobial potency against a microorganism is considered significant if the MIC is $\leq 100\ \mu g/mL$ (Eloff, 2004).

Medicinal plants with antimicrobial properties

In the following sections, we highlight 22 species within the Balkan region for their strong antimicrobial properties. A comprehensive summary of 130 Balkan species and their medicinal uses related to infection is provided in Table 3.1.

Amaryllidaceae

Allium sativum L.

Allium sativum L. (garlic) is a bulbous perennial native to Central Asia, though there is evidence of its medicinal use in Ancient Sumeria, Egypt Greece, and Rome (Harris et al., 2001). Ethanolic extracts of the cloves of *A. sativum* exhibited inhibitory effects against *Escherichia coli* (MIC = 65.50 μg/mL), *Pseudomonas aeruginosa* (MIC = 58.50 μg/mL), and *Bacillus subtilis* (MIC = 80.10 μg/mL), and the essential oil inhibited *Staphylococcus aureus* (MIC = 24 μg/mL) (Karuppiah & Rajaram, 2012; Tsao & Yin, 2001). Allicin, a compound metabolized by crushing or cutting the bulb of *A. sativum*, as well as diallyl sulfides, is primarily responsible for the antibacterial activity of garlic (Tsao & Yin, 2001; Block, 1985). An in vivo study showed that the bulbs of *A. sativum* exhibited significant antibacterial activity against penicillin-sensitive *Staphylococcus aureus* injected into the tissue of rats (Venâncio et al., 2017).

TABLE 3.1 Medicinal plants from the Balkans with antimicrobial properties. Botanical taxa are reported here as found in the literature and may include synonyms rather than current accepted names per http://www.worldfloraonline.org/.

Botanical taxon	Part(s) used	Preparation and/or administration	Treated disease(s) or folk medical uses(s)	Reference/Sources
Abies alba Mill.	Resin	Mixed with fat	Antifungal	Mustafa et al. (2015)
		Boiled in oil	Eczema	Mustafa et al. (2012)
		Mixed and boiled with milk butter	Skin infections	Mustafa et al. (2012)
		Topically applied	Skin infections	Mustafa et al. (2012); Pieroni et al. (2013)
		Balm	Skin injuries	Šarić-Kundalić et al. (2010)
		Mixture	Eye inflammations	Šarić-Kundalić et al. (2011)
	Needles	Mixture and single component	Pulmonary ailments	Šarić-Kundalić et al. (2011)
Achillea millefolium L.	Leaves	Tea	Respiratory problems	Jarić et al. (2015)
			Wounds	Zlatković et al. (2014)
			Hemorrhoids	Rexhepi et al. (2013)
			Eczema	Mustafa et al. (2012)
			Urogenital disorders	Hajdari et al. (2018)
			Antimicrobial and to treat influenza	Mustafa et al. (2015)
		Tincture, topical use	Antibacterial, wounds	Mustafa et al. (2012)
		Fresh leaves, topically applied	Infected wounds caused by dog bite	Hajdari et al. (2018)
			Wound healing (ulcers)	Mustafa et al. (2012), Mustafa et al. (2020), Rexhepi et al. (2013)
			Inflammation of the skin and mucous membranes	Menković et al. (2011)
		Mixed with *Plantago major* leaves and topically applied to the eye area	Conjunctivitis	Hajdari et al. (2018)
		Squeezed juice of leaves and drunk	Pulmonary disorders	Hajdari et al. (2018)
		Dried and ground plant applied directly on wound	Wounds, ulcers, and hemorrhoids	Jarić et al. (2007)
		Mixed with milk fat (*mehlem*), topically applied	Acne	Mustafa et al. (2020)

(*Continued*)

TABLE 3.1 (Continued)

Botanical taxon	Part(s) used	Preparation and/or administration	Treated disease(s) or folk medical uses(s)	Reference/Sources
		Cataplasm from the leaves	Reduce joint inflammation and accelerate wound healing	Jarić et al. (2015)
	Flowers	Decoction, externally applied	Skin irritations and acne	Mustafa et al. (2012)
Aconitum divergens Pančić	Aerial parts	Infusion	Antiseptic for the oral cavity	Mustafa et al. (2012)
	Leaves	Squeezed and topically applied to the wound	Skin infections (antibacterial)	Mustafa et al. (2012)
Agrimonia eupatoria L.	Aerial parts	Infusion	Antiinflammatory and to treat earache	Mustafa et al. (2015)
		Decoctions or macerated in *raki* and drunk	Skin diseases	Pieroni (2008)
		Used externally	Skin, mouth, and pharyngeal inflammations	Menković et al. (2011)
		Tea	As "natural penicillin" for all manner of ailments, colds, laryngitis	Jarić et al. (2015)
			For blood purification and to treat wounds	Šarić-Kundalić et al. (2011)
	Leaves	Applied directly on the wound	Wounds and cuts	Jarić et al. (2007)
Alchemilla vulgaris L.	Whole plant	Tea	Wounds and ulcers	Jarić et al. (2015)
	Aerial parts	Tea, externally used	Ulcers, eczema, skin rashes	Menković et al. (2011)
Allium ampeloprasum Thunb. (*Allium porrum*)	Whole plant	Leaf juice instilled in ear	Ear infection	Mustafa et al. (2020), Pieroni et al. (2014), Pieroni et al. (2015), Pieroni et al. (2013); Pieroni, Sõukand, Quave, Hajdari, and Mustafa (2017)
	Leaves	Externally applied	Wounds (suppurative)	Pieroni et al. (2015)
Allium cepa L.	Bulb	Epidermis is removed and it is directly applied, or the onion is chopped up and applied as a cataplasm, or lightly baked onions are placed on the affected area under a bandage (externally used)	Injuries, for draining pus from infected areas (antibacterial)	Jarić et al. (2015), Mustafa et al. (2015)

(Continued)

TABLE 3.1 (Continued)

Botanical taxon	Part(s) used	Preparation and/or administration	Treated disease(s) or folk medical uses(s)	Reference/Sources
		Fresh sliced bulb	Eye inflammations	Pieroni et al. (2014)
		Boiled with soap and after cooling applied on the nail	Nail infections	Mustafa et al. (2012), Mustafa et al. (2015)
Allium sativum L.	Bulb	Eaten	Antifungal, to treat urinary tract infections, bronchitis	Mustafa et al. (2012), Mustafa et al. (2015)
			Eye inflammations	Pieroni et al. (2017)
		Mixed with honey	Respiratory system disorders (bronchitis)	Mustafa et al. (2015)
		Tincture	Antibacterial	Mustafa et al. (2012)
		Boiled in milk (four to 5 cloves) and drunk as tea	To "disinfect" the intestine	Mustafa et al. (2012)
		Tea	Inflammations and infections of the urogenital tract, cystitis	Jarić et al. (2015)
		Equal quantities of chopped garlic and honey are mixed together—one teaspoon to be taken three times a day with warm water	Pneumonia	Jarić et al. (2015)
		Decoction	Skin problems	Jarić et al. (2015)
		Cataplasm from chopped garlic	External ulcers	Jarić et al. (2015)
		Internally applied	Tuberculosis, purification of lungs, tooth inflammation, throat inflammation	Šarić-Kundalić et al. (2011)
		Externally applied	Ulcers and rheumatism, for skin ailments	Šarić-Kundalić et al. (2011)
Alnus glutinosa (L.) Gaertn.	Bark	Externally applied	Inflammation of the mouth and pharynx	Menković et al. (2011)
Althaea officinalis L.	Leaves, flowers, roots	Internally applied	Urinary tract inflammation and bronchitis	Šarić-Kundalić et al. (2011)
	Roots	Decoction	Lung disorders and oral cavity antiseptic	Mustafa et al. (2012)
Arctium lappa Willd.	Leaves	Applied directly on wound	Ulcers and to promote wound healing	Jarić et al. (2007)
		Boiled in milk (used externally)	Skin inflammation and ulcers	Mustafa et al. (2012)

(Continued)

TABLE 3.1 (Continued)

Botanical taxon	Part(s) used	Preparation and/or administration	Treated disease(s) or folk medical uses(s)	Reference/Sources
	Roots, leaves, seeds	Tea	Mouth infections, digestive ailments, and internal ulcers	Šarić-Kundalić et al. (2010)
			Stomach inflammation, stomach ulcers, and digestive disorders; wounds and skin rash	Šarić-Kundalić et al. (2011)
Arctostaphylos uva-ursi (L.) Spreng.	Leaves	Infusion	Urinary tract inflammations	Jarić et al. (2015), Menković et al. (2011), Mustafa et al. (2015), Šarić-Kundalić et al. (2010), Šarić-Kundalić *et al.* (2011), Šavikin et al. (2013)
	Aerial parts, flowers, leaves	Tea	Bronchitis, urinary system inflammation, and inflamed tonsils	Mustafa et al. (2020)
	Aerial parts	Tea	Antiseptic	Jarić et al. (2007)
Aristolochia clematitis L.	Fruits	Tea	Antibacterial properties. Externally used for skin infections and wounds (for rinsing and poultice tea)	Jarić et al. (2007)
Artemisia absinthium L.	Aerial parts	Mixed with honey	Promote wound healing	Pieroni et al. (2014)
Bellis perennis L.	Whole plants	Decoction	Skin infection	Mustafa et al. (2012)
	Aerial parts	Externally used	Skin diseases and promote wound healing	Menković et al. (2011)
	Aerial parts	Fresh juice of leaves is applied to the eyes	Eye inflammations	Pieroni et al. (2015)
Betula pendula Roth.	Bark, leaves		Bladder infections, urinary tract infections, purification of urinary bladder, renal inflammations, blood purification, purification of lungs, common cold, and fever	Šarić-Kundalić et al. (2011)
Betula verrucosa Ehrh.	Cortex	Decoction	Kidney infections	Menković et al. (2011)
	Bark	Burned, the vapors are exposed to the skin	Skin inflammations	Pieroni et al. (2013)

(*Continued*)

TABLE 3.1 (Continued)

Botanical taxon	Part(s) used	Preparation and/or administration	Treated disease(s) or folk medical uses(s)	Reference/Sources
Brassica oleracea L.	Leaves	Topically applied	Burn injuries	Hajdari et al. (2018)
		Fermented leaves topically applied	Antibacterial	Mustafa et al. (2012)
		Fresh leaf as compress	Skin inflammation	Šavikin et al. (2013)
Calendula officinalis L.	Flowers	Extracted with cold milk	Stomach ulcers	Mustafa et al. (2012)
		Internally applied	Urinary tract inflammations, cervical wounds, eye ailments, hemorrhoids, different injuries, wounds, and external ulcers	Šarić-Kundalić et al. (2011)
		Fat-based ointments	Skin complaints, burns, and wounds	Šavikin et al. (2013)
		Tea and tincture	Gastric and duodenal ulcers	Šavikin et al. (2013)
		Mixed with pig or other fats to create an ointment "*mehlem*," which is topically applied	Eczema and other inflammatory skin disorders, the skin for burns	Hajdari et al. (2018)
		Extracted in oil in the sun for 40 days and stored for use when needed; topically applied	Burns of the skin	Hajdari et al. (2018)
		Flowers combined with flowering aerial parts of *Hypericum perforatum* and extracted in olive or sunflower oil for 40 days in the sun to yield a blood-red oleolite; topical application	Skin injuries and promote wound healing	Hajdari et al. (2018)
	Aerial parts	*Mehlem* (mixed with pig's fat)	Laceration and skin infections	Mustafa et al. (2020)
		Extracted with different oils	Antibacterial, antifungal	Jarić et al. (2007), Mustafa et al. (2015)
Calluna vulgaris (L.) Hull	Aerial parts	Used internally	Urinary tract infections	Šarić-Kundalić et al. (2011)
Capsella bursa-pastoris (L.) Medik	Aerial parts	Used externally	Superficial skin injuries, wounds, and burns	Menković et al. (2011)

(Continued)

TABLE 3.1 (Continued)

Botanical taxon	Part(s) used	Preparation and/or administration	Treated disease(s) or folk medical uses(s)	Reference/Sources
Carlina acaulis L.	Roots	Used externally	Dermatosis and to rinse wounds and ulcers	Menković et al. (2011)
	Flowers, roots, stems	Decoction	Eczema and acne	Rexhepi et al. (2013)
Castanea sativa Mill.	Fruits	Tea	Skin ailments	Šarić-Kundalić et al. (2010)
Centaurea cyanus L.	Flowers	Decoction	Eye infections	Mustafa et al. (2012)
		Externally used	Eye inflammations	Šarić-Kundalić et al. (2011)
		Tea	Respiratory disorders	Mustafa et al. (2015)
Centaurium erythraea Rafn.	Aerial parts	Tea, internally used	Stomach ulcers, laryngitis, cold remedies	Jarić et al. (2015)
Chelidonium majus L.	Aerial parts	Infusion	Bronchitis and stomach ulcers	Mustafa et al. (2012)
		Externally	Skin conditions such as blister rashes and scabies	Menković et al. (2011)
		Extract applied directly to wound	Skin infections	Rexhepi et al. (2013)
	Latex	Leaves topically applied	Skin inflammations and eczema	Hajdari et al. (2018), Mustafa et al. (2015), Mustafa et al. (2020)
		Externally applied and used	Skin complaints (skin eruption, psoriasis, eczema)	Jarić et al. (2007)
Cichorium intybus L.	Seed oil	Topically applied	Skin infections (in children)	Hajdari et al. (2018)
	Roots	Decoction	Urinary system infections and bronchitis	Mustafa et al. (2012)
Citrullus lanatus (Thunb.) Matsum. & Nakai	Fruits	Juice is expressed and dropped into the ear canal to treat ear	Ear infection	Hajdari et al. (2018)
		Internally applied to treat	Gastrointestinal ailments, bacterial infections, common cold, pneumonia, throat ache, for influenza, blood purification	Šarić-Kundalić et al. (2011)
		Juice mixed with honey	Cough	Mustafa et al. (2012)

(*Continued*)

TABLE 3.1 (Continued)

Botanical taxon	Part(s) used	Preparation and/or administration	Treated disease(s) or folk medical uses(s)	Reference/Sources
Conium maculatum L.	Roots	Boiled in milk and used	Ulcers	Mustafa et al. (2012)
Cucumis sativus L.	Fruits	Juice	Skin inflammations	Šarić-Kundalić et al. (2010)
Cucurbita pepo L.	Leaves	Tea	Sore throat, intestinal infections, kidney inflammations	Mustafa et al. (2012)
	Fruit, seed	Internally used	Bacterial infections and prostate inflammations.	Šarić-Kundalić et al. (2011)
Cydonia oblonga Mill	Leaves, seeds	Internally used	Stomach ulcers	Šarić-Kundalić et al. (2011)
	Leaves	Infusion	Respiratory inflammations	Mustafa et al. (2012)
Equisetum arvense L.	Stem and leaves	Infusion	Urinary system infections	Mustafa et al. (2012)
	Whole plant	Fresh plant mixed with honey and milk cream, extracted for 1 week	General infections	Hajdari et al. (2018), Jarić et al. (2007), Menković et al. (2011), Mustafa et al. (2012), Mustafa et al. (2020), Seifi, Abbasalizadeh, Mohammad-Alizadeh-Charandabi, Khodaie, and Mirghafourvand (2018)
	Aerial parts	Externally applied	Wounds and burns	Menković et al. (2011), Šarić-Kundalić et al. (2011)
		Internally applied	Urinary tract infections, renal ailments, renal inflammations, urinary bladder inflammations	Šarić-Kundalić et al. (2011)
Euphrasia officinalis L.	Leaves	Fresh leaves topically applied	Eye inflammation	Hajdari et al. (2018), Menković et al. (2011), Mustafa et al. (2020)
Euphrasia rostkoviana Hayne	Aerial parts, leaves	Internally applied	Bronchitis and gingivitis	Šarić-Kundalić et al. (2011)
		Externally applied	Eye ailments, inflammations, and injuries	Šarić-Kundalić et al. (2011)
Fragaria vesca L.	Fruits, leaves	Internally applied	Throat inflammations	Šarić-Kundalić et al. (2011)
Fraxinus angustifolia Vahl	Leaf	Directly to wound	Wound healing	Rexhepi et al. (2013)

(*Continued*)

TABLE 3.1 (Continued)

Botanical taxon	Part(s) used	Preparation and/or administration	Treated disease(s) or folk medical uses(s)	Reference/Sources
Fraxinus ornus L.	Bark	Mixed with beeswax and topically applied	Infected wounds	Hajdari et al. (2018)
Galium verum L.	Aerial parts	Tea taken externally	Skin ailments, wounds, ulcers, and acne	Jarić et al. (2007)
		Externally applied	Poorly healing wounds	Menković et al. (2011)
	Flowers	Tea	Urinary system infections	Mustafa et al. (2012)
Gentiana cruciata L.	Whole plant	Fresh plant mixed with honey and milk cream, extracted for 1 week	Nail infections	Mustafa et al. (2012)
Gentiana lutea L.	Leaves	Infusion, drunk	Urinary tract infections	Hajdari et al. (2018)
	Root	Macerate in *rakija* (40 days of maceration), to be drunk in the morning before eating	Stomach ulcer	Mustafa et al. (2020), Pieroni, Giusti, and Quave (2011)
Glechoma hederacea L.	Aerial parts, leaves	Internally applied and used	Bronchial purification, bronchitis, laryngitis, nasal congestion, throat inflammations, tuberculosis	Šarić-Kundalić et al. (2011)
	Aerial parts	Tea (drunk)	Astringent, diuretic, tonic effects on the bronchial and urinary system, for bronchitis, diarrhea, and to improve appetite	Jarić et al. (2007)
Hedera helix L.	Leaves, fruits, flowers	Internally applied and used	Bladder inflammations, bronchitis	Šarić-Kundalić et al. (2011)
		Externally applied and used	Wounds and swollen legs	Šarić-Kundalić et al. (2011)
Helianthus annuus L.	Seed	Extracted with animal fat	Skin infections	Mustafa et al. (2015)
	Seed oil	Topically applied	Skin infections (in children)	Hajdari et al. (2018)
Hypericum maculatum Crantz	Aerial parts	Externally applied	Skin inflammation, blunt injuries, wounds, burns	Menković et al. (2011)
	Flowers	Decoction used	Urinary tract inflammations	Šarić-Kundalić et al. (2010)
	Aerial parts, flowers	Internally applied and used	Throat inflammations	Šarić-Kundalić et al. (2011)
		Externally applied and used (Oleolite)	Ovarian inflammations, cervical wounds, burns, hemorrhoids, eye injuries, and throat inflammations.	Šarić-Kundalić et al. (2011)

(Continued)

TABLE 3.1 (Continued)

Botanical taxon	Part(s) used	Preparation and/or administration	Treated disease(s) or folk medical uses(s)	Reference/Sources
Hypericum perforatum L.	Aerial parts	Tea	Wound healing	Mustafa et al. (2015), Seifi et al. (2018), Zlatković et al. (2014)
		Extracted with olive oil, topically applied	Skin infections, skin after sunburn or thermal burn, eczemas, wound healing	Hajdari et al. (2018), Jarić et al. (2007), Mustafa et al. (2012), Mustafa et al. (2020), Seifi et al. (2018), Zlatković et al. (2014)
			Variola, athlete's foot, postpartum infection (applied to genitals), to treat acne	Hajdari et al. (2018)
		Oleolite mixed with iodine and topically applied	Skin infections	Hajdari et al. (2018)
		Externally applied	Skin inflammation, injuries, wounds, burns	Menković et al. (2011)
	Flowers, aerial parts	Tea	Sore throat	Pieroni et al. (2014)
		Decoction topically applied	Wounds and skin inflammations	Pieroni et al. (2013), Pieroni et al. (2014)
	Flowers	Infusion	Stomach disorders, genital infections	Mustafa et al. (2012), Pieroni et al. (2015)
Hyssopus officinalis L.	Aerial parts	Tea internally applied	Bronchitis, colds	Jarić et al. (2015)
Juglans regia L.	Fruits and leaves	Tea, directly into the wound	Eczema, shingles, skin inflammation	Rexhepi et al. (2013)
	Leaves, fruits (unripe)	Cataplasm	Gangrene, wounds, ulcers	Jarić et al. (2015)
	Unripe fruits	Extracted with oil 30–40 days exposed to sun	Skin burns and protecting skin from sunburn	Mustafa et al. (2020)
	Fruits	Honey (1 kg) mixed with fruits (1 kg) extracted for 1 month	Lung inflammations	Mustafa et al. (2012)
	Roots	Extracted for 1 month with sunflower oil and then liquid mixed with honey	Lung inflammations and bronchitis	Mustafa et al. (2012)

(*Continued*)

TABLE 3.1 (Continued)

Botanical taxon	Part(s) used	Preparation and/or administration	Treated disease(s) or folk medical uses(s)	Reference/Sources
Juniperus communis L.	Fruits	Extracted for 10 days in cold water mixed with lemons	Kidney inflammations	Mustafa et al. (2012)
		Decoction	Respiratory inflammations	Mustafa et al. (2012)
		Berries are used for making and flavoring the brandy known as "*klekovaca*"	Used for disinfection, owing to antibacterial properties	Jarić et al. (2007)
		Infusion	Urinary tract ailments	Šavikin et al. (2013)
	Fruits, needles	Internally applied	Urinary tract inflammations, tuberculosis, common cold, inflammations of mucosa	Šarić-Kundalić et al. (2011)
		Externally applied	Skin rash and throat inflammation	Šarić-Kundalić et al. (2011)
Juniperus oxycedrus L.	Fruits	Tea, tincture, oil	Skin infections	Rexhepi et al. (2013)
Larix decidua Mill.	Bark, fruits, needles	Externally applied	Wounds, ulcers, and restlessness	Šarić-Kundalić et al. (2011)
Lavandula angustifolia Mill.	Aerial parts	Tea, used to rinse the eyes	Inflammation and conjunctivitis	Hajdari et al. (2018)
Linum hirsutum L.	Seeds	Decoction	Urinary system inflammations	Mustafa et al. (2012)
	Leaves	Infusion	Respiratory inflammations	Mustafa et al. (2012)
Linum usitatissimum L.	Seed	Internally applied	Gastrointestinal ulcers, urinary tract infections, bronchitis, and stomach ulcers	Šarić-Kundalić et al. (2011)
Lycopersicon esculentum Mill.	Fruits	Topically applied	Skin inflammation and ulcers	Mustafa et al. (2012)
		Baked fruits mixed with sugar topically applied in wound	Wound infections	Mustafa et al. (2012)
	Aerial parts	Topically applied to skin	Antimicrobial	Mustafa et al. (2015)
Malva sylvestris L.	Leaf	Tea	Bronchitis	Rexhepi et al. (2013)
		Fresh, external application	Wounds	Pieroni et al. (2015)
	Flower	Tea	Sore throat	Pieroni et al. (2013)

(*Continued*)

TABLE 3.1 (Continued)

Botanical taxon	Part(s) used	Preparation and/or administration	Treated disease(s) or folk medical uses(s)	Reference/Sources
		Infusion	Respiratory tract infections	Hajdari et al. (2018)
		Internally applied	Lung ailments, bronchitis, throat infections	Jarić et al. (2007)
	Aerial parts	Extracted with fat (*melhem*)	Wound healing	Mustafa et al. (2015)
Matricaria chamomilla L.	Aerial parts	Decoction	Sinusitis	Mustafa et al. (2012)
		Tea internally used	Antiseptic, antiinflammatory	Hajdari et al. (2018), Jarić et al. (2007)
		Tea externally used	Skin and mucous complaints, burns, wounds, ulcers, for vaginal douche	Jarić et al. (2007)
		Steam inhalation	Sinusitis	Jarić et al. (2007)
		Tea drunk	Respiratory tract infections	Hajdari et al. (2018), Menković et al. (2011)
		Tea	For rinsing the oral cavity	Hajdari et al. (2018), Jarić et al. (2015)
	Flowers	External use	Inflammations of the skin, mouth and pharynx, wounds and burns	Menković et al. (2011)
		Tea used to rinse/cleanse eyes	Conjunctivitis	Hajdari et al. (2018), Mustafa et al. (2020)
		Decoction	Sinusitis	Mustafa et al. (2012)
Matricaria recutita L.	Aerial parts	Tea	Oral cavity inflammations, gingivitis, urinary system infections	Mustafa et al. (2012)
			Eye inflammations	Pieroni et al. (2011)
			Wound healing (ulcers of the skin and soft tissues)	Rexhepi et al. (2013)
Mentha longifolia (L.) L.	Leaves	Tea	Bronchitis, lungs inflammation	Mustafa et al. (2020)
	Aerial parts	Tea	Respiratory infections	Hajdari et al. (2018), Mustafa et al. (2015)
Mentha x piperita L.	Aerial parts	Infusion	Respiratory disorders	Hajdari et al. (2018)
			Antiseptic, used for gastric ulcers, indigestion, influenza, and colds	Jarić et al. (2007)

(Continued)

TABLE 3.1 (Continued)

Botanical taxon	Part(s) used	Preparation and/or administration	Treated disease(s) or folk medical uses(s)	Reference/Sources
Mentha pulegium L.	Aerial parts	Infusion	Respiratory system infections	Mustafa et al. (2015)
Momordica charantia L.	Fruits	Mixed with oil—internal use	Wound healing	Mustafa et al. (2015)
Morus nigra L.	Fruits	Eaten fresh	Infections of upper respiratory system	Mustafa et al. (2015)
Nicotiana tabacum L.	Leaves (dried)	Crushed leaves, topically applied	Wounds	Pieroni et al. (2013)
	Leaves	Leaf material removed from cigarette, mixed with sugar, and topically applied	Fungal infections of hands or feet	Hajdari et al. (2018)
Ocimum basilicum L.	Aerial parts	Inhalation	To clear the bronchial passages	Jarić et al. (2015)
		Externally used	Nipple inflammation during lactation and greasy skin	Šarić-Kundalić et al. (2011)
		Infusion	Kidney infections, tuberculosis	Mustafa et al. (2015)
		Topically applied to eye	Conjunctivitis	Hajdari et al. (2018)
		Tea mixed with sugar and used	Genital wash for feminine hygiene	Hajdari et al. (2018)
	Leaves	Tea	Ear inflammations and common cold	Šarić-Kundalić et al. (2010)
			Colds/flu	Pieroni et al. (2017)
Olea europaea L.	Fruits	Oil topically applied	Skin rashes and inflammations	Hajdari et al. (2018)
		Eaten fresh	Tuberculosis	Mustafa et al. (2015)
Ononis spinosa L.	Flowers	Tea	Gastritis and gastric ulcers	Rexhepi et al. (2013)
Orchis morio L	Root	Herbal tea	Cleaning wounds	Menković et al. (2011)
Origanum vulgare L.	Aerial parts	Tea	Sore throats, colds, flu, and bronchitis	Hajdari et al. (2018), Jarić et al. (2015), Jarić et al. (2007), Mustafa et al. (2012), Mustafa et al. (2015), Mustafa et al. (2020), Pieroni (2008), Pieroni et al. (2015), Pieroni et al. (2014), Pieroni et al. (2013), Pieroni et al. (2017), Šarić-Kundalić et al. (2010)
			Skin inflammations in infants	Hajdari et al. (2018)

(Continued)

TABLE 3.1 (Continued)

Botanical taxon	Part(s) used	Preparation and/or administration	Treated disease(s) or folk medical uses(s)	Reference/Sources
			Inflammation of the urinary tract, respiratory disorders, digestive disorders	Menković et al. (2011), Šarić-Kundalić et al. (2010)
	Leaves	Infusion	Urinary tract infection	Šavikin et al. (2013)
Petroselinum crispum (Mill.) Fuss	Leaves, roots	Tea	Urinary tract infection	Šarić-Kundalić et al. (2011), Šavikin et al. (2013)
	Leaves	Boiled with garlic and carrot	Stomach infections	Mustafa et al. (2012)
Phaseolus vulgaris L.	Seeds	Burned and topically applied	Dog bite wounds	Hajdari et al. (2018), Mustafa et al. (2020), Pieroni et al. (2014)
		Burned, mixed with oil, and externally applied with a hen's feather	Skin inflammations in babies and kids	Pieroni et al. (2015)
Physalis alkekengi L.	Fruits	Fresh fruits are eaten	Used to treat urinary tract infections	Jarić et al. (2015)
Picea abies (L.) H.Karst.	Resin	Topically applied	Skin ailments, throat inflammation, and eye ailments	Šarić-Kundalić et al. (2011)
			Wounds	Pieroni et al. (2011)
Pinus heldreichii Christ	Needles	Syrup	Pulmonary ailments	Šarić-Kundalić et al. (2011)
Pinus peuce Griseb.	Cones	Tea	Inflammation of the urinary tract	Menković et al. (2011)
Pinus mugo Turra	Needles, shoots	Inhalation	Common colds and bronchitis	Menković et al. (2011)
	Needles	Syrup	Pulmonary ailments	Šarić-Kundalić et al. (2011)
Pinus nigra J.F. Arnold	Resin	Extracted with oil	Skin infections	Mustafa et al. (2015)
	Young branches	Boiled for 4 h slowly, added some sugar, lemon, and honey, then drunk	Bronchitis and respiratory disease	Hajdari et al. (2018)
	Needles	Syrup for pulmonary ailments	Pulmonary ailments	Šarić-Kundalić et al. (2011)
	Resin	Topically applied	Skin infections	Hajdari et al. (2018)

(Continued)

TABLE 3.1 (Continued)

Botanical taxon	Part(s) used	Preparation and/or administration	Treated disease(s) or folk medical uses(s)	Reference/Sources
Pinus sylvestris L.	Bark, needles	Syrup (100 g of buds in 0.5 L of boiled water, strained after 2 h and mixed with 1 kg of honey-dose: four to five tablespoons a day)	Chronic bronchitis	Jarić et al. (2007)
	Cones	Cones 40 cones mixed with honey (1 kg) eaten after 1 month	Bronchitis	Mustafa et al. (2012)
	Needles	Syrup for pulmonary ailments	Pulmonary ailments	Šarić-Kundalić et al. (2011)
		Tea	Chronic bronchitis	Rexhepi et al. (2013)
	Needles, resin	Balm (the sap of Scots pine, kajmak/a creamy dairy product similar to clotted cream/and homemade wine vinegar are mixed together)	Eczema	Jarić et al. (2015)
Plantago lanceolata L.	Leaves	750 g honey is mixed with 35 ground leaves: one teaspoon, three times a day, before a meal	Stomach ulcers	Jarić et al. (2015)
		Cataplasm	Wounds, ulcers, and for draining pus	Jarić et al. (2015), Menković et al. (2011)
		Internally applied	Tuberculosis, bronchitis, throat inflammation; for internal healing	Šarić-Kundalić et al. (2011)
		Externally applied	Skin ailments, external ulcers, inflamed wounds	Šarić-Kundalić et al. (2011)
		Compress	Wounds, cuts, ulcers, inflammation of the skin	Šavikin et al. (2013)
		Freshly crushed, external application, sometimes with salt	Wounds, internal hemorrhages	Pieroni et al. (2015)
		Fresh leaves applied topically in wound	Wound infections	Mustafa et al. (2012)
		Externally as a poultice to be applied (fresh, dried, and oily leaf can be applied externally onto sore area)	Wounds, cuts, festering wounds, ulcers	Jarić et al. (2007)
		Tea	Intestinal and stomach ulcers	Jarić et al. (2007), Menković et al. (2011)

(*Continued*)

TABLE 3.1 (Continued)

Botanical taxon	Part(s) used	Preparation and/or administration	Treated disease(s) or folk medical uses(s)	Reference/Sources
	Aerial parts (dried)	Tea	Skin burn injuries	Rexhepi et al. (2013)
Plantago major L.	Leaves	Tea	Urogenital infections	Mustafa et al. (2012)
		Fresh, topically applied	Skin infections, wounds, furuncle, skin cut, skin burns, skin eczema	Hajdari et al. (2018), Jarić et al. (2015), Jarić et al. (2007), Menković et al. (2011), Mustafa et al. (2012), Mustafa et al. (2012), Mustafa et al. (2020), Pieroni et al. (2011), Pieroni et al. (2015), Šarić-Kundalić et al. (2011), Šavikin et al. (2013), Zlatković et al. (2014)
		Topically applied to eye area	Conjunctivitis	Hajdari et al. (2018)
		Tea	Eczema	Rexhepi et al. (2013)
			Earache (antimicrobial)	Mustafa et al. (2020)
			Intestinal and stomach ulcers	Jarić et al. (2007), Šarić-Kundalić et al. (2011), Zlatković et al. (2014)
			Wound healing	Mustafa et al. (2015)
		Internally applied	Tuberculosis, bronchitis, throat inflammation, for internal healing	Šarić-Kundalić et al. (2011)
		750 g honey is mixed with 35 ground leaves: one teaspoon, three times a day, before a meal	Stomach ulcers	Jarić et al. (2015)
	Leaves, roots	Infusion, poultice	Bronchitis, peptic ulcers. Externally, for purulent wounds, skin diseases, skin ulcers, wound healing	Zlatković et al. (2014)
Polygonum bistorta L.	Rhizomes, roots	Macerated roots (200–300 g) mixed honey (1 kg)	Respiratory infections	Mustafa et al. (2012)
		Tea (tea prepared and applied as a compress)	Skin complaints, festering, wounds, and hemorrhoids	Jarić et al. (2007)
		Externally applied	Skin inflammations	Menković et al. (2011)
Populus alba L.	Aerial parts	Topically uses	Wound healing	Mustafa et al. (2015)

(*Continued*)

TABLE 3.1 (Continued)

Botanical taxon	Part(s) used	Preparation and/or administration	Treated disease(s) or folk medical uses(s)	Reference/Sources
Populus nigra L.	Cortex	Decoction	Urinary system inflammations	Mustafa et al. (2012)
	Leaves	Decoction	Tuberculosis, bronchitis	Mustafa et al. (2012)
Primula acaulis (L.) Hill	Rhizomes	Taken as a syrup and tea	Bronchitis	Jarić et al. (2007)
Primula veris L.	Whole plant	Taken as a syrup and tea	Bronchitis and colds	Jarić et al. (2015)
	Aerial parts	Infusion	Respiratory system disorders (bronchitis)	Hajdari et al. (2018), Mustafa et al. (2015)
	Roots	Decoction	Bronchitis	Menković et al. (2011), Šavikin et al. (2013)
	Flowers	Tea	Respiratory disease	Mustafa et al. (2020)
				Menković et al. (2011)
Primula vulgaris Huds.	Flowers, leaves, roots	Internally applied	Influenza, bronchitis, pulmonary ailments, pneumonia	Šarić-Kundalić et al. (2011)
Prunus avium (L.) L.	Fruits	Tea	Respiratory inflammations	Mustafa et al. (2012)
	Fruits	Infusion	Urinary infections	Zlatković et al. (2014)
Prunus cerasus L.	Resin	Externally applied	Skin inflammations	Pieroni et al. (2013)
	Fruits	Whole fresh leaf applied	Conjunctivitis (antibacterial)	Hajdari et al. (2018), Mustafa et al. (2015)
Prunus domestica L.	Fruits	Fruits (fermented 1–2 months and then resulting must be distilled) *raki*, topically applied	Antiseptic on wounds	Pieroni (2008); Pieroni et al. (2014), Pieroni et al. (2011), Pieroni et al. (2015), Pieroni et al. (2014), Pieroni et al. (2013)
			Earaches (instilled in the ear)	Pieroni et al. (2013)
		Hot *rakia*, in fumigations	Sinusitis	Pieroni et al. (2011)
Prunus spinosa L.	Flowers, fruits	Externally applied	Inflammation of the mouth and pharynx	Menković et al. (2011)
Pteridium aquilinum (L.) Kuhn.	Leaves	Extracted with oil	Wound healing	Mustafa et al. (2015)
		Decoction	Antibacterial	Mustafa et al. (2012)

(Continued)

TABLE 3.1 (Continued)

Botanical taxon	Part(s) used	Preparation and/or administration	Treated disease(s) or folk medical uses(s)	Reference/Sources
Pulmonaria officinalis L.	Aerial parts	Infusion	Bronchitis	Mustafa et al. (2015), Šarić-Kundalić et al. (2011)
Robinia pseudoacacia L.	Flowers	Tea	Skin infections	Mustafa et al. (2015)
		Decoction	Respiratory inflammations	Mustafa et al. (2012), Mustafa et al. (2012)
Rosa canina L.	Fruits	Tea	Sore throat	Mustafa et al. (2012)
			Sore throat, flu	Hajdari et al. (2018), Jarić et al. (2007), Mustafa et al. (2012), Mustafa et al. (2015), Pieroni et al. (2011), Pieroni et al. (2015), Pieroni et al. (2014), Pieroni et al. (2017)
	Flowers and fruits	Tea	Respiratory problems (cough, bronchitis, and cold)	Rexhepi et al. (2013)
		Topically applied	Wound healing	Rexhepi et al. (2013)
Rubus fruticosus L.	Fruits, leaves, roots	Internally used	Skin ailments, throat inflammations	Šarić-Kundalić et al. (2011)
	Aerial parts	Infusion	Wound healing	Mustafa et al. (2015)
	Fruits	Infusion	Kidney infections, oral cavity infections	Mustafa et al. (2015)
	Leaves	Fresh leaves applied topically in wound	Skin infection	Mustafa et al. (2012)
		Externally applied	Inflammation of the mouth and pharynx	Menković et al. (2011)
		Tea	Sore throat	Pieroni et al. (2017)
	Roots	Decoction	Appendicitis	Mustafa et al. (2012)
Rubus idaeus L.	Leaves	Infusion	Throat inflammations	Šarić-Kundalić et al. (2011)
			Oral cavity infections	Mustafa et al. (2015)
		Decoction	Sore throat, influenza	Mustafa et al. (2012)
	Leaves, fruits	Externally applied	Inflammation of the mouth and pharynx	Menković et al. (2011)
Rubus ulmifolius Schott	Leaves	Crushed and mixed with clarified butter (*tëlynë*), topically applied	Skin infections, wounds	Pieroni et al. (2014)

(*Continued*)

TABLE 3.1 (Continued)

Botanical taxon	Part(s) used	Preparation and/or administration	Treated disease(s) or folk medical uses(s)	Reference/Sources
Rumex patientia L.	Leaves	Crushed and mixed with animal fat	Wound healing	Pieroni et al. (2014)
Salvia officinalis L.	Herb	Infusion	Mouth and throat infections	Šavikin et al. (2013)
	Aerial parts	Tea	Disinfection of the oral cavity and teeth	Jarić et al. (2015), Šarić-Kundalić et al. (2011)
		Infusion, then added honey	Tonsillitis and other infections of respiratory system	Mustafa et al. (2015)
		Apply topically	Acne, psoriasis	Hajdari et al. (2018)
	Leaves	Tea	Skin ailments, urinary tract infections, bronchitis, throat inflammations, influenza	Šarić-Kundalić et al. (2011)
			Sore throat, cough, liver protective	Pieroni et al. (2017)
			Respiratory inflammations	Mustafa et al. (2020)
		Decoction	Sore throats, flu, tonsillitis	Pieroni et al. (2005)
Salvia pratensis L.	Leaves	Internally applied	Gingivitis, skin ailments, urinary tract infections, bronchitis, throat inflammations, for influenza inflammations	Šarić-Kundalić et al. (2011)
		Externally applied	Sore throat, skin rash, wounds, and greasy skin	Šarić-Kundalić et al. (2011)
Salvia verticillata L.	Leaves	Cataplasm	Wounds	Jarić et al. (2015)
	Aerial parts	Fresh, crushed, or the fresh:	Wound healing (humans), snake bites, and skin inflammations (animals)	Pieroni et al. (2014)
	Flowers	Visiting the plant is considered very effective	Bronchitis	Pieroni et al. (2013)
Sambucus ebulus L.	Leaves	Crushed and applied	Wounds (in animals and humans)	Pieroni et al. (2014)
	Flowers	Tincture	Urinary inflammations	Mustafa et al. (2012)
Sambucus nigra L.	Flowers	Infusion	Respiratory inflammations (bronchitis), sore throats	Hajdari et al. (2018), Mustafa et al. (2012), Mustafa et al. (2020), Pieroni et al. (2011), Rexhepi et al. (2013), Zlatković et al. (2014)

(*Continued*)

TABLE 3.1 (Continued)

Botanical taxon	Part(s) used	Preparation and/or administration	Treated disease(s) or folk medical uses(s)	Reference/Sources
		Tea	Antiseptic to reduce inflammation, influenza	Jarić et al. (2015)
		Tea mixed with lemon and sugar	Bronchitis	Mustafa et al. (2012), Mustafa et al. (2015)
		Extracted with oil—topically used	Skin infections	Mustafa et al. (2015)
		Balm	Wounds	Jarić et al. (2007)
	Stems	Boiled with milk cream	Skin inflammations, eczema	Mustafa et al. (2012)
	Bark	Decoction mixed with sheep fat or bee wax to create poultice	Wounds	Pieroni et al. (2014)
		Mixed with butter or cream, resin, in poultice (*mehlem*)	Wounds, bruises	Pieroni et al. (2011)
	Fruits	Tea	Drunk for asthma, cough, bronchitis, fever	Hajdari et al. (2018), Rexhepi et al. (2013), Zlatković et al. (2014)
		250 g sugar and 1 L of fruit juice are boiled until reduced to 1 L, then drunk	Respiratory infections and bronchitis	Hajdari et al. (2018)
	Cambium	Fresh, externally applied with honey	Skin inflammations	Pieroni et al. (2015)
Sedum spectabile Boreau	Leaves	Fresh leaves are chewed	Stomach ulcer	Jarić et al. (2007)
		Cataplasm	Wounds and draining pus from wounds	Jarić et al. (2007)
		Top epidermis of leaf is removed to reveal a fleshy tissue that is topically applied	Conjunctivitis	Hajdari et al. (2018)
Sempervivum hirtum L.	Leaves	Internally applied	Ear infection, stomach ulcers, throat inflammation	Šarić-Kundalić et al. (2011)
		Externally applied	Skin ailments: warts, ulcers, corns, and skin rash	Šarić-Kundalić et al. (2011)
Sempervivum tectorum L.	Leaves	Liquid is expressed from the leaf and topically applied in ear	Ear infection	Šarić-Kundalić et al. (2011)
		One leaf a day	Stomach and intestinal ulcers	Šarić-Kundalić et al. (2011)

(Continued)

TABLE 3.1 (Continued)

Botanical taxon	Part(s) used	Preparation and/or administration	Treated disease(s) or folk medical uses(s)	Reference/Sources
	Leaves	Decoction after cooled applied in ear	Earache and ear infections	Mustafa et al. (2012)
		Extracted with fat (cow or pig fat)—topically applied	Wound healing	Mustafa et al. (2015), Pieroni et al. (2015)
Sisymbrium officinale (L.) Scop.	Stems, leaves, fruits	Tea	Respiratory system problems (mostly to protect from tuberculosis)	Rexhepi et al. (2013)
Solanum tuberosum L.	Bulb	Topically applied	External ulcers	Šarić-Kundalić et al. (2011)
		Externally applied (in slices)	Eye inflammations	Pieroni et al. (2014)
Stachys officinalis (L.) Trevis.	Whole plant	Tea	Skin complaints (wounds, burns), for bronchitis, cough, asthma (tea)	Šarić-Kundalić et al. (2011)
	Leaves	Fresh leaves are topically applied	Skin infection	Mustafa et al. (2012)
		Two to three drops applied in the ear	Earache	Mustafa et al. (2012)
		Infusion, topically applied	Wounds	Mustafa et al. (2012)
Symphytum officinale L.	Leaves	Mixed with pork fat (poultice)	Wounds	Pieroni et al. (2011)
Tanacetum balsamita L.	Aerial parts	Mixed with wax, incense, pot marigold, and resin, in a poultice (*mehlem*)	Every skin disease	Pieroni et al. (2011)
Taraxacum officinale (L.) eber ex F.H. Wigg.	Flowers	Tea	Urinary system inflammations and respiratory inflammation	Mustafa et al. (2012), Mustafa et al. (2020)
		Topically applied	Wounds	Pieroni et al. (2011)
		Decoction mixed with lemon fruits	Bronchitis	Mustafa et al. (2012)
		"Honey": 150 flowers are covered with 3 L water, sugar is added and this is cooked, one teaspoon before a meal	Bronchitis	Jarić et al. (2015)
	Leaves	Compress	Eczema, acne, and wounds	Šavikin et al. (2013)
Teucrium chamaedrys L.	Flowering aerial parts	Tea	Respiratory inflammation	Mustafa et al. (2012)

(*Continued*)

TABLE 3.1 (Continued)

Botanical taxon	Part(s) used	Preparation and/or administration	Treated disease(s) or folk medical uses(s)	Reference/Sources
Teucrium montanum L.	Leaves	Tea (topically applied)	Skin problems	Mustafa et al. (2020)
Teucrium polium L.	Aerial parts	Mixed with fat	Tuberculosis, "*Saraxha*" (cutaneous tuberculosis)	Mustafa et al. (2015)
Thymus serpyllum L.	Aerial parts	Infusion	Respiratory inflammations (bronchitis)	Hajdari et al. (2018), Mustafa et al. (2012), Mustafa et al. (2012), Mustafa et al. (2015), Mustafa et al. (2020), Rexhepi et al. (2013)
		Decoction	Respiratory inflammations	Mustafa et al. (2012)
Tilia cordata Mill.	Flowers	Tea	Respiratory inflammations (bronchitis)	Jarić et al. (2007), Mustafa et al. (2012), Mustafa et al. (2015), Pieroni et al. (2005), Pieroni et al. (2017), Rexhepi et al. (2013), Šavikin et al. (2013), Zlatković et al. (2014)
			Colds and sore throats	Mustafa et al. (2015)
		Decoction	Sore throat, lung inflammations	Mustafa et al. (2015)
Tilia platyphyllos Scop.	Flowers	Tea	Bronchitis, flu	Mustafa et al. (2012), Mustafa et al. (2020)
		Decoction	Sore throat, lung inflammations	Mustafa et al. (2012)
Trifolium pratense L.	Flowers	Infusion	Oral cavity antiseptic	Mustafa et al. (2012)
Triticum vulgare Vill.	Flour	Mixed with hot water—topically used	Skin inflammation and ulcers	Mustafa et al. (2015)
Tussilago farfara L.	Leaves, flowers	Tea	Ulcers	Rexhepi et al. (2013)
		Wounds are to be bandaged with fresh leaf	Festering wounds and ulcers	Jarić et al. (2007)
Ulmus minor Mill.	Leaves	Extracted with fat	Antimycotic, antibacterial, "*Saraxha*" (cutaneous tuberculosis)	Mustafa et al. (2015)
	Bark	Decoction, externally applied	Wounds, burns	Pieroni et al. (2015)

(*Continued*)

TABLE 3.1 (Continued)

Botanical taxon	Part(s) used	Preparation and/or administration	Treated disease(s) or folk medical uses(s)	Reference/Sources
Urtica dioica L.	Aerial parts	Infusion	Bronchitis, antibacterial, urinary disorders	Mustafa et al. (2015)
		Tea	Urinary tract infections	Hajdari et al. (2018)
		Tea used to wash and treat hands and feet	Fungal skin infection	Hajdari et al. (2018)
	Leaves	Externally applied	Skin complaints, neuralgia, hemorrhoids, hair problems	Menković et al. (2011)
Vaccinium myrtillus L.	Fruits	Also eaten as dried fruit	Sore throats, digestive troubles	Pieroni (2008)
	Fruits, leaves	Juice of fresh fruits	Digestive tract infections, eye inflammations, urinary disorders	Mustafa et al. (2015)
		Tea	Urinary tract infections	Jarić et al. (2015)
			Inflammation of the mouth and throat.	Menković et al. (2011)
		Tea, syrup (gargle as a throat wash)	Viral infection	Rexhepi et al. (2013)
		Externally applied	Skin rash, inflamed ulcers	Šarić-Kundalić et al. (2011)
Vaccinium vitis-idaea L.	Leaves	Infusion	Urinary inflammations	Mustafa et al. (2015)
	Fruits	Infusion	Urinary tract infections	Mustafa et al. (2015), Šavikin et al. (2013)
	Fruits and leaves	Infusion	Wound healing	Mustafa et al. (2015)
	Areal parts		Respiratory ailments, influenza, nipple inflammation during lactation	Šarić-Kundalić et al. (2011)
		Tea and fresh fruit as part of a person's diet	Urinary tract infections	Jarić et al. (2015)
Valeriana officinalis L.	Leaves	Macerated leaves are mixed with yogurt and topically applied	Breast inflammations	Mustafa et al. (2012)
Verbascum phlomoides L.	Flowers	Tea	Antiseptic properties, respiratory ailments (bronchitis, laryngitis, asthma, influenza, tuberculosis)	Jarić et al. (2007), Rexhepi et al. (2013)

(Continued)

TABLE 3.1 (Continued)

Botanical taxon	Part(s) used	Preparation and/or administration	Treated disease(s) or folk medical uses(s)	Reference/Sources
Veronica officinalis L.	Leaves	Infusion	Respiratory system inflammations, wound healing	Mustafa et al. (2015)
	Aerial parts	Externally applied	Skin diseases, wounds	Menković et al. (2011)
Vitis labrusca L.	Young shoots	Squeezed to extract juice that is topically applied to wounds	Antiseptic	Pieroni et al. (2014)
Zea mays L.	Fruits	Infusion	Urinary tract inflammations	Mustafa et al. (2015)

Apiaceae

Petroselinum crispum (Mill.) Fuss

Petroselinum crispum (Mill.) Fuss (parsley) is an herb native to the Mediterranean region. The aerial parts of *P. crispum* showed antibacterial activity against several *Vibrio* spp. strains with MIC values ranging from 19 to 39 μg/mL (Snoussi et al., 2016). A recent clinical trial tested the efficacy of using *P. crispum* powder as an herb-based antimicrobial treatment for urinary tract infections (UTIs). With a sample size of 37 patients with UTIs, patients indicated a significant decline in indicators such as frequency, dysuria, suprapubic pain, and loin pain upon completion of the study. Furthermore, the general urine exam saw a significant decrease in terms of acidity, pus cells, crystals, and epithelial cells (Nashtar & Al-Attar, 2018). In vitro studies similarly demonstrated strong antibacterial activity of *P. crispum* against UTI clinical isolates (Petrolini et al., 2013). Studies such as these show promising results for the use of parsley as an antimicrobial agent against human urinary infections.

Asteraceae

Achillea millefolium L.

Achillea millefolium L. (yarrow) is a flowering plant native to eastern Turkey. Ethanolic extracts of *A. millefolium* exhibit high antibacterial activity against *Bacillus cereus* (MIC = 0.85 μg/mL), *Enterococcus faecalis* (MIC = 1.71 μg/mL), *Serratia rubidaea* (MIC = 1.19 μg/mL), *Escherichia coli* (MIC = 13.59 μg/mL), *Lactobacillus brevis* (MIC = 0.85 μg/mL), and *Lactobacillus hilgardii* (MIC = 0.59 μg/mL). Constituents such as terpinolene, 1,8-cineole, thujone, camphor, and borneol may be responsible for the antibacterial activity of *A. millefolium* (Ali et al., 2017). Although some clinical trials investigate the efficacy of *A. millefolium* as a pain reliever, antiinflammatory agent, and wound healing aid, further clinical studies must be conducted to test the antibacterial potential of the species (Hajhashemi et al., 2018; Jenabi & Fereidoony, 2015).

Artemisia absinthium L.

Artemisia absinthium L. (wormwood) is a flowering plant with origins in eastern Turkey. The chloroform extract from the leaves of *A. absinthium* exhibited MIC values ranging from 128 to 256 μg/mL against *Staphylococcus aureus, Enterococcus faecalis,* and *Bacillus cereus* (Fiamegos et al., 2011). The compounds thujone, linalool, and β-caryophyllene are associated with the antibacterial activity found in *A. absinthium*. An in vivo study analyzing surgical rat wounds infected with *Staphylococcus aureus* showed a significant reduction in bacterial count at the wound site upon topical application of *A. absinthium* (Moslemi et al., 2012).

Calendula officinalis L.

Calendula officinalis L. (pot marigold) is an annual herbaceous plant found throughout Europe. Leaves and flowers of *C. officinalis* extracted with methanol had MICs against several Gram-positive and Gram-negative bacteria, including *Staphylococcus epidermidis* (MIC = 10 μg/mL), *Listeria monocytogenes* (MIC = 15 μg/mL), *Bacillus megaterium* (MIC = 35 μg/mL), *Escherichia coli* (MIC = 35 μg/mL), and *Shigella flexneri* (MIC = 50 μg/mL) (Szakiel et al., 2008). Major flavonoid-based components such as rutin, gallic acid, and quercetin-3-O-glucoside may be responsible for the antibacterial activity of the methanolic flower extracts (Rigane et al., 2013). Furthermore, the antibacterial activity of the essential oil of *C. officinalis* shown in other studies may be a result of its major chemical constituents, including citral, geraniol, eugenol, menthol, and cinnamic aldehyde (Chaleshtori et al., 2016). A clinical trial conducted on 18 patients assessed the efficacy of a mouthwash containing *C. officinalis* against microorganisms adhering to sutures upon molar extraction. The treatment group saw a reduction in microorganism count compared to baseline, thus suggesting that *C. officinalis* showed antibacterial activity against microorganisms adhering to dental sutures (Faria et al., 2011).

Matricaria chamomilla L.

Matricaria chamomilla L. (chamomile) is an annual herb native to Europe and Western Asia. Essential oils showed inhibitory activity with MIC values ranging from 0.011 to 4 μg/mL against *Staphylococcus aureus, Bacillus cereus, Bacillus subtilis, Shigella shiga, Shigella sonnei,* and *Pseudomonas aeruginosa* (Kazemi, 2015). Ethanolic and methanolic extracts also showed MICs of 12.5–15 μg/mL against *Escherichia coli, Bacillus cereus, Staphylococcus aureus,* and *Salmonella typhi* (Roby et al., 2013). Compounds isolated from the essential oil, such as bisabolol, bisabolol oxide, limonene, camphene, and camphor demonstrate high antibacterial activity and may be responsible for the antimicrobial properties seen in *M. chamomilla* (Kazemi, 2015). For more information on the antibacterial properties and clinical trials conducted with *M. chamomilla,* refer to a comprehensive literature review (Chassagne et al., 2021).

Betulaceae

Alnus glutinosa (L.) Gaertn.

Alnus glutinosa (L.) Gaertn. (black alder) is a tree species found throughout Europe, southwest Asia, and northern Africa. The ethanolic and methanolic extracts from the leaves of *A. glutinosa* showed antibacterial against *Staphylococcus aureus, Bacillus subtilis, Escherichia coli, Klebsiella aerogenes,* and *Pseudomonas aeruginosa* with MIC values ranging from 125 to 250 μg/mL (Altinyay et al., 2015; Middleton et al., 2010). The compound oregonin produced from the methanolic extract of *A. glutinosa* shows extremely high antibacterial properties against both Gram-negative and Gram-positive bacteria (*Bacillus subtilis*—15.6 μg/mL, *Staphylococcus aureus*—15.6 μg/mL, *Proteus vulgaris*—31.2 μg/mL) and may be responsible for the antibacterial effects seen in *A. glutinosa* (Abedini et al., 2016). The flavonoid, genkwanin, also extracted from the methanol, may also be responsible for some antibacterial activity, with an MIC value of 0.5 μg/mL against *Bacillus cereus* (Kumarasamy et al., 2006).

Lamiaceae

Lavandula angustifolia Mill.

Lavandula angustifolia Mill. (lavender) is a flowering plant native to the Mediterranean region. Its essential oil has strong antibacterial activities against *Escherichia coli, Pseudomonas aeruginosa, Proteus mirabilis, Klebsiella pneumoniae, Acinetobacter baumannii, Staphylococcus aureus, Enterococcus faecalis*, and *Bacillus subtilis* with MIC values ranging from 2 to 8 μg/mL (Erdoğan Orhan et al., 2012). Hydroxycinnamic acids such as rosmarinic acid, chlorogenic acid, and caffeic acid, as well as flavonoids like rutin and quercetin are a few of the constituents primarily responsible for the antibacterial activity of *L. angustifolia* (Zenão et al., 2017). An in vivo study was conducted to assess the effect of the topical antibacterial treatment of *L. angustifolia* and *Thymus vulgaris* against bovine staphylococcal and streptococcal mastitis. The results indicated that intramammary application of a solution of *Lavandula angustifolia* and *Thymus vulgaris* significantly decreased the bacterial count of staphylococci and streptococci in the milk produced by the cows in the study (Abboud et al., 2015).

Mentha longifolia (L.) L.

Mentha longifolia (L.) L. (wild mint) is an herbaceous perennial found in Europe, western and central Asia, and parts of Africa. The essential oil of *M. longifolia* showed antibacterial activity with MIC's ranging from 15.62 to 62.50 μg/mL against *Bacillus macerans, Acinetobacter baumannii, Escherichia coli, Klebsiella pneumoniae, Bacillus subtilis*, and *Enterococcus faecalis* (Gulluce et al., 2007). Flavonoids such as quercetin-3-O-glycoside and terpenoids such as menthol isolated from the leaves of *M. longifolia* may be responsible for the antibacterial activity demonstrated (Al-Bayati, 2009). Furthermore, active constituents in the essential oil of *M. longifolia*, such as *cis*-piperitone epoxide, piperitenone oxide, pulegone, and menthone, also contribute greatly to the antibacterial activity of the essential oil.

Mentha x piperita L.

Mentha x piperita L. (peppermint) is a hybrid cross between *Mentha spicata* (spearmint) and *Mentha aquatica* L. (watermint) and is a small herbaceous species native to the Mediterranean region. The essential oil showed antibacterial activity against *Streptococcus pneumonia, Staphylococcus aureus, Pseudomonas aeruginosa, Escherichia coli, Salmonella typhi, Listeria monocytogenes*, and *Klebsiella pneumoniae* with MIC values ranging from 0.5 to 8 μg/mL

(Abolfazl et al., 2014). Further information about *Mentha x piperita* can be found in a prior review by our group (Chassagne et al., 2021).

***Ocimum basilicum* L.**

Ocimum basilicum L. (basil) is an herb found in subtropical regions of Asia, Africa, and South America. *O. basilicum* showed antibacterial activity against several *Vibrio* spp. strains with MIC values ranging from 19 to 39 μg/mL (Snoussi et al., 2016). Methanol extracts from the leaves of *O. basilicum* also exhibited MICs ranging from 62.5 to 125 μg/mL against *Bacillus cereus, Bacillus megaterium, Staphylococcus. aureus, Bacillus subtilis, Listeria monocytogenes*, and *Escherichia coli* (Hossain et al., 2010). Chemical constituents such as rosmarinic acid may be responsible for the antibacterial activity seen in the essential oils of *O. basilicum* (Bais et al., 2002).

***Origanum vulgare* L.**

Origanum vulgare L. (oregano) is an aromatic perennial herb found in the Mediterranean region. Coccimiglio et al. reported that the ethanolic extract showed MIC values ranging from 6.3 to 12.5 μg/mL against *Pseudomonas aeruginosa, Escherichia coli, Bordetella bronchiseptica, Burkholderia cenocepacia, Acinetobacter baumannii*, and *Bacillus subtilis* (Coccimiglio et al., 2016). The essential oils show the highest antibacterial activity, with MIC's ranging from 0.3 to 100 μg/mL (Helal et al., 2019; Santos et al., 2017). For a more comprehensive review of *Origanum vulgare*, refer to previous work by our group (Chassagne et al., 2021).

Malvaceae

***Althaea officinalis* L.**

Althaea officinalis L. (marshmallow) is a perennial species native to Europe, Western Asia, and Northern Africa. The methanolic extract has antibacterial activity against *Bacillus psychrosaccharolyticus* (MIC = 31.25 μg/mL), *Exiguobacterium acetylicum* (MIC = 125 μg/mL), *Escherichia coli* (MIC = 31.25 μg/mL, *Staphylococcus aureus* (MIC = 15.63 μg/mL), and *Pseudomonas aeruginosa* (MIC = 62.5 μg/mL) (Mehreen et al., 2016; Ozturk & Ercisli, 2007). A rat model in vivo study assessed the wound healing properties of *Althaea officinalis*. The flavonoids in *A. officinalis* may be responsible for the antibacterial activity exhibited, through further phytochemical studies must be conducted to assess the specific compounds and their inhibitory activity (Mehreen et al., 2016). The topically administered hydroethanolic extract was found to significantly reduce the wound size compared to the control, and the results of the in vitro analyses demonstrated the antibacterial properties of *A. officinalis* against several Gram-positive bacteria (Rezaei et al., 2015).

***Malva sylvestris* L.**

Malva sylvestris L. (common mallow) is an annual plant native to regions in Europe, Northern Africa, and Southwest Asia. Methanolic extracts from the leaves of *M. sylvestris* exhibited antibacterial activity against *Staphylococcus aureus, Enterococcus faecalis, Streptococcus agalactiae*, and *Erwinia carotovora* with MICs ranging from 35 to 37 μg/mL, and the flowers showed activity against *Escherichia coli* with a MIC of 31 μg/mL (Razavi et al., 2011).

Although specific chemical constituents in methanolic extracts of *M. sylvestris* have not been identified to explain its antibacterial activity against the listed microbes, the major constituent of the oil, eugenol, was shown to exhibit inhibitory activity (Cecotti et al., 2016). Furthermore, the phytoalexin, malvone A (2-methyl-3-methoxy-5,6-dihydroxy-1,4-naphthoquinone), showed inhibitory activity against the pathogen *Verticillium dahliae* (Veshkurova et al., 2006). The mechanism behind this important antibacterial agent may provide insight into future research on the pharmacology of *M. sylvestris*. Two different in vivo studies conducted on Wistar rats exhibited both the wound healing and burn healing potential of the diethyl ester extract of *Malva sylvestris*, with the *M. sylvestris*-treated group showing a significant reduction in wound size and burn size compared to the control groups (Pirbalouti et al., 2009; Pirbalouti et al., 2010).

Pinaceae

***Larix decidua* Mill.**

Larix decidua Mill. (European larch) is a deciduous conifer found in the Alps, the Sudetes, and the Carpathian mountains. Methanolic extracts from the bark showed antibacterial activity against *Listeria monocytogenes, Bacillus cereus, Staphylococcus aureus, Dickeya solani, Pectobacterium atrospecticum,* and *Micrococcus flavus* with MIC values ranging from 150 to 220 μg/mL (Salem et al., 2016). Compounds such as abietic acid, oleanolic acid, larixol, 2,9-dihydroxyverrucosane, nonacosane are responsible for the antibacterial properties of the methanolic extract of *L. decidua* (Salem et al., 2016). Animal studies must be conducted to assess the in vivo efficacy of this plant species as an antibacterial agent.

***Picea abies* (L.) H.Karst.**

Picea abies (L.) H.Karst. (Norway spruce) is the most common conifer in Europe and is native to regions spanning from Scandinavia to Northern Russia. Methanolic extracts from the wood and bark showed antibacterial activity against *Escherichia coli* (MIC = 60 μg/mL), *Staphylococcus aureus* (MIC = 130 μg/mL), *Listeria monocytogenes* (160 μg/mL), and *Bacillus cereus* (MIC = 140 μg/mL) (Salem et al., 2016). Several constituents of *P. abies* are responsible for its antibacterial activity including the piperidine alkaloid, epidihydropinidine, as well as flavonols like kaempferol, quercetin, and myricetin, and stilbene glucosides such as piceid (Fyhrquist et al., 2018; Metsämuuronen & Sirén, 2019). Resin salve extracted from *P. abies* was used in a pilot clinical trial to assess its effectiveness in surgical wound healing. The trial included 23 patients whose wound healing postsurgery was delayed. The resin salve treatment group saw a significant reduction in wound size and an improved healing time compared to the control, despite the presence of microbes at the infection site (Sipponen et al., 2012).

Rosaceae

***Agrimonia eupatoria* L.**

Agrimonia eupatoria L. (agrimony) is a perennial herb native to Europe and Southwest Asia. The acetone extract from the aerial parts showed antibacterial activity with MIC values ranging from 120 to 310 μg/mL against *Staphylococcus aureus, Bacillus subtilis, Bacillus cereus, Lactobacillus rhamnosus*, and *Bifidobacterium animalis* subsp. *lactis* (Muruzović et al., 2016). An ointment prepared from the ethanolic extract of *A. eupatoria* was shown to have significant wound healing properties in rats, with the ethanolic extract exhibiting faster wound healing times compared to the control group. The researchers concluded that *A. euphoria* tannin and flavonoid constituents contributed to the observed accelerated wound healing process due to their antibacterial properties (Ghaima, 2013).

Prunus spinosa L.

Prunus spinosa L. (blackthorn) is a flowering plant native to Europe, western Asia, and northwest Africa. The methanolic extract from the fruits of *P. spinosa* showed MIC against *Staphylococcus aureus* (MIC = 7.8 μg/mL), *Citrobacter freundii* (MIC = 31.25 μg/mL), *Listeria innocua* (MIC = 31.2 μg/mL), and *Sarcina lutea* (62.5 μg/mL) (Kumarasamy et al., 2004; Radovanović et al., 2013).

Rosa canina L.

Rosa canina L. (dog rose) is a deciduous shrub native to Europe, western Asia, and northwest Africa. A geometric isometric mixture (2:1) extracted with methanol from the seeds of *R. canina* exhibited high antibacterial activity against *Lactobacillus plantarum* (MIC = 0.1 μg/mL), *Staphylococcus epidermidis* (0.1 μg/mL), and *Proteus mirabilis* (10 μg/mL). The compounds responsible for such antibacterial properties were identified as the flavonoid glycoside isomers, kaempferol 3-*O*-(6″-*O*-*E*-*p*-coumaroyl)-β-D-glucopyranoside and kaempferol 3-*O*-(6″-*O*-*Z*-*p*-coumaroyl)-β-D-glucopyranoside (Kumarasamy et al., 2003). A triple-blind randomized clinical trial was conducted to assess the effect of the fruits of *R. canina* in preventing UTIs in women following a cesarean section. The results showed that the incidence of UTI was significantly lower in the group that received 500 mg of *R. canina* compared to the placebo group. Because UTIs are characterized by inflammation due to bacterial invasion, it can be concluded that the antibacterial activity exhibited in *R. canina* may be partially responsible for the results of this clinical trial (Seifi et al., 2018).

Rubus fruticosus L.

Rubus fruticosus L. (blackberry) is a perennial shrub native to regions throughout Europe. The methanolic extract from the stem of *R. fruticosus* showed an inhibitory concentration of 20 μg/mL against *Escherichia coli, Salmonella typhi, Staphylococcus aureus, Proteus mirabilis, Micrococcus luteus, Bacillus subtilis*, and *Pseudomonas aeruginosa* (Riaz et al., 2011).

Urticaceae

Urtica dioica L.

Urtica dioica L. (nettle) is a perennial flowering plant native to Europe, temperate Asia, western North Africa but cultivated throughout the world. The hexane extract showed inhibitory activity against *Salmonella typhi, Escherichia coli, Klebsiella pneumoniae, Staphylococcus aureus, Enterococcus faecalis, Shigella flexneri*, and *Pseudomonas aeruginosa* with MIC values ranging from 7.81 to 250 μg/mL. Constituents isolated from the hexane extracts, such as neophytadiene, heptadecyl ester, hexyl octyl ester, butyl tetradecyl ester, and 1,2 benzenedicarboxylic acid, may be responsible for the antibacterial activity exhibited in *U. dioica* (Dar et al., 2013). The wound-healing potential of hydroethanolic extract from the leaves of *U. dioica* was assessed in a rat model in vivo study. The topical treatment of *U. dioica* on the surgical wound showed faster healing rates compared to the control group. An in vitro quantification of the inhibitory potential of *U. dioica* in this study showed strong antibacterial activity against *Pseudomonas aeruginosa* and *Enterococcus faecalis*, two microorganisms commonly associated with open wounds (Zouari Bouassida et al., 2017).

Conclusions

The Balkan region contains a reservoir of TEK, which serves as a predominant form of healthcare especially in isolated mountainous and rural areas where access to Western medicine is limited. The geographic isolation, rich biodiversity, and complex biocultural variety have shaped specific knowledge on the traditional uses of botanical species to treat infectious diseases. Thus, TEK applications in regions can be used not only to preserve cultural traditions and local natural resources but also for drug discovery initiatives and public health issues, as not all-natural remedies are safe.

The Balkans also possess a wealth of diverse antibacterial species that may show promising clinical applications for bacterial diseases. Of the species described herein, *Allium sativum, Achillea millefolium, Calendula officinalis, Matricaria chamomilla, Origanum vulgare, Lavandula angustifolia, Mentha x piperita, Malva sylvestris, Rosa canina*, and *Urtica dioica* showed the most potent antibacterial activity, with MICs $\leq 100\ \mu g/mL$ against a range of microorganisms. Several of these species, including *A. sativum, M. chamomilla, Mentha x piperita, O. vulgare, M. sylvestris*, and *U. dioica* exhibited wound or burn healing activity in vivo. However, the safety of these extracts needs to be studied more in depth prior to their use in human clinical trials. Despite this limitation, some pilot human-based clinical trials have been conducted to test species such as *C. officinalis* and *M. chamomilla* for their wound-healing potential against oral pathogens, as well as *R. canina* against UTIs. Future research should focus on conducting more in vivo studies and human clinical trials that highlight and explain the diverse antibacterial properties of the above Balkan species, as well as those listed in Table 3.1.

References

Abboud, M., Rammouz., Jammal, B., & Sleiman, M. (2015). In vitro and in vivo antimicrobial activity of two essential oils *Thymus vulgaris* and *Lavandula angustifolia* against bovine *Staphylococcus* and *Streptococcus mastitis* pathogen. *Middle East Journal of Agriculture Research, 4*, 975–983.

Abedini, A., Chollet, S., Angelis, A., Borie, N., Nuzillard, J. M., Skaltsounis, A. L., ... Hubert, J. (2016). Bioactivity-guided identification of antimicrobial metabolites in *Alnus glutinosa* bark and optimization of oregonin purification by centrifugal partition chromatography. *Journal of Chromatography B: Analytical Technologies in the Biomedical and Life Sciences, 1029–1030*, 121–127. Available from https://doi.org/10.1016/j.jchromb.2016.07.021.

Abolfazl, M., Hadi, A., Frhad, M., & Hossein, N. (2014). In vitro antibacterial activity and phytochemical analysis of some medicinal plants. *Journal of Medicinal Plants Research, 8*, 186–194.

Al-Bayati, F. A. (2009). Isolation and identification of antimicrobial compound from *Mentha longifolia* L. leaves grown wild in Iraq. *Annals of Clinical Microbiology and Antimicrobials, 8*, 20. Available from https://doi.org/10.1186/1476-0711-8-20.

Ali, S. I., Gopalakrishnan, B., & Venkatesalu, V. (2017). Pharmacognosy, phytochemistry and pharmacological properties of *Achillea millefolium* L.: A review. *Phytotherapy Research, 31*, 1140–1161. Available from https://doi.org/10.1002/ptr.5840.

Altinyay, M., Eryilmaz, A. N., Yazgan, B. S., & Yilmaz, M. L. (2015). Altun, antimicrobial activity of some *Alnus* species. *European Review for Medical and Pharmacological Sciences, 19*, 4671–4674. Available from http://www.europeanreview.org/.

Bais, H. P., Walker, T. S., Schweizer, H. P., & Vivanco, J. M. (2002). Root specific elicitation and antimicrobial activity of rosmarinic acid in hairy root cultures of *Ocimum basilicum*. *Plant Physiology and Biochemistry, 40*, 983–995. Available from https://doi.org/10.1016/S0981-9428(02)01460-2.

Block, E. (1985). The chemistry of garlic and onions. *Scientific American, 252*, 114–119. Available from https://doi.org/10.1038/scientificamerican0385-114.

CLSI. (2012). Methods for dilution antimicrobial susceptibiity tests for bacteria that grow aerobically; approved standard; CLSI document M07-A9.

Cecotti, R., Bergomi, P., Carpana, E., & Tava, A. (2016). Chemical characterization of the volatiles of leaves and flowers from cultivated *Malva sylvestris* var. mauritiana and their antimicrobial activity against the aetiological agents of the European and American Foulbrood of Honeybees (*Apis mellifera*). *Natural Product Communications, 11*(10), 1527–1530. Available from https://doi.org/10.1177/1934578x1601101026, 1934578X1601101.

Chaleshtori, S. H., Kachoie, M. A., & Pirbalouti, A. G. (2016). Phytochemical analysis and antibacterial effects of Calendula of cinalis essential oil. *Bioscience Biotechnology Research Communications, 9*(3), 517–522. Available from https://doi.org/10.21786/bbrc/9.3/26.

Chassagne, F., Samarakoon, T., Porras, G., Lyles, J. T., Dettweiler, M., Marquez, L., ... Systematic, A. (2021). Review of plants with antibacterial activities: A taxonomic and phylogenetic perspective. *Frontiers in Pharmacology, 11*, 586548. Available from https://doi.org/10.3389/fphar.2020.586548.

Coccimiglio, J., Alipour, M., Jiang, Z. H., Gottardo, C., & Suntres, Z. (2016). Antioxidant, antibacterial, and cytotoxic activities of the ethanolic origanum vulgare extract and its major constituents. *Oxidative Medicine and Cellular Longevity, 2016*, 1404505. Available from https://doi.org/10.1155/2016/1404505.

Dar, S. A., Ganai, F. A., Yousuf, A. R., Balkhi, M. U. H., Bhat, T. M., & Sharma, P. (2013). Pharmacological and toxicological evaluation of *Urtica dioica*. *Pharmaceutical Biology, 51*, 170–180. Available from https://doi.org/10.3109/13880209.2012.715172.

Eloff, J. N. (2004). Quantification the bioactivity of plant extracts during screening and bioassay guided fractionation. *Phytomedicine: International Journal of Phytotherapy and Phytopharmacology, 11*, 370–371. Available from https://doi.org/10.1078/0944711041495218.

Erdoğan Orhan, I., Özçelik, B., Kartal, M., & Kan, Y. (2012). Seçilmiş Umbelliferae ve Labiatae bitkilerinin uçucu yağları ve tek uçucu yağ bileşiklerinin antimikrobiyal ve antiviral etkileri. *Turkish Journal of Biology, 36*, 239–246. Available from https://doi.org/10.3906/biy-0912-30.

Faria, R. L., Cardoso, L. M. L., Akisue, G., Pereira, C. A., Junqueira, J. C., Jorge, A. O. C., & Santos, P. V. (2011). Antimicrobial activity of *Calendula officinalis, Camellia sinensis* and chlorhexidine against the adherence of microorganisms to sutures after extraction of unerupted third molars. *Journal of Applied Oral Science, 19*, 476–482. Available from https://doi.org/10.1590/S1678-77572011000500007.

Fiamegos, Y. C., Kastritis, P. L., Exarchou, V., Han, H., Bonvin, A. M. J. J., Vervoort, J., ... Tegos, G. P. (2011). Antimicrobial and efflux pump inhibitory activity of caffeoylquinic acids from *Artemisia absinthium* against Gram-positive pathogenic bacteria. *PLoS One, 6*, e18127. Available from https://doi.org/10.1371/journal.pone.0018127.

Fyhrquist, P., Virjamo, V., Hiltunen, E., & Julkunen-Tiitto, R. (2018). Epidihydropinidine, the main piperidine alkaloid compound of Norway spruce (*Picea abies*) shows antibacterial and anti-Candida activity. *Fitoterapia, 134*, 503–511. Available from https://doi.org/10.1016/j.fitote.2018.12.015.

Ghaima, K. K. (2013). Antibacterial and wound healing activity of some *Agrimonia eupatoria* extracts. *Baghdad Science Journal, 10*, 152–160.

Gulluce, M., Sahin, F., Sokmen, M., Ozer, H., Daferera, D., Sokmen, A., ... Ozkan, H. (2007). Antimicrobial and antioxidant properties of the essential oils and methanol extract from *Mentha longifolia* L. ssp. longifolia. *Food Chemistry, 103*, 1449–1456. Available from https://doi.org/10.1016/j.foodchem.2006.10.061.

Hajdari, A., Pieroni, A., Jhaveri, M., Mustafa, B., & Quave, C. L. (2018). Ethnomedical knowledge among Slavic speaking people in South Kosovo. *Ethnobiology and Conservation, 7*, 6. Available from https://doi.org/10.15451/ec2018-03-07.06-1-42.

Hajhashemi, M., Ghanbari, Z., Movahedi, M., Rafieian, M., Keivani, A., & Haghollahi, F. (2018). The effect of *Achillea millefolium* and *Hypericum perforatum* ointments on episiotomy wound healing in primiparous women. *Journal of Maternal-Fetal and Neonatal Medicine, 31*, 63–69. Available from https://doi.org/10.1080/14767058.2016.1275549.

Harris, J. C., Cottrell, S. L., Plummer, S., & Lloyd, D. (2001). Antimicrobial properties of *Allium sativum* (garlic). *Applied Microbiology and Biotechnology, 57*, 282–286. Available from https://doi.org/10.1007/s002530100722.

Helal, I. M., El-Bessoumy, A., Al-Bataineh, E., Joseph, M. R. P., Rajagopalan, P., Chandramoorthy, H. C., & Ben Hadj Ahmed, S. (2019). Antimicrobial efficiency of essential oils from traditional medicinal plants of asir region, Saudi Arabia, over drug resistant isolates. *BioMed Research International, 2019*, 8928306. Available from https://doi.org/10.1155/2019/8928306.

Hossain, M. A., Kabir, M. J., Salehuddin, S. M., Rahman, S. M. M., Das, A. K., Singha, S. K., ... Rahman, A. (2010). Antibacterial properties of essential oils and methanol extracts of sweet basil *Ocimum basilicum* occurring in Bangladesh. *Pharmaceutical Biology, 48*, 504–511. Available from https://doi.org/10.3109/13880200903190977.

Jarić, S., Popović, Z., Mačukanović-Jocić, M., Djurdjević, L., Mijatović, M., Karadžić, B., ... Pavlović, P. (2007). An ethnobotanical study on the usage of wild medicinal herbs from Kopaonik Mountain (Central Serbia). *Journal of Ethnopharmacology, 111*, 160–175. Available from https://doi.org/10.1016/j.jep.2006.11.007.

Jarić, S., MačUkanović-Jocić, M., Djurdjević, L., Mitrović, M., Kostić, O., Karadžić, B., & Pavlović, P. (2015). An ethnobotanical survey of traditionally used plants on *Suva planina* mountain (south-eastern Serbia). *Journal of Ethnopharmacology, 175*, 93–108. Available from https://doi.org/10.1016/j.jep.2015.09.002.

Jenabi, E., & Fereidoony, B. (2015). Effect of *Achillea millefolium* on relief of primary dysmenorrhea: A double-blind randomized clinical trial. *Journal of Pediatric and Adolescent Gynecology, 28*, 402–404. Available from https://doi.org/10.1016/j.jpag.2014.12.008.

Karuppiah, P., & Rajaram, S. (2012). Antibacterial effect of *Allium sativum* cloves and *Zingiber officinale* rhizomes against multiple-drug resistant clinical pathogens. *Asian Pacific Journal of Tropical Biomedicine, 2*, 597–601. Available from https://doi.org/10.1016/S2221-1691(12)60104-X.

Kathe, W., Honnef, S., & Heym, A. (2003). Croatia and Romania: A study of the collection of and trade in medicinal and aromatic plants (MAPs), relevant legislation and the potential of MAP use for financing nature conservation and protected areas.

Kazemi, M. (2015). Chemical composition and antimicrobial activity of essential oil of *Matricaria recutita*. *International Journal of Food Properties, 18*, 1784–1792. Available from https://doi.org/10.1080/10942912.2014.939660.

Kumarasamy, Y., Cox, P. J., Jaspars, M., Rashid, M. A., & Sarker, S. D. (2003). Bioactive flavonoid glycosides from the seeds of *Rosa canina*. *Pharmaceutical Biology, 41*, 237–242. Available from https://doi.org/10.1076/phbi.41.4.237.15663.

Kumarasamy, Y., Cox, P. J., Jaspars, M., Nahar, L., & Sarker, S. D. (2004). Comparative studies on biological activities of *Prunus padus* and *P. spinosa*. *Fitoterapia, 75*, 77–80. Available from https://doi.org/10.1016/j.fitote.2003.08.011.

Kumarasamy, Y., Cox, P. J., Jaspars, M., Nahar, L., & Sarker, S. D. (2006). Bioactivity of hirsutanolol, oregonin and genkwanin, isolated from the seeds of *Alnus glutinosa* (Betulaceae). *Natural Product Communications, 1*, 641–644. Available from https://doi.org/10.1177/1934578x0600100808.

Mehreen, A., Waheed, M., Liaqat, I., & Arshad, N. (2016). Phytochemical, antimicrobial, and toxicological evaluation of traditional herbs used to treat sore throat. *BioMed Research International, 2016*, 8503426. Available from https://doi.org/10.1155/2016/8503426.

Menković, N., Šavikin, K., Tasić, S., Zdunić, G., Stešević, D., Milosavljević, S., & Vincek, D. (2011). Ethnobotanical study on traditional uses of wild medicinal plants in Prokletije Mountains (Montenegro). *Journal of Ethnopharmacology, 133*, 97–107. Available from https://doi.org/10.1016/j.jep.2010.09.008.

Metsämuuronen, S., & Sirén, H. (2019). Bioactive phenolic compounds, metabolism and properties: A review on valuable chemical compounds in Scots pine and Norway spruce. *Phytochemistry Reviews, 18*, 623–664. Available from https://doi.org/10.1007/s11101-019-09630-2.

Middleton, P., Stewart, F., Al-Qahtani, S., Egan, P., O'Rourke, C., Abdulrahman, A., & Sarker, S. D. (2010). Antioxidant, antibacterial activities and general toxicity of *Alnus glutinosa*, *Fraxinus excelsior* and *Papaver rhoeas*. *Iranian Journal of Pharmaceutical Research, 4*, 101–103. Available from https://doi.org/10.22037/ijpr.2010.620.

Moslemi, H. R., Hoseinzadeh, H., Badouei, M. A., Kafshdouzan, K., & Fard, R. M. N. (2012). Antimicrobial activity of *Artemisia absinthium* against surgical wounds infected by *Staphylococcus aureus* in a rat model. *Indian Journal of Microbiology, 52*, 601–604. Available from https://doi.org/10.1007/s12088-012-0283-x.

Muruzović, M., Mladenović, K. G., Stefanović, O. D., Vasić, S. M., & Čomić, L. R. (2016). Extracts of *Agrimonia eupatoria* L. as sources of biologically active compounds and evaluation of their antioxidant, antimicrobial, and antibiofilm activities. *Journal of Food and Drug Analysis, 24*, 539–547. Available from https://doi.org/10.1016/j.jfda.2016.02.007.

Mustafa, B., Hajdari, A., Pajazita, Q., Syla, B., Quave, C. L., & Pieroni, A. (2012). An ethnobotanical survey of the Gollak region, Kosovo. *Genetic Resources and Crop Evolution, 59*, 739–754. Available from https://doi.org/10.1007/s10722-011-9715-4.

Mustafa, B., Hajdari, A., Krasniqi, F., Hoxha, E., Ademi, H., Quave, C. L., & Pieroni, A. (2012). Medical ethnobotany of the Albanian Alps in Kosovo. *Journal of Ethnobiology and Ethnomedicine, 8*, 6. Available from https://doi.org/10.1186/1746-4269-8-6.

Mustafa, B., Hajdari, A., Pieroni, A., Pulaj, B., Koro, X., & Quave, C. L. (2015). A cross-cultural comparison of folk plant uses among Albanians, Bosniaks, Gorani and Turks living in south Kosovo. *Journal of Ethnobiology and Ethnomedicine, 11*, 39. Available from https://doi.org/10.1186/s13002-015-0023-5.

Mustafa, B., Hajdari, A., Pulaj, B., Quave, C. L., & Pieroni, A. (2020). Medical and food ethnobotany among Albanians and Serbs living in the Shtërpcë/Štrpce area, South Kosovo. *Journal of Herbal Medicine, 22*, 100344. Available from https://doi.org/10.1016/j.hermed.2020.100344.

Nashtar, S. B., & Al-Attar, Z. (2018). The effect of parsley in the treatment of UTI in Iraqi patients. *International Journal of Medical Research & Health Sciences, 7*, 1–7. Available from https://doi.org/10.1016/j.hermed.2020.100344.

Ozturk, S., & Ercisli, S. (2007). Antibacterial activity of aqueous and methanol extracts of *Althaea officinalis* and *Althaea cannabina* from Turkey. *Pharmaceutical Biology, 45*, 235–240. Available from https://doi.org/10.1080/13880200701213179.

Petrolini, F. V. B., Lucarini, R., de Souza, M. G. M., Pires, R. H., Cunha, W. R., & Martins, C. H. G. (2013). Evaluation of the antibacterial potential of *Petroselinum crispum* and *Rosmarinus officinalis* against bacteria that cause urinary tract infections. *Brazilian Journal of Microbiology, 44*, 829–834. Available from https://doi.org/10.1590/S1517-83822013005000061.

Pieroni, A. (2008). Local plant resources in the ethnobotany of Theth, a village in the Northern Albanian Alps. *Genetic Resources and Crop Evolution, 55*, 1197–1214. Available from https://doi.org/10.1007/s10722-008-9320-3.

Pieroni, A., Dibra, B., Grishaj, G., Grishaj, I., & Gjon Maçai, S. (2005). Traditional phytotherapy of the Albanians of Lepushe, Northern Albanian Alps. *Fitoterapia, 76*, 379–399. Available from https://doi.org/10.1016/j.fitote.2005.03.015.

Pieroni, A., Giusti, M. E., & Quave, C. L. (2011). Cross-cultural ethnobiology in the Western Balkans: Medical ethnobotany and ethnozoology among Albanians and Serbs in the Pešter Plateau, Sandžak, South-Western Serbia. *Human Ecology, 39*, 333–349. Available from https://doi.org/10.1007/s10745-011-9401-3.

Pieroni, A., Rexhepi, B., Nedelcheva, A., Hajdari, A., Mustafa, B., Kolosova, V., ... Quave, C. L. (2013). One century later: The folk botanical knowledge of the last remaining Albanians of the upper Reka Valley, Mount Korab, Western Macedonia. *Journal of Ethnobiology and Ethnomedicine, 9*, 22. Available from https://doi.org/10.1186/1746-4269-9-22.

Pieroni, A., Cianfaglione, K., Nedelcheva, A., Hajdari, A., Mustafa, B., & Quave, C. L. (2014). Resilience at the border: Traditional botanical knowledge among Macedonians and Albanians living in Gollobordo, Eastern Albania. *Journal of Ethnobiology and Ethnomedicine, 10*, 31. Available from https://doi.org/10.1186/1746-4269-10-31.

Pieroni, A., Nedelcheva, A., Hajdari, A., Mustafa, B., Scaltriti, B., Cianfaglione, K., & Quave, C. L. (2014). Local knowledge on plants and domestic remedies in the mountain villages of Peshkopia (Eastern Albania). *Journal of Mountain Science, 11*, 180–193. Available from https://doi.org/10.1007/s11629-013-2651-3.

Pieroni, A., Ibraliu, A., Abbasi, A. M., & Papajani-Toska, V. (2015). An ethnobotanical study among Albanians and Aromanians living in the Rraicë and Mokra areas of Eastern Albania. *Genetic Resources and Crop Evolution, 62*, 477–500. Available from https://doi.org/10.1007/s10722-014-0174-6.

Pieroni, A., Sõukand, R., Quave, C. L., Hajdari, A., & Mustafa, B. (2017). Traditional food uses of wild plants among the Gorani of South Kosovo. *Appetite, 108*, 83–92. Available from https://doi.org/10.1016/j.appet.2016.09.024.

Pirbalouti, A. G., Yousefi, M., Nazari, H., Karimi, I., & Koohpayeh, A. (2009). Evaluation of burn healing properties of *Arnebia euchroma* and *Malva sylvestris*. *Electronic Journal of Biology, 5*, 62–66.

Pirbalouti, A. G., Shahrzad, A., Abed, K., & Hamedi, B. (2010). Wound healing activity of *Malva sylvestris* and *Punica granatum* in alloxan-induced diabetic rats. *Acta Poloniae Pharmaceutica - Drug Research, 67*, 511–516. Available from http://www.ptfarm.pl/pub/File/acta_pol_2010/5_2010/511-516.pdf.

Radovanović, B., Anđelković, S., Radovanović, A., & Anđelković, M. (2013). Antioxidant and antimicrobial activity of polyphenol extracts from wild berry fruits grown in Southeast Serbia. *Tropical Journal of Pharmaceutical Research, 12*(5), 813–819. Available from https://doi.org/10.4314/tjpr.v12i5.23.

Razavi, S. M., Zarrini, G., Molavi, G., & Ghasemi, G. (2011). Bioactivity of *Malva sylvestris* L., a medicinal plant from Iran. *Iranian Journal of Basic Medical Sciences, 14*, 574–579. Available from http://www.mums.ac.ir/shares/basic_medical/basicmedjou/2011/nov/a12.pdf.

Rexhepi, B., Mustafa, B., Hajdari, A., Rushidi-Rexhepi, J., Quave, C. L., & Pieroni, A. (2013). Traditional medicinal plant knowledge among Albanians, Macedonians and Gorani in the Sharr Mountains (Republic of Macedonia). *Genetic Resources and Crop Evolution, 60*, 2055–2080. Available from https://doi.org/10.1007/s10722-013-9974-3.

Rezaei, Z., Dadgar, A., Noori-Zadeh, S. A., Mesbah-Namin., Pakzad, I., & Davodian, E. (2015). Evaluation of the antibacterial activity of the *Althaea officinalis* L. leaf extract and its wound healing potency in the rat model of excision wound creation. *Avicenna Journal of Phytomedicine, 5*, 105–112.

Riaz, M., Ahmad, M., & Rahman, N. (2011). Antimicrobial screening of fruit, leaves, root and stem of rubus fruticosus. *Journal of Medicinal Plant Research, 5*, 5920–5924. Available from http://www.academicjournals.org/JMPR/PDF/pdf2011/30%20Oct/Riaz%20et%20al.pdf.

Rigane, G., Ben Younes, S., Ghazghazi, H., & Ben Salem, R. (2013). Investigation into the biological activities and chemical composition of *Calendula officinalis* L. Growing in Tunisia. *International Food Research Journal, 20*, 3001–3007. Available from http://www.ifrj.upm.edu.my/20%20(06)%202013/4%20IFRJ%2020%20(06)%202013%20Rigane%20129.pdf.

Roby, M. H. H., Sarhan, M. A., Selim, K. A. H., & Khalel, K. I. (2013). Antioxidant and antimicrobial activities of essential oil and extracts of fennel (*Foeniculum vulgare* L.) and chamomile (*Matricaria chamomilla* L.). *Industrial Crops and Products, 44*, 437–445. Available from https://doi.org/10.1016/j.indcrop.2012.10.012.

Salem, M. Z. M., Elansary, H. O., Elkelish, A. A., Zeidler, A., Ali, H. M., Hefny, M. E. L., & Yessoufou, K. (2016). In vitro bioactivity and antimicrobial activity of *Picea abies* and *Larix decidua* wood and bark extracts. *BioResources., 11*, 9421–9437. Available from https://doi.org/10.15376/biores.11.4.9421-9437.

Santos, M. I. S., Martins, S. R., Veríssimo, C. S. C., Nunes, M. J. C., Lima, A. I. G., Ferreira, R. M. S. B., … Ferreira, M. A. S. S. (2017). Essential oils as antibacterial agents against food-borne pathogens: Are they really as useful as they are claimed to be? *Journal of Food Science and Technology, 54*, 4344–4352. Available from https://doi.org/10.1007/s13197-017-2905-0.

Šarić-Kundalić, B., Dobeš, C., Klatte-Asselmeyer, V., & Saukel, J. (2010). Ethnobotanical study on medicinal use of wild and cultivated plants in middle, south and west Bosnia and Herzegovina. *Journal of Ethnopharmacology, 131*, 33–55. Available from https://doi.org/10.1016/j.jep.2010.05.061.

Šarić-Kundalić, B., Dobeš, C., Klatte-Asselmeyer, V., & Saukel, J. (2011). Ethnobotanical survey of traditionally used plants in human therapy of east, north and north-east Bosnia and Herzegovina. *Journal of Ethnopharmacology, 133*, 1051–1076. Available from https://doi.org/10.1016/j.jep.2010.11.033.

Šavikin, K., Zdunić, G., Menković, N., Živković, J., Ćujić, N., Tereščenko, M., & Bigović, D. (2013). Ethnobotanical study on traditional use of medicinal plants in South-Western Serbia, Zlatibor district. *Journal of Ethnopharmacology, 146*, 803–810. Available from https://doi.org/10.1016/j.jep.2013.02.006.

Seifi, M., Abbasalizadeh, S., Mohammad-Alizadeh-Charandabi, S., Khodaie, L., & Mirghafourvand, M. (2018). The effect of Rosa (*L. Rosa canina*) on the incidence of urinary tract infection in the puerperium: A randomized placebo-controlled trial. *Phytotherapy Research*, *32*, 76–83. Available from https://doi.org/10.1002/ptr.5950.

Sipponen, A., Kuokkanen, O., Tiihonen, R., Kauppinen, H., & Jokinen, J. J. (2012). Natural coniferous resin salve used to treat complicated surgical wounds: Pilot clinical trial on healing and costs. *International Journal of Dermatology*, *51*, 726–732. Available from https://doi.org/10.1111/j.1365-4632.2011.05397.x.

Snoussi, M., Dehmani, A., Noumi, E., Flamini, G., & Papetti, A. (2016). Chemical composition and antibiofilm activity of *Petroselinum crispum* and *Ocimum basilicum* essential oils against *Vibrio* spp. strains. *Microbial Pathogenesis*, *90*, 13–21. Available from https://doi.org/10.1016/j.micpath.2015.11.004.

Szakiel, A., Ruszkowski, D., Grudniak, A., Kurek, A., Wolska, K. I., Doligalska, M., & Janiszowska, W. (2008). Antibacterial and antiparasitic activity of oleanolic acid and its glycosides isolated from marigold (*Calendula officinalis*). *Planta Medica*, *74*, 1709–1715. Available from https://doi.org/10.1055/s-0028-1088315.

Tsao, S. M., & Yin, M. C. (2001). In-vitro antimicrobial activity of four diallyl sulphides occurring naturally in garlic and Chinese leek oils. *Journal of Medical Microbiology*, *50*, 646–649. Available from https://doi.org/10.1099/0022-1317-50-7-646.

Venâncio, P. C., Figueroba, S. R., Nani, B. D., Nunes Ferreira, L. E., Muniz, B. V., Del Fiol, F. d. S., ... Groppo, F. C. (2017). Antimicrobial activity of two garlic species (*Allium sativum* and *A. tuberosum*) against staphylococci infection: in vivo study in rats. *Advanced Pharmaceutical Bulletin*, *7*, 115–121. Available from https://doi.org/10.15171/apb.2017.015.

Veshkurova, O., Golubenko, Z., Pshenichnov, E., Arzanova, I., Uzbekov, V., Sultanova, E., ... Stipanovic, R. D. (2006). Malvone A, a phytoalexin found in *Malva sylvestris* (family Malvaceae). *Phytochemistry*, *67*, 2376–2379. Available from https://doi.org/10.1016/j.phytochem.2006.08.010.

Zenão, S., Aires, A., Dias, C., Saavedra, M. J., & Fernandes, C. (2017). Antibacterial potential of *Urtica dioica* and *Lavandula angustifolia* extracts against methicillin resistant *Staphylococcus aureus* isolated from diabetic foot ulcers. *Journal of Herbal Medicine*, *10*, 53–58. Available from https://doi.org/10.1016/j.hermed.2017.05.003.

Zlatković, B. K., Bogosavljević, S. S., Radivojević, A. R., & Pavlović, M. A. (2014). Traditional use of the native medicinal plant resource of Mt. Rtanj (Eastern Serbia): Ethnobotanical evaluation and comparison. *Journal of Ethnopharmacology*, *151*, 704–713. Available from https://doi.org/10.1016/j.jep.2013.11.037.

Zouari Bouassida, K., Bardaa, S., Khimiri, M., Rebaii, T., Tounsi, S., Jlaiel, L., & Trigui, M. (2017). Exploring the *Urtica dioica* leaves hemostatic and wound-healing potential. *BioMed Research International*, *2017*, 1047523. Available from https://doi.org/10.1155/2017/1047523.

CHAPTER

4

Medicinal plants used in South Africa as antibacterial agents for wound healing

Samantha Rae Loggenberg[1], *Danielle Twilley*[1], *Marco Nuno De Canha*[1] *and Namrita Lall*[1,2,3]

[1]Department of Plant and Soil Sciences, Faculty of Natural and Agricultural Sciences, University of Pretoria, Pretoria, South Africa [2]School of Natural Resources, University of Missouri, Columbia, MO, United States [3]College of Pharmacy, JSS Academy of Higher Education and Research, Mysuru, India

Introduction

Wounds account for one of the major reasons for hospital visits in Africa each year, equating to approximately 30%–42% of hospital attendance, relating to an estimated 9% death rate each year (Builders & Builders, 2016). There is a concerning lack of published information regarding the burden of wounds on South African healthcare systems; however, a study by Lotz (2019) aimed at determining the wound burden in a South African hospital that had limited resources. On the day of the study, 518 inpatients were admitted of which 179 patients had wounds, accounting for a wound burden of 34.6%, of which acute surgical wounds and burn wounds were the most predominant, with a wound burden of 45% and 18%, respectively. Chronic wounds were also reported, with pressure injuries (9%) and leg ulcers (9%) accounting for the highest burden, followed by fungating wounds (3%). It was further found that 13% of all wounds presented on the day were infected. When comparing this to patients visiting clinics for outpatient visits, on the day of the study it was found that 63 of the 333 outpatients presented with wounds, accounting for an additional 18.9% wound burden (Lotz, 2019).

In South Africa, it has been estimated that approximately 72% of the population rely on traditional medicine as a source of health care, equating to an estimated R 2.9 million and R 520 million per year in trade of traditional medicinal products and raw plant material,

Medicinal Plants as Anti-infectives
DOI: https://doi.org/10.1016/B978-0-323-90999-0.00018-5

respectively (Mander, Ntuli, Diederichs, & Mavundla, 2019), which includes the use of medicinal plants for the treatment of dermatological disorders and ailments. An ethnobotanical literature review by Mabona and VanVuuren (2013) summarized that over 100 plant species in southern African were used for dermatological disorders of which 41% of the plants were used for wound treatment. Additionally, the treatment of infectious skin diseases, which included the treatment of bacterial infections, comprised 32% of the plant species. The treatment of sores and ulcers was categorized separately, where approximately 25% of medicinal plants were used for these treatments, whereas treatment of burns only comprised 10%. Furthermore, Mabona and VanVuuren (2013) stated that medicinal plants could potentially be used for multiple disorders, such as the use of a plant for its wound-healing, antiinfective, and antiinflammatory properties.

According to data analysis using Web of Science, the number of publications (including articles, literature reviews, and book chapters) within the last 5–10 years that include the topics of "medicinal plants" and "wound healing" equate to 611 (Fig. 4.1). No publications of this nature were published between the years 2011 and 2016. However, there were under a hundred publications that met the set requirements published in 2017, with a steady increase in the number of publications for years thereafter until 2021. Along with the growing trend in publications between 2017 and 2021, an exponential increase in citations was observed during this period. These data may indicate a shift towards research aimed at investigating medicinal plants in wound healing within the last 5 years.

The general shift in interest for investigating traditionally used plants for their medicinal value in South Africa has greatly increased within the last 5 years. There are various foundations in South Africa that provide funding to all fields of research, with specific focus on the investigation of South African medicinal plants. These organizations include but are not limited to the South African Medical Research Council and the South African

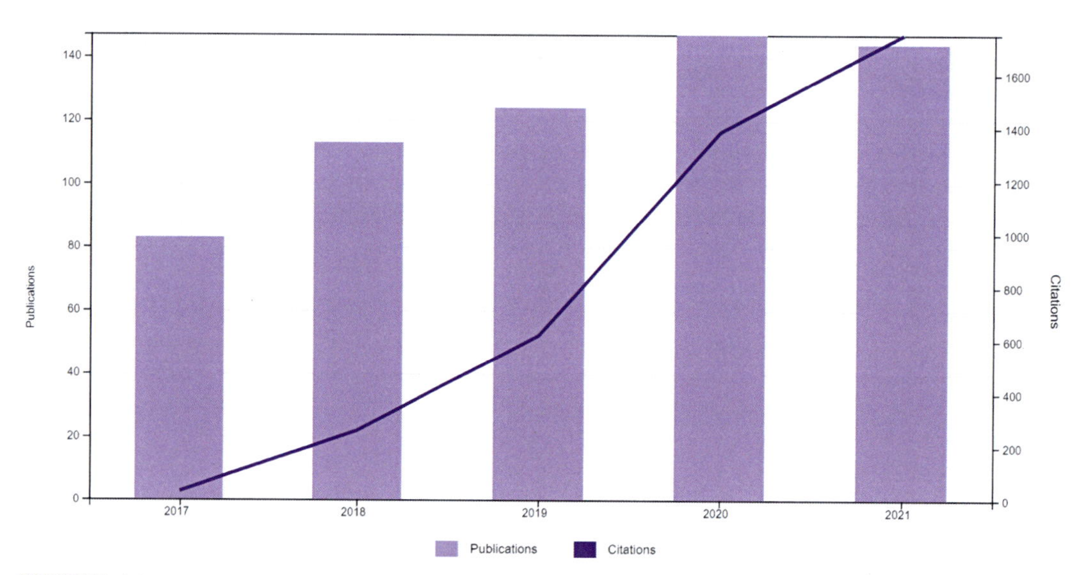

FIGURE 4.1 Publications and citations relating to the use of medicinal plants for wound healing from 2017 to 2021.

Research Chairs Initiative, which focus on research developments in South Africa, as well as foundations such as the National Research Foundation, Technology Innovation Agency, and Department of Science and Innovation which aid in funding research outputs conducted by universities and scientific institutions nation-wide. All of these foundations aim to further the development of science, technology, medicine, and indigenous knowledge within South Africa.

Pathophysiology of wound healing

The epidermal layer of the skin acts as the main protective barrier for the body. Lesions and breaks in the skin may expose the internal tissues to pathogens and microbes from the external environment, leaving soft tissue vulnerable to pathogenic infections that may lead to adverse health conditions which may be life-threatening (Aldridge, 2015).

The process of wound healing is a dynamic system consisting of multiple phases which involve several factors and cellular components including hemostasis, inflammation, proliferation/granulation, and remodeling/maturation (Fig. 4.2).

Upon damage to the epidermal layer, cellular membranes are disrupted causing leakage of cellular contents and the release of collagen into the extracellular fluid between cells. Inactive blood plasma platelets present in the extracellular fluid interact with the exposed collagen and become active. The activated platelets recruit red blood cells and actin filaments as well as stimulate blood clot formation by facilitating the coagulation of these molecules to the wound site (Lynch, Colvin, & Antoniades, 1989). The formation of a blood clot prevents further bleeding and exposure of the soft tissues to the external environment; however, pathogens may have already entered the wound site. White blood cells and neutrophils, also known as natural killer cells, near the wound, recognize and destroy pathogens whilst simultaneously removing cellular debris. The identification of pathogenic proteins activates the white blood cells to secrete fibrogenic cytokines which facilitate the recruitment of macrophages and neutrophils to the wound site by initiating the inflammatory response, leading to swelling around the wound. Macrophages continue to recruit immune cells and secrete cytokines which initiate the formation of granulation tissue within the wound which facilitates tissue repair (Almine, Wise, & Weiss, 2012).

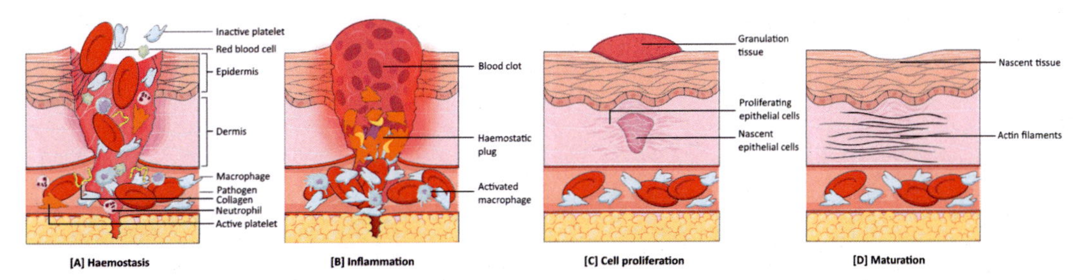

FIGURE 4.2 The phases of wound healing; (A) hemostasis, (B) inflammation, (C) cell proliferation, and (D) maturation. Source: *From Servier. (2019).* The phases of wound healing. *Retrieved from https://smart.servier.com/category/anatomy-and-the-human-body/cardiovascular-system/blood/.*

The granulation tissue is filled with collagen and actin filaments that dynamically contract to promote closure of the wound and allow for the migration of newly formed tissue cells to the center of the wound. Once the wound has closed completely, tissue remodeling and degradation of granulation tissue around the site of injury take place to increase the tensile strength of the tissue and allow for the formation of other structures such as hair follicles and cell junctions (Li et al., 2013).

Wound infection

The duration of the wound-healing process may vary depending on the size of the wound and the severity of the tissue damage. The damaged tissue becomes vulnerable and susceptible to pathogenic infection during this period since the moist environment of wounds provides optimal conditions for bacterial growth, which can lead to bacterial infection. The body's immune response limits pathogenic infection in a healthy patient; however, immunocompromised patients are at a higher risk of infections resulting in severe health complications (Aldridge, 2015). If left untreated, infected wounds may lead to bacterial overgrowth which allows the pathogen to spread into deeper tissues within the body or disseminate into the blood, causing bacteremia (bacterial infection of the blood). Bacteremia can lead to the spreading and infection of various tissues and organs within the body which can overstimulate the immune response leading to septicemia (Zielińska-Borkowska, 2015). Septicemia involves rapid widespread inflammation throughout infected tissue and organs to combat pathogenic infection. This abrupt inflammatory response may cause several health complications such as sepsis, which involves the tearing of tissues and in severe cases, organ failure. Depending on the preexisting health condition of a patient when organ failure occurs, septicemia may result in mortality. Even with early diagnosis and treatment, septicemia is difficult to treat due to the causative pathogens often developing antibiotic resistance (Churpek et al., 2017). It is therefore important to identify antibacterial treatments suitable for dermal wounds which aid in the prevention of bacterial infection.

Currently available treatments and products

Topical creams

Topical antibacterial treatments are produced in the form of creams, ointments, gels, and serums which can be directly applied to dermal wounds, allowing for the diffusion of the active compounds into the superficial layers of the skin and providing direct antibacterial treatment. *Aloe* extracts, produced from *Aloe vera* and *Aloe ferox*, are commonly utilized in many of these products due to their wound-healing properties and their antibacterial effects (Hashemi, Madani, & Abediankenari, 2015; Jia, Zhao, & Jia, 2008). Additionally, transdermal drug delivery systems, such as adhesive patches, which are placed directly onto the skin are designed to deliver an effective dose of a drug across the epidermal and dermal layers of the skin to reach the blood circulatory system, can be used. This method of drug delivery allows for systemic circulation of the drug whilst avoiding first-pass metabolism which maximizes the therapeutic effects of the drug (Ashok Kumar, Pullakandam, Lakshmana Prabu, & Gopal, 2010).

Transdermal drug delivery systems

Transdermal drug delivery systems provide controlled and continuous administration of a particular drug dosage which allows for constant plasma levels of a drug, irrespective of its biological half-life (Prausnitz & Langer, 2008). Transdermal patches are generally made up of a hydrogel which is a network of polymeric chains which can retain both water and the active compound(s) within the matrix. The polymeric matrix acts as a reservoir and allows for the steady and continuous release of the drug for extended periods of time (Girard, Teferra, & Awika, 2019). Research has shown that the use of plant-derived polymers in transdermal patches provided more efficient controlled release and drug permeation in comparison to synthetic polymers (Saidin, Anuar, & Affandi, 2018). Nanocarriers, a form of transdermal drug delivery, have provided an efficient method of administration for both lipophilic and hydrophilic drugs. Different types of nanocarriers that have been developed for transdermal drug delivery include liposomes, transfersomes, ethosomes, niosomes, dendrimers, lipid-based nanoparticles, polymer nanoparticles, and nanoemulsions (Escobar-Chavez, Diaz-Torres, & Rodriguez-Cruz, 2012). Transdermal drug delivery patches consisting of nanoparticles that exhibit antibacterial activity have also been developed (Ashok Kumar et al., 2010). The use of nanocarrier technology may provide efficient drug delivery systems in the development of antimicrobial treatments against wound infection (Escobar-Chavez et al., 2012). It is therefore important to consider research conducted on the optimization of drug uptake and bioavailability of antibacterial compounds and plant extracts.

Bacteria associated with infections of dermal wounds

Numerous bacteria are involved in the infection of dermal wounds; however, common pathogens that cause wound infections include species such as *Staphylococcus aureus* and *Pseudomonas aeruginosa* (Bessa, Fazii, Di Giulio, & Cellini, 2015; Manzuoerh, Farahpour, Oryan, & Sonboli, 2019). In a case study conducted by Bessa et al. (2015), 312 swabs were collected from infected wounds of 213 patients of which *S. aureus* (37%) and *P. aeruginosa* (17%) were the most common pathogens detected.

Sepsis is commonly caused by adverse infections of postsurgical wounds known as nosocomial infections. Nosocomial infections, also known as healthcare-associated infections, may occur in patients, generally with preexisting health complications, which are exposed to unsanitary healthcare facilities or equipment. The main cause of nosocomial infections is bacterial biofilm growth on indwelling medical devices which can disseminate into the bloodstream (Otto, 2009). Bacterial infections in the bloodstream, leading to sepsis, may be lethal to both healthy and immunocompromised patients due to the difficulty in treatment and management (Churpek et al., 2017). Nosocomial infections in immunocompromised patients may even occur due to secondary infection by commensal bacteria. Several studies conducted identified commensal bacteria such as *Bacillus subtilis* and *Staphylococcus epidermidis* to be the cause of septicemia by nosocomial infection (Christensen et al., 1982; Saleh, Kheirandish, Azizi, & Azizi, 2014).

The most common cause of infections in dermal or postsurgical wounds is due to the colonization of these wounds by bacteria from either the environment or normal flora of the human body which have pathogenic potential (Singal & Grover, 2016). The normal human flora consists of an aggregate of all microbiota that reside in and on the human body. These microbes may compete and synergistically interact to modulate each other and prevent the colonization of pathogenic microbes. However, imbalances in the skin can lead to infection of dermal wounds or even cause nosocomial infections. Breaks in the skin can cause numerous changes, such as a shift in pH, which may favor the growth of one microbe to another, thus altering the balanced dynamic between microbiota of the skin, therefore, perpetuating the growth of certain bacteria (Saleh et al., 2014).

Bacillus subtilis

Bacillus subtilis is a Gram-positive rod-shaped bacterium found largely in soil and decomposing plant material and is essential in the degradation of waste products. This spore-forming pathogen is a facultative aerobe and has an optimal growth temperature ranging from 25°C to 35°C (Turnbull, 1996). The bacterium is considered nonpathogenic to humans; however, it can cause secondary nosocomial infections in immunocompromised patients which may be lethal. In addition to infections of dermal wounds and burns, *B. subtilis* also causes infections of the respiratory and urinary tracts (Saleh et al., 2014).

Bacillus subtilis secretes tissue-damaging toxins and enzymes such as penicillinase, which inactivate penicillin, thereby interfering with certain penicillin-based antibiotic treatments. This microorganism also has the ability to produce antibiotic compounds such as ribosomal antibiotics (i.e., subtilosin and sublancin) (Saleh et al., 2014), which play an important role in the first step of infection through the process of niche establishment and microbial monopolization. The secretion of antibiotics eliminates competing bacteria and allows for the bacterium to thrive since this pathogen has acquired immunity against the self-produced antibiotics (Turnbull, 1996). For example, *B. subtilis* may secrete penicillin whilst also producing penicillinase enzyme to provide the bacterium with immunity against the antibiotic (Saleh et al., 2014).

Treating infections caused by *B. subtilis* has several challenges including the development of antibiotic resistance, and antagonistic interaction between self-produced antibiotics and administered antibiotics (Hashimoto et al., 2017).

Staphylococcus aureus

Staphylococcus aureus is a Gram-positive, cocci-shaped bacterium that forms part of the normal human flora of the skin and mucous. This opportunistic pathogen is a highly resilient facultative anaerobe that can survive extreme conditions, thriving in moderately acidic environments (pH 4–6) and at a temperature of approximately 37°C. This bacterium is highly resistant to antibiotics, such as methicillin and β-lactam antibiotics, and can cause a vast range of infections including those of the respiratory tract, soft tissue, and epidermis (Cohen & Kurzrock, 2004).

Staphylococcus aureus infections of the epidermis are considered nonlethal if treated accordingly; however, *S. aureus* infections may spread to deeper tissues within the body if the soft tissues are exposed. Dermal lesions and injured tissue maintain a slightly acidic environment which creates optimal conditions for cellular functions and the wound-healing process to occur (Schneider, Korber, Grabbe, & Dissemond, 2007). These conditions, however, also provide a favorable environment in which *S. aureus* may proliferate (Cohen & Kurzrock, 2004). *Staphylococcus aureus* secretes toxins that facilitate pathogenic attachment to the host surface or cause harm to the host tissue aiding the spread of infection. The secreted toxins adversely affect the host tissue by damaging cell membranes or exerting enzymatic degradation of essential host molecules. Hemolysins and leukotoxins are pathogenic factors that target and cause lysis of red and white blood cells, respectively. The lysis of cells causes the release of cellular contents containing nutrients that can sustain *S. aureus* and further progress the infection. This cycle continues as more red and white blood cells are continuously transported to the site of infection during the inflammatory response (Zhang, Hu, & Rao, 2017). *Staphylococcus aureus* damages the host tissue upon secretion of α-toxins which facilitate the degradation of cellular adhesion molecules, such as E-cadherin, in epithelial cells. The degradation of tissue structures and induced cellular lysis inhibits the wound-healing process allowing for an increased infection (Otto, 2014). Deep tissue infections of *S. aureus* are difficult to treat with antibiotics due to the multidrug resistance of the pathogen (Shajari & Khorshidi, 2002).

Staphylococcus epidermidis

Staphylococcus epidermidis is a Gram-positive, cocci-shaped bacterium that forms part of the normal microflora of the skin. This bacterium is found mainly on the skin or sweat glands and is responsible for the production of body odor through perspiration. This facultative anaerobe can cause opportunistic infections such as nosocomial infections in immunocompromised patients, especially through infection of surgical wounds (Rupp, Fey, Heilmann, & Götz, 2001). Forming part of the human microbiome of the skin, *S. epidermidis* may surpass other competing bacteria leading to the infection of dermal wounds.

Infections by *S. epidermidis* are often treatable due to their susceptibility to antibiotics such as penicillinase-resistant penicillin and cephalosporin antibiotics, which disrupt the formation of the peptidoglycan layer within the bacterial cell membrane. However, the pathogen has been reported to easily acquire antibiotic resistance through horizontal gene transfer of antibiotic resistance genes between other species of *Staphylococcus* such as *Staphylococcus aureus* (Forbes & Schaberg, 1983).

Pseudomonas aeruginosa

Pseudomonas aeruginosa is a Gram-negative, bacillus-shaped bacterium which has an optimal growth temperature of 37°C which is able to infect soft tissues such as the urinary and respiratory tract, and dermal wounds. Chronic wound infections by *P. aeruginosa* may develop due to the ability of this pathogen to avoid phagocytosis through the downregulation and alteration of flagella proteins on the pathogens surface, which play a role in pathogen detection and phagocyte activation (Amiel, Lovewell, O'Toole, Hogan, & Berwin, 2010).

Pseudomonas aeruginosa is highly resistant to beta-lactam antibiotics as well as first- and second-generation cephalosporins, due to decreased bacterial cell wall permeability (Micek et al., 2005). Although combinatorial antibiotic therapy has been reported to be an effective treatment against this pathogen, infections by *P. aeruginosa* are difficult to treat due to the intrinsic antibiotic resistance (Rahal, 2006).

South African medicinal plant species with activity against wound-associated bacteria

Several plants are traditionally used in South Africa for their wound-healing activity, which are discussed below (Fig. 4.3). These plants were further evaluated for their antibacterial activity against wound-associated bacteria, as summarized in Table 4.1. According to Rahal (2006) antimicrobial thresholds have been defined for medicinal plants with an MIC of $<100\ \mu g/mL$ as having strong/significant antimicrobial activity, while those with MICs of 100–625 μg/mL have moderate activity. For compounds, MICs $<10\ \mu g/mL$ are considered noteworthy, while those with MICs between 64 and 100 μg/mL are considered clinically relevant.

In addition, the in vitro and in vivo wound-healing activity and toxic potential of the traditionally used plants were summarized in Tables 4.2 and 4.3

Aloe barberae Dyer

Aloe barberae Dyer is the largest tree-forming *Aloe* species that is indigenous to South Africa and is found in the Eastern Cape, KwaZulu-Natal, and northwards to Mozambique (Ndhlala, Amoo, Stafford, Finnie, & Van Staden, 2009a). This slow-growing tree succulent belongs to the Asphodelaceae family and can reach an average height of 9 m but can reach heights of 18 m (Succulent Plant Site, 2004). The *A. barberae* tree has dichotomous branching and is covered in a smooth gray bark. Each branch forms succulent rosettes of curved dark-green leaves with light-green spines lining the leaf margins. This *Aloe* species has a multibranched inflorescence that does not grow much higher than the leaf rosettes. The flower racemes appear swollen or round in shape and salmon-pink in color. *Aloe barberae* is highly resilient and long-living once established with a flowering period throughout May–July, over early to midwinter (Ndhlala et al., 2009a).

Traditional usage

Aloe barberae Dyer has been used by the natives of the Eastern Cape and KwaZulu-Natal to treat wounds and skin irritations based on its hydrating and antiinflammatory properties. Similarly, to other aloe species, the leaf sap of *A. barberae* possesses hydrating

FIGURE 4.3 Plants traditionally used in South Africa for wound healing (A) *Aloe barberae* Dyer, (B) *Aloe excelsa* Berger, (C) *Aloe ferox* Miller, (D) *Elephantorrhiza elephantina* (Burch.) Skeel leaves and pods, (E) *Elephantorrhiza elephantina* (Burch.) Skeel flowers, (F) *Erythrina lysistemon* Hutch., (G) *Galenia africana* L., (H) *Grewia occidentalis* L., (I) *Melianthus comosus* Vahl. leaves, (J) *Melianthus comosus* Vahl. flowers, (K) Plectranthus fruticosus L' Hér, (L) *Sutherlandia frutescens* (L.) R.Br., and (M) *Urtica urens* L.

properties which can be used to treat skin conditions such as eczema. However, traditional medicinal uses and preparations of *A. barberae* have not been well documented apart from deduction of knowledge pertaining to the use of other *Aloe* species (Ndhlala, Amoo, Stafford, Finnie, & Van Staden, 2009b).

Aloe excelsa Berger

Aloe excelsa, also known as the "Zimbabwe Aloe," is an unbranched tree-forming *Aloe* species belonging to the "Asphodelaceae" family that is native to the Northern and Western Cape provinces of South Africa. It grows to an average height of 4–6 m with old, dried leaves running along the trunk of the plant. *Aloe excelsa* leaves grow in rosette formation and can reach a length of 1 m each. The rosettes sprout a branched inflorescence with

TABLE 4.1 Antibacterial activity of plants used in South Africa for the treatment of wounds.

Plant species/family	Extraction solvent	Plant part	Antibacterial activity	Reference
Aloe barberae Dyer/ Asphodelaceae	Petroleum ether	Whole leaf	MIC: 0.78 mg/mL (*B. subtilis*) MIC: 1.5 mg/mL (*S. aureus*)	Ndhlala et al. (2009a)
	Dichloromethane		MIC: 0.39 mg/mL (*B. subtilis*) MIC: 0.78 mg/mL (*S. aureus*)	
	Ethanol		MIC: 0.78 mg/mL (*B. subtilis* and *S. aureus*)	
	Water		MIC: 1.56 mg/mL (*B. subtilis*) MIC: 6.25 mg/mL (*S. aureus*)	
	Petroleum ether	Upper stem	MIC: 3.125 mg/mL (*B. subtilis* and *S. aureus*)	
	Dichloromethane		MIC: 1.56 mg/mL (*B. subtilis* and *S. aureus*)	
	Ethanol		MIC: 3.125 mg/mL (*B. subtilis* and *S. aureus*)	
	Water		MIC: 6.25 mg/mL (*B. subtilis* and *S. aureus*)	
	Petroleum ether	Young bark	MIC: 1.56 mg/mL (*B. subtilis* and *S. aureus*)	
	Dichloromethane		MIC: 1.56 mg/mL (*B. subtilis* and *S. aureus*)	
	Ethanol		MIC: 1.56 mg/mL (*B. subtilis* and *S. aureus*)	
	Water		MIC: 3.125 mg/mL (*B. subtilis* and *S. aureus*)	
	Petroleum ether	Mature bark	MIC: 0.78 mg/mL (*B. subtilis* and *S. aureus*)	
	Dichloromethane		MIC: 0.39 mg/mL (*B. subtilis* and *S. aureus*)	
	Ethanol		MIC: 1.56 mg/mL (*B. subtilis*) MIC: 3.125 mg/mL (*S. aureus*)	
	Water		MIC: 3.125 mg/mL (*B. subtilis* and *S. aureus*)	
	Petroleum ether	Roots	MIC: 1.56 mg/mL (*B. subtilis*) MIC: 0.78 mg/mL (*S. aureus*)	
	Dichloromethane		MIC: 0.78 mg/mL (*B. subtilis*) MIC: 0.39 mg/mL (*S. aureus*)	
	Ethanol		MIC: 1.56 mg/mL (*B. subtilis* and *S. aureus*)	
	Water		MIC: 1.56 mg/mL (*B. subtilis*) MIC: 6.25 mg/mL (*S. aureus*)	

Aloe excelsa Berger/Asphodelaceae	Hot water (decoction)	Leaves	MIC: 5.0 mg/mL (*B. subtilis*) MIC: 6.0 mg/mL (*S. aureus*)	Coopoosamy and Magwa (2007)
	Ethyl acetate		MIC: 3.0 mg/mL (*B. subtilis*) MIC: 1.0 mg/mL (*S. aureus*)	
	Acetone		MIC: 2.0 mg/mL (*B. subtilis*) MIC: 1.0 mg/mL (*S. aureus* and *S. epidermidis*)	Coopoosamy and Magwa (2007)
	Ethanol		MIC: 2.0 mg/mL (*B. subtilis*) MIC: 1.0 mg/mL (*S. aureus*)	Coopoosamy and Naidoo (2012)
Aloe ferox Miller/Xanthorrhoeaceae	Leaf material was homogenized and filtered	Whole leaf	A 1:1 ratio of whole leaf juice and nutrient broth showed complete growth inhibition of *P. aeruginosa* and *S. aureus* after 24 h	Jia, Zhao, and Jia (2008)
	Water	Leaves	No inhibition at a concentration of 50 mg/mL against *S. aureus* or *P. aeruginosa.*	Obi et al., 2003
Elephantorrhiza elephantina (Burch.) Skeels/Fabaceae	Ethanol: water (7:3) and n-butanol	Rhizome	Bioautography using TLC showed inhibition of *B. subtilis*, *P. aeruginosa*, and *S. aureus* at loading capacities lower than 15 μg.	Maroyi (2017)
	Methanol	Stem rhizome	Zone of inhibition (ZI): 23.3 mm at 100 mg/mL (*S. aureus*). MIC: 0.156 mg/mL (*S. aureus*)	Mathabe, Nikolova, Lall, and Nyazema (2006)
	Ethanol		ZI: 23.7 mm at 100 mg/mL (*S. aureus*) MIC: 0.156 mg/mL (*S. aureus*)	
	Acetone		ZI: 24.0 mm at 100 mg/mL (*S. aureus*) MIC: 0.132 mg/mL (*S. aureus*)	
	Water		ZI: 25.0 mm at 100 mg/mL (*S. aureus*) MIC: 0.156 mg/mL (*S. aureus*)	
	Ethanol	Root	ZI: 1.0 mm (*B. subtilis*) ZI: 3.0 mm (*S. aureus*) ZI: 2.5 mm (*P. aeruginosa*)	Mukanganyama, Ntumy, Maher, Muzila, and Andrae-Marobela (2011)
	Dichloromethane: methanol (1:1)	Leaf	MIC: 0.5 mg/mL [*S. aureus* and Gentamycin methicillin-resistant *Staphylococcus aureus* (GMRSA)] MIC: 1.0 mg/mL [Methicillin-resistant *S. aureus* (MRSA)] MIC: 0.38 mg/mL (*S. epidermidis*) MIC: 1.0 mg/mL (*P. aeruginosa*)	Mabona, Viljoen, Shikanga, Marston, and Van Vuuren (2013)
		Root and rhizome	MIC: 0.5 mg/mL (*S. aureus*, MRSA and GMRSA) and MIC: 1.0 mg/mL (*S. epidermidis*) MIC: 2.0 mg/mL (*P. aeruginosa*)	

(*Continued*)

TABLE 4.1 (Continued)

Plant species/family	Extraction solvent	Plant part	Antibacterial activity	Reference
	Water	Leaf	MIC: 16 mg/mL (*S. aureus* and *S. epidermidis*) MIC: 8.0 mg/mL (MRSA and GMRSA) MIC: 12 mg/mL (*P. aeruginosa*)	
		Root and rhizome	MIC: 2.0 mg/ mL (*S. aureus* and GMRSA) MIC: 1.0 mg/mL (MRSA) MIC: 4.0 mg/mL (*S. epidermidis* and *P. aeruginosa*)	
	Dichloromethane: methanol	Rhizome	MIC: 2000 μg/mL (GMRSA) MIC: 4000 μg/mL (MRSA) MIC: 1000 μg/mL (*S. aureus* and *S. epidermidis*) MIC: 500 μg/mL (*P. aeruginosa*)	Nciki, Vuuren, Van Eyk, and De Wet (2016)
	Water		MIC: 4000 μg/mL (GMRSA, MRSA and *S. epidermidis*) MIC: 6000 μg/mL (*S. aureus*) MIC: 2000 μg/mL (*P. aeruginosa*)	
	70% ethanol		Antibacterial activity at a loading capacity of 10, 13, and 15 μg against *B. subtilis*, *S. aureus*, and *P. aeruginosa*, respectively	Aaku et al. (1998)
	n-butanol fraction		Antibacterial activity at a loading capacity of 4, 6, and 8 μg against *B. subtilis*, *S. aureus*, and *P. aeruginosa*, respectively	
Erythrina lysistemon Hutch./Fabaceae	Ethyl acetate	Leaves	ZI (extract at 1 mg/mL)/ZI (neomycin 200–500 μg/mL): no activity on *S. aureus*, *S. epidermidis*, *B. subtilis*, and *P. aeruginosa*	Pillay, Jäger, Mulholland, and Van Staden (2001)
		Bark	ZI (extract at 1 mg/mL)/ZI (neomycin 200–500 μg/mL): 0.9 (*S. aureus*) and 0.4 (*B. subtilis*); no activity on *S. epidermidis* or *P. aeruginosa*	
	Ethanol	Leaves	ZI (extract at 1 mg/mL)/ZI (neomycin 200–500 μg/mL): no activity on *S. aureus*, *S. epidermidis*, *B. subtilis*, and *P. aeruginosa*	
		Bark	ZI (extract at 1 mg/mL)/ZI (neomycin 200–500 μg/mL): 0.7 (*S. aureus*) and 0.4 (*B. subtilis*); no activity on *S. epidermidis* or *P. aeruginosa*	
	Water	Leaves	ZI (extract at 1 mg/mL)/ZI (neomycin 200–500 μg/mL): no activity on *S. aureus*, *S. epidermidis*, *B. subtilis*, and *P. aeruginosa*	
		Bark	ZI (extract at 1 mg/mL)/ZI (neomycin 200–500 μg/mL): 0.8 (*S. aureus*) and 0.5 (*B. subtilis*); no activity on *S. epidermidis* or *P. aeruginosa* ZI: 0 mm against *P. aeruginosa* and *S. aureus*	Nsele (2012), Pillay et al. (2001)
	60% Ethanol tincture	Bark	ZI: 1 mm (*P. aeruginosa*), 60% ethanol alone showed ZI of 2 mm ZI: 4 mm (*S. aureus*), 60% ethanol alone showed ZI of 1 mm	Nsele (2012)

	Dichloromethane	Stem bark	MIC: 104 μg/mL (*S. aureus*) MIC: 5 μg/mL (*S. epidermidis*) MIC: 500 μg/mL (*P. aeruginosa*)	Sadgrove, Oliveira, Khumalo, van Vuuren, and van Wyk (2020)
		Leaves	MIC: 0.313 (1 h) and 0.156 mg/mL (24 h) (*P. aeruginosa*) MIC: 0.313 (1 and 24 h) (*S. aureus*)	Mukandiwa, Naidoo, and Eloff (2012)
	Methanol	Stem bark	MIC: 125 μg/mL (*S. aureus* and *S. epidermidis*) MIC: 830 μg/mL (*P. aeruginosa*)	Mukandiwa et al. (2012), Sadgrove et al. (2020)
		Leaves	MIC: 0.313 (1 h) and 0.156 mg/mL (24 h) (*P. aeruginosa*) MIC: 0.313 (1 and 24 h) (*S. aureus*)	Mukandiwa et al. (2012)
	Acetone	Leaves	MIC: 0.313 (1 h) and 0.156 mg/mL (24 h) (*P. aeruginosa*) MIC: 0.156 (1 h) and 0.078 mg/mL (24 h) (*S. aureus*)	
	Hexane		MIC: 0.625 (1 and 24 h) (*P. aeruginosa*) MIC: 1.25 (1 and 24 h) (*S. aureus*)	
Galenia africana L./Aizoaceae	80% Ethanol	Unspecified	MIC: 3.125 mg/mL (*S. aureus*, MRSA) and methicillin-sensitive strain (MSSA). Minimum bactericidal concentration (MBC): 3.125 mg/mL (MSSA) MBC: 6.25 mg/mL (MRSA)	Ng'uni et al (2018)
	Hot water (decoction)	Fresh aerial	MIC: >0.48 mg/mL (*S. aureus*, *S. epidermidis*, and *P. aeruginosa*)	Elbagory, Meyer, Cupido, and Hussein (2017)
	Gold nanoparticles prepared from hot water decoction		MIC: >32 nM (*S. aureus* and *S. epidermidis*) MIC: 32 nM (*P. aeruginosa*)	
Grewia occidentalis L./Malvaceae	Methanol	Shoots	MIC: 1.0 mg/mL (*S. aureus*) MIC: 4.0 mg/mL (*P. aeruginosa* and *B. subtilis*)	Grierson and Afolayan (1999)
	Acetone		MIC: >5.0 mg/mL (*S. aureus, P. aeruginosa*, and *B. subtilis*)	
	Water		MIC: 1.0 mg/mL (*S. aureus*) MIC: >5.0 mg/mL (*P. aeruginosa* and *B. subtilis*)	
		Roots	MIC: >12.5 mg/mL (*B. subtilis*) MIC: 12.5 mg/mL (*S. aureus*)	Mabona (2013)
	Petroleum ether	Roots	MIC: 3.125 mg/mL (*B. subtilis*) MIC: 12.5 mg/mL (*S. aureus*)	
	Dichloromethane		MIC: 3.125 mg/mL (*B. subtilis* and *S. aureus*)	
	Ethanol		MIC: 0.78 mg/mL (*S. aureus*) MIC: 3.125 mg/mL (*B. subtilis*)	

(*Continued*)

TABLE 4.1 (Continued)

Plant species/family	Extraction solvent	Plant part	Antibacterial activity	Reference
Melianthus comosus Vahl/Melianthaceae	Acetone	Leaves	MIC: >6.3 mg/mL (*P. aeruginosa*) MIC: 0.78 (*S. aureus*) MIC: 0.00078 mg/mL (*S. aureus*)	McGaw and Eloff (2005)
	Ethanol		MIC: 0.500 mg/mL; MBC: 2.00 mg/mL (MSSA) MIC: 0.391 mg/mL; MBC: 1.562 mg/mL (MRSA)	Heyman, Hussein, Meyer, and Lall (2009)
	Dichloromethane: methanol (1:1)		MIC: 0.4 mg/mL (*S. aureus*) MIC: 0.5 mg/mL (MRSA) MIC: 0.25 mg/mL (GMRSA) MIC: 0.25 mg/mL (*S. epidermidis*) MIC: 0.10 mg/mL (*P. aeruginosa*)	Mabona et al. (2013)
	Water		MIC: 1.60 (*S. aureus*) MIC: 0.25 mg/mL (MRSA) MIC: 0.25 mg/mL (GMRSA) MIC: 0.25 mg/mL (*S. epidermidis*) MIC: 2.00 mg/mL (*P. aeruginosa*)	
	Methanol		MIC: 2.00 mg/mL (*S. aureus*) MIC: 4.00 mg/mL (*S. epidermidis*) MIC: 8.00 mg/mL (*B. subtilis*) MIC: > 8.00 mg/mL (*P. aeruginosa*)	Kelmanson, Jäger, and Van Staden (2000)
Plectranthus fruticosus/ Lamiaceae	Essential oil from hydrodistillation	Leaves	No activity was observed against *S. aureus*, *S. epidermidis*, or *P. aeruginosa*	Maistry (2003)
Polystichum pungens (Kaulf.) C. Presl/ Dryopteridaceae	Methanol	Shoots	MIC: 5.0 mg/mL (*P. aeruginosa*) MIC: 0.5 mg/mL (*S. aureus*) MIC: 1.0 mg/mL (*B. subtilis*)	Grierson and Afolayan (1999)
	Acetone		MIC: >5.0 mg/mL (*P. aeruginosa*) MIC: 1.0 mg/mL (*S. aureus*) MIC: 0.5 mg/mL (*B. subtilis*)	
	Water		MIC: >5.0 mg/mL (*P. aeruginosa*) MIC: 0.5 mg/mL (*S. aureus*) MIC: 5.0 mg/mL (*B. subtilis*)	
Sutherlandia frutescens (L.) R.Br/Fabaceae	Hexane	Leaves	MIC: 0.31 mg/mL (*S. aureus*) MIC: >10 mg/mL (*P. aeruginosa*)	Katerere and Eloff (2005)
	Acetone (sequential extraction)		MIC: >10 mg/mL (*S. aureus*) MIC: 2.5 mg/mL (*P. aeruginosa*)	
	Dichloromethane (sequential extraction)		MIC: 2.5 mg/mL (*S. aureus*) MIC: 5.0 mg/mL (*P. aeruginosa*)	
	Ethyl acetate (sequential extraction)		MIC: 1.25 mg/mL (*S. aureus*) MIC: 5.0 mg/mL (*P. aeruginosa*)	

	Acetone		MIC: 10 mg/mL (*S. aureus*) MIC: 1.25 mg/mL (*P. aeruginosa*)	
	Ethanol		MIC: 10 mg/mL (*S. aureus* and *P. aeruginosa*)	
	Water		MIC: 10 mg/mL (*S. aureus* and *P. aeruginosa*)	
	Gold and silver nanoparticles prepared from a water extract	Leaf	MIC of water extract: >50 mg/mL (*S. epidermidis* and *P. aeruginosa*) MIC of AgNPs: 0.075 mg/mL (*S. epidermidis* and *P. aeruginosa*)	Dube, Meyer, Madiehe, and Meyer (2020)
Urtica urens L./ Urticaceae	Petroleum ether	Leaf	MIC: 6.125 mg/mL (*B. subtilis*) MIC: 3.125 mg/mL (*S. aureus*) MIC: 6.25 mg/mL (*S. epidermidis*)	Thibane, Ndhlala, Abdelgadir, Finnie, and Van Staden (2019)
	Dichloromethane		MIC: 1.562 mg/mL (*B. subtilis*) MIC: 1.562 mg/mL (*S. aureus*) MIC: 6.25 mg/mL (*S. epidermidis*)	
	70% Ethanol		MIC: 0.098 mg/mL (*B. subtilis*) MIC: 0.098 mg/mL (*S. aureus*) MIC: 1.562 mg/mL (*S. epidermidis*)	
	Water		MIC: 0.098 mg/mL (*B. subtilis*) MIC: 3.125 mg/mL (*S. aureus*) MIC: 6.25 mg/mL (*S. epidermidis*)	
	Chloroform	Roots	MIC: 25 mg/mL (*P. aeruginosa*) MIC: 12.5 mg/mL (*B. subtilis*) MIC: >100 mg/mL (*S. aureus*)	Rajput, Choudhary, and Sharma (2019)
		Stem	MIC: >100 mg/mL (*P. aeruginosa*) MIC: >100 mg/mL (*B. subtilis*) MIC: 25 mg/mL (*S. aureus*)	
		Leaves	MIC: 12.5 mg/mL (*P. aeruginosa*) MIC: >100 mg/mL (*B. subtilis*) MIC: 12.5 mg/mL (*S. aureus*)	
	Methanol	Roots	MIC: >100 mg/mL (*P. aeruginosa*) MIC: 12.5 mg/mL (*B. subtilis*) MIC: >100 mg/mL (*S. aureus*)	
		Stem	MIC: 25 mg/mL (*P. aeruginosa*) MIC: >100 mg/mL (*B. subtilis*) MIC: >100 mg/mL (*S. aureus*)	
		Leaves	MIC: >100 mg/mL (*P. aeruginosa*) MIC: >100 mg/mL (*B. subtilis*) MIC: 12.5 mg/mL *S. aureus*	
	Ethanol	Aerial parts	MIC: 150 μg/mL (*B. subtilis, S. aureus, S. epidermidis*, and *P. aeruginosa*)	Mzid, Khedir, Salem, Regaieg, and Rebai (2017)

TABLE 4.2 In vitro and in vivo wound-healing activity of plants traditionally used in South Africa.

Plant species/ family	Extraction solvent	Plant part	Wound-healing activity	Reference
Aloe ferox Miller/ Xanthorrhoeaceae	Leaf material was homogenized and filtered	Whole leaf	Wound severity scores were significantly lower in rats treated with *A. ferox* and rats treated with *A. ferox* followed by *S. aureus*, when compared to the group that did not receive treatment. Showed 51.77 ± 0.75% wound closure when compared to the untreated control which showed 42.88 ± 1.23% closure after 32 h exposure	Fox et al. (2017), Jia et al. (2008)
	Leaf gel was liquidized	Leaf gel	Showed 82.41 ± 16.30% wound closure when compared to the untreated control which showed 52.04 ± 3.39% closure after 32 h exposure	Fox et al. (2017)
	Hot water	Leaves	At 10 and 100 μg/mL there was a significant increase in human primary epidermal keratinocytes (HPEK) cell migration, proliferation, and differentiation	Moriyama et al. (2016)
Urtica urens L./ Urticaceae	Ethanol	Leaves, stems, bark, and roots	Showed an increase in wound contraction in rabbits inoculated with *P. aeruginosa* (in vivo) when tested at 25% in an ointment formulation	Taqa, Mustafa, & Al-Haliem (2014)
	Water	Leaves, stems, bark, and roots	Did not show an increase in wound-healing contraction in rabbits	

red-orange compact racemes. The succulent leaves are dark green in color with dark red thorns along the leaf margins. Leaves of juvenile plants of this species have spines running along the entire leaf epidermis to deter herbivores but often contribute to the misidentification of this species as *Aloe ferox*. However, the racemes of this species are far shorter and slightly curved in comparison to that of *Aloe ferox*. Mature *A. excelsa* plants are established after 3–6 years of germination and flower over August and September during the late winter and early spring period (Coopoosamy & Magwa, 2007).

Traditional usage

Aloe excelsa is used by traditional healers throughout South Africa but is particularly popular amongst Sotho-speaking indigenous people, to treat a variety of ailments. Traditional healers prepare the leaves of *A. excelsa* in various ways depending on the intended medicinal use or route of administration. Decoctions of the whole leaf are prepared by boiling the leaves in water; the mixture is subsequently strained and administered as a tea. Tea made from *A. excelsa* is said to have purgative, blood purifying, and immune-boosting effects (Coopoosamy & Magwa, 2006). Gel exudates of the leaves are used to treat skin irritations such as rashes and burns by direct application of the sap to

TABLE 4.3 Toxicity of South African medicinal plants used for wound healing.

Plant species/family	Extraction solvent	Plant part	Toxicity potential	Reference
Aloe barberae Dyer/ Asphodelaceae	Petroleum ether	Upper stem	Nonmutagenic at 50, 500, and 5000 μg/mL using *Salmonella typhimurium* strain TA98	Ndhlala et al. (2009a)
	Dichloromethane			
	Ethanol			
	Water			
	Petroleum ether	Young bark	Nonmutagenic at 50, 500, and 5000 μg/mL using *Salmonella typhimurium* strain TA98	
	Dichloromethane			
	Ethanol			
	Water			
	Petroleum ether	Mature bark	Nonmutagenic at 50, 500, and 5000 μg/mL using *Salmonella typhimurium* strain TA98	
	Dichloromethane			
	Ethanol			
	Water			
	Petroleum ether	Leaves	Nonmutagenic at 50, 500, and 5000 μg/mL using *Salmonella typhimurium* strain TA98	
	Dichloromethane			
	Ethanol			
	Water			
	Petroleum ether	Roots	Nonmutagenic at 50, 500, and 5000 μg/mL using *Salmonella typhimurium* strain TA98	
	Dichloromethane			
	Ethanol			
	Water			

(*Continued*)

TABLE 4.3 (Continued)

Plant species/family	Extraction solvent	Plant part	Toxicity potential	Reference
Aloe ferox Miller/ Xanthorrhoeaceae	Leaf material was homogenized and filtered	Whole leaf	No erythema or edema was observed in Guinea pigs after a single dose application of *A. ferox* up to 72 h and no erythema or swelling was observed in Guinea pigs after multiple applications of *A. ferox* over a 10-day course	Andersen (2007), Fox et al. (2017), Jia et al. (2008)
			No skin irritation was observed in mice with healthy skin or rats with damaged skin over a 14-day period	
			Negligible cytotoxicity observed against human keratinocytes (HaCat) at concentrations of 0.4, 0.6, and 1.3 mg/mL	
			Considered a nonirritant when tested on damaged and nondamaged skin of New Zealand rabbits	
	Water		Did not show toxicity in rats in acute, subacute, and chronic toxicity studies and no behavioral or physiological changes were observed at doses of 50, 100, 200, and 400 mg/kg.	Mwale and Masika (2012)
	Oil		Application of a 0.2% concentration showed no dermal phototoxicity on both irradiated and nonirradiated skin	Andersen (2007)
	Leaf gel was liquidized	Leaf gel	Negligible cytotoxicity observed against human keratinocytes (HaCat) at concentrations of 0.4, 0.6, and 1.3 mg/mL	Fox et al. (2017)
	Evaporation of latex which drains from the leaves	Resin	Single oral dose of the resin at 5.0 g/kg in Wistar rats did not show signs of toxicity or death ($LD_{50} > 5.0$ g/kg).	Celestino et al. (2013)
	Fresh	Nectar of flowers	Persistent ingestion of the nectar has caused joint weakness and partial paralysis	Watt and Breyer-Brandwijk (1962)
Elephantorrhiza elephantina (Burch.) Skeels/Fabaceae	Aqueous extract	Rhizome	Acute toxicity: decreased respiratory rate after 10 min of administration in rats at 1600 mg/kg	Maphosa, Masika, and Moyo (2010)
			Subacute toxicity: increased white blood cell and monocyte counts at 400 and 800 mg/kg and increased the serum levels of creatine at 400 and 800 mg/kg	
	Acetone	Leaves	LC_{50}: 416.40 μg/mL on Vero cells	Kudumela, McGaw, and Masoko (2018)

	Water	Rhizomes	LC_{50}: 5.8 ppm against Brine Shrimp	Mpofu, Msagati, and Krause (2014)
	Methanol		LC_{50}: 1.8 ppm against Brine Shrimp	
	Chloroform fraction		LC_{50}: 10 ppm against Brine Shrimp (KwaZulu-Natal cultivar)	
	Ethyl acetate fraction		LC_{50}: 7.9 ppm against Brine Shrimp (KwaZulu-Natal cultivar)	
	Methanol fraction		LC_{50}: 15 ppm against Brine Shrimp (KwaZulu-Natal cultivar)	
	Fresh	Seed/bean	Roasted beans have been used as a substitute for coffee. The seed is toxic to rabbits, sheep, and guinea pigs, causing gastroenteritis	Watt and Breyer-Brandwijk (1962)
	Aqueous	Seeds	An equivalent dose of 0.75 g caused necrosis at the injection point in guinea pigs and caused gastroenteritis and pulmonary edema upon subcutaneous injection	
			Oral LD_{50} in rabbits is between 5 and 7.5 g, causing death within 24 h. Ingestion caused apathy, loss of appetite, diarrhea, and extreme exhaustion	
			Oral dose of 250 g in sheep caused death within 24 h causing gastroenteritis, hemorrhaging, and degeneration of the liver	
Erythrina lysistemon Hutch./Fabaceae			The plant and seeds have been reported to contain alkaloids which are highly toxic	Mbambezeli and Notten (2002)
	Dichloromethane	Stem bark	Mortality rate of 10.2% and 16.7% at 24 and 48 h, respectively, in brine shrimp	Sadgrove et al. (2020)
	Methanol	Stem bark	Mortality rate of 23.5% and 37.1% at 24 and 48 h, respectively, in brine shrimp	
Galenia africana L./Aizaoceae	80% Ethanol	Unspecified	Acute oral toxicity: oral administration of	Ng'uni, Klaasen, and Fielding (2018)
			300 and 2000 mg/kg doses to Sprague–Dawley rats showed no adverse reactions and no deaths were recorded	

(*Continued*)

TABLE 4.3 (Continued)

Plant species/family	Extraction solvent	Plant part	Toxicity potential	Reference
			Acute dermal toxicity: 2000 mg/kg topical application on Sprague–Dawley rats showed no systemic signs of toxicity and no deaths were recorded ($LD_{50} > 2000$ mg/kg). Necroscopy abnormalities were noted, which have been reported to be common background finding in Sprague–Dawley rats	
			Skin irritation: 20% wt./vol. concentration and 1% vol./vol. dilution showed a viability of 84.75 ± 11.26 and 92.97 ± 14.63% on the Episkin, respectively. Therefore, considered a nonirritant	
			Dermal sensitization: application of 50, 100, and 200 mg/mL/kg doses did not cause signs of systemic toxicity and did not cause dermal sensitization	
	Gold nanoparticles prepared from hot water (decoction)	Fresh aerial	Cell viability >80 and >70% at concentrations of 8 and 32 nM, respectively, against human fibroblasts (KMST-6)	Elbagory et al. (2017)
	Fresh	Whole plant	Causes waterpens in small ruminants such as sheep and goats. Waterpens are characterized by extensive ascites and liver cirrhosis	Watt and Breyer-Brandwijk (1962)
Grewia occidentalis L./Malvaceae			Listed as nonpoisonous and the fruits are used traditionally to make beer and milkshake (boiled in milk)	Turner (2008), Hepplewhite and Witkoppen Wildflower Nursery (2018)
			The yellow berry is eaten by both the Xhosa and Zulu communities	Watt and Breyer-Brandwijk (1962)
Melianthus comosus Vahl/Melianthaceae	Ethanol	Leaves	IC_{50}: 51.40 μg/mL on Vero cells	Heyman et al. (2009)
	Decoction	Leaves and stems	A 5% decoction caused vomiting and gastrointestinal irritation	Watt and Breyer-Brandwijk (1962)

	Aqueous infusion	Roots	Subcutaneous injection in animals caused lassitude, loss of appetite, vomiting, and resulted in death. Necrosis was observed at the injection site and organs showed signs of congestion	
	Dried plant parts	Leaves, flowers, and young fruit	A sheep drenched with 80 g of leaves, flowers, and young fruit, died within 4–5 h after administration by causing hemorrhaging and inflammation of the duodenum, jejunum, and other organs	
	Acetone-enriched extract	Leaves	No toxicity observed against brine shrimps at a concentration of 0.1, 0.2, and 0.5 mg/mL. At 1.0 and 2.0 mg/mL, 9.76% and 21.65% mortality rate was recorded, respectively ($LC_{50} > 2.0$ mg/mL)	Angeh (2006)
			IC_{50}: 0.0442 mg/mL on Vero cells	
Plectranthus fruticosus/ Lamiaceae	Distillation water extracted with diethyl ether	Leaves	The plant is embryo- and foetotoxic to rodents	Chamorro, Salazar, Fournier, and Pages (1991)
Sutherlandia frutescens (L.) R.Br/Fabaceae	Ethanol	Leaves	LD_{50}: 40.54 μg/mL (zebrafish larvae)	Chen et al. (2018), Zonyane et al. (2019)
			At 5 μg/mL the hatching rate of zebrafish was not affected	
			There was a 0% hatch rate and 100% mortality rate at 300 μg/mL	
			Mortality rate increased from 22.2% to 95.5% at concentrations ranging from 10 to 30 μg/mL	
			Concentrations >30 μg/mL led to abdominal excess fluid and yolk sac edema. At 200 μg/mL, there was a 38% frequency of abnormal morphological features	
			Decreased heart rate (28 ± 1 beats/10 s to 27 ± 3 beats/10 s) at 5 μg/mL	
	Water		LD_{50}: 297.57 μg/mL (zebrafish larvae)	
			At 5 μg/mL the hatching rate of zebrafish was not affected	

(*Continued*)

TABLE 4.3 (Continued)

Plant species/family	Extraction solvent	Plant part	Toxicity potential	Reference
			There was a 0% hatch rate and 100% mortality rate at 300 μg/mL	
			Mortality rate increased from 0% to 33.3% at concentrations ranging from 0 to 200 μg/mL	
			Concentrations >30 μg/mL led to abdominal excess fluid and yolk sac edema	
			No decrease in heart rate at 5 μg/mL, however, a decrease was noted at 10 and 30 μg/mL	
	Ethyl acetate	Whole plant	Antimutagenic activity at a concentration of 5%, 10%, and 20% per plate using *Salmonella typhimurium* strains TA97a, TA98, and TA100	
	50% Methanol		At 50% the methanolic extract showed pro-mutagenic effects in comparison with the diagnostic mutagen 2-acetamidofluorene for *S. typhimurium* strain TA98. The same extract also exhibited pro-mutagenic effects in comparison with Aflatoxin B1 using the *S. typhimurium* strain TA100. In both cases the revertant colonies were higher than that of the mutagenic compounds in the presence of the S9 rat liver enzyme fraction	Ntuli, Gelderblom, and Katerere (2018)
	800 mg/day leaf powder capsules for 3 months		No significant difference was noted in the physical, vital, blood, and biomarker indices when compared to the placebo group. No general adverse events were observed	Johnson, Syce, Nell, Rudeen, and Folk (2007)
Urtica urens L./ Urticaceae	50% Aqueous ethanol	Aerial parts	No cytotoxic effects exhibited on mouse macrophages (RAW 264.7) or hepatocyte (HepG2) cells	Carvalho et al. (2017)
	The plant is used as a relish by African communities			Watt and Breyer-Brandwijk (1962)
	Aqueous	Leaves	No harmful effects were noted against skin fibroblasts (HSF) at 5 and 10 μg/mL	Al Doghaither, Omar, Rahimulddin, & Al-Ghafari (2016)

the affected area. The gel exudates are also administered orally to act as an immune booster, improve digestion, and alleviate constipation. Similar to other *Aloe* species, the gel of *Aloe excelsa* is directly applied to dermal wounds to accelerate wound closure and to moisturize the skin (Coopoosamy & Magwa, 2007).

Aloe ferox Miller

Aloe ferox Miller, also known as the bitter aloe, is an arborescent species belonging to the Xanthorrhoeaceae family. It is indigenous to southern Africa but is found mainly in KwaZulu-Natal. The name "ferox," which translates to "ferocious," describes the morphology of the large spiny leaves. It is single-stemmed, grows to an average height of 3 m, and forms rosettes of large fleshy leaves. The leaves are thick and dull green in color with red-brown spines along the edges of each leaf and smaller spines along the entire leaf epidermis. The flowers are uniform, red-orange in color, and grow between 0.5 and 1.22 m above the leaves into multibranched inflorescences with tightly compacted conical racemes. It takes 2–4 years to reach maturity and flowers between May and August over the winter period (O'Brien, Van Wyk, & Van Heerden, 2011).

Traditional usage

There are ancient depictions of *Aloe ferox* in cave paintings in the Aliwal North area of South Africa (over 250 years old) and scriptures by the Greek physician Dioscorides (CE AD) indicating that the plant has been used for its medicinal properties for centuries. The gel is known for its antiinflammatory and hydrating properties and is therefore used in wound healing and cosmetic products that treat unfavorable skin conditions such as eczema, conjunctivitis, or sunburn (O'Brien et al., 2011).

Aloe ferox is commonly used in the production of purgative medication such as "aloe bitters" or "aloe lump." Aloe bitters are made from the dark sap of the aloe, the sap between the leaf epidermis and the gel of the leaf, which is extracted and dried to make a dark crystalline substance. The crystals are used in the Eastern and Western Cape provinces of South Africa to alleviate symptoms of arthritis but are mainly used to treat constipation. The purgative effects of the dark sap are attributed to the action of reducing water reabsorption into the intestine and by directly stimulating the smooth muscle of the gut (Chen, Van Wyk, Vermaak, & Viljoen, 2012). The dark sap and gel of *Aloe ferox* are used by the Xhosa in the treatment of dermal wounds and are believed to stimulate wound closure as well as provide antimicrobial and antiinflammatory activity. Slices of the leaves are placed in the drinking water of livestock to provide immune-boosting properties. Traditional healers of the Eastern Cape use it to treat wounds developed from sexually transmitted infections such as gonorrhea and syphilis. Decoctions of the fresh and dried leaves are administered orally or directly applied as a gel to wounds to prevent bacterial infection and promote wound healing (Kambizi, Sultana, Afolayan, & Afolayan, 2008). Additionally, the Zulu's prepare a decoction from the leaves, which is applied to venereal sores (Watt & Breyer-brandwijk, 1962).

Elephantorrhiza elephantina (Burch.) Skeel

Elephantorrhiza elephantina, which forms part of the Fabaceae family, is a species of a small genus, consisting of nine species that are distributed over the African continent. In southern Africa specifically, *E. burkhei*, *E. elephantina*, *E. goetzei*, and *E. suffruticosa* are used for their medicinal properties. This plant has a wide distribution throughout southern Africa including South Africa, Namibia, Botswana, Zimbabwe, Mozambique, Swaziland (now Eswatini), and Lesotho. It thrives in hot and dry grasslands and open scrub. This species is described as a perennial suffrutex or low shrub, producing annual stems growing up to 90 cm tall, with a thickened 8 m long rhizome. The bipinnately compound leaves are arranged alternately, described as almost glabrous with a petiole of up to 8 cm long. The raceme inflorescence is axillary and commonly confined to the lower stem occurring solitary or clustered. The yellow-white bisexual flowers have reddish-brown glands at the base, with linear-oblong petals that are connate at the base. The fruit is described as a compressed-oblong, straight, or slightly curved pod that is reddish-brown in color and swollen over the seeds (Maroyi, 2017).

Traditional usage

The species is used to treat various gastrointestinal and dermatological ailments in South Africa, including diarrhea, dysentery, stomach pain, ulcers, acne, male pattern baldness, and hemorrhoids. Some nonmedicinal uses include the use of the rhizome and/or bark as a dye to tan animal hides. Young shoots are used as livestock feed and the seeds are often roasted and used as a substitute for coffee. In Lesotho the plant is used to prevent bleeds, syphilis, and disorders of the intestinal tract. Maroyi (2017) has recorded numerous ethnomedicinal uses, including its use for dermatological ailments. An infusion of the rhizomes and roots is used in South Africa for acne and to treat shingles in combination with other species. Rhizome and root decoctions are administered orally, both alone and in combination with other species to treat sores. A rhizome decoction is used to relieve itching. The underground parts are used to treat sunburn. In Lesotho, a root decoction is applied directly to bleeding wounds and is also taken orally as a blood purifier.

Erythrina lysistemon Hutch

Erythrina lysistemon, commonly known as the coral tree, forms part of the Fabaceae family. It is a medium-sized tree, reaching heights of up to 10 m, and is well known for its bright red flowers which bloom during spring (August to September). The leaves, which can reach a size of up to 17 × 18 cm, are compound and have three leaflets. Long and slender fruit pods, which can be up to 15 cm long contain bright red seeds, which are easy to germinate for propagation. *Erythrina lysistemon* is found throughout South Africa as well as in southern Africa, such as Botswana, Angola, and Zimbabwe. It can be found in the Eastern Cape, North West, KwaZulu-Natal, Gauteng, Limpopo, and Mpumalanga, where it can be found growing in various habitats such as rocky slopes, dry woodlands, and savannah, as well as forest and coastal areas (Mbambezeli & Notten, 2002).

Traditional usage

The leaves are crushed and used as a dressing for wounds that have been infested with maggots. A poultice prepared from the bark is used to treat abscesses, wounds, and sores as well as arthritis. Other uses include infusions prepared from the leaves which are applied as ear drops to alleviate earaches and root decoctions which are used on sprains (Mbambezeli & Notten, 2002).

Galenia africana L

Galenia africana, known as the kraalbos, forms part of the Aizoaceae family. It is a shrub that can reach a meter in height. It is most often found in the Namaqualand and Karoo regions of South Africa, growing in areas that have been disturbed, such as roads, old lands, and kraals. The leaves are oppositely arranged, hairless, and can reach a length of up to 5 cm. As the shrub starts to mature, the leaves start to turn yellow (Vries et al., 2005).

Traditional usage

A study performed in the south-eastern Karoo reported that the species is boiled in salt-water and used to wash wounds. The stems, which have been steeped in water are applied topically to treat rough skin. Other medicinal properties include bathing in a weak infusion for the treatment of rheumatism (van Wyk, 2008). Watt and Breyer-brandwijk (1962) reported that a decoction of *G. africana* is used to prepare a lotion for the treatment of wounds in humans and animals. It was also reported that people in the Western Cape, used a decoction to treat skin diseases. An ointment to dress leg wounds, more specifically in women, is made by frying *G. africana* in butter together with the plant species *Cyanella lutea, Lobostemon fruticosus, Melianthus major, Melianthus comosus* "Tiendaegeneesblare" and "Jakkalsoorblare" has been reported. External lesions associated with syphilis are washed using a decoction prepared from *G. africana, L. fruticosus, M. major*, and *M. comosus* (Watt & Breyer-brandwijk, 1962).

Grewia occidentalis L

Grewia occidentalis, commonly known as cross-berry or four-corner, belonging to the Malvaceae family, is an indigenous South African shrub, which can reach heights of up to 3 m. It can be found in a variety of habitats such as wooded grasslands, arid Karoo and Highveld, coastal and mountain slopes, or forest areas. It is widely distributed throughout South Africa ranging from the Western Cape, throughout the Eastern Cape, Gauteng, Limpopo, Free State, and North West, Mpumalanga and KwaZulu-Natal, and is also found in Zimbabwe and Mozambique (Foden & Potter, 2005). During the summer months (October–January), purple star-shaped flowers appear, which in turn develop into four-lobed red-brown to purple fruits, from which the common name is derived. The simple, alternate leaves are dark green and shiny, with hairs that appear on both sides (Grierson & Afolayan, 1999; Turner, 2008).

Traditional usage

In the Eastern Cape, the Zulu, Tswana, and Xhosa communities soak the bark and twigs in hot water which is used as a wash, lotion, or dressing for the treatment of wounds (Grierson & Afolayan, 1999; Watt & Breyer-brandwijk, 1962). Other medicinal properties include the use of the bark and roots to treat bladder ailments (Pitso & Lebese, 2014).

Melianthus comosus Vahl.

Melianthus comosus commonly known as the honey flower or the "touch me not" (kruidjie-roer-my-nie) belongs to the Melianthaceae family. This erect shrub can grow up to 1−2 m tall with multiple branches covered in stellate hairs. The pinnately compound leaves can grow up to 20 cm long and are arranged alternately with four to seven pairs of opposite leaflets. The leaflets are oblong to lanceolate with toothed margins. The raceme inflorescences originate from the leaf axils and can grow up to 10 cm long. Flowers have four, unequal petals which are red in color. Fruits are bladder shaped, winged, and have a membranous capsule. The seeds are ovoid, black, or dark brown and grow between 3 and 6 mm (Weber, 2017). This species is distributed throughout Namibia and in the dry interior regions of seven South African provinces and in neighboring Lesotho (Harris, 2004).

Traditional usage

A poultice and a decoction prepared from the leaves of *Melianthus comosus* are used to treat bed sores, septic wounds, and to reduce swelling (Mabona, Viljoen, Shikanga, Marston, & Van Vuuren, 2013). Harris (2004) also reported the use of the leaf poultice and decoction for bruising, backache, and rheumatoid arthritis in the joints. A study in the south-eastern Karoo reported that the species is applied topically to reduce inflammation in the leg and for various skin ailments. Other uses include applying the leaves as a poultice, and bathing or soaking in water in which the plant has been boiled, to treat pain associated with legs, knees, and the back (van Wyk, 2008). The Xhosa's make a paste prepared from the leaves or a tincture prepared from the root bark or the leaves to treat wounds associated with snake bites. Additionally, a decoction is used to treat slow-healing wounds, and a dressing prepared from the leaf paste is used to treat sores and reduce swelling caused by bruises (Watt & Breyer-brandwijk, 1962).

Plectranthus fruticosus L'Hér

Plectranthus fruticosus otherwise known as the skunk-leaf is a 1−1.5 m high, semiwoody, branched perennial belonging to the Lamiaceae family. This species can be found along rocky shaded areas and forests occurring from the Western Cape, Limpopo and Mpumalanga and Swaziland borders, particularly at altitudes higher than 1000 m. Cross-sections of the branches are characteristically square with leaves arranged oppositely. The dark green leaves are broadly ovate, toothed along the margin, and slightly velvety, reaching 120−150 mm in diameter. The leaves have a characteristic dark green venation. The leaf petiole is green and grows to lengths of 60−80 mm. The inflorescence is an erect panicle reaching up to 300 mm in length and made up of around 600 mauve, blue-purple flowers, generally occurring as six flowers per node. Flowering occurs from mid-March to early May. The calyx is green with a purple mouth and the seeds or nutlets are 1.5 mm long and brown in color occurring at the base of the calyx (Harrower, 2010; Jodamus & Notten, 2002).

Traditional usage

Plectranthus fruticosus is used in Romanian traditional medicine for its healing and soothing properties, particularly for burns of the skin. The stems are used as a repellent for flies and are often rubbed along windows to deter entry. In South Africa, this species is often used for its ornamental or esthetic properties (Lukhoba, Simmonds, & Paton, 2006). It has further been reported that the Khoi-San and Cape Dutch used the fresh leaves for the treatment of open wounds (van Wyk, 2008).

Polystichum pungens (Kaulf.) C. Presl

Polystichum pungens, known as the forest shield fern or the prickly shield fern, belonging to the Dryopteridaceae family, is found distributed throughout South Africa, such as in the Eastern and Western Cape, Limpopo, Mpumalanga and KwaZulu-Natal, and is found in Swaziland. It is confined to forest areas with elevations ranging from 600 to 1350 m, such as Table Mountain and the foothills of the Drakensberg. The rhizomes lie along the surface of the ground resulting in the formation of a clonal stand and therefore a shrubby growth appearance. The leaves can reach a height of up to 1 m long and are dark green in appearance (Foden & Potter, 2005; Roux, 2004).

Traditional usage

In the Eastern Cape, it was reported that the dried fronds were pulverized into a powder and applied on wounds. Similarly, the fresh fronds are ground and used as a poultice (Grierson & Afolayan, 1999). The southern Sotho use a decoction prepared from the rhizome as an enema for intestinal worms in humans and to treat bot fly infestations in horses (Watt & Breyer-brandwijk, 1962).

Sutherlandia frutescens (L.) R.Br.

Sutherlandia frutescens, better known by its common name, cancer bush, forms part of the Fabaceae family and has recently been transferred to the genus *Lassertia*, thereby it has been renamed to *Lasseria frutescens* (L.) It is classified as a shrub that is soft-wooded and can reach heights of up to 1 m. The gray-green colored leaves are pinnately compound with the leaflets reaching lengths of between 4 and 10 mm. During spring to mid-summer (September-December), bright orange flowers, up to 35 mm long, form, which in turn develop into large bladder-like fruits. It is widely distributed throughout the drier regions of southern Africa, such as Namibia and Botswana. In South Africa it can be found in the Western and Eastern Cape, including the Karoo, as well as KwaZulu-Natal and Mpumalanga (Xaba & Notten, 2003).

Traditional usage

It has been reported that the Khoisan and the Nama people, the original Cape inhabitants, used a decoction of the species as a wash to treat wounds. A decoction has also been used as a wash to treat ailments relating to the eyes. It is extensively used for numerous other health-related problems and diseases such as cancer (which the common name is derived from), colds, bronchitis, TB, arthritis, liver ailments, hemorrhoids, and several other disorders (Xaba & Notten, 2003).

Urtica urens L.

Urtica urens more commonly known as the small nettle belongs to the Urticaceae family. This plant is considered a weed and is often found in cultivated fields, gardens, and orchards. This species has a wide distribution spanning from the temperate regions of Europe, California in the United States of America, the tropics in Africa, and is widespread throughout South Africa. The species is an erect, branched, annual shrub growing up to 65 cm high. The four-angled stem is sparsely covered in stinging hairs. The olive-green leaves are in an opposite arrangement, with a 3–5 cm long petiole. The dense, monecious flowers occur as a cyme inflorescence, blooming in March to May months. The nut is compressed, ovate, smooth, and often shiny and is between 1–4.2 mm × 1–1.4 mm (Aswal, 1972).

Traditional usage

In the south-eastern Karoo, the powdered leaves are applied topically to burn wounds (van Wyk, 2008). The Zulus use this species as an aphrodisiac, whilst in the south-western Cape a bark infusion is ingested to treat inflammation, pain in the bladder, and internal bleeds as well as to stimulate lactation. An infusion is applied externally for the treatment of burns. The ground plant powder is used as a snuff to treat nose bleeds and made into a sirup using brown sugar for the treatment of whooping cough. A warm compress prepared using an infusion is applied to different tumors (Watt & Breyer-brandwijk, 1962). A study by Al-Bakri and Afifi (2007) recorded the use of the leaves to have antirheumatic, antispasmodic properties and is used as a diuretic.

Compounds present in plants traditionally used for wound healing in South Africa

Although several of the tested plant extracts (Table 4.1) have not shown significant antibacterial activity against the wound-associated pathogens, isolation of compounds from these plants and other species has shown noteworthy activity and may be a new source of antibacterial agents. Compounds isolated from the species discussed throughout the chapter have been summarized below.

Aloe species

Anthraquinone compounds isolated from both *A. ferox* and *A. excelsa* exhibited antibacterial activity against wound-associated bacteria (Fig. 4.4). Three compounds, aloe-emodin,

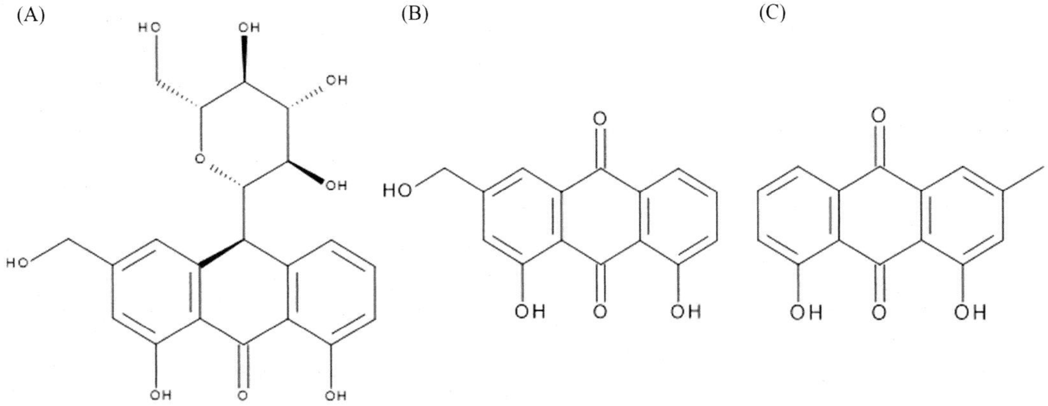

FIGURE 4.4 Major compounds isolated from Aloe species (A) aloin A, (B) aloe-emodin, and (C) chrysophanol.

chrysophanol, and aloin A, which showed antibacterial activity were isolated from the hexane extract of the dried plant material. The MICs of aloe-emodin, chrysophanol, and aloin A were 125, 250, and 62.5 μg/mL, respectively, against *Bacillus subtilis*. Against *Staphylococcus aureus* the MICs were 125, >250, and 62.5 μg/mL, respectively, and MICs of 250, 31.25, and 125 μg/mL were obtained against *Staphylococcus epidermidis*, respectively (Aaku, Dharani, Majinda, & Motswaiedi, 1998; Coopoosamy & Magwa, 2006; Kambizi et al., 2008). In addition, aloin A has been found to significantly accelerate wound healing in rats and enhance the proliferation of human skin fibroblasts and endothelial cells in vitro.

The isolation and identification of these active compounds from *A. barberae* were conducted using chromatographic and ultraviolet spectroscopic methods. The ethanolic and dichloromethane leaf extracts were subjected to thin-layer chromatography (TLC) and the UV spectra were obtained to identify possible compounds present. The results from the TLC confirmed that phenolic compounds were present within the leaf extracts of *A. barberae* which are hypothesized to be the antibacterial compounds, such as aloin derivatives and chrysophanol, commonly found in aloe species. However, further isolation and characterization of these phenolic compounds within *A. barberae* is needed to confirm the exact chemical structure of these compounds (Ndhlala et al., 2009a).

Elephantorrhiza elephantina

A study by Aaku et al. (1998) isolated dihydrokaempferol, kaempferol, (-)-catechin, ethyl gallate, gallic acid, 2-(3,4-dihydroxyphenyl) ethanol, 4-hydroxybenzoic acid, ethyl-1-O-β-D-galactopyranoside, and quercetin-3-O-β-D-glucopyranoside from the 70% ethanolic rhizome extract of *E. elephantina* (Fig. 4.5). A TLC bioautography method was used to determine the antibacterial activity of the isolated compounds against *B. subtilis*, *S. aureus*, and *P. aeruginosa*. Each of the compounds was found to have antibacterial activity against the tested pathogens only at a loading capacity of above 50 μg, except for gallic acid and ethyl gallate which showed activity at loading capacities of 15 and 25 μg against *B. subtilis*

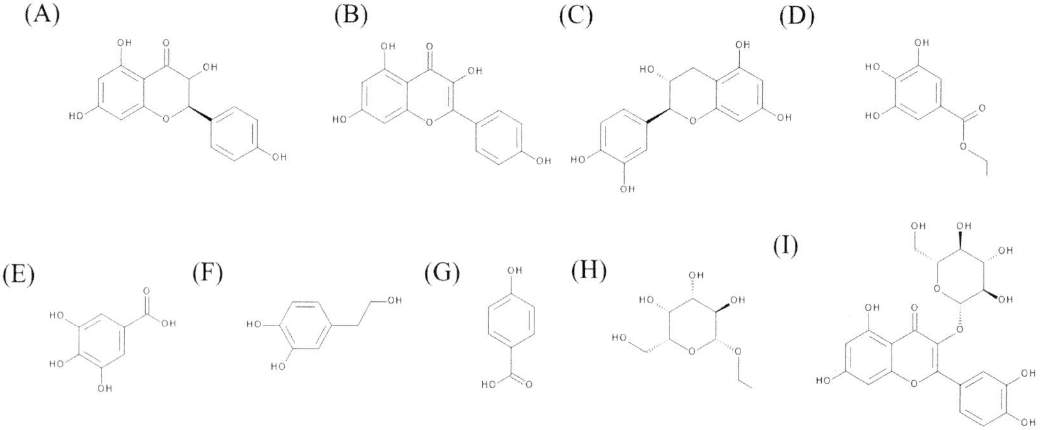

FIGURE 4.5 Compounds isolated from the 70% ethanolic rhizome extract of *Elephantorrhiza elephantina* (A) dihydrokaempferol, (B) kaempferol, (C) (-)-catechin, (D) ethyl gallate, (E) gallic acid, (F) 2-(3,4-dihydroxyphenyl) ethanol, (G) 4-hydroxybenzoic acid, (H) ethyl-1-O-β-D-galactopyranoside, and (I) quercetin-3-O-β-D-glucopyranoside.

and 20 and 35 μg against *S. aureus*, respectively. However, the authors concluded that the constituents may have a synergistic effect, as the 70% ethanolic rhizome extract and the n-butanol semipure fraction showed higher activity than the compounds (Table 4.1).

In a study by Ming et al. (2017), kaempferol was found to inhibit 80% of *S. aureus* biofilm formation at a concentration of 64 μg/mL, however, did not show antibacterial activity, which correlates with the study by Aaku et al. (1998). This further corresponds to a study by Adamczak, Ożarowski, and Karpiński (2020), where kaempferol showed an MIC > 1000 μg/mL against both *S. aureus* and *P. aeruginosa*. However, in a study by Tajuddeen, Sani Sallau, Muhammad Musa, James Habila, and Muhammad Yahaya (2014), kaempferol was reported to have an MIC of 6.25 μg/mL against *S. aureus*, which was also the MIC obtained for dihydrokaempferol. Against *B. subtilis*, dihydrokaempferol showed an MIC of 1 mg/mL (Zhou, Li, Wang, Liu, & Wu, 2007). Kaempferol has also been reported to have wound-healing activity in both nondiabetic and diabetic wounds in rats after a 14-day topical treatment of 1% (wt./wt.) kaempferol ointment (Özay et al., 2019).

(-)-Catechin has been tested against three strains of *S. aureus* (BB568, EMRSA-15, and EMRSA-16). The compound exhibited no activity against these strains with MICs >256 mg/L. The compound was also tested in combination with the positive control oxacillin and showed no potentiation of the antibiotic against the three strains (Stapleton et al., 2004). Ethyl gallate showed activity against *B. subtilis* with an MIC of 1000 μg/mL. Using a macrodilution and microdilution methods the compound was only active using the macrodilution method with MICs of 600–1200 and 600–2400 μg/mL against *P. aeruginosa* and *S. aureus*, respectively (Mazurova et al., 2015). Vandal, Abou-Zaid, Ferroni, and Leduc (2015) reported an MIC of 1000 μg/mL against *S. aureus* and no activity against *P. aeruginosa* (> 1000 μg/mL). Sato et al. (1997) reported the antibacterial activity of this compound against *S. epidermidis* and *P. aeruginosa* with MICs of 1000 and 500 μg/mL, respectively. In a study by Mazurova et al. (2015) the MIC of gallic acid was determined using macrodilution and microdilution methods. The macrodilution MIC was determined as 600–1200 μg/mL and the microdilution MIC was 300–600 μg/mL against *P. aeruginosa*. The macrodilution MIC and microdilution MIC against *S. aureus* were 2400–4800 and >4800 μg/mL, respectively. Borges, Ferreira, Saavedra, and Simões (2013) reported similar MICs with 500 and 1750 μg/mL against *P. aeruginosa* and *S. aureus*, respectively. The compound 4-hydroxybenzoic acid has been tested for its antibacterial activity against *S. aureus* (ATCC 5838), *S. epidermidis* (12228), *B. subtilis* (ATCC 6633), and *P. aeruginosa* (KCTC 1628) with an IC_{50} of 926, 355, 956, and 619 μg/mL, respectively (Cho, Moon, Seong, & Park, 1998). A recent study by Liu, Du, Beaman, Beth, and Monroe (2020) tested the compound against *S. epidermidis* and it exhibited an IC_{50} of 3.2 mg/mL after 4 h and 4 mg/mL after 24 h. The compound also exhibited antibacterial activity against drug-resistant *S. epidermidis* with an IC_{50} of 4 mg/mL after 4 h and 3.2 mg/mL after 24 h. When tested against *S. aureus* the IC_{50} was 5.2 mg/mL after 4 h and 4.5 mg/mL after 24 h. Against drug-resistant *S. aureus* the IC_{50} was 2.9 mg/mL after 4 h and 2.3 mg/mL after 24 h (Liu et al., 2020). There is no reported antibacterial activity for the compounds 2-(3,4-dihydroxyphenyl) ethanol and ethyl-1-O-β-D-galactopyranoside. Quercetin-3-O-β-D-glucopyranoside isolated from *Halostachys caspica* was investigated for its antibacterial activity against *S. aureus* and *B. subtilis* and exhibited MICs of 200 and 100 μg/mL, respectively. The compound also exhibited an IC_{50} of 167.61 and 56.00 μg/mL, respectively (Liu et al., 2010).

Erythrina lysistemon

In a study by Sadgrove, Oliveira, Khumalo, van Vuuren, and van Wyk (2020), seven isoflavone derivatives were isolated from the bark of *E. lysistemon*, namely, three pterocarpans (erybraedin A, phaseollidin, and cristacarpin), one flavonoid (abyssinone V-4′ methyl ether), one isoflavan (eryzerin C) and two isoflavonoids (alpumisoflavone and lysisteisoflavone), which were tested for antibacterial activity against various skin pathogens (Fig. 4.6). Erybraedin A was found to be the most active compound against the tested

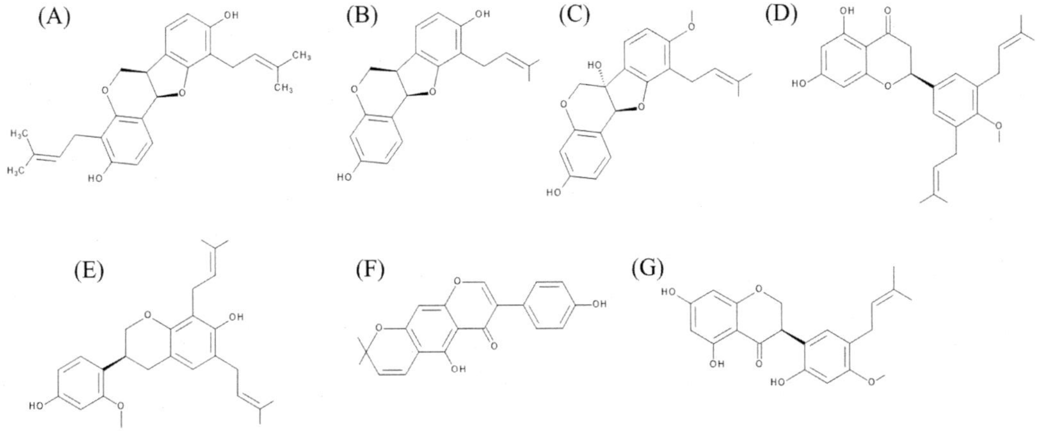

FIGURE 4.6 Compounds isolated from the bark of *Erythrina lysistemon* (A) erybraedin A, (B) phaseollidin, (C) cristacarpin, (D) abyssinone V-4′ methyl ether, (E) eryzerin C, (F) alpumisoflavone, and (G) lysisteisoflavone.

pathogens with an MIC value of 2 μg/mL against both *S. aureus* and *S. epidermidis*, whereas an MIC of 20 μg/mL was obtained against *P. aeruginosa*. Eryserin C was found to be the second most active compound, with an MIC value of 5 μg/mL against both *S. aureus* and *P. aeruginosa*, and an MIC value of 2 μg/mL against *S. epidermidis*. This was followed by phaseollidin, which showed MIC values of 5, 10, and 20 μg/mL against *S. epidermidis*, *S. aureus*, and *P. aeruginosa* respectively. Abyssinone V-4′ methyl ether, alpumisoflavone, cristacarpin, and lysisteisoflavone showed less activity against the tested skin pathogens with MIC values of 59, 31, 156, and 62 μg/mL against *S. aureus*, MIC values of 117, 125, 412, and 26 μg/mL against *S. epidermidis* and MIC values of 260, 20, 78, and 31 μg/mL against *P. aeruginosa* (Sadgrove et al., 2020). In a study by Tanaka et al. (2002) cristacarpin showed similar activity to that reported by Sadgrove et al. (2020) with an MIC of 100 μg/mL against methicillin-resistant *S. aureus*.

Galenia africana

Ticha (2015) determined the antimicrobial activity of several flavanones and chalcones isolated from *G. africana* (Fig. 4.7). (E)-2′,4′-dihydroxylchalcone exhibited activity against both *S. aureus* and methicillin-resistant *S. aureus* using the disk diffusion assay. The zone of inhibition was 11 and 7 mm against *S. aureus* and MRSA, respectively, at a concentration of 0.5 mg/mL.

Melianthus comosus

A study by Bedane et al. (2020) reported the isolation of six bufadienolides, 16β-formyloxymelianthugenin, 2β-acetoxymelianthusigenin, 2β-hydroxy-3β,5β-di-O-acetylhellebrigenin,

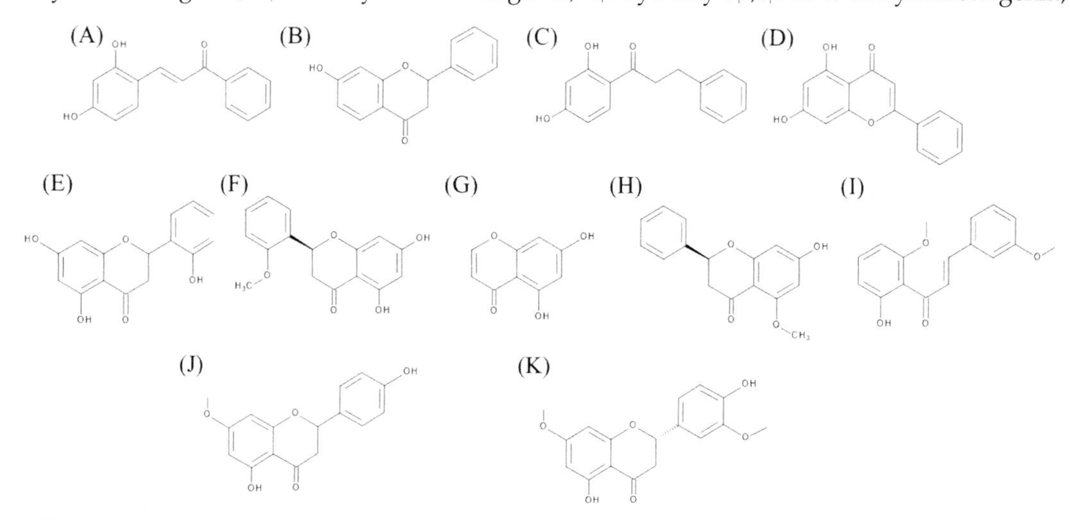

FIGURE 4.7 Flavanone and chalcone compounds isolated from *Galenia africana*, (A) (E)-2′,4′-dihydroxylchalcone, (B) 7-hydroxyflavanone, (C) 2′,4′-dihydroxydihydrochalcone, (D) (S)-5,7-dihydroxy flavone, (E) 2′,5,7-trihydroxyflavanone, (F) (S)-5,7-dihydroxy-2′-methoxy flavanone, (G) chromenone, (H) alpinetin, (I) (S)-2′-hydroxy-3,6′-dimethoxydihydrochalcone, (J) (S)-4′,5-dihydroxy-7-methoxy flavanone, and (K) (S)-4′,5-dihydroxy-3′,7-dimethoxy flavanone.

2β-acetoxy-5β-O-acetylhellebrigenin, melianthusigenin, and 16β-hydroxybersaldegenin 1,3,5-orthoacetate from a dichloromethane:methanol (1:1) extract prepared from the leaves of *M. comosus* (Fig. 4.8). While these compounds have not been tested for their antimicrobial activity, several bufadienolide compounds have been tested against *S. aureus*, *P. aeruginosa*, *S. epidermidis*, and *B. subtilis*. These compounds have also been isolated from the glandular skin secretions of toad species and several species of Kalanchoe (Cunha Filho et al., 2005; Rodriguez, Ibáñez, Rollins-Smith, Gutiérrez, & Durant-Archibold, 2020; Stefanowicz-Hajduk et al., 2020).

Plectranthus fruticosus

Several diterpenoids including 10(14)-aromadendrene-4β,15-diol, *ent*-3β-acetoxy-labda-8(17),12Z,14-trien-2α-ol, *ent*-12β-acetoxy-15β-hydroxykaur-16-en-19-oic acid, *ent*-12β-acetoxy-7β-hydroxykaur-16-en-19-oic acid, *ent*-7β-hydroxykaur-15-en-19-oic acid, *ent*-12β-acetoxy-17-oxokaur-15-en-19-oic acid, *ent*-7β-hydroxy-15β,16β-epoxykauran-19-oic acid, *ent*-labda-8(17),12Z,14-triene-2α,3β-diol, methyl *ent*-12β-hydroxykaur-16-en-19-oate, methyl *ent*-12β-acetoxy-7β-hydroxykaur-15-en-19-oate, *ent*-labda8(17),12Z,14-triene-2α,3β-dibenzoate, *ent*-12β-acetoxy-15β-hydroxykaur-16-en-19-oate, and methyl *ent*-12β-acetoxy-17-oxokaur-15-en-19-oate have been isolated from this species which have been tested against *P. aeruginosa* and *S. aureus*. The diterpene kaurane was identified as the most active compound with an MIC of 62.5 μg/mL against *S. aureus*. None of the compounds exhibited antibacterial activity against *P. aeruginosa* (Gaspar-Marques, Fá Tima Simões, & Rodríguez, 2004) (Fig. 4.9).

Sutherlandia frutescens

There have been four cycloartane glycosides (sutherlandioside A, B, C, and D) isolated from this species which have been tested for their antimicrobial activity against *S. aureus*,

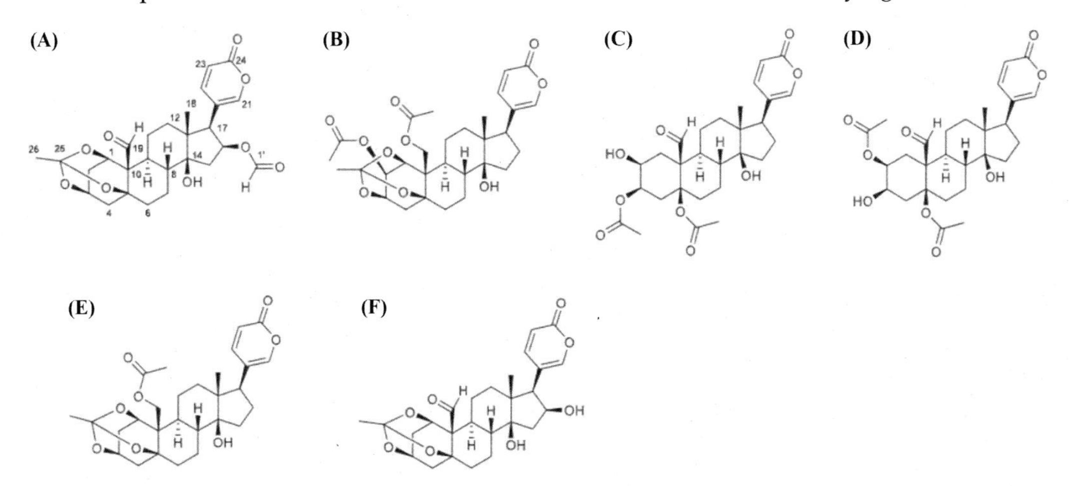

FIGURE 4.8 Bufadienolides isolated from *Melianthus comosus*, (A) 16β-formyloxymelianthugenin, (B) 2β-acetoxymelianthusigenin, (C) β-hydroxy-3β,5β-di-O-acetylhellebrigenin, (D) β-acetoxy-5β-O-acetylhellebrigenin, (E) melianthusigenin, and (F) 6β-hydroxybersaldegenin 1,3,5-orthoacetate.

FIGURE 4.9 Kaurane, an active diterpenoid isolated from the acetone extract of *P. fruticosus*.

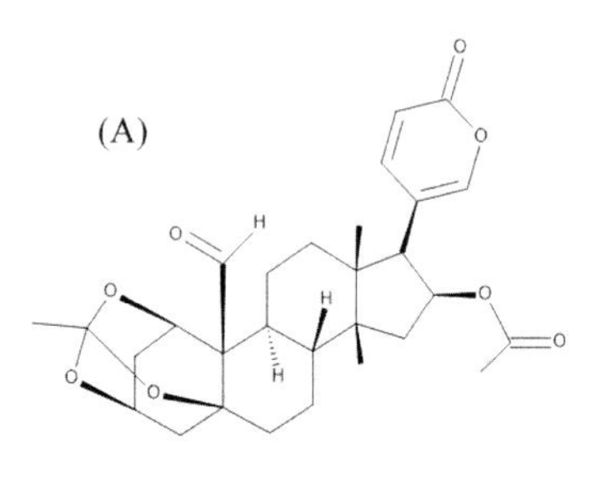

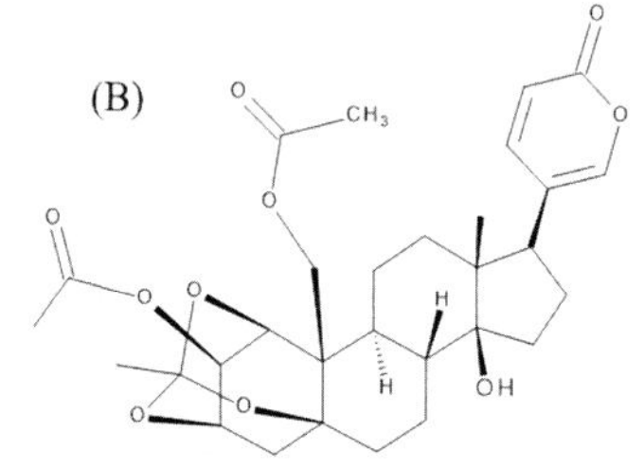

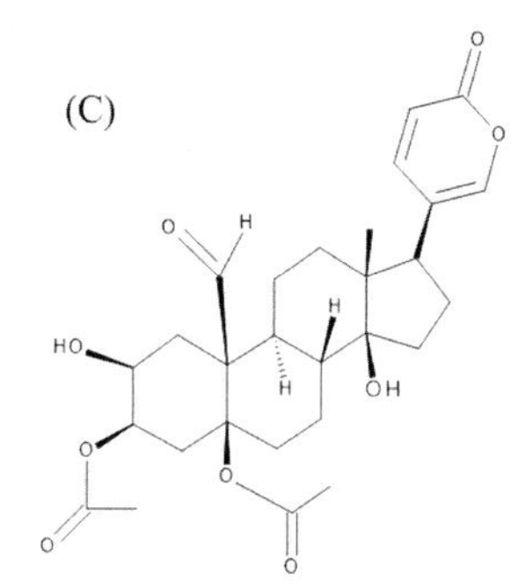

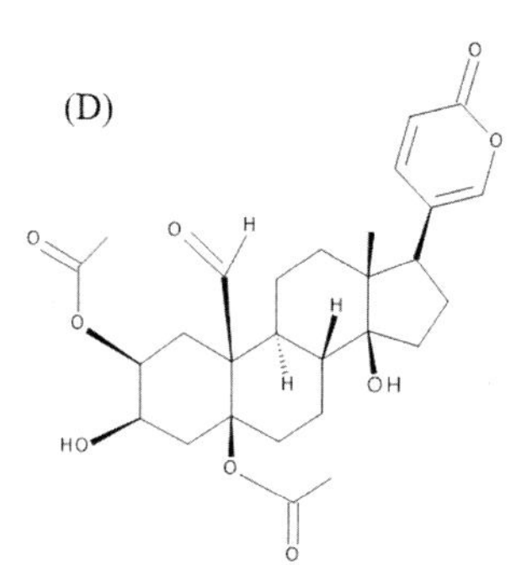

FIGURE 4.10 Cycloartane glycosides isolated from *Sutherlandia frutescens*, namely, (A) sutherlandioside A, (B) sutherlandioside B, (C) sutherlandioside C, and (D) sutherlandioside D.

methicillin-resistant *S. aureus*, *P. aeruginosa*; however, these showed no antimicrobial activity at the highest tested concentration of 20 μg/mL (Fu et al., 2008) (Fig. 4.10).

Discussion

Africa is amongst one of the richest continents with regards to species richness and plant biodiversity. It is estimated to have between 40,000–45,000 higher plant species of which approximately 10% are used medicinally for various diseases. This is largely due to the tropical and subtropical climatic regions, allowing these species to flourish (Agyare et al., 2016). Many southern African species have been identified through empirical evidence and traditional knowledge as potential wound-healing agents, largely based on their phytochemical composition. Various plant extracts and their secondary metabolites, which have antibacterial activity, have been identified as potential leads for the development of wound care products (Chingwaru, Bagar, Maroyi, Kapewangolo, & Chingwaru, 2019).

Of the plant species evaluated in this chapter, all the species were evaluated for their antibacterial activity against at least one wound-associated pathogen; however, *Aloe barberae, Aloe excelsa, Elephantorrhiza elephantina, Erythrina lysistemon, Galenia africana, Grewia occidentalis, Melianthus comosus, Plectranthus fruticosus, Polystichum pungens*, and *Sutherlandia frutescens* had no reports of in vitro or in vivo wound-healing activity.

South African *Aloe* species have shown potential for medicinal use in wound healing and preventative action against microbial growth against certain strains of bacteria associated with wound infections (Coopoosamy & Magwa, 2007; Ndhlala et al., 2009a; O'Brien et al., 2011). Research conducted on the antimicrobial activity of certain *Aloe* species has shown a general trend in which Gram-positive bacteria are more susceptible in comparison to Gram-negative bacteria (Pellizzoni, Ruzickova, Kalhotka, & Lucini, 1975). The mechanism of antibacterial action of the compounds isolated from the various *Aloe* species is hypothesized to be as a result of destabilization of the bacterial cell wall or oxidative stress. However, further studies are required to determine the specific mode of action for each compound. Although the anthraquinone compounds isolated from *A. ferox* and *A. excelsa* displayed similar antibacterial activity, the activity of the crude leaf extracts of *A. barberae, A. excelsa*, and *A. ferox* against *Bacillus subtilis, Staphylococcus aureus*, and *Staphylococcus epidermidis* varied (Coopoosamy & Magwa, 2007; Jia et al., 2008; Ndhlala et al., 2009a), which may be due to chemical and physiological differences amongst the *Aloe* species as well as their growth conditions which may alter the accumulation of metabolites within the plant tissue (Malmir, Serrano, & Silva, 2017). However, limited research has been conducted on the differences in chemical composition of anthraquinone compounds isolated from the above-mentioned *Aloe* species. Toxicity studies conducted using *A. ferox* have shown that leaf extracts prepared from *A. ferox* have shown no adverse dermatological effects and no in vitro or in vivo toxicity (Andersen, 2007). There are currently no toxicity studies that have been conducted using *A. excelsa;* however, *A. barberae* has been reported to be nonmutagenic.

Despite reports showing that anthraquinone compounds, isolated from the *Aloe* species, displayed promising antibacterial activity, there is limited research conducted on the activity and use of these compounds against *Pseudomonas aeruginosa*. Whilst the anthraquinone compounds isolated from the *Aloe* species have been tested for their antibacterial activity, there are no reports on the in vitro or in vivo wound-healing activity of aloe-emodin and chrysophanol. These compounds have been identified as the major constituents in Aloe species, however, other compounds could potentially be responsible for the activity, therefore, additional compound isolation and elucidation should be considered. Furthermore, confirmation of whether these are also major constituents present in *A. barberae* should be evaluated.

Compounds have been isolated from most of the species described in this work, with the exception of *Grewia occidentalis*. The compound kaempferitrin isolated from the aerial parts of *Urtica urens* has not been reported for its antibacterial or wound-healing activity and should be considered for further evaluation. Similarly, compounds isolated from *M. comosus* have not been evaluated for their antibacterial or wound-healing activity, even though various extract preparations of *M. comosus* have shown promising antibacterial activity against wound-associated pathogens. Furthermore, several other compounds which have been reported for their antibacterial activity should be further tested. Kaempferol, which was isolated from *E. elephantina*, should be reassessed for its antibacterial activity against wound-associated pathogens as there are contradicting reports describing its activity; however, it has been shown to have in vivo wound-healing activity. In addition, dihydrokaempferol, from *E. elephantina*, and kaurane from *P. fruticosus*, have not been fully evaluated for their antibacterial activity and should be considered for further evaluation. Compounds isolated from *Erythrina lysistemon* are of particular interest since they showed the lowest MIC values against wound-associated bacteria and are yet to be tested for their wound-healing activity. It should also be noted that although several of the plant extracts did not show significant antibacterial activity, these should also be evaluated for antibiofilm activity. The formation of biofilms is a mechanism pathogens have developed to form resistance, thereby making them difficult to treat (Famuyide, Aro, Fasina, Eloff, & McGaw, 2019).

Several models have been developed for the determination of both in vitro and in vivo wound-healing activity. In vitro models are advantageous since they require less ethical considerations while providing valuable information on biochemical and physiological processes that are induced by a test substance and also allow for testing of multiple test substances concurrently. These models also allow for the detection of cell–cell interactions as well as cell–matrix interaction and mimic events of cell migration during the wound-healing process. Single-cell systems, three-dimensional cell systems, multicell systems, and organ cultures are all used in in vitro wound-healing models. To accurately assess the wound-healing activity of new leads, dermal fibroblasts and human fibroblasts (primary or secondary cell lines) should be used, and it would therefore be beneficial to have cytotoxicity data on these cell lines. Models that assess the in vivo wound-healing activity of test substances include both artificial and tissue models and are highly dependent on which aspect of wound healing is being evaluated. Subcutaneous chambers/sponges and subcutaneous tubes can be used as in vivo models, as well as excision, incision, superficial dead space, and burn wounds. These can be used to directly determine the effects of a test substance on re-epithelialization, collagenation, neovascularization, and tensile breaking strength. Other animal models include the use of rabbit ear chambers, hamster cheek pouches, rabbit corneal pockets, and the chick chorioallantoic membranes (Agyare et al., 2016). The compounds and extracts that have not yet been tested using these models should be investigated to determine which stage of wound-healing process is enhanced after treatment with these agents.

No toxicity studies have been reported for *Aloe excelsa* and *Polystichum pungens*. Both safety and efficacy are important factors to consider when developing antibacterial and wound-healing agents. The antibacterial activity and wound-healing properties should outweigh any cytotoxic effects, and compounds that show activity against wound-associated bacteria and in vitro and in vivo wound-healing models need to be tested in the appropriate toxicity models. Often the extracts that have shown significant antibacterial or wound-healing activity are not always the same extracts that have been evaluated for their toxic potential. Therefore, extracts that have shown significant biological activity should also be evaluated for their toxicity, in order to determine the overall efficacy and therapeutic effect of a plant extract.

Conclusion

This chapter has demonstrated that medicinal plants with ethnobotanical usage for the treatment of wounds have been studied for their antibacterial activity against some pathogenic microbes linked to wound infections. A review of the literature has shown that many of the extracts lack data for both in vitro and in vivo activity of these extracts on wound-healing models. While some biomarker compounds have been isolated from these species, many have not yet been tested for either their antimicrobial activity or their wound-healing activity. While this chapter focuses on the antibacterial activity and wound-healing activity of the selected species, it is also imperative to highlight the lack of information on which stages of the wound-healing process the extracts and compounds act upon as well as the potential toxicity of the plant extracts or compounds. More studies can be conceived in order to determine the wound-healing mechanism of action for natural products, to emphasize the impact of natural products in treatment development. In addition, this chapter emphasizes the importance of using traditional knowledge as a tool for selecting plants for further biological evaluation as it may provide insight into potential lead candidates for the development of wound care and antibacterial products.

Index

(-)-catechin; (E)-2′,4′-dihydroxychalcone; 2-(3,4-dihydroxyphenyl) ethanol; 4-hydroxybenzoic acid; abyssinone V-4′ methyl ether; *Aloe barberae; Aloe excelsa; Aloe ferox*; aloe-emodin; aloin A; alpumisoflavone; anthraquinone; antibacterial; *Bacillus subtilis*; chrysophanol; cristacarpin; *Elephantorrhiza elephantina*; epidermal layer; erybraedin A; *Erythrina lysistemon*; eryzerin C; ethyl gallate; ethyl-1-O-β-D-galactopyranoside; *Galenia Africana*; gallic acid; *Grewia occidentalis*; dihydrokaempferol; infections; isolated compounds; kaempferol; lysisteisoflavone; *Melianthus comosus*; 16β-formyloxymelianthugenin; 2β-acetoxymelianthusigenin; 2β-hydroxy-3β,5β-di-O-acetylhellebrigenin; 2β-acetoxy-5β-O-acetylhellebrigenin; melianthusigenin; 16β-hydroxybersaldegenin 1,3,5-orthoacetate; microbes; pathogens; phaseollidin; *Plectranthus fruticosus*; 10(14)-aromadendrene-4β,15-diol; ent-3β-acetoxy-labda-8(17),12Z,14-trien-2α-ol; ent-12β-acetoxy-15β-hydroxykaur-16-en-19-oic acid; ent-12β-acetoxy-7β-hydroxykaur-16-en-19-oic acid; ent-7β-hydroxykaur-15-en-19-oic acid; ent-12β-acetoxy-17-oxokaur-15-en-19-oic acid; ent-7β-hydroxy-15β,16β-epoxykauran-19-oic acid; ent-labda-8(17),12Z,14-triene-2α,3β-diol; methyl ent-12β-hydroxykaur-16-en-19-oate; methyl ent-12β-acetoxy-7β-hydroxykaur-15-en-19-oate; ent-labda8 (17),12Z,14-triene-2α,3β-dibenzoate; ent-12β-acetoxy-15β-hydroxykaur-16-en-19-oate; methyl ent-12β-acetoxy-17-oxokaur-15-en-19-oate; *Polystichum pungens*; *Pseudomonas aeruginosa*; quercetin-3-O-β-D-glucopyranoside; South Africa; *Staphylococcus aureus*; *Staphylococcus epidermidis*; *Sutherlandia frutescens*; sutherlandioside A; sutherlandioside B; sutherlandioside C; sutherlandioside D; toxicity; traditional medicine; *Urtica urens*; wound healing

Glossary

IC_{50} The inhibitory concentration of a test substance required to inhibit 50% of a specific biological or biochemical function

LC_{50} The lethal concentration of a test substance required to kill 50% of a population in a given period of time
Minimum biocidal concentration the lowest concentration of test sample required to cause bacterial cell death
Minimum inhibitory concentration (MIC) the lowest concentration of test sample required to inhibit the visible growth of a microorganism
Nosocomial infection an infection often occurring within 48 h after admission to a healthcare facility, 3 days after being discharged or 30 days after an operation
Zone of inhibition (ZI) an area of growth media where no bacterial growth is observed, due to the presence of a test sample that impedes bacterial growth

References

Aaku, E., Dharani, S. P., Majinda, R. R. T., & Motswaiedi, M. S. (1998). Chemical and antimicrobial studies on *Elephantorrhiza elephantina*. *Fitoterapia (Milano)*, *69*(5), 464–465.
Adamczak, A., Ożarowski, M., & Karpiński, T. M. (2020). Antibacterial activity of some flavonoids and organic acids widely distributed in plants. *Journal of Clinical Medicine*, *9*, 109. Available from https://doi.org/10.3390/jcm9010109.
Agyare, C., Boakye, Y. D., Bekoe, E. O., Hensel, A., Dapaah, S. O., & Appiah, T. (2016). Review: African medicinal plants with wound healing properties. *Journal of Ethnopharmacology*, *177*, 85–100. Available from https://doi.org/10.1016/j.jep.2015.11.008.
Al Doghaither, H. A., Omar, U. M., Rahimulddin, S. A., & Al-Ghafari, A. B. (2016). Cytotoxic effects of aqueous extracts of *Urtica urens* on most common cancer types in Saudi Arabia. *Journal of Biological Sciences*, *16*(6), 242–246. Available from https://doi.org/10.3923/jbs.2016.242.246.
Al-Bakri, A. G., & Afifi, F. U. (2007). Evaluation of antimicrobial activity of selected plant extracts by rapid XTT colorimetry and bacterial enumeration. *Journal of Microbiological Methods*, *68*, 19–25. Available from https://doi.org/10.1016/j.mimet.2006.05.013.
Aldridge, P. (2015). Complications of wound healing: Causes and prevention. *Companion Animal*, *20*, 453–459. Available from https://doi.org/10.12968/coan.2015.20.8.453.
Almine, J. F., Wise, S. G., & Weiss, A. S. (2012). Elastin signaling in wound repair. *Birth Defects Research Part C: Embryo Today: Reviews*, *96*, 248–257. Available from https://doi.org/10.1002/bdrc.21016.
Amiel, E., Lovewell, R. R., O'Toole, G. A., Hogan, D. A., & Berwin, B. (2010). *Pseudomonas aeruginosa* evasion of phagocytosis is mediated by loss of swimming motility and is independent of flagellum expression. *Infection and Immunity*, *78*, 2937–2945. Available from https://doi.org/10.1128/IAI.00144-10.
Andersen, F. A. (2007). Final report on the safety assessment of *Aloe andongensis* extract, *Aloe andongensis* leaf juice, *Aloe arborescens* leaf extract, *Aloe arborescens* leaf juice, *Aloe arborescens* leaf protoplasts, *Aloe barbadensis* flower extract, *Aloe barbadensis* leaf, *Aloe barbadensis* leaf extract, *Aloe barbadensis* leaf juice, *Aloe barbadensis* leaf polysaccharides, *Aloe barbadensis* leaf water, *Aloe ferox* leaf extract, *Aloe ferox* leaf juice, and *Aloe ferox* leaf juice extract. *International Journal of Toxicology*, *26*, 1–50. Available from https://doi.org/10.1080/10915810701351186.
Angeh, I. E. (2006). *Potentising and application of an extract of Melianthus comosus against plant fungal pathogens Declaration* (p. 105) University of Pretoria. Available from https://repository.up.ac.za/handle/2263/30601.
Ashok Kumar, J., Pullakandam, N., Lakshmana Prabu, S., & Gopal, V. (2010). Transdermal drug delivery system: An overview. *International Journal of Pharmaceutical Sciences Review and Research*, *3*. Available from http://www.globalresearchonline.net.
Aswal, B. S. (1972). A note on *Urtica urens* Linn. (Urticaceae). *Nelumbo*, *14*, 169–170.
Bedane, K. G., Brieger, L., Strohmann, C., Seo, E. J., Efferth, T., & Spiteller, M. (2020). Cytotoxic bufadienolides from the leaves of a medicinal plant *Melianthus comosus* collected in South Africa. *Bioorganic Chemistry*, *102*, 104102. Available from https://doi.org/10.1016/j.bioorg.2020.104102.
Bessa, L. J., Fazii, P., Di Giulio, M., & Cellini, L. (2015). Bacterial isolates from infected wounds and their antibiotic susceptibility pattern: Some remarks about wound infection. *International Wound Journal*, *12*, 47–52. Available from https://doi.org/10.1111/iwj.12049.
Borges, A., Ferreira, C., Saavedra, M. J., & Simões, M. (2013). Antibacterial activity and mode of action of ferulic and gallic acids against pathogenic bacteria. *Microbial Drug Resistance*, *19*, 256–265. Available from https://doi.org/10.1089/mdr.2012.0244.

Builders, P. F., & Builders, M. I. (2016). *Wound care: Traditional African medicine approach. Worldwide wound healing - Innovation in natural and conventional methods.* InTech. Available from https://doi.org/10.5772/65521.

Carvalho, A. R., Costa, G., Figueirinha, A., Liberal, J., Prior, J. A. V., Lopes, M. C., ... Batista, M. T. (2017). *Urtica* spp.: Phenolic composition, safety, antioxidant and anti-inflammatory activities. *Food Research International, 99*, 485–494. Available from https://doi.org/10.1016/j.foodres.2017.06.008.

Celestino, V. R. L., Maranhão, H. M. L., Vasconcelos, C. F. B., Lima, C. R., Medeiros, G. C. R., Araújo, A. V., & Wanderley, A. G. (2013). Acute toxicity and laxative activity of *Aloe ferox* resin. *Brazilian Journal of Pharmacognosy, 23*(2), 279–283. Available from https://doi.org/10.1590/S0102-695X2013005000009.

Chamorro, G., Salazar, M., Fournier, G., & Pages, N. (1991). Anti-implantation effects of various extracts of *Plectranthus fruticosus* on pregnant rats [1]. *Planta Medica, 57*(1), 81. Available from https://ipn.elsevierpure.com/en/publications/anti-implantation-effects-of-various-extracts-of-plectranthus-fru.

Chen, L., Xu, M., Gong, Z., Zonyane, S., Xu, S., & Makunga, N. P. (2018). Comparative cardio and developmental toxicity induced by the popular medicinal extract of *Sutherlandia frutescens* (L.) R.Br. detected using a zebrafish Tuebingen embryo model. *BMC Complementary and Alternative Medicine, 18*(1), 273. Available from https://doi.org/10.1186/s12906-018-2303-9.

Chen, W., Van Wyk, B. E., Vermaak, I., & Viljoen, A. M. (2012). Cape aloes - A review of the phytochemistry, pharmacology and commercialisation of *Aloe ferox*. *Phytochemistry Letters, 5*, 1–12. Available from https://doi.org/10.1016/j.phytol.2011.09.001.

Chingwaru, C., Bagar, T., Maroyi, A., Kapewangolo, P. T., & Chingwaru, W. (2019). Wound healing potential of selected Southern African medicinal plants: A review. *Journal of Herbal Medicine, 17–18*, 100263. Available from https://doi.org/10.1016/j.hermed.2019.100263.

Cho, J.-Y., Moon, J.-H., Seong, K.-Y., & Park, K.-H. (1998). Antimicrobial activity of 4-hydroxybenzoic acid and trans 4-hydroxycinnamic acid isolated and identified from rice hull. *Biotechnology and Biochemistry, 62*, 2273–2276. Available from https://doi.org/10.1271/bbb.62.2273.

Christensen, G. D., Bisno, A. L., Parisi, J. T., McLaughlin, B., Hester, M. G., & Luther, R. W. (1982). Nosocomial septicemia due to multiply antibiotic-resistant *Staphylococcus epidermidis*. *Annals of Internal Medicine, 96*, 1–10. Available from https://doi.org/10.7326/0003-4819-96-1-1.

Churpek, M. M., Snyder, A., Han, X., Sokol, S., Pettit, N., Howell, M. D., & Edelson, D. P. (2017). Quick sepsis-related organ failure assessment, systemic inflammatory response syndrome, and early warning scores for detecting clinical deterioration in infected patients outside theintensive care unit. *American Journal of Respiratory and Critical Care Medicine, 195*, 906–911. Available from https://doi.org/10.1164/rccm.201604-0854OC.

Cohen, P. R., & Kurzrock, R. (2004). Community-acquired methicillin-resistant *Staphylococcus aureus* skin infection: An emerging clinical problem. *Journal of the American Academy of Dermatology, 50*, 277–280. Available from https://doi.org/10.1016/j.jaad.2003.06.005.

Coopoosamy, R. M., & Magwa, M. L. (2006). Antibacterial activity of aloe emodin and aloin A isolated from *Aloe excelsa*. *African Journal of Biotechnology, 5*, 1092–1094. Available from http://www.academicjournals.org/AJB.

Coopoosamy, R. M., & Magwa, M. L. (2007). Traditional use, antibacterial activity and antifungal activity of crude extract of *Aloe excelsa*. *African Journal of Biotechnology, 6*(20), 2406–2410. Available from http://www.academicjournals.org/AJB.

Coopoosamy, R. M., & Naidoo, K. K. (2012). A comparative study of three aloe species used to treat skin diseases in South African rural communities. *The Journal of Alternative and Complementary Medicine, 19*(5), 425–428. Available from https://doi.org/10.1089/acm.2012.0087.

Cunha Filho, G. A., Schwartz, C. A., Resck, I. S., Murta, M. M., Lemos, S. S., Castro, M. S., ... Schwartz, E. F. (2005). Antimicrobial activity of the bufadienolides marinobufagin and telocinobufagin isolated as major components from skin secretion of the toad *Bufo rubescens*. *Toxicon, 45*, 777–782. Available from https://doi.org/10.1016/j.toxicon.2005.01.017.

Dube, P., Meyer, S., Madiehe, A., & Meyer, M. (2020). Antibacterial activity of biogenic silver and gold nanoparticles synthesized from *Salvia africana*-lutea and *Sutherlandia frutescens*. *Nanotechnology, 31*(50), 505607. Available from https://doi.org/10.1088/1361-6528/abb6a8.

Elbagory, A. M., Meyer, M., Cupido, C. N., & Hussein, A. A. (2017). Inhibition of bacteria associated with wound infection by biocompatible green synthesized gold nanoparticles from South African plant extracts. *Nanomaterials, 7*(417), 1–22. Available from https://doi.org/10.3390/nano7120417.

Escobar-Chavez, J., Diaz-Torres, R., Rodriguez-Cruz, I. M., Domnguez-Delgado., Sampere-Morales., Angeles-Anguiano., & Melgoza-Contreras. (2012). Nanocarriers for transdermal drug delivery. *Research and Reports in Transdermal Drug Delivery, 2021*, 3–17. Available from https://doi.org/10.2147/rrtd.s32621.

Famuyide, I. M., Aro, A. O., Fasina, F. O., Eloff, J. N., & McGaw, L. J. (2019). Antibacterial and antibiofilm activity of acetone leaf extracts of nine under-investigated South African Eugenia and Syzygium (Myrtaceae) species and their selectivity indices. *BMC Complementary and Alternative Medicine, 19*, 1–13. Available from https://doi.org/10.1186/s12906-019-2547-z.

Foden, W., & Potter, L. (2005). *Grewia occidentalis* L. var. occidentalis. National Assessment: Red list of South African Plants. Retrieved from http://redlist.sanbi.org/species.php?species = 933-26. (Accessed 27 October 2020).

Foden, W., & Potter, L. (2005). *Polystichum pungens* (Kaulf.) C. Presl. National Assessment: Red List of South African Plants. Retrieved from http://redlist.sanbi.org/species.php?species = 3734-8. (Accessed 27 October 2020).

Forbes, B. A., & Schaberg, D. R. (1983). Transfer of resistance plasmids from *Staphylococcus epidermidis* to *Staphylococcus aureus*: Evidence for conjugative exchange of resistance. *Journal of Bacteriology, 153*(2), 627–634.

Fox, L. T., Mazumder, A., Dwivedi, A., Gerber, M., du Plessis, J., & Hamman, J. H. (2017). In vitro wound healing and cytotoxic activity of the gel and whole-leaf materials from selected aloe species. *Journal of Ethnopharmacology, 200*, 1–7. Available from https://doi.org/10.1016/j.jep.2017.02.017.

Fu, X., Li, X.-C., Smillie, T. J., Carvalho, P., Mabusela, W., Syce, J., … Khan, I. A. (2008). Cycloartane glycosides from *Sutherlandia frutescens*. *Jpurnal of Natural Products, 71*, 1749–1753. Available from https://doi.org/10.1021/np800328r.

Gaspar-Marques, C., Fá Tima Simões, M., & Rodríguez, B. (2004). Further labdane and kaurane diterpenoids and other constituents from *Plectranthus fruticosus*. *Journal of Natural Products, 67*, 614–621. Available from https://doi.org/10.1021/np030490j.

Girard, A. L., Teferra, T., & Awika, J. M. (2019). Effects of condensed vs hydrolysable tannins on gluten film strength and stability. *Food Hydrocolloids, 89*, 36–43. Available from https://doi.org/10.1016/j.foodhyd.2018.10.018.

Grierson, D. S., & Afolayan, A. J. (1999). Antibacterial activity of some indigenous plants used for the treatment of wounds in the Eastern Cape, South Africa. *Journal of Ethnopharmacology, 66*, 103–106.

Harris, S. (2004). *Melianthus comosus (CC BY 2.0)*. Retrieved from http://pza.sanbi.org/melianthus-comosus. (Accessed 29 October 2020).

Harrower, A. (2010). *Plectranthus fruticosus "Liana,"*. Retrieved from http://opus.sanbi.org/bitstream/20.500.12143/3730/1/Plectranthusfruticosus%27Liana%27_PlantzAfrica.pdf. (Accessed 30 October 2020).

Hashemi, S. A., Madani, S. A., & Abediankenari, S. (2015). The review on properties of aloe vera in healing of cutaneous wounds. *BioMed Research International, 2015*, 714216. Available from https://doi.org/10.1155/2015/714216.

Hashimoto, T., Hayakawa, K., Mezaki, K., Kutsuna, S., Takeshita, N., Yamamoto, K., … Ohmagari, N. (2017). Bacteremia due to *Bacillus subtilis*: A case report and clinical evaluation of 10 cases. *Kansenshogaku Zasshi. the Journal of the Japanese Association for Infectious Diseases, 91*, 151–154. Available from http://europepmc.org/abstract/MED/30277700.

Heyman, H. M., Hussein, A. A., Meyer, J. J. M., & Lall, N. (2009). Antibacterial activity of South African medicinal plants against methicillin resistant *Staphylococcus aureus*. *Pharmaceutical Biology, 47*(1), 67–71. Available from https://doi.org/10.1080/13880200802434096.

Hepplewhite, M.D., & Witkoppen Wildflower Nursery. (2018). *Grewia occidentalis*. Retrieved from Available from https://witkoppenwildflower.co.za/grewia-occidentalis/#:~:text = Poisonous%3A%20Not%20poisonous., walking%20sticks%20and%20assegai%20handles.

Jia, Y., Zhao, G., & Jia, J. (2008). Preliminary evaluation: The effects of *Aloe ferox* Miller and *Aloe arborescens* Miller on wound healing. *Journal of Ethnopharmacology, 120*(2), 181–189. Available from https://doi.org/10.1016/j.jep.2008.08.008.

Jodamus, N., & Notten, A. (2002). *Plectranthus fruticosus L'Hérit. "James"*. Retrieved from http://pza.sanbi.org/plectranthus-fruticosus-james#:~:text = This%20plectranthus%20is%20a%20particularly,succulent%20leaves%20and%20pink%20flowers. (Accessed 30 October 2020).

Johnson, Q., Syce, J., Nell, H., Rudeen, K., & Folk, W. R. (2007). A randomized, double-blind, placebo-controlled trial of *Lessertia frutescens* in healthy adults. *PLoS Clinical Trials, 2*(4), e16. Available from https://doi.org/10.1371/journal.pctr.0020016.

Kambizi, L., Sultana, N., Afolayan, A. J., & Afolayan, A. J. (2008). A plant traditionally used for the treatment of sexually transmitted infections in the Eastern Cape. *Pharmaceutical Biology, 42*, 636–639. Available from https://doi.org/10.1080/13880200490902581.

Katerere, D. R., & Eloff, J. N. (2005). Antibacterial and antioxidant activity of *Sutherlandia frutescens* (Fabaceae), a reputed Anti-HIV/AIDS phytomedicine. *Phytotherapy Research*, *19*(9), 779–781. Available from https://doi.org/10.1002/ptr.1719.

Kelmanson, J. E., Jäger, A. K., & Van Staden, J. (2000). Zulu medicinal plants with antibacterial activity. *Journal of Ethnopharmacology*, *69*(3), 241–246, Vol. Available from http://www.elsevier.com/locate/jethpharm.

Kudumela, R. G., McGaw, L. J., & Masoko, P. (2018). Antibacterial interactions, anti-inflammatory and cytotoxic effects of four medicinal plant species. *BMC Complementary and Alternative Medicine*, *18*(1), 199. Available from https://doi.org/10.1186/s12906-018-2264-z.

Li, J.-F., Duan, H.-F., Wu, C.-T., Zhang, D.-J., Deng, Y., Yin, H.-L., ... Wang, Y.-L. (2013). HGF accelerates wound healing by promoting the dedifferentiation of epidermal cells through B-1-Integrin/ILK pathway. *BioMed Research International*, *2013*, 470418. Available from https://doi.org/10.1155/2013/470418.

Liu, H., Mou, Y., Zhao, J., Wang, J., Zhou, L., Wang, M., ... Yang, F. (2010). Flavonoids from *Halostachys caspica* and their antimicrobial and antioxidant activities. *Molecules (Basel, Switzerland)*, *15*, 7933–7945. Available from https://doi.org/10.3390/molecules15117933.

Liu, J., Du, C., Beaman, H. T., Beth, M., & Monroe, B. (2020). Characterization of phenolic acid antimicrobial and antioxidant structure-property relationships. *Pharmaceutics*, *12*, 1–17. Available from https://doi.org/10.3390/pharmaceutics12050419.

Lotz, M. E. (2019). The burden of wounds in a resource-constrained tertiary hospital: A cross-sectional study. *Wound Healing Southern Africa*, *12*, 29–33. Available from https://journals.co.za/content/journal/10520/EJC-17b0876853.

Lukhoba, C. W., Simmonds, M. S. J., & Paton, A. J. (2006). Plectranthus: A review of ethnobotanical uses. *Journal of Ethnopharmacology*, *103*, 1–24. Available from https://doi.org/10.1016/j.jep.2005.09.011.

Lynch, S. E., Colvin, R. B., & Antoniades, H. N. (1989). Growth factors in wound healing. Single and synergistic effects on partial thickness porcine skin wounds. *Journal of Clinical Investigation*, *84*, 640–646. Available from https://doi.org/10.1172/JCI114210.

Mabona, U. (2013). Antimicrobial activity of southern African medicinal plants with dermatological relevance. *Journal of Ethnopharmacology*, *148*(1), 45–55.

Mabona, U., & VanVuuren, S. F. (2013). Southern African medicinal plants used to treat skin diseases. *South African Journal of Botany*, *87*, 175–193. Available from https://doi.org/10.1016/j.sajb.2013.04.002.

Mabona, U., Viljoen, A., Shikanga, E., Marston, A., & Van Vuuren, S. (2013). Antimicrobial activity of southern African medicinal plants with dermatological relevance: From an ethnopharmacological screening approach, to combination studies and the isolation of a bioactive compound. *Journal of Ethnopharmacology Journal*, *148*, 45–55. Available from https://doi.org/10.1016/j.jep.2013.03.056.

Maistry, K. (2003). *The antimicrobial properties and chemical composition of leaf essential oils of indigenous Plectranthus (Lamiaceae) species* (p. 71).

Malmir, M., Serrano, R., & Silva, O. (2017). *Anthraquinones as potential antimicrobial agents - A review*. Retrieved from https://www.researchgate.net/profile/Maryam_Malmir/publication/319620317_Anthraquinones_as_potential_antimicrobial_agents-A_review/links/59b567a4458515a5b4939841/Anthraquinones-as-potential-antimicrobial-agents-A-review.pdf.

Mander, M., Ntuli, L., Diederichs, N., & Mavundla, K. (2019). *Economics of the traditional medicine trade in South Africa. South African health review, health systems trust*. Westville.

Manzuoerh, R., Farahpour, M. R., Oryan, A., & Sonboli, A. (2019). Effectiveness of topical administration of *Anethum graveolens* essential oil on MRSA-infected wounds. *Biomedicine and Pharmacotherapy*, *109*, 1650–1658. Available from https://doi.org/10.1016/j.biopha.2018.10.117.

Maphosa, V., Masika, P. J., & Moyo, B. (2010). Toxicity evaluation of the aqueous extract of the rhizome of *Elephantorrhiza elephantina* (Burch.) Skeels. (Fabaceae), in rats. *Food and Chemical Toxicology*, *48*(1), 196–201. Available from https://doi.org/10.1016/j.fct.2009.09.040.

Maroyi, A. (2017). *Elephantorrhiza elephantina*: Traditional uses, phytochemistry, and pharmacology of an important medicinal plant species in Southern Africa. *Evidence-Based Complementary and Alternative Medicine*, *2017*, 6403905. Available from https://doi.org/10.1155/2017/6403905.

Mathabe, M. C., Nikolova, R. V., Lall, N., & Nyazema, N. Z. (2006). Antibacterial activities of medicinal plants used for the treatment of diarrhoea in Limpopo Province, South Africa. *Journal of Ethnopharmacology*, *105*(1–2), 286–293. Available from https://doi.org/10.1016/j.jep.2006.01.029.

Mazurova, J., Kukla, R., Rozkot, M., Lustykova, A., Slehova, E., Sleha, R., ... Opletal, L. (2015). Use of natural substances for boar semen decontamination. *Veterinarni Medicina*, *60*, 235–247. Available from https://doi.org/10.17221/8175-VETMED.

Mbambezeli, G., & Notten, A. (2002). *Erythrina lysistemon*. Retrieved from http://pza.sanbi.org/erythrina-lysistemon. (Accessed 28 October 2020).

Mbambezeli, G., & Notten, A. (2002). *Erythrina lysistemon*. Retrieved from http://pza.sanbi.org/erythrina-lysistemon.

McGaw, L. J., & Eloff, J. N. (2005). Screening of 16 poisonous plants for antibacterial, anthelmintic and cytotoxic activity in vitro. *South African Journal of Botany*, *71*(3–4), 302–306. Available from https://doi.org/10.1016/S0254-6299(15)30102-2.

Micek, S. T., Lloyd, A. E., Ritchie, D. J., Reichley, R. M., Fraser, V. J., & Kollef, M. H. (2005). *Pseudomonas aeruginosa* bloodstream infection: Importance of appropriate initial antimicrobial treatment. *Antimicrobial Agents and Chemotherapy*, *49*, 1306–1311. Available from https://doi.org/10.1128/AAC.49.4.1306-1311.2005.

Ming, D., Wang, D., Cao, F., Xiang, H., Mu, D., Cao, J., ... Wang, T. (2017). Kaempferol inhibits the primary attachment phase of biofilm formation in *Staphylococcus aureus*. *Frontiers in Microbiology*, *8*, 2263. Available from https://www.frontiersin.org/article/10.3389/fmicb.2017.02263.

Moriyama, M., Moriyama, H., Uda, J., Kubo, H., Nakajima, Y., Goto, A., ... Hayakawa, T. (2016). Beneficial effects of the genus aloe on wound healing, cell proliferation, and differentiation of epidermal keratinocytes. *PLoS One*, *11*(10), e0164799. Available from https://doi.org/10.1371/journal.pone.0164799.

Mpofu, S. J., Msagati, T. A. M., & Krause, R. W. M. (2014). Cytotoxicity, phytochemical analysis and antioxidant activity of crude extracts from rhizomes of *Elephantorrhiza elephantina* and *Pentanisia prunelloides*. *African Journal of Traditional, Complementary, and Alternative Medicines: AJTCAM/African Networks on Ethnomedicines*, *11*(1), 34–52.

Mukandiwa, L., Naidoo, V., & Eloff, J. N. (2012). In vitro antibacterial activity of seven plants used traditionally to treat wound myiasis in animals in Southern Africa. *Journal of Medicinal Plants Research*, *6*(27), 4379–4388. Available from https://doi.org/10.5897/JMPR11.1130.

Mukanganyama, S., Ntumy, A. N., Maher, F., Muzila, M., & Andrae-Marobela, K. (2011). Anti-infective properties of selected medicinal plants from Botswana. *The African Journal of Plant Science and Biotechnology*, *5*(1), 1–7.

Mwale, M., & Masika, P. J. (2012). Scientific Research and Essay. *Toxicological studies on the leaf extract of Aloe ferox Mill. (Aloaceae)*, *7*(15), 1605–1613. Available from https://doi.org/10.5897/SRE09.334.

Mzid, M., Khedir, B., Salem, B., Regaieg, W., & Rebai, T. (2017). Antioxidant and antimicrobial activities of ethanol and aqueous extracts from *Urtica urens* antioxidant and antimicrobial activities of ethanol and aqueous extracts from *Urtica urens*. *Pharmaceutical Biology*, *55*(1), 775–781. Available from https://doi.org/10.1080/13880209.2016.1275025.

Nciki, S., Vuuren, S., Van Eyk, A., & De Wet, H. (2016). Plants used to treat skin diseases in northern Maputaland, South Africa: Antimicrobial activity and in vitro permeability studies. *Pharmaceutical Biology*, *54*(11), 2420–2436. Available from https://doi.org/10.3109/13880209.2016.1158287.

Ndhlala, A. R., Amoo, S. O., Stafford, G. I., Finnie, J. F., & Van Staden, J. (2009a). Antimicrobial, anti-inflammatory and mutagenic investigation of the South African tree aloe (*Aloe barberae*). *Journal of Ethnopharmacology*, *124*(3), 404–408. Available from https://doi.org/10.1016/j.jep.2009.05.037.

Ndhlala, A. R., Amoo, S. O., Stafford, G. I., Finnie, J. F., & Van Staden, J. (2009b). Pharmacological activities of the South African tree aloe (*Aloe barberae*). *South African Journal of Botany*, *75*, 413. Available from https://doi.org/10.1016/j.sajb.2009.02.085.

Ng'uni, T., Klaasen, J. A., & Fielding, B. C. (2018). Acute toxicity studies of the South African medicinal plant *Galenia africana*. *Toxicology Reports*, *5*, 813–818. Available from https://doi.org/10.1016/j.toxrep.2018.08.008.

Nsele, N.W. (2012). *Assessment of the antibacterial activity of Artemisia afra, Erythrina lysistemon and Psidium guajava*. Retrieved from https://openscholar.dut.ac.za/handle/10321/938.

Ntuli, S. S. B. N., Gelderblom, W. C. A., & Katerere, D. R. (2018). The mutagenic and antimutagenic activity of *Sutherlandia frutescens* extracts and marker compounds. *BMC Complementary and Alternative Medicine*, *18*(1), 1–10. Available from https://doi.org/10.1186/s12906-018-2159-z.

Obi, C. L., Potgieter, N., Bessong, P. O., Masebe, T., Mathebula, H., & Molobela, P. (2003). In vitro antibacterial activity of Venda medicinal plants. *South African Journal of Botany*, *69*(2), 199–203. Available from https://doi.org/10.1016/S0254-6299(15)30346-X.

Otto, M. (2009). *Staphylococcus epidermidis*—The "accidental" pathogen. *Nature Reviews Microbiology*, *7*, 555–567. Available from https://doi.org/10.1038/nrmicro2182.

Otto, M. (2014). *Staphylococcus aureus* toxins. *Current Opinion in Microbiology, 17*, 32–37. Available from https://doi.org/10.1016/j.mib.2013.11.004.

O'Brien, C., Van Wyk, B. E., & Van Heerden, F. R. (2011). Physical and chemical characteristics of *Aloe ferox* leaf gel. *South African Journal of Botany, 77*, 988–995. Available from https://doi.org/10.1016/j.sajb.2011.08.004.

Özay, Y., Güzel, S., Yumrutaş, Ö., Pehlivanoğlu, B., Erdoğdu, İ. H., Yildirim, Z., ... Darcan, S. (2019). Wound healing effect of kaempferol in diabetic and nondiabetic rats. *Journal of Surgical Research, 233*, 284–296. Available from https://doi.org/10.1016/j.jss.2018.08.009.

Pellizzoni, M., Ruzickova, G., Kalhotka, L., & Lucini, L. (1975). Antimicrobial activity of different *Aloe barbadensis* Mill. and *Aloe arborescens* Mill. leaf fractions. *Journal of Medicinal Plants Research, 6*(10), 1975–1981. Available from https://doi.org/10.5897/JMPR011.1680.

Pillay, C. C. N., Jäger, A. K., Mulholland, D. A., & Van Staden, J. (2001). Cyclooxygenase inhibiting and antibacterial activities of South African Erythrina species. *Journal of Ethnopharmacology, 74*(3), 231–237. Available from https://doi.org/10.1016/S0378-8741(00)00366-4.

Pitso, F. S., & Lebese, M. R. (2014). Traditional uses of wild edible plants in arid areas of South Africa. *Journal of Human Ecology, 48*, 23–31. Available from https://doi.org/10.1080/09709274.2014.11906771.

Prausnitz, M. R., & Langer, R. (2008). Transdermal drug delivery. *Nature Biotechnology, 26*, 1261–1268. Available from https://doi.org/10.1038/nbt.1504.

Rahal, J. J. (2006). Novel antibiotic combinations against infections with almost completely resistant *Pseudomonas aeruginosa* and *Acinetobacter* species. *Clinical Infectious Diseases, 43*, S95–S104. Available from https://academic.oup.com/cid/article/43/Supplement_2/S95/333943.

Rajput, P., Choudhary, M., & Sharma, R. A. (2019). Comparing antibacterial potential and phytochemical constituents of two species of genus *Urtica*. *International Journal of Life science and Pharma Research, 9*(4), 90–102. Available from https://doi.org/10.22376/ijpbs/lpr.2019.9.4.P90-102.

Rodriguez, C., Ibáñez, R., Rollins-Smith, L. A., Gutiérrez, M., & Durant-Archibold, A. A. (2020). Antimicrobial Secretions of Toads (Anura, Bufonidae): Bioactive Extracts and Isolated Compounds against Human Pathogens. *Antibiotics, 9*, 843. Available from https://doi.org/10.3390/antibiotics9120843.

Roux, K. (2004). *Polystichum*. Retrieved from http://pza.sanbi.org/polystichum. (Accessed 27 October 2020).

Rupp, M. E., Fey, P. D., Heilmann, C., & Götz, F. (2001). Characterization of the importance of *Staphylococcus epidermidis* autolysin and polysaccharide intercellular adhesin in the pathogenesis of intravascular catheter-associated infection in a rat model. *The Journal of Infectious Diseases, 183*, 1038–1042. Available from https://academic.oup.com/jid/article/183/7/1038/859058.

Sadgrove, N. J., Oliveira, T. B., Khumalo, G. P., van Vuuren, S. F., & van Wyk, B.-E. (2020). Antimicrobial isoflavones and derivatives from *Erythrina* (Fabaceae): Structure activity perspective (Sar & Qsar) on experimental and mined values against *Staphylococcus aureus*. *Antibiotics, 9*(5), 223. Available from https://doi.org/10.3390/antibiotics9050223.

Saidin, N. M., Anuar, N. K., & Affandi, M. M. R. M. M. (2018). Roles of polysaccharides in transdermal drug delivery system and future prospects. *Journal of Applied Pharmaceutical Science, 8*, 141–157. Available from https://doi.org/10.7324/JAPS.2018.8320.

Saleh, F., Kheirandish, F., Azizi, H., & Azizi, M. (2014). Molecular diagnosis and characterization of *Bacillus subtilis* isolated from burn wound in Iran. *Research in Molecular Medicine., 2*, 40. Available from https://doi.org/10.18869/acadpub.rmm.2.2.40.

Sato, Y., Oketani, H., Singyouchi, K., Ohtsubo, T., Kihara, M., Shibata, H., & Higuti, T. (1997). Extraction and purification of effective antimicrobial constituents of *Terminalia chebula* RETS against methicillin-resistant *Staphylococcus aureus*. *Biological & Pharmaceutical Bulletin, 20*, 401–404. Available from https://doi.org/10.1248/bpb.20.401.

Schneider, L. A., Korber, A., Grabbe, S., & Dissemond, J. (2007). Influence of pH on wound-healing: A new perspective for wound-therapy? *Archives of Dermatological Research, 298*, 413–420. Available from https://doi.org/10.1007/s00403-006-0713-x.

Shajari, G., Khorshidi, A., & Moosavi, G. (2002). Vancomycin resistance in *Staphylococcus aureus* strains short communication. *Archives of Razi Institute, 54*(1), 107–110. Available from http://www.SID.ir.

Singal, A., & Grover, C. (2016). *Normal flora of the skin. Comprehensive approach to infections in dermatology*. New Delhi, India: Jaypee Brothers Medical Publishers (P) Ltd. Available from https://books.google.co.za/books?hl = en&lr = &id = QbGfCwAAQBAJ&oi = fnd&pg = PR1&dq = Comprehensive + Approach + to + Infections +

in + Dermatology&ots = j0s-hhQ19J&sig = _dHIVGu1tOGWOd_VGT8Let1iu9g&redir_esc = y#v = onepage&q = ComprehensiveApproach to Infections in Dermatology&f = false.

Stapleton, P. D., Shah, S., Anderson, J. C., Hara, Y., Hamilton-Miller, J. M. T., & Taylor, P. W. (2004). Modulation of β-lactam resistance in *Staphylococcus aureus* by catechins and gallates. *International Journal of Antimicrobial Agents, 23*, 462–467. Available from https://doi.org/10.1016/j.ijantimicag.2003.09.027.

Stefanowicz-Hajduk, J., Hering, A., Gucwa, M., Hałasa, R., Soluch, A., Kowalczyk, M., ... Ochocka, R. (2020). Biological activities of leaf extracts from selected Kalanchoe species and their relationship with bufadienolides content. *Pharmaceutical Biology, 58*, 732–740. Available from https://doi.org/10.1080/13880209.2020.1795208.

Succulent Plant Site. (2004). *Aloe barberae (Aloe bainesii)*. Retrieved from https://www.succulents.co.za/aloes/tree-aloes/aloe-barberae.php. (Accessed 14 December 2020).

Tajuddeen, N., Sani Sallau, M., Muhammad Musa, A., James Habila, D., & Muhammad Yahaya, S. (2014). Flavonoids with antimicrobial activity from the stem bark of *Commiphora pedunculata* (Kotschy & Peyr.) Engl. *Natural Product Research, 28*, 1915–1918. Available from https://doi.org/10.1080/14786419.2014.947488.

Tanaka, H., Sato, M., Fujiwara, S., Hirata, M., Etoh, H., & Takeuchi, H. (2002). Antibacterial activity of isoflavonoids isolated from *Erythrina variegata* against methicillin-resistant *Staphylococcus aureus*. *Letters in Applied Microbiology, 35*, 494–498. Available from https://doi.org/10.1046/j.1472-765X.2002.01222.x.

Taqa, G. A., Mustafa, E. A., & Al-Haliem, S. M. (2014). Evaluation of antibacterial and efficacy of plant extract (*Urtica urens*) on skin wound healing in rabbits. *International Journal of Enhanced Research in Science Technology & Engineering, 3*(1), 64–70.

Thibane, V. S., Ndhlala, A. R., Abdelgadir, H. A., Finnie, J. F., & Van Staden, J. (2019). The cosmetic potential of plants from the Eastern Cape Province traditionally used for skincare and beauty. *South African Journal of Botany, 122*, 475–483. Available from https://doi.org/10.1016/j.sajb.2018.05.003.

Ticha, L.A. (2015). *Phytochemical and biological activity studies of Galenia africana*. Retrieved from https://doclinika.ru/wp-content/uploads/2015/10/Ticha.pdf. (Accessed 14 December 2020).

Turnbull, P. C. B. (1996). *Medical microbiology* (4th (ed.)). Galveston: University of Texas Medical Branch.

Turner, S. (2008). *Grewia occidentalis*. Retrieved from http://pza.sanbi.org/grewia-occidentalis. (Accessed 27 October 2020).

Turner, S. (2008). *Grewia occidentalis*. Retrieved from http://pza.sanbi.org/grewia-occidentalis.

Vandal, J., Abou-Zaid, M. M., Ferroni, G., & Leduc, L. G. (2015). Antimicrobial activity of natural products from the flora of Northern Ontario, Canada. *Pharmaceutical Biology, 53*, 800–806. Available from https://doi.org/10.3109/13880209.2014.942867.

van Wyk, B. E. (2008). A review of Khoi-San and Cape Dutch medical ethnobotany. *Journal of Ethnopharmacology, 119*, 331–341. Available from https://doi.org/10.1016/j.jep.2008.07.021.

Vries, F.A., El Bitar, H., Green, I.R., Klaasen, J.A., Mabusela, W.T., Bodo, B., & Johnson, Q. (2005). An antifungal active extract from the aerial parts of Galenia africana structure of cyclopeptides view project phytochemical studies of extracts from *Aloe succotrina* view project. In: *11th NAPRECA Symposium* (pp. 123–131). Retrieved from https://www.researchgate.net/publication/228835536.

Watt, J. M., & Breyer-brandwijk, M. G. (1962). The medicinal and poisonous plants of Southern and Eastern Africa: Being an account of their medicinal and other uses, chemical composition, pharmacological effects and toxicology in man and animal. *Nature, 132*, 336.

Weber, E. (2017). *Invasive plant species of the world: A reference guide to environmental weeds*. UK: CABI. Available from https://doi.org/10.1079/9781780643861.0000.

Xaba, P.M.A., & Notten, A. (2003). *Lessertia frutescens*. Retrieved from http://pza.sanbi.org/lessertia-frutescens. (Accessed 31 October 2020).

Zhang, X., Hu, X., & Rao, X. (2017). Apoptosis induced by *Staphylococcus aureus* toxins. *Microbiological Research, 205*, 19–24. Available from https://doi.org/10.1016/j.micres.2017.08.006.

Zhou, L., Li, D., Wang, J., Liu, Y., & Wu, J. (2007). Antibacterial phenolic compounds from the spines of undefined Lam. *Natural Product Research, 21*, 283–291. Available from https://doi.org/10.1080/14786410701192637.

Zielińska-Borkowska, U. (2015). Sepsis – 2015. *Pielęgniarstwo w Anestezjologii i Intensywnej Opiece, 1*, 177–180. Available from https://doi.org/10.15374/pwaiio2015010.

Zonyane, S., Chen, L., Xu, M. J., Gong, Z. N., Xu, S., & Makunga, N. P. (2019). Geographic-based metabolomic variation and toxicity analysis of *Sutherlandia frutescens* L. R.Br. – An emerging medicinal crop in South Africa. *Industrial Crops and Products, 133*, 414–423. Available from https://doi.org/10.1016/j.indcrop.2019.03.010.

CHAPTER

5

The use of South African medicinal plants in the pursuit to treat gonorrhea and other sexually transmitted diseases

Tanyaradzwa Tiandra Dembetembe, Namrita Lall and Quenton Kritzinger

Department of Plant and Soil Sciences, Faculty of Natural and Agricultural Sciences, University of Pretoria, Pretoria, South Africa

Introduction

Sexually transmitted diseases (STDs) are a major problem internationally in that they have adverse effects on people's quality of life. According to the World Health Organization (WHO) (2016a), there are about one million new infections of STDs recorded daily worldwide. Chlamydia caused by *Chlamydia trachomatis*, gonorrhea caused by *Neisseria gonorrhoeae*, syphilis caused by *Treponema pallidum*, and trichomoniasis caused by *Trichomonas vaginalis*, which is a parasitic disease, each account for 25% of all the new infections excluding the human immunodeficiency virus, acquired immunodeficiency syndrome (HIV/AIDS) (World Health Organization, 2016a). Across the globe, there are approximately 350 million cases reported yearly of the four major STDs: chlamydia (131 million), gonorrhea (78 million), syphilis (6 million), and trichomoniasis (142 million) (WHO Regional Office for Africa, 2018). These venereal diseases are commonly associated with HIV/AIDS. South Africa has a relatively high incidence of HIV/AIDS therefore STDs are highly prevalent (Johnson, Coetzee, & Dorrington, 2005; Kharsany et al., 2020).

STDs are communicable diseases that are transmitted through sexual contact in humans via skin lesions in genitalia or bodily fluids (Belcher & Dawson, 2017). Clinical symptoms include abdominal pains, vaginal or penile discharge, dysuria, venereal sores, and lesions on the skin and mucosal tissues (Achakazai et al., 2017). Some of the diseases are latent

Medicinal Plants as Anti-infectives
DOI: https://doi.org/10.1016/B978-0-323-90999-0.00001-X

infections (e.g., syphilis) whereby the patient is asymptomatic for years and symptoms appear in the late stages of the disease (Centers for Disease Control and Prevention, 2017). If STDs are not treated they may manifest as pelvic inflammatory disease (PID), infertility, pregnancy complications, birth defects, and neurological damage. Prolonged exposure to STDs has been reported to increase the chances of HIV/AIDS (Cates & Wasserheit, 1991; Maier & Katsufrakis, 2015).

Venereal diseases have been attributed to several pathogens including bacteria, viruses, and fungi (Table 5.1). In South Africa, *N. gonorrhoeae*, *C. trachomatis*, and *T. pallidum* are the top bacterial causal agents of STDs (Johnson, Coetzee, & Dorrington, 2005; Kharsany et al., 2020). *Candida albicans* is the most common fungus causing vaginal yeast infections. *Candida albicans* is part of the normal human microbiota, however, if the immune system is suppressed the fungus may be pathogenic (Anwar, Malik, & Subhan, 2012; Chin, Lee, Rusliza, & Chong, 2016). *Haemophilus ducreyi* is another relatively common bacterial pathogen that causes chancroid, whilst *Gardnerella vaginalis* that causes bacterial vaginosis is most prevalent in females than in males (Johnson et al., 2005).

TABLE 5.1 Microorganisms associated with sexually transmitted diseases.

Disease	Type of organism	Microorganism	References
Vaginal yeast infections	Fungi	*Candida albicans*	Cavalcanti Filho et al. (2017), Mayer, Wilson, and Hube (2013)
Chlamydia	Bacteria	*Chlamydia trachomatis*	Achakazai et al. (2017), Geisler and Stamm (2007)
Lymphogranuloma venereum (LGV)	Bacteria	*Chlamydia trachomatis serovars (1–3)*	Geisler and Stamm (2007), Johnson et al. (2005)
Bacterial vaginosis	Bacteria	*Gardnerella vaginalis*	Machado, Castro, Palmeira-de-Oliveira, Martinez-de-Oliveira, and Cerca (2016)
Chancroid	Bacteria	*Haemophilus ducreyi*	Belcher and Dawson (2017), Johnson et al. (2005)
Herpes	Virus	Herpes simplex virus (HSV)	Talwar et al. (2000)
Human immunodeficiency virus (HIV)	Virus	Human immunodeficiency virus	Simon, Ho, and Karim (2006)
Human papillomavirus (HPV)	Virus	Human papillomavirus	Boxman, Hogewoning, Mulder, Bouwes Bavinck, and Ter Schegget (1999)
Granuloma inguinale	Bacteria	*Klebsiella granulomatis*	Kibbi, Bahhady, and El-Shareef (2012)
Gonorrhea	Bacteria	*Neisseria gonorrhoeae*	Tshikalange, Mamba, and Adebayo (2016), Van Vuuren and Holl (2017)
Shigellosis	Bacteria	*Shigella sonnei, Shigella flexneri*	Centers for Disease Control and Prevention (2017)
Syphilis	Bacteria	*Treponema pallidum*	Achakazai et al. (2017)
Trichomoniasis	Parasite	*Trichomonas vaginalis*	Centers for Disease Control and Prevention (2017)

The high prevalence of STDs has heightened the burden on the available market therapies as there is a rampant increase in drug-resistant STD pathogens. Therefore this has brought medicinal plants in the limelight as candidates for new alternative therapeutics for STDs (Palmeira-de-Oliveira et al., 2013). This review will focus primarily on one of the prevalent STDs in South Africa, namely, gonorrhea.

Background on gonorrhea

Gonorrhea is one of the most problematic STDs in the world. The global incidence rate for 2016 was estimated to be at 20 and 26 new reported gonococcal infections per 1000 women and men, respectively (Rowley et al., 2019). According to the WHO, sub-Saharan Africa is the most affected region in the world (WHO Regional Office for Africa, 2018; World Health Organisation, 2019). In South Africa, gonorrhea is the second most prevalent bacterial STD affecting humans (Kularatne, Niit, et al., 2018). In 2017, it was estimated that of the 4.5 million South Africans that were infected with gonorrhea, 6.6% were females and 3.5% were males (Kularatne, Kufa-Chakezha, Maseko, & Gumede, 2018). Gonorrhea is prevalent in sexually active individuals mostly below 30 years (Shim, 2011). The high incidence of gonorrhea has been due to several factors which include low socioeconomic status, the onset of sexual activity, safe sex awareness, and risky sexual behaviors (Barnes & Holmes, 1984; Oller, Wood, & St Luke, 1970).

Gonorrhea is a disease transmitted via sexual contact in individuals through lesions and bodily fluids (Belcher & Dawson, 2017). The symptoms of the disease include lower abdominal pains, vaginal or penial discharge, pain when urinating, itching of genitalia, venereal sores, and ulcerations (lesions on the skin and mucosal tissue) (Achakazai et al., 2017; Centers for Disease Control and Prevention, 2017). In some cases, the disease may cause urethritis and PID. If symptoms are seen they are usually mild and nonspecific (Curran et al., 1975; McCormack, Stumacher, Johnson, & Donner, 1977). Gonorrhea can also be asymptomatic in both males and females and most cases are diagnosed in women (Platt, Rice, & McCormack, 1983; Wallin, 1974). Untreated gonococcal infections could result in infertility, birth defects, and life-threatening ectopic pregnancies (Platt et al., 1983). In some cases, the infection can spread to the blood resulting in disseminated gonococcal infection, which can lead to death (Centers for Disease Control and Prevention, 2017).

The causal agent: *Neisseria gonorrhoeae*

Gonorrhea is caused by *N. gonorrhoeae*, a pathogen affecting humans, belonging to the proteobacteria group and the Neisseriaceae family. It is nonspore-forming diplococcus, which means that when viewed under the light microscope the bacteria exist as two bacterial cells joined together (Shim, 2011; Westling-Haggstrom, Elmros, Normark, & Winblad, 1977) (Fig. 5.1). *Neisseria gonorrhoeae* mostly infects genital mucosa but has been found to affect rectal, pharyngeal, oral and conjunctiva mucosa (Quillin & Seifert, 2018). The pathogen cannot survive outside of the host as it relies on the host to acquire nutrients.

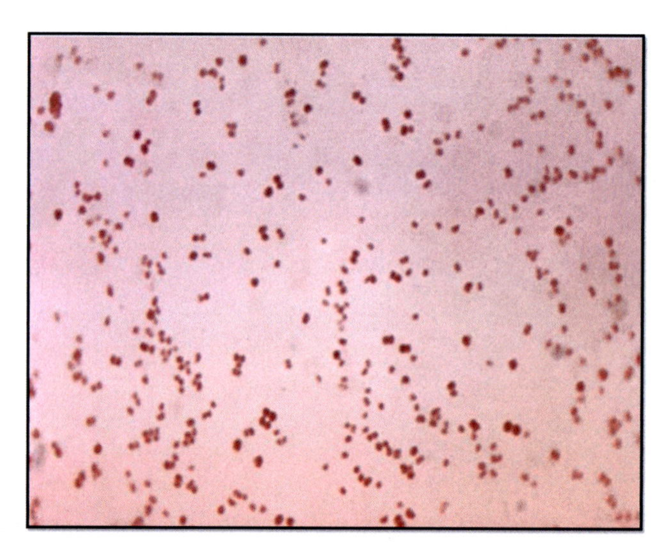

FIGURE 5.1 Causative agent diplococcus *Neisseria gonorrhoeae* viewed under the light microscope (CDC/Renelle Woodall, 1969). Source: *Woodall, R. (1969).* Photomicrograph of a Gram-stained specimen, revealed the presence of numerous Gram-negative, diplococcal bacteria that were identified as *Neisseria gonorrhoeae. ID#14855. https://phil.cdc.gov/Details.aspx?pid = 14855 (Accessed 2 February 2021).*

N. gonorrhoeae is a Gram-negative obligate anaerobic bacterium (Knapp & Clark, 1984). It, therefore, does not require oxygen to grow but rather carbon dioxide. The pathogen has a lipooligosaccharide (LOS) as the major component of the outer membrane with a lipid A inner core (Arenas, 2012; Christodoulides, 2019). The LOS lacks the repetitive O-polysaccharide antigen which is found in lipopolysaccharide of other Gram-negative bacteria (McSheffrey and Gray-Owen, 2015) (Fig. 5.2). The gonococcal LOS is an immunostimulatory molecule which results in inflammation consequently increasing pathogenicity of the bacteria (Zhou et al., 2014). Changes in the fatty acid chains of lipid A alters the pathogenesis and antibiotic susceptibility of *N. gonorrhoeae* (Lewis et al., 2009; Liu, John, & Jarvis, 2010).

N. gonorrhoeae has a relatively small genome of ~2 Mb compared to *Escherichia coli* with a genome DNA of ~4 Mb (Chung et al., 2008; McSheffrey & Gray-Owen, 2015). The gonococci express *pilin* and *Opa* (Opacity-associated protein) genes. Subsequently, these genes translate to surface proteins that allow adhesion between bacterial and human cells which heightens the virulence of *N. gonorrhoeae* (Ball & Criss, 2013; Griffiss, Lammel, Wang, Dekker, & Brooks, 1999). The *Opa* and *pilin* genes are subject to phase variation which is the switching on and off of genes triggered by DNA polymerase slippage while replicating tandem repeats. The slippage results in the addition or deletion of repeats which can cause reading frame shifts affecting the resultant protein. Thus *N. gonorrhoeae* has antigenic variation that allows the bacteria to evade the host immune systems enabling bacterial persistence and reinfection in a host (Murphy, Connell, Barritt, Koomey, & Cannon, 1989; Sadarangani, Pollard, & Gray-Owen, 2011; Yu et al., 2013).

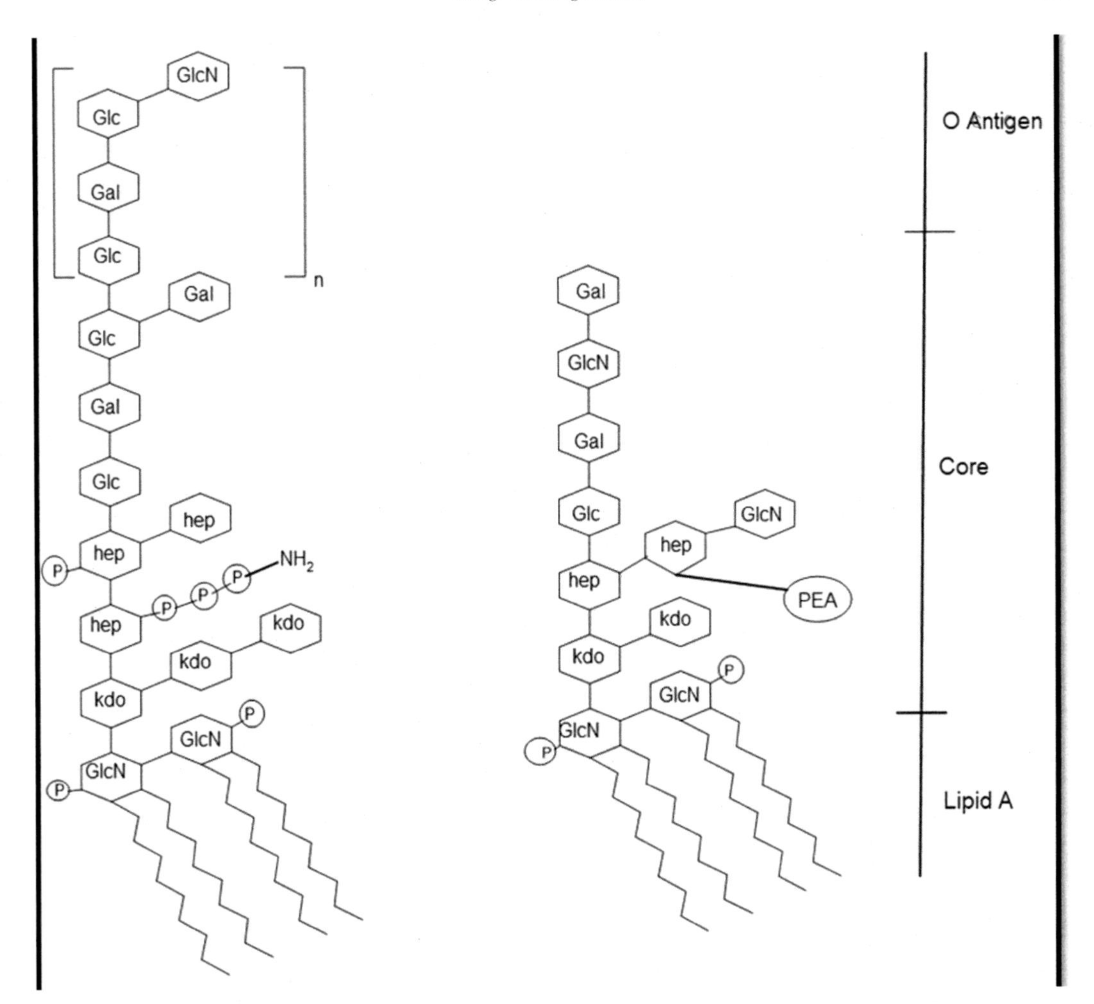

FIGURE 5.2 Virulence factor of *Neisseria gonorrhoeae*. Difference between lipooligosaccharide (LOS) and lipopolysaccharide (LPS) membrane structures in Gram-negative bacteria showing the polysaccharide core and lipid A found in both molecules. *Glu*, glucose; *GluN*, *N*-glucosamine; *kdo*, 2-keto-3-deoxyoctinic acid; *PEA*, diphosphoethanolamine; *P*, phosphate group.

Pathogenesis of Neisseria gonorrhoeae *and evasion of host immune system*

Gonococci infections result in prolonged inflammation allowing for persistence and reinfection of the disease. The initial step in the infection of *N. gonorrhoeae* involves the tethering of the bacteria to the genital epithelial cells via pathogen-associated molecular patterns (PAMPs). The PAMPs involved include pili, LOS, and Opa proteins. The pili attach to host CD46/complement receptor 3 whilst the LOS binds to asialoglycoprotein receptor (McSheffrey & Gray-Owen, 2015; Sadarangani et al., 2011). Additionally, Opa proteins have an affinity for the carcinoembryonic antigen-related cell adhesion molecule family (CEACAM) (Lenz & Dillard, 2018; Sadarangani et al., 2011). The close interaction of either of these PAMPs and the epithelial cell receptors trigger the endocytosis of the bacterium, which elicits an immune response (McSheffrey & Gray-Owen, 2015; Wang, Gray-Owen, Knorre, Meyer, & Dehio, 1998). *N. gonorrhoeae* triggers cytokine release which results in the overzealous recruitment of neutrophils to the site of infection (Criss & Seifert, 2012; Edwards & Butler, 2011). Cytokines such as nuclear factor kappa B (NF-κB) activate signal transduction pathways that induce IL-17 that drives the T_h17 response, which recruits neutrophils (Feinen, Jerse, Gaffen, & Russell, 2010). The gonococci can evade the host immune system as T_h17 response that favors the production of IL-10; suppresses T cells in adaptive immunity which allows the infection to persist (Liu, Liu, & Russell, 2014). The Opa proteins, within infected tissue, can bind to CEACAM 1 of dendritic cells which suppresses maturation of T_1, T_2, and B cells (Yu et al., 2013; Zhu et al., 2012). The T_1 helper cells are responsible for the killing and clearing of infected cells whilst T_2 helper cells work in combination with B cells to produce memory cells and antibodies. The inactivation of this adaptive immunity results in prolonged inflammation, thus gonococci infections are not cleared and the lack of antibodies permits reinfection in individuals (Hedges, Mayo, Mestecky, Hook, & Russell, 1999; Hedges, Sibley, Mayo, Hook, & Russell, 1998). Opa proteins can also interact with neutrophils via CEACAM 3 resulting in stimulation of oxidative burst of the bacteria. The fragments of the bacteria can elicit immune responses in other noninfected cells within the tissue (Sadarangani et al., 2011; Sarantis & Gray-Owen, 2007).

Evasion of host immune system via nutrition immunity

The small genome of *N. gonorrhoeae* limits the metabolic capacity of the gonococci. Thus the bacteria rely on host mucosal membranes for nutrition (Cornelissen, 2018; McSheffrey & Gray-Owen, 2015). Naturally, as a host defense mechanism, the body limits nutrients available to microbes. The available iron in the body is bound to transferrin and lactoferrin after which is transported and stored in the liver (Parrow, Fleming, & Minnick, 2013). *N. gonorrhoeae* has transferrin (Tbps) and lactoferrin (Lbps) receptors which interact with transferrin and lactoferrin that have bound iron (McSheffrey & Gray-Owen, 2015; Parrow et al., 2013). This mechanism allows the gonococcal bacteria to acquire iron for metabolism. It has been shown that a defective gonococcal bacterium without both of these receptors eliminates virulence (Anderson, Hobbs, Biswas, & Sparling, 2003).

Coinfections of **Neisseria gonorrhoeae**

STDs are acquired via sexual contact and there is a possibility of contracting two infections simultaneously. Gonorrhea and chlamydia are the most common coinfections in individuals (Guy et al., 2015; Leonard, Schoborg, Low, Unemo, & Borel, 2019; Seo, Choi, & Lee, 2019). Approximately 50% of gonococcal infections are coupled with chlamydia (Creighton, Tenant-Flowers, Taylor, Miller, & Low, 2003; McSheffrey & Gray-Owen, 2015). Individuals are usually treated for both diseases even though they have contracted one of the infections. The mechanistic interaction between the two is, however, unknown (Leonard et al., 2019).

N. gonorrhoeae can interact with HIV by suppressing the expression of HIV memory response. The Opa_{CEA} protein of *N. gonorrhoeae* suppresses the maturation of dendritic cells, which are antigen-presenting cells (APCs) (Yu et al., 2013). These cells are responsible for activation of the humoral immune response which is activated by APCs, such as dendritic lymphocytes, which are recognized by T_2 ($CD4^+$) helper cells. Consequently, triggering B lymphocytes maturation into plasma cells that produce antibodies (Martin-Gayo & Yu, 2019; Steinman, 1991).

N. gonorrhoeae increases replication of HIV-1 in T cells by activation of 5′ HIV long terminal repeats. This is because *N. gonorrhoeae* causes genital epithelial cells to increase the production of pro-inflammatory cytokines tumor necrosis factor α and interleukins (IL-6 & 8) (Ferreira et al., 2011). The gonococci pilin protein activates NF-κB, which is known to also escalate HIV replication (Acchioni et al., 2019; Dietrich et al., 2011). The occurrence of gonococcal infections increases the likelihood of individuals to seroconvert to HIV positive (Bernstein, Marcus, Nieri, Philip, & Klausner, 2010; McSheffrey & Gray-Owen, 2015; Mlisana et al., 2012).

Status of available treatments for gonorrhea

At present, for treatment of gonorrhea a combination of first-line antibiotics, azithromycin, and ceftriaxone/cefixime is recommended (Ryan, 2017; Wi et al., 2017). The treatment is given as a single oral/intramuscular dose of ceftriaxone/cefixime in combination with a single oral dose of azithromycin (World Health Organization, 2016b). The therapy is also used in the treatment of other STDs such as chlamydia that normally exists as coinfections with gonorrhea (Belcher & Dawson, 2017).

Ceftriaxone and cefixime are antibiotics classified as third-generation extended-spectrum cephalosporins (ESC) (Fig. 5.3). These antibiotics inhibit *N. gonorrhoeae* by inhibiting the penicillin-binding proteins (PBP) which are used to make peptidoglycan cross-links in the bacterial cell wall (Unemo, Del Rio, & Shafer, 2016). Azithromycin is a macrolide antibiotic that inhibits translation by binding to the 50S ribosomal subunit (Douthwaite and Champney, 2001).

Azithromycin was first introduced for the treatment of gonorrhea in the 1980s and the third-generation cephalosporins in the 1990s (Suay-García & Pérez-Gracia, 2018; Unemo et al., 2016). These antibiotics were introduced as *N. gonorrhoeae* had gained resistance to the then used tetracycline, penicillin, amoxicillin, and fluoroquinolones like ciprofloxacin (Ohnishi, Golparian, et al., 2011; Ohnishi, Saika, et al., 2011; Unemo & Nicholas, 2012). In the 1980s ciprofloxacin and azithromycin were used as the first-line treatments, however, *N. gonorrhoeae* gained resistance (Lewis et al., 2008; Lynagh et al., 2015; Stevens et al., 2014; Unemo et al., 2016). The mode of action used by the ciprofloxacin-resistant strains is by reducing the affinity of DNA gyrase whilst in azithromycin-resistant strains by altering the 50S ribosomal subunit target (Unemo & Shafer, 2011, 2014). The escalation in ciprofloxacin resistance resulted in the abandonment of the regiment for gonorrhea treatment (Lewis et al., 2008; Unemo et al., 2016). As the years progressed azithromycin was discouraged as a monotherapy due to the resistance that was observed and it was therefore recommended to be used in combination with cefixime/ceftriaxone (Bignell et al., 2013). Azithromycin-resistant strains have been detected in many countries including Australia, Ireland, and Canada (Kularatne, Kufa-Chakezha, et al., 2018; Lynagh et al., 2015; Martin

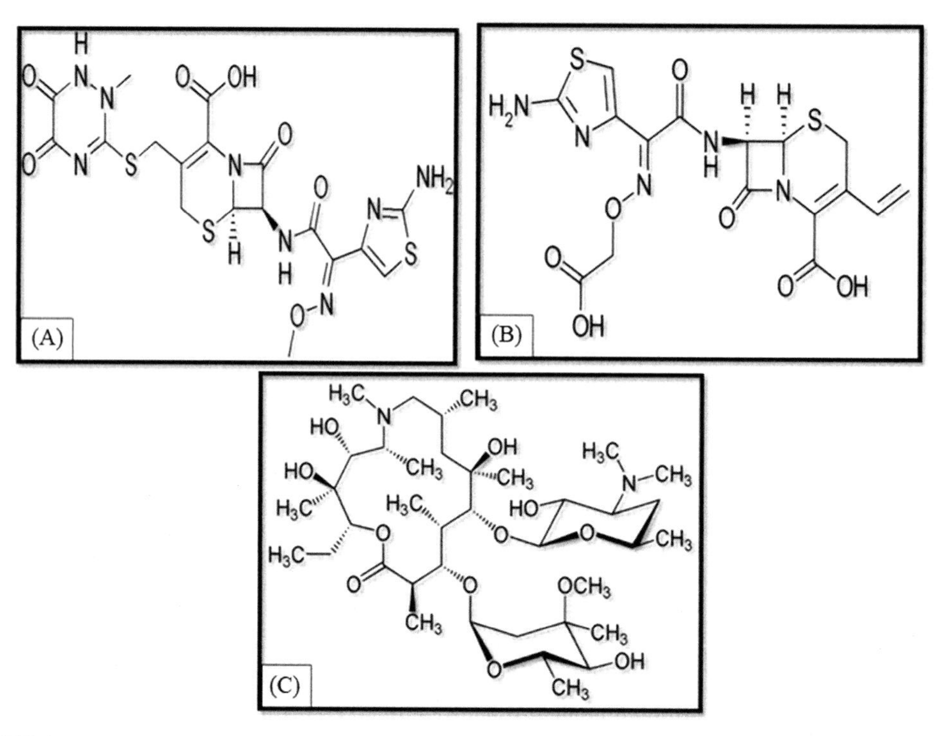

FIGURE 5.3 First-line drugs. Structure of (A) ceftriaxone; (B) cefixime; and (C) azithromycin.

et al., 2019; Stevens et al., 2014; World Health Organisation, 2019). Even though ceftriaxone and cefixime are used as current regiments, antibiotic resistance has been reported in various parts of the world (Unemo et al., 2016; Wi et al., 2017). According to the National Institute for Communicable Diseases (NICD) of South Africa, cefixime resistance isolates have been detected in Cape Town, Gauteng, and the Eastern Cape province (Kularatne, Maseko, Gumede, Radebe, & Chakezha, 2017). Resistance of ESCs in *N. gonorrhoeae* is by mutations to the penicillin-binding proteins and also increasing efflux of the ESCs in the bacterium (Unemo & Shafer, 2014). The rise of cephalosporin antibiotic resistance has resulted in some countries opting to alternatively use spectinomycin or gentamicin in combination with azithromycin (Australasia Sexual Alliance, 2019; Bignell et al., 2013; Public Health Canada, 2017; Suay-García & Pérez-Gracia, 2018; World Health Organisation, 2019).

The upsurge of ESC resistance has been a concern as there is now an emergence of multidrug resistant (MDR) and extensively drug-resistant (XDR) *N. gonorrhoeae* (Alirol et al., 2017; Martin et al., 2019). Bacterial isolates that have resistance to at least two of the currently recommended therapeutics, as well as resistance to at least two of penicillin, tetracycline, erythromycin, or ciprofloxacin, are categorized as XDR strains (Martin et al., 2019). On the other hand, MDR bacterial strains have resistance to at least one current recommended therapy and resistance to at least two other antibiotics (penicillin, tetracycline, erythromycin, or ciprofloxacin). Treatment failures of both azithromycin and ESCs toward XDR *N. gonorrhoeae* have been reported (Martin et al., 2019). The first of such cases were reported in Australia and the United Kingdom (Australian Government Department of Health, 2018; European Centre for Disease Prevention and Control, 2018; Public Health England, 2018). Recently some XDR isolates have been detected in Canada as well (Martin et al., 2019). The heightened antibiotic resistance observed in *N. gonorrhoeae* globally has resulted in the WHO placing it on the global priority list level 2 (World Health Organization, 2017). The emergence of XDR *N. gonorrhoeae* isolates thus puts a burden on the available antibiotics by reducing their efficiency (Crowther-Gibson et al., 2011). This, therefore, prompts the need for

TABLE 5.2 Selected South African medicinal plants used traditionally for the treatment of sexually transmitted diseases.

Plant species	Family	Parts used	References
*Tabernaemontana elegans**	Apocynaceae	Leaves/roots	De Wet et al. (2012)
*Catharanthus roseus**	Apocynaceae	Roots	Mongalo and Makhafola (2018), Semenya, Potgieter, Johannes, and Erasmus (2013)
*Aloe ferox**	Asphodelaceae	Leaves	Van Wyk, Van Oudtshoorn, and Gericke (2017)
Aloe marlothii	Asphodelaceae	Leaves	De Wet et al. (2012)
*Helichrysum caespitatum**	Asteraceae	Whole plant	Semenya, Potgieter, and Erasmus (2013)
*Senecio serratuloides**	Asteraceae	Whole plant	De Wet et al. (2012), Semenya, Potgieter, and Erasmus (2013), Semenya, Potgieter, Johannes et al. (2013)
Helichrysum populifolium	Asteraceae	Leaves	De Wet et al. (2012)
Elaeodendron transvaalense	Celastraceae	Roots	Mabogo (1990), Semenya, Potgieter, and Erasmus (2013)
Terminalia sericea	Combretaceae	Roots/stem bark	Mongalo and Makhafola (2018)
*Combretum molle**	Combretaceae	Leaves/roots	De Wet et al. (2012), Fyhrquist et al. (2002)
*Dioscorea sylvatica**	Dioscoreaceae	Bulb	Semenya, Potgieter, and Erasmus (2013), Semenya, Potgieter, Johannes et al. (2013)
*Diospyros lycioides**	Ebenaceae	Roots	Semenya, Potgieter, Johannes et al. (2013)
*Jatropha zeyheri**	Euphorbiaceae	Roots	Mongalo and Makhafola (2018), Semenya, Potgieter, Johannes et al. (2013)
*Cassia abbreviata**	Fabaceae	Roots/stem bark	Mongalo and Makhafola (2018)
*Senna italica**	Fabaceae	Root	Chauke, Shai, Mogale, Tshisikhawe, and Mokgotho (2015)
*Albizia adianthifolia**	Fabaceae	Leaves	De Wet et al. (2012)

(*Continued*)

TABLE 5.2 (Continued)

Plant species	Family	Parts used	References
Elephantorrhiza elephantina	Fabaceae	Roots	Mongalo and Makhafola (2018), Semenya, Potgieter, Johannes et al. (2013)
Burkea africana	Fabaceae	Roots	Semenya, Potgieter, Johannes et al. (2013)
Pelargonium spp.	Geraniaceae	Roots	Semenya, Potgieter, and Erasmus (2013), Semenya, Potgieter, Johannes et al. (2013)
*Hypoxis hemerocallidea**	Hypoxidaceae	Bulb	Mongalo and Makhafola (2018), Semenya, Potgieter, and Erasmus (2013)
*Hypoxis obtusa**	Hypoxidaceae	Roots	Semenya, Potgieter, and Erasmus (2013)
*Ximenia caffra**	Olacaceae	Roots	Chauke et al. (2015), De Wet et al. (2012), Mongalo and Makhafola (2018)
*Peltophorum africanum**	Rhamnaceae	Roots/stem bark	Mongalo and Makhafola (2018), Semenya, Potgieter, and Erasmus (2013)
*Ziziphus mucronata**	Rhamnaceae	Roots	Semenya, Potgieter, and Erasmus (2013), Semenya, Potgieter, Johannes et al. (2013)

Note: *Plants used to treat gonorrhea and have been reported to have antigonococcal activity.

drug discovery to produce new and alternative therapies for the treatment of gonorrhea.

Selected South African plants used in traditional medicine for the treatment of sexually transmitted diseases and their bioactivity

The antibiotic resistance burden has consequently brought plants into the limelight for drug development for gonorrhea and other STDs (Palmeira-de-Oliveira, Silva, Palmeira-de-Oliveira, Martinez-de-Oliveira, & Salgueiro, 2013). Plants have been reported to have a vast diversity in chemistry and they have been used throughout history in folk medicine by traditional healers worldwide (De Wet, Nzama, & Van Vuuren, 2012; Palmeira-de-Oliveira, Silva, Palmeira-de-Oliveira, Martinez-de-Oliveira, & Salgueiro, 2013; Yang et al., 2012). Table 5.2 provides a summary of some of the South African medicinal plants used in traditional medicine for the treatment of venereal diseases including gonorrhoea. In the section below selected plants are discussed further in detail with emphasis on their bioactivity and in vitro studies that have been conducted. In general, for antimicrobial studies, a plant extract is considered to have noteworthy activity if the minimum inhibitory concentration (MIC) is below 1 mg/ml (Ndhlala et al., 2013; van Vuuren, 2008).

Aloe ferox

Aloe ferox Mill. (bitter aloe) belongs to the Asphodelaceae family of plants (Semenya, Potgieter, & Erasmus, 2013; Semenya, Potgieter, Johannes, et al., 2013; Van Wyk et al., 2017) (Fig. 5.4). It is

FIGURE 5.4 Four South African plants used to treat gonorrhea: (A) *Aloe ferox*, (B) *Cassia abbreviata*, (C) *Combretum molle*, (D) *Hypoxis hemerocallidea*. Source: *From BotBin. (2005).* Habitus, leaves and inflorescence. Species: Hypoxis hemerocallidea Fisch. & C.A. Mey. Genus: Hypoxis. Family: Hypoxidaceae. Location: Berlin Botanical Gardens Berlin-Dahlem. *https://commons.wikimedia.org/wiki/File:Hypoxis_hemerocallidea_BotGardBln1105InflorescenceHabitus.JPG. Gregory, D. (2007).* Aloe ferox on R61 route between Cofimvaba and Ngcobo. *https://commons.wikimedia.org/wiki/File:Aloe_Ferox_between_Cofimvaba_and_Ngcobo.jpg. Jeppetown. (2010). Sjambok Pod (Cassia abbreviata). https://commons.wikimedia.org/wiki/File:Cassia_abbreviata_1.jpg. Rotational. (2007).* Combretum molle R. Br. ex G. Don. Fruit and foliage of a Velvet bushwillow, at Hamerkop Kloof, Magaliesberg, South Africa. *https://commons.wikimedia.org/wiki/File:Combretum_molle00.jpg.*

used in South African traditional medicine for the treatment of arthritis, conjunctivitis, eczema, hypertension, STDs, stress, and venereal sores. Related species *Aloe marlothii*, A. Berger, and *Aloe vera* (L.) Burm.f. are also used in phytomedicine for treatment of STDs including HIV (Semenya et al., 2013). The fleshy leaves are made into a decoction for the treatment of STDs such as gonorrhoea and syphilis (Van Wyk et al., 2017).

Aloe ferox has been reported to have antimicrobial activity against *Candida albicans*, *Gardnerella vaginalis*, *Shigella sonnei*, and *N. gonorrhoeae* (Kambizi & Afolayan, 2008; van Vuuren & Naidoo, 2010). Dichloromethane:methanol (DCM:MeOH) extracts were found to have a MIC of 8 mg/mL compared to the 0.04 mg/mL of the ciprofloxacin control (van Vuuren & Naidoo, 2010). Kambizi and Afolayan (2008) found that methanol extracts of *A. ferox* inhibit *N. gonorrhoeae* at MIC of 0.5 mg/mL. The DCM:MeOH extracts are thus less active than the polar methanol extracts (Kambizi & Afolayan, 2008). The variation seen in bioactivity could be due to the solvents used for extracting compounds differently.

Bioactive compounds including aloin, aloe-emodin, and chrysophanol have been isolated from *A. ferox* extracts (Kambizi, Sultana, & Afolayan, 2005). Aloin exhibits antimicrobial activity against *C. albicans* and *N. gonorrhoeae* with MICs of 5 and 0.1 mg/mL, respectively (Kambizi & Afolayan, 2008). This bioactive compound has antiviral activity against Herpes Simplex Virus (HSV-1). Potent activity was seen at 63 μg/mL with 25%–50% cytopathic effect (CPE) being observed 120 h after infection with HSV-1. The positive control had 75% CPE noticed 24 h after infection. Aloin was, therefore, able to delay infection of HSV-1 in the Vero cells more than the positive control (Kambizi, Goosen, Taylor, & Afolayan, 2007).

Cassia abbreviata

Cassia abbreviata Oliv., wild senna is part of the Fabaceae family (Sobeh et al., 2018). In traditional medicine, it is used as a treatment for all STDs. The usual preparation of the prescription is a decoction of the stem/stem bark of the plant. This decoction is taken orally by the patient (Chauke et al., 2015) (Fig. 5.4).

Water stem bark extracts of *C. abbreviata* have antifungal activity against *C. albicans* with a MIC of 0.1 mg/mL (Mongalo, McGaw, Finnie, & Van Staden, 2017). Ethanolic extracts of the plant have exhibited antigonococcal activity with a MIC of 46.88 μg/mL. The ethyl acetate and acetone extracts have been seen to inhibit HIV-1 RT with an IC_{50} of 1.25 mg/mL (Chauke, Shai, Mogale, & Mokgotho, 2016). Leteane et al. (2012) also tested the inhibitory effect of ethanolic root extracts on the production of antigen p24 produced by HIV-1. It was seen that 150 μg/mL of the extract resulted in 55.1% inhibition (Leteane et al., 2012). A relative of the plant, *Cassia sieberiana* DC., had better inhibition activity of 98.1% at the same test concentration. *Cassia abbreviata* also exhibits some antiviral activity against HSV (Viol, Chagonda, Moyo, & Mericli, 2016).

Palmitic acid has been isolated from *C. abbreviata* (Dangarembizi et al., 2015). Palmitic acid is an inhibitor of HIV-1 by inhibiting entry and fusion of the virus. These effects were noted when concentrations between 22 and 100 μM were used resulting in inhibition of HIV infection of between 70% and 98% (Lee et al., 2009; Lin, Paskaleva, Chang, Shekhtman, & Canki, 2011). Palmitoleic acid, an unsaturated fatty acid, that can be biosynthesized from palmitic acid has activity against *N. gonorrhoeae* (Bergsson, Steingrímsson, & Thormar, 1999).

Combretum molle

Combretum molle R. Br. ex G. Don, the velvet bushwillow is a member of the Combretaceae family (Hedberg et al., 1982). The whole plant is used traditionally in the treatment of syphilis and gonorrhoea (De Wet, Nzama, & Van Vuuren, 2012; Fyhrquist et al., 2002). Oral decoctions are the preferred treatment of STDs (Bryant, 1966) (Fig. 5.4).

There have been reports on the antimicrobial activity of the plant. Methanol and acetone extracts of *C. molle* have the same activity with MICs of 40 μg/mL against *C. albicans*. Dichloromethane extracts inhibited the yeast at 320 μg/mL (Masoko, Picard, & Eloff, 2007). Methanol plant extracts inhibited HIV-1 ribonuclease (RNase H) and RNA-dependent DNA-polymerase (RDDP) with IC_{50s} of 9.7 and 9.5 μg/mL, respectively (Bessong et al., 2005). This makes the plant a potential candidate to develop novel anti-HIV drugs. Other plants in the Combretaceae family, namely, *Combretum adenogonium* Steud. ex A.Rich. and *Terminalia sericea* Burch. ex DC., have exhibited anti-HIV activity with 79% and 98% inhibition at 100 μg/mL (Bessong et al., 2005; Mushi, Mbwambo, Innocent, & Tewtrakul, 2012).

Terpenoids combretene A & B have been isolated from the plant as well as punicalagin and sericoside (Ahmed, Al-Howiriny, Passreiter, & Mossa, 2004; Asres et al., 2001). However, little is known about the antimicrobial properties of these compounds isolated from *C. molle*.

Elaeodendron transvaalense

Elaeodendron transvaalense (Burtt Davy) R.H. Archer, Transvaal saffronwood, is a medicinal plant in the Celastraceae family. It is used traditionally in the management of STDs (including gonorrhea and herpes) and HIV infections (Mabogo, 1990; Maroyi & Semenya, 2019; Samie et al., 2010). It is used sometimes in combination with *Elephantorrhiza elephantina* (Burch.) Skeels in the treatment of these diseases (Semenya, Potgieter, Johannes, et al., 2013). The root/bark decoctions of the plant are mostly preferred for treatment of STDs (Maroyi & Semenya, 2019; Semenya, Potgieter, Johannes, et al., 2013).

Mamba, Adebayo, and Tshikalange (2016) found that the ethanolic plant extracts have antimicrobial activity against *C. albicans, N. gonorrhoeae* and *G. vaginalis* with MICs of 3.1, 1.6, and 12.5 mg/mL, respectively, compared to the ciprofloxacin control (<0.01 mg/mL) (Mamba et al., 2016). They also found that *Elaeodendron croceum* (Thunb.) DC., a related species used in the treatment of HIV, has the same activity on the bacterial species and better activity against *C. albicans* with a MIC of 1.6 mg/mL. The plant has exhibited anti-fungal activity against other *Candida* spp. including *C. krusei* and *C. neoformans* (Samie et al., 2010). The water and ethanol plant extracts, at 100 μg/mL, have moderate reverse transcriptase (RT) inhibition of ~40% when compared nevirapine (~80%) and doxorubicin (100%) controls (Mamba et al., 2016; Sigidi et al., 2017).

Three compounds, isolated from the plant, lup-20(30)-ene-3α,29-diol, lup-20(29)-ene-30-hydroxy-3-one and 4′-O-methyl-epigallocatechin have been tested on STD pathogens (Mamba et al., 2016). 4′-O-methyl-epigallocatechin inhibited *N. gonorrhoeae* at a MIC of 6.3 mg/mL (Mamba et al., 2016). In the same study, it was seen that the active compound had antiviral properties inhibiting HIV-1 RT by 63.7% compared to the doxorubicin control that had total inhibition of the enzyme.

Hypoxis hemerocallidea

Hypoxis hemerocallidea Fisch., C.A. Mey. & Avé-Lall., previously known as *Hypoxis rooperi*, is part of the Hypoxidaceae family. It is known as the yellow star 'African potato'

(Van Wyk et al., 2017). It is a very popular plant by traditional healers to treat gonorrhoea and HIV. The plant tubers are usually made into a decoction and taken orally (De Wet et al., 2012; S. Semenya et al., 2013). The plant extracts are commercially available in pharmacies as tonics, tinctures, or capsules (Fig. 5.4).

Naidoo, Van Vuuren, Van Zyl, and De Wet (2013) showed that aqueous extracts have antigonococcal properties with a MIC of 500 μg/mL compared to the 0.04 μg/mL of the ciprofloxacin control. In the same study, DCM:MeOH extracts had less activity with a MIC of 8 mg/mL (Naidoo et al., 2013). The difference could have resulted from the different compounds eluted by the solvents since they differ in polarity. The increased antigonococcal activity in the aqueous extracts could have been due to polar compounds. The plant leaf aqueous, ethanolic and DCM extracts have activity against *C. albicans* with a MIC of 0.8 mg/mL. The leaf extracts are more bioactive against *Candida* than those of the corms (Ncube, Finnie, & Staden, 2011). *Hypoxis hemerocallidea* has been reported to have anti-HIV activity. The plant has been shown to maintain CD4 cells in HIV patients who were given capsules containing methanolic extracts. This validates the plant's use as a dietary supplement in individuals with HIV (Albrecht, 1996; Matyanga, Morse, Gundidza, & Nhachi, 2020). A related species, *Hypoxis sobolifera* Jacq., aqueous and ethanolic extracts inhibit HIV-1 RT ~80% and ~55% at 200 μg/mL, respectively (Klos et al., 2009).

Hypoxoside is a compound that has been isolated from the plants (Drewes, Hall, Learmonth, & Upfold, 1984; Matyanga et al., 2020). This inactive compound is converted to rooperol which improves the immune system (Drewes, Elliot, Khan, Dhlamini, & Gcumisa, 2008; Mills, Cooper, Seely, & Kanfer, 2005). Phytosterol, β-sitosterol (BSS), isolated from the plant has also been found to be an immune enhancer. Clinical trials have been done on humans and Bouic et al. (1996) showed that β-sitosterol glucoside (BSSG) can increase the proliferation of T cells by increasing the expression of CD25 in the cells thus boosting the immune system (Bouic et al., 1996). Furthermore, people with HIV who were not using the antiretroviral therapy were given BSS/BSSG, maintained their CD4 count after 12 months of drinking the prescription. Individuals that had >500 CD4 cells at the start of the trial maintained their CD4 count and had decreased viral loads of HIV (Bouic et al., 2001).

Peltophorum africanum

Peltophorum africanum Sond., weeping wattle, belonging to the Fabaceae family, is utilised traditionally to treat STDs such as HIV and gonorrhoea (Samie et al., 2010; Semenya, Potgieter, Johannes, & Erasmus, 2013; Tshikalange, Mamba, & Adebayo, 2016). The root and stem bark of this plant are usually made into a decoction that is orally taken to treat STDs (Mongalo, 2013; Mongalo & Makhafola, 2018) (Fig. 5.5).

Ethanolic plant extracts of the plant have antigonococcal activity with a MIC of 1.6 mg/mL and activity against *C. albicans* of 3.1 mg/mL compared to ciprofloxacin (<0.01 mg/mL) (Mamba et al., 2016). Furthermore, aqueous root extracts have been shown to have antibacterial activity against *G. vaginalis* and *N. gonorrhoeae* with MICs of 0.5 mg/mL (Naidoo et al., 2013). This plant has been studied for the potential treatment of HIV and has been reported to inhibit HIV-1 RT (Mamba et al., 2016; Tshikalange et al., 2008). The methanolic stem/bark extracts inhibit RDDP of RT with IC_{50} of 3.5 μg/mL. It also inhibits RNase H of RT with an IC_{50} of 10.6 μg/mL (Bessong et al., 2005).

FIGURE 5.5 Three South African plants used to treat gonorrhea. (A) *Peltophorum africanum*, (B) *Tabernaemontana elegans*, (C) *Terminalia sericea*. Source: *From JMK. (2014).* Peltophorum africanum. *https://commons.wikimedia.org/wiki/File:Peltophorum_africanum,_habitus,_c,_Zoutpan.jpg. Dupount, B. (2014).* Terminalia sericea S40 Road West of Satara, Kruger NP, South Africa. *https://commons.wikimedia.org/wiki/File:Silver_Clusterleaf_(Terminalia_sericea)_(13927744861).jpg. SA plants. (2017).* Tabernaemontana elegans, paired fruit (derived from a single flower); cultivated in garden, Pretoria. Gauteng, South Africa. *https://commons.wikimedia.org/wiki/File:Tabernaemontana_elegans_5Dsr_5370.jpg.*

Bessong et al. (2005) have isolated gallotannin and catechin from methanolic extracts of stem/bark of the plant which have anti-HIV properties. They also showed that gallotannin can inhibit the RDDP and RNase H functions of RT with IC_{50} values of 6 and 5 μM, respectively, compared to the 0.5 μM of ODN 93 (positive control). The compound also stopped 3′-end processing activity of HIV-1 integrase at 100 μM whilst catechin had moderate activity inhibiting 65% of the enzyme at 100 μM (Bessong et al., 2005). Epigallocatechin-3-O-gallate (EGCG) was isolated from methanolic extracts of the plant (Ebada, Ayoub, Singab, & Al-Azizi, 2008). EGCG has antigonococcal activity with a MIC of 32 μg/mL whilst an EGCG derivate containing acyl group palmitoleate (C16) had a MIC of 16 μg/mL. The compounds had less activity than the positive controls amoxicillin (0.25 μg/mL) and cefazolin (0.5 μg/mL). The two compounds C16 and EGCG showed anti-*Candida* activity with a MIC of 16 and 64 μg/mL, respectively. Fluconazole control had a MIC of 0.25 μg/mL (Matsumoto et al., 2012).

Tabernaemontana elegans

Tabernaemontana elegans Stapf, the toad tree, is a South African medicinal plant belonging to the Apocynaceae family (De Wet et al., 2012). The roots and the leaves are commonly used to treat STDs. The plant is finely chopped and made into a decoction with either *H. hemerocallidea* corms or *Ipomoea batata* (L.) Lam. (sweet potato) leaves to treat gonorrhoea (De Wet et al., 2012) (Fig. 5.5).

Tabernaemontana elegans has some pharmacological effects which include: antibacterial, antifungal, and antiprotozoal activity. Naidoo et al. (2013) showed that DCM:MeOH extract has antibacterial activity against *G. vaginalis* and *N. gonorrhoeae* with MIC of 0.25 and 1 mg/mL, respectively. The extracts did not have better activity than the positive control (ciprofloxacin) (0.39 μg/mL) against *G. vaginalis* whilst the extract was not better than the positive control (0.04 μg/mL) for *N. gonorrhoeae*. The aqueous extracts had antifungal activity of 0.25 mg/mL compared to the 2.5 μg/mL amphotericin B control. The DCM: MeOH extracts had antiprotozoan properties inhibiting *T. vaginalis* at a MIC of 1 mg/mL (Naidoo et al., 2013).

Terminalia sericea

Terminalia sericea Burch. ex DC., the silver cluster leaf, is a medicinal plant in the Combretaceae family (Van Wyk et al., 2017). The stem bark or roots are used as a decoction to treat STDs (syphilis & gonorrhoea) and other opportunistic illnesses related to HIV (Chinsembu, 2016; Hutchings, 1989; Mongalo & Makhafola, 2018; Watt & Breyer-Brandwijk, 1962) (Fig. 5.5).

The DCM:MeOH extracts have been reported to have antimicrobial properties against *N. gonorrhoeae* with a MIC of 1 mg/mL whilst *T. vaginalis* and *C. albicans* both with MICs of 2 mg/mL. Aqueous extracts of the plant inhibited *G. vaginalis* at the lowest concentration of 2 mg/mL (van Vuuren & Naidoo, 2010).

Chauke et al. (2016) revealed that the plant has anti-HIV properties, where acetone and water extracts had IC_{50} values of 0.08 mg/mL (Chauke et al., 2016). Tshikalange et al. (2016) revealed that the ethanolic plant extracts of *T. sericea* can inhibit 100% HIV-1 RT at 100 μg/mL (Machado et al., 2016). Bessong et al. (2005) showed that the methanolic leaf extracts of the plant inhibited the RDDP functions of HIV-1 RT by 98% at 100 μg/mL (Bessong et al., 2005). A related species of the plant, *Terminalia paniculata* Roth, has also exhibited anti-HIV-1 activity with both acetone and methanol extracts having an IC_{50} $\leq$10.3 μg/mL at 100 μg/mL (Durge et al., 2017). These results suggest that *T. sericea* has the potential to be used for drug development for the treatment of HIV.

Resveratrol has been isolated from ethanol extracts of the plant and has antimicrobial activity (Joseph, Moshi, Innocent, & Nkunya, 2007; Mongalo, McGaw, Segapelo, Finnie, & Van Staden, 2016). The compound inhibits *N. gonorrhoeae* with a MIC of 25 μg/mL (Docherty, Fu, & Tsai, 2001). Furthermore, the compound has exhibited antifungal activity against *Candida* spp. including *C. albicans* (Houillé et al., 2014; Weber, Schulz, & Ruhnke, 2011). The bioactive compound has antibacterial properties against *C. trachomatis* with pathogenesis seen at concentrations below 75 μM (Petyaev et al., 2017). Resveratrol has antiviral activity against HSV with inhibition of 99% at 100 μg/mL (Annunziata et al., 2018; Docherty et al., 1999).

Conclusion

The high incidence of STDs worldwide results in a burden on our available therapies. This is a consequence of the emergence of resistant STD strains. The decline in the efficient therapeutics for the treatment of these diseases makes it important to look for new candidates for drug discovery.

Most of the research on South Africans plants and STDs is mainly in vitro studies. There is limited research on this subject area, as few scientists focus on indigenous plants and venereal diseases. There are generally more studies on HIV and gonorrhea. Although few plants, such as *H. hemerocallidea*, have moved on to clinical studies, from the literature study conducted there is certainly evidence that plants can be utilized in developing novel therapeutics. Plants such as *A. ferox, C. abbreviata*, and *H. hemerocallidea* have potential to be used for the development of novel gonorrhea treatments due to their good antigonococcal properties. More studies have to be done on phytochemicals isolated from plants for the treatment of STDs. This is because research shows that bioactivity against pathogens increases when isolated compounds are used. For example, resveratrol, EGCG, and aloin all have better antigonococcal activity than the crude extracts from which they are isolated. The in vitro studies that have been conducted thus far validate the use of plants in treatment against STDs by traditional healers.

References

Acchioni, C., Remoli, A. L., Marsili, G., Acchioni, M., Nardolillo, I., Orsatti, R., ... Sgarbanti, M. (2019). Alternate NF-κB-independent signaling reactivation of latent HIV-1 provirus. *Journal of Virology, 93*. Available from https://doi.org/10.1128/jvi.00495-19, e00495-19.

Achakazai, B., Abbas, F., Samad, A., Achakazai, H., Malghani, A., Achakazai, H., ... Taj, M. K. (2017). Prevalence of syphilis and chlamydia in married females of Quetta. *Pure and Applied Biology, 6*(3), 1037–1043. Available from https://doi.org/10.19045/bspab.2017.600110.

Ahmed, B., Al-Howiriny, T., Passreiter, C., & Mossa, J. (2004). Pharmaceutical biology combretene-A and B: Two new triterpenes from Combretum molle. *Pharmaceutical Biology, 42*, 109–113. Available from https://doi.org/10.1080/13880200490510883.

Albrecht, C. F. (1996). Hypoxoside: A putative, non-toxic prodrug for the possible treatment of certain malignancies, HIV-infection and inflammatory conditions. In: *Chemistry, biological and pharmacological properties of African medicinal plants: Proceedings of the first international IOCD-symposium*, Victoria Falls, Zimbabwe, February 25–28, (pp. 303–309).Harare: University of Zimbabwe Publications.

Alirol, E., Wi, T. E., Bala, M., Bazzo, M. L., Chen, X., Deal, C., ... Balasegaram, M. (2017). Multidrug-resistant gonorrhea: A research and development roadmap to discover new medicines. *PLoS Medicine, 14*, e1002366. Available from https://doi.org/10.1371/journal.pmed.1002366.

Anderson, J. E., Hobbs, M. M., Biswas, G. D., & Sparling, P. F. (2003). Opposing selective forces for expression of the gonococcal lactoferrin receptor. *Molecular Microbiology, 48*, 1325–1337. Available from https://doi.org/10.1046/j.1365-2958.2003.03496.x.

Annunziata, G., Maisto, M., Schisano, C., Ciampaglia, R., Narciso, V., Tenore, G. C., & Novellino, E. (2018). Resveratrol as a novel anti-Herpes Simplex Virus nutraceutical agent: An overview. *Viruses, 10*, 473. Available from https://doi.org/10.3390/v10090473.

Anwar, K. P., Malik, A., & Subhan, K. H. (2012). Profile of candidiasis in HIV infected patients. *Iranian Journal of Microbiology, 4*, 204–209. Available from http://ijm.tums.ac.ir.

Arenas, J. (2012). The role of bacterial lipopolysaccharides as immune modulator in vaccine and drug development. *Endocrine, Metabolic & Immune Disorders Drug Targets, 12*(3), 221–235. Available from https://doi.org/10.2174/187153012802002884.

Asres, K., Bucar, F., Knauder, E., Yardley, V., Kendrick, H., & Croft, S. L. (2001). *In vitro* antiprotozoal activity of extract and compounds from the stem bark of *Combretum molle*. *Phytotherapy Research, 15*, 613–617. Available from https://doi.org/10.1002/ptr.897.

Australasia Sexual Alliance. (2019). *Australian STI Management Guidelines for use in primary care*. Retrieved from http://www.sti.guidelines.org.au/sexually-transmissible-infections/gonorrhoea.

Australian Government Department of Health. (2018). Multi-drug resistant gonorrhoea | Australian Government Department of Health. Retrieved from https://www.health.gov.au/news/multi-drug-resistant-gonorrhoea.

Ball, L. M., & Criss, A. K. (2013). Constitutively opa-expressing and opa-deficient *Neisseria gonorrhoeae* strains differentially stimulate and survive exposure to human neutrophils. *Journal of Bacteriology, 195*, 2982–2990. Available from https://doi.org/10.1128/JB.00171-13.

Barnes, R. C., & Holmes, K. K. (1984). Epidemiology of gonorrhea: Current perspectives. *Epidemiologic Reviews*, *6*, 1–30. Available from https://doi.org/10.1093/oxfordjournals.epirev.a036267.

Belcher, C., & Dawson, M. (2017). Infectious disease emergencies. In C. K. Stone, & R. L. Humphries (Eds.), *Current diagnosis & treatment; treatment: Emergency medicine, 8e*. New York, NY: McGraw-Hill Education. Available from http://accessmedicine.mhmedical.com/content.aspx?aid = 1144320961.

Bergsson, G., Steingrímsson, Ó., & Thormar, H. (1999). *In vitro* susceptibilities of *Neisseria gonorrhoeae* to fatty acids and monoglycerides. *Antimicrobial Agents and Chemotherapy*, *43*, 2790. Available from https://www.ncbi.nlm.nih.gov/pmc/articles/PMC89562/.

Bernstein, K. T., Marcus, J. L., Nieri, G., Philip, S. S., & Klausner, J. D. (2010). Rectal gonorrhea and chlamydia reinfection is associated with increased risk of HIV seroconversion. *Journal of Acquired Immune Deficiency Syndromes*, *53*, 537–543. Available from https://doi.org/10.1097/QAI.0b013e3181c3ef29.

Bessong, P. O., Obi, C. L., Andréola, M. L., Rojas, L. B., Pouységu, L., Igumbor, E., ... Litvak, S. (2005). Evaluation of selected South African medicinal plants for inhibitory properties against human immunodeficiency virus type 1 reverse transcriptase and integrase. *Journal of Ethnopharmacology*, *99*, 83–91. Available from https://doi.org/10.1016/j.jep.2005.01.056.

Bignell, C., Unemo, M., Radcliffe, K., Jensen, J. S., Babayan, K., Barton, S., ... Van De Laar, M. (2013). 2012 European guideline on the diagnosis and treatment of gonorrhoea in adults. *International Journal of STD & AIDS*, *24*, 85–92. Available from https://doi.org/10.1177/0956462412472837.

Bouic, P. J., Clark, A., Brittle, W., Lamprecht, J. H., Freestone, M., & Liebenberg, R. W. (2001). Plant sterol/sterolin supplement use in a cohort of South African HIV-infected patients-effects on immunological and virological surrogate markers. *South African Medical Journal*, *91*, 848–850.

Bouic, P. J. D., Etsebeth, S., Liebenberg, R. W., Albrecht, C. F., Pegel, K., & Van Jaarsveld, P. P. (1996). Beta-sitosterol and beta-sitosterol glucoside stimulate human peripheral blood lymphocyte proliferation: Implications for their use as an immunomodulatory vitamin combination. *International Journal of Immunopharmacology*, *18*, 693–700. Available from https://doi.org/10.1016/S0192-0561(97)85551-8.

Boxman, I. A., Hogewoning, A., Mulder, L. H., Bouwes Bavinck, J. N., & Ter Schegget, J. (1999). Detection of human papillomavirus types 6 and 11 in pubic and perianal hair from patients with genital warts. *Journal of Clinical Microbiology*, *37*(7), 2270–2273.

Bryant, A. T. (1966). *Zulu medicine and medicine-men* (2nd ed.). *Cape Town: Struik.*

Cates, W., & Wasserheit, J. N. (1991). Genital chlamydial infections: Epidemiology and reproductive sequelae. *American Journal of Obstetrics and Gynecology*, *164*, 1771–1781. Available from https://doi.org/10.1016/0002-9378(91)90559-A.

Cavalcanti Filho, J. R., Silva, T. F., Nobre, W. Q., Oliveira de Souza, L. I., Silva e Silva Figueiredo, C. S., Figueiredo, R. C., & Correia, M. T. D. S. (2017). Antimicrobial activity of *Buchenavia tetraphylla* against *Candida albicans* strains isolated from vaginal secretions. *Pharmaceutical Biology*, *55*(1), 1521–1527. Available from https://doi.org/10.1080/13880209.2017.1304427.

Centers for Disease Control and Prevention. (2017). *Gonorrhea - CDC Fact Sheet*, 1–2. Retrieved from https://wwwn.cdc.gov/dcs/.

Chauke, M. A., Shai, L. J., Mogale, M. A., & Mokgotho, M. P. (2016). Antibacterial and anti HIV-1 reverse transcriptase activity of selected medicinal plants from Phalaborwa, South Africa. *Research Journal of Medicinal Plant*, *10*, 88–395. Available from https://doi.org/10.3923/rjmp.2016.388.395.

Chauke, M. A., Shai, L. J., Mogale, M. A., Tshisikhawe, M. P., & Mokgotho, M. P. (2015). Medicinal plant use of villagers in the Mopani district, Limpopo province, South Africa. *African Journal of Traditional, Complementary and Alternative Medicines*, *12*(3), 9–26. Available from https://doi.org/10.4314/ajtcam.v12i3.2.

Chin, V. K., Lee, T. Y., Rusliza, B., & Chong, P. P. (2016). Dissecting *Candida albicans* infection from the perspective of *C. albicans* virulence and omics approaches on host–pathogen interaction: A review. *International Journal of Molecular Sciences*, *17*, 1643. Available from https://doi.org/10.3390/ijms17101643.

Chinsembu, K. C. (2016). Ethnobotanical study of medicinal flora utilised by traditional healers in the management of sexually transmitted infections in Sesheke District, Western Province, Zambia. *Brazilian Journal of Pharmacognosy*, *26*, 268–274. Available from https://doi.org/10.1016/j.bjp.2015.07.030.

Christodoulides, M. (2019). *Preparation of lipooligosaccharide (LOS) from* Neisseria gonorrhoeae. *Neisseria Gonorrhoeae: Methods in molecular biology* (pp. 87–96). New York, NY: Humana. Available from https://doi.org/10.1007/978-1-4939-9496-0_6.

Chung, G. T., Yoo, J. S., Oh, H. B., Lee, Y. S., Cha, S. H., Kim, S. J., & Yoo, C. K. (2008). Complete genome sequence of *Neisseria gonorrhoeae* NCCP11945. *Journal of Bacteriology, 190*, 6035–6036. Available from https://doi.org/10.1128/JB.00566-08.

Cornelissen, C. N. (2018). Subversion of nutritional immunity by the pathogenic Neisseriae. *Pathogens and Disease, 76*, 1–14. Available from https://doi.org/10.1093/femspd/ftx112.

Creighton, S., Tenant-Flowers, M., Taylor, C. B., Miller, R., & Low, N. (2003). Co-infection with gonorrhoea and chlamydia: how much is there and what does it mean? *International Journal of STD & AIDS, 14*, 109–113. Available from https://doi.org/10.1258/095646203321156872.

Criss, A. K., & Seifert, H. S. (2012). A bacterial siren song: intimate interactions between *Neisseria* and neutrophils. *Nature Reviews Microbiology, 10*, 178–190. Available from https://doi.org/10.1038/nrmicro2713.

Crowther-Gibson, P., Govender, N., Lewis, D. A., Bamford, C., Brink, A., von Gottberg, A., ... Botha, M. (2011). Part IV. GARP: Human infections and antibiotic resistance. *South African Medical Journal, 101*, 567–578. Available from http://www.samj.org.za/index.php/samj/article/view/5102/3367.

Curran, J., Rendtorff, R., Chandler, R., Wiser, W., Robinson, H., Rendtorff, R., ... Robinson, H. (1975). Female gonorrhea: Its relation to abnormal uterine bleeding, urinary tract symptoms, and cervicitis. *Obstetrics and Gynecology, 45*, 195–198.

Dangarembizi, R., Chivandi, E., Dawood, S., Erlwanger, K., Gundidza, V., Magwa, M., ... Samie, A. (2015). The fatty acid composition and physicochemical properties of the underutilised *Cassia abbreviata* seed oil. *Pakistan Journal of Pharmaceutical Sciences, 28*, 1005–1008. Available from https://pubmed.ncbi.nlm.nih.gov/26004707/.

De Wet, H., Nzama, V. N., & Van Vuuren, S. F. (2012). Medicinal plants used for the treatment of sexually transmitted infections by lay people in northern Maputaland, KwaZulu-Natal Province, South Africa. *South African Journal of Botany, 78*, 12–20. Available from https://doi.org/10.1016/j.sajb.2011.04.002.

Dietrich, M., Bartfeld, S., Munke, R., Lange, C., Ogilvie, L. A., Friedrich, A., & Meyer, T. F. (2011). Activation of NF-kappaB by *Neisseria gonorrhoeae* is associated with microcolony formation and type IV pilus retraction. *Cellular Microbiology, 13*, 1168–1182. Available from https://doi.org/10.1111/j.1462-5822.2011.01607.x.

Docherty, J. J., Fu, M. M., Stiffler, B. S., Limperos, R. J., Pokabla, C. M., & Delucia, A. L. (1999). Resveratrol inhibition of Herpes Simplex Virus replication. *Antiviral Research, 43*, 145–155. Available from https://doi.org/10.1016/S0166-3542(99)00042-X.

Docherty, J. J., Fu, M. M., & Tsai, M. (2001). Resveratrol selectively inhibits *Neisseria gonorrhoeae* and *Neisseria meningitidis*. *Journal of Antimicrobial Chemotherapy, 47*, 243–244. Available from https://doi.org/10.1093/jac/47.2.243.

Douthwaite, S., & Champney, W. S. (2001). Structures of ketolides and macrolides determine their mode of interaction with the ribosomal target site. *Journal of Antimicrobial Chemotherapy, 48*, 1–8. Available from https://doi.org/10.1093/jac/48.suppl_2.1.

Drewes, S. E., Elliot, E., Khan, F., Dhlamini, J. T. B., & Gcumisa, M. S. S. (2008). Hypoxis hemerocallidea-Not merely a cure for benign prostate hyperplasia. *Journal of Ethnopharmacology, 119*, 593–598. Available from https://doi.org/10.1016/j.jep.2008.05.027.

Drewes, S. E., Hall, A., Learmonth, R., & Upfold, U. (1984). Isolation of hypoxoside from *Hypoxis rooperi* and synthesis of (E)-1,5-bis(3′,4′-dimethoxyphenyl)pent-4-en-1-yne. *Phytochemistry, 23*, 1313–1316. Available from https://doi.org/10.1016/S0031-9422(00)80449-5.

Durge, A., Jadaun, P., Wadhwani, A., Chinchansure, A. A., Said, M., Thulasiram, H. V., ... Kulkarni, S. S. (2017). Acetone and methanol fruit extracts of *Terminalia paniculata* inhibit HIV-1 infection *in vitro*. *Natural Product Research, 31*, 1468–1471. Available from https://doi.org/10.1080/14786419.2016.1258561.

Ebada, S. S., Ayoub, N. A., Singab, A. N. B., & Al-Azizi, M. M. (2008). Phytophenolics from *Peltophorum africanum* Sond. (Fabaceae) with promising hepatoprotective activity. *Pharmacognosy Magazine, 4*, 287–293.

Edwards, J. L., & Butler, E. K. (2011). The pathobiology of *Neisseria gonorrhoeae* lower female genital tract infection. *Frontiers in Microbiology, 2*, 102. Available from https://doi.org/10.3389/fmicb.2011.00102.

European Centre for Disease Prevention and Control. (2018). *Extensively drug-resistant (XDR) Neisseria gonorrhoeae in the United Kingdom and Australia. Health Protection Report Advanced Access Report.* 12, 1–11. Retrieved from https://ecdc.europa.eu/sites/portal/files/documents/RRA-Gonorrhoea%2CAntimicrobialresistance-UnitedKingdom%2CAustralia.pdf.

Feinen, B., Jerse, A. E., Gaffen, S. L., & Russell, M. W. (2010). Critical role of Th17 responses in a murine model of Neisseria gonorrhoeae genital infection. *Mucosal Immunology., 3*, 312–321. Available from https://doi.org/10.1038/mi.2009.139.

Ferreira, V. H., Nazli, A., Khan, G., Mian, M. F., Ashkar, A. A., Gray-Owen, S., ... Kaushic, C. (2011). Endometrial epithelial cell responses to coinfecting viral and bacterial pathogens in the genital tract can activate the HIV-1 LTR in an NFκB-and AP-1–dependent manner. *The Journal of Infectious Diseases, 204*, 299–308. Available from https://doi.org/10.1093/infdis/jir260.

Fyhrquist, P., Mwasumbi, L., Hæggström, C. A., Vuorela, H., Hiltunen, R., & Vuorela, P. (2002). Ethnobotanical and antimicrobial investigation on some species of Terminalia and Combretum (Combretaceae) growing in Tanzania. *Journal of Ethnopharmacology, 79*(2), 169–177. Available from https://doi.org/10.1016/S0378-8741(01)00375-0.

Geisler, W. M., & Stamm, W. E. (2007). Chapter 13: Genital chlamydial infections. In J. D. Klausner, & E. W. Hook (Eds.), *Current diagnosis & treatment; sexually transmitted diseases*. New York, NY: The McGraw-Hill Companies. Available from http://accessmedicine.mhmedical.com/content.aspx?aid = 3025206.

Griffiss, J. M. L., Lammel, C. J., Wang, J., Dekker, N. P., & Brooks, G. F. (1999). *Neisseria gonorrhoeae* coordinately uses pili and Opa to activate HEC-1-B cell microvilli, which causes engulfment of the gonococci. *Infection and Immunity, 67*, 3469–3480. Available from https://doi.org/10.1128/iai.67.7.3469-3480.1999.

Guy, R., Ward, J., Wand, H., Rumbold, A., Garton, L., Hengel, B., ... Group, S. I. (2015). Coinfection with *Chlamydia trachomatis*, *Neisseria gonorrhoeae* and *Trichomonas vaginalis*: A cross-sectional analysis of positivity and risk factors in remote Australian Aboriginal communities. *Sexually Transmitted Infections, 91*, 201–206. Available from https://doi.org/10.1136/sextrans-2014-051535.

Hedberg, I., Hedberg, O., Madati, P. J., Mshigeni, K. E., Mshiu, E. N., & Samuelsson, G. (1982). Inventory of plants used in traditional medicine in Tanzania. I. Plants of the families acanthaceae-cucurbitaceae. *Journal of Ethnopharmacology, 6*, 29–60. Available from https://doi.org/10.1016/0378-8741(82)90070-8.

Hedges, S. R., Mayo, M. S., Mestecky, J., Hook, E. W., & Russell, M. W. (1999). Limited local and systemic antibody responses to *Neisseria gonorrhoeae* during uncomplicated genital infections. *Infection and Immunity, 67*, 3937–3946. Available from https://doi.org/10.1128/IAI.67.8.3937-3946.1999.

Hedges, S. R., Sibley, D. A., Mayo, M. S., Hook, E. W., 3rd, & Russell, M. W. (1998). Cytokine and antibody responses in women infected with *Neisseria gonorrhoeae*: Effects of concomitant infections. *The Journal of Infectious Diseases, 178*, 742–751. Available from https://doi.org/10.1086/515372.

Houillé, B., Papon, N., Boudesocque, L., Bourdeaud, E., Besseau, S., Courdavault, V., ... Lanoue, A. (2014). Antifungal activity of resveratrol derivatives against Candida Species. *Journal of Natural Products, 77*, 1658–1662. Available from https://doi.org/10.1021/np5002576.

Hutchings, A. (1989). A survey and analysis of traditional medicinal plants as used by the Zulu; Xhosa and Sotho. *Bothalia., 19*, 111–123. Available from https://doi.org/10.4102/abc.v19i1.947.

Johnson, L. F., Coetzee, D. J., & Dorrington, R. E. (2005). Sentinel surveillance of sexually transmitted infections in South Africa: A review. *Sexually Transmitted Infections, 81*(4), 287–293. Available from https://doi.org/10.1136/sti.2004.013904.

Joseph, C. C., Moshi, M. J., Innocent, E., & Nkunya, M. H. H. (2007). Isolation of a stilbene glycoside and other constituents of *Terminalia sericea*. *African Journal of Traditional, Complementary and Alternative Medicines, 4*, 383–386. Available from https://doi.org/10.4314/ajtcam.v4i4.31231.

Kambizi, L., & Afolayan, A. J. (2008). Extracts from Aloe ferox and *Withania somnifera* inhibit *Candida albicans* and *Neisseria gonorrhoea*. *African Journal of Biotechnology, 7*, 12–15. Available from https://doi.org/10.4314/ajb.v7i1.58301.

Kambizi, L., Goosen, B. M., Taylor, M. B., & Afolayan, A. J. (2007). Anti-viral effects of aqueous extracts of *Aloe ferox* and *Withania somnifera* on Herpes Simplex Virus type 1 in cell culture. *South African Journal of Science, 103*, 359–360.

Kambizi, L., Sultana, N., & Afolayan, A. J. (2005). Bioactive compounds isolated from *Aloe ferox*.: A plant traditionally used for the treatment of sexually transmitted infections in the Eastern Cape, South Africa. *Pharmaceutical Biology, 42*, 636–639. Available from https://doi.org/10.1080/13880200490902581.

Kharsany, A. B., McKinnon, L., Lewis, L., Cawood, C., Khanyile, D., Maseko, D. V., ... Toledo, C. (2020). Population prevalence of sexually transmitted infections in a high HIV burden district in KwaZulu-Natal, South Africa: Implications for HIV epidemic control. *International Journal of Infectious Diseases, 98*, 130–137. Available from https://doi.org/10.1016/j.ijid.2020.06.046.

Kibbi, A., Bahhady, R., & El-Shareef, M. (2012). Granuloma inguinale. In L. A. Goldsmith, S. I. Katz, B. A. Gilchrest, A. S. Paller, D. J. Leffell, & K. Wolff (Eds.), *Fitzpatrick's dermatology in general medicine, 8e*. New York, NY: The McGraw-Hill Companies. Available from http://accessmedicine.mhmedical.com/content.aspx?aid = 56091772.

Klos, M., van de Venter, M., Milne, P. J., Traore, H. N., Meyer, D., & Oosthuizen, V. (2009). *In vitro* anti-HIV activity of five selected South African medicinal plant extracts. *Journal of Ethnopharmacology, 124*, 182–188. Available from https://doi.org/10.1016/j.jep.2009.04.043.

Knapp, J. S., & Clark, V. L. (1984). Anaerobic growth of *Neisseria gonorrhoeae* coupled to nitrite reduction. *Infection and Immunity, 46*, 176–181. Available from https://doi.org/10.1128/iai.46.1.176-181.1984.

Kularatne, R., Kufa-Chakezha, T., Maseko, V., & Gumede, F. (2018). *Neisseria gonorrhoeae* antimicrobial resistance surveilance: NICD GERMS-SA 2017. *Communicable Diseases Surveillance Bullettin, 16*, 131–139.

Kularatne, R., Maseko, V., Gumede, L., Radebe, F., & Chakezha, T. K. (2017). *Neisseria gonorrhoeae* antimicrobial surveillance in Gauteng province, South Africa. *Communicable Diseases Surveillance Bullettin, 14*, 56–64.

Kularatne, R., Niit, R., Rowley, J., Kufa-Chakezha, T., Peters, R. P., Taylor, M. M., ... Korenromp, E. L. (2018). Adult gonorrhea, chlamydia and syphilis prevalence, incidence, treatment and syndromic case reporting in South Africa: Estimates using the Spectrum-STI model, 1990-2017. *PLoS One, 13*, e0205863. Available from https://doi.org/10.1371/journal.pone.0205863.

Lee, D. Y.-W., Lin, X., Paskaleva, E. E., Liu, Y., Puttamadappa, S. S., Thornber, C., ... Canki, M. (2009). Palmitic acid is a novel CD4 fusion inhibitor that blocks HIV entry and infection. *AIDS Research and Human Retroviruses, 25*, 1231–1241. Available from https://doi.org/10.1089/aid.2009.0019.

Lenz, J. D., & Dillard, J. (2018). Pathogenesis of *Neisseria gonorrhoeae* and the host defense in ascending infections of human fallopian tube. *Frontiers in Immunology, 9*, 2710. Available from https://doi.org/10.3389/fimmu.2018.02710.

Leonard, C. A., Schoborg, R. V., Low, N., Unemo, M., & Borel, N. (2019). Pathogenic interplay between *Chlamydia trachomatis* and *Neisseria gonorrhoeae* that influences management and control efforts-more questions than answers? *Current Clinical Microbiology Reports, 6*, 182–191. Available from https://doi.org/10.1007/s40588-019-00125-4.

Leteane, M. M., Ngwenya, B. N., Muzila, M., Namushe, A., Mwinga, J., Musonda, R., ... Andrae-Marobela, K. (2012). Old plants newly discovered: *Cassia sieberiana* D.C. and *Cassia abbreviata* Oliv. Oliv. root extracts inhibit in vitro HIV-1c replication in peripheral blood mononuclear cells (PBMCs) by different modes of action. *Journal of Ethnopharmacology, 141*, 48–56. Available from https://doi.org/10.1016/j.jep.2012.01.044.

Lewis, D. A., Scott, L., Slabbert, M., Mhlongo, S., van Zijl, A., Sello, M., ... Wasserman, E. (2008). Escalation in the relative prevalence of ciprofloxacin-resistant gonorrhoea among men with urethral discharge in two South African cities: association with HIV seropositivity. *Sexually Transmitted Infections, 84*, 352–355. Available from https://doi.org/10.1136/sti.2007.029611.

Lewis, L. A., Choudhury, B., Balthazar, J. T., Martin, L. E., Ram, S., Rice, P. A., ... Shafer, W. M. (2009). Phosphoethanolamine substitution of lipid A and resistance of *Neisseria gonorrhoeae* to cationic antimicrobial peptides and complement-mediated killing by normal human serum. *Infection and Immunity, 77*, 1112–1120. Available from https://doi.org/10.1128/IAI.01280-08.

Lin, X., Paskaleva, E. E., Chang, W., Shekhtman, A., & Canki, M. (2011). Inhibition of HIV-1 infection in ex vivo cervical tissue model of human vagina by palmitic acid; implications for a microbicide development. *PLoS One, 6*, e24803. Available from https://doi.org/10.1371/journal.pone.0024803.

Liu, M., John, C. M., & Jarvis, G. A. (2010). Phosphoryl moieties of lipid a from *Neisseria meningitidis* and *N. gonorrhoeae* lipooligosaccharides play an important role in activation of both MyD88-and TRIF-dependent TLR4-MD-2 signaling pathways. *Journal of Immunology, 185*, 6974–6984. Available from https://doi.org/10.4049/jimmunol.1000953.

Liu, Y., Liu, W., & Russell, M. W. (2014). Suppression of host adaptive immune responses by *Neisseria gonorrhoeae*: role of interleukin 10 and type 1 regulatory T cells. *Mucosal Immunology., 7*, 165–176. Available from https://doi.org/10.1038/mi.2013.36.

Lynagh, Y., Mac Aogáin, M., Walsh, A., Rogers, T. R., Unemo, M., & Crowley, B. (2015). Detailed characterization of the first high-level azithromycin-resistant *Neisseria gonorrhoeae* cases in Ireland. *Journal of Antimicrobial Chemotherapy, 70*, 2411–2413. Available from https://doi.org/10.1093/jac/dkv106.

Mabogo, D. E. N. (1990). *The ethnobotany of the Vhavenda*. University of Pretoria.

Machado, D., Castro, J., Palmeira-de-Oliveira, A., Martinez-de-Oliveira, J., & Cerca, N. (2016). Bacterial vaginosis biofilms: Challenges to current therapies and emerging solutions. *Frontiers in Microbiology, 6*, 1528. Available from https://doi.org/10.3389/fmicb.2015.01528.

Maier, R., & Katsufrakis, P. J. (2015). *Chapter 14: Sexually transmitted diseases. Current diagnosis & treatment: family medicine*. New York, NY: McGraw-Hill Education. Available from http://accessmedicine.mhmedical.com/content.aspx?aid = 1106847551.

Mamba, P., Adebayo, S. A., & Tshikalange, T. E. (2016). Anti-microbial, anti-inflammatory and HIV-1 reverse transcriptase activity of selected South African plants used to treat sexually transmitted diseases. *International Journal of Pharmacognosy and Phytochemical Research*, *8*, 1870–1876.

Maroyi, A., & Semenya, S. S. (2019). Medicinal uses, phytochemistry and pharmacological properties of Elaeodendron transvaalense. *Nutrients.*, *11*, 545. Available from https://doi.org/10.3390/nu11030545.

Martin, I., Sawatzky, P., Allen, V., Lefebvre, B., Hoang, L., Naidu, P., ... Mulvey, M. (2019). Multidrug-resistant and extensively drug-resistant *Neisseria gonorrhoeae* in Canada, 2012–2016. *Canada Communicable Disease Report*, *45*, 45–53. Available from https://doi.org/10.14745/ccdr.v45i23a01.

Martin-Gayo, E., & Yu, X. G. (2019). Role of dendritic cells in natural immune control of HIV-1 infection. *Frontiers in Immunology*, *10*, 1306. Available from https://doi.org/10.3389/fimmu.2019.01306.

Masoko, P., Picard, J., & Eloff, J. N. (2007). The antifungal activity of twenty-four southern African *Combretum* species (Combretaceae). *South African Journal of Botany*, *73*, 173–183. Available from https://doi.org/10.1016/j.sajb.2006.09.010.

Matsumoto, Y., Kaihatsu, K., Nishino, K., Ogawa, M., Kato, N., & Yamaguchi, A. (2012). Antibacterial and antifungal activities of new acylated derivatives of epigallocatechin gallate. *Frontiers in Microbiology*, *3*, 53. Available from https://doi.org/10.3389/fmicb.2012.00053.

Matyanga, C. M. J., Morse, G. D., Gundidza, M., & Nhachi, C. F. B. (2020). African potato (*Hypoxis hemerocallidea*): A systematic review of its chemistry, pharmacology and ethno medicinal properties. *BMC Complementary Medicine and Therapies*, *20*, 1–12. Available from https://doi.org/10.1186/s12906-020-02956-x.

Mayer, F. L., Wilson, D., & Hube, B. (2013). *Candida albicans* pathogenicity mechanisms. *Virulence*, *4*(2), 119–128. Available from https://doi.org/10.4161/viru.22913.

McCormack, W. M., Stumacher, R. J., Johnson, K., & Donner, A. (1977). Clinical spectrum of gonococcal infection in women. *Lancet (London, England)*, *1*, 1182–1185. Available from https://doi.org/10.1016/s0140-6736(77)92720-9.

McSheffrey, G. G., & Gray-Owen, S. D. (2015). *Neisseria gonorrhoeae. Molecular medical microbiology* (2nd (ed.), pp. 1471–1485). Boston: Academic Press. Available from https://doi.org/10.1016/B978-0-12-397169-2.00082-2.

Mills, E., Cooper, C., Seely, D., & Kanfer, I. (2005). African herbal medicines in the treatment of HIV: Hypoxis and Sutherlandia. An overview of evidence and pharmacology. *Nutrition Journal*, *4*, 19. Available from https://doi.org/10.1186/1475-2891-4-19.

Mlisana, K., Naicker, N., Werner, L., Roberts, L., van Loggerenberg, F., Baxter, C., ... Abdool Karim, S. S. (2012). Symptomatic vaginal discharge is a poor predictor of sexually transmitted infections and genital tract inflammation in high-risk women in South Africa. *The Journal of Infectious Diseases*, *206*, 6–14. Available from https://doi.org/10.1093/infdis/jis298.

Mongalo, N. I. (2013). Peltophorum africanum Sond [Mosetlha]: A review of its ethnomedicinal uses, toxicology, phytochemistry and pharmacological activities. *Journal of Medicinal Plants Research*, *7*, 3484–3491. Available from https://doi.org/10.5897/JMPR2013.5302.

Mongalo, N. I., & Makhafola, T. J. (2018). Ethnobotanical knowledge of the lay people of Blouberg area (Pedi tribe), Limpopo Province, South Africa. *Journal of Ethnobiology and Ethnomedicine*, *14*(1), 46. Available from https://doi.org/10.1186/s13002-018-0245-4.

Mongalo, N. I., McGaw, L. J., Finnie, J. F., & Van Staden, J. (2017). Pharmacological properties of extracts from six South African medicinal plants used to treat sexually transmitted infections (STIs) and related infections. *South African Journal of Botany*, *112*, 290–295. Available from https://doi.org/10.1016/j.sajb.2017.05.031.

Mongalo, N. I., McGaw, L. J., Segapelo, T. V., Finnie, J. M., & Van Staden, J. (2016). Ethnobotany, phytochemistry, toxicology and pharmacological properties of Terminalia sericea Burch. ex DC. (Combretaceae) – A review. *Journal of Ethnopharmacology*, *194*, 789–802. Available from https://doi.org/10.1016/j.jep.2016.10.072.

Murphy, G. L., Connell, T. D., Barritt, D. S., Koomey, M., & Cannon, J. G. (1989). Phase variation of gonococcal protein II: Regulation of gene expression by slipped-strand mispairing of a repetitive DNA sequence. *Cell.*, *56*, 539–547. Available from https://doi.org/10.1016/0092-8674(89)90577-1.

Mushi, N. F., Mbwambo, Z. H., Innocent, E., & Tewtrakul, S. (2012). Antibacterial, anti-HIV-1 protease and cytotoxic activities of aqueous ethanolic extracts from *Combretum adenogonium* Steud. Ex A. Rich (Combretaceae). *BMC Complementary and Alternative Medicine*, *12*, 163. Available from https://doi.org/10.1186/1472-6882-12-163.

Naidoo, D., Van Vuuren, S. F., Van Zyl, R. L., & De Wet, H. (2013). Plants traditionally used individually and in combination to treat sexually transmitted infections in northern Maputaland, South Africa: Antimicrobial activity and cytotoxicity. *Journal of Ethnopharmacology, 149*, 656–667. Available from https://doi.org/10.1016/j.jep.2013.07.018.

Ncube, B., Finnie, J. F., & Staden, J. V. (2011). *In vitro* antimicrobial synergism within plant extract combinations from three South African medicinal bulbs. *Journal of Ethnopharmacology, 139*, 81–89. Available from https://doi.org/10.1016/j.jep.2011.10.025.

Ndhlala, A. R., Amoo, S. O., Ncube, B., Moyo, M., Nair, J. J., & Van Staden, J. (2013). *Antibacterial, antifungal, and antiviral activities of African medicinal plants. Medicinal plant research in Africa: Pharmacology and chemistry* (pp. 621–659). Elsevier Inc. Available from https://doi.org/10.1016/B978-0-12-405927-6.00016-3.

Ohnishi, M., Golparian, D., Shimuta, K., Saika, T., Hoshina, S., Iwasaku, K., ... Unemo, M. (2011). Is *Neisseria gonorrhoeae* initiating a future era of untreatable gon *Neisseria gonorrhoeae* initiating a future era of untreatable gonorrhea?: Detailed characterization of the first strain with high-level resistance to ceftriaxone. *Antimicrobial Agents and Chemotherapy, 55*, 3538–3545. Available from https://doi.org/10.1128/AAC.00325-11.

Ohnishi, M., Saika, T., Hoshina, S., Iwasaku, K., Nakayama, S. I., Watanabe, H., & Kitawaki, J. (2011). Ceftriaxone-resistant *Neisseria gonorrhoeae*, Japan. *Emerging Infectious Diseases, 17*, 148–149. Available from https://doi.org/10.3201/eid1701.100397.

Oller, L. Z., Wood, T., & St Luke, S. (1970). Factors influencing the incidence of gonorrhoea and non-gonococcal urethritis in men in an industrial city. *British Journal of Venereal Diseases, 46*, 96. Available from https://doi.org/10.1136/sti.46.2.96.

Palmeira-de-Oliveira, A., Silva, B. M., Palmeira-de-Oliveira, R., Martinez-de-Oliveira, J., & Salgueiro, L. (2013). Are plant extracts a potential therapeutic approach for genital infections? *Current Medicinal Chemistry, 20*, 2914–2928. Available from https://doi.org/10.2174/0929867311320990007.

Parrow, N., Fleming, R. E., & Minnick, M. F. (2013). Sequestration and scavenging of iron in infection. *Infection and Immunity, 81*, 3503–3514. Available from https://doi.org/10.1128/IAI.00602-13.

Petyaev, I., Zigangirova, N., Morgunova, E., Kyle, N., Fedina, E., & Bashmakov, Y. (2017). Resveratrol inhibits propagation of *Chlamydia trachomatis* in McCoy cells. *BioMed Research International, 2017*, 1–7. Available from https://doi.org/10.1155/2017/4064071.

Platt, R., Rice, P. A., & McCormack, W. M. (1983). Risk of acquiring gonorrhea and prevalence of abnormal adnexal findings among women recently exposed to gonorrhea. *Journal of the American Medical Association, 250*, 3205–3209. Available from https://doi.org/10.1001/jama.1983.03340230057031.

Public Health Canada. (2017). Treatment of *N. gonorrhoeae* in response to the discontinuation of spectinomycin: Alternative treatment guidance statement. *Canadian Guidelines on Sexually Transmitted Infections* (pp. 1–4). Retrieved from https://doi.org/10.1186/s13063-016-1683-8.

Public Health England. (2018). *UK case of Neisseria gonorrhoeae with high-level resistance to azithromycin and resistance to ceftriaxone acquired abroad* (pp. 1–4). Retrieved from https://assets.publishing.service.gov.uk/government/uploads/system/uploads/attachment_data/file/694655/hpr1118_MDRGC.pdf.

Quillin, S. J., & Seifert, H. S. (2018). *Neisseria gonorrhoeae* host adaptation and pathogenesis. *Nature Reviews Microbiology, 16*, 226–240. Available from https://doi.org/10.1038/nrmicro.2017.169.

Rowley, J., Hoorn, S. V., Korenromp, E., Low, N., Unemo, M., Abu-Raddad, L. J., ... Taylor, M. M. (2019). Global and regional estimates of the prevalence and incidence of four curable sexually transmitted infections in 2016. *WHO Bulletin, 97*, 548–562P. Available from https://doi.org/10.2471/BLT.18.228486.

Ryan, K. J. (2017). *Neisseria. Sherris medical microbiology, 7e*. New York, NY: McGraw-Hill Education. Available from http://accessmedicine.mhmedical.com/content.aspx?aid = 1148675051.

Sadarangani, M., Pollard, A. J., & Gray-Owen, S. D. (2011). Opa proteins and CEACAMs: Pathways of immune engagement for pathogenic *Neisseria*. *FEMS Microbiology Reviews, 35*, 498–514. Available from https://doi.org/10.1111/j.1574-6976.2010.00260.x.

Samie, A., Tambani, T., Harshfield, E., Green, E., Ramalivhana, J. N., & Bessong, P. O. (2010). Antifungal activities of selected venda medicinal plants against *Candida albicans, Candida krusei* and *Cryptococcus neoformans* isolated from South African AIDS patients. *African Journal of Biotechnology, 9*, 2965–2976. Available from https://doi.org/10.5897/AJB09.1521.

Sarantis, H., & Gray-Owen, S. D. (2007). The specific innate immune receptor CEACAM3 triggers neutrophil bactericidal activities via a Syk kinase-dependent pathway. *Cellular Microbiology, 9*, 2167–2180. Available from https://doi.org/10.1111/j.1462-5822.2007.00947.x.

Semenya, S., Potgieter, M. J., & Erasmus, L. J. C. (2013). Indigenous plant species used by Bapedi healers to treat sexually transmitted infections: Their distribution, harvesting, conservation and threats. *South African Journal of Botany*, *87*, 66–75. Available from https://doi.org/10.1016/j.sajb.2013.03.001.

Semenya, S. S., Potgieter, M. J., Johannes, L., & Erasmus, C. (2013). Bapedi phytomedicine and their use in the treatment of sexually transmitted infections in Limpopo Province, South Africa. *African Journal of Pharmacy and Pharmacology*, *7*(6), 250–262. Available from https://doi.org/10.5897/AJPP12.608.

Seo, Y., Choi, K. H., & Lee, G. (2019). Characterization and trend of co-infection with *Neisseria gonorrhoeae* and *Chlamydia trachomatis* from the Korean National infectious diseases surveillance database. *World Journal of Men's Health*, *37*, 1–9. Available from https://doi.org/10.5534/WJMH.190116.

Shim, B. S. (2011). Current concepts in bacterial sexually transmitted diseases. *Korean Journal of Urology*, *52*, 589–597. Available from https://doi.org/10.4111/kju.2011.52.9.589.

Sigidi, M. T., Ndama Traoré, A., Boukandou, M., Tshisikhawe, M. P., Ntuli, S. S., & Potgieter, N. (2017). Anti-HIV, pro-inflammatory and cytotoxicity properties of selected Venda plants. *Indian Journal of Traditional Knowledge*, *16*, 545–552.

Simon, V., Ho, D. D., & Karim, A. Q. (2006). HIV/AIDS epidemiology, pathogenesis, prevention, and treatment. *Lancet*, *368*(9534), 489–504. Available from https://doi.org/10.1016/S0140-6736(06)69157-5.

Sobeh, M., Mahmoud, M. F., Abdelfattah, M. A., Cheng, H., El-Shazly, A. M., & Wink, M. (2018). A proanthocyanidin-rich extract from *Cassia abbreviata* exhibits antioxidant and hepatoprotective activities in vivo. *Journal of Ethnopharmacology*, *213*, 38–47. Available from https://doi.org/10.1016/j.jep.2017.11.007.

Steinman, R. M. (1991). The dendritic cell system and its role in immunogenicity. *Annual Review of Immunology*, *9*, 271–296. Available from https://doi.org/10.1146/annurev.iy.09.040191.001415.

Stevens, K., Zaia, A., Tawil, S., Bates, J., Hicks, V., Whiley, D., … Howden, B. P. (2014). *Neisseria gonorrhoeae* isolates with high-level resistance to azithromycin in Australia. *Journal of Antimicrobial Chemotherapy*, *70*, 1267–1268. Available from https://doi.org/10.1093/jac/dku490.

Suay-García, B., & Pérez-Gracia, M. (2018). Future prospects for *Neisseria gonorrhoeae* treatment. *Antibiotics.*, *7*, 49. Available from https://doi.org/10.3390/antibiotics7020049.

Talwar, G. P., Raghuvanshi, P., Mishra, R., Banerjee, U., Rattan, A., Whaley, K. J., … Doncel, G. F. (2000). Polyherbal formulations with wide spectrum antimicrobial activity against reproductive tract infections and sexually transmitted pathogens. *American Journal of Reproductive Immunology*, *43*(3), 144–151. Available from https://doi.org/10.1111/j.8755-8920.2000.430303.x.

Tshikalange, T. E., Mamba, P., & Adebayo, S. A. (2016). Antimicrobial, antioxidant and cytotoxicity studies of medicinal plants used in the treatment of sexually transmitted diseases. *International Journal of Pharmacognosy and Phytochemical Research*, *8*(11), 1891–1895.

Tshikalange, T. E., Meyer, J. J. M., Lall, N., Muñoz, E., Sancho, R., Van de Venter, M., & Oosthuizen, V. (2008). *In vitro* anti-HIV-1 properties of ethnobotanically selected South African plants used in the treatment of sexually transmitted diseases. *Journal of Ethnopharmacology*, *119*, 478–481. Available from https://doi.org/10.1016/j.jep.2008.08.027.

Unemo, M., Del Rio, C., & Shafer, W. M. (2016). Antimicrobial resistance expressed by *Neisseria gonorrhoeae*: A major global public health problem in the 21st century. *Microbiology Spectrum.*, *4*. Available from https://doi.org/10.1128/microbiolspec.EI10-0009-2015.

Unemo, M., & Nicholas, R. A. (2012). Emergence of multidrug-resistant, extensively drug-resistant and untreatable gonorrhea. *Future Microbiology*, *7*, 1401–1422. Available from https://doi.org/10.2217/fmb.12.117.

Unemo, M., & Shafer, W. M. (2011). Antibiotic resistance in *Neisseria gonorrhoeae*: Origin, evolution, and lessons learned for the future. *Annals of the New York Academy of Sciences*, *1230*, 19–28. Available from https://doi.org/10.1111/j.1749-6632.2011.06215.x.

Unemo, M., & Shafer, W. M. (2014). Antimicrobial resistance in *Neisseria gonorrhoeae* in the 21st century: Past, evolution, and future. *Clinical Microbiology Reviews*, *27*, 587–613. Available from https://doi.org/10.1128/CMR.00010-14.

Van Vuuren, S., & Holl, D. (2017). Antimicrobial natural product research: A review from a South African perspective for the years 2009–2016. *Journal of Ethnopharmacology*, *208*, 236–252. Available from https://doi.org/10.1016/j.jep.2017.07.011.

van Vuuren, S. F. (2008). Antimicrobial activity of South African medicinal plants. *Journal of Ethnopharmacology*, *119*, 462–472. Available from https://doi.org/10.1016/j.jep.2008.05.038.

van Vuuren, S. F., & Naidoo, D. (2010). An antimicrobial investigation of plants used traditionally in southern Africa to treat sexually transmitted infections. *Journal of Ethnopharmacology, 130*, 552–558. Available from https://doi.org/10.1016/j.jep.2010.05.045.

Van Wyk, B., Van Oudtshoorn, B., Gericke, N. (2017). *Medicinal plants of South Africa* (2nd ed.). *Pretoria: Briza.*

Viol, D. I., Chagonda, L. S., Moyo, S. R., & Mericli, A. H. (2016). Toxicity and antiviral activities of some medicinal plants used by traditional medical practitioners in Zimbabwe. *American Journal of Plant Sciences, 7*, 1538–1544. Available from https://doi.org/10.4236/ajps.2016.711145.

Wallin, J. (1974). Gonorrhoea in 1972 A 1-year study of patients attending the VD Unit in Uppsala. *British Journal of Venereal Diseases, 51*, 41–47. Available from https://www.ncbi.nlm.nih.gov/pmc/articles/PMC1045109/pdf/brjvendis00055-0047.pdf.

Wang, J., Gray-Owen, S. D., Knorre, A., Meyer, T. F., & Dehio, C. (1998). Opa binding to cellular CD66 receptors mediates the transcellular traversal of *Neisseria gonorrhoeae* across polarized T84 epithelial cell monolayers. *Molecular Microbiology, 30*, 657–671. Available from https://doi.org/10.1046/j.1365-2958.1998.01102.x.

Watt, J. M., & Breyer-Brandwijk, M. G. (1962). *The medicinal and poisonous plants of southern and eastern Africa*. E. & S. Livingstone. Available from https://books.google.co.za/books?id = 2ZjwAAAAMAAJ.

Weber, K., Schulz, B., & Ruhnke, M. (2011). Resveratrol and its antifungal activity against *Candida* species. *Mycoses, 54*, 30–33. Available from https://doi.org/10.1111/j.1439-0507.2009.01763.x.

Westling-Haggstrom, B., Elmros, T., Normark, S., & Winblad, B. (1977). Growth pattern and cell division in *Neisseria gonorrhoeae. Journal of Bacteriology, 129*, 333–342. Available from https://doi.org/10.1128/jb.129.1.333-342.1977.

WHO Regional Office for Africa. (2018). *Global health sector stratergy on sexually transmitted infections 2016-2021 implementation framework for the African region*. Retrieved from http://apps.who.int/bookorders.

Wi, T., Lahra, M. M., Ndowa, F., Bala, M., Dillon, J. R., Ramon-Pardo, P., ... Unemo, M. (2017). Antimicrobial resistance in *Neisseria gonorrhoeae*: Global surveillance and a call for international collaborative action. *PLoS Medicine, 14*, e1002344. Available from https://doi.org/10.1371/journal.pmed.1002344.

World Health Organization. (2016a). *Sexually transmitted infections (STIs)*, WHO. Retrieved from http://www.who.int/mediacentre/factsheets/fs110/en/.

World Health Organization. (2016b). *WHO Guidelines for the treatment of Neisseria gonorrhoeae.*

World Health Organization. (2017). *Global priority list of antibiotic-resistant bacteria to guide research, discovery and development of new antibiotics* (pp. 1–7). Retrieved from http://www.cdc.gov/drugresistance/threat-report-2013/.

World Health Organisation. (2019). *WHO report on global sexually transmitted infection surveillance 2018*, WHO. Retrieved from https://www.who.int/reproductivehealth/publications/stis-surveillance-2018/en/.

Yang S., Wu T., Zheng J., Huang Y., Chen X.Y., & Wu H. (2012). Traditional Chinese medicinal herbs for *Condyloma acuminatum, Cochrane Database of Systematic Reviews*. Retrieved from https://doi.org/10.1002/14651858.CD010234.

Yu, Q., Chow, E. M., McCaw, S. E., Hu, N., Byrd, D., Amet, T., ... Gray-Owen, S. (2013). Association of *Neisseria gonorrhoeae* OpaCEA with dendritic cells suppresses their ability to elicit an HIV-1-specific T cell memory response. *PLoS One, 8*, e56705. Available from https://doi.org/10.1371/journal.pone.0056705.

Zhou, X., Gao, X., Broglie, P. M., Kebaier, C., Anderson, J. E., Thom, N., ... Duncan, J. A. (2014). Hexa-acylated lipid A is required for host inflammatory response to *Neisseria gonorrhoeae* in experimental gonorrhea. *Infection and Immunity, 82*, 184–192. Available from https://doi.org/10.1128/IAI.00890-13.

Zhu, W., Ventevogel, M. S., Knilans, K. J., Anderson, J. E., Oldach, L. M., McKinnon, K. P., ... Duncan, J. A. (2012). *Neisseria gonorrhoeae* suppresses dendritic cell-induced, antigen-dependent CD4 T cell proliferation. *PLoS One, 7*, e41260. Available from https://doi.org/10.1371/journal.pone.0041260.

CHAPTER

6

Antibacterial activity of some selected medicinal plants of Pakistan

Zia Ur Rehman Mashwani, Rahmat Wali, Muhammad Faraz Khan, Fozia Abasi, Nadia Khalid and Naveed Iqbal Raja

Department of Botany, PMAS - Arid Agriculture University, Rawalpindi, Pakistan

Introduction

Plants have been used for therapeutic purposes by humans since ancient times. Almost every resident area on the planet has developed a traditional therapeutic system based on information about medicinal plants. Two-thirds of the world's population, or approximately 6.8 billion people, use medicinal plants as a treatment for diseases ranging from the common cold to cancer (Ali, Faizi, & Kazmi, 2011). Medicines obtained from plants, animals, or minerals fall under the category of "Ethnomedicine." These medicines are dependent on local pharmacopeia and are source of healing for a wide range of diseases.

According to the World Health Organization (WHO), medicinal plants are classified as those plants in which one or more part has a therapeutic potential and can be used for drug synthesis (Zahoor, Shah, Gul, & Amin, 2018). Medicinal plants show an important role in both herbal medicine and healthcare systems (Rahimullah, Shah, Mujaddad-ur-Rehman, & Hayat, 2019). A growing number of researches have shown that the plants are rich in phytochemicals required for the synthesis and development of different drugs (Kumar, Karthik, & Rao, 2010). Different species of Gram-positive and Gram-negative bacteria are responsible for causing infections in a large number of human populations (Ahameethunisa & Hopper, 2010; Bibi, Nisa, Chaudhary, & Zia, 2011). Diverse types of diseases are caused by bacteria comprising bloodstream infections, urinary tract infections, wound infections, skin infections, pneumonia, asthma, and so on. Certain bacterial strains are dangerous enough to cause death in humans such as 3–4 million people die each year all over the globe from diarrhea as a result of intestinal infection (Ahameethunisa & Hopper, 2010; Munazir, Qureshi, Arshad, & Gulfraz, 2012).

Medicinal Plants as Anti-infectives
DOI: https://doi.org/10.1016/B978-0-323-90999-0.00007-0

In the world as well as in developing countries, most humans die due to infectious bacterial diseases (Nathan, Ahameethunisa, & Hoper, 2004). The causative bacterial organisms include Gram-positive and Gram-negative such as different species of *Bacillus*, *Staphylococcus*, *Salmonella*, and *Pseudomonas*, which are the main source of severe infections in humans. Because these organisms have the ability to survive in harsh conditions due to their multiple environmental habitats (Ahameethunisa & Hopper, 2010). The synthetic antibiotics have the following limitation: First, these are costly and are out of range from the patient belonging to developing countries. Second, with the passage of time, microorganisms develop resistance against antibiotics. Therefore, after some time these antibiotics are not effective against the microbes (Alder, 2005; Walsh, 2003). Furthermore, the antibiotics may be associated with adverse effects on the host, including hypersensitivity, immune suppression, and also allergic reactions. On the other hand, natural products have got incredible success in serving as a guidepost for new antibacterial drug discovery. Moreover, antibiotics obtained in this way have biological friendliness nature (Walsh, 2003; Koehn & Carter, 2005). Also, it is well known that the bioactive plant extracts are a promising source of majority of drugs (Nathan et al., 2004). For example, quinine (*Cinchona*) and berberine (*Berberis*) are the antibiotics obtained from plants that are highly effective against microbes (*Staphylococcus aureus*, *Escherichia coli*) (Ahmad, Farman, Najmi, Mian, & Hasan, 2008).

A wide range of bioactive plants grow naturally in Pakistan. In this study, a selection of 108 medicinal species from Pakistan were investigated, including six naturally growing plants: *Aesculus indica* Linn., *Arisaema flavum* (Forssk.) Schott, *Carissa opaca* Stapf ex Haines, *Debregeasia salicifolia* (D. Don) Rendle, *Pistacia integerrima* Stew. ex Brand, and *Toona ciliata* M. Roem (Abbasi et al., 2009; Badoni, 2000; Chakraborthy, 2009; Shah & Khan, 2006). Their distribution, traditional use, and properties are described in Table 6.1.

Antibacterial properties of different medicinal plants from Pakistan

Roots and fruits extracts of *Leptadenia pyrotechnica* (Forssk.) Decne. were analyzed for antibacterial properties. Plant material was obtained from Thal desert of Pakistan. Extracts were made by using eight solvents such as n-hexane, chloroform, acetone, ethyl acetate, butanol, methanol, ethanol, and water and were investigated against *Staphylococcus epidermidis* and *Staphylococcus aureus*. All solvents inhibited the growth of *S. aureus*. Comparison suggested that *S. aureus* was more profoundly inhibited by root extracts, whereas growth of *S. epidermidis* was more largely inhibited by fruit extracts (Munazir et al., 2012).

Antimicrobial properties of three medicinal plants, that is, *Artemisia indica* Willd., *Medicago falcata* L. and *Tecoma stans* (L.) Juss. ex. Kunth were studied against four diseases caused by bacterial strains, that is, *Escherichia coli*, *Pseudomonas aeruginosa*, *Salmonella typhi*, and *Staphylococcus aureus*. High inhibitory properties were exhibited by butanol, chloroform, and ethyl acetate extracts of *A. indica*, *M. falcate*, and *T. stans* ranging between 15 and 20 mm against *E. coli*, *P. aeruginosa*, and *S. aureus*. *A. indica* revealed inhibitory activity against *S. typhi* ranging between 12 and 14 mm, for all extracts tested (Javid et al., 2015).

Antibacterial activities of four significant medicinal herbs found in Balochistan such as *Grewia erythraea* Schwein f., *Hymenocrater sessilifolius* Fisch. and C.A. Mey, *Vincetoxicum stocksii* Ali and Khatoon, and *Zygophyllum fabago* L. were investigated against 12 bacterial

TABLE 6.1 List of various plants from Pakistan showing antibacterial activity.

Species number	Botanical name	Common name	Family	Plant part	Solvent used	Concentration	Bacterial strain	Zone of inhibition (mm)	References
1.	*Aesculus indica* (Wall. ex Cambess.) Hook.	Jawaz	Sapindaceae	Leaf	Crude	20 mg/mL	*Bacillus subtilis*/ *Micrococcus luteus*/ *Salmonella setubal*/ *Staphylococcus aureus*/ *Pseudomonas pickettii*	12 ± 0/14 ± 0.5/ 13.5 ± 0.5/14.5 ± 1/ 13 ± 0.5	Bibi et al. (2011)
					Aqueous			16 ± 1/14 ± 0.5/ 15 ± 0.5/13 ± 0.5/ 13 ± 0.2	
					Hexane			NA/NA/NA/NA/ NA	
					Chloroform			10.5 ± 0.1/12 ± 0/ 11.5 ± 0.5/13 ± 0.5/ 14.5 ± 0.1	
					Ethyl acetate			12 ± 0.5/NA/ 13 ± 0.5/12 ± 0.2/ 12 ± 0.5	
					Methanol			10 ± 0.5/11 ± 0.1/ 10 ± 0.1/11 ± 0.2/ 10 ± 1	
2.	*Ajuga integrifolia* Buch.-Ham.	Bugleweed or ground pine	Lamiaceae	Leaf	n-hexane methanol	100 mg/mL 50 mg/mL	*Bacillus cereus* /*Salmonella* Typhi	14.0 ± 0.2/14.0 ± 0.5	Rahman et al. (2015)
3.	*Alpinia galanga* (L.) Willd.	Siamese ginger	Zingiberaceae		Ethanol		*Salmonella* Typhi	11	Khattak, Saeed-ur-Rehman, Shah, Ahmad, an Ahmad (2005)
4.	*Alpinia galanga* (L.) Willd.	Siamese ginger	Zingiberaceae		Ethanol		*Staphylococcus aureus*	10	Khattak et al. (2005)
5.	*Althaea officinalis* L.	Marshmallow, Khatmi	Malvaceae	Root, leaf, and flower	Methanol	15 mg/mL	*Staphylococcus aureus*	2.7	Walter, Shinwari, Afzal, and Malik (2011)

(*Continued*)

TABLE 6.1 (Continued)

Species number	Botanical name	Common name	Family	Plant part	Solvent used	Concentration	Bacterial strain	Zone of inhibition (mm)	References
6.	*Arisaema flavum* (Forssk.) Schott	Marjarai	Araceae	Rhizome	Crude	20 mg/mL	*Bacillus subtilis*/ *Micrococcus luteus*/ *Salmonella setubal*/ *Staphylococcus aureus*/ *Pseudomonas pickettii*	10.3 ± 0.17/ 10.6 ± 0.07/ 10.6 ± 0.05/NA/ 13.7 ± 0.05	Bibi et al. (2011)
					Aqueous			NA/NA/NA/NA/ NA	
					Hexane			NA/NA/NA/NA/ NA	
					Chloroform			11.2 ± 0.08/12 ± 0/ NA/10.3 ± 0.42/NA	
					Ethyl acetate			NA/9.6 ± 0.61/NA/ NA/NA	
					Methanol			12 ± 0.5/NA/ 12.6 ± 0.02/ 13.6 ± 0.23/NA	
7.	*Artemisia dubia* L. ex B.D.Jacks.	Tarkha, Valati afsanthin	Asteraceae	Leaf	Methanol	15 mg/mL	*Escherichia coli* ATCC 15224/ *Bacillus subtilis* ATCC 6633/ *Staphylococcus aureus* ATCC 6538/*Micrococcus luteus* ATCC 10240	9.5 ± 0.03/ 11.5 ± 0.07/ 12.0 ± 0.10/ 9.5 ± 0.03	Mannan, Ahmed, Hussain, Jamil, and Miza (2012)
				Flower	Chloroform		*Escherichia coli* ATCC 15224	10 ± 0.10	
				Leaf	Chloroform		*Staphylococcus aureus* ATCC 6538/*Micrococcus luteus* ATCC 10240	10.5 ± 0.10/ 9.0 ± 0.05	

8.	*Artemisia indica* Willd.	Indian Wormwood	Asteraceae	Whole plant	Chloroform butanol ethyl acetate n-hexane	200 μL	*Pseudomonas aeruginosa*/ *Salmonella Typhi*/ *Staphylococcus aureus*/*Escherichia coli*	17.33 ± 1.15/ 13.66 ± 0.57/ 15.33 ± 1.15/ 18.66 ± 1.15	Javid et al. (2015)
9.	*A. maritime*	Tarakh	Asteraceae	Aerial part	Ethanol	100 mg/mL	*Klebsiella pneumoniae*	16	Malik, Mirza, Riaz, Hameed, and Hussain (2010)
10.	*Asphodelus tenuifolius* Cav.	Onionweed, White asphodel, or Piazi	Asphodelaceae	Seed	Ethanol	100 mg/mL	*Vibrio cholerae*	15	Malik et al. (2010)
11.	*Azadirachta indica* A.Juss.	Neem tree or margosa tree	Meliaceae	Leaf	Ethanol	100 mg/mL	*Micrococcus pyogenes*	19	Malik et al. (2010)
12.	*Berberis aristata* DC.	Indian barberry, chutro, or tree turmeric	Berberidaceae	Fruit	Ethanol	100 mg/mL	*Shigella dysenteriae*	13	Malik et al. (2010)
13.	*Bergenia ciliata* (Haw.) Sternb.	Fringed bergenia	Saxifragaceae	Rhizome	Aqueous	100 mg/mL	*Salmonella* Typhi	20	Malik et al. (2010)
14.	*Calligonum polygonoides* L.	Phok	Polygonaceae	Stem, leaf, fruit, flower	Methanol	10 mg/mL	*Escherichia coli*	10.5 ± 0.9	Mustafa, Ahmed, Ahmed, and Jamil (2016)
15.	*Calotropis procera* (Aiton) Dryand.	Apple of sodom or sodom apple	Apocynaceae	Leaf	Ethanol	100 mg/mL	*Vibrio cholerae*	14	Malik et al. (2010)
16.	*Calotropis procera* (Aiton) Dryand.	Apple of sodom or sodom apple	Apocynaceae	Leaf	Methanol	100 mg/mL	*Bacillus subtilis*	15.0 ± 0.2	Rahman et al. (2015)

(*Continued*)

TABLE 6.1 (Continued)

Species number	Botanical name	Common name	Family	Plant part	Solvent used	Concentration	Bacterial strain	Zone of inhibition (mm)	References
17.	*Calotropis procera* (Aiton) Dryand.	Auricula tree, Dead sea apple	Apocynaceae	Leaf	Methanol	10 μg	*Proteus mirabilis/Pseudomonas aeruginosa/Bacillus cereus*	17 ± 2/19 ± 2/16 ± 2	Bilal et al. (2020)
18.	*Cannabis sativa* L.	Marijuana, Hemp, or Gallow grass	Cannabaceae	Leaf	n-hexane	100 mg/mL	*Bacillus subtilis/Bacillus cereus/Escherichia coli*	15 ± 0.5/11 ± 0.2/10 ± 0.3	Nasrullah, Rahman, Ikram, Nisar, and Khan (2012)
					Methanol	50 mg/mL	*Pseudomonas aeruginosa/Salmonella* Typhi	11.4 ± 0.2/14 ± 0.2	
19.	*Cannabis sativa* L.	Marijuana, Hemp, or Gallow grass	Cannabaceae	Leaf	Methanol	100 mg/mL	*Escherichia coli*	23.3 ± 0.5	Hazrat, Nisar, and Zaman (2013)
20.	*Carissa carandas* L.	Kerenda	Apocynaceae	Fruit	Ethanol	5 mg/mL	*Staphylococcus aureus*	4.50 ± 0.4082	Naqvi, Azhar, Jabeen, and Hasan (2012)
21.	*Carissa spinarum* L.	Granda	Apocynaceae	Leaf	Crude	20 mg/mL	*Bacillus subtilis/Micrococcus luteus/Salmonella setubal/Staphylococcus aureus/Pseudomonas pickettii*	10.2 ± 0.07/NA/NA/NA/NA	Bibi et al. (2011)
					Aqueous			NA/NA/NA/NA/NA	
					Hexane			NA/NA/NA/NA/NA	
					Chloroform			11.3 ± 0.03/NA/NA/10.2 ± 0.05/NA	
					Ethyl acetate			12.3 ± 0.04/NA/NA/NA/NA	
					Methanol			NA/NA/NA/NA/NA	

22.	*Cerastium glomeratum* Thuill.	Clammy chickweed, Sticky mouse-ear chickweed	Caryophyllaceae	Whole plant	Aqueous	20 mg/mL	*Escherichia coli*/*Staphylococcus aureus*	26.03 ± 0.1/28.06 ± 0.02	Ullah et al. (2020)
					n-hexane			8 ± 0/8 ± 0	
					Dichloromethane (DCM)			10 ± 0/10 ± 0	
					Ethyl acetate			8 ± 0/8 ± 0	
23.	*Cichorium intybus* L.	Chicory, Kasni	Asteraceae	Whole plant	Aqueous	250 g/L	*Escherichia coli*/*Staphylococcus aureus*	13 ± 0.3/16 ± 0.5	Rahimullah et al. (2019)
					Ethanol			10 ± 0.5/11 ± 0.5	
					Chloroform			10 ± 0.5/11 ± 0.5	
					Hexane			11 ± 0.5/14 ± 0.5	
24.	*Cordia latifolia* Roxb.	Latifolia, Sebestan, Bara lasura	Boraginaceae	Flower and fruit	Methanol	15 mg/mL	*Staphylococcus aureus*	2.53	Walter et al. (2011)
25.	*Coriandrum sativum* L.	Coriander, dhania	Apiaceae	Fruit	Ethanol		*Bacillus cereus*/*Staphylococcus aureus*/*Escherichia coli*/*Pseudomonas aeruginosa*	NA	Khan et al. (2013)
26.	*Cucumis sativus* L.	Cucumber	Cucurbitaceae	Seed	Ethanol		*Bacillus cereus*/*Staphylococcus aureus*/*Escherichia coli*/*Pseudomonas aeruginosa*	NA	Khan et al. (2013)
27.	*Curcuma longa* L.	Common turmeric	Zingiberaceae		Ethanol		*Staphylococcus aureus*	10	Khattak et al. (2005)
28.	*Alpinia galanga* (L.) Willd.	Haldi	Zingiberaceae		Ethanol		*Staphylococcus aureus*	10	Khattak et al. (2005)
29.	*Datura innoxia* Mill.	Pricklyburr or recurved thorn-apple	Solanaceae	Leaf/seed/stem/root	Methanol	10 μL	*Escherichia coli*	24/21/21/23	Hussain et al. (2016)

(Continued)

TABLE 6.1 (Continued)

Species number	Botanical name	Common name	Family	Plant part	Solvent used	Concentration	Bacterial strain	Zone of inhibition (mm)	References
30.	*Debregeasia saeneb* (Forssk.) Hepper & J.R.I. Wood	Chewr, Ajlai	Urticaceae	Stem	Crude	20 mg/mL	*Bacillus subtilis*/ *Micrococcus luteus*/ *Salmonella setubal*/ *Staphylococcus aureus*/ *Pseudomonas pickettii*	14 ± 0.03/NA/NA/ NA/13.2 ± 0.05	Bibi et al. (2011)
					Aqueous			NA/NA/NA/ 11.6 ± 0.08/NA	
					Hexane			NA/NA/NA/NA/ NA	
					Chloroform			14.1 ± 0.12/ 15.1 ± 0.14/ 13.3 ± 0.09/ 12.1 ± 0.21/ 17.2 ± 0.07	
					Ethyl acetate			13.7 ± 0.11/NA/ NA/12.3 ± 0.13/ 13.7 ± 0.01	
					Methanol			NA/NA/NA/NA/ NA	
31.	*Diospyros kaki* L. f.	Kaki, Japanese persimmon, or Oriental persimmon	Ebenaceae	Bark	Methanol	100 mg/mL	*Escherichia coli*	27.8 ± 0.5	Hazrat et al. (2013)
32.	*Dodonaea viscosa* (L.) Jacq.	Sand olive, hopbush	Sapindaceae	Aerial part	Crude ethanol	3.2 mg/mL	*Staphylococcus aureus*/*Micrococcus luteus*/*Bacillus subtilis*/*Bacillus cereus*/*Escherichia coli*/*Pseudomonas aeruginosa*/ *Salmonella* Typhi	12.0 ± 0.3/ 12.9 ± 0.2/ 13.3 ± 0.2/0/ 11.0 ± 0.2/ 12.4 ± 0.4/0	Khurram et al. (2009)
33.	*Dodonaea viscosa* (L.) Jacq.	Broad leaf hopbush, candlewood, or giant hopbush	Sapindaceae	Leaf	Methanol n-hexane	100 mg/mL	*Bacillus subtilis*/*P. aeruginosa*/*B. cereus*/*Escherichia coli*/*S.* Typhi	10 ± 0.4/17.3 ± 0.1/ 13.2 ± 0.1/18 ± 0.1/ 15 ± 0.2	Nasrullah et al. (2012)

34.	*Dodonaea viscosa* (L.) Jacq.	Broad leaf hopbush, candlewood, or giant hopbush	Sapindaceae	Leaf	Methanol	100 mg/mL	*Escherichia coli*	18.6 ± 1.1	Hazrat et al. (2013)
35.	*Ephedra gerardiana* Wall. ex Stapf	Jointfir, Soma kalpa	Ephedraceae	Whole plant	Methanol	15 mg/mL	*Escherichia coli*	2.57	Walter et al. (2011)
36.	*Eucalyptus globulus* Labill.	Southern blue gum	Myrtaceae	Leaf	Ethanol	100 mg/mL	*Salmonella* Typhi	24	Malik et al. (2010)
37.	*Euphorbia prostrata* Aiton	Prostrate spurge or prostrate sandmat	Euphorbiaceae	Whole plant	Methanol	3 μg/mL	*Micrococcus luteus*	24	Ahmad et al. (2011)
						5 μg/mL	*Staphylococcus aureus/Escherichia coli/Pseudomonas aeruginosa*	24/20/26	
38.	*Fagonia indica* Burm.f.	Dramaaho	Zygophyllaceae	Whole plant	Methanol	10 mg/mL	*Escherichia coli*	12.3 ± 0.7	Mustafa et al. (2016)
39.	*Ficus sarmentosa* Buch.-Ham. ex Sm.	Nepal Fig	Moraceae	Whole plant	n-hexane ethyl acetate n-hexane		*Klebsiella pneumonia/ Salmonella* Typhimurium/ *Bacillus stearothermophilus*	16/15/15	Rauf et al. (2012)
40.	*Glycyrrhiza glabra* L.	Liquorice	Fabaceae	Root	Methanol	15 mg/mL	*Escherichia coli*	3.6	Walter et al. (2011)
41.	*Grewia erythraea* Schweinf.	Gaddeim, gaddein, or godem	Malvaceae	Whole plant	Methanol	200 μg/mL	*Bacillus subtilis/ Escherichia coli/ Pseudomonas aeruginosa/ Staphylococcus aureus/ Streptococcus pyogenes/Klebsiella pneumoniae*	12 ± 2/7 ± 1/ 24 ± 2/8 ± 1/7 ± 2/ 8 ± 1	Zaidi and Crow (2005)

(*Continued*)

TABLE 6.1 (Continued)

Species number	Botanical name	Common name	Family	Plant part	Solvent used	Concentration	Bacterial strain	Zone of inhibition (mm)	References
42.	*Helianthus annuus* L.	Common sunflower	Asteraceae	Leaf	Methanol	100 mg/mL	*Escherichia coli*	36.6 ± 0.1	Hazrat et al. (2013)
43.	*Heliotropium strigosum* Willd.	Gorakh pam, Kharsan	Boraginaceae	Leaf	Methanol	10 mg/mL	*Escherichia coli*	9.3 ± 1.3	Mustafa et al. (2016)
44.	*Himalaiella heteromalla* (D. Don) Raab-Straube	Kaliziri	Asteraceae	Whole plant	Methanol	30 μg/mL 60 μg/mL 90 μg/mL	*Escherichia coli*/ *Pseudomonas aeruginosa*/ *Salmonella* Typhi/ *Stenotrophomonas maltophilia*/*Serratia marcescens*/*Bacillus subtilis*/ *Staphylococcus aureus*/*Micrococcus luteus*	11.5/11.75/10.95/ 8.67/13.87/6.45/ 6.78/9.57 12.27/13.13/12.67/ 9.11/16.2/6.96/ 7.21/9.75 13.57/11.29/12.87/ 15.23/20.65/7.83/ 9.78/10.92	Batool, Miana, Muddassir, Khan, and Zafar (2019)
45.	*Hymenocrater sessilifolius* Benth.		Lamiaceae	Whole plant	Methanol	200 μg/mL	*Bacillus subtilis*/ *Escherichia coli*/ *Pseudomonas aeruginosa*	22 ± 1/8 ± 1/16 ± 2	Zaidi and Crow (2005)
46.	*Hyssopus officinalis* L.	Hyssopus, Zoofa	Lamiaceae	Leaf	Methanol	15 mg/mL	*Staphylococcus aureus*	3.37	Walter et al. (2011)
47.	*Isodon rugosus* (Wall. ex Benth.) Codd	Wrinkled Leaf Isodon	Lamiaceae	Whole plant	Ethyl acetate Methanol		*Salmonella* Typhimurium/ *Bacillus stearothermophilus*/ *Staphylococcus aureus*	14/12/14	Rauf et al. (2012)
48.	*Justicia adhatoda* L.	Vasica, Vasaka, Baikar basuti	Acanthaceae	Whole plant	Methanol	15 mg/mL	*Pseudomonas aeruginosa*	2.67 ± 0.06	Walter et al. (2011)
49.	*Lavandula angustifolia* Mill.	Lavender	Lamiaceae	Flower	Essential oil	2 μL	*Staphylococcus aureus*/ *Streptococcus* spp./ *Escherichia coli*/ *Pseudomonas aeruginosa*	7.5/7/2/0.5	Sohoo et al. (2019)

50.	*Lawsonia inermis* L.	Hina, The henna tree, The mignonette tree or The Egyptian privet	Lythraceae	Leaf/stem	Aqueous	100 mg/mL	*Bacillus subtilis*	23	Malik et al. (2010)
51.	*Lawsonia inermis* L.	Henna, Mehndi	Lythraceae	Leaf	Ethanol		*Bacillus cereus*/*Staphylococcus aureus*/*Escherichia coli*/*Pseudomonas aeruginosa*	7.10 ± 0.03/7.10 ± 0.03/7.10 ± 0.03/8.67 ± 0.52	Khan et al. (2013)
52.	*Lens culinaris* Medik.	Lentil, Masur	Fabaceae	Seed	Ethanol		*Bacillus cereus*/*Staphylococcus aureus*/*Escherichia coli*/*Pseudomonas aeruginosa*	NA	Khan et al. (2013)
53.	*Litchi chinensis* Sonn.	Maida Sak	Lauraceae	Bark	n-hexane	100 mg/mL	*Saccharomyces cerevisiae*	18	Malik et al. (2010)
54.	*Leptadenia pyrotechnica* (Forssk.) Decne.	Khimp, Kheep	Asclepiadaceae	Fruit	Ethanol	50 mg/mL	*Staphylococcus aureus*	16	Munazir et al. (2012)
					Methanol			19	
					Ethyl acetate		*Staphylococcus epidermidis*	10	
				Root	Methanol			15	
55.	*Malva sylvestris* L.	Common mallow, Tree mallow, Khabazi	Malvaceae	Dried leaf, root, and flower	Methanol	15 mg/mL	*Staphylococcus aureus*	3.1	Walter et al. (2011)
56.	*Mangifera indica* L.	Sindhri variety	Anacardiaceae	Peel	Ethanol	4 mg/mL	*Bacillus subtilis*/*Micrococcus luteus*/*Pseudomonas septica*/*Staphylococcus aureus*/*Enterobacter aerogenes*	10 ± 1/10 ± 1/13 ± 0.25/8 ± 1/15 ± 0.5	Suleman, Ali, Haq, Nisa, and Zia (2019)
				Pulp				7 ± 1/10 ± 1/12 ± 1/9 ± 0.5/19 ± 1	
				Kernel				12 ± 0.25/14 ± 0.75/10 ± 1/10 ± 0.5/13 ± 1	

(Continued)

TABLE 6.1 (Continued)

Species number	Botanical name	Common name	Family	Plant part	Solvent used	Concentration	Bacterial strain	Zone of inhibition (mm)	References
57.	*Mangifera indica* L.	Hujra variety	Anacardiaceae	Peel	Ethanol	4 mg/mL	*Bacillus subtilis*/*Micrococcus luteus*/*Pseudomonas septica*/*Staphylococcus aureus*/*Enterobacter aerogenes*	12 ± 0.5/15 ± 1/10 ± 1/10 ± 0.4/15 ± 0.5	Suleman et al. (2019)
				Pulp				10 ± 0.3/9 ± 0.1/12 ± 0.2/8 ± 0.4/10 ± 1	
				Kernel				15 ± 0.5/11 ± 1/9 ± 1/10 ± 1/15 ± 0.5	
58.	*Mangifera indica* L.	Langra variety	Anacardiaceae	Peel	Ethanol	4 mg/mL	*Bacillus subtilis*/*Micrococcus luteus*/*Pseudomonas septica*/*Staphylococcus aureus*/*Enterobacter aerogenes*	17 ± 1/8 ± 0.5/10 ± 0.2/22 ± 0.5/20 ± 0.3	Suleman et al. (2019)
				Pulp				14 ± 1/10 ± 0.25/7 ± 0.7/9 ± 0.25/15 ± 1	
				Kernel				15 ± 1/14 ± 0.75/13 ± 0.5/10 ± 0.5/9 ± 0.2	
59.	*Mangifera indica* L.	Chaunsa variety	Anacardiaceae	Peel	Ethanol	4 mg/mL	*Bacillus subtilis*/*Micrococcus luteus*/*Pseudomonas septica*/*Staphylococcus aureus*/*Enterobacter aerogenes*	12 ± 1/16 ± 0.5/8 ± 0.3/17 ± 0.5/20 ± 0.7	Suleman et al. (2019)
				Pulp				10 ± 0.6/10 ± 1/10 ± 1/12 ± 1/19 ± 0.8	
				Kernel				14 ± 0.9/13 ± 0.5/9 ± 1/19 ± 0.4/14 ± 1	
60.	*Mangifera indica* L.	Dosehri variety	Anacardiaceae	Peel	Ethanol	4 mg/mL	*Bacillus subtilis*/*Micrococcus luteus*/*Pseudomonas septica*/*Staphylococcus aureus*/*Enterobacter aerogenes*	15 ± 0.4/10 ± 1/11 ± 1/15 ± 0.9/17 ± 0.10	Suleman et al. (2019)
				Pulp				9 ± 1/8 ± 0.7/12 ± 0.75/14 ± 0.4/19 ± 0.7	
				Kernel				16 ± 1/18 ± 0.9/9 ± 0.75/8 ± 1/8 ± 1	

61.	*Mangifera indica* L.	Almashil variety	Anacardiaceae	Peel	Ethanol	4 mg/mL	*Bacillus subtilis*/ *Micrococcus luteus*/ *Pseudomonas septica*/ *Staphylococcus aureus*/*Enterobacter aerogenes*	11 ± 0.9/15 ± 0.3/ 7 ± 0.5/9 ± 0.5/ 9 ± 0.10	Suleman et al. (2019)
				Pulp				13 ± 0.8/9 ± 1/ 11 ± 1/10 ± 0.7/ 11 ± 0.8	
				Kernel				8 ± 1/10 ± 0.5/ 15 ± 0.2/7 ± 0.6/ 9 ± 0.5	
62.	*Mangifera indica* L.	Dalasi variety	Anacardiaceae	Peel	Ethanol	4 mg/mL	*Bacillus subtilis*/ *Micrococcus luteus*/ *Pseudomonas septica*/ *Staphylococcus aureus*/*Enterobacter aerogenes*	10 ± 1/13 ± 0.75/ 10 ± 0.7/10 ± 0.6/ 9 ± 1	Suleman et al. (2019)
				Pulp				9 ± 0.7/12 ± 0.75/ 18 ± 0.5/11 ± 1/ 10 ± 0.4	
				Kernel				14 ± 0.75/10 ± 1/ 15 ± 0.6/12 ± 0.9/ 13 ± 1	
63.	*Medicago falcata* L.	Yellow lucerne, sickle alfalfa, or yellow-flowered alfalfa	Fabaceae	Whole plant	n-hexane Ethyl acetate Chloroform Butanol	200 μL	*Salmonella* Typhi	15.33 ± 1.15/ 11.33 ± 0.57/ 13.33 ± 1.15/ 13.33 ± 0.57	Javid et al. (2015)
64.	*Melaleuca alternifolia* (Maiden & Betche) Cheel	Tea tree	Myrtaceae	Leaf	Essential oil	2 μL	*Staphylococcus aureus*/ *Streptococcus* spp./ *Escherichia coli*/ *Pseudomonas aeruginosa*	4.5/3.7/4/1.7	Sohoo et al. (2019)
65.	*Melia azedarach* L.	Chinaberry tree or Pride of India	Meliaceae	Leaf	Aqueous	100 mg/mL	*Bacillus subtilis*	18	Malik et al. (2010)
66.	*Mentha longifolia* (L.) L.	Wild mint	Lamiaceae	Leaf	Methanol	100 mg/mL	*Escherichia coli*	26.1 ± 0.5	Hazrat et al. (2013)

(*Continued*)

TABLE 6.1 (Continued)

Species number	Botanical name	Common name	Family	Plant part	Solvent used	Concentration	Bacterial strain	Zone of inhibition (mm)	References
67.	*Mentha × piperita* L.	Breena	Lamiaceae	Leaf	Ethanol, methanol	0.1 mL	*E. coli, Staphylococcus, Pseudomonas, Salmonella, Streptobacillus*	15	Sabahat and Perween (2005)
68.	*Momordica charantia* L.	Karela	Cucurbitaceae	Skin, pulp	Ethanol, methanol	0.1 mL	*Staphylococcus, Pseudomonas, Salmonella, Streptobacillus*	15	Sabahat and Perween (2005)
69.	*Moringa oleifera* Lam.	Suhanjna	Moringaceae	Root bark	Ethanol	100 mg	*Staphylococcus aureus*	13	Nikkon, Saud, Rahman, and Haque (2003)
70.	*Myrtus communis* L.	Myrtle	Myrtaceae	Fruit	Aqueous	100 mg/mL	*Shigella dysenteriae*	16	Malik et al. (2010)
71.	*Nigella sativa* L.	Black caraway	Ranunculaceae	Seed	Ethanol	100 mg/mL	*Salmonella* Typhi	20	Malik et al. (2010)
72.	*Nigella sativa* L.	Black seed	Ranunculaceae	Seed	Essential oil	2 μL	*Staphylococcus aureus*/ *Streptococcus* spp./ *Escherichia coli*/ *Pseudomonas aeruginosa*	10/10/6/11.7	Sohoo et al. (2019)
73.	*Ocimum basilicum* L.	Great basil	Lamiaceae	Leaf	Methanol	100 mg/mL	*Escherichia coli*	24.0 ± 0.0	Hazrat et al. (2013)
74.	*Onosma bracteatum* Wall.	Gao zuban	Boraginaceae	Leaf	Methanol	15 mg/mL	*Staphylococcus aureus*	2.67	Walter et al. (2011)
75.	*Periploca aphylla* Decne.	Leafless Silkflower Shrub	Asclepiadaceae	Whole plant	Methanol chloroform		*Klebsiella pneumoniae*/ *Bacillus stearothermophilus*	12/14	Rauf et al. (2012)

76.	*Polygonum bistorta* L.	Bistorta, Anjbar	Polygonaceae	Root	Methanol	20 μL	*Bacillus subtilis/ Enterococcus faecalis/ Staphylococcus aureus/ Pseudomonas aeruginosa/ Salmonella* Typhi	6.5/8/0/6/0	Khalid et al. (2011)
						30 μL		11/13/8/ 9/0	
					Aqueous (Cold)	20 μL		8/0/0/6/0	
						30 μL		12/9/NA/10/8	
					Aqueous (Hot)	20 μL		7/0/0/0/0	
						30 μL		10/10/9/8/8	
77.	*Phyllanthus emblica* L.	Indian gooseberry or amla	Phyllanthaceae	Fruit	Aqueous	100 mg/mL	*Shigella dysenteriae*	24	Malik et al. (2010)
78.	*Phyllanthus emblica* L.	Indian gooseberry, amla	Phyllanthaceae	Fruit	Ethanol		*Bacillus cereus/ Staphylococcus aureus/Escherichia coli/Pseudomonas aeruginosa*	22.03 ± 0.10/ 20.31 ± 0.43/ 23.68 ± 0.36/ 20.39 ± 0.17	Khan et al. (2013)
79.	*Pistacia chinensis* Bunge	Pistachio or Chinese pistache	Anacardiaceae	Bark	Methanol	100 mg/mL	*Escherichia coli*	26.3 ± 0.5	Hazrat et al. (2013)
80.	*Pistacia chinensis* subsp. *integerrima* (J. L. Stewart ex Brandis) Rech. f.	Pistachio, Kakar singi, Kangar	Anacardiaceae	Stem	Crude	20 mg/mL	*Bacillus subtilis/ Micrococcus luteus/ Salmonella setubal/ Staphylococcus aureus/ Pseudomonas pickettii*	20.33 ± 0.41/17 ± 0/ 17.33 ± 0.34/ 23 ± 0.34/ 19.66 ± 0.05	Bibi et al. (2011)
					Aqueous			19.66 ± 0.05/NA/ NA/NA/NA	
					Hexane			14.6 ± 0.11/ 13.3 ± 0.05/ 13.3 ± 0.2/12 ± 0.2/ 10.3 ± 0.05	
					Chloroform			13 ± 0/12 ± 0/ 12 ± 0/11.66 ± 0/ 12 ± 0	
					Ethyl acetate			15 ± 0/12.66 ± 0.05/ 14.3 ± 0.37/ 13.3 ± 0.2/NA	
					Methanol			NA/NA/10 ± 0/ 11.5 ± 0/11 ± 0.5	

(*Continued*)

TABLE 6.1 (Continued)

Species number	Botanical name	Common name	Family	Plant part	Solvent used	Concentration	Bacterial strain	Zone of inhibition (mm)	References
81.	*Pistacia chinensis* subsp. *integerrima* (J. L. Stewart ex Brandis) Rech. f.	Pistachio, Kakar singi, Kangar	Anacardiaceae	Gall	Methanol	20 μL	*Bacillus subtilis*/ *Enterococcus faecalis*/ *Staphylococcus aureus*/ *Pseudomonas aeruginosa*/ *Salmonella* Typhi	8/6/6/6.5/7	Khalid et al. (2011)
						30 μL		12/10/10.5/10/11	
					Aqueous (Cold)	30 μL		8/0/9/0/8	
					Aqueous (Hot)	30 μL		9/8/8/8/9	
82.	*Pisum sativum* L.	Matar	Fabaceae	Skin, seed	Ethanol, methanol	0.1 mL	*E. coli*, *Staphylococcus*, *Pseudomonas*, *Salmonella*, *Streptobacillus*	16	Sabahat and Perween (2005)
83.	*Polygonum plebeium* R. Br.	Common knotweed	Polygonaceae	Whole plant	Methanol	100 mg/mL	*Salmonella* Typhi	15.00 ± 0.0	Hazrat et al. (2013)
84.	*Populus ciliata* Wall. ex. Royle	Himalayan popla, Paloch, Phals, Chalun	Salicaceae	Leaf	Aqueous		*Klebsiella pneumonia*/*Serratia marcescens*/ *Pseudomonas pseudoalcaligenes*/ *Staphylococcus epidermidis*/ *Streptococcus pyogenes*	12.6 ± 9.355/ 20.8 ± 7.73/ 18.0 ± 4.728/ 17.8 ± 5.148/ 19.2 ± 5.216	Hafeez et al. (2021)
85.	*Punica granatum* L.	Pomegranate	Punicaceae	Rind of fruit	Ethanol	100 mg/mL	*Salmonella* Typhi	23	Malik et al. (2010)
86.	*Punica granatum* L.	Pomegranate	Lythraceae	Pericarp	Methanol	100 mg/mL	*Pseudomonas aeruginosa*	23.00 ± 0.4	Hazrat et al. (2013)
87.	*Punica granatum* L.	Pomegranate, Anar	Lythraceae	Peel	Aqueous	250 g/L	*Escherichia coli*/ *Staphylococcus aureus*/ *Acinetobacter baumannii*/ *Pseudomonas aeruginosa*	13 ± 0.3/0/15 ± 0.5/ 14 ± 0.6	Khan et al. (2017)
					Hexane			12 ± 0.7/11 ± 0.3/0/ 10 ± 0	
					Ethanol			0/0/14 ± 0.5/0	

88.	*Rosmarinus officinalis* L.	Rosemary	Lamiaceae	Leaf	Methanol	100 mg/mL	*Bacillus subtilis*	26.66 ± 0.2	Hazrat et al. (2013)
89.	*Rumex hastatus* D. Don	Khatimber	Polygonaceae	Leaf	Methanol	100 mg/mL	*Escherichia coli*	25.4 ± 0.3	Hazrat et al. (2013)
90.	*Sideroxylon mascatense* (A. DC.) T.D.Penn.	Gargole	Sapotaceae	Fruit	Methanol	100 mg/mL	*Escherichia coli*	30.1 ± 0.5	Hazrat et al. (2013)
91.	*Solanum nigrum* L.	Black night shade	Solanaceae	Fruit	Methanol	100 mg/mL	*Bacillus cereus*	16.3 ± 0.4	Nasrullah et al. (2012)
					Methanol	50 mg/mL	*Escherichia coli*/ *Pseudomonas aeruginosa*/ *Salmonella* Typhi	14 ± 0.4/17.2 ± 0.4/ 15.2 ± 0.3	
92.	*Solanum nigrum* L.	Black night shade	Solanaceae	Fruit	Methanol	100 mg/mL	*Escherichia coli*	20.0 ± 0.0	Hazrat et al. (2013)
93.	*Solanum surattense* Burm. f.	Kundiari, Momoli	Solanaceae	Whole plant	Methanol	10 mg/mL	*Escherichia coli*	14.8 ± 0.5	Mustafa et al. (2016)
94.	*Sphaeranthus indicus* L.	East Indian globe thistle	Asteraceae	Leaf/stem	n-hexane	100 mg/mL	*Micrococcus pyogenes*	13	Malik et al. (2010)
95.	*Suaeda vermiculata* Forssk. ex J.F. Gmel.	Khaari, Boi booti	Amaranthaceae	Stem, leaf	Methanol	10 mg/mL	*Escherichia coli*	19.5 ± 0.3	Mustafa et al. (2016)
96.	*Swertia chirata* Buch.-Ham. ex Wall.	Chirayata, Chiretta	Gentianaceae	Stem	Methanol	20 μL	*Bacillus subtilis*/ *Enterococcus faecalis*/ *Staphylococcus aureus*/ *Pseudomonas aeruginosa*/ *Salmonella* Typhi	7.5/0/0/0/0	Khalid et al. (2011)
						30 μL		12/0/0/8/9	
					Aqueous (Cold)	20 μL		9/0/6/0/0	
						30 μL		15/8/10/0/9	
					Aqueous (Hot)	20 μL		0/0/0/0/0	
						30 μL		9/8/8/8/10	

(Continued)

TABLE 6.1 (Continued)

Species number	Botanical name	Common name	Family	Plant part	Solvent used	Concentration	Bacterial strain	Zone of inhibition (mm)	References
97.	*Syzygium aromaticum* (L.) Merr. & L.M. Perry	Clove	Myrtaceae	Flower bud	Essential oil	2 μL	*Staphylococcus aureus*/ *Streptococcus* spp./ *Escherichia coli*/ *Pseudomonas aeruginosa*	8.6/9.3/5.2/2	Sohoo et al. (2019)
98.	*Syzygium aromaticum* (L.) Merr. & L.M. Perry	Clove	Myrtaceae	Bud	Ethanol		Vancomycin-resistant *Staphylococcus aureus*	5.5 ± 0.41	Asghar, Yousuf, Shoaib, and Asghar (2020)
							Methicillin-resistant *Staphylococcus aureus*	10.3 ± 0.61	
99.	*Tecoma stans* (L.) Juss. ex Kunth	Yellow trumpetbush or yellow bells	Bignoniaceae	Whole plant	n-hexane	200 μL	*Escherichia coli*/ *Pseudomonas aeruginosa*/ *Salmonella* Typhi/ *Staphylococcus aureus*	15.66 ± 0.57/ 13.66 ± 0.57/15 ± 1/ 17.33 ± 1.15	Javid et al. (2015)
100.	*Thymus vulgaris* L.	Common thyme, German thyme, or Garden thyme	Lamiaceae	Whole plant	Methanol	100 mg/mL	*Pseudomonas aeruginosa*	22.0 ± 0.0	Hazrat et al. (2013)
101.	*Tinospora sinensis* (Lour.) Merr.	Gurjo, Heart-leaved moonseed, or guduchi	Menispermaceae	Stem	Ethanol	100 mg/mL	*Escherichia coli*	16	Malik et al. (2010)
102.	*Toona ciliata* M. Roem.	Cedrela, Red cedar, or Toon tree	Meliaceae	Leaf	Aqueous	100 mg/mL	*Staphylococcus*/ *Escherichia coli*/ *Bacillus subtilis*	NA/NA/NA	Malik et al. (2010)
					Ethanol			16/14/13	
					n-hexane			NA/NA/NA	

103.	*Toona ciliata* M. Roem.	Mahanim	Meliaceae	Leaf	Crude	20 mg/mL	*Bacillus subtilis*/ *Micrococcus luteus*/ *Salmonella setubal*/ *Staphylococcus aureus*/ *Pseudomonas pickettii*	11.2 ± 0.06/ 17 ± 0.41/ 17.1 ± 0.06/NA/ 12.3 ± 0.01	Bibi et al. (2011)
					Aqueous			NA/NA/NA/NA/ NA	
					Hexane			10.6 ± 0.21/NA/ NA/NA/NA	
					Chloroform			13.6 ± 0.06/ 15.8 ± 0.12/ 14.5 ± 0.04/ 13 ± 0.08/NA	
					Ethyl acetate			15 ± 0.06/ 16.2 ± 0.09/NA/ NA/11.2 ± 0.11	
					Methanol			NA/12.7 ± 0.07/ NA/NA/NA	
104.	*Vincetoxicum stocksii* Ali & Khatoon		Asclepiadaceae	Whole plant	Methanol	200 µg/mL	*Bacillus subtilis*/ *Escherichia coli*/ *Proteus mirabilis*/ *Pseudomonas aeruginosa*/*Shigella dysenteriae*/ *Klebsiella pneumoniae*	25 ± 2/12 ± 1/ 15 ± 2/6 ± 1/ 13 ± 1/15 ± 2	Zaidi and Crow (2005)
105.	*Zingiber officinale* Roscoe	Ginger, Adrak	Zingiberaceae	Root	Methanol	10 µL	*Bacillus subtilis*/ *Enterococcus faecalis*/ *Staphylococcus aureus*/ *Pseudomonas aeruginosa*/ *Salmonella* Typhi	6/0/7/6/0	
						20 µL		11/0/12/9/6	
						30 µL		17/8/19/16/10	
					Aqueous (Cold)	10 µL		0/0/0/0/0	
						20 µL		0/0/7/10/0	
						30 µL		9/8/10/15/8	
					Aqueous (Hot)	10 µL		0/0/0/0/0	
						20 µL		6/0/0/0/0	
						30 µL		9/0/8/9/8	

(*Continued*)

TABLE 6.1 (Continued)

Species number	Botanical name	Common name	Family	Plant part	Solvent used	Concentration	Bacterial strain	Zone of inhibition (mm)	References
106.	*Ziziphus sativa* Geartn.	Chinese jujube	Rhamnaceae	Leaf	Methanol	50 mg/mL	*Bacillus cereus*	12.5 ± 0.2	Rahman et al. (2015)
107.	*Ziziphus vulgaris* L.	Jujube, Anab, Unnab, Singli	Rhamnaceae	Fruit	Methanol	15 mg/mL	*Staphylococcus aureus*	2.17	Walter et al. (2011)
108.	*Zygophyllum fabago* L.	Syrian bean-caper	Zygophyllaceae	Whole plant	Methanol	200 μg/mL	*Bacillus subtilis*/*Bacillus cereus*/*Escherichia coli*/*Pseudomonas aeruginosa*	16 ± 1/16 ± 2/11 ± 2/8 ± 1	Zaidi and Crow (2005)

Legend: NA = No Activity.

strains. *Z. fabago* extract showed high antibacterial activity against *Escherichia coli. V. stocksii* extract exhibited great antibacterial activity against *Bacillus subtilis* and *Bacillus cereus*. However, extracts of *H. sessilifolius* and *G. erythraea* were only active against *Pseudomonas aeruginosa* (Zaidi & Crow, 2005).

Methanolic and n-hexane extracts of fruits of *Solanum nigrum* L., leaves of *Dodonaea viscosa* Jacq and *Cannabis sativa* L. were explored for antibacterial properties. It was observed that methanolic extracts of all plants excluding *Solanum nigrum* L. exhibited inhibitory activity against all the tested bacterial strains, whereas n-hexane extracts of plants showed inactive properties against *Pseudomonas aeruginosa* (Nasrullah et al., 2012).

Antibacterial properties of 16 medicinal plant species from Dir Kohistan Valley KPK, Pakistan were tested. Six bacterial strains, that is, *Staphylococcus aureus, Escherichia coli, Pseudomonas aeruginosa, Salmonella typhi, Bacillus subtilis,* and *Bacillus cereus* were investigated. Highest antibacterial activity was shown by methanolic extracts of 10 plant species *Helianthus annuus* L. (36.6 ± 0.1), *Sideroxylon mascatense* (A.DC.) T.D. Penn. (30.1 ± 0.5), *Diospyros kaki* L.f. (27.8 ± 0.5), *Pistacia chinensis* Bunge (26.3 ± 0.5), *Mentha longifolia* (L.) L. (26.1 ± 0.5), *Rumex hastatus* D. Don (25.4 ± 0.3), *Ocimum basilicum* L. (24.0 ± 0.0), *Cannabis sativa* L. (23.3 ± 0.5), *Punica granatum* L. (23.00 ± 0.4), and *Thymus vulgaris* L. (22.0 ± 0.0) (Hazrat et al., 2013).

Indigenous medicinal plants *Curcuma longa* L. and *Alpinia galanga* (L.) Willd. were investigated for antibacterial activities against *Escherichia coli, Bacillus subtilis, Shigella flexneri, Staphylococcus aureus, Pseudomonas aeruginosa,* and *Salmonella typhi*. Inhibition zone of 11 mm was shown by ethanolic extract of *A. galanga* against *S. typhi*, whereas ethanolic extracts of *C. longa* and *A. galanga* showed inhibition zone of 10 mm against *S. aureus* (Khattak et al., 2005).

Methanolic extracts of leaves, stem, root, and seeds of *Datura innoxia* Mill. were analyzed for antibacterial properties against *Staphylococcus aureus, Pseudomonas aeruginosa, Escherichia coli, Klebsiella pneumoniae, Streptococcus pneumonia, Proteus* spp., and *Salmonella typhi* with the help of agar well diffusion method. Methanolic extracts of *D. inoxia* showed significant zones of inhibition between 6 and 24 mm (Mustafa et al., 2016).

Antibacterial properties of 17 medicinal plants from Pakistan were analyzed against 10 Gram-positive and Gram-negative bacterial strains. Aqueous, ethanolic, and n-hexane extracts were utilized. High antibacterial activity was shown by *Eucalyptus globulus* Labill., *Phyllanthus emblica* L., and *Sphaeranthus indicus* L. extracts against all 10 bacterial strains. Significant activity was exhibited by ethanolic extracts of *Azadirachta indica* A. Juss., *Toona ciliata* M. Roem., *Punica granatum* L., *Bergenia ciliata* (Haw.) Sternb., and *Lawsonia inermis* L. (Malik et al., 2010).

Mustafa et al. (2016) carried out an ethnobotanical survey in order to explore the phytochemical and antibacterial activities of the plants of Cholistan Desert. Five medicinal plants (i.e., *Calligonum polygonoides* L., *Fagonia indica* Burm.f., *Heliotropium strigosum* Willd., *Solanum surattense* Burm. f., and *Suaeda vermiculata* Forssk. ex J.F. Gmel.) were collected and their methanolic extracts were evaluated for their antibacterial properties against *Escherichia coli*. The methanolic extract of *S. fruticosa* showed a maximum inhibition zone of 19.5 ± 0.3, while *H. strigosum* extract showed minimum inhibition of 9.3 ± 1.3 (Mustafa et al., 2016).

The methanolic, cold and hot water extracts of *Swertia chirata* Buch.-Ham. ex Wall., *Pistacia chinensis* subsp. *integerrima* (J. L. Stewart ex Brandis) Rech. f., *Persicaria bistorta* (L.)

Samp., and *Zingiber officinale* Roscoe were evaluated against both Gram-negative (*Salmonella typhi, Pseudomonas aeruginosa*) and Gram-positive (*Bacillus subtilis, Enterococcus faecalis, Staphylococcus aureus*) bacteria. The maximum inhibition was reported by the *Z. officinale* methanol extract against *S. aureus*, with a diameter of 19 mm (Khalid et al., 2011).

Sohoo et al. (2019) investigated the antibacterial efficacy of essential oils (EO) of *Syzygium aromaticum* (L.) Merr. & L.M. Perry, *Lavandula angustifolia* Mill., *Melaleuca alternifolia* (Maiden & Betche) Cheel, and *Nigella sativa* L. against the pathogenic bacteria responsible for causing mastitis such as *Escherichia coli, Pseudomonas aeruginosa, Streptococcus* spp., and *Staphylococcus aureus*, with the help of disc diffusion technique. *N. sativa* EO showed more inhibitory action against *P. aeruginosa* in comparison to *M. alternifolia* and *L. angustifolia* EO while *S. aromaticum* EO exhibited higher inhibitory activity than *M. alternifolia* and *L. angustifolia* EO against all bacterial strains (Sohoo et al., 2019).

The aqueous and crude extracts of *Dodonaea viscosa* (L.) Jacq. were investigated for their bactericidal potential against three Gram-negative bacteria (*Salmonella typhi, Escherichia coli, Pseudomonas aeruginosa*) and four Gram-positive bacteria (*Bacillus cereus, Bacillus subtilis, Staphylococcus aureus, Micrococcus luteus*). The average zone of inhibition indicated that the crude extract showed inhibition against almost all of the bacterial strains except *B. cereus* and *S. typhi*, while aqueous extract showed no inhibition against any of the bacterial strains (Khurram et al., 2009).

Asghar et al. (2020) carried out the green synthesis of silver nanoparticles (AgNPs) using the extract of *Populus ciliata* Wall. ex. Royle in order to assess its antimicrobial potential. The synthesized AgNPs exhibited inhibitory properties against selected Gram-negative (*Serratia marcescens, Klebsiella pneumoniae*, and *Pseudomonas pseudoalcaligenes*) and Gram-positive (*Staphylococcus epidermidis* and *Streptococcus pyogenes*) bacterial strains. The highest mean antibacterial activity was observed against *S. marcescens* (20.8 mm $\pm$ 7.7) and *S. pyogenes* (19.2 mm $\pm$ 5.2) showing bactericidal potential (Asghar et al., 2020).

Khan et al. (2013) conducted a study for demonstrating the antibacterial properties of five medicinal plants of Pakistan such as *Coriandrum sativum* L., *Cucumis sativus* L., *Lens culinaris* Medik., *Lawsonia inermis* L., and *Phyllanthus emblica* L. Antibacterial assay was performed with the help of bacterial strains of *B. cereus, E. coli, P. aeruginosa*, and *S. aureus*. The extract of *C. sativum, C. sativus, L. culinaris* showed no activity on any of the clinical isolates while both *L. alba* and *P. emblica* showed a range of inhibitory activity with maximum inhibition zones against *P. aeruginosa* (8.67 mm $\pm$ 0.52) and *E. coli* (23.68 mm $\pm$ 0.36), respectively (Khan et al., 2013).

Mannan et al. (2012) examined the antibacterial activities of different extracts of *Artemisia dubia* L. ex B.D. Jacks, that is, leaf methanol extract, flower methanol extract, leaf chloroform extract, and flower chloroform extract, against the clinical isolates of three Gram-positive (*Bacillus subtilis* ATCC 6633, *Micrococcus luteus* ATCC 10240, *Staphylococcus aureus* ATCC 6538) and five Gram-negative (*Escherichia coli* ATCC 15224, *Salmonella setubal* ATCC 19196, *Pseudomonas pickettii* ATCC 49129, *Enterobacter aerogenes* ATCC 13048, *Bordetella bronchiseptica* ATCC 4617) bacteria. The results of bioassay showed that while flower chloroform extract was not effective against any of the bacterial strains, leaf methanol extract was effective against *E. coli, B. subtilis, S. aureus*, and *M. luteus*, leaf chloroform extract was effective against *S. aureus* and *M. luteus*, and flower methanol extract showed antibacterial activity against *E. coli* (Ihsan-ul-haq et al., 2012).

Asghar (2020) used the methanolic extract of leaves of *Calotropis procera* (Aiton) Dryand. to check its antibacterial activity against *Bacillus cereus, Enterococcus faecalis, Escherichia coli, Klebsiella pneumonia, Proteus mirabilis, Pseudomonas aeruginosa,* and *Salmonella typhi* using the disk diffusion method. The leaves extract of *C. procera* was found to be effective against *B. cereus, P. mirabilis,* and *P. aeruginosa* while it showed no activity against *E. coli, E. faecalis, K. pneumonia,* and *S. typhi* (Asghar et al., 2020).

The green synthesis of chitosan functionalized silver nanoparticles (CS-AgNPs) was reported using the ethanolic extract of the buds of *Syzygium aromaticum* (L.) Merr. & L.M. Perry. The CS-AgNPs were then tested against vancomycin-resistant *Staphylococcus aureus* and methicillin-resistant *Staphylococcus aureus* and the zone of inhibition were 5.5 ± 0.4 and 10.3 ± 0.6 mm, respectively showing their potential as an antibacterial agent (Asghar et al., 2020).

Bibi et al. (2011) determined the antibacterial activity of crude methanolic extract of nine medicinal plants of Pakistan, that is, *Althaea officinalis* L., *Cordia dichotoma* G. Forst., *Ephedra gerardiana* Wall. ex Stapf, *Glycyrrhiza glabra* L., *Hyssopus officinalis* L., *Malva sylvestris* L., *Justicia adhatoda* L., *Onosma bracteatum* Wall., and *Ziziphus vulgaris* L. using the agar well diffusion method. Different concentrations of extracts were screened against two Gram-negative bacteria (*Escherichia coli* and *Pseudomonas aeruginosa*) and one Gram-positive bacterial strain (*Staphylococcus aureus*). Maximum inhibitory effect of all plants was observed at 15 mg/mL concentration of extract. The maximum antibacterial activity was seen in *J. adhatoda* against *P. aeruginosa, H. officinalis* against *S. aureus, G. glabra* against *E. coli* having the diameter of 2.67 ± 0.06, 3.37 ± 0.05, and 3.6 ± 0.3 mm, respectively (Bibi et al., 2011).

Khan et al. (2017) designed an experiment to analyze the antibacterial activity of HPLC fractions of aqueous, hexane, chloroform, and methanol-based peel extracts of *Punica granatum* L. The extracts were then screened against multidrug-resistant (MDR) pathogenic bacteria (*Escherichia coli, Acinetobacter baumannii, Staphylococcus aureus,* and *Pseudomonas aeruginosa*). The HPLC fractions collected from aqueous peel extract showed maximum inhibitory activity against *P. aeruginosa,* hexane fraction exhibited activity against three pathogens, while the ethanol fraction showed activity against *A. baumannii.* The results of the study showed that HPLC fractions of peel extract of *P. granatum* exhibited potential inhibitory activity against MDR bacterial human pathogens (Khan et al., 2017).

Extracts of six medicinal plants, that is, *Aesculus indica* (Wall. ex Cambess.) Hook., *Arisaema flavum* (Forssk.) Schott, *Carissa spinarum* L., *Debregeasia saeneb* (Forssk.) Hepper & J.R.I. Wood, *Pistacia chinensis* subsp. *integerrima* (J. L. Stewart ex Brandis) Rech. f., and *Toona ciliata* M. Roem. were tested for their bactericidal potential against two Gram-negative (*Pseudomonas pickettii, Salmonella setubal*) and three Gram-positive (*Bacillus subtilis, Micrococcus luteus, Staphylococcus aureus*) bacterial strains. Other than hexane, all fractions of *A. indica* showed antibacterial potential with significantly more activity by aqueous extract against *B. subtilis* with zone of inhibition of 16 ± 1 mm. The crude leaf extract of *T. ciliata* was significantly active against all bacterial strains except *S. aureus* and showed maximum inhibition zone against *S. setubal,* that is, 17.1 mm. Among all the plants, the maximum antibacterial activity was shown by aqueous and crude extracts of *P. integerrima* against *B. subtilis* (19.66 ± 0.05 mm) and *S. aureus* (23 ± 0.34 mm) (Bibi et al., 2011).

The phytochemical and antibacterial potential of the seeds of *Cichorium intybus* L. used traditionally as a medicine in Pakistan was analyzed by agar well diffusion method.

Aqueous, ethanol, hexane, and chloroform extracts were tested against *Escherichia coli* and *Staphylococcus aureus*. Among all the extracts, aqueous seeds were the most active ones and exhibited a wide range of inhibition zones (Rahimullah et al., 2019).

Batool et al. (2019) carried out an in vitro study to screen the extract of *Himalaiella heteromalla* (D. Don) Raab-Straube against different bacterial strains. Gram-positive strains were *Bacillus subtilis* ATCC 5230, *Micrococcus luteus* ATCC 9341s, *Staphylococcus aureus* ATCC 6538, while Gram-negative strains were *Pseudomonas aeruginosa* ATCC 9027, *Salmonella typhi* ATCC 14028, *Serratia marcescens* ATCC 13880, *Escherichia coli* ATCC 8739, and *Stenotrophomonas maltophilia* ATCC 13637. Three concentrations of methanolic extract were used and maximum antibacterial activity was exhibited against *S. marcescens* with inhibitory zones of 13.87, 16.20, and 20.65 mm at 30, 60, and 90 μg/mL concentration, respectively (Batool et al., 2019).

Suleman et al. (2019) analyzed seven cultivars of *Mangifera indica* L. (Mango) such as Almashil, Chaunsa, Dalasi, Dosehri, Hujra, Langra, and Sindhri for their potential antioxidant and antibacterial properties. Extracts were prepared from peel, kernel, and pulp and tested against *Bacillus subtilis*, *Enterobacter aerogenes*, *Micrococcus luteus*, *Pseudomonas septica*, and *Staphylococcus aureus*, using the disk diffusion method. It was observed that the bacterial growth was inhibited most effectively by Langra peel extract with inhibition zone of 22 mm against *S. aureus* which was 95% more effective than the positive control (Suleman et al., 2019).

Ullah et al. (2020) carried out an experiment for evaluating the antibacterial potential of different fractional extracts of *Cerastium glomeratum* Thuill. The aqueous extracts of *Cerastium glomeratum* showed maximum inhibitory activity against both *Escherichia coli* and *Staphylococcus aureus* having inhibition zones of 26.03 ± 0.1 and 28.06 ± 0.02, respectively. On the other hand, n-hexane, dichloromethane, and ethyl acetate extracts comparably showed less inhibition (Ullah et al., 2020).

Conclusion

Different compounds such as ascorbic acid, curcumin, vasicine, piperine, quercetin, myricetin, and gallic acid being reportedly isolated from these plants possess antibacterial potential. Pakistan has a variety of ethnomedicinal plants used to treat different bacterial diseases; however, studies on in vivo activity, toxicology, and mechanism of action are very limited. Hence, a detailed investigation on these aspects needs to be carried out for the development of novel antibacterial drugs from the studied plant species.

References

Abbasi, A. M., Khan, M. A., Ahmad, M., Zafar, M., Khan, H., Muhammad, N., ... Sultana, S. (2009). Medicinal plants used for the treatment of jaundice and hepatitis based on socio-economic documentation. *African Journal of Biotechnology*, *8*, 1643–1650. Available from http://www.academicjournals.org/AJB/PDF/pdf2009/20Apr/Abbasi%20et%20al.pdf.

Ahameethunisa, A. R., & Hopper, W. (2010). Antibacterial activity of *Artemisia nilagirica* leaf extracts against clinical and phytopathogenic bacteria. *BMC Complementary and Alternative Medicine*, *10*, 6. Available from https://doi.org/10.1186/1472-6882-10-6.

Ahmad, N. S., Farman, M., Najmi, M. H., Mian, K. B., & Hasan, A. (2008). Pharmacological basis for use of *Pistacia integerrima* leaves in hyperuricemia and gout. *Journal of Ethnopharmacology, 117*, 478–482. Available from https://doi.org/10.1016/j.jep.2008.02.031.

Ahmad, M., Khan, R. A., Khan, F. U., Khan, N. A., Shah, M. S., & Khan, M. R. (2011). Antioxidant and antibacterial activity of crude methanolic extract of Euphorbia prostrata collected from District Bannu (Pakistan). *African Journal of Pharmacy and Pharmacology, 5*(8), 1175–1178.

Alder, J. D. (2005). Daptomycin, a new drug class for the treatment of Gram-positive infections. *Drugs of Today, 41*, 81–90. Available from https://doi.org/10.1358/dot.2005.41.2.882660.

Ali, N. H., Faizi, S., & Kazmi, S. U. (2011). Antibacterial activity in spices and local medicinal plants against clinical isolates of Karachi, Pakistan. *Pharmaceutical Biology, 49*, 833–839. Available from https://doi.org/10.3109/13880209.2010.551136.

Asghar, M. A., Yousuf, R. I., Shoaib, M. H., & Asghar, M. A. (2020). Antibacterial, anticoagulant and cytotoxic evaluation of biocompatible nanocomposite of chitosan loaded green synthesized bioinspired silver nanoparticles. *International Journal of Biological Macromolecules, 160*, 934–943. Available from https://doi.org/10.1016/j.ijbiomac.2020.05.197.

Badoni, A. (2000). An ethnobotanical survey of Pinswari community – A preliminary survey. *Bulletin of the Botanical Survey of India, 32*, 110–115.

Batool, A., Miana, G. A., Muddassir, M., Khan, M. A., & Zafar, S. (2019). In vitro cytotoxic, antioxidant, antibacterial and antifungal activity of *Saussurea heteromalla* indigenous to Pakistan. *Pakistan Journal of Pharmaceutical Sciences, 32*, 2771–2777.

Bibi, Y., Nisa, S., Chaudhary, F. M., & Zia, M. (2011). Antibacterial activity of some selected medicinal plants of Pakistan. *BMC Complementary and Alternative Medicine, 11*, 52. Available from https://doi.org/10.1186/1472-6882-11-52.

Bilal, H., Ali, I., Uddin, S., Khan, I., Said, A., Ur Rahman, M., … Khan, A. A. (2020). Biological evaluation of antimicrobial activity of *Calotropis procera* against a range of bacteria. *Journal of Pharmacognosy and Phytochemistry, 9*(1), 31–35.

Chakraborthy, G. S. (2009). Evaluation of immunomodulatory activity of *Aesculus indica*. *International Journal of PharmTech Research, 1*, 132–134. Available from http://sphinxsai.com/pdf/jpt_Ap_Ju_09/PT=4%20Chakraborthy%20(132-134).pdf.

Hafeez, M., Zeb, M., Khan, A., Akram, B., Abdin, Z. U., Haq, S., … Ali, S. (2021). Populus ciliata mediated synthesis of silver nanoparticles and their antibacterial activity. *Microscopy Research and Technique, 84*(3), 480–488.

Hazrat, A., Nisar, M., & Zaman, S. (2013). Antibacterial activities of sixteen species of medicinal plants reported from dir kohistan valley KPK, Pakistan. *Pakistan Journal of Botany, 45*, 1369–1374. Available from http://www.pakbs.org/pjbot/PDFs/45%284%29/33.pdf.

Hussain, F., Kalim, M., Ali, H., Ali, T., Khan, M., Xiao, S., … Ashraf, A. (2016). Antibacterial activities of methanolic extracts of Datura inoxia. *PSM Microbiology, 1*(1), 33–35.

Ihsan-ul-haq., Mannan, A., Ahmed, I., Hussain, I., Jamil, M., & Mirza, B. (2012). Antibacterial activity and brine shrimp toxicity of artemisia dubia extract. *Pakistan Journal of Botany, 44*, 1487–1490. Available from http://www.pakbs.org/pjbot/PDFs/44%284%29/49.pdf.

Javid, T., Adnan, M., Tariq, A., Akhtar, B., Ullah, R., & AbdElsalam, N. M. (2015). Antimicrobial activity of three medicinal plants (*Artemisia indica, Medicago falcata* and *Tecoma stans*). *African Journal of Traditional, Complementary and Alternative Medicines, 12*, 91–96. Available from https://doi.org/10.4314/ajtcam.v12i3.11.

Khalid, A., Waseem, A., Saadullah, M., Uzair-Ur-Rehman, S., Khiljee, A., Sethi, M. H. H. B., … Murtaza. (2011). Antibacterial activity analysis of extracts of various plants against Gram -positive and -negative bacteria. *African Journal of Pharmacy and Pharmacology, 5*, 887–893. Available from http://www.academicjournals.org/AJPP/PDF/pdf2011/July/Khalid%20et%20al.pdf.

Khan, D. A., Hassan, F., Ullah, H., Karim, S., Baseer, A., Abid, M. A., … Murtaza, G. (2013). Antibacterial activity of *Phyllantus emblica, Coriandrum sativum, Culinaris medic, Lawsonia alba* and *Cucumis sativus*. *Acta Poloniae Pharmaceutica, 70*, 855–859.

Khan, I., Rahman, H., Abd El-Salam, N. M., Tawab, A., Hussain, A., Khan, T. A., … Ullah, R. (2017). *Punica granatum* peel extracts: HPLC fractionation and LC MS analysis to quest compounds having activity against multidrug resistant bacteria. *BMC Complementary and Alternative Medicine, 17*, 247. Available from https://doi.org/10.1186/s12906-017-1766-4.

Khattak, S., Saeed-ur-Rehman., Shah, H. U., Ahmad, W., & Ahmad, M. (2005). Biological effects of indigenous medicinal plants *Curcuma longa* and *Alpinia galanga*. *Fitoterapia, 76*, 254–257. Available from https://doi.org/10.1016/j.fitote.2004.12.012.

Khurram, M., Khan, M. A., Hameed, A., Abbas, N., Qayum, A., & Inayat, H. (2009). Antibacterial activities of *Dodonaea viscosa* using contact bioautography technique. *Molecules (Basel, Switzerland), 14*, 1332–1341. Available from https://doi.org/10.3390/molecules14031332.

Koehn, F., & Carter. (2005). The evolving role of natural products in drug discovery, origins of plant derived medicines. *Ethnobotanical Leaflets*, *4*, 373–387.

Kumar, G., Karthik, L., & Rao, K. V. B. (2010). Antimicrobial activity of latex of *Calotropis gigantea* against pathogenic microorganisms - An in vitro study. *Pharmacologyonline*, *3*, 155–163. Available from http://www.unisa.it/download/1966_11225_424168271_17.Kumar.pdf.

Malik, F., Mirza, T., Riaz, H., Hameed, A., & Hussain, S. (2010). Biological screening of seventeen medicinal plants used in the traditional systems of medicine in Pakistan for antimicrobial activities. *African Journal of Pharmacy and Pharmacology*, *4*, 335–340. Available from http://www.academicjournals.org/AJPP/PDF/pdf2010/June/Malik%20et%20al.pdf.

Mannan, H. A., Ahmed, I. B. R. A. R., Hussain, I. Z. H. A. R., Jamil, M., & Miza, B. (2012). Antibacterial activity and brine shrimp toxicity of Artemisia dubia extract. *Pakistan Journal of Botany*, *44*, 1487–1490.

Munazir, M., Qureshi, R., Arshad, M., & Gulfraz, M. (2012). Antibacterial activity of root and fruit extracts of *Leptadenia pyrotechnica* (Asclepiadaceae) from Pakistan. *Pakistan Journal of Botany*, *44*, 1209–1213. Available from http://www.pakbs.org/pjbot/PDFs/44(4)/06.pdf.

Mustafa, G., Ahmed, S., Ahmed, N., & Jamil, A. (2016). Phytochemical and antibacterial activity of some unexplored medicinal plants of Cholistan Desert. *Pakistan Journal of Botany*, *48*, 2057–2062. Available from http://www.pakbs.org/pjbot/PDFs/48(5)/36.pdf.

Naqvi, S. B. S., Azhar, I., Jabeen, S., & Hasan, S. F. (2012). Report: Studies on antibacterial activity of some traditional medicinal plants used in folk medicine. *Pakistan Journal of Pharmaceutical Sciences*, *25*(3), 669–674.

Nasrullah, S., Rahman, K., Ikram, M., Nisar, M., & Khan, I. (2012). Screening of antibacterial activity of medicinal plants. *International Journal of Pharmaceutical Sciences Review and Research*, *14*, 25–29. Available from http://globalresearchonline.net/journalcontents/v14-2/03.pdf.

Nathan, C., Ahameethunisa, A. R., & Hoper, W. (2004). Antibacterial activity of *Artemisia nilagirica* leaf extract against clinical and phytopathogenic bacteria. *BMC Complementry and Alternative Medicines*, *431*.

Nikkon, F., Saud, Z. A., Rahman, H., & Haque, E. (2003). In vitro antimicrobial activity of the compound isolated from chloroform extract of *Moringa oleifera* Lam. *Pakistan Journal of Biological Sciences*, *6*(22), 1888–1890.

Rahimullah, G., Shah, T., Mujaddad-ur-Rehman, S. T., & Hayat, A. (2019). Phytochemical and antibacterial screening of *Cichorium intybus* seeds use in traditional medicine systems in Pakistan. *International Journal of Basic Medical Sciences and Pharmacy (IJBMSP)*, *8*.

Rahman, K., Nisar, M., Jan, A. U., Suliman, M., Iqbal, A., Ahmad, A., … Ghaffar, R. (2015). Antibacterial activity of important medicinal plants on human pathogenic bacteria. *International Journal of Agronomy and Agricultural Research*, *6*(06), 106–111.

Rauf, A., Muhammad, N., Khan, A., Uddin, N., Atif, M., & Barkatullah. (2012). Antibacterial and phytotoxic profile of selected Pakistani medicinal plants. *World Applied Sciences Journal*, *20*(4), 540–544. Available from https://doi.org/10.5829/idosi.wasj.2012.20.04.1718.

Sabahat, S., & Perween, T. (2005). Antibacterial activities of *Mentha piperita*, *Pisum sativum* and *Momordica charantia*. *Pakistan Journal of Botany*, *37*(4), 997–1001.

Shah M., & Khan, M. (2006). Common medicinal folk recipes of Siran Valley.

Sohoo, A. B., Kamboh, A. A., Leghari, R. A., Abro, S. H., Korejo, N. A., & Soomro, J. (2019). Individual and combined antibacterial activity of plant essential oils and antibiotics against bacterial isolates of mastitis. *International Journal of Applied Research in Veterinary Medicine*, *17*, 22–28.

Suleman, M., Ali, J. S., Haq, I. U., Nisa, S., & Zia, M. (2019). Antioxidative, protein kinase inhibition and antibacterial potential of seven mango varieties cultivated in Pakistan. *Pakistan Journal of Pharmaceutical Sciences*, *32*, 1687–1695. Available from http://www.pjps.pk/wp-content/uploads/pdfs/32/4/Paper-30.pdf.

Ullah, H., Hubaib, M., Israr, M., Mushtaq, M., Zeeshan, M., & Mustafa, M. (2020). Antibacterial activity of different fractional extracts of *Cerastium glomeratum*. *Journal of Tropical Pharmacy and Chemistry*, *5*(2), 57–62.

Walsh, C. (2003). Where will new antibiotics come from? *Nature Reviews Microbiology*, *1*, 65–70. Available from https://doi.org/10.1038/nrmicro727.

Walter, C., Shinwari, Z. K., Afzal, I., & Malik, R. N. (2011). Antibacterial activity in herbal products used in Pakistan. *Pakistan Journal of Botany*, *43*, 155–162.

Zahoor, M., Shah, A. B., Gul, S., & Amin, S. (2018). HPLC-UV analysis of antioxidants in *Citrus sinensis* stem and root extracts. *Journal of the Chemical Society of Pakistan*, *40*, 595–601. Available from http://www.jcsp.org.pk/PublishedVersion/187a8a7e-d1f5-43a3-b8b6-c9ce9015ed8eManuscript%20no%2024,%20Final%20Gally%20Proof%20of%2011333%20(Muhammad%20%20Zahoor).pdf.

Zaidi, M. A., & Crow, S. A. (2005). Biologically active traditional medicinal herbs from Balochistan, Pakistan. *Journal of Ethnopharmacology*, *96*, 331–334. Available from https://doi.org/10.1016/j.jep.2004.07.023.

CHAPTER

7

Medicinal plants used as antidiarrheal agents in the lower Mekong basin

François Chassagne

UMR 152 PharmaDev, IRD, UPS, Université de Toulouse, Toulouse, France

Introduction

The Mekong is the 10th largest river in the world, it originates in the Tibetan Plateau and runs through six different countries: China, Myanmar, Lao PDR, Thailand, Cambodia, and Vietnam before emptying into the South China Sea (Adamson, Rutherfurd, Peel, & Conlan, 2009). The lower part of the Mekong, also known as the lower Mekong basin, flows from Lao PDR to Vietnam, and is home to about 67 million people including more than 100 ethnic groups (Laws & Semone, 2009; Mainuddin, Kirby, & Hoanh, 2011) (Fig. 7.1). This area is also highly biodiverse with about 20,000 plant species, 430 mammals, 1200 bird species, 800 reptile and amphibian species, and 1148 fish species (The Mekong River Commission, 2019). This rich biodiversity is supporting by a large variety of landscapes with the highlands of northern Thailand and northern Lao PDR, the Annamites mountains of eastern Lao PDR and western Vietnam, the plains in Lao PDR, Thailand and Cambodia, and the Mekong delta and costal mangroves (Douglas, 2005). Different forest formations are encountered in the area such as wet evergreen forests, semievergreen forests, mixed deciduous forests, deciduous dipterocarp woodlands, montane forests, wetlands, and swamp forests (Rundel, 2009). This unique biodiversity provides a wide range of ecosystem services which benefit human well-being and human health. It represents a significant source of food for rural and indigenous communities living in the area, and it also plays an important role in health care systems by providing medicinal resources such as medicinal plants (Walther et al., 2016).

In the lower Mekong basin, diarrheal diseases are frequent and remain one of the leading causes of morbidity and mortality, especially among children under five (Fischer Walker, Perin, Aryee, Boschi-Pinto, & Black, 2012; Troeger et al., 2016). From 2010 to 2015, 147, 90, and 27 diarrhea outbreaks have been reported in Cambodia, Lao PDR, and Vietnam, respectively (Lawpoolsri et al., 2018). Moreover, global change (i.e., climate change, rapid population growth, industrialization, and intensive agricultural development) in the area affect

Medicinal Plants as Anti-infectives
DOI: https://doi.org/10.1016/B978-0-323-90999-0.00015-X

Upper Mekong Basin
GREATER MEKONG
CHINA
MYANMAR
Hanoi
LAO P.D.R
Vientiane
Lower
Mekong
Basin
Mekong River
THAILAND
VIET NAM
Bangkok
Phnom Penh
CAMBODIA
N

water quality and can increase the number of diarrheal events (Boithias et al., 2016; Chea, Grenouillet, & Lek, 2016; McIver et al., 2014; Phung, Huang, Rutherford, Chu, Wang, & Nguyen, 2015; Phung, Huang, Rutherford, Chu, Wang, Nguyen, . . . Nguyen, 2015).

The WHO defines diarrhea as the passage of three or more loose or liquid stools per 24 h, or more frequently than is normal for the individual (WHO, 2020). Various clinical manifestations of diarrhea have been described including acute (<7 days), prolonged (7–13 days), persistent (14–19 days), and chronic (14–29 days) diarrhea (Shane et al., 2017). Two main forms of acute diarrhea are recognized: acute watery diarrhea (includes cholera), and acute bloody diarrhea, also called dysentery, which manifests as frequent scant stools with blood and mucus (Shane et al., 2017).

Acute diarrhea is mainly caused by bacteria and viruses and to a lesser extent by parasites (Marcos & DuPont, 2007). In the lower Mekong basin, the main causative agents are rotaviruses and diarrheagenic *Escherichia coli* strains including enterotoxigenic *E. coli* (ETEC), enteroaggregative *E. coli*, enteropathogenic *E. coli* (EPEC), and enteroinvasive *E. coli* (EIEC) (Meng et al., 2011; Thompson et al., 2015; Vu Nguyen, Le Van, Le Huy, Nguyen Gia, & Weintraub, 2006). Other bacterial agents have also been reported in the area: *Salmonella* spp., *Shigella* spp., *Vibrio cholerae*, and *Campylobacter jejuni/coli* (Mahapatra et al., 2014; Ngoc, 2017; Phetsouvanh et al., 2008; Poramathikul et al., 2016; Von Seidlein et al., 2006).

The first-line treatment of acute diarrhea is oral rehydration solution for all cases, while antibiotics are mostly used for severe cases of diarrhea, at-risk populations (e.g., child, elderly person, immunocompromised patients), and traveler's diarrhea (Riddle, Dupont, & Connor, 2016).

However, these bacterial agents are becoming increasingly resistant to antibiotics, causing problems in the management of patients with infectious diarrhea. Antibiotic-resistant bacteria include strains of *Campylobacter*, ETEC, *Salmonella*, *Shigella*, and *V. cholerae* in Cambodia (Meng et al., 2011; Poramathikul et al., 2016); strains of *Campylobacter*, *Salmonella*, *Shigella*, and all types of *E. coli* in Vietnam (Thompson et al., 2015; Vu Nguyen et al., 2006); and strains of nontyphi *Salmonella*, Shigella, and *V. cholerae* in Lao PDR (Iwanaga et al., 2004; Phuong et al., 2017; Yamashiro et al., 1998).

Traditional medicine for diarrheal diseases in the Mekong Basin

The role of traditional medicine in the management of diarrhea

Traditional practices, especially medicinal plants, have been reported to treat diarrhea in various continents including Africa, Europe, India, and South America (Palombo, 2006). In the lower Mekong region, they continue to play an important role in primary health care, especially for rural populations due to their availability, accessibility, low cost, and effectiveness (Maneenoon et al., 2015; Nguyen, Bun, Ollivier, & Dang, 2020). Sydara et al. (2005) interviewed 592 villagers of the Champasak province in Lao PDR and found that 77% had already

FIGURE 7.1 Map showing the four countries from the lower Mekong basin. Source: *From Adamson, P.T., Rutherfurd, I.D., Peel, M.C., & Conlan, I.A. (2009). The hydrology of the Mekong River. In: The Mekong (pp. 53-76). Australia: Elsevier Inc. https://doi.org/10.1016/B978-0-12-374026-7.00004-8. Mekong River Commission. (2018). An introduction to MRC procedural rules for Mekong water cooperation. Retrieved from https://www.mrcmekong.org/assets/Publications/MRC-procedures-EN-V.7-JUL-18.pdf. (Accessed December 2021).*

used traditional medicine, and 59% had used traditional medicine in the last 6 months (Sydara et al., 2005). Among the interviewees using traditional medicine, almost all of them reported using traditional medicine for treating diarrhea (Sydara et al., 2005). In a Kho-Cil community from the South of Vietnam (Lam Dong province), 79% of the 124 informants interviewed reported to use medicinal plants for their health problems including diarrhea (Nguyen et al., 2020). In northern Thailand, Panyadee, Balslev, Wangpakapattanawong, and Inta (2019) inventoried medicinal plants found in 195 home gardens from four villages and showed that diarrhea was one of the most common ailment treated with these plants (Panyadee et al., 2019). In another study, Tangjitman, Wongsawad, Kamwong, Sukkho, and Trisonthi (2015) studied the medicinal plants used by Karen people of northern Thailand for digestive disorders and showed that diarrhea was associated with the highest number of plant species and uses recorded (Tangjitman et al., 2015). In a Vietnamese rural district, mothers mentioned traditional herbs as being the most used therapies for treating diarrhea in children under the age of five (Le, Ottosson, Nguyen, Kim, & Allebeck, 2011). In northeastern Cambodia, Chassagne, Hul, Deharo, and Bourdy (2016) reported that 87% of Bunong people used traditional medicine as a first step for treating diarrhea (Chassagne et al., 2016).

The cultural belief system of people living in the Mekong area

The use of traditional medicine for diarrhea is also associated with local perceptions and beliefs that can influence the management of the disease by villagers. Shawyer, Gani, Punufimana, and Seuseu (1996) identified four main factors in the belief system influencing decision making and behavior including diagnostic categories, explanatory models, perceived ill-effects, and appropriate treatments (Shawyer et al., 1996). Here, I present some examples of beliefs and practices related to these four factors in the Mekong population.

In the Khon Kaen province of Thailand, diarrheal illnesses are classified into different categories which have their own perceived causes and their own therapeutic practices (Shawyer et al., 1996). While most of these categories present similar features than those from the biomedical classification, one hold "spirits" responsible for the manifestation of diarrhea. In this case, the only treatment recommended by villagers is to ask a spirit medium for ceremony to apologize to spirits. Also, one of the most frequently cited causes of diarrhea is centered around food taboos. This widespread belief involves the eating of wrong food such as sour foods and fermented foods. Regarding therapeutic practices, herbal medicine was reported in the treatment of some categories of diarrhea especially the type called "tong sia" or "tong ruang" also described as a common watery diarrhea.

In rural Vietnam, mothers consider that diarrhea without fever is a minor illness and thus can be treated by self-medication such as traditional therapies (Le et al., 2011). In another study, Vietnamese people living in urban area mentioned that traditional medicine is "much safer" than Western medicine, and they often used it for mild symptoms (Hoa, Öhman, Lundborg, & Chuc, 2007). Similar perceptions were noted among four ethnic groups in northern Vietnam. In this study, it was shown that ethnic minority caregivers recognized the danger signs of child diarrhea. They administer home remedies such as herbal mixtures for minor episodes of diarrhea lasting 2–4 days, but they consult the communal health station if the diarrhea lasts more than 4 days, reemerged shortly after recovery, or include high fever, very frequent defecation, loss of weight, and lethargy (Rheinländer, Samuelsen, Dalsgaard, & Konradsen, 2011).

An important medical concept to consider when studying the management of diarrheal illnesses is the hot-cold explanatory model for illnesses. In Vietnam, Kaljee et al. (2004) reported that diarrhea and dysentery are perceived as "hot" diseases that should be treated by "cold" medicine (Kaljee et al., 2004). In this study, herbs are considered by Vietnamese people as "cold," while modern medicine is considered to be "hot" (Kaljee et al., 2004). For this reason, some informants reported to be reluctant to use Western medicine for diarrhea and dysentery.

In Thai and Cambodian medicine, "wind illness" has been reported as another explanatory model for diarrheal illnesses (Chassagne et al., 2016; Muecke, 1979). The "hot-cold" balance and "wind illness" are linked to the humoral theory which states that the human body is composed of four basic elements, namely, earth, water, fire, and wind. In this theory, health is maintained through equilibrium, and a dysfunction of one of these elements can cause illnesses (Chassagne, Deharo, Punley, & Bourdy, 2017; Shawyer et al., 1996).

Organoleptic properties (i.e., taste, smell, color, shape) have also been reported as major selection criteria of plants for diarrhea (Ankli, Sticher, & Heinrich, 1999; Leonti, Sticher, & Heinrich, 2002). In traditional Thai medicine, the taste of drug is of utmost importance to identify antidiarrheal plant species, and Thai people consider that astringent tasting plants have healing properties for diarrhea (Neamsuvan, Phumchareon, Bunphan, & Kaosaeng, 2016). In Cambodian medicine, astringent taste is also a criterion to select the plant species used for treating diarrhea (Chassagne et al., 2016). Odor can also play a role in the selection of plants for diarrhea. For example, *Paederia foetida* L., an herbaceous plant native to Asia, has a strong sulfurous odor resembling that of a flatus or feces. This plant is widely used in Asia for intestinal disorders such as diarrhea and dysentery because native doctors think that the odor given off by the leaves identifies them as a specific treatment (Steinmetz, 1961). In Vietnam, the plant has been reported for the treatment of diarrhea and dysentery (van Sam, Baas, & Keßler, 2008).

Pharmacological validation of plants used for diarrhea

The use of medicinal plants for diarrhea is based on empirical knowledge of traditional healers and villagers and influenced by sociocultural factors. In a scientific context, the ethnomedical use of plants can be validated by pharmacological assays to provide rationalization of their presumed clinical effect. In this approach, in vitro models are used as a first step due to their low cost, simplicity, lack of ethical constraints, and the small amount of material required (Brusotti, Cesari, Dentamaro, Caccialanza, & Massolini, 2014; Butterweck & Nahrstedt, 2012). However, in vitro tests are less clinically relevant than in vivo assays which must be considered as a second step in the validation process of medicinal plants (Houghton, Howes, Lee, & Steventon, 2007). Furthermore, clinical trials are the last step in the process to verify or refute empirical knowledge, and only well-designed randomized double-blind placebo-controlled clinical studies should be used (Gertsch, 2009).

In the case of plants used for diarrhea, several pharmacological actions can be involved in the observed clinical effect (Fig. 7.2). First, plants can act on the signs and symptoms of diarrhea (e.g., frequent loose stools, abdominal cramps), and thus exhibit anti-diarrheal and/or antispasmodic effect. Second, plants can act on the pathophysiology of diarrhea (e.g., alteration in intestinal motility, abnormal water adsorption or electrolyte secretion into the intestinal lumen) and display antimotility or antisecretory activity. Third, plants

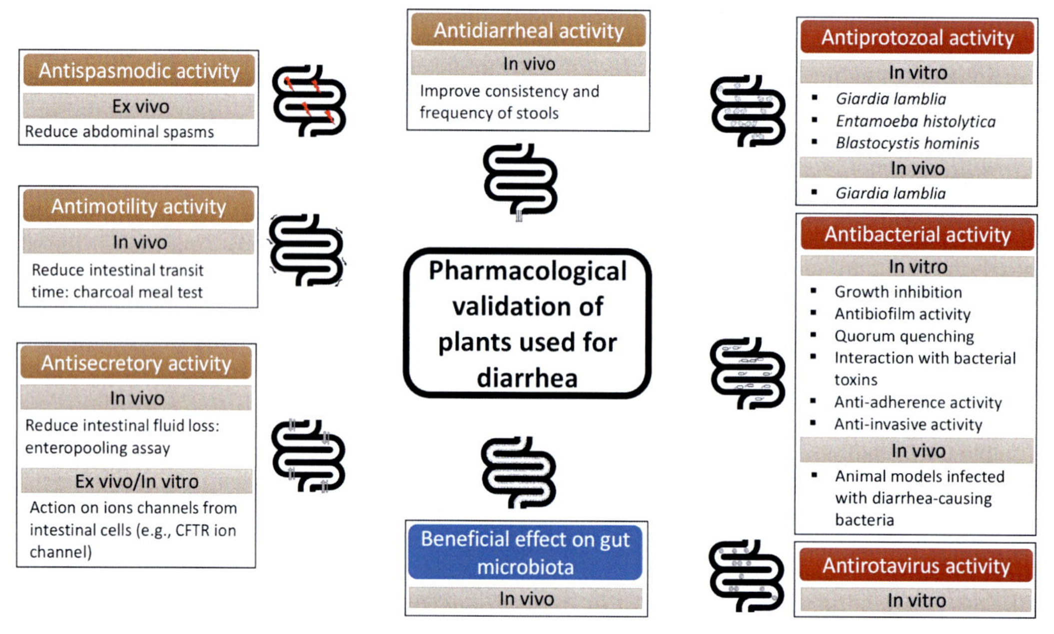

FIGURE 7.2 Current pharmacological models used to evaluate the effect of plants treating diarrhea. In red: pharmacological models targeting microbes causing diarrhea. In brown: pharmacological models acting on the diarrheal symptoms and pathophysiology. In blue: pharmacological model studying the change in intestinal microbiota.

can target the specific agents causing diarrhea (e.g., viruses, bacteria, and parasites), and show antibacterial, antiviral or antiparasitic activity. This leads to the analysis of the main pharmacological models currently employed for evaluating plants used for diarrhea.

Models assessing the effect of plants on the signs and symptoms of diarrhea

Antidiarrheal effect

One of the most popular ways to evaluate the antidiarrheal effect of plants is to induce a diarrhea in an animal model (e.g., mice or rats) and to monitor changes in feces consistency, mass, and frequency with and without administration of plant extracts. The in vivo castor oil-induced diarrhea model is the most employed one and consists of administering a castor oil (a vegetable oil obtained from the bean of *Ricinus communis* L.) treatment to mice or rats (Kinuthia, Muriithi, & Mwangi, 2016; Palla et al., 2015). Other agents are also used to induce diarrhea such as magnesium sulfate, and lactose (Boakye, Brierley, Pasilis, & Balemba, 2012; Uddin et al., 2005).

Spasmolytic activity

To evaluate the antispasmodic effect of plants, *ex vivo* models (i.e., isolated intestinal tissues taken from a living animal) are usually employed. Isolated rabbit jejunum is widely used due to its spontaneous rhythmic contractions in controlled experimental conditions, avoiding the use of an agonist to test the relaxant activity (Kinuthia et al., 2016; Saqib et al., 2015). In a lesser manner, isolated guinea ileum is also employed, but this model requires the use of an agonist (e.g., acetylcholine) to induce ileum contraction (Ghayur & Gilani, 2005).

By using the same *ex vivo* models, additional experiments can be performed to assess the mechanism behind the spasmolytic activity. For example, a high concentration of K^+ can be added to the *ex vivo* preparation to test a potential calcium channel blocking activity. Indeed, a high concentration of K^+ induces a contraction of smooth muscle through influx of Ca^{2+}, and extracts or compounds inhibiting this contraction are known to block calcium channels (Palla et al., 2015; Saqib et al., 2015). Also, standard antagonists such as naloxone (μ opioid blocker), propranolol (β-adrenergic blocker), tamsulosin (α1-adrenergic blocker), and yohimbine (α2-adrenergic blocker) can be used to determine the type of receptor involved in the spasmolytic activity (Kinuthia et al., 2016).

Models assessing the antimotility and antisecretory activities

Antimotility activity

Inhibition of intestinal motility by plants can be assessed by using the gut meal travel test. In this model, castor oil is used as a pharmacological agent to increase peristaltic activity, and the transit of a charcoal meal (5%–10% activated charcoal in an aqueous suspension) along the small intestine is measured. A peristaltic index is usually calculated by comparing the distance traveled by the charcoal plug from pylorus to the cecum to the total length of the small intestine (Kinuthia et al., 2016; Palla et al., 2015).

Antisecretory activity

Under physiological conditions, the small intestine (mainly intestinal crypts) secretes fluid and electrolytes to prepare for digestion, and this process is balanced by fluid absorption by the intestinal villi (Farthing, 2006). In diarrheal diseases, particularly during cholera, enterotoxigenic bacterial infection, and rotavirus infection, the normal balance between absorption and secretion in the crypt-villus axis is disrupted. A hypersecretion across the intestinal lumen occurs and induces an excess in intestinal fluid losses. This secretory process results predominantly from the active secretion of chloride ions through cystic fibrosis transmembrane conductance regulator (CFTR) channels or calcium-activated chloride channels (CaCCs) (Farthing, 2006; Thiagarajah, Donowitz, & Verkman, 2015). In secretory diarrhea, CFTR channels are mostly activated by the bacterial enterotoxins released in cholera and ETEC infection, while CaCCs channels seem to be the main target of rotavirus infection and drug-induced secretory diarrhea (Thiagarajah & Verkman, 2013). Other pathways are also involved in the intestinal fluid secretion and thus could serve as alternative therapeutic targets such as Na/K/Cl transporters, calcium-sensing receptors, and K^+ channels (Thiagarajah et al., 2015).

To date, the main pharmacological assays used to evaluate the antisecretory activity of plant extracts are (1) animal-based models aiming to measure the fluid accumulation in the intestine, using ligated intestinal loop, after administration of toxins (e.g., cholera toxin or heat-stable enterotoxin of ETEC), castor oil, or prostaglandin E2 (Kinuthia et al., 2016; Palla et al., 2015; Velázquez et al., 2012). These experiments are also called enteropooling assays; (2) *ex vivo* models aiming to evaluate the action on ions transports (Na^+ absorption, Cl^- secretion) across the intestinal epithelium by using intestinal pieces of mice mounted in Ussing chambers (Makrane et al., 2019); (3) in vitro tests aiming to assess the CFTR channel inhibitory activity in intestinal cells (e.g., T84, Caco-2, HT29-CL19A) or other types of cells (e.g., recombinant FRT) (Akrimajirachoote, Satitsri, Sommart, Rukachaisirikul, & Muanprasat, 2020; Dawurung et al., 2020; Gabriel et al., 1999; Ma et al., 2002; Ren et al., 2012; Zhang, Fujii, & Naren, 2012; Zhang et al., 2014).

Models assessing the antiinfective properties

Antibacterial activity

In vitro models are commonly used to assess the antibacterial activity of plant extracts by targeting diarrhea-causing bacteria (e.g., *E. coli*, *Salmonella* spp., *Shigella boydii*, *Shigella flexneri*, *Shigella sonnei*, and *V. cholerae*). Traditional strategies aim at killing or preventing the growth of bacteria (Chassagne et al., 2020; Porras et al., 2020). For this purpose, different methods have been described such as the disk diffusion assay (or agar diffusion method), the agar dilution method, and the broth microdilution test (Alanís, Calzada, Cervantes, Torres, & Ceballos, 2005; De Esparza, Bye, Meckes, López, & Jiménez-Estrada, 2007; Madikizela, Ndhlala, Finnie, & Van Staden, 2012). Due to its simplicity, the disk diffusion method is widely employed, but the diffusion of the substance will depend mainly on its polarity and molecular weight, and thus prevent any comparison of different samples (Tan & Lim, 2015). For quantitative analysis, agar dilution and the broth microdilution methods are more appropriate (Wiegand, Hilpert, & Hancock, 2008).

Because killing or preventing the growth of bacteria leads to strong selective pressure, and thus development of resistance, other antibacterial approaches have been developed including targeting virulence factors (Rasko & Sperandio, 2010). Among the latter, we can mention the interaction of plant extracts with bacterial toxins. For example, the binding properties of natural substances to heat-labile enterotoxin from ETEC or to cholera toxin have been studied using an ELISA assay and an SDS-PAGE analysis, respectively (Chen et al., 2007; Velázquez et al., 2012). More recently, antibiofilm and quorum quenching assays have been used to target biofilm produced by bacteria and to interfere with bacterial signaling molecules (e.g., AI-2) (Castillo, Heredia, Arechiga-Carvajal, & García, 2014; Oh et al., 2017).

Although less common, other in vitro tests have been reported to be used for assessing the effect of plant extracts on the adherence and invasiveness of bacterial enteric pathogens to epithelial cells (Brijesh et al., 2006).

Finally, animal-based studies have also been employed to evaluate the antibacterial activity of plant extracts. In this case, a bacteria causing diarrhea is administered to animals and different physiological, biochemical, hematological, and histopathological parameters are measured (Choi et al., 2011; Fokam Tagne, Noubissi, Fankem, & Kamgang, 2018; Mukherjee et al., 2013; Noubissi et al., 2019; Thakurta et al., 2007).

Antiviral and antiparasitic activity

While bacteria are the most common pathogens used in pharmacological studies, viruses and parasites are also employed to explain the clinical effect of plant extracts used for diarrhea.

Rotaviruses are the main viruses to be tested as they represent the major causes of infantile gastroenteritis worldwide (Marcos & DuPont, 2007). Several rotaviruses strains have been reported to be used in pharmacological studies, including Human (HCR3), simian (RRV or SA-11 strains), porcine (KJ205−1), and bovine (KJ56−1) rotaviruses. These strains were multiplied in monkey kidney cells (i.e., MA-104 or TF-104 a cloned derivative of MA-104), then different strategies were used to determine the antirotavirus effect such as blocking effect of the viral infection, virucidal activity, or reduction of infection particles produced postinfection (Gonçalves et al., 2005; Kwon et al., 2010; Téllez, Téllez, Vélez, & Ulloa, 2015).

Regarding the antiparasitic effect, most of the studies focused on the antiprotozoal activity of natural substances either in vitro or in vivo. In vitro cytotoxicity activity of these substances were evaluated against trophozoites of *Giardia lamblia* (Brandelli, Giordani, De Carli, & Tasca, 2009), trophozoites of *Entamoeba histolytica* (Calzada, Bautista, Yépez-Mulia, García-Hernandez, & Ortega, 2015), and the intestinal protozoa *Blastocystis hominis* (Sawangjaroen & Sawangjaroen, 2005). Infection with trophozoites of *G. lamblia* in animal model was also reported (Barbosa, Calzada, & Campos, 2007).

Other models

Last but not least, new pharmacological models have been developed to evaluate the interaction of plants with gut microbes. Indeed, there is growing evidence showing that plants interact with the gut microbiota and exhibit beneficial health effect on gastrointestinal disorders (Thumann, Pferschy-Wenzig, Moissl-Eichinger, & Bauer, 2019). With the emergence of new technologies such as metagenomics, metabolomics, and metatranscriptomics, researchers are now able to study the complex relationship between gut microbiota, genes, and metabolites (Feng, Ao, Peng, & Yan, 2019). For example, Li et al. (2019) studied the effect of *Panax ginseng* polysaccharides on the diversity and composition of the gut microbiota in mice with antibiotic-associated diarrhea. The authors showed that *P. ginseng* polysaccharides positively affected the gut microbiota by increasing beneficial bacteria such as *Lactobacillus* or *Lactococcus* species, and also promoted the recovery of the mucosa by reversing carbohydrate, amino acid, and energy metabolism to normal levels. Altogether, these effects helped to alleviate the symptoms of diarrhea (Li et al., 2019).

This emerging field of study opens the way for a better understanding of the interaction of plant with gut microbiota in diarrheal diseases, and more researchers should integrate these experiments into their laboratory workflow when studying plants used for diarrhea.

Medicinal plants used for diarrhea in the lower Mekong basin

Literature search methodology

A review of the scientific literature was performed to identify medicinal plants used in traditional medicine for the treatment of diarrheal illnesses. Electronic databases (i.e.,

Google Scholar and Pubmed) were consulted by using specific keywords such as "plant" and "diarrhea," combined with countries' names from the lower Mekong basin: "Thailand" or "Lao PDR" or "Vietnam" or "Cambodia." Libraries from France, Cambodia, and Lao PDR were also consulted to gather books, thesis, and reports by focusing on ethnobotanical data from Southeast Asia.

From these documents, information on plant species names, botanical tissues, therapeutic indications, and countries of origin were extracted. All botanical names (genus and species) were checked on The Plant List website (http://www.theplantlist.org/).

A score was then calculated to rank the plant species by order of ethnobotanical importance, and thus generated a list of the most frequently used plants in the area. For each plant species, citation frequency (number of bibliographic references citing the plant species) and citation distribution (number of countries where the plant was reported to be used) were obtained, and these two categories were multiplied to give a score. Only plants with the highest score were considered in the discussion below.

Overview of the dataset

Overall, 431 plant species belonging to 307 botanical genera and 101 botanical families were identified from 51 bibliographic references. Among the latter, 35 references were original ethnobotanical surveys, while 16 were review articles or documents presenting anecdotal ethnomedical uses. Twenty-three documents reported ethnobotanical uses from Thailand, 14 from Lao PDR, 12 from Cambodia, and 11 from Vietnam.

Of the 431 plant species reported, the most represented botanical families were Fabaceae (48 plant species, 11.1%), Zingiberaceae (18, 4.2%), Rubiaceae (16, 3.7%), Asteraceae (15, 3.5%), and Malvaceae (14, 3.2%). *Terminalia* was the most represented botanical genus with seven plant species (1.6%), followed by *Bauhinia*, *Calamus*, and *Ficus* with five plant species each (1.2%), and *Alpinia*, *Amomum*, *Dillenia*, *Dioscorea*, *Euphorbia*, *Lagerstroemia*, *Melastoma*, *Rubus*, *Smilax*, *Syzygium*, *Zingiber*, and *Ziziphus* with four plant species each (0.9%). The most represented plant tissues were root (148 plant species, 34.3%), bark (124, 28.7%), leaf (118, 27.3%), stem (58, 13.4%), and fruit (54, 12.5%).

Regarding ethnobotanical uses, diarrhea and dysentery were mentioned for 351 and 185 plant species respectively. A total of 20 plants were mentioned for treating diarrhea and vomiting (e.g., *Psidium guajava* L., *Chromolaena odorata* (L.) R.M.King & H.Rob., and *Careya arborea* Roxb.), 11 for diarrhea in children (e.g., *P. guajava*, *Oroxylum indicum* (L.) Kurz, and *Saccharum officinarum* L.), and 9 for cholera (e.g., *Holarrhena pubescens* Wall. ex G.Don).

Only 6 plant species (i.e., *P. guajava*, *C. odorata*, *Alstonia scholaris* (L.) R. Br., *Centella asiatica* (L.) Urb., *Allium sativum* L., *Caesalpinia sappan* L.) were used in the four countries from the lower Mekong area, 10 species (i.e., *Punica granatum* L., *Mangifera indica* L., *H. pubescens*, *O. indicum*, *Piper nigrum* L., *Coscinium fenestratum* (Goetgh.) Colebr., *Rhodomyrtus tomentosa* (Aiton) Hassk., *Cyperus rotundus* L., *Ageratum conyzoides* (L.) L., and *Curcuma aromatica* Salisb.) were used in three different countries, 49 were used in two countries, and 366 plant species were used in only one country from the lower Mekong basin. Of the 431 plant species, 203 (47.1%) plant species were reported in Cambodia for the treatment of diarrheal illnesses, 113 (26.2%) in Thailand, 112 (26.0%) in Vietnam, and 75 (17.4%) in Lao PDR.

The top 24 plant species with the highest score are shown in Table 7.1

TABLE 7.1 Most used plant species for diarrheal illnesses in the lower Mekong basin.

Plant species	Botanical family	Geographical distribution	Tissue	Ethnomedical uses	Countries and ethnic groups[a]	Citation distribution[b]	Citation frequency[c]	Score[d]
Psidium guajava L.	Myrtaceae	Five continents (native to Central and South America)	Bark, bud, fruit, leaf, leaf (young), stem, root, trunk, wood	Diarrhea, diarrhea (infant), diarrhea (mucus), diarrhea (vomiting), dysentery	Cambodia (Bunong, Khmer), Lao PDR, (Brou, Hmong, Kry, Saek), Thailand (Karen, Lahu, Lisu, other, Thai Yuan, Yao), Vietnam	4	15	60
Chromolaena odorata (L.) R.M.King & H. Rob.	Asteraceae	Five continents (native to America)	Leaf, leaf (young), root	Diarrhea, diarrhea (vomiting), dysentery	Cambodia (Bunong), Lao PDR (Brou, Hmong, Kry, Saek), Thailand (Karen), Vietnam (Van Kieu, other)	4	8	32
Alstonia scholaris (L.) R. Br.	Apocynaceae	Asia, Australia	Bark, leaf, stem bark	Diarrhea, dysentery	Cambodia, Lao PDR, Thailand, Vietnam	4	7	28
Allium sativum L.	Amaryllidaceae	Five continents (native to Central Asia)	Bulb	Diarrhea, dysentery	Cambodia, Lao PDR, Thailand, Vietnam	4	6	24
Centella asiatica (L.) Urb.	Apiaceae	Five continents (native to Africa, Asia and Oceania)	Leaf, whole plant	Diarrhea, dysentery	Cambodia, Lao PDR (Kry), Thailand (Karen, Lahu), Vietnam	4	6	24
Punica granatum L.	Lythraceae	Five continents (native to Central Asia)	Flower, fruit (pericarp), leaf, leaf (young), root	Diarrhea, dysentery	Cambodia, Thailand (Karen, other), Vietnam	3	7	21
Caesalpinia sappan L.	Fabaceae	Southeast Asia	Bark, leaf, root, stem wood	Diarrhea, dysentery	Cambodia, Lao PDR, Thailand (Karen, Yao), Vietnam	4	5	20
Mangifera indica L.	Anacardiaceae	Five continents (native to South Asia)	Bark, root, seed	Diarrhea, dysentery	Cambodia (Bunong, Khmer), Thailand (Lahu, Lisu), Vietnam (Van Kieu)	3	6	18

(*Continued*)

TABLE 7.1 (Continued)

Plant species	Botanical family	Geographical distribution	Tissue	Ethnomedical uses	Countries and ethnic groups[a]	Citation distribution[b]	Citation frequency[c]	Score[d]
Holarrhena pubescens Wall. ex G.Don	Apocynaceae	Asia, Africa	Bark, root, stem	Cholera, diarrhea, dysentery	Cambodia, Lao PDR, Thailand (Tai Yai)	3	6	18
Oroxylum indicum (L.) Kurz	Bignoniaceae	Asia	Bark, root	Diarrhea, diarrhea (infant), dysentery	Cambodia, Lao PDR (Brou, Saek), Vietnam	3	5	15
Piper nigrum L.	Piperaceae	Africa, America, Asia (native to South India)	Fruit	Diarrhea, dysentery	Cambodia, Thailand, Vietnam	3	5	15
Careya arborea Roxb.	Lecythidaceae	Asia	Bark, leaf, leaf (young), root, stem, trunk	Diarrhea, diarrhea (mucus), diarrhea (simple), diarrhea (vomiting), dysentery	Cambodia (Bunong, Khmer), Vietnam	2	6	12
Coscinium fenestratum (Goetgh.) Colebr.	Menispermaceae	Asia	Liana, root, stem	Diarrhea, dysentery	Cambodia, Lao PDR (Brou, Kry), Vietnam	3	4	12
Rhodomyrtus tomentosa (Aiton) Hassk.	Myrtaceae	America, Asia (native to South Asia)	Fruit, leaf, root, stem, wood	Diarrhea, dysentery	Cambodia (Bunong), Thailand, Vietnam	3	4	12
Cyperus rotundus L.	Cyperaceae	Five continents (native to Africa, Asia, and Europe)	Root	Diarrhea, dysentery	Lao PDR, Thailand, Vietnam	3	4	12
Cratoxylum formosum (Jacq.) Benth. & Hook.f. ex Dyer	Hypericaceae	Southeast Asia	Bark, leaf, root, stem, trunk	Diarrhea (mucus), diarrhea (vomiting), dysentery	Cambodia (Bunong), Thailand (Lahu, Lawa)	2	5	10
Eurycoma longifolia Jack	Simaroubaceae	Southeast Asia	Bark, fruit, leaf, root	Diarrhea, dysentery	Cambodia, Vietnam	2	5	10

Ageratum conyzoides (L.) L.	Asteraceae	Five continents (native to Central America)	Leaf, root	Diarrhea, dysentery	Cambodia (Bunong), Lao PDR (Brou), Vietnam	3	3	9
Curcuma aromatica Salisb.	Zingiberaceae	Asia	Rhizome	Diarrhea, diarrhea bloating, diarrhea piles	Cambodia, Lao PDR, Thailand	3	3	9
Musa × paradisiaca L.	Musaceae	Five continents (native to Malaysia)	Flower, fruit, trunk (young)	Diarrhea, dysentery	Cambodia (Bunong, Khmer), Thailand (Karen, Yuan)	2	4	8
Zingiber officinale Roscoe	Zingiberaceae	Five continents (native to India)	Rhizome	Diarrhea	Cambodia (Bunong), Thailand (Karen, Other)	2	4	8
Areca catechu L.	Arecaceae	Africa, Asia (native to Philippines)	Leaf, fruit, root, seed	Diarrhea, dysentery	Cambodia, Vietnam	2	4	8
Shorea obtusa Wall. ex Blume	Dipterocarpaceae	Southeast Asia	Bark, root, sap, stem, trunk, wood	Diarrhea, diarrhea (mucus)	Cambodia (Bunong), Thailand (Tai Yai)	2	4	8
Xylia xylocarpa (Roxb.) Taub.	Fabaceae	Africa, Asia (native to South Asia)	Bark, fruit, root, stem, trunk, wood	Diarrhea (mucus), diarrhea (vomiting), dysentery	Cambodia (Bunong), Thailand (Yuan)	2	4	8

[a]*Countries and ethnic groups in which the plant species were reported to be used. When the ethnic group was not mentioned in the reference, the term "other" was employed.*
[b]*Number of countries reporting the plant in the treatment of diarrhea illnesses.*
[c]*Number of bibliographic references reporting the plant in the treatment of diarrheal illnesses.*
[d]*Calculated by multiplying "citation frequency" and "citation distribution" numbers.*

Of these 24 plant species, four are endemic to Southeast Asia (*C. sappan* L., *Cratoxylum formosum* (Jacq.) Benth. & Hook.f. ex Dyer; *Eurycoma longifolia* Jack; *Shorea obtusa* Wall. ex Blume), 21 are native to Asia and 10 are distributed throughout the five continents. The following section focuses on the first 10 most represented plant species.

Discussion of some selected plant species

Psidium guajava

Psidium guajava (Myrtaceae) (Fig. 7.3), also known as guava tree, is a shrub or small tree native to Central and South America, and widely cultivated throughout the world (Sitther et al., 2014). According to the bibliographic review, a total of 13 ethnic groups distributed throughout the four countries from the lower Mekong area mentioned the use of guava for the treatment of diarrheal illnesses. Various parts of the tree were reported to be used as antidiarrheic, with leaves and bark being the most used ones. Also, guava tree was mentioned for the treatment of different diarrheal illnesses such as simple diarrhea, infant diarrhea, diarrhea with mucus, diarrhea with vomiting, and dysentery. *P. guajava* was already reported to be used for diarrheal illnesses in other countries such as Brazil, China, Congo, Fiji, Mexico, Mozambique, Peru, Philippines, Senegal, South Africa, Trinidad, and USA (Gutiérrez, Mitchell, & Solis, 2008). From a pharmacological perspective, a wide range of in vitro and in vivo assays validated the use of *P. guajava* for infectious diarrhea. Its in vitro antibacterial activity was confirmed using a hot aqueous extract of dried leaves of *P. guajava* on *S. flexneri* and *V. cholerae*. The same extract also demonstrated an antiadherence effect on

FIGURE 7.3 Photographs of the three most used antidiarrheal plant species from the lower Mekong basin: (A) *Psidium guajava*; (B) *Chromolaena odorata*; (C) *Alstonia scholaris*.

EPEC and an antiinvasive effect on both EIEC and *S. flexneri*, and finally decreased the production of *E. coli* heat-labile toxin and cholera toxin (Birdi et al., 2010). Antirotavirus activity was also evaluated with a methanolic extract of *P. guajava* leaves which showed activity against a simian rotavirus (SA-11) at a concentration of 8 μg/mL (Gonçalves et al., 2005). Regarding in vivo studies, *P. guajava* leaf extract (300 mg/kg per day) showed quicker clearance of infection after 19 days in an infectious diarrhea model using *Citrobacter rodentium*-infected mice (Gupta & Birdi, 2015). In another in vivo model using *S. flexneri*-infected rats, *P. guajava* leaf extract at 200 mg/kg reduced the number and weight of stools collected, and the density of *S. flexneri* after 5 days of treatment (Hirudkar et al., 2020a, 2020b). Similar results were obtained in another study performed by the same authors in an EPEC-induced diarrhea rat model (Hirudkar et al., 2020). Clinical trials have also been performed to confirm the antidiarrheal effect of guava tree. In Mexico, a phytodrug containing guava leaves administered every 8 h during 3 days to 50 patients with acute diarrheic disease significantly reduced the duration of abdominal pain in these patients (Lozoya et al., 2002). In India, an oral guava leaf decoction administered to 109 patients suffering from acute diarrhea significantly decreased stool frequency after 24 h and reduced the intensity of abdominal pain after 48 h (Birdi, Krishnan, Kataria, Gholkar, & Daswani, 2020). All these studies demonstrate that *P. guajava* is of great interest in the treatment of diarrheal illnesses. Some bioactive compounds (e.g., quercetin and quercetin derivatives) have been isolated, but none of them seem to be solely responsible for the antidiarrheal effect observed (Birdi et al., 2010). Thus the development of guava-based phytomedicines should be encouraged, and then be added to the therapeutics used in the management of infectious diarrhea.

Chromolaena odorata

Chromolaena odorata (Asteraceae) (Fig. 7.3) is a small herbaceous plant native to the Americas and is considered one of the worst terrestrial invasive species in the Old World tropics (Yu, He, Zhao, & Li, 2014). In the literature review, *C. odorata* was cited as antidiarrheal agents by eight ethnic groups from the lower Mekong basin. Leaf and root of the plant are used for treating simple diarrhea, diarrhea with vomiting, and dysentery. *C. odorata* is also used for treating diarrheal illnesses in Bangladesh and Nigeria (Aba et al., 2015; Jahan et al., 2019). Antidiarrheal effect (reduction in number of feces) of a methanolic extract of *C. odorata* leaves at 50, 100, and 200 mg/kg was demonstrated on a castor oil-induced diarrhea in mice. In the same study, the extract significantly reduced the intestinal motility in mice at the same concentrations (Taiwo, Olajide, Soyannwo, & Makinde, 2000). The antidiarrheal activity (reduction in the frequency and wetness of stools) of *C. odorata* was later confirmed in a rat model of castor oil-induced diarrhea using an ethanolic leaf extract of *C. odorata* at 200 and 400 mg/kg (Aba et al., 2015). A dichloromethane extract of *C. odorata* demonstrated antibacterial activity against *V. cholerae* with a MIC value of 156 μg/mL. Two flavonoid compounds (i.e., scutellarein tetramethyl ether, sinensetin) were identified as being responsible for this antibacterial activity (Atindehou et al., 2013). Despite its efficacy, the use of *C. odorata* is controversial due to the presence of pyrrolizidine alkaloids which exhibit hepatotoxicity and carcinogenicity (Anyanwu et al., 2017). Therefore more research is needed to investigate the dosage range that is safe for humans (Omokhua, McGaw, Finnie, & Van Staden, 2016).

Alstonia scholaris

Alstonia scholaris (Apocynaceae) (Fig. 7.3) is a tree native to tropical and subtropical Asia and North Australia (Pandey et al., 2020). In the literature analysis, a total of seven references reported the use of *A. scholaris* as antidiarrheal and antidysenteric agents all over the four countries of the lower Mekong basin. Bark and leaf were the two most cited plant tissues. Other studies from India, Indonesia, Papua New Guinea, and the Philippines also reported the use of *A. scholaris* for treating diarrheal illnesses (Khyade et al., 2014). Regarding its pharmacological properties, a methanolic extract of *A. scholaris* showed antidiarrheal effect (reduction in the frequency of defecation) in the castor-oil induced diarrhea in mice model and its spasmolytic activity was confirmed using an isolated rabbit jejunum preparation (Shah et al., 2010). The methanolic extract was also studied for its acute and sub-acute toxicity, and showed liver damage (slight degeneration and centrilobular necrosis) after 28 days at 500 and 1000 mg/kg but no toxicity was observed after 14 days (Bello et al., 2016). It was previously noted that the subacute toxicity might be due to the presence of echitamine (Baliga et al., 2004). Moreover, teratogenic effect was also observed in mice using a hydroalcoholic extract of *A. scholaris* at doses above 240 mg/kg (Jagetia and Baliga, 2003). Therefore, the long-term use of *A. scholaris* should be avoided, and more studies should be performed to evaluate its safety in humans.

Allium sativum

Allium sativum (Amaryllidaceae) is a bulb crop native to Central Asia that has been cultivated all over the world for thousands of years. In the lower Mekong region, the bulb was reported to be used for treating diarrhea and dysentery in Cambodia, Lao PDR, Thailand, and Vietnam. In Palestine, the raw bulb is mixed with yogurt then eaten thrice a day to treat diarrhea (Jaradat, Ayesh, & Anderson, 2016). The bulb of *A. sativum* was also cited as an antidysenteric agent in Mexico (Alanís et al., 2005). Most of the studies validating the use of garlic in the treatment of diarrhea have focused on its antiinfective properties, especially its antibacterial and antiprotozoal activities. A methanolic extract of *A. sativum* was tested on *E. coli*, *Salmonella* sp., *S. flexneri*, and *S. sonnei*, but failed to inhibit their growth at 8 mg/mL using the agar dilution method (Alanís et al., 2005). In another study, a garlic concentrate along with its organosulfur compounds demonstrate a bactericidal activity on *C. jejuni*. This activity was due to cell membrane damages (Lu et al., 2011). Its antibacterial activity was also confirmed on *Salmonella typhi* using in vitro and in vivo assays. In the latter experiment, the consumption of garlic extract in infected mice caused a significant reduction in *S. typhi* load in the feces and reduced the duration of infection (Adebolu, Adeoye, & Oyetayo, 2011). Other studies focused on its antiprotozoal activity. A methanolic extract of *A. sativum* showed moderate in vitro antiprotozoal activity against *E. histolytica* ($IC_{50} = 61.8$ μg/mL) and *G. lamblia* ($IC_{50} = 64.9$ μg/mL) (Calzada, Yépez-Mulia, & Aguilar, 2006). In another study, some compounds (i.e., allyl alcohol, allyl mercaptan) were identified as being responsible for the antiprotozoal activity against *Giardia intestinalis* (Harris, Plummer, Turner, & Lloyd, 2000). Also, an aqueous extract of garlic was administered to mice infected with *Blastocystis* spp. at 20 mg/kg per day and showed a reduction in shedding of cysts (Abdel-Hafeez, Ahmad, Kamal, Abdellatif, &

Abdelgelil, 2015). Garlic also presents some adverse effects and can induce contact dermatitis along with bleeding events. This explains why garlic should not be used by patients using anticoagulant therapy (Kuete, 2017). Besides the numerous studies aiming to evaluate its antiinfective properties, there is a lack of studies validating its use for diarrhea. Therefore more effort should be done to confirm its antidiarrheal potential.

Centella asiatica

Centella asiatica (Apiaceae) is a small herbaceous plant species native to Asia, Africa, and Oceania and widely distributed throughout the five continents. According to the literature review, its leaf and the whole plant have been used for treating diarrhea and dysentery in Cambodia, Lao PDR, Thailand, and Vietnam. In India, the whole plant is ground to extract the juice which is used orally to relieve diarrhea and dysentery (Laloo & Hemalatha, 2011). The antibacterial activity of a dichloromethane/methanol extract (1:1, vol./vol.) was tested on bacteria responsible for diarrhea such as *E. coli*, *S. typhi*, and *S. sonnei*, but the very high MIC values found (MIC > 50 mg/mL) did not confirm its activity (Sieberi, Omwenga, Wambua, Samoei, & Ngugi, 2020). While *C. asiatica* is widely studied, especially for its neuroprotective activity, antidiabetic effect, or wound healing activity, its effect on diarrhea has been poorly investigated (Sun et al., 2020). Thus further pharmacological studies should be performed to validate the use of *C. asiatica* for diarrheal illnesses.

Punica granatum

Punica granatum (Lythraceae), also known as pomegranate, is a shrub native to Central Asia and now cultivated in most regions of the five continents (Shaygannia, Bahmani, Zamanzad, & Rafieian-Kopaei, 2015). Regarding its use in the lower Mekong region, the flower, fruit, leaf, and root of *P. granatum* were mentioned for treating diarrheal illnesses (i.e., diarrhea and dysentery) in Cambodia, Thailand, and Vietnam. Pomegranate was also reported to be used for diarrheal illnesses in Algeria, China, Iran, Mexico, Pakistan, and Turkey (Bouasla & Bouasla, 2017; Ghorbani, 2005; Lee, Xiao, & Pei, 2008; Navarro, Villarreal, Rojas, & Lozoya, 1996; Rashid et al., 2015; Rose, Özünel, & Bennett, 2013). The effect of an aqueous extract of *P. granatum* peel at 100, 200, 300, and 400 mg/kg was evaluated for its antisecretory, antimotility, and antidiarrheal activities using the enteropooling assay in isolated rat ileum, the charcoal meal test in rats, and the castor oil-induced diarrhea in rats, respectively. The results revealed that *P. granatum* reduced diarrhea in a dose-dependent manner by inhibiting intestinal motility and fluid accumulation (Qnais, Elokda, Ghalyun, & Abdulla, 2007). Another study confirmed the antidiarrheal activity of pomegranate, by testing a methanolic extract of *P. granatum* fruit at 800 mg/kg on a castor oil-induced diarrhea rat model (Souli et al., 2015). Punicalagin, corilagin, and ellagic acid were identified as responsible for the antidiarrheal effect of an ethyl acetate fraction of *P. granatum* peels (Zhao et al., 2018). Regarding its antimicrobial effect, a methanolic extract of *P. granatum* dried fruit peel showed a significant antibacterial activity against a multidrug-resistant *S. typhi* with a MIC value of 32 μg/mL (Rani & Khullar, 2004). A mouse *Salmonella typhimurium* infection model was used to study the in vivo antibacterial activity of an ethanolic extract of *P. granatum* peel and resulted in significant reduction of mouse mortality after

6 days. In the extract used, the major compounds were gallic acid, ellagic acid, and punicalagin (Choi et al., 2011). Also, an aqueous extract of *P. granatum* leaf was tested on human (HCR3) and simian (SA-11) rotaviruses but did not show any inhibition activity (Gonçalves et al., 2005). In a randomized controlled clinical trial, 62 patients with ulcerative colitis were treated with an aqueous extract of pomegranate peel (6 g of dry peel per day) and resulted in a reduction of antidiarrheal medication need after 4 weeks (Kamali et al., 2015). Overall, the effect of *P. granatum* fruit has been validated by various pharmacological models which encourage the development of pomegranate-based phytomedicines for treating diarrheal illnesses. However, some concerns regarding its long-term toxicity have been raised and this should incite further exploration of its safety (Ismail, Sestili, & Akhtar, 2012).

Caesalpinia sappan

Caesalpinia sappan (Fabaceae) is a tree found in Southeast Asia and is mainly used as a dyeing plant (Nirmal, Rajput, Prasad, & Ahmad, 2015). According to the literature analysis, the bark, leaf, root, and stem wood of *C. sappan* are used to treat diarrhea and dysentery in the four countries from the lower Mekong region. Its antibacterial activity has been widely studied using in vitro models. An ethanolic and an aqueous extract of *C. sappan* exhibited good growth inhibition activity against *S. typhi* and *E. coli* using the disk diffusion method (Srinivasan et al., 2012). 5-Hydroxy-1,4-naphthoquinone isolated from *C. sappan* heartwood showed growth inhibition activity on *Clostridium perfringens*, a bacteria causing food poisoning (Lim, Jeon, Jeong, Lee, & Lee, 2007). Brazilin, another compound from *C. sappan*, also showed growth inhibition on *S. typhimurium* (MIC = 64 μg/mL) and *E. coli* (MIC = 256 μg/mL) (Xu & Lee, 2004). Besides its antibacterial activity, no studies have been yet performed to validate its full antidiarrheal potential, and thus a better investigation of its pharmacological activities is needed.

Mangifera indica

Mangifera indica (Anacardiaceae), commonly known as mango, is a tree native to Asia and widely cultivated in tropical regions (Ediriweera, Tennekoon, & Samarakoon, 2017). In the lower Mekong area, the bark, root, and seed of *M. indica* are used for treating diarrhea and dysentery in Cambodia, Thailand, and Vietnam. Other studies from Bangladesh, Canary Islands, Guyana, India, Senegal, and Sri Lanka also reported the use of mango in the treatment of diarrheal illnesses (Ediriweera et al., 2017). Several in vivo studies were performed on mango to test for its antidiarrheal effect. Two studies focused on the antidiarrheal potential of mango seeds. Sairam et al. (2003) demonstrated that a methanolic and an aqueous extract of *M. indica* seeds administered orally at 250 mg/kg to mice using a castor oil-induced diarrhea model significantly reduced the number of feces excreted. In the same study, the authors also showed a significant reduction in intestinal transit time with the methanolic extract of *M. indica* seeds administered orally at 250 mg/kg (Sairam et al., 2003). Rajan, Suganya, Thirunalasundari, and Jeeva (2012) showed similar results by testing *M. indica* seed kernels in a castor oil-induced diarrhea model and in a charcoal meal test (Rajan et al., 2012). Another in vivo study focused on the antidiarrheal potential of mango stem bark. In this work, Tchoumba Tchoumi et al. (2020) demonstrated an antidiarrheal (i.e., reduction in the frequency of stools), antisecretory, and antimotility effect of an aqueous and a methanolic extract of *M. indica* stem bark by testing concentrations ranging from 300 to 500 mg/kg (Tchoumba Tchoumi et al., 2020). In Yakubu and Salimon (2015), an aqueous extract of *M. indica* leaves was tested at 25, 50, and 100 mg/kg using a castor oil-induced diarrhea model, an enteropooling assay, and a charcoal meal test. In this work, *M. indica* leaves reduced by half the total number of wet feces, reduced the mass and volume of intestinal fluid, and reduced the intestinal motility in a dose-dependent manner (Yakubu & Salimon, 2015). Not only various parts of *M. indica* have proven their antidiarrheal activity but the plant has also demonstrated a good antibacterial activity against enteric pathogens. For example, an organic extract of *M. indica* leaves has shown to inhibit the growth of *S. flexneri* with an MIC value of 250 μg/mL (van Vuuren, Nkwanyana, & de Wet, 2015), and an aqueous and an ethanolic extract of *M. indica* seeds have demonstrated an inhibition of *Shigella dysenteriae* growth with MIC values of 380 and 190 μg/mL, respectively (Rajan, Thirunalasundari, & Jeeva, 2011). Regarding in vivo studies, an aqueous and a methanolic extract of *M. indica* stem bark at concentrations ranging from 300 to 500 mg/kg showed a significant reduction of bacterial load in feces after 14 days using an EPEC-infected mice model (Tchoumba Tchoumi et al., 2020). Several compounds might be involved in the antibacterial activity of *M. indica* including gallic acid, kaempferol, linalool, mangiferin, methyl gallate, and quercetin (Ediriweera et al., 2017). In a recent clinical trial, a mango juice by-product comprising peel and pulp of mango significantly reduced the severity of diarrheal events in children of 6–8 years old over a period of 2 months. This effect was attributed to the presence of gallotannins (Anaya-Loyola et al., 2020). The safety of mango was also studied using a subchronic (28 days) toxicity assay, and it was shown that the aqueous extract of *M. indica* leaves does not present genotoxic, clastogenic, and cytotoxic effect at doses of 150, 250, 500, and 1000 mg/kg (Villas Boas et al., 2019). Overall, *M. indica* present a good safety profile and multiple pharmacological actions which make it a good candidate for further development as an antidiarrheal phytomedicine.

Holarrhena pubescens

Holarrhena pubescens (syn. *H. antidysenterica* (Roth) Wall. ex A.DC., Apocynaceae) is a flowering plant native to southern Africa, India, Southeast Asia, and southern China (Zahara, Panda, Swain, & Luyten, 2020). In the literature analysis, *H. pubescens* was mentioned in six bibliographic references originating from Cambodia, Lao PDR, and Thailand. In these countries, the bark, root, and stem of *H. pubescens* have been used for treating cholera, diarrhea, and dysentery. This plant is also used in other parts of the world such as India where the bark is used to treat dysentery (Gairola, Sharma, Gaur, Siddiqi, & Painuli, 2013), and Pakistan where the beans are mixed with yogurt to treat diarrhea and cramps (Ahmad et al., 2018). The antidiarrheal activity of an ethanolic extract of *H. pubescens* seeds was evaluated in rats using a castor oil-induced diarrhea model, and an oral administration of this extract was shown to reduce the severity of diarrhea at 200 and 400 mg/kg. In the same study, this extract was administered orally to rats using an ETEC-infected model, and it showed a reduction in body weight change comparable to the positive control gentamicin (Sharma et al., 2015). Further investigation of the antibacterial activity of *H. pubescens* on diarrheagenic *E. coli* showed that its alkaloids reduce initial bacterial adhesion of EPEC to intact epithelial cells (Kavitha & Niranjali, 2009). In another study, a methanolic extract of *H. pubescens* seeds exhibited a growth inhibition of *S. typhimurium* with a MIC value of 256 μg/mL (Rani & Khullar, 2004). However, the growth inhibitory activity of *H. pubescens* leaf and bark was not confirmed on other enteropathogenic bacteria (i.e., *Enterobacter aerogenes*, *Salmonella paratyphi*, *S. typhi*, *S. dysenteriae*, *S. sonnei*, and *V. cholerae*) as the MIC values were found to be superior to 1 mg/mL (Rath & Padhy, 2015). Gilani et al. (2010) demonstrated that *H. pubescens* also present spasmogenic and spasmolytic effects, mediated through activation of histamine receptors and Ca^{++} channel blockade, which might explain its action on diarrhea and constipation (Gilani et al., 2010). In a randomized controlled trial, a monoherbal formulation containing *H. pubescens* administered to chronic ulcerative colitis patients induced a reduction in diarrhea, abdominal pain and a beneficial effect on stool consistency after 30 days compared to mesalamine (Johari & Gandhi, 2016). Despite its apparent efficacy, *H. pubescens* possess pyrrolizidine alkaloids which are responsible for an hepatotoxic effect (Arseculeratne, Gunatilaka, & Panabokke, 1981). Therefore further investigation of its short-term and long-term safety is necessary.

Oroxylum indicum

Oroxylum indicum (Bignoniaceae) is a tree native to India, Southeast Asia, and China (Dinda et al., 2015). In the lower Mekong area, the bark and root of *O. indicum* are used to treat dysentery and diarrhea in children and adults. Different parts of *O. indicum* were also mentioned to treat diarrheal illnesses in Bangladesh, India, Malaysia, Myanmar, Nepal, Philippines, and Sri Lanka (Dinda et al., 2015). A flavonoid-rich fraction of *O. indicum* root bark administered orally at 100 mg/kg to mice with diarrhea (induced by castor oil or by magnesium sulfate) resulted in a reduction of diarrheal episodes. In the same study, this extract also demonstrated antisecretory and antimotility assay in a dose-dependent manner (Joshi et al., 2012). Regarding its antibacterial activity, *O. indicum* showed a weak growth inhibitory activity against *S. typhimurium* and *E. coli* with MIC values ranging from 640 to 1280 μg/mL (Yossathera et al., 2016). *O. indicum* is a plant rich in flavonoids such as baicalein, chrysin, kaempferol, oroxylin A, quercetin, and scutellarein (Dinda et al., 2015). It also contains ellagic acid, an antidiarrheal compound (Chen et al., 2020). Overall, only one study demonstrated the antidiarrheal effect of *O. indicum*, and there is a lack of rigorous toxicological studies. Therefore more research should be performed to confirm the efficacy and safety of *O. indicum*.

Conclusion

The countries from the lower Mekong basin offer a rich flora that is still used by local people to treat diarrheal illnesses. A total of 431 plants have been reported to be used in the region for this purpose demonstrating the extensive traditional knowledge of people from the lower Mekong basin.

Interestingly, the most cited plants are introduced and cultivable species. This is the case of *A. sativum*, *C. odorata*, *P. guajava*, and *P. granatum*. Since remedies that spread between cultures are likely proven effective cures, it can indicate a high efficacy of this plant for treating diarrheal illnesses.

However, for most of the plants, there is a lack of well-designed clinical trials that are essential to clearly demonstrate their efficacy and safety. Moreover, new pharmacological models should be developed to fully understand the antidiarrheal mechanism of these plants (e.g., effect on the gut microbiota, effect on specific receptor).

In addition, safety concerns have been raised for some plants due to the presence of pyrrolizidine alkaloids (i.e., *C. odorata*, *H. pubescens*), due to adverse effects (i.e., *A. sativum*), or due to teratogenic and hepatotoxic effects (i.e., *A. scholaris*). Within this regard, safety guidelines are needed to limit the health risk associated with the use of antidiarrheal plants.

Still, some plants (e.g., *M. indica*, *P. guajava*) present good safety profiles and various pharmacological activities validating their use as antidiarrheal. As they are also easy to cultivate, these plants could be used to develop a standardized phytomedicine. Such endeavor could help control the variable quality of homemade remedies and serve a population in need of local and good quality health products. To succeed, academic scientists, private companies, public hospitals, and local traditional medicine centers should work together and make sure their products benefit the people who provided the knowledge.

Another solution could be to isolate the compound or mixture of compounds involved in the antidiarrheal activity and markets a botanical drug. A successful antidiarrheal product from plant is crofelemer, an FDA-approved drug for the treatment of diarrhea associated with anti-HIV drugs. Crofelemer is a purified oligomeric proanthocyanidin from the sap of *Croton lechleri* Müll.Arg. native to South America. Regarding the high number of antidiarrheal plants, it is likely that new antidiarrheal compounds could be isolated. Therefore more research is needed to unveil these new medicines.

In conclusion, antidiarrheal plants represent a rich source of medicine that needs to be further studied to strengthen their proper use, efficacy, safety, and quality as recommended by the WHO in its report on traditional medicine for 2014–23. This will help to integrate these products into national health care programs from countries of the lower Mekong basin.

References

Aba, P. E., Joshua, P. E., Ezeonuogu, F. C., Ezeja, M. I., Omoja, V. U., & Umeakuana, P. U. (2015). Possible antidiarrhoeal potential of ethanol leaf extract of *Chromolaena odorata* in castor oil-induced rats. *Journal of Complementary and Integrative Medicine, 12*, 301–306. Available from https://doi.org/10.1515/jcim-2014-0033.

Abdel-Hafeez, E. H., Ahmad, A. K., Kamal, A. M., Abdellatif, M. Z. M., & Abdelgelil, N. H. (2015). In vivo antiprotozoan effects of garlic (*Allium sativum*) and ginger (*Zingiber officinale*) extracts on experimentally infected mice with *Blastocystis* spp. *Parasitology Research, 114*, 3439–3444. Available from https://doi.org/10.1007/s00436-015-4569-x.

Adamson, P. T., Rutherfurd, I. D., Peel, M. C., & Conlan, I. A. (2009). *The hydrology of the Mekong River. The Mekong* (pp. 53–76). Australia: Elsevier Inc. Available from https://doi.org/10.1016/B978-0-12-374026-7.00004-8.

Adebolu, T. T., Adeoye, O. O., & Oyetayo, V. O. (2011). Effect of garlic (*Allium sativum*) on *Salmonella typhi* infection, gastrointestinal flora and hematological parameters of albino rats. *African Journal of Biotechnology, 10*, 6804–6808. Available from http://www.academicjournals.org/AJB/PDF/pdf2011/13Jul/Adebolu%20et%20al.pdf.

Ahmad, M., Zafar, M., Shahzadi, N., Yaseen, G., Murphey, T. M., & Sultana, S. (2018). Ethnobotanical importance of medicinal plants traded in Herbal markets of Rawalpindi - Pakistan. *Journal of Herbal Medicine, 11*, 78–89. Available from https://doi.org/10.1016/j.hermed.2017.10.001.

Akrimajirachoote, N., Satitsri, S., Sommart, U., Rukachaisirikul, V., & Muanprasat, C. (2020). Inhibition of CFTR-mediated intestinal chloride secretion by a fungus-derived arthropsolide A: Mechanism of action and anti-diarrheal efficacy. *European Journal of Pharmacology, 885*, 173393. Available from https://doi.org/10.1016/j.ejphar.2020.173393.

Alanís, A. D., Calzada, F., Cervantes, J. A., Torres, J., & Ceballos, G. M. (2005). Antibacterial properties of some plants used in Mexican traditional medicine for the treatment of gastrointestinal disorders. *Journal of Ethnopharmacology, 100*, 153–157. Available from https://doi.org/10.1016/j.jep.2005.02.022.

Anaya-Loyola, M. A., García-Marín, G., García-Gutiérrez, D. G., Castaño-Tostado, E., Reynoso-Camacho, R., López-Ramos, J. E., ... Pérez-Ramírez, I. F. (2020). A mango (*Mangifera indica* L.) juice by-product reduces gastrointestinal and upper respiratory tract infection symptoms in children. *Food Research International, 136*, 109492. Available from https://doi.org/10.1016/j.foodres.2020.109492.

Ankli, A., Sticher, O., & Heinrich, M. (1999). Yucatec Maya medicinal plants vs nonmedicinal plants: Indigenous characterization and selection. *Human Ecology, 27*, 557–580. Available from https://doi.org/10.1023/A:1018791927215.

Anyanwu, S., Inyang, I. J., Asemota, E. A., Obioma, O. O., Okpokam, D. C., & Agu, V. O. (2017). Effect of ethanolic extract of *Chromolaena odorata* on the kidneys and intestines of healthy albino rats. *Integrative Medicine Research, 6*(3), 292–299. Available from https://doi.org/10.1016/j.imr.2017.06.004.

Arseculeratne, S. N., Gunatilaka, A. A. L., & Panabokke, R. G. (1981). Studies on medicinal plants of Sri Lanka: Occurrence of pyrrolizidine alkaloids and hepatotoxic properties in some traditional medicinal herbs. *Journal of Ethnopharmacology, 4*, 159–177. Available from https://doi.org/10.1016/0378-8741(81)90033-7.

Atindehou, M., Lagnika, L., Guérold, B., Strub, J. M., Zhao, M., Van Dorsselaer, A., ... Metz-Boutigue, M.-H. (2013). Isolation and identification of two antibacterial agents from *Chromolaena odorata* L. active against four diarrheal strains. *Advances in Microbiology, 3*(1), 115–121. Available from https://doi.org/10.4236/aim.2013.31018.

Baliga, M. S., Jagetia, G. C., Ulloor, J. N., Baliga, M. P., Venkatesh, P., Reddy, R., ... Bairy, K. L. (2004). The evaluation of the acute toxicity and long term safety of hydroalcoholic extract of Sapthaparna (*Alstonia scholaris*) in mice and rats. *Toxicology Letters, 151*, 317–326. Available from https://doi.org/10.1016/j.toxlet.2004.01.015.

Barbosa, E., Calzada, F., & Campos, R. (2007). In vivo antigiardial activity of three flavonoids isolated of some medicinal plants used in Mexican traditional medicine for the treatment of diarrhea. *Journal of Ethnopharmacology, 109*, 552–554. Available from https://doi.org/10.1016/j.jep.2006.09.009.

Bello, I., Bakkouri, A., Tabana, Y., Al-Hindi, B., Al-Mansoub, M., Mahmud, R., & Asmawi, M. (2016). Acute and sub-acute toxicity evaluation of the methanolic extract of *Alstonia scholaris* stem bark. *Medical Sciences, 4*(1), 5. Available from https://doi.org/10.3390/medsci4010004.

Birdi, T., Daswani, P., Brijesh, S., Tetali, P., Natu, A., & Antia, N. (2010). Newer insights into the mechanism of action of *Psidium guajava* L. leaves in infectious diarrhoea. *BMC Complementary and Alternative Medicine, 10*, 33. Available from https://doi.org/10.1186/1472-6882-10-33.

Birdi, T., Krishnan, G. G., Kataria, S., Gholkar, M., & Daswani, P. (2020). A randomized open label efficacy clinical trial of oral guava leaf decoction in patients with acute infectious diarrhoea. *Journal of Ayurveda and Integrative Medicine, 11*, 163–172. Available from https://doi.org/10.1016/j.jaim.2020.04.001.

Boakye, P. A., Brierley, S. M., Pasilis, S. P., & Balemba, O. B. (2012). *Garcinia buchananii* bark extract is an effective anti-diarrheal remedy for lactose-induced diarrhea. *Journal of Ethnopharmacology, 142*, 539–547. Available from https://doi.org/10.1016/j.jep.2012.05.034.

Boithias, L., Choisy, M., Souliyaseng, N., Jourdren, M., Quet, F., Buisson, Y., ... Ribolzi, O. (2016). Hydrological regime and water shortage as drivers of the seasonal incidence of diarrheal diseases in a tropical Montane environment. *PLoS Neglected Tropical Diseases, 10*(12), e0005195. Available from https://doi.org/10.1371/journal.pntd.0005195.

Bouasla, A., & Bouasla, I. (2017). Ethnobotanical survey of medicinal plants in northeastern of Algeria. *Phytomedicine: International Journal of Phytotherapy and Phytopharmacology, 36*, 68–81. Available from https://doi.org/10.1016/j.phymed.2017.09.007.

Brandelli, C. L. C., Giordani, R. B., De Carli, G. A., & Tasca, T. (2009). Indigenous traditional medicine: In vitro anti-giardial activity of plants used in the treatment of diarrhea. *Parasitology Research, 104*, 1345–1349. Available from https://doi.org/10.1007/s00436-009-1330-3.

Brijesh, S., Daswani, P. G., Tetali, P., Rojatkar, S. R., Antia, N. H., & Birdi, T. J. (2006). Studies on *Pongamia pinnata* (L.) Pierre leaves: Understanding the mechanism(s) of action in infectious diarrhea. *Journal of Zhejiang University. Science B, 7*, 665–674. Available from https://doi.org/10.1631/jzus.2006.B0665.

Brusotti, G., Cesari, I., Dentamaro, A., Caccialanza, G., & Massolini, G. (2014). Isolation and characterization of bioactive compounds from plant resources: The role of analysis in the ethnopharmacological approach. *Journal of Pharmaceutical and Biomedical Analysis, 87*, 218–228. Available from https://doi.org/10.1016/j.jpba.2013.03.007.

Butterweck, V., & Nahrstedt, A. (2012). What is the best strategy for preclinical testing of botanicals? A critical perspective. *Planta Medica, 78*, 747–754. Available from https://doi.org/10.1055/s-0031-1298434.

Calzada, F., Bautista, E., Yépez-Mulia, L., García-Hernandez, N., & Ortega, A. (2015). Antiamoebic and antigiardial activity of clerodane diterpenes from Mexican Salvia species used for the treatment of diarrhea. *Phytotherapy Research, 29*, 1600–1604. Available from https://doi.org/10.1002/ptr.5421.

Calzada, F., Yépez-Mulia, L., & Aguilar, A. (2006). In vitro susceptibility of *Entamoeba histolytica* and *Giardia lamblia* to plants used in Mexican traditional medicine for the treatment of gastrointestinal disorders. *Journal of Ethnopharmacology, 108*, 367–370. Available from https://doi.org/10.1016/j.jep.2006.05.025.

Castillo, S., Heredia, N., Arechiga-Carvajal, E., & García, S. (2014). Citrus extracts as inhibitors of quorum sensing, biofilm formation and motility of *Campylobacter jejuni. Food Biotechnology, 28*, 106–122. Available from https://doi.org/10.1080/08905436.2014.895947.

Chassagne, F., Deharo, E., Punley, H., & Bourdy, G. (2017). Treatment and management of liver diseases by Khmer traditional healers practicing in Phnom Penh area, Cambodia. *Journal of Ethnopharmacology, 202*, 38–53. Available from https://doi.org/10.1016/j.jep.2017.03.002.

Chassagne, F., Hul, S., Deharo, E., & Bourdy, G. (2016). Natural remedies used by Bunong people in Mondulkiri province (Northeast Cambodia) with special reference to the treatment of 11 most common ailments. *Journal of Ethnopharmacology, 191*, 41–70. Available from https://doi.org/10.1016/j.jep.2016.06.003.

Chassagne, F., Samarakoon, T., Porras, G., Lyles, J. T., Dettweiler, M., Marquez, L., ... Quave, C. L. (2020). A systematic review of plants with antibacterial activities: A taxonomic and phylogenetic perspective. *Frontiers in Pharmacology, 11*, 586548. Available from https://doi.org/10.3389/fphar.2020.586548.

Chea, R., Grenouillet, G., & Lek, S. (2016). Evidence of water quality degradation in lower Mekong basin revealed by self-organizing map. *PLoS One, 11*, e0145527. Available from https://doi.org/10.1371/journal.pone.0145527.

Chen, J., Yang, H., & Sheng, Z. (2020). Ellagic acid activated PPAR signaling pathway to protect ileums against castor oil-induced diarrhea in mice: Application of Transcriptome Analysis in Drug Screening. *Frontiers in Pharmacology*, 10. Available from https://doi.org/10.3389/fphar.2019.01681.

Chen, J. C., Chang, Y. S., Wu, S. L., Chao, D. C., Chang, C. S., Li, C. C., ... Hsiang, C. Y. (2007). Inhibition of *Escherichia coli* heat-labile enterotoxin-induced diarrhea by Chaenomeles speciosa. *Journal of Ethnopharmacology, 113*, 233–239. Available from https://doi.org/10.1016/j.jep.2007.05.031.

Choi, J. G., Kang, O. H., Lee, Y. S., Chae, H. S., Oh, Y. C., Brice, O. O., ... Kwon, D. Y. (2011). In vitro and in vivo antibacterial activity of *Punica granatum* peel ethanol extract against *Salmonella. Evidence-Based Complementary and Alternative Medicine, 2011*, 690518. Available from https://doi.org/10.1093/ecam/nep105.

Dawurung, C. J., Noitem, R., Rattanajak, R., Bunyong, R., Richardson, C., Willis, A. C., ... Pyne, S. G. (2020). Isolation of CFTR and TMEM16A inhibitors from *Neorautanenia mitis* (A. Rich) Verdcourt: Potential lead compounds for treatment of secretory diarrhea. *Phytochemistry, 179*, 112464. Available from https://doi.org/10.1016/j.phytochem.2020.112464.

De Esparza, R. R., Bye, R., Meckes, M., López, J. T., & Jiménez-Estrada, M. (2007). Antibacterial activity of *Piqueria trinervia*, a Mexican medicinal plant used to treat diarrhea. *Pharmaceutical Biology, 45*, 446–452. Available from https://doi.org/10.1080/13880200701389011.

Dinda, B., SilSarma, I., Dinda, M., & Rudrapaul, P. (2015). *Oroxylum indicum* (L.) Kurz, an important Asian traditional medicine: From traditional uses to scientific data for its commercial exploitation. *The Journal of Ethnopharmacology, 161*, 255–278. Available from https://doi.org/10.1016/j.jep.2014.12.027-.

Douglas, I. (2005). The physical geography of Southeast Asia. In A. Gupta (Ed.), *The Mekong river basin* (pp. 193–218). Oxford: Oxford University Press.

Ediriweera, M. K., Tennekoon, K. H., & Samarakoon, S. R. (2017). A review on ethnopharmacological applications, pharmacological activities, and bioactive compounds of *Mangifera indica* (Mango). *Evidence-Based Complementary and Alternative Medicine, 2017*, 6949835. Available from https://doi.org/10.1155/2017/6949835.

Farthing, M. J. G. (2006). Antisecretory drugs for diarrheal disease. *Digestive Diseases, 24*, 47–58. Available from https://doi.org/10.1159/000090308.

Feng, W., Ao, H., Peng, C., & Yan, D. (2019). Gut microbiota, a new frontier to understand traditional Chinese medicines. *Pharmacological Research*, *142*, 176–191. Available from https://doi.org/10.1016/j.phrs.2019.02.024.

Fischer Walker, C. L., Perin, J., Aryee, M. J., Boschi-Pinto, C., & Black, R. E. (2012). Diarrhea incidence in low- and middle-income countries in 1990 and 2010: A systematic review. *BMC Public Health*, *12*, 220. Available from https://doi.org/10.1186/1471-2458-12-220.

Fokam Tagne, M. A., Noubissi, P. A., Fankem, G. O., & Kamgang, R. (2018). Effects of *Oxalis barrelieri* L. (Oxalidaceae) aqueous extract on diarrhea induced by *Shigella dysenteriae* type 1 in rats. *Health Science Reports*, *1*(2), e20. Available from https://doi.org/10.1002/hsr2.20.

Gabriel, S. E., Davenport, S. E., Steagall, R. J., Vimal, V., Carlson, T., & Rozhon, E. J. (1999). A novel plant-derived inhibitor of cAMP-mediated fluid and chloride secretion. *American Journal of Physiology-Gastrointestinal and Liver Physiology*, *276*(1), G58–G63. Available from https://doi.org/10.1152/ajpgi.1999.276.1.G58.

Gairola, S., Sharma, J., Gaur, R. D., Siddiqi, T. O., & Painuli, R. M. (2013). Plants used for treatment of dysentery and diarrhoea by the Bhoxa community of district Dehradun, Uttarakhand, India. *Journal of Ethnopharmacology*, *150*, 989–1006. Available from https://doi.org/10.1016/j.jep.2013.10.007.

Gertsch, J. (2009). How scientific is the science in ethnopharmacology? Historical perspectives and epistemological problems. *Journal of Ethnopharmacology*, *122*, 177–183. Available from https://doi.org/10.1016/j.jep.2009.01.010.

Ghayur, M. N., & Gilani, A. H. (2005). Pharmacological basis for the medicinal use of ginger in gastrointestinal disorders. *Digestive Diseases and Sciences*, *50*, 1889–1897. Available from https://doi.org/10.1007/s10620-005-2957-2.

Ghorbani, A. (2005). Studies on pharmaceutical ethnobotany in the region of Turkmen Sahra, north of Iran (Part 1): General results. *Journal of Ethnopharmacology*, *102*, 58–68. Available from https://doi.org/10.1016/j.jep.2005.05.035.

Gilani, A. H., Khan, A., Khan, A. U., Bashir, S., Rehman, N. U., & Mandukhail, S. U. R. (2010). Pharmacological basis for the medicinal use of *Holarrhena antidysenterica* in gut motility disorders. *Pharmaceutical Biology*, *48*, 1240–1246. Available from https://doi.org/10.3109/13880201003727960.

Gonçalves, J. L. S., Lopes, R. C., Oliveira, D. B., Costa, S. S., Miranda, M. M. F. S., Romanos, M. T. V., ... Wigg, M. D. (2005). In vitro anti-rotavirus activity of some medicinal plants used in Brazil against diarrhea. *Journal of Ethnopharmacology*, *99*, 403–407. Available from https://doi.org/10.1016/j.jep.2005.01.032.

Gupta, P., & Birdi, T. (2015). *Psidium guajava* leaf extract prevents intestinal colonization of *Citrobacter rodentium* in the mouse model. *Journal of Ayurveda and Integrative Medicine*, *6*, 50–52. Available from https://doi.org/10.4103/0975-9476.146557.

Gutiérrez, R. M. P., Mitchell, S., & Solis, R. V. (2008). *Psidium guajava*: A review of its traditional uses, phytochemistry and pharmacology. *Journal of Ethnopharmacology*, *117*, 1–27. Available from https://doi.org/10.1016/j.jep.2008.01.025.

Harris, J. C., Plummer, S., Turner, M. P., & Lloyd, D. (2000). The microaerophilic flagellate *Giardia intestinalis*: *Allium sativum* (garlic) is an effective antigiardial. *Microbiology (Reading, England)*, *146*, 3119–3127. Available from https://doi.org/10.1099/00221287-146-12-3119.

Hirudkar, J. R., Parmar, K. M., Prasad, R. S., Sinha, S. K., Jogi, M. S., Itankar, P. R., & Prasad, S. K. (2020a). Quercetin a major biomarker of *Psidium guajava* L. inhibits SepA protease activity of *Shigella flexneri* in treatment of infectious diarrhoea. *Microbial Pathogenesis*, *138*, 103807. Available from https://doi.org/10.1016/j.micpath.2019.103807.

Hirudkar, J. R., Parmar, K. M., Prasad, R. S., Sinha, S. K., Lomte, A. D., Itankar, P. R., & Prasad, S. K. (2020b). The antidiarrhoeal evaluation of *Psidium guajava* L. against enteropathogenic *Escherichia coli* induced infectious diarrhoea. *Journal of Ethnopharmacology*, *251*, 112567. Available from https://doi.org/10.1016/j.jep.2020.112561.

Hoa, N. Q., Öhman, A., Lundborg, C. S., & Chuc, N. T. K. (2007). Drug use and health-seeking behavior for childhood illness in Vietnam-A qualitative study. *Health Policy (Amsterdam, Netherlands)*, *82*, 320–329. Available from https://doi.org/10.1016/j.healthpol.2006.10.005.

Houghton, P. J., Howes, M. J., Lee, C. C., & Steventon, G. (2007). Uses and abuses of in vitro tests in ethnopharmacology: Visualizing an elephant. *Journal of Ethnopharmacology*, *110*, 391–400. Available from https://doi.org/10.1016/j.jep.2007.01.032.

Ismail, T., Sestili, P., & Akhtar, S. (2012). Pomegranate peel and fruit extracts: A review of potential anti-inflammatory and anti-infective effects. *Journal of Ethnopharmacology*, *143*, 397–405. Available from https://doi.org/10.1016/j.jep.2012.07.004.

Iwanaga, M., Toma, C., Miyazato, T., Insisiengmay, S., Nakasone, N., & Ehara, M. (2004). Antibiotic resistance conferred by a class I integron and SXT constin in *Vibrio cholerae* O1 strains isolated in Laos. *Antimicrobial Agents and Chemotherapy*, *48*, 2364–2369. Available from https://doi.org/10.1128/AAC.48.7.2364-2369.2004.

Jagetia, G. C., & Baliga, M. S. (2003). Induction of developmental toxicity in mice treated with *Alstonia scholaris* (Sapthaparna) in Utero. *Birth Defects Research Part B - Developmental and Reproductive Toxicology, 68*, 472–478. Available from https://doi.org/10.1002/bdrb.10047.

Jahan, R., Jannat, K., Shoma, J. F., Arif Khan, M., Shekhar, H. U., & Rahmatullah, M. (2019). *Drug discovery and herbal drug development: A special focus on the anti-diarrheal plants of Bangladesh. Herbal medicine in India: Indigenous knowledge, practice, innovation and its value* (pp. 363–400). Bangladesh: Springer Singapore. Available from https://doi.org/10.1007/978-981-13-7248-3_23.

Jaradat, N. A., Ayesh, O. I., & Anderson, C. (2016). Ethnopharmacological survey about medicinal plants utilized by herbalists and traditional practitioner healers for treatments of diarrhea in the West Bank/Palestine. *Journal of Ethnopharmacology, 182*, 57–66. Available from https://doi.org/10.1016/j.jep.2016.02.013.

Johari, S., & Gandhi, T. (2016). A randomized single blind parallel group study comparing monoherbal formulation containing *Holarrhena antidysenterica* extract with mesalamine in chronic ulcerative colitis patients. *Ancient Science of Life, 36*(1), 19–27. Available from https://doi.org/10.4103/0257-7941.195409.

Joshi, S. V., Gandhi, T. R., Vyas, B. A., Shah, P. D., Patel, P. K., & Vyas, H. G. (2012). Effect of *Oroxylum indicum* on intestinal motility in rodents. *Oriental Pharmacy and Experimental Medicine, 12*, 279–285. Available from https://doi.org/10.1007/s13596-012-0082-2-.

Kaljee, L. M., Thiem, V. D., von Seidlein, L., Genberg, B. L., Canh, D. G., Tho, L. H., ... Trach, D. D. (2004). Healthcare use for diarrhoea and dysentery in actual and hypothetical cases, Nha Trang, Viet Nam. *Journal of Health, Population, and Nutrition, 22*, 139–149.

Kamali, M., Tavakoli, H., Khodadoost, M., Daghaghzadeh, H., Kamalinejad, M., Gachkar, L., ... Adibi, P. (2015). Efficacy of the *Punica granatum* peels aqueous extract for symptom management in ulcerative colitis patients. A randomized, placebo-controlled, clinical trial. *Complementary Therapies in Clinical Practice, 21*, 141–146. Available from https://doi.org/10.1016/j.ctcp.2015.03.001.

Kavitha, D., & Niranjali, S. (2009). Inhibition of enteropathogenic *Escherichia coli* adhesion on host epithelial cells by *Holarrhena antidysenterica* (L.) WALL. *Phytotherapy Research, 23*(9), 1229–1236. Available from https://doi.org/10.1002/ptr.2520.

Khyade, M. S., Kasote, D. M., & Vaikos, N. P. (2014). *Alstonia scholaris* (L.) R. Br. and *Alstonia macrophylla* Wall. ex G. Don: A comparative review on traditional uses, phytochemistry and pharmacology. *Journal of Ethnopharmacology, 153*, 1–18. Available from https://doi.org/10.1016/j.jep.2014.01.025.

Kinuthia, D. G., Muriithi, A. W., & Mwangi, P. W. (2016). Freeze dried extracts of *Bidens biternata* (Lour.) Merr. and Sheriff. show significant antidiarrheal activity in in-vivo models of diarrhea. *Journal of Ethnopharmacology, 193*, 416–422. Available from https://doi.org/10.1016/j.jep.2016.09.041.

Kuete, V. (2017). *Allium sativum. Medicinal spices and vegetables from Africa: Therapeutic potential against metabolic, inflammatory, infectious and systemic diseases* (pp. 363–377). Cameroon: Elsevier Inc. Available from https://doi.org/10.1016/B978-0-12-809286-6.00015-7.

Kwon, H. J., Kim, H. H., Ryu, Y. B., Kim, J. H., Jeong, H. J., Lee, S. W., ... Lee, W. S. (2010). In vitro anti-rotavirus activity of polyphenol compounds isolated from the roots of *Glycyrrhiza uralensis*. *Bioorganic and Medicinal Chemistry, 18*, 7668–7674. Available from https://doi.org/10.1016/j.bmc.2010.07.073.

Laloo, D., & Hemalatha, S. (2011). Ethnomedicinal plants used for diarrhea by tribals of Meghalaya, Northeast India. *Pharmacognosy Reviews, 5*, 147–154. Available from https://doi.org/10.4103/0973-7847.91108.

Lawpoolsri, S., Kaewkungwal, J., Khamsiriwatchara, A., Sovann, L., Sreng, B., Phommasack, B., ... Ko Oo, M. (2018). Data quality and timeliness of outbreak reporting system among countries in Greater Mekong subregion: Challenges for international data sharing. *PLoS Neglected Tropical Diseases, 12*(4), e0006425. Available from https://doi.org/10.1371/journal.pntd.0006425.

Laws, E., & Semone, P. (2009). *The Mekong: Developing a new tourism region. River tourism* (pp. 55–73). Australia: CABI Publishing. Available from http://bookshop.cabi.org/.

Le, T. H., Ottosson, E., Nguyen, T. K. C., Kim, B. G., & Allebeck, P. (2011). Drug use and self-medication among children with respiratory illness or diarrhea in a rural district in Vietnam: A qualitative study. *Journal of Multidisciplinary Healthcare., 4*, 329–336. Available from https://doi.org/10.2147/JMDH.S22769.

Lee, S., Xiao, C., & Pei, S. (2008). Ethnobotanical survey of medicinal plants at periodic markets of Honghe Prefecture in Yunnan Province, SW China. *Journal of Ethnopharmacology, 117*, 362–377. Available from https://doi.org/10.1016/j.jep.2008.02.001.

Leonti, M., Sticher, O., & Heinrich, M. (2002). Medicinal plants of the Popoluca, México: Organoleptic properties as indigenous selection criteria. *Journal of Ethnopharmacology, 81*, 307–315. Available from https://doi.org/10.1016/S0378-8741(02)00078-8.

Li, S., Qi, Y., Chen, L., Qu, D., Li, Z., Gao, K., … Sun, Y. (2019). Effects of *Panax ginseng* polysaccharides on the gut microbiota in mice with antibiotic-associated diarrhea. *International Journal of Biological Macromolecules, 124*, 931–937. Available from https://doi.org/10.1016/j.ijbiomac.2018.11.271.

Lim, M. Y., Jeon, J. H., Jeong, E. Y., Lee, C. H., & Lee, H. S. (2007). Antimicrobial activity of 5-hydroxy-1,4-naphthoquinone isolated from *Caesalpinia sappan* toward intestinal bacteria. *Food Chemistry, 100*, 1254–1258. Available from https://doi.org/10.1016/j.foodchem.2005.12.009.

Lozoya, X., Reyes-Morales, H., Chávez-Soto, M. A., Martínez-García, M. D. C., Soto-González, Y., & Doubova, S. V. (2002). Intestinal anti-spasmodic effect of a phytodrug of *Psidium guajava* folia in the treatment of acute diarrheic disease. *Journal of Ethnopharmacology, 83*, 19–24. Available from https://doi.org/10.1016/S0378-8741(02)00185-X.

Lu, X., Rasco, B. A., Jabal, J. M. F., Eric Aston, D., Lin, M., & Konkel, M. E. (2011). Investigating antibacterial effects of garlic (*Allium sativum*) concentrate and garlic-derived organosulfur compounds on *Campylobacter jejuni* by using Fourier transform infrared spectroscopy, Raman spectroscopy, and electron microscopy. *Applied and Environmental Microbiology, 77*, 5257–5269. Available from https://doi.org/10.1128/AEM.02845-10.

Ma, T., Thiagarajah, J. R., Yang, H., Sonawane, N. D., Folli, C., Galietta, L. J. V., & Verkman, A. S. (2002). Thiazolidinone CFTR inhibitor identified by high-throughput screening blocks cholera toxin-induced intestinal fluid secretion. *Journal of Clinical Investigation, 110*, 1651–1658. Available from https://doi.org/10.1172/JCI0216112.

Madikizela, B., Ndhlala, A. R., Finnie, J. F., & Van Staden, J. (2012). Ethnopharmacological study of plants from Pondoland used against diarrhoea. *Journal of Ethnopharmacology, 141*, 61–71. Available from https://doi.org/10.1016/j.jep.2012.01.053.

Mahapatra, T., Mahapatra, S., Babu, G. R., Tang, W., Banerjee, B., Mahapatra, U., & Das, A. (2014). Cholera outbreaks in south and Southeast Asia: Descriptive analysis, 2003-2012. *Japanese Journal of Infectious Diseases, 67*, 145–156. Available from https://doi.org/10.7883/yoken.67.145.

Mainuddin, M., Kirby, M., & Hoanh, C. T. (2011). Adaptation to climate change for food security in the lower Mekong Basin. *Food Security., 3*, 433–450. Available from https://doi.org/10.1007/s12571-011-0154-z.

Makrane, H., Aziz, M., Mekhfi, H., Ziyyat, A., Legssyer, A., Melhaoui, A., … Eto, B. (2019). *Origanum majorana* L. extract exhibit positive cooperative effects on the main mechanisms involved in acute infectious diarrhea. *Journal of Ethnopharmacology, 239*, 111503. Available from https://doi.org/10.1016/j.jep.2018.09.005.

Maneenoon, K., Khuniad, C., Teanuan, Y., Saedan, N., Prom-in, S., Rukleng, N., … Wongwiwat, W. (2015). Ethnomedicinal plants used by traditional healers in Phatthalung Province, Peninsular Thailand. *Journal of Ethnobiology and Ethnomedicine, 11*, 43. Available from https://doi.org/10.1186/s13002-015-0031-5.

Marcos, L. A., & DuPont, H. L. (2007). Advances in defining etiology and new therapeutic approaches in acute diarrhea. *Journal of Infection., 55*, 385–393. Available from https://doi.org/10.1016/j.jinf.2007.07.016.

McIver, L. J., Chan, V. S., Bowen, K. J., Iddings, S. N., Hero, K., & Raingsey, P. P. (2014). Review of climate change and water-related diseases in Cambodia and findings from stakeholder knowledge assessments. *Asia-Pacific Journal of Public Health., 28*, 49–58. Available from https://doi.org/10.1177/1010539514558059.

Meng, C. Y., Smith, B. L., Bodhidatta, L., Richard, S. A., Vansith, K., Thy, B., … Mason, C. J. (2011). Etiology of diarrhea in young children and patterns of antibiotic resistance in Cambodia. *Pediatric Infectious Disease Journal., 30*, 331–335. Available from https://doi.org/10.1097/INF.0b013e3181fb6f82.

Muecke, M. A. (1979). An explication of "wind illness" in Northern Thailand. *Culture, Medicine and Psychiatry, 3*, 267–300. Available from https://doi.org/10.1007/BF00114614.

Mukherjee, S., Koley, H., Barman, S., Mitra, S., Datta, S., Ghosh, S., … Dhar, P. (2013). *Oxalis corniculata* (Oxalidaceae) leaf extract exerts in vitro antimicrobial and in vivo anticolonizing activities against *Shigella dysenteriae* 1 (NT4907) and *Shigella flexneri* 2a (2457T) in induced diarrhea in suckling mice. *Journal of Medicinal Food, 16*, 801–809. Available from https://doi.org/10.1089/jmf.2012.2710.

Navarro, V., Villarreal, M. L., Rojas, G., & Lozoya, X. (1996). Antimicrobial evaluation of some plants used in Mexican traditional medicine for the treatment of infectious diseases. *Journal of Ethnopharmacology, 53*, 143–147. Available from https://doi.org/10.1016/0378-8741(96)01429-8.

Neamsuvan, O., Phumchareon, T., Bunphan, W., & Kaosaeng, W. (2016). Plant materials for gastrointestinal diseases used in Chawang District, Nakhon Si Thammarat Province, Thailand. *Journal of Ethnopharmacology, 194*, 179–187. Available from https://doi.org/10.1016/j.jep.2016.09.001.

Ngoc, M. N. T. (2017). Thermophilic Campylobacter - Neglected Foodborne Pathogens in Cambodia, Laos and Vietnam. *Gastroenterology & Hepatology, 8*(3), 00279. Available from https://doi.org/10.15406/ghoa.2017.08.00279.

Nguyen, X. M. A., Bun, S. S., Ollivier, E., & Dang, T. P. T. (2020). Ethnobotanical study of medicinal plants used by K'Ho-Cil people for treatment of diarrhea in Lam Dong Province, Vietnam. *Journal of Herbal Medicine., 19*, 100320. Available from https://doi.org/10.1016/j.hermed.2019.100320.

Nirmal, N. P., Rajput, M. S., Prasad, R. G. S. V., & Ahmad, M. (2015). Brazilin from *Caesalpinia sappan* heartwood and its pharmacological activities: A review. *Asian Pacific Journal of Tropical Medicine, 8*, 421–430. Available from https://doi.org/10.1016/j.apjtm.2015.05.014.

Noubissi, P. A., Fokam Tagne, M. A., Fankem, G. O., Ngakou Mukam, J., Wambe, H., & Kamgang, R. (2019). Effects of *Crinum jagus* water/ethanol extract on *Shigella flexneri*-induced diarrhea in rats. *Evidence-Based Complementary and Alternative Medicine, 2019*, 9537603. Available from https://doi.org/10.1155/2019/9537603.

Oh, S. Y., Yun, W., Lee, J. H., Lee, C. H., Kwak, W. K., & Cho, J. H. (2017). Effects of essential oil (blended and single essential oils) on anti-biofilm formation of *Salmonella* and *Escherichia coli*. *Journal of Animal Science and Technology, 59*, 4. Available from https://doi.org/10.1186/s40781-017-0127-7.

Omokhua, A. G., McGaw, L. J., Finnie, J. F., & Van Staden, J. (2016). *Chromolaena odorata* (L.) R.M. King & H. Rob. (Asteraceae) in sub-Saharan Africa: A synthesis and review of its medicinal potential. *Journal of Ethnopharmacology, 183*, 112–122. Available from https://doi.org/10.1016/j.jep.2015.04.057.

Palla, A. H., Khan, N. A., Bashir, S., Ur-Rehman, N., Iqbal, J., & Gilani, A. H. (2015). Pharmacological basis for the medicinal use of *Linum usitatissimum* (Flaxseed) in infectious and non-infectious diarrhea. *Journal of Ethnopharmacology, 160*, 61–68. Available from https://doi.org/10.1016/j.jep.2014.11.030.

Palombo, E. A. (2006). Phytochemicals from traditional medicinal plants used in the treatment of diarrhoea: Modes of action and effects on intestinal function. *Phytotherapy Research., 20*, 717–724. Available from https://doi.org/10.1002/ptr.1907.

Pandey, K., Shevkar, C., Bairwa, K., & Kate, A. S. (2020). Pharmaceutical perspective on bioactives from *Alstonia scholaris*: Ethnomedicinal knowledge, phytochemistry, clinical status, patent space, and future directions. *Phytochemistry Reviews, 19*, 191–233. Available from https://doi.org/10.1007/s11101-020-09662-z.

Panyadee, P., Balslev, H., Wangpakapattanawong, P., & Inta, A. (2019). Medicinal plants in homegardens of four ethnic groups in Thailand. *Journal of Ethnopharmacology, 239*, 111927. Available from https://doi.org/10.1016/j.jep.2019.111927.

Phetsouvanh, R., Nakatsu, M., Arakawa, E., Davong, V., Vongsouvath, M., Lattana, O., ... Newton, P. N. (2008). Fatal bacteremia due to immotile *Vibrio cholerae* serogroup O21 in Vientiane, Laos - A case report. *Annals of Clinical Microbiology and Antimicrobials, 7*, 10. Available from https://doi.org/10.1186/1476-0711-7-10.

Phung, D., Huang, C., Rutherford, S., Chu, C., Wang, X., & Nguyen, M. (2015). Climate change, water quality, and water-related diseases in the Mekong Delta Basin: A systematic review. *Asia-Pacific Journal of Public Health., 27*, 265–276. Available from https://doi.org/10.1177/1010539514565448.

Phung, D., Huang, C., Rutherford, S., Chu, C., Wang, X., Nguyen, M., ... Nguyen, T. H. (2015). Association between climate factors and diarrhoea in a Mekong Delta area. *International Journal of Biometeorology, 59*, 1321–1331. Available from https://doi.org/10.1007/s00484-014-0942-1.

Phuong, T. L. T., Rattanavong, S., Vongsouvath, M., Davong, V., Lan, N. P. H., Campbell, J. I., ... Baker, S. (2017). Non-typhoidal *Salmonella* serovars associated with invasive and non-invasive disease in the Lao People's Democratic Republic. *Transactions of the Royal Society of Tropical Medicine and Hygiene, 111*, 418–424. Available from https://doi.org/10.1093/trstmh/trx076.

Poramathikul, K., Bodhidatta, L., Chiek, S., Oransathid, W., Ruekit, S., Nobthai, P., ... Swierczewski, B. (2016). Multidrug-resistant *Shigella* infections in patients with diarrhea, Cambodia, 2014–2015. *Emerging Infectious Diseases, 22*, 1640–1643. Available from https://doi.org/10.3201/eid2209.152058.

Porras, G., Chassagne, F., Lyles, J. T., Marquez, L., Dettweiler, M., Salam, A. M., ... Quave, C. L. (2020). Ethnobotany and the role of plant natural products in antibiotic drug discovery. *Chemical Reviews, 121*(6), 3495–3560. Available from https://doi.org/10.1021/acs.chemrev.0c00922.

Qnais, E. Y., Elokda, A. S., Ghalyun, Y. Y. A., & Abdulla, F. A. (2007). Antidiarrheal activity of the aqueous extract of *Punica granatum* (pomegranate) peels. *Pharmaceutical Biology, 45*, 715–720. Available from https://doi.org/10.1080/13880200701575304.

Rajan, S., Suganya, H., Thirunalasundari, T., & Jeeva, S. (2012). Antidiarrhoeal efficacy of *Mangifera indica* seed kernel on Swiss albino mice. *Asian Pacific Journal of Tropical Medicine, 5*, 630–633. Available from https://doi.org/10.1016/S1995-7645(12)60129-1.

Rajan, S., Thirunalasundari, T., & Jeeva, S. (2011). Anti-enteric bacterial activity and phytochemical analysis of the seed kernel extract of *Mangifera indica* Linnaeus against *Shigella dysenteriae* (Shiga, corrig.) Castellani and Chalmers. *Asian Pacific Journal of Tropical Medicine, 4*, 294–300. Available from https://doi.org/10.1016/S1995-7645(11)60089-8.

Rani, P., & Khullar, N. (2004). Antimicrobial evaluation of some medicinal plants for their anti-enteric potential against multi-drug resistant *Salmonella typhi*. *Phytotherapy Research, 18*, 670–673. Available from https://doi.org/10.1002/ptr.1522.

Rashid, S., Ahmad, M., Zafar, M., Sultana, S., Ayub, M., Khan, M. A., & Yaseen, G. (2015). Ethnobotanical survey of medicinally important shrubs and trees of Himalayan region of Azad Jammu and Kashmir, Pakistan. *Journal of Ethnopharmacology, 166*, 340–351. Available from https://doi.org/10.1016/j.jep.2015.03.042.

Rasko, D. A., & Sperandio, V. (2010). Anti-virulence strategies to combat bacteria-mediated disease. *Nature Reviews Drug Discovery, 9*, 117–128. Available from https://doi.org/10.1038/nrd3013.

Rath, S., & Padhy, R. N. (2015). Antibacterial efficacy of five medicinal plants against multidrug-resistant enteropathogenic bacteria infecting under-5 hospitalized children. *Journal of Integrative Medicine, 13*, 45–57. Available from https://doi.org/10.1016/S2095-4964(15)60154-6.

Ren, A., Zhang, W., Thomas, H. G., Barish, A., Berry, S., Kiel, J. S., & Naren, A. P. (2012). A tannic acid-based medical food, cesinex ®, exhibits broad-spectrum antidiarrheal properties: A mechanistic and clinical study. *Digestive Diseases and Sciences, 57*, 99–108. Available from https://doi.org/10.1007/s10620-011-1821-9.

Rheinländer, T., Samuelsen, H., Dalsgaard, A., & Konradsen, F. (2011). Perspectives on child diarrhoea management and health service use among ethnic minority caregivers in Vietnam. *BMC Public Health, 11*, 690. Available from https://doi.org/10.1186/1471-2458-11-690.

Riddle, M. S., Dupont, H. L., & Connor, B. A. (2016). ACG clinical guideline: Diagnosis, treatment, and prevention of acute diarrheal infections in adults. *American Journal of Gastroenterology., 111*, 602–622. Available from https://doi.org/10.1038/ajg.2016.126.

Rose, J. L., Özünel, E. O., & Bennett, B. C. (2013). Ethnobotanical remedies for acute diarrhea in central Anatolian villages. *Economic Botany, 67*, 137–146. Available from https://doi.org/10.1007/s12231-013-9233-8.

Rundel, P. W. (2009). *Vegetation in the Mekong Basin. The Mekong* (pp. 143–160). United States: Elsevier Inc. Available from https://doi.org/10.1016/B978-0-12-374026-7.00007-3.

Sairam, K., Hemalatha, S., Kumar, A., Srinivasan, T., Ganesh, J., Shankar, M., & Venkataraman, S. (2003). Evaluation of anti-diarrhoeal activity in seed extracts of *Mangifera indica*. *Journal of Ethnopharmacology, 84*, 11–15. Available from https://doi.org/10.1016/S0378-8741(02)00250-7.

Saqib, F., Ahmed, M. G., Janbaz, K. H., Dewanjee, S., Jaafar, H. Z. E., & Zia-Ul-Haq, M. (2015). Validation of ethnopharmacological uses of *Murraya paniculata* in disorders of diarrhea, asthma and hypertension. *BMC Complementary and Alternative Medicine, 15*, 319. Available from https://doi.org/10.1186/s12906-015-0837-7.

Sawangjaroen, N., & Sawangjaroen, K. (2005). The effects of extracts from anti-diarrheic Thai medicinal plants on the in vitro growth of the intestinal protozoa parasite: *Blastocystis hominis*. *Journal of Ethnopharmacology, 98*, 67–72. Available from https://doi.org/10.1016/j.jep.2004.12.024.

Shah, A. J., Gowani, S. A., Zuberi, A. J., Ghayur, M. N., & Gilani, A. H. (2010). Antidiarrhoeal and spasmolytic activities of the methanolic crude extract of *Alstonia scholaris* L. are mediated through calcium channel blockade. *Phytotherapy Research, 24*, 28–32. Available from https://doi.org/10.1002/ptr.2859.

Shane, A. L., Mody, R. K., Crump, J. A., Tarr, P. I., Steiner, T. S., Kotloff, K., . . . Pickering, L. K. (2017). Infectious diseases society of America clinical practice guidelines for the diagnosis and management of infectious diarrhea. *Clinical Infectious Diseases., 65*, e45–e80. Available from https://doi.org/10.1093/cid/cix669.

Sharma, D. K., Gupta, V. K., Kumar, S., Joshi, V., Mandal, R. S. K., Bhanu Prakash, A. G., & Singh, M. (2015). Evaluation of antidiarrheal activity of ethanolic extract of *Holarrhena antidysenterica* seeds in rats. *Veterinary World, 8*, 1392–1395. Available from https://doi.org/10.14202/vetworld.2015.1392-1395.

Shawyer, R. J., Gani, A. S. B., Punufimana, A. N., & Seuseu, N. K. F. (1996). The role of clinical vignettes in rapid ethnographic research: A folk taxonomy of diarrhoea in Thailand. *Social Science & Medicine, 42*, 75. Available from https://doi.org/10.1016/0277-9536(95.

Shaygannia, E., Bahmani, M., Zamanzad, B., & Rafieian-Kopaei, M. (2015). A review study on *Punica granatum* L. *Journal of Evidence-Based Complementary and Alternative Medicine*, *21*, 221–227. Available from https://doi.org/10.1177/2156587215598039.

Sieberi, B. M., Omwenga, G. I., Wambua, R. K., Samoei, J. C., & Ngugi, M. P. (2020). Screening of the dichloromethane: Methanolic extract of *Centella asiatica* for antibacterial activities against *Salmonella typhi, Escherichia coli, Shigella sonnei, Bacillus subtilis,* and *Staphylococcus aureus*. *Scientific World Journal*, *2020*, 6378712. Available from https://doi.org/10.1155/2020/6378712.

Sitther, V., Zhang, D., Harris, D. L., Yadav, A. K., Zee, F. T., Meinhardt, L. W., & Dhekney, S. A. (2014). Genetic characterization of guava (*Psidium guajava* L.) germplasm in the United States using microsatellite markers. *Genetic Resources and Crop Evolution*, *61*, 829–839. Available from https://doi.org/10.1007/s10722-014-0078-5.

Souli, A., Sebai, H., Rtibi, K., Chehimi, L., Sakly, M., Amri, M., ... Marzouki, L. (2015). Inhibitory effects of two varieties of Tunisian pomegranate (*Punica granatum* L.) extracts on gastrointestinal transit in rat. *Journal of Medicinal Food*, *18*, 1007–1012. Available from https://doi.org/10.1089/jmf.2014.0110.

Srinivasan, R., Selvam, G. G., Karthik, S., Mathivanan, K., Baskaran, R., Karthikeyan, M., ... Govindasamy, C. (2012). In vitro antimicrobial activity of *Caesalpinia sappan* L. *Asian Pacific Journal of Tropical Biomedicine*, *2*, S136–S139. Available from https://doi.org/10.1016/S2221-1691(12)60144-0.

Steinmetz, E. F. (1961). Pæderia foetida. *Pharmaceutical Biology*, *1*, 133–144. Available from https://doi.org/10.3109/13880206109087314.

Sun, B., Wu, L., Wu, Y., Zhang, C., Qin, L., Hayashi, M., ... Liu, T. (2020). Therapeutic potential of *Centella asiatica* and its triterpenes: A review. *Frontiers in Pharmacology*, *11*, 568032. Available from https://doi.org/10.3389/fphar.2020.568032.

Sydara, K., Gneunphonsavath, S., Wahlström, R., Freudenthal, S., Houamboun, K., Tomson, G., & Falkenberg, T. (2005). Use of traditional medicine in Lao PDR. *Complementary Therapies in Medicine*, *13*, 199–205. Available from https://doi.org/10.1016/j.ctim.2005.05.004.

Taiwo, O. B., Olajide, O. A., Soyannwo, O. O., & Makinde, J. M. (2000). Anti-inflammatory, antipyretic and antispasmodic: Properties of *Chromolaena odorata*. *Pharmaceutical Biology*, *38*, 367–370. Available from https://doi.org/10.1076/phbi.38.5.367.5970.

Tan, J. B. L., & Lim, Y. Y. (2015). Critical analysis of current methods for assessing the in vitro antioxidant and antibacterial activity of plant extracts. *Food Chemistry*, *172*, 814–822. Available from https://doi.org/10.1016/j.foodchem.2014.09.141.

Tangjitman, K., Wongsawad, C., Kamwong, K., Sukkho, T., & Trisonthi, C. (2015). Ethnomedicinal plants used for digestive system disorders by the Karen of northern Thailand. *Journal of Ethnobiology and Ethnomedicine*, *11*, 27. Available from https://doi.org/10.1186/s13002-015-0011-9.

Tchoumba Tchoumi, L. M., Nchouwet, M. L., Poualeu Kamani, S. L., Yousseu Nana, W., Douho Djimeli, R. C., Kamanyi, A., & Ngnokam, S. L. W. (2020). Antimicrobial and antidiarrhoeal activities of aqueous and methanolic extracts of *Mangifera indica* Linn stem bark (Anarcadiaceae) in Wistar rats. *Advances in Traditional Medicine*, *21*, 485–498. Available from https://doi.org/10.1007/s13596-020-00470-6.

Téllez, M. A., Téllez, A. N., Vélez, F., & Ulloa, J. C. (2015). In vitro antiviral activity against rotavirus and astrovirus infection exerted by substances obtained from *Achyrocline bogotensis* (Kunth) DC. (Compositae). *BMC Complementary and Alternative Medicine*, *15*, 428. Available from https://doi.org/10.1186/s12906-015-0949-0.

Thakurta, P., Bhowmik, P., Mukherjee, S., Hajra, T. K., Patra, A., & Bag, P. K. (2007). Antibacterial, antisecretory and antihemorrhagic activity of *Azadirachta indica* used to treat cholera and diarrhea in India. *Journal of Ethnopharmacology*, *111*(3), 607–612. Available from https://doi.org/10.1016/j.jep.2007.01.022.

The Mekong River Commission. (2019). *Summary, State of the Basin Report 2018*, The Mekong River Commission.

Thiagarajah, J. R., Donowitz, M., & Verkman, A. S. (2015). Secretory diarrhoea: Mechanisms and emerging therapies. *Nature Reviews Gastroenterology and Hepatology*, *12*, 446–457. Available from https://doi.org/10.1038/nrgastro.2015.111.

Thiagarajah, J. R., & Verkman, A. S. (2013). Chloride channel-targeted therapy for secretory diarrheas. *Current Opinion in Pharmacology*, *13*, 888–894. Available from https://doi.org/10.1016/j.coph.2013.08.005.

Thompson, C. N., Phan, M. V. T., Van Minh Hoang, N., Van Minh, P., Vinh, N. T., Thuy, C. T., ... Baker, S. (2015). A prospective multi-center observational study of children hospitalized with diarrhea in Ho Chi Minh City, Vietnam. *American Journal of Tropical Medicine and Hygiene*, *92*, 1045–1052. Available from https://doi.org/10.4269/ajtmh.14-0655.

Thumann, T. A., Pferschy-Wenzig, E. M., Moissl-Eichinger, C., & Bauer, R. (2019). The role of gut microbiota for the activity of medicinal plants traditionally used in the European Union for gastrointestinal disorders. *Journal of Ethnopharmacology, 245*, 112153. Available from https://doi.org/10.1016/j.jep.2019.112153.

Troeger, C., Blacker, B. F., Khalil, I. A., Rao, P. C., Cao, S., Zimsen, S. R., ... Reiner, R. C. (2016). *The Lancet Infectious Diseases, 18*. Available from https://doi.org/10.1016/S1473-3099, 30362–1.

Uddin, S. J., Shilpi, J. A., Alam, S. M. S., Alamgir, M., Rahman, M. T., & Sarker, S. D. (2005). Antidiarrhoeal activity of the methanol extract of the barks of *Xylocarpus moluccensis* in castor oil- and magnesium sulphate-induced diarrhoea models in mice. *Journal of Ethnopharmacology, 101*, 139–143. Available from https://doi.org/10.1016/j.jep.2005.04.006.

van Sam, H., Baas, P., & Keßler, P. J. A. (2008). Traditional medicinal plants in Ben En National Park, Vietnam, Blumea. *Journal of Plant Taxonomy and Plant Geography, 53*, 569–601. Available from https://doi.org/10.3767/000651908X607521.

van Vuuren, S. F., Nkwanyana, M. N., & de Wet, H. (2015). Antimicrobial evaluation of plants used for the treatment of diarrhoea in a rural community in northern Maputaland, KwaZulu-Natal, South Africa. *BMC Complementary and Alternative Medicine, 15*, 53. Available from https://doi.org/10.1186/s12906-015-0570-2.

Velázquez, C., Correa-Basurto, J., Garcia-Hernandez, N., Barbosa, E., Tesoro-Cruz, E., Calzada, S., & Calzada, F. (2012). Anti-diarrheal activity of (-)-Epicatechin from *Chiranthodendron pentadactylon* Larreat: Experimental and computational studies. *Journal of Ethnopharmacology, 143*, 716–719. Available from https://doi.org/10.1016/j.jep.2012.07.039.

Von Seidlein, L., Deok, R. K., Ali, M., Lee, H., Wang, X. Y., Vu, D. T., ... Clemens, J. (2006). A multicentre study of *Shigella diarrhoea* in six Asian countries: Disease burden, clinical manifestations, and microbiology. *PLoS Medicine, 3*, 1556–1569. Available from https://doi.org/10.1371/journal.pmed.0030353.

Villas Boas, G. R., Rodrigues Lemos, J. M., de Oliveira, M. W., dos Santos, R. C., Stefanello da Silveira, A. P., Bacha, F. B., ... Oesterreich, S. A. (2019). Preclinical safety evaluation of the aqueous extract from *Mangifera indica* Linn. (Anacardiaceae): genotoxic, clastogenic and cytotoxic assessment in experimental models of genotoxicity in rats to predict potential human risks. *Journal of Ethnopharmacology, 243*, 112086. Available from https://doi.org/10.1016/j.jep.2019.112086.

Vu Nguyen, T., Le Van, P., Le Huy, C., Nguyen Gia, K., & Weintraub, A. (2006). Etiology and epidemiology of diarrhea in children in Hanoi, Vietnam. *International Journal of Infectious Diseases., 10*, 298–308. Available from https://doi.org/10.1016/j.ijid.2005.05.009.

Walther, B. A., Boëte, C., Binot, A., By, Y., Cappelle, J., Carrique-Mas, J., ... Morand, S. (2016). Biodiversity and health: Lessons and recommendations from an interdisciplinary conference to advise Southeast Asian research, society and policy. *Infection, Genetics and Evolution, 40*, 29–46. Available from https://doi.org/10.1016/j.meegid.2016.02.003.

Wiegand, I., Hilpert, K., & Hancock, R. E. W. (2008). Agar and broth dilution methods to determine the minimal inhibitory concentration (MIC) of antimicrobial substances. *Nature Protocols, 3*, 163–175. Available from https://doi.org/10.1038/nprot.2007.521.

Xu, H. X., & Lee, S. F. (2004). The antibacterial principle of *Caesalpina sappan*. *Phytotherapy Research, 18*, 647–651. Available from https://doi.org/10.1002/ptr.1524.

Yakubu, M. T., & Salimon, S. S. (2015). Antidiarrhoeal activity of aqueous extract of *Mangifera indica* L. leaves in female albino rats. *Journal of Ethnopharmacology, 163*, 135–141. Available from https://doi.org/10.1016/j.jep.2014.12.060.

Yamashiro, T., Nakasone, N., Higa, N., Iwanaga, M., Insisiengmay, S., Phounane, T., ... Vongsanith, P. (1998). Etiological study of diarrheal patients in Vientiane, Lao People's Democratic Republic. *Journal of Clinical Microbiology, 36*, 2195–2199. Available from https://doi.org/10.1128/jcm.36.8.2195-2199.1998.

Yossathera, K., Sriprang, S., Suteerapataranon, S., & Deachathai, S. (2016). Antibacterial and antioxidative compounds from *Oroxylum indicum*. *Chemistry of Natural Compounds, 52*, 311–313. Available from https://doi.org/10.1007/s10600-016-1625-4-.

Yu, X., He, T., Zhao, J., & Li, Q. (2014). Invasion genetics of *Chromolaena odorata* (Asteraceae): extremely low diversity across Asia. *Biological Invasions, 16*, 2351–2366. Available from https://doi.org/10.1007/s10530-014-0669-2.

Zahara, K., Panda, S. K., Swain, S. S., & Luyten, W. (2020). Metabolic diversity and therapeutic potential of holarrhena pubescens: An important ethnomedicinal plant. *Biomolecules, 10*, 1–28. Available from https://doi.org/10.3390/biom10091341.

Zhang, W., Fujii, N., & Naren, A. P. (2012). Recent advances and new perspectives in targeting CFTR for therapy of cystic fibrosis and enterotoxin-induced secretory diarrheas. *Future Medicinal Chemistry, 4*, 329–345. Available from https://doi.org/10.4155/fmc.12.1.

Zhang, Y., Yu, B., Sui, Y., Gao, X., Yang, H., & Ma, T. (2014). Identification of resveratrol oligomers as inhibitors of cystic fibrosis transmembrane conductance regulator by high-throughput screening of natural products from Chinese medicinal plants. *PLoS One, 9*, e94302. Available from https://doi.org/10.1371/journal.pone.0094302.

Zhao, S.-S., Ma, D.-X., Zhu, Y., Zhao, J.-H., Zhang, Y., Chen, J.-Q., & Sheng, Z.-L. (2018). Antidiarrheal effect of bioactivity-guided fractions and bioactive components of pomegranate (*Punica granatum* L.) peels. *Neurogastroenterology & Motility, 30*(7), e13364. Available from https://doi.org/10.1111/nmo.13364.

WHO. Diarrhoea, (2020). <https://www.who.int/news-room/fact-sheets/detail/diarrhoeal-disease> Accessed 12.03.21).

CHAPTER

8

Medicinal plants from West Africa used as antimalarial agents: an overview

Agnès Aubouy[1], *Aissata Camara*[2] *and Mohamed Haddad*[1]

[1]UMR 152 PharmaDev, Université de Toulouse, IRD, UPS, Toulouse, France [2]Institute for Research and Development of Medicinal and Food Plants of Guinea (IRDPMAG), Dubréka, Guinea

Introduction

Despite a decrease in malaria mortality over the last 10 years due to extensive malaria control through insecticide-impregnated bednets and increased use of artemisinin derivatives, malaria prevalence and burden in terms of morbidity and mortality is still extremely high. In 2019 the World Health Organization (WHO) reported 229 million cases of malaria and 409,000 deaths worldwide, the vast majority of which occurred in sub-Saharan African region (94% of malaria cases and deaths). African children pay the heaviest price, since they accounted for 67% (274,000) of all malaria deaths worldwide (WHO, 2019).

Numerous factors contribute to the complexity of this pathology and the difficulty of eradicating it, an objective that has already been claimed several times by the WHO in the past through the "roll back malaria" initiative. These factors include the *Plasmodium* parasite, the *Anopheles* vector, the human host, and the socio-geographical environment. *Plasmodium* is an extremely complex parasite with a high genetic diversity, in particular, due to its sexual multiplication. *P. falciparum*, the far most widespread and most dangerous species in sub-Saharan Africa, has a tremendous antigenic variability, which complicates vaccine research. Such genetic diversity is also a key factor in the emergence of parasites resistant to the antimalarial drugs used in a given area. Vector control is similarly hampered by the development of insecticide-resistant mosquitoes. The human host also offers a genetic and immunological complexity that contributes to a range of various clinical expressions of *P. falciparum* malaria. Finally, the geographical, economic, and social

Medicinal Plants as Anti-infectives
DOI: https://doi.org/10.1016/B978-0-323-90999-0.00014-8

environment of a given area contributes to the level of transmission, which in turn affects the level of immunity of the population, the effectiveness of the health system in the management of malaria, the means of control deployed, including the type of antimalarial drugs and insecticide used, the level of access to health care for the population, and so on.

In West African countries, the main characteristics of the malaria problem are a high level of transmission, parasite resistance to most antimalarial drugs available on the pharmaceutical market (except for artemisinin-based combination therapies at the moment), the resistance of *Anopheles* to insecticides, and limited access to health care. Natural products and more particularly plant-derived products constitute an enormous reservoir of bioactive molecules with novel biological targets and mode-of-action. The malaria field has already benefited from such biodiversity with quinine extracted from *Cinchona* bark, and artemisinin from *Artemisia annua* leaves, two molecules that are extremely active on *Plasmodium falciparum*, still in use today, and which have resulted in several derivative molecules by synthesis or hemisynthesis. Plant-derived compounds are therefore one of the answers to face the problem of the scarcity of effective antimalarial molecules due to the selection of resistant parasites. However, herbal products are not only molecules but also preparations based on whole plants or parts of plants (herbal teas, decoctions, etc.). The use of such preparations is commonly linked to strong cultural practices and offers the advantage of public acceptance and easy access (low cost, availability in markets and small shops). Thus, there are two ways to deal with the issue of medicinal plants traditionally used against malaria:

- the search for new active molecules to feed the antimalarial drug pipeline. This way is long and expensive, but it is how two of the most widely used molecules for the management of malaria were discovered: quinine and artemisinin.
- the evaluation and validation of the use of whole plant or plant part preparations. However, the use of this type of preparation raises the problem of the quality of the remedy, linked to the origin of the plant, the storage method of the plant, the method of preparation, the dosage used, and so on. Indeed, the chemical composition of a plant is influenced by genetic, agricultural, and environmental factors. If the content of the main active molecules is altered, the efficacy may be impacted. To circumvent this difficulty, the standardization of these remedies may be a solution. However, this would have an impact on the cost of such products, thereby limiting their value in terms of lower cost and wider access for the poorest populations. Thus, given that malaria is a fatal disease and that effective drugs exist, the recommendation of plant-based preparations that are difficult to standardize raises obvious ethical questions.

This chapter will focus on the medicinal plants from West Africa used as antimalarial agents. West African countries explored here are malaria-endemic areas, and for which data are available on the use of antimalarial plants. It includes, from the West to the East: Senegal, Cape Verde, The Gambia, Mali, Guinea-Bissau, Guinea, Sierra Leone, Liberia, Ivory Coast, Burkina Faso, Ghana, Togo, Benin, Niger, and Nigeria. In this chapter, we will discuss the traditional use of plants against malaria, and present the plants scientifically validated for their antimalarial activity, in the form of plant extracts (from traditional preparations or chemical solvents), or isolated molecules. Finally, we will discuss the special case of *Artemisia afra* and *A. annua*, two plants of Chinese origin and not endemic to West Africa, whose use as an antimalarial is beginning to spread to this part of the world.

Traditional use of medicinal plants in West Africa

Traditional Medicine is defined by the sum total of knowledge or practices whether explicable or inexplicable, used in diagnosing, preventing, or eliminating a physical, mental, or social disease which may rely exclusively on past experience or observations handed down from generation to generation, verbally or in writing. It also comprises therapeutic practices that have been in existence often for hundreds of years before the development of modern scientific medicine and are still in use today without any documented evidence of adverse effects.

The rich flora of West Africa has been known in Europe since the first contact with the populations of the Gulf of Guinea by the Portuguese in the 15th century and represents an inexhaustible source of remedies available to the population. This richness of West African flora is notably linked to two different environments: the tropical forest along the coast, and the savannah in the hinterland (Oguakwa, 1980). Thus, it is estimated that there are 300,000 plant species in the world, of which more than 200,000 are found in the intertropical zone, including West Africa (Abayomi, 2010). In the absence of modern western medicine, traditional African medicine provided most of the health needs of the population during the precolonial period. In Africa, a large majority of the population uses only local plants to treat themselves for various reasons: lack of essential medicines, lack of access to so-called "modern" medicines, inadequate health care, lack of proximity to health centers, and high cost of medicines. Moreover, sociocultural practices are also to be taken into account in the success of the widespread use of traditional medicine, since it is an integral part of the culture of the people who rely on it. The use of plants for treatment is an ancestral practice, particularly in Africa, and frequently there is an oral transmission of knowledge from generation to generation to certain categories of initiates such as traditional practitioners and herbalists.

According to 2008 world statistics, West Africa has about 1.5 doctors for every 10,000 inhabitants, whereas the ratio in France, for example, is 6.6 for the same number of inhabitants (WHO, 2008) (Fig. 8.1). Furthermore, it is estimated that there are more traditional healers (100 times more) than conventional doctors or nurses in the region, and a high proportion of the 300 million West Africans exposed to malaria prefer to use inexpensive and handy traditional medicines when they have malaria symptoms rather than seeking treatment from the formal health system (Bodeker & Kronenberg, 2002; Soh & Benoit-Vical, 2007). In terms of health expenditures, a West African spends between $10 and $40 on medical care per year, whereas in France this amount is $4719 (more than 100 times). In this context, the contribution of traditional medicine in the fight against certain pathologies, in particular malaria represents an important asset for the well-being of populations.

Regarding malaria, the treatments available are mainly based on the use of traditional herbal remedies. However, there is a lack of data on their efficacy and safety, while the validation of traditional practices could lead to innovative strategies in malaria control. Moreover, natural products from plants or other organisms represent an almost inexhaustible reservoir of molecules, most of which are poorly explored and may constitute lead molecules for new antimalarial drugs, such as artemisinin, initially isolated from *Artemisia annua* (Kayser & Kiderlen, 2003). The West African region has incredible biodiversity and many indigenous plants are used as antimalarial agents, although few formal studies have examined the safest and most effective dosages and delivery mechanisms (Soh & Benoit-Vical, 2007).

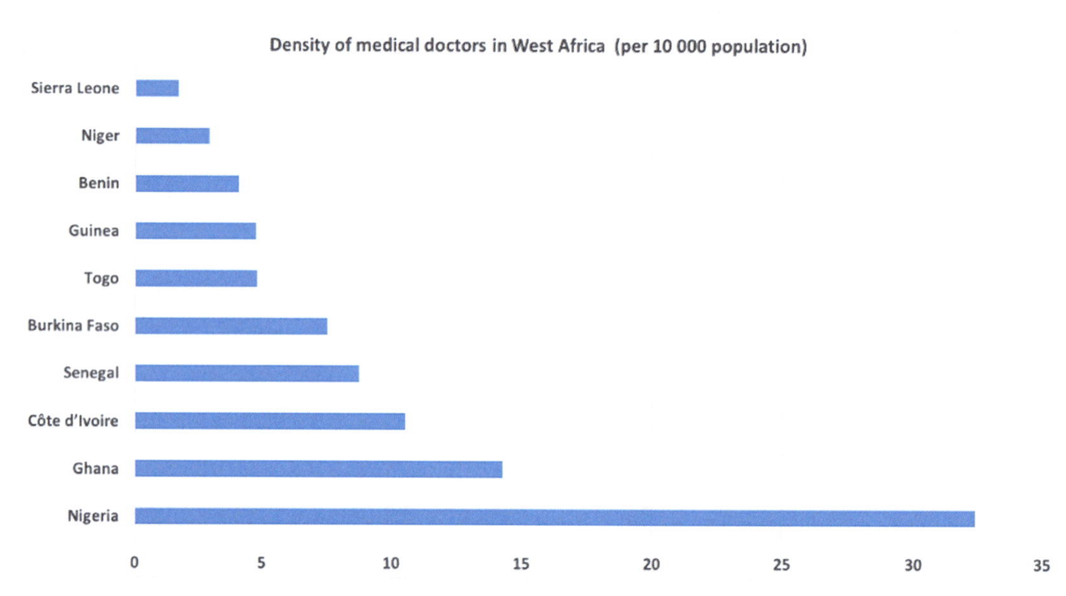

FIGURE 8.1 Density of medical doctors in West Africa. Data are from the latest available year ranging from 2004 to 2019 (WHO,2008).

Through ethnopharmacology, it is possible to validate traditional uses, especially for antimalarial plants. In West Africa, four families of plants are widely used as antimalarial remedies: plants of the Combretaceae, Euphorbiaceae, Meliaceae, and Rubiaceae families (Soh & Benoit-Vical, 2007). The Euphorbiaceae family includes species such as *Jatropha* sp. or *Ricinus* sp., used against fever or malaria by populations in three continents: Africa, South America, and Asia; while Meliaceae and Rubiaceae are used in Africa and South America. Fig. 8.2 shows the main antimalarial plants listed in the West African pharmacopoeia.

However, despite the richness of the West African pharmacopoeia, the majority of plant extracts used in traditional medicine are still produced using ancient traditional methods that do not always guarantee the efficacy, stability, and safety of the remedy. One of the most difficult but crucial tasks in the search for new antimalarial drugs from traditional medicine is the selection of plants or compounds that have the best chance to produce safe and effective antimalarial drugs for use in phytomedicine. For this purpose, it is therefore essential to evaluate the efficacy of the isolated plants and compounds by in vitro and in vivo approaches.

Plant extracts and plant compounds validated by in vitro and/or in vivo approach

To confirm the antimalarial effect of a plant, several factors should be considered:

1. From an ethnopharmacologic perspective, the traditional way of use needs to be known, particularly for the following points:

   The part of plant used: the whole plant may be used, but usually a specific part of the plant is used: roots, all aerial parts together, the stem, the leaves, the flowers, and the root or stem bark.

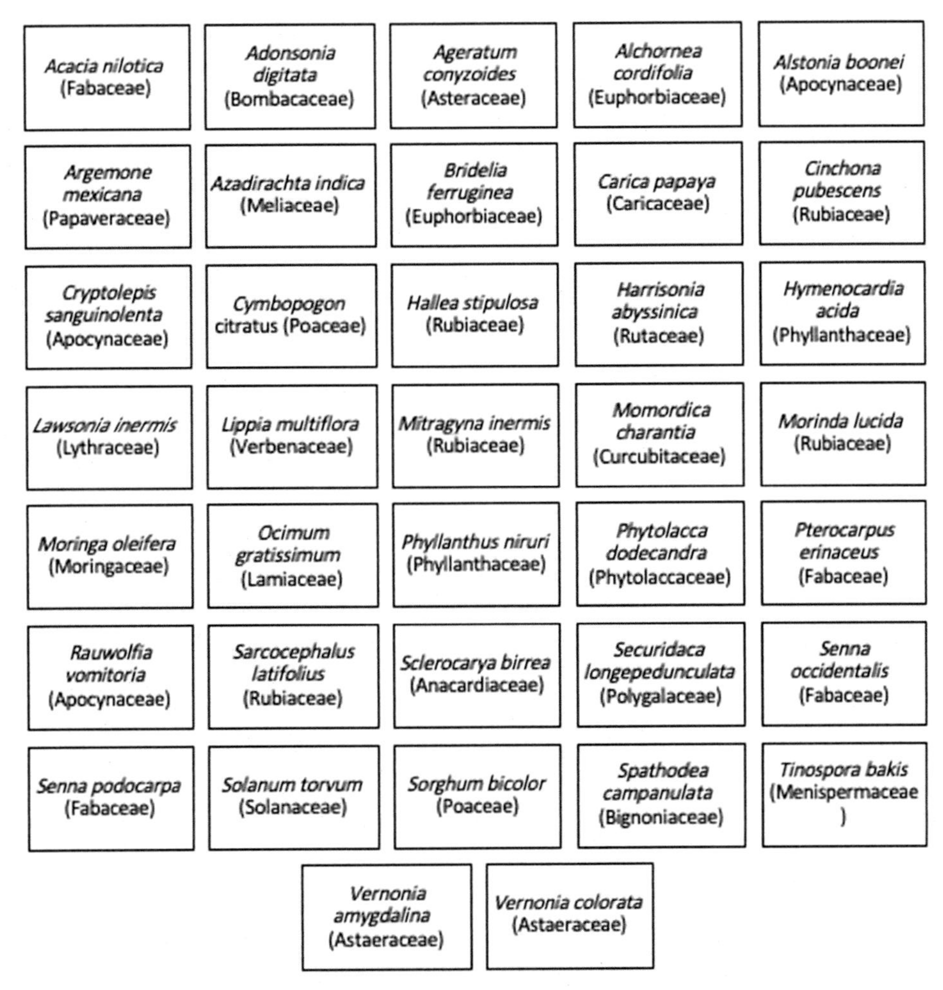

FIGURE 8.2 Main antimalarial plants listed in the West African Pharmacopoeia.

The mode of preparation: a plant-based preparation can be an infusion, a decoction, a maceration, or an extraction of juice. This point is of high importance for validating the traditional preparation. The nature of the extracted molecules present in the preparation depends on the method of preparation. The preparation may also involve a phase of shaking or crushing the plant.

Infusion: simmering or boiling water (or brought to an appropriate temperature) is poured over the herb, before an appropriate incubation time in the liquid. *Decoction*: the plant is immersed in cold water and then the mixture is boiled for an appropriate time, before being left to infuse for a defined period of time. *Maceration*: room temperature liquid (usually water or alcohol) is poured over the herb and left to macerate during a defined period of time.

The mode of absorption: it can also vary and includes absorption by oral route, body bath, drop in eyes or on the skin, and so on. In this chapter, only orally absorbed plants will be discussed.

The use of the plant alone or mixed with other plants: plants are also often mixed with others to combine effects against several symptoms, like fever and malaria, or to potentialize the activity.

2. From a research perspective, the method used to test the plant has to be known to get an idea of the comparability of the results. Thus, the main elements to know are the following:

The method of extraction used: depending on the purpose of the study, the researchers will test the traditional preparation or other modes of plant extraction. The traditional preparation will be studied to validate a traditional use. To isolate and identify the bioactive molecules in a complex mixture of compounds, specific plant extracts will be prepared and bioguided fractionation will be carried out. This involves testing the activity of less and less complex fractions until a single molecule is identified. The nature of the solvent chosen for extraction determines the type of molecule that will be present in the solute. A solvent will extract molecules whose polarity is close to its own polarity, according to the law of similarity and intermiscibility (like dissolves like). Water, alcohol (ethanol, methanol), chloroform, pentane, hexane, and methylene chloride are the main solvents used for extraction and isolation of plant products. Please, see the review proposed by Zhang, Lin, and Ye (2018) for more details about the different types of extraction processes. More recent techniques, such as dereplication, metabolomics, and molecular networking, allow also us to identify bioactive molecules from a complex extract (Chassagne et al., 2018; Vial et al., 2020).

The biological test used: unfortunately, research teams do not follow a unique method to evaluate antimalarial activity of plants. Each team adapts his method according to its means, material, and biological facilities. Three types of antimalarial activities can be measured: in vitro, in animal models, or in humans through clinical assays.

In vitro antimalarial evaluation of plant extracts

In vitro, antimalarial activity is measured by incubating the plant preparation with *P. falciparum* strains. Parasite strains used are the most often lab strains, but some teams also test *P. falciparum* isolates from patients. In vitro culture conditions may also differ from one lab to the other. The main used *P. falciparum* lab strains are the chloroquino-resistant K1, FcB1, Dd2, and W2, and the chloroquino-sensitive 3D7, FCM29, and D6. These lab strains differ by their geographical origin and their resistance phenotype toward antimalarial reference molecules as chloroquine. Finally, the level of antimalarial efficacy is assessed by measuring the concentration that inhibits the growth of the parasite by 50% called IC_{50}. In addition, laboratories frequently measure the toxicity of the extract or molecule on human cells, in order to calculate the selectivity index, which is equal to the cytotoxic dose over the antimalarial dose. Again, the methods differ between laboratories in the choice of cells used for the cytotoxicity measurement, either primary cells or cell lines.

Among the plants we have listed in Table 8.1, six families of plants have been the most worked on in vitro, with four to nine species per family tested for their antiplasmodial activity *in vitro*: Asteraceae, Combretaceae, Euphorbiaceae, Fabaceae, Meliaceae, and Rubiaceae. Twelve species presented interesting IC_{50}s for their crude extracts, below 0.1 μg/mL. However, these results are interesting if the selectivity index is relatively high (>25), which reduces the number of species to six, as follows: *Enantia polycarpa* (Annonaceae), *Acalypha wilkesiana* (Euphorbiaceae), *Phyllanthus fraternus* (Euphorbiaceae), *Tectona grandis* (Lamiaceae), and *Bambusa vulgaris* (Poaceae). Unfortunately, many studies do not report the selectivity index. These species were tested on the chloroquine-sensitive strain 3D7, except for *Enantia polycarpa* which was tested on the chloroquine-resistant strain K1. It is unfortunate that the teams do not systematically test the extracts on two strains, one chloroquine-sensitive and the other chloroquine-resistant. It will also be interesting in the future to work with artemisinin-resistant strains.

In vivo antimalarial evaluation of plant extracts

In vivo, the most often rodent models are used, either mice or rats, infected by murine *Plasmodium* strains. Depending on both the genetic background of the mouse and on the *Plasmodium* species, the infection will be more or less severe, and parasitemia more or less high (Li, Seixas, & Langhorne, 2001; Wykes & Good, 2009). There are four species of rodent malaria: *P. chabaudi*, *P. vinckei*, *P. berghei*, and *P. yoelii* and various strains (e.g., *P. yoelii* YM, *P. yoelii* 17XNL, *P. chabaudi chabaudi* AS, *P. chabaudi adami*, *P. berghei* ANKA) with different parasite biology and pathogenicity. *P. berghei* ANKA has the specificity to cause cerebral malaria characterized by low parasitemia and by the onset of neurological symptoms followed by death 7–10 days after infection in the BALB/c, C57BL/6, CBA mice, whereas its infection resolves in DBA/2j mice. Conversely, *P. berghei* K173 is lethal in all four species of mice but without neurological symptoms. *P. chabaudi chabaudi* (Pcc) is nonlethal in BALB/c, C57BL/6, CBA mice, causing hyperparasitemia, while Pcc AS is lethal in A/J and DBA/2 mice. *P. yoelii* 17XL and YM are lethal in the main species of mice (BALB/c, C57BL/6, CBA, DBA/2j), whereas *P. yoelii* 17XNL is nonlethal. Finally, *P. vinckei vinckei* is lethal in BALB/c mice, but not in the other species of mice.

In murine models (Table 8.2), 12 plant species were reported for their highly active crude extracts with parasite inhibition ≥80%. Of these plants, five were toxic to mice or the toxicity was not documented, reducing the list to *Terminalia albida* (Combretaceae), *Annickia polycarpa* (Annonaceae), *Vernonia amygdalina* (Asteraceae), *Carica papaya* (Caricaceae), *Ficus thonningii* (Moraceae), *Quassia amara* (Simaroubaceae), and *Triumfetta cordifolia* (Tiliaceae). Unfortunately, none of these plants achieved complete inhibition of parasitemia in the model used.

In vitro and in vivo evaluation of antimalarial compounds

In Table 8.3, molecules isolated from plant extracts were tested for their antiplasmodial or antimalarial activity. The most interesting molecules presenting with IC_{50} values ≤1 μM were the cryptolepine analog (from *Cryptolepis sanguinolenta*, Periplocaceae), dioconpeltine A, habropetaline A, and dioncophylline A and C (from plants of the

TABLE 8.1 In vitro antimalarial activity of the main West African plants studied.

Plant family	Botanical name	Used part	Type of extract	Plasmodium strain	IC50* (μg/mL)	Selectivity index§	References
Annonaceae	*Enantia polycarpa*	Stem bark	Ethanol 90%	K1	0.126	616	Kamanzi Atindehou, Schmid, Brun, Koné, and Traore (2004)
	Monanthotaxis caffra	Leaves and twigs	Ethanol 70%	FcB1	5.86	15.12	Laryea and Borquaye (2019)
				W2	18.94	4.68	
				CAM06	18.54	4.78	
	Polyalthia longifolia	Leaves	Aqueous	NF54	24.00	>4.17	Kwansa-Bentum, Agyeman, Larbi-Akor, Anyigba, and Appiah-Opong (2019)
			Ethanol 70%		22.46	>4.45	
			Ethyl acetate		9.50	>10.53	
Apocynaceae	*Funtumia elastic*	Stem bark	Ethanol	FcB1	3.3	Nd	Zirihi, Mambu, Guédé-Guina, Bodo, and Grellier (2005)
	Rauvolfia vomitoria	Root bark	Ethanol	FcB1	2.5	>100	Zirihi et al. (2005)
Amaranthaceae	*Gomphrena celosioides*	Aerial parts	Methylene chloride	3D7	14.41	Nd	Weniger et al. (2004)
			Methanol	K1	6.83	Nd	
Arecaceae	*Borassus aethiopum*	Roots	Dichloromethane	FCR3	25	Nd	Gruca et al. (2015)
	Cocos nucifera	Husk fibers	Ethyl acetate	W2	10.94	30.3	Adebayo et al. (2013)

Asteraceae	*Acanthospermum hispidum*	Whole plant	Aqueous	Dd2	3.70	Nd	Adukpo et al. (2020)
				3D7	3.66	Nd	
		Aerial parts	Dichloromethane	3D7	7.5	4.7	Bero et al. (2009)
				W2	4.8	4.7	
	Bidens pilosa	Leaves and twigs	Ethanol 70%	FcB1	23.48	4.34	Laryea and Borquaye (2019)
				W2	4.60	22.17	
				CAM06	21.43	4.76	
	Dicoma tomentosa	Whole plant	Dichloromethane	3D7	3.4	3.3	Jansen et al. (2012)
				W2	1.9	6.1	
			Diethyl ether	3D7	3.9	1.6	
				W2	4.8	1.3	
			Ethyl acetate	3D7	4.4	1.3	
				W2	4.6	1.3	
			Methanol	3D7	5.8	3.5	
				W2	3	6.7	
	Erigeron floribundus	Leaves	Pentane	Nigerian patient isolate	6	Nd	Ménan et al. (2006)
				FcM29	4	4.6	
	Launaea taraxacifolia	Leaves	Chloroform	D6	21	>2.1	Bello et al. (2017)
				W2	18	>2.6	

(*Continued*)

TABLE 8.1 (Continued)

Plant family	Botanical name	Used part	Type of extract	Plasmodium strain	IC50* (μg/mL)	Selectivity index§	References
Cochlospermaceae	*Cochlospermum planchonii*	Rhizome and leaves	Dichlorométhane	3D7	2.4	Nd	Lamien-Meda et al. (2015)
			Ethyl acetate	3D7	11.5	Nd	
		Roots	Methylene chloride	K1	4.4	15	Vonthron-Sénécheau et al. (2003)
	Cochlospermum tinctorium	Roots	Ethanol	3D7	2.3	Nd	Togola, Diallo, Dembélé, Barsett, and Paulsen (2005)
				Dd2	3.8	Nd	
Combretaceae	*Anogeissus leiocarpus*	Leaves	Methylene chloride	K1	3.8	19	Vonthron-Sénécheau et al. (2003)
			Methanol	FcB1	6.3	Nd	Okpekon et al. (2004)
			Methylene chloride	FcB1	9.3	Nd	
		Roots	Methanol	FcB1	2.6	Nd	
			Methylene chloride	FcB1	2.6	Nd	
		Bark	Methanol	FcB1	4.8	Nd	
			Methylene chloride	FcB1	19.5	Nd	
	Combretum collinum	Leaves	Dichloromethane	K1	4	6.5	Ouattara et al. (2014)
			Total alkaloids	K1	4	9	
	Combretum fragrans	Leaves	Dichloromethane	K1	5	9	
			Methanol	K1	9	4.6	

		$MeOH/H_2O$ (1/1)	K1	12	>4.2	
		Aqueous	K1	11	1.7	
		Total alkaloids	K1	3	>8.3	
Combretum micranthum	Leaves	Aqueous	W2	0.8	Nd	Benoit et al. (1996)
Terminalia albida	Stem bark	Aqueous	K1	1.5	Nd	Camara et al. (2019)
Terminalia macroptera	Leaves	Ethanol 90%	FcB1	1.2	Nd	Haidara et al. (2018)
	Roots		FcB1	1.6	Nd	
	Root bark	Aqueous	W2	1	Nd	Sanon et al. (2003)
Terminalia ivorensis	Leaves	Aqueous	3D7	0.64	9.77	Komlaga et al. (2016)
			W2	10.52	0.5	
		Petroleum ether	3D7	14.8	>14	
		Ethyl acetate	3D7	16	6	
Terminalia avicennioides	Leaves	Methanol	K1	6.5	1.9	Ouattara et al. (2014)
		$MeOH/H_2O$ (1/1)	K1	3.5	9.3	
		Ethyl acetate	K1	7	3	
Terminalia glaucescens	Stem, leaves	Aqueous	FcM29	2.34	Nd	Mustofa et al. (2000)
			FcB1	4.21	Nd	
			Nigerian patient isolate	2.36	Nd	
		Ethanol	FcM29	0.41	Nd	
			FcB1	0.50	Nd	

(*Continued*)

TABLE 8.1 (Continued)

Plant family	Botanical name	Used part	Type of extract	Plasmodium strain	IC50* (μg/mL)	Selectivity index§	References
				Nigerian patient isolate	0.55	Nd	
			Pentane	FcM29	4.60	Nd	
				FcB1	8.08	Nd	
				Nigerian patient isolate	26.8	Nd	
		Bark	Methanol	FcB1	2	Nd	Okpekon et al. (2004)
			Methylene chloride	FcB1	1.8	Nd	
	Terminalia schimperiana	Young Leaves	Ethanol 90%	K1	2.4	33	Kamanzi Atindehou et al. (2004)
Cucurbitaceae	*Momordica balsamina*	Nd	Heptane + EtOAc	F32	4	Nd	Benoit-Vical et al. (2006)
				FcM29	3	Nd	
Euphorbiaceae	*Acalypha wilkesiana*	Leaves	Methanol 70%	3D7	<0.39	>100	Amlabu et al. (2018)
	Alchornea cordifolia	Stem, leaves	Aqueous	FcM29	3.51	Nd	Valentin et al. (2000)
				FcBI	4.01	Nd	
				Nigerian patient isolate	2.47	Nd	
			Ethanol	FcM30	3.20	Nd	
				FcBI	3.95	Nd	
				Nigerian patient isolate	2.90	Nd	

Chrozophora senegalensis	Leaves	Aqueous	FcM29	1.85	8	Garcia-Alvarez et al. (2013)
			FcB1	1.9		
			F32	0.8		
			W2	0.33		
Croton lobatus	Roots	Methylene chloride	3D7	4.42	Nd	Weniger et al. (2004)
			K1	2.80	Nd	
		Methanol	3D7	6.56	Nd	
			K1	4.91	Nd	
	Aerial parts	Methylene chloride	3D7	3.74	Nd	
			K1	3.64	Nd	
		Methanol	3D7	0.38	Nd	
			K1	>20	Nd	
Croton penduliflorus	Leaves and twigs	Ethanol 70%	FCB	5.37	50.65	Laryea and Borquaye (2019)
			W2	14.03	19.39	
			CAM06	14.66	18.55	
Phyllanthus fraternus	Whole plant	Aqueous	3D7	4.07	7.64	Komlaga et al. (2016)
		Methanol	3D7	0.44	410	
Sebastiana chamaelea	Whole plant	Aqueous	FcM29	6.6	7.5	Garcia-Alvarez et al. (2013)
			FcB1	10		
			F32	14		
			W2	12		

(*Continued*)

TABLE 8.1 (Continued)

Plant family	Botanical name	Used part	Type of extract	Plasmodium strain	IC50* (μg/mL)	Selectivity index§	References
Fabaceae	*Cassia alata*	Leaves	Dichloromethane/ methane (1:1)	D10	7.02	Nd	Da et al. (2016)
	Erythrina senegalensis	Stem bark	Ethanol 90%	K1	1.8	Nd	Kamanzi Atindehou et al. (2004)
	Mezoneuron benthamianum	Leaves	Ethanol 70% (Precipitate)	3D7	6.4	Nd	Jansen et al. (2017)
	Pterocarpus erinaceus	Leaves	Ethanol 70%	Patient isolate	14.63	Nd	Karou, Dicko, Sanon, Simpore, and Traore (2003)
			Ether		7.38	Nd	
			Chloroform		1.93	Nd	
	Pericopsis laxiflora	Bark	Methanol	NF54	11.5	Nd	Koffi, Silué, Tano, Dable, and Yavo (2020)
				K1	7.4	Nd	
	Senna siamea	Roots	Aqueous	Dd2	4.47	Nd	Adukpo et al. (2020)
				3D7	3.95	Nd	
	Tetrapleura tetraptera	Bark	Dichloromethane	3D7	13.0	6.15	Lekana-Douki et al. (2011)
				FcB1	10.1	7.91	
Hypericaceae	*Harungana madagascariensis*	Bark	Aqueous	NF54	6.16	Nd	Koffi et al. (2020)
				K1	7.3	Nd	
			Ethanol	NF54	20.3	Nd	
				K1	22.2	Nd	
			Methanol	NF54	23.9	Nd	
				K1	22.8	Nd	

Lamiaceae	*Tectona grandis*	Leaves	Aqueous	3D7	7.18	>13.93	Komlaga et al. (2016)
			Petroleum ether	3D7	4.48	41	
			Ethyl acetate	3D7	14.15	12	
			Methanol	3D7	0.92	>217	
Loganiaceae	*Anthocleista djalonensis*	Stem bark	Ethanol 70%	ANKTC023	9.94	Nd	Attemene et al. (2018)
				ANKTC024	10	Nd	
				ANKTC024	9.69	Nd	
				ANKTC024	5.36	Nd	
				K1	15.94	Nd	
	Strychnos spinosa	Leaves	Dichloromethane	3D7	15.6	>6.4	Bero et al. (2009)
				W2	8.9	>6.4	
Malvaceae	*Clappertonia ficifolia*	Leaves	Ethanol 70%	FcB1	4.43	61.7	Laryea and Borquaye (2019)
				W2	7.94	34.4	
				CAM06	6.56	41.7	
	Sida acuta	whole plant	Ethanol 70%	Patient isolate	4.37	Nd	Karou et al. (2003)
			Chloroform		0.87	Nd	
			Aqueous		0.92	Nd	
Maranthaceae	*Thalia geniculata*	Roots	Methylene chloride	K1	14.33	Nd	Weniger et al. (2004)
			Methanol	3D7	2.83	Nd	
				K1	6.38	Nd	

(*Continued*)

TABLE 8.1 (Continued)

Plant family	Botanical name	Used part	Type of extract	Plasmodium strain	IC50* (μg/mL)	Selectivity index§	References
Meliaceae	*Azadirachta indica*	Leaves	Ethanol	D6	2.5	Nd	MacKinnon et al. (1997)
				W2	2.48	Nd	
	Cedrela odorata	Wood	Ethanol	D6	9.29	Nd	MacKinnon et al. (1997)
				W2	2.77	Nd	
	Trichilia emetica	Root bark	Ethanol 90%	K1	3.91	2.14	Kamanzi Atindehou et al. (2004); Togola et al. (2005)
		Leaves	Methanol	Dd2	2.5	3.34	
	Trichilia monadelpha	Stem bark	Ethanol 90%	K1	3.6	Nd	Kamanzi Atindehou et al. (2004)
Mimosaceae	*Schrankia leptocarpa*	Aerial parts	Methylene chloride	3D7	16.58	Nd	Weniger et al. (2004)
				K1	3.38	Nd	
			Methanol	3D7	8.00	Nd	
	Cylicodiscus gabunensis	Bark	Ethanol 70%	Dd2	20.8	2.9	Aldulaimi et al. (2017)
			Ethanol/n-butanol		10.4	4.8	
Moraceae	*Ficus thonningii*	Leaves	Methanol	NF54	5.3	>3.8	Falade et al. (2014)
				K1	21.1		
			Hexane	NF54	2.7	>7.4	
				K1	10.4		
			Ethyl acetate	NF54	5.3	>3.8	
				K1	15.3		

Myrtaceae	*Syzygium guineense*	Leaves	Ethanol 70%	FcB1	14.9	77.9	Laryea and Borquaye (2019)
				W2	4.6	5.21	
				CAM06	5.5	16.9	
Ochnaceae	*Lophira alata*	Leaves	Methanol	NF54	11.3	>1.7	Falade et al. (2014)
				K1	5.3		
			Hexane	NF54	2.5	>8.0	
				K1	2.5		
			Ethyl acetate	NF54	9.7	>2.0	
Oleaceae	*Ximenia americana*	Nd	Aqueous	F32	0.6–2.6	Nd	Benoit et al. (1996)
				FcB1	1.05–1.83	Nd	
Oxalidaceae	*Biophytum umbraculum*	Aerial parts	Dichloromethane	K1	7.4	>13.6	Austarheim et al. (2016)
			Ethyl acetate	NF54	6.7	19.1	
				K1	5.6		
Papilionaceae	*Afrormosia laxiflora*	Leaves	Methylene chloride	FcB1	14.5	Nd	Okpekon et al. (2004)
		Roots	Methanol	FcB1	7.5	Nd	
			Methylene chloride	FcB1	1.5	Nd	
Phyllanthaceae	*Phyllanthus niruri*	Whole plant	Aqueous	Dd2	5.7	Nd	Adukpo et al. (2020)
				3D7	5.5	Nd	
Piperaceae	*Pothomorphe umbellata*	Leaves	Ethanol 90%	K1	3.7	Nd	Kamanzi Atindehou et al. (2004)

(*Continued*)

TABLE 8.1 (Continued)

Plant family	Botanical name	Used part	Type of extract	Plasmodium strain	IC50* (μg/mL)	Selectivity index§	References
Poaceae	*Bambusa vulgaris*	Leaves	Aqueous	3D7	7.5	>13.3	Komlaga et al. (2016)
			Petroleum ether	3D7	0.7	>267	
			Ethyl acetate	3D7	0.5	>408	
Polygalaceae	*Carpolobia lutea*	Aerial parts	Dichloromethane	3D7	19.4	3.4	Bero et al. (2009)
				W2	8.1	3.4	
Proteaceae	*Faurea speciosa*	Leaves and twigs	Ethanol 70%	FcB1	14.8	10.4	Laryea and Borquaye (2019)
				W2	9.3	16.6	
				CAM06	6.9	22.3	
Rhamnaceae	*Ziziphus mauritiana*	Leaves	Ethanol 70%	ANKTC023	9.7	Nd	Attemene et al. (2018)
				ANKTC024	10.2	Nd	
				ANKTC024	13.6	Nd	
				ANKTC024	15.4	Nd	
				K1	20	Nd	
Rubiaceae	*Canthium setosum*	Aerial parts	Methylene chloride	3D7	2.8	Nd	Weniger et al. (2004)
			Methanol	K1	4.8	Nd	
	Keetia leucantha	Twigs	Dichloromethane	3D7	11.3	Nd	Bero et al. (2009)
	Mitragyna inermis	Leaves	Chloroform	W2	4.36	Nd	Traore-Keita et al. (2000)
				3D7	4.82	Nd	
		Roots	Chloroform	W2	22.26	Nd	
				3D7	22.21	Nd	
	Morinda morindoides	Leaves	Ethanol 90%	K1	3.54		Kamanzi Atindehou et al. (2004)

	Nauclea latifolia	Stem	Aqueous	Nigerian patient isolate	1.7	Nd	Benoit-Vical et al. (1998)
				FcB1	2.2	Nd	
		Root		Nigerian patient isolate	0.7	Nd	
				FcB1	0.6	Nd	
	Pavetta crassipes	Leaves	Total alkaloids	W2	1.2	Nd	Sanon, Azas, et al. (2003)
				D6	1	Nd	
Rutaceae	*Fagara macrophylla*	Stem bark	Ethanol	FcB1	2.3	12	Zirihi et al. (2005)
Verbenaceae	*Lantana rhodesiensis*	Leaves	methanol 50%	3D7	12.5	Nd	Nea et al. (2021)
Violaceae	*Hybanthus enneaspermus*	Aerial parts	Methylene chloride	K1	2.57	Nd	Weniger et al. (2004)

IC50*: half maximal inhibitory concentration; Selectivity index§: ratio of the toxic concentration of a sample against its effective bioactive concentration.

TABLE 8.2 In vivo antimalarial activity of the main West African plants studied.

Plant family	Botanical name	Used part	Type of extract	Model used	Inhibition of parasitemia (%)	Toxicity in the model	References
Anacardiaceae	*Antrocaryon micraster*	Stem bark	Ethanol	*ICR mice, P. berghei ANKA*	46.1% at 400 mg/kg per day	Nontoxic	Kumatia et al. (2021)
Annonaceae	*Annickia polycarpa*	Leaves	Ethanol 96%	*ICR mice, P. berghei ANKA*	95.5% at 400 mg/kg per day	Nontoxic	Kumatia et al. (2021)
	Enantia polycarpa	Stem bark	Ethanol	*Swiss albino mice, P. berghei berghei NK65*	75.8% at 600 mg/kg per day	Nontoxic	Anosa, Udegbunam, Okoro, and Okoroafor (2014)
	Polyalthia longifolia	Leaves	Aqueous	*Swiss albino mice, P. berghei ANKA*	53% at 800 mg/kg per day	Nd	Bankole et al. (2016)
	Uvaria chamae	Roots	Ethanol 70%	Swiss albino mice, *P. berghei berghei*	75.9% schizonticide at 900 mg/kg per day	Toxic	Okokon, Ita, and Udokpoh (2006)
Amaranthaceae	*Amaranthus spinosus*	Red barks of the stems	Aqueous	NMRI mice, *P. berghei berghei*	53% at 900 mg/kg per day	Nontoxic	Hilou, Nacoulma, and Guiguemde (2006)
Arecaceae	*Cocos nucifera*	Husk fibers	Ethyl acetate	*Swiss albino mice, P. berghei NK65*	86.3% at 125 mg/kg per day	Nd	Adebayo et al. (2013)
Asteraceae	*Bidens pilosa*	Leaves and twigs	Ethanol 70%	*BALB/C mice, P. berghei ANKA*	74.7% at 400 mg/kg per day	Nontoxic	Laryea and Borquaye (2019)
	Dicoma tomentosa	Whole plant	Methanol and ethanol/water	Swiss mice, *P. berghei* NK173	40%–60% at 100 mg/kg per day	Toxic at 200 mg/kg	Jansen et al. (2012)
	Launaea taraxacifolia	Fresh leaves	Methanol	Swiss mice, *P. berghei*	73.5% at 200 mg/kg per day	Nd	Adetutu, Olorunnisola, Owoade, and Adegbola (2016)

	Tithonia diversifolia	Aerial part	Ethanol 70%	*Swiss albino mice, P. berghei* ANKA	74.97% at 200 mg/kg per day	Toxic at 400 mg/kg	Elufioye and Agbedahunsi (2004)
	Vernonia amygdalina	Fresh leaves	Aqueous	Unidentified mice species, *P. berghei* NK65	80% at 350 mg/kg per day	Nontoxic	Okpe et al. (2016)
Caricaceae	*Carica papaya*	Fresh leaves	Aqueous	Unidentified mice species, *P. berghei* NK65	93% at 350 mg/kg per day	Nontoxic	Okpe et al. (2016)
	Terminalia albida	Stem bark	Aqueous	C57BL 6 mice, *P. berghei ANKA*	100% at 100 mg/kg per day (intraperitoneal)	Nontoxic	Camara et al. (2019)
Combretaceae	*Terminalia macroptera*	Leaves or roots	Ethanol 90%	Swiss albino mice, *P. chabaudi*	37.2% and 46.4% at 100 mg/kg per day for leaves and roots, respectively	Nontoxic	Haidara et al. (2018)
Cucurbitaceae	*Momordica balsamina*	Nd	Methanol	Unidentified mice species, *P. vinckei petteri*	52% at 100 mg/kg per day	Nontoxic	Benoit-Vical et al. (2006)
Euphorbiaceae	*Chrozophora senegalensis*	Stems	Aqueous	Swiss mice, *P. vinckei petteri*	75.2% at 25 mg/kg per day (intraperitoneal)	Nd	Garcia-Alvarez et al. (2013)
Fabaceae	*Bauhinia rufescens*	Leaves	Aqueous	NMRI mice, *P. berghei ANKA*	50% at 100 mg/kg per day	Nd	Bonkian et al. (2018)
	Cassia alata	Leaves	Dichloromethane/methane 1:1	NMRI mice, *P. berghei ANKA*	45.2% at 100 mg/kg per day	Nontoxic	Da et al. (2016)
	Cassia singueana	Roots	Methanol	Wistar rats and Swiss albino mice, *P. berghei*	80% at 200 mg/kg per day (subcutaneous)	Nontoxic	Adzu et al. (2003)
Hippocrateaceae	*Hippocratea africana*	Roots	Ethanol 70%	Swiss albino mice. *P. berghei berghei*	90.9% schizonticidal at 600 mg/kg per day	Toxic	Okokon et al. (2006)
Icacinaceae	*Icacina senegalensis*	Leaves	Methanol	Swiss albino mice, *P. berghei* NK65	80% at 100 mg/kg per day	Toxic	David-Oku, Ifeoma, Christian, and Dick (2014)
		Root bark	Ethanol	Swiss albino mice, *P. berghei* NK65	92% at 200 mg/kg per day	Toxic	Akuodor et al. (2017)

(Continued)

TABLE 8.2 (Continued)

Plant family	Botanical name	Used part	Type of extract	Model used	Inhibition of parasitemia (%)	Toxicity in the model	References
Loganiaceae	*Anthocleista djalonensis*	Stem bark	Ethanol 70%	Swiss albino mice, *P. berghei*	70.5% at 600 mg/kg per day	Nd	Attemene et al. (2018)
Loranthaceae	*Tapinanthus sessilifolius*	Whole plant	Methanol	Unidentified mice species, *P. berghei* ANKA	51.3% at 400 mg/kg per day	Nd	Okpako and Ajaiyeoba (2004)
Malvaceae	*Clappertonia ficifolia*	Leaves	Ethanol 70%	Swiss albino mice, *P. berghei*	62.6% at 400 mg/kg per day	Nontoxic	Laryea and Borquaye (2019)
Meliaceae	*Azadirachta indica*	Whole plant	Aqueous	Swiss albino mice, *P. berghei* NK65	92% at 1240 mg/kg per day	Nd	Alaribe et al. (2021)
		Unripe fruit kernel	Methanol	C57BL/6 mice, *P. berghei* ANKA	30% at 150 mg/kg per day	Nd	Habluetzel et al. (2019)
		Leaves	Ethanol 70%	BALB/C mice, *P. berghei* NK65	68 to 69.3% at 300 mg/kg per day	Nontoxic	Tepongning et al. (2018)
Moraceae	*Artocarpus altilis*	Stem bark	Ethanol 70%	Wistar albino mice, *P. berghei berghei* NK65	55.5% at 200 mg/kg per day	Nontoxic	Adebajo et al. (2014)
	Ficus thonningii	Leaves	Hexane	Swiss albino mice, *P. berghei* NK65	84.5% at 500 mg/kg per day	Nontoxic	Falade et al. (2014)
Ochnaceae	*Lophira alata*	Leaves	Hexane	Swiss albino mice, *P. berghei* NK65	74.4% at 500 mg/kg per day	Nontoxic	Falade et al. (2014)
Papilionaceae	*Erythrina senegalensis*	Bark	Aqueous	Wistar rats and Swiss albino mice, *P. yoelii nigeriensis*	23% at 100 mg/kg per day	Nontoxic	Saidu et al. (2000)
Rhamnaceae	*Ziziphus mauritiana*	Leaves	Ethanol 70%	Swiss albino mice, *P. berghei*	88.9% at 600 mg/kg per day	Nd	Attemene et al. (2018)

Rubiaceae	*Crossopteryx febrifuga*	Stem bark	Ethanol 70%	*unidentified mice type, P. berghei ANKA*	71% at 400 mg/kg per day	Nd	Elufioye and Agbedahunsi (2004)
	Keetia leucantha	Twigs	Dichloromethane or aqueous	Swiss mice, *P. berghei* NK173	56.8% at 53% at 200 mg/kg per day for CH_2Cl_2 and H_2O extracts respectively	Nd	Bero et al. (2009)
Scrophulariaceae	*Striga hermonthica*	Whole plant	Methanol	Unidentified mice type, *P. berghei* ANKA	68.5% at 400 mg/kg per day	Nd	Okpako and Ajaiyeoba (2004)
Simaroubaceae	*Quassia amara*	Stem	Hexane	Albino mice, *P. berghei ANKA*	98% at 100 mg/kg per day (intraperitoneal)	Nontoxic	Ajaiyeoba et al. (1999)
Tiliaceae	*Triumfetta cordifolia*	Leaves	Ethanol 70%	Swiss albino, *P. berghei ANKA*	98% at 400 mg/kg per day	Nontoxic	Ezenyi, Verma, Singh, Okhale, and Adzu (2020)

TABLE 8.3 List of molecules isolated from West African plants.

Plant family	Botanical name	Used part	Type of extract	Isolated metabolite	In vitro			In vivo			References
					Plasmodium strain	IC50*	Selectivity index§	Model used	Inhibition of parasitemia (%)	Toxicity	
Ancistrocladaceae	*Ancistrocladus abbreviatus*	Root bark	Methanol/Water	Ancistrobrevine E	NF54	0.213 μM	604				Fayez et al. (2018)
				5-epi-Ancistrobrevine E	NF54	0.663 μM	225				
				Ancistrobrevine F	NF54	0.928 μM	175				
				5-epi-Ancistrobrevine F	NF54	0.846 μM	138				
				Ancistrobrevine G	NF54	0.134 μM	183				
Annonaceae	*Polyalthia longifolia*	Stem bark	Ethyl acetate	16-Hydroxycleroda-3,13-dién-16,15-olide	K1	16.76 μM	Nd				Gbedema, Bayor, Annan, and Wright (2015)
				Acide 16-oxocléroda-3,13 (14) E-dién-15-oique	K1	9.59 μM	Nd				
				3,16-dihydroxycleroda-4 (18), 13 (14) Z-diène-15,16-olide	K1	18.41 μM	Nd				
				Bisclerodane imide	3D7	4.53 μM	Nd				Annan et al. (2015)
				Cleroda-3-ène	3D7	4.76 μM	Nd				
				Pyrrole-15,16-dione; cleroda-3-ène	3D7	112.14 μM	Nd				
				Pyrrolidine-15,16-dione; cleroda-3,13 (14)-diène-15,16-diamide	3D7	67.12 μM	Nd				
				Cleroda-3-ène-15,16-diamide	3D7	10.17 μM	Nd				

	Xylopia aethiopica	Fruits	Petroleum ether	Xylopic acid				*ICR mice, P. berghei NK65*	99.6% suppression at 100 mg/kg per day	Not available	Boampong et al. (2013)
Apocynaceae	*Picralima nitida*	Fruit peels	Total alkaloids	Akuammicine	D6/W2	0.45/ 0.73 μg/mL	Nd				Okunji, Iwu, Ito, and Smith (2005)
				Akuammine	D6/W2	0.95/ 0.66 μg/mL	Nd				
				Alstonine	D6/W2	0.017/ 0.038 μg/ mL	Nd				
				Picraline	D6/W2	0.44/ 0.53 μg/mL	Nd				
				Picratidine	D6/W2	0.80/ 0.92 μg/mL	Nd				
				Picranitidine	D6/W2	0.04/ 0.03 μg/mL	Nd				
				c-Akuammigine	D6/W2	0.42/ 0.10 μg/mL	Nd				
Asteraceae	*Artemisia gorgonum*	Leaves and flowers	Ethanol	Epimagnolin A	FcB1	5.7 μg/mL	Nd				Ortet et al. (2011)
				Aschantin	FcB1	5.7 μg/mL	Nd				
				Kobusin	FcB1	7.67 μg/mL	Nd				
				Sesamin	FcB1	3.37 μg/mL	Nd				
				Artemetin	FcB1	3.5 μg/mL	Nd				

(*Continued*)

TABLE 8.3 (Continued)

Plant family	Botanical name	Used part	Type of extract	Isolated metabolite	In vitro			In vivo			References
Clusiaceae	*Garcinia kola*	Fresh seeds	Ethyl acetate	Kolaviron				*Swiss albino mice, P. berghei*	92% at 200 mg/kg per day	Nontoxic in murine model	Oluwatosin et al. (2014)
Cochlospermaceae	*Cochlospermum planchonii*	Rhizomes	Dichlorométhane	Cochloxanthine	3D7	6.8 μg/mL	Nd				Lamien-Meda et al. (2015)
				Dihydrocochloxanthine	3D7	6.9 μg/mL	Nd				
	Cochlospermum tinctorium	Rhizomes	Ethanol	3-0-E-P coumaroylalphitolic acid	3D7	2.3 μM	18.7				Ballin et al. (2002)
					Dd2	3.8 μM	11.3				
Combretaceae	*Guiera senegalensis*	Roots	Chloroform	Harman	D6	2.2 μg/mL	Nd				Ancolio et al. (2002)
					W2	1.3 μg/mL	Nd				
				Tetreahydroharmine	D6	3.9 μg/mL	Nd				
					W2	1.4 μg/mL	Nd				
Dioncophylaceae	*Triphyophyllum peltatum*	Roots and stem bark	Dichlorométhane	Dioncophylline A				OF1 mice, *P. berghei* ANKA	99% at Day 4 at 50 mg/kg per day		François et al. (1997)
				Dioncophylline B					47% at Day 4 at 50 mg/kg per day		
				Dioncophylline C					100% at Day 4 at 50 mg/kg per day		
	Nd	Nd	Synthesis	Dioncopeltine A	NF54	0.008 μM	4891				Moyo et al. (2020)
					W2	0.304 μM					

				Habropetaline A	NF54	0.015 μM	2923				
					W2	0.084 μM					
				Dioncophylline C	NF54	0.038 μM	552.6				
					W2	0.112 μM					
Euphorbiaceae	*Alchornea cordifolia*	Leaves	Ethanol	Ellagic acid	FcM29	0.08 μg/mL	77.5				Banzouzi et al. (2002)
					Nigerian	0.14 μg/mL	44.29				
Fabaceae	*Mezoneuron Benthamianum*	Leaves	Ethanol 70% (Precipitate)	Ethyl gallate	3D7	6.2 μg/mL	Nd				Jansen et al. (2017)
				Quercetin	3D7	9.5 μg/mL	Nd				
				13b-OH-pheophorbide a	3D7	5.1 μg/mL	Nd				
Lauraceae	*Persea americana*	Seeds	Methanol	1,2,4-Trihydroxyheptadec-16-ene	D6	1.6 μg/mL	>1.6				Falodun et al. (2014)
					W2	2.1 μg/mL	>1.4				
				1,2,4-Tetrahydroxyheptadecane-6, 16-diene	D6	1.4 μg/mL	>2.1				
					W2	1.4 μg/mL	>1.4				
Meliaceae	*Cedrela odorata*	Leaves	Ethanol	Gedunin	D6	0.039 μg/mL	58.98				MacKinnon et al. (1997)
					W2	0.02 μg/mL	115				
Papaveraceae	*Argemone mexicana*	Leaves	Aqueous	Protopine	K1	0.32 μg/mL	Nd	NMRI mice, *P. berghei*	No reduction of parasitemia	Nd	Simoes-Pires et al. (2014)

(*Continued*)

TABLE 8.3 (Continued)

Plant family	Botanical name	Used part	Type of extract	Isolated metabolite	In vitro			In vivo		References
				Allocryptopine	K1	1.46 μg/mL	Nd	No reduction of parasitemia	Nd	
				Berberine	K1	0.32 μg/mL	Nd	No reduction of parasitemia	Nd	
Periplocaceae	*Cryptolepis sanguinolenta*	Roots	Synthesis	Cryptolepine triflate	K1	0.8 μM	11	43% at Day 5 at 50 mg/kg per day	Nd	Rocha e Silva et al. (2012)
					3D7	0.91 μM	10			
				Cryptolepine analog	K1	0.1 μM	330	55% at Day 5 at 50 mg/kg per day	Nd	
					3D7	0.087 μM	390			
Rutaceae	*Zanthoxylum zanthoxyloides*	Root bark	Dichlorométhane	Bis-dihydrochelerythrinyl ether	3D7	6 μM	Nd			Goodman et al. (2019)
				Buesgenine	3D7	5.7 μM	Nd			
				Chelerythrine	3D7	1.1 μM	Nd			
				γ-Fagarine	3D7	9.6 μM	Nd			
				Skimmianine	3D7	2.7 μM	Nd			
				Pellitorine	3D7	8.8 μM	Nd			

Dioncophylaceae family), all presenting very high selectivity index (range 330–4891). All were tested on both sensitive and resistant *falciparum* strains. Interestingly, *P. berghei* ANKA-infected mice treated with dioncophylline C were totally free of parasites 4 days after infection, while dioconphylline A treatment led to 99% parasite inhibition. In addition, the root extract of *Cryptolepsis sanguinolenta* was evaluated for its clinical efficacy for the treatment of uncomplicated malaria in Ghana and resulted in total parasite elimination in all patients ($n = 44$) by Day 7.

Clinical trials in humans for the evaluation of antimalarial plants and compounds

Clinical assays in humans: clinical trials evaluating antimalarial efficacy with herbal medicine are scarce, but some have been carried out (Table 8.4). Clinical trials differ according to the question raised about tolerability or efficacy, which determines the design of the study, specifying the primary and secondary endpoints, the duration of the follow-up, and so on. Then, the study population may be different, children or adults, and the type of clinical malaria, asymptomatic, uncomplicated, or severe. The number of patients included is also of high importance; when larger samples are used, the confidence level is increased and the margin of error decreased. Then, the experimental tools used to follow primary and secondary endpoints may be different from a lab to another, as the use of PCR to verify the occurrence of reinfection. Molecular approach is indeed necessary to distinguish recrudescent parasites from new infections during the follow-up, and/or can be used to search for submicroscopic parasitemia. Finally, the way to interpret the results is of high importance. Results can be mostly based on parasite densities during the follow-up, or on a clinical assessment proposed by the WHO and defined as "adequate clinical response" (ACR), "early treatment failure" (ETF), or "late treatment failure" (LTF). ACR is defined by the absence of parasitemia on Day 14 irrespective of axillary temperature, or absence of fever irrespective of the presence of parasitemia, without previously meeting any of the criteria of treatment failure. ETF is defined as development of danger signs on Days 1, 2, or 3 in the presence of parasitemia; fever on Day 2 with parasitemia greater than at baseline; or fever on Day 3 with parasitemia. LTF is defined as development of any danger signs or signs of severe malaria in the presence of parasitemia on any day from Days 4 to 14, without previously meeting any of the criteria of ETF; or fever with parasitemia on any day from Days 4 to 14, without previously meeting any of the criteria of ETF. In addition, "fever" is usually defined as history of fever in the previous 24 h or axillary temperature $\geq 37.5°C$.

In general, few herbal preparations have been tested in patients via clinical trials. Table 8.4 shows that five plants of West African origin have been tested by a clinical approach in humans by three different teams. The evaluation of the therapeutic efficacy against malaria of any product raises an obvious ethical problem since malaria is a fatal disease and effective treatments exist, although the need to complement the therapeutic arsenal is a reality. As children are the population most affected by severe malaria, most studies have been done in adults to limit the risks, except for the two studies carried out in Mali with *Argemone mexicana* (Graz et al., 2010; Willcox et al., 2007). Results of efficacy

TABLE 8.4 Main West African plants evaluated alone in human clinical trials for their antimalarial activity during *P. falciparum* uncomplicated malaria.

Plant family	Plant species	Preparation	Controlled intake	Molecular biology	Treatment duration	Control arm	Nb of participants	Age in year	Follow-up duration	Efficacy	References
Asteraceae	*Vernonia amygdalina*	Tea from leaves	No	No	7 days	No	41	Mean: 28.7 (Range: 13–60)	28	ACR/ETF/LTF D14: 67/15/12%; complete parasite clearance by D14: 32%	Challand and Willcox (2009)
Cochlospermaceae	*Cochlospermum planchonii*	Tuberous roots decoction	No	No	6 days	CQ	CQ $n = 21$; Cp $n = 46$	Mean: 23 (Range: 12–45)	6	By D5, 52% had a PD = 0 versus 57% for CQ; PD range: 0–12,000 versus 0–127 for CQ	Benoit-Vical et al. (2003)
N. D.	N. D.	Tea from leaves	No	No	Mean 8.4 days (1–35 days)	No	19	Mean: 17 (Range: 0.9–50)	14	Parasites were not cleared, except for 1 patient by D7. D14: 95% CI for PD = 25–5330/μL	Willcox (1999)
Papaveraceae	*Argemone mexicana*	Decoction with aerial parts	No	Yes for positive cases at D28	A: 3; B:7; C: 8 days	No	A $n = 23$; B $n = 40$; C $n = 17$	Medians A: 2.3; B: 1.9; C: 2.0	28	ACR/ETF/LTF D14: A: 35/26/35%, B: 73/15/13%, C: 65/12/24 ACR/ETF/LTF D28: A: 22/26/39%, B: 63/15/23%, C: 59/12/23%	Willcox et al. (2007)
		Decoction with aerial parts	No	No	7 days	Art-AQ	Am: 199 Art-AQ: 102	Medians Am: 5; Art-AQ: 5	28	ACR D14: Am 65.7%, Art-AQ: 100%	Graz et al. (2010)
Periplocaceae	*Cryptolepsis sanguinolenta*	Tea from roots	No	No	5 days	No	44	Mean: 9.2	28	By Day 4, 50% had a PD = 0; By Day 7, 100% had a PD = 0	Bugyei, Boye, and Andy (2010)

Rubiaceae	*Nauclea pobeguinii*	Capsules of 80% ethanolic quantified extract from NP stem bark	Yes for the first 3 days	No	7 days	No	11	Mean: 25.7 (Range: 20–33)	14	ACR/ETF/LTF D14: 91/0/9%	Mesia et al. (2012a)
		Capsules of 80% ethanolic quantified extract from NP stem bark	Yes for the first 3 days	No	7 days	Art-AQ	Np: 33; Art-AQ: 32	Means: Np: 22.8; Art-AQ: 25.7	14	APCR D14: Np: 87.9, Art-AQ: 96.9%	Mesia et al. (2012)

were interesting for *A. mexicana*, *Vernonia amygdalina*, and *Nauclea pobeguinii* with ACRs ranging from 67% for *V. amygdalina* to 87.9% for *N. pobeguinii*, although well below the rate of artesunate-amodiaquine, the reference association used in two of the studies. However, these studies suffer from design weaknesses. The lack of supervision for treatment administration in most of the studies is a source of bias in the analysis of the results. Follow-up of 14 days (or less) does not comply with WHO recommended protocols and the lack of molecular methods does not allow to distinguish true failures due to parasite recrudescences from failures due to new infections. In addition, WHO guidelines for the treatment of malaria recommend that first-line treatment should be changed if the total failure rate exceeds 10% (WHO, 2009). Finally, none of these studies presented a chemical analysis of the administered preparation. Knowing the chemical composition, especially the amount of the main active metabolites, would give an idea of the composition needed for a certain activity, or provide a basis for comparison when evaluating the efficacy of the same preparation in other contexts.

Surprisingly, the problem of malaria prevention with plant preparations has been little studied so far, except with *Artemisia* which will be discussed in the next paragraph. Yet, preventive use, as opposed to curative use, is less subject to ethical questions and could help reduce malaria transmission.

The case of Artemisia in West Africa

Artemisia annua and *A. afra* are two plant species described for their antimalarial activities. Both plants are not endemic in West Africa. *A. annua* originates from China and grows all over Europe and in Asia. It is also found in Canada and the United States, following its settlement in North America from northern Asia. *A. annua* is well known for its historic use in China to treat or prevent malaria and for the research work of Chinese researchers that led to the identification of artemisinin as a bioactive antimalarial molecule, work that was awarded by the 2015 Nobel Prize in Medicine (Tu, 2011). Semisynthetic artemisinin derivatives (artesunate, artemether, arteether, etc.) are also highly active against *Plasmodium*. Most often used in combination with another antimalarial molecule with a different mechanism of action and half-life, these molecules constitute the last effective bulwark against malaria. The antimalarial activity of the traditional remedy, an infusion prepared with the fresh leaves, has been validated in vitro on both asexual (De Donno et al., 2012; Liu et al., 2010; Rocha e Silva et al., 2012) and sexual forms of *P. falciparum* (Snider & Weathers, 2021). In vitro, the antiparasite activity of the infusion preparation on chloroquine-sensitive and resistant *falciparum* strains, but also on field isolates, is characterized by IC_{50}s values between 0.11 and 1.11 μg/mL, very good values for crude extracts. Concerning its clinical activity in humans, *A. annua* tea in a curative scheme has not shown sufficient efficacy to be recommended (Mueller et al., 2004). But some efficacy in the prevention of malaria has been demonstrated through a clinical trial which reported 40% of more than one clinical malaria attacks during the 9-months follow-up in the control group, compared to 17.9% in the *A. annua* group (Ogwang et al., 2012). *A. afra*, a plant of the same genus, is found naturally in South and East Africa; it is one of the most popular and commonly used herbal medicines in

Southern Africa. This plant is used in decoction or infusion of fresh leaves, to treat a wide range of diseases, including malaria. Interestingly, its antimalarial activity in vitro was confirmed despite the absence of artemisinin, both on asexual and sexual stages of *P. falciparum* (Moyo et al., 2019; Snider & Weathers, 2021). However, IC_{50}s of asexual stages are much lower than those of *A. annua* with IC_{50}s values between 8.9 and 15.3 μg/mL (Kraft et al., 2003).

At this time, *A. annua* tea is at the heart of a political-scientific debate whose objective is to promote or not its use in malaria-endemic areas. As herbal teas derived from traditional medicine to prevent or treat fevers are widely used by local populations in malaria-endemic areas, it is a potentially powerful public health tool. The WHO reports very poor indicators of access to care in malaria-endemic countries, particularly in Africa where only 20% of pregnant women have access to malaria preventive treatment during pregnancy. Traditional and/or complementary medicine may constitute a way to improve access to care. The WHO Strategy for Traditional Medicine 2014–23 highlights that herbal medicines and traditional treatments are the main health care for millions of people around the world. It is also argued that this care is close to the people, easily accessible, and affordable (WHO, 2013). However, the WHO's main warnings regarding the use of this tea are the variable quality of homemade remedies and the promotion of parasite resistance to artemisinin, an extremely serious threat in the fight against malaria (WHO, 2019). However, the multiplicity of antiplasmodial molecules in the leaves of *A. annua* may limit the risk of resistance promotion, as previously shown in a murine model (Elfawal, Towler, Reich, Weathers, & Rich, 2015). Although additional scientific data are needed to confirm or not the relevance and the absence of risk regarding artemisinin resistance of using such a tea, a small group of people have been very active in promoting the use of *A. annua* teas since 2017, by, for example, writing a note for the WHO (Weathers, Cornet-Vernet, Hassanali, & Schul, 2017), producing the film "malaria business" promoting the benefits of the tea for eradicating malaria and accusing the pharmaceutical industry of hiding information, and creating an association to promote the production, sale, and use of this plant against malaria ("La Maison de l'Artemisia"). This association already has 91 sites in 24 countries, including 39 sites in West African countries. In France, a group of scientists and clinical doctors expressed their concern about the use of *A. annua* tea in various media and reported cases of imported severe malaria following the use of the tea for prophylaxis during a stay in sub-Saharan Africa (Argemi et al., 2019; Lagarce et al., 2016). The French journal "Prescrire," intended for doctors, also published a warning in 2021 about the consumption of this plant, which is associated with a prolongation of the QT interval of the electrocardiogram (Anonymous, 2021).

Meanwhile, the use of *A. annua* teas seems to have become very popular in West Africa, although, to date, no ethnopharmacological studies have established such facts. Since March 2020, the Covid-19 pandemic seems to have increased the popularity of antimalarial treatments, including *A. annua*-based teas, partly due to the controversy surrounding chloroquine. In Madagascar in particular, the use of a preparation containing several plants, the Covid Organics, including *A. annua*, is recommended by the government as a preventive treatment for COVID-19. Again, ethnopharmacological studies are needed to confirm and understand this increasing use of *A. annua* in West Africa.

Conclusion

The use of plants against malaria remains an important way of treatment for local populations in West Africa. Ethnopharmacological studies depicting local uses of specific plants have enabled the identification of active plant species against malaria and of isolated compounds that still need research and development to be used as a medicine. Numerous West African plants and compounds are thus promising candidates for drug development, but this requires a significant investment of time and money. So far, only quinine and artemisinin are natural molecules from plants that have led to antimalarial drugs. Another antimalarial drug, atovaquone, is also a molecule derived from a plant compound. The use of whole plants or parts of plants as in traditional use can also be of significant interest in Public Health, due to its accessibility to the most vulnerable populations. Scientific validation of the preventive or curative activity and safety of these products remains an essential step.

References

Abayomi, S. (2010). *Plantes medicinales et medecine traditionnelle d'Afrique*. Academie Suisse des Sciences Naturelles.

Adebajo, A., Odediran, S., Aliyu, F., Nwafor, P., Nwoko, N., & Umana, U. (2014). In vivo antiplasmodial potentials of the combinations of four Nigerian antimalarial plants. *Molecules (Basel, Switzerland), 19*(9), 13136–13146. Available from https://doi.org/10.3390/molecules190913136.

Adebayo, J. O., Balogun, E. A., Malomo, S. O., Soladoye, A. O., Olatunji, L. A., Kolawole, O. M., … Krettli, A. U. (2013). Antimalarial activity of *Cocos nucifera* husk fibre: Further studies. *Evidence-Based Complementary and Alternative Medicine, 2013*, 742476. Available from https://doi.org/10.1155/2013/742476.

Adetutu, A., Olorunnisola, O. S., Owoade, A. O., & Adegbola, P. (2016). Inhibition of in vivo growth of *Plasmodium berghei* by *Launaea taraxacifolia* and *Amaranthus viridis* in Mice. *Malaria Research and Treatment, 2016*, 9248024. Available from https://doi.org/10.1155/2016/9248024.

Adukpo, S., Elewosi, D., Asmah, R. H., Nyarko, A. K., Ekpe, P. K., Edoh, D. A., & Ofori, M. F. (2020). Antiplasmodial and genotoxic study of selected ghanaian medicinal plants. *Evidence-Based Complementary and Alternative Medicine, 2020*, 1582724. Available from https://doi.org/10.1155/2020/1582724.

Adzu, B., Abbah, J., Vongtau, H., et, K., & Gamaniel. (2003). Studies on the use of *Cassia singueana* in malaria ethnopharmacy. *Journal of Ethnopharmacology, 88*(2), 257. Available from https://doi.org/10.1016/S0378-8741.

Ajaiyeoba, E. O., Abalogu, U., Krebs, H., & Oduola, A. M. (1999). In vivo antimalarial activities of *Quassia amara* and *Quassia undulata* plant extracts in mice. *Journal of Ethnopharmacology, 67*(3), 73. Available from https://doi.org/10.1016/S0378-8741(99)00073-2.

Akuodor, G. C., Essien, D.-O., Nkorroh, J. A., Essien, A. D., Nkanor, E. E., Ezeunala, M. N., & Chilaka, K. C. (2017). Antiplasmodial activity of the ethanolic root bark extract of *Icacina senegalensis* in mice infected by *Plasmodium berghei*. *Journal of Basic and Clinical Physiology and Pharmacology, 28*(2). Available from https://doi.org/10.1515/jbcpp-2016-0109.

Alaribe, S. C., Oladipupo, A. R., Uche, G. C., Onumba, M. U., Ota, D., Awodele, O., & Oyibo, W. A. (2021). Suppressive, curative, and prophylactic potentials of an antimalarial polyherbal mixture and its individual components in *Plasmodium berghei*-infected mice. *Journal of Ethnopharmacology, 277*, 114105. Available from https://doi.org/10.1016/j.jep.2021.114105.

Aldulaimi, O., Uche, F. I., Hameed, H., Mbye, H., Ullah, I., Drijfhout, F., … Li, W.-W. (2017). A characterization of the antimalarial activity of the bark of *Cylicodiscus gabunensis* Harms. *Journal of Ethnopharmacology, 198*, 221–225. Available from https://doi.org/10.1016/j.jep.2017.01.014.

Amlabu, W. E., Nock, I. H., Kaushik, N. K., Mohanakrishnan, D., Tiwary, J., Audu, P. A., … Sahal, D. (2018). Exploration of antiplasmodial activity in *Acalypha wilkesiana* Müller Argoviensis, 1866 (family: Euphorbiaceae) and its GC-MS fingerprint. *Parasitology Research, 117*(5), 1473–1484. Available from https://doi.org/10.1007/s00436-018-5802-1.

Ancolio, C., Azas, N., Mahiou, V., Ollivier, E., Di Giorgio, C., Keita, A., ... Balansard, G. (2002). Antimalarial activity of extracts and alkaloids isolated from six plants used in traditional medicine in Mali and Sao Tome. *Phytotherapy Research, 16*(7), 646–649. Available from https://doi.org/10.1002/ptr.1025.

Annan, K., Ekuadzi, E., Asare, C., Sarpong, K., Pistorius, D., Oberer, L., ... Ofori, M. (2015). Antiplasmodial constituents from the stem bark of *Polyalthia longifolia* var pendula. *Phytochemistry Letters, 11*, 28–31. Available from https://doi.org/10.1016/j.phytol.2014.10.028.

Anonymous. (2021). Extraits de la plante Artemisia annua: Allongements de l'intervalle QT de l'électrocardiogramme. *Prescrire, 41*, 114–115. Available from https://www.prescrire.org/fr/3/31/60637/0/NewsDetails.aspx.

Anosa, G. N., Udegbunam, R. I., Okoro, J. O., & Okoroafor, O. N. (2014). In vivo antimalarial activities of *Enantia polycarpa* stem bark against *Plasmodium berghei* berghei in mice. *Journal of Ethnopharmacology, 153*(2), 531–534. Available from https://doi.org/10.1016/j.jep.2014.02.022.

Argemi, X., Houze, S., Noel, H., Broca, O., Chidiac, C., & Rapp, C. (2019). Imported *Plasmodium falciparum* malaria following non-pharmaceutical forms of artemisia annua prophylaxis. *Journal of Travel Medicine, 26*, taz073. Available from https://doi.org/10.1093/jtm/taz073.

Attemene, S. D. D., Beourou, S., Tuo, K., Gnondjui, A. A., Konate, A., Toure, A. O., ... Djaman, J. A. (2018). Antiplasmodial activity of two medicinal plants against clinical isolates of *Plasmodium falciparum* and *Plasmodium berghei* infected mice. *Journal of Parasitic Diseases, 42*(1), 68–76. Available from https://doi.org/10.1007/s12639-017-0966-7.

Austarheim, I., Pham, A. T., Nguyen, C., Zou, Y.-F., Diallo, D., Malterud, K. E., & Wangensteen, H. (2016). Antiplasmodial, anti-complement and anti-inflammatory in vitro effects of *Biophytum umbraculum* Welw. traditionally used against cerebral malaria in Mali. *Journal of Ethnopharmacology, 190*, 159–164. Available from https://doi.org/10.1016/j.jep.2016.05.058.

Ballin, N. Z., Traore, M., Tinto, H., Sittie, A., Mølgaard, P., Olsen, C. E., ... Christensen, S. B. (2002). Antiplasmodial compounds from *Cochlospermum tinctorium*. *Journal of Natural Products, 65*(9), 1325–1327. Available from https://doi.org/10.1021/np020008h.

Bankole, A. E., Adekunle, A. A., Sowemimo, A. A., Umebese, C. E., Abiodun, O., & Gbotosho, G. O. (2016). Phytochemical screening and in vivo antimalarial activity of extracts from three medicinal plants used in malaria treatment in Nigeria. *Parasitology Research, 115*(1), 299–305. Available from https://doi.org/10.1007/s00436-015-4747-x.

Banzouzi, J.-T., Prado, M., Valentin, C., Roumestan., Mallie, Y., Pelissier., & Blache. (2002). In vitro antiplasmodial activity of extracts of *Alchornea cordifolia* and identification of an active constituent: Ellagic acid. *Journal of Ethnopharmacology, 81*(3), 121–126. Available from https://doi.org/10.1016/s0378-8741(02)00121-6.

Bello, O. M., Zaki, A. A., Khan, S. I., Fasinu, P. S., Ali, Z., Khan, I. A., ... Oguntoye, O. S. (2017). Assessment of selected medicinal plants indigenous to West Africa for antiprotozoal activity. *South African Journal of Botany, 113*, 200–211. Available from https://doi.org/10.1016/j.sajb.2017.08.002.

Benoit, F., Valentin, A., Pelissier, Y., Diafouka, F., Marion, C., Kone-Bamba, D., ... Bastide, J.-M. (1996). In vitro antimalarial activity of vegetal extracts used in West African traditional medicine. *The American Journal of Tropical Medicine and Hygiene, 54*(1), 67–71. Available from https://doi.org/10.4269/ajtmh.1996.54.67.

Benoit-Vical, F., Valentin, A., Cournac, V., Pélissier, Y., Mallié, M., & Bastide, J.-M. (1998). *In vitro antiplasmodial activity of stem and root extracts of* Nauclea latifolia *S.M. (Rubiaceae). Journal of Ethnopharmacology, 61*(3), 173–178. Available from https://doi.org/10.1016/S0378-8741(98)00036-1.

Benoit-Vical, F., Valentin, A., Da, B., Dakuyo, Z., Descamps, L., & Mallié, M. (2003). N'Dribala (*Cochlospermum planchonii*) versus chloroquine for treatment of uncomplicated *Plasmodium falciparum* malaria. *Journal of Ethnopharmacology, 89*(1), 111–114. Available from https://doi.org/10.1016/S0378-8741(03)00277-0.

Benoit-Vical, F., Grellier, P., Abdoulaye, A., Moussa, I., Ousmane, A., Berry, A., ... Poupat, C. (2006). In vitro and in vivo antiplasmodial activity of *Momordica balsamina* alone or in a traditional mixture. *Chemotherapy, 52*(6), 288–292. Available from https://doi.org/10.1159/000095960.

Bero, J., Ganfon, H., Jonville, M.-C., Frédérich, M., Gbaguidi, F., DeMol, P., ... Quetin-Leclercq, J. (2009). In vitro antiplasmodial activity of plants used in Benin in traditional medicine to treat malaria. *Journal of Ethnopharmacology, 122*(3), 439–444. Available from https://doi.org/10.1016/j.jep.2009.02.004.

Boampong, J. N., Ameyaw, E. O., Aboagye, B., Asare, K., Kyei, S., Donfack, J. H., & Woode, E. (2013). The curative and prophylactic effects of xylopic acid on *Plasmodium berghei* infection in mice. *Journal of Parasitology Research, 2013*, 356107. Available from https://doi.org/10.1155/2013/356107.

Bodeker, G., & Kronenberg, F. (2002). A public health agenda for traditional, complementary, and alternative medicine. *American Journal of Public Health, 92*, 1582–1591. Available from https://doi.org/10.2105/AJPH.92.10.1582.

Bonkian, L. N., Yerbanga, R. S., Koama, B., Soma, A., Cisse, M., Valea, I., ... Traore/Coulibaly, M. (2018). In vivo antiplasmodial activity of two Sahelian plant extracts on *Plasmodium berghei* ANKA infected NMRI mice. *Evidence-Based Complementary and Alternative Medicine, 2018*, 6859632. Available from https://doi.org/10.1155/2018/6859632.

Bugyei, K., Boye, G., & Andy, M. (2010). Clinical efficacy of a tea-bag formulation of *Cryptolepis sanguinolenta* root in the treatment of acute uncomplicated *Falciparum malaria*. *Ghana Medical Journal, 44*(1), 3–9. Available from http://10.4314/gmj.v44i1.68849.

Camara, A., Haddad, M., Reybier, K., Traoré, M. S., Baldé, M. A., Royo, J., ... Aubouy, A. (2019). Terminalia albida treatment improves survival in experimental cerebral malaria through reactive oxygen species scavenging and anti-inflammatory properties. *Malaria Journal, 18*, 431. Available from https://doi.org/10.1186/s12936-019-3071-9.

Challand, S., & Willcox, M. (2009). A clinical trial of the traditional medicine *Vernonia amygdalina* in the treatment of uncomplicated malaria. *Journal of Alternative and Complementary Medicine, 15*(11), 1231–1237. Available from https://doi.org/10.1089/acm.2009.0098.

Chassagne, F., Haddad, M., Amiel, A., Phakeovilay, C., Manithip, C., Bourdy, G., ... Marti, G. (2018). A metabolomic approach to identify anti-hepatocarcinogenic compounds from plants used traditionally in the treatment of liver diseases. *Fitoterapia, 127*, 226–236. Available from https://doi.org/10.1016/j.fitote.2018.02.021.

Da, O., Yerbanga, R. S., Traore/Coulibaly, M., Koama, B. K., Kabre, Z., Tamboura, S., ... Ouedraogo, G. A. (2016). Evaluation of the antiplasmodial activity and lethality of the leaf extract of *Cassia alata* L. (Fabaceae). *Pakistan Journal of Biological Sciences, 19*(4), 171–178. Available from https://doi.org/10.3923/pjbs.2016.171.178.

David-Oku, E., Ifeoma, O.-O. J., Christian, A. G., & Dick, E. A. (2014). Evaluation of the antimalarial potential of Icacina senegalensis Juss (Icacinaceae). *Asian Pacific Journal of Tropical Medicine, 7*(1), S469–S472. Available from https://doi.org/10.1016/S1995-7645(14)60276-5.

De Donno, A., Grassi, T., Idolo, A., Guido, M., Papadia, P., Caccioppola, A., ... Fanizzi, F. P. (2012). First-time comparison of the in vitro antimalarial activity of *Artemisia annua* herbal tea and artemisinin. *Transactions of the Royal Society of Tropical Medicine and Hygiene, 106*, 696–700. Available from https://doi.org/10.1016/j.trstmh.2012.07.008.

Elfawal, M. A., Towler, M. J., Reich, N. G., Weathers, P. J., & Rich, S. M. (2015). Dried whole-plant *Artemisia annua* slows evolution of malaria drug resistance and overcomes resistance to artemisinin. *Proceedings of the National Academy of Sciences of the United States of America, 112*, 821–826. Available from https://doi.org/10.1073/pnas.1413127112.

Elufioye, T. O., & Agbedahunsi, J. M. (2004). Antimalarial activities of *Tithonia diversifolia* (Asteraceae) and *Crossopteryx febrifuga* (Rubiaceae) on mice in vivo. *Journal of Ethnopharmacology, 93*(2–3), 167–171. Available from https://doi.org/10.1016/j.jep.2004.01.009.

Ezenyi, I. C., Verma, V., Singh, S., Okhale, S. E., & Adzu, B. (2020). Ethnopharmacology-aided antiplasmodial evaluation of six selected plants used for malaria treatment in Nigeria. *Journal of Ethnopharmacology, 254*, 112694. Available from https://doi.org/10.1016/j.jep.2020.112694.

Falade, M. O., Akinboye, D. O., Gbotosho, G. O., Ajaiyeoba, E. O., Happi, T. C., Abiodun, O. O., & Oduola, A. M. J. (2014). In vitro and in vivo antimalarial activity of *Ficus thonningii* Blume (Moraceae) and *Lophira alata* Banks (Ochnaceae), identified from the ethnomedicine of the Nigerian Middle Belt. *Journal of Parasitology Research, 2014*, 972853. Available from https://doi.org/10.1155/2014/972853.

Falodun, A., Imieje, V., Erharuyi, O., Ahomafor, J., Jacob, M., Khan, S., & Hamann, M. (2014). Evaluation of three medicinal plant extracts against *Plasmodium falciparum* and selected microganisms. *African Journal of Traditional, Complementary and Alternative Medicines, 11*(4), 142–146. Available from http://10.4314/ajtcam.v11i4.22.

Fayez, S., Feineis, D., Aké Assi, L., Kaiser, M., Brun, R., Awale, S., & Bringmann, G. (2018). Ancistrobrevines E-J and related naphthylisoquinoline alkaloids from the West African liana *Ancistrocladus abbreviatus* with inhibitory activities against *Plasmodium falciparum* and PANC-1 human pancreatic cancer cells. *Fitoterapia, 131*, 245–259. Available from https://doi.org/10.1016/j.fitote.2018.11.006.

François, G., Timperman, G., Eling, W., Assi, L. A., Holenz, J., & Bringmann, G. (1997). Naphthylisoquinoline alkaloids against malaria: Evaluation of the curative potentials of dioncophylline C and dioncopeltine A against *Plasmodium berghei* in vivo. *Antimicrobial Agents and Chemotherapy, 41*(11), 2533–2539. Available from https://doi.org/10.1128/AAC.41.11.2533.

Garcia-Alvarez, M. C., Moussa, I., Njomnang Soh, P., Nongonierma, R., Abdoulaye, A., Nicolau-Travers, M. L., … Benoit-Vical, F. (2013). Both plants *Sebastiania chamaelea* from Niger and *Chrozophora senegalensis* from Senegal used in African traditional medicine in malaria treatment share a same active principle. *Journal of Ethnopharmacology, 149*(3), 676–684. Available from https://doi.org/10.1016/j.jep.2013.07.024.

Gbedema, S. Y., Bayor, M. T., Annan, K., & Wright, C. W. (2015). Clerodane diterpenes from *Polyalthia longifolia* (Sonn) Thw. var. pendula: Potential antimalarial agents for drug resistant *Plasmodium falciparum* infection. *Journal of Ethnopharmacology, 169*, 176–182. Available from https://doi.org/10.1016/j.jep.2015.04.014.

Goodman, C. D., Hoang, A. T., Diallo, D., Malterud, K. E., McFadden, G. I., & Wangensteen, H. (2019). Anti-plasmodial effects of *Zanthoxylum zanthoxyloides*. *Planta Medica, 85*(13), 1073–1079. Available from https://doi.org/10.1055/a-0973-0067.

Graz, B., Willcox, M. L., Diakite, C., Falquet, J., Dackuo, F., Sidibe, O., … Diallo, D. (2010). *Argemone mexicana* decoction versus artesunate-amodiaquine for the management of malaria in Mali: Policy and public-health implications. *Transactions of the Royal Society of Tropical Medicine and Hygiene, 104*(1), 33–41. Available from https://doi.org/10.1016/j.trstmh.2009.07.005.

Gruca, M., Yu, W., Amoateng, P., Nielsen, M. A., Poulsen, T. B., & Balslev, H. (2015). Ethnomedicinal survey and in vitro anti-plasmodial activity of the palm *Borassus aethiopum* Mart. *Journal of Ethnopharmacology, 175*, 356–369. Available from https://doi.org/10.1016/j.jep.2015.09.010.

Habluetzel, A., Pinto, B., Tapanelli, S., Nkouangang, J., Saviozzi, M., Chianese, G., … Bruschi, F. (2019). Effects of *Azadirachta indica* seed kernel extracts on early erythrocytic schizogony of *Plasmodium berghei* and pro-inflammatory response in inbred mice. *Malaria Journal, 18*(1), 35. Available from https://doi.org/10.1186/s12936-019-2671-8.

Haidara, M., Haddad, M., Denou, A., Marti, G., Bourgeade-Delmas, S., Sanogo, R., … Aubouy, A. (2018). In vivo validation of anti-malarial activity of crude extracts of *Terminalia macroptera*, a Malian medicinal plant. *Malaria Journal, 17*(1), 68. Available from https://doi.org/10.1186/s12936-018-2223-7.

Hilou, A., Nacoulma, O. G., & Guiguemde, T. R. (2006). In vivo antimalarial activities of extracts from *Amaranthus spinosus* L. and *Boerhaavia erecta* L. in mice. *Journal of Ethnopharmacology, 103*(2), 236–240. Available from https://doi.org/10.1016/j.jep.2005.08.006.

Jansen, O., Tits, M., Angenot, L., Nicolas, J.-P., De Mol, P., Nikiema, J.-B., & Frédérich, M. (2012). Anti-plasmodial activity of *Dicoma tomentosa* (Asteraceae) and identification of urospermal A-15-O-acetate as the main active compound. *Malaria Journal, 11*, 289. Available from https://doi.org/10.1186/1475-2875-11-289.

Jansen, O., Tchinda, A. T., Loua, J., Esters, V., Cieckiewicz, E., Ledoux, A., … Frédérich, M. (2017). Antiplasmodial activity of *Mezoneuron benthamianum* leaves and identification of its active constituents. *Journal of Ethnopharmacology, 203*, 20–26. Available from https://doi.org/10.1016/j.jep.2017.03.021.

Kamanzi Atindehou, K., Schmid, C., Brun, R., Koné, M. W., & Traore, D. (2004). Antitrypanosomal and antiplasmodial activity of medicinal plants from Côte d'Ivoire. *Journal of Ethnopharmacology, 90*(2–3), 221–227. Available from https://doi.org/10.1016/j.jep.2003.09.032.

Karou, D., Dicko, M. H., Sanon, S., Simpore, J., & Traore, A. S. (2003). Antimalarial activity of *Sida acuta* Burm. f. (Malvaceae) and *Pterocarpus erinaceus* Poir. (Fabaceae). *Journal of Ethnopharmacology, 89*(2–3), 291–294. Available from https://doi.org/10.1016/j.jep.2003.09.010.

Kayser, O., & Kiderlen, A. F. (2003). Nicht-viraler gentransfer und gentherapie. *Deutsche Apotheker Zeitung, 143*, 55–62.

Koffi, J. A., Silué, K. D., Tano, D. K., Dable, T. M., & Yavo, W. (2020). Evaluation of antiplasmodial activity of extracts from endemic medicinal plants used to treat malaria in Côte d'Ivoire. *BioImpacts, 10*(3), 151–157. Available from https://doi.org/10.34172/bi.2020.19.

Komlaga, G., Cojean, S., Dickson, R. A., Beniddir, M. A., Suyyagh-Albouz, S., Mensah, M. L. K., … Loiseau, P. M. (2016). Antiplasmodial activity of selected medicinal plants used to treat malaria in Ghana. *Parasitology Research, 115*(8), 3185–3195. Available from https://doi.org/10.1007/s00436-016-5080-8.

Kraft, C., Jenett-Siems, K., Siems, K., Jakupovic, J., Mavi, S., Bienzle, U., & Eich, E. (2003). In vitro antiplasmodial evaluation of medicinal plants from Zimbabwe. *Phytotherapy Research, 17*, 123–128. Available from https://doi.org/10.1002/ptr.1066.

Kumatia, E. K., Ayertey, F., Appiah-Opong, R., Bagyour, G. K., Asare, K. O., Mbatcho, V. C., & Dabo, J. (2021). Intervention of standardized ethanol leaf extract of Annickia polycarpa, (DC.) Setten and Maas ex I.M. Turner. (Annonaceae), in *Plasmodium berghei* infested mice produced anti-malaria action and normalized gross hematological indices. *Journal of Ethnopharmacology, 267*, 113449. Available from https://doi.org/10.1016/j.jep.2020.113449.

Kumatia, E. K., Ayertey, F., Appiah-Opong, R., Bolah, P., Ehun, E., & Dabo, J. (2021). *Antrocaryon micraster* (A. Chev. And Guillaumin) stem bark extract demonstrated anti-malaria action and normalized hematological indices in *Plasmodium berghei* infested mice in the Rane's test. *Journal of Ethnopharmacology, 266*, 113427. Available from https://doi.org/10.1016/j.jep.2020.113427.

Kwansa-Bentum, B., Agyeman, K., Larbi-Akor, J., Anyigba, C., & Appiah-Opong, R. (2019). In vitro assessment of antiplasmodial activity and cytotoxicity of *Polyalthia longifolia* leaf extracts on *Plasmodium falciparum* Strain NF54. *Malaria Research and Treatment, 2019*, 6976298. Available from https://doi.org/10.1155/2019/6976298.

Lagarce, L., Lerolle, N., Asfar, P., Le Govic, Y., Lainé-Cessac, P., & de Gentile, L. (2016). A non-pharmaceutical form of *Artemisia annua* is not effective in preventing *Plasmodium falciparum* malaria. *Journal of Travel Medicine, 23*, taw049. Available from https://doi.org/10.1093/jtm/taw049.

Lamien-Meda, A., Kiendrebeogo, M., Compaoré, M., Meda, R. N. T., Bacher, M., Koenig, K., ... Novak, J. (2015). Quality assessment and antiplasmodial activity of West African Cochlospermum species. *Phytochemistry, 119*, 51–61. Available from https://doi.org/10.1016/j.phytochem.2015.09.006.

Laryea, M. K., & Borquaye, L. S. (2019). Antimalarial efficacy and toxicological assessment of extracts of some Ghanaian medicinal plants. *Journal of Parasitology Research, 2019*, 1630405. Available from https://doi.org/10.1155/2019/1630405.

Lekana-Douki, J. B., Oyegue Liabagui, S. L., Bongui, J. B., Zatra, R., Lebibi, J., & Toure-Ndouo, F. S. (2011). In vitro antiplasmodial activity of crude extracts of *Tetrapleura tetraptera* and *Copaifera religiosa*. *BMC Research Notes, 4*, 506. Available from https://doi.org/10.1186/1756-0500-4-506.

Li, C., Seixas, E., & Langhorne, J. (2001). Rodent malaria: The mouse as a model for understanding immune responses and pathology induced by the erythrocytic stages of the parasite. *Medical Microbiology and Immunology, 189*, 115–126. Available from https://doi.org/10.1007/s430-001-8017-8.

Liu, N. Q., Cao, M., Frédérich, M., Choi, Y. H., Verpoorte, R., & van der Kooy, F. (2010). Metabolomic investigation of the ethnopharmacological use of *Artemisia afra* with NMR spectroscopy and multivariate data analysis. *Journal of Ethnopharmacology, 128*, 230–235. Available from https://doi.org/10.1016/j.jep.2010.01.020.

MacKinnon, S., Durst, T., Arnason, J. T., Angerhofer, C., Pezzuto, J., Sanchez-Vindas, P. E., ... Gbeassor, M. (1997). Antimalarial activity of tropical meliaceae extracts and gedunin derivatives. *Journal of Natural Products, 60*(4), 336–341. Available from https://doi.org/10.1021/np9605394.

Ménan, H., Banzouzi, J.-T., Hocquette, A., Pélissier, Y., Blache, Y., Koné, M., ... Valentin, A. (2006). Antiplasmodial activity and cytotoxicity of plants used in West African traditional medicine for the treatment of malaria. *Journal of Ethnopharmacology, 105*(1–2), 131–136. Available from https://doi.org/10.1016/j.jep.2005.10.027.

Mesia, K., Tona, L., Mampunza, M. M., Ntamabyaliro, N., Muanda, T., Muyembe, T., ... Vlietinck, A. J. (2012). Antimalarial efficacy of a quantified extract of *Nauclea pobeguinii* stem bark in human adult volunteers with diagnosed uncomplicated falciparum malaria. Part 2: A clinical phase IIB trial. *Planta Medica, 78*(9), 853–860. Available from https://doi.org/10.1055/s-0031-1298488.

Mesia, K., Tona, L., Mampunza, M. M., Ntamabyaliro, N., Muanda, T., Muyembe, T., ... Vlietinck, A. J. (2012a). Antimalarial efficacy of a quantified extract of *Nauclea pobeguinii* stem bark in human adult volunteers with diagnosed uncomplicated falciparum malaria. Part 1: a clinical phase IIA trial.Planta. *Medica, 78*(3), 211–218. Available from https://doi.org/10.1055/s-0031-1280359.

Moyo, P., Kunyane, P., Selepe, M. A., Eloff, J. N., Niemand, J., Louw, A. I., ... Birkholtz, L. M. (2019). Bioassay-guided isolation and identification of gametocytocidal compounds from *Artemisia afra* (Asteraceae). *Malaria Journal, 18*, 65. Available from https://doi.org/10.1186/s12936-019-2694-1.

Moyo, P., Shamburger, W., van der Watt, M. E., Reader, J., de Sousa, A. C. C., Egan, T. J., ... Birkholtz, L.-M. (2020). Naphthylisoquinoline alkaloids, validated as hit multistage antiplasmodial natural products. *International Journal for Parasitology: Drugs and Drug Resistance, 13*, 51–58. Available from https://doi.org/10.1016/j.ijpddr.2020.05.003.

Mueller, M. S., Runyambo, N., Wagner, I., Borrmann, S., Dietz, K., & Heide, L. (2004). Randomized controlled trial of a traditional preparation of *Artemisia annua* L. (Annual Wormwood) in the treatment of malaria. *Transactions of the Royal Society of Tropical Medicine and Hygiene, 98*, 318–321. Available from https://doi.org/10.1016/j.trstmh.2003.09.001.

Mustofa., Valentin, A., Benoit-Vical, F., Pélissier, Y., Koné-Bamba, D., & Mallié, M. (2000). Antiplasmodial activity of plant extracts used in West African traditional medicine. *Journal of Ethnopharmacology, 73*(1), 296–298. Available from https://doi.org/10.1016/s0378-8741(00)00296-8.

Nea, F., Bitchi, M. B., Genva, M., Ledoux, A., Tchinda, A. T., Damblon, C., … Fauconnier, M.-L. (2021). Phytochemical investigation and biological activities of *Lantana rhodesiensis*. *Molecules (Basel, Switzerland), 26*(4), 846. Available from https://doi.org/10.3390/molecules26040846.

Oguakwa, J. U. (1980). Plants used in traditional medicine in West Africa. *Journal of Ethnopharmacology, 2*, 29–31. Available from https://doi.org/10.1016/0378-8741(80)90025-2.

Ogwang, P. E., Ogwal, J. O., Kasasa, S., Olila, D., Ejobi, F., Kabasa, D., & Obua, C. (2012). *Artemisia annua* L. infusion consumed once a week reduces risk of multiple episodes of malaria: A randomised trial in a Ugandan community. *Tropical Journal of Pharmaceutical Research, 11*, 445–453. Available from https://doi.org/10.4314/tjpr.v11i3.14.

Okokon, J. E., Ita, B. N., & Udokpoh, A. E. (2006). The in-vivo antimalarial activities of *Uvaria chamae* and *Hippocratea africana*. *Annals of Tropical Medicine and Parasitology, 100*(7), 585–590. Available from https://doi.org/10.1179/136485906X118512.

Okpako, L., & Ajaiyeoba, E. (2004). In vitro and in vivo antimalarial studies of *Striga hermonthica* and *Tapinanthus sessilifolius* extracts. *African Journal of Medicine and Medical Sciences, 33*(1), 73–75. Available from http://europepmc.org/abstract/MED/15490799.

Okpe, O., Habila, N., Ikwebe, J., Upev, V. A., Okoduwa, S. I. R., & Isaac, O. T. (2016). Antimalarial potential of *Carica papaya* and *Vernonia amygdalina* in mice infected with *Plasmodium berghei*. *Journal of Tropical Medicine, 2016*, 8738972. Available from https://doi.org/10.1155/2016/8738972.

Okpekon, T., Yolou, S., Gleye, C., Roblot, F., Loiseau, P., Bories, C., … Hocquemiller, R. (2004). Antiparasitic activities of medicinal plants used in Ivory Coast. *Journal of Ethnopharmacology, 90*(1), 91–97. Available from https://doi.org/10.1016/j.jep.2003.09.029.

Okunji, C. O., Iwu, M. M., Ito, Y., & Smith, P. L. (2005). Preparative separation of indole alkaloids from the rind of *Picralima nitida* (Stapf) T. Durand & H. Durand by pH-zone-refining countercurrent chromatography. *Journal of Liquid Chromatography and Related Technologies, 28*(5), 775–783. Available from https://doi.org/10.1081/JLC-200048915.

Oluwatosin, A., Tolulope, A., Ayokulehin, K., Patricia, O., Aderemi, K., Catherine, F., & Olusegun, A. (2014). Antimalarial potential of kolaviron, a biflavonoid from *Garcinia kola* seeds, against *Plasmodium berghei* infection in Swiss albino mice. *Asian Pacific Journal of Tropical Medicine, 7*(2), 97–104. Available from https://doi.org/10.1016/S1995-7645(14)60003-1.

Ortet, R., Prado, S., Regalado, E. L., Valeriote, F. A., Media, J., Mendiola, J., & Thomas, O. P. (2011). *Furfuran lignans* and a flavone from *Artemisia gorgonum* Webb and their in vitro activity against *Plasmodium falciparum*. *Journal of Ethnopharmacology, 138*(2), 637–640. Available from https://doi.org/10.1016/j.jep.2011.09.039.

Ouattara, L. P., Sanon, S., Mahiou-Leddet, V., Gansané, A., Baghdikian, B., Traoré, A., … Sirima, S. B. (2014). In vitro antiplasmodial activity of some medicinal plants of Burkina Faso. *Parasitology Research, 113*(1), 405–416. Available from https://doi.org/10.1007/s00436-013-3669-8.

Rocha e Silva, L. F., de Magalhães, P. M., Costa, M. R. F., das, M., Alecrim, G. C., Chaves, F. C. M., … Vieira, P. P. R. (2012). In vitro susceptibility of *Plasmodium falciparum* Welch field isolates to infusions prepared from *Artemisia annua* L. cultivated in the Brazilian Amazon. *Memórias Do Instituto Oswaldo Cruz, 107*, 859–866. Available from https://doi.org/10.1590/s0074-02762012000700004.

Rocha e Silva, L. F., Montoia, A., Amorim, R. C. N., Melo, M. R., Henrique, M. C., Nunomura, S. M., … Pohlit, A. M. (2012). Comparative in vitro and in vivo antimalarial activity of the indole alkaloids ellipticine, olivacine, cryptolepine and a synthetic cryptolepine analog. *Phytomedicine: International Journal of Phytotherapy and Phytopharmacology, 20*(1), 71–76. Available from https://doi.org/10.1016/j.phymed.2012.09.008.

Saidu, K., Onah, J., Orisadipe, A., Olusola, A., Wambebe, C., & Gamaniel, K. (2000). Antiplasmodial, analgesic, and anti-inflammatory activities of the aqueous extract of the stem bark of *Erythrina senegalensis*. *Journal of Ethnopharmacology, 71*(1–2), 275–280. Available from https://doi.org/10.1016/S0378-8741(00)00188-4.

Sanon, O., Azas, M., Gasquet, C., Ouattara, N., Nebie, I., Traore, A. S., Esposito, F., Balansard, G., Timon-David, P., & Fumoux, F. (2003). Ethnobotanical survey and in vitro antiplasmodial activity of plants used in traditional medicine in Burkina Faso. *Journal of Ethnopharmacology, 86*(2), 381. Available from https://doi.org/10.1016/s0378-8741(02)00381-1.

Simoes-Pires, C., Hostettmann, K., Haouala, A., Cuendet, M., Falquet, J., Graz, B., & Christen, P. (2014). Reverse pharmacology for developing an anti-malarial phytomedicine. The example of *Argemone mexicana*. *International*

Journal for Parasitology: Drugs and Drug Resistance, 4(3), 338–346. Available from https://doi.org/10.1016/j.ijpddr.2014.07.001.

Snider, D., & Weathers, P. J. (2021). In vitro reduction of *Plasmodium falciparum* gametocytes: *Artemisia* spp. tea infusions vs. artemisinin. *Journal of Ethnopharmacology, 268*, 113638. Available from https://doi.org/10.1016/j.jep.2020.113638.

Soh, P. N., & Benoit-Vical, F. (2007). Are West African plants a source of future antimalarial drugs? *Journal of Ethnopharmacology, 114*, 130–140. Available from https://doi.org/10.1016/j.jep.2007.08.012.

Tepongning, R. N., Mbah, J. N., Avoulou, F. L., Jerme, M. M., Ndanga, E.-K. K., & Fekam, F. B. (2018). Hydroethanolic extracts of *Erigeron floribundus* and *Azadirachta indica* reduced *Plasmodium berghei* parasitemia in Balb/c Mice. *Evidence-Based Complementary and Alternative Medicine, 2018*, 5156710. Available from https://doi.org/10.1155/2018/5156710.

Togola, A., Diallo, D., Dembélé, S., Barsett, H., & Paulsen, B. S. (2005). Ethnopharmacological survey of different uses of seven medicinal plants from Mali, (West Africa) in the regions Doila, Kolokani and Siby. *Journal of Ethnobiology and Ethnomedicine, 1*, 7. Available from https://doi.org/10.1186/1746-4269-1-7.

Traore-Keita, F., Gasquet, M., Di Giorgio, C., Ollivier, E., Delmas, F., Keita, A., … Timon-David, P. (2000). Antimalarial activity of four plants used in traditional medicine in Mali. *Phytotherapy Research, 14*(1), 45–47, https://doi.org/10.1002/(SICI)1099-1573(200002)14:1<45::AID-PTR544>3.0.CO;2-C.

Tu, Y. (2011). The discovery of artemisinin (qinghaosu) and gifts from Chinese medicine. *Nature Medicine, 17*, 1217–1220. Available from https://doi.org/10.1038/nm.2471.

Vial, T., Tan, W. L., Deharo, E., Missé, D., Marti, G., & Pompon, J. (2020). Mosquito metabolomics reveal that dengue virus replication requires phospholipid reconfiguration via the remodeling cycle. *Proceedings of the National Academy of Sciences of the United States of America, 117*, 27627–27636. Available from https://doi.org/10.1073/pnas.2015095117.

Vonthron-Sénécheau, C., Weniger, B., Ouattara, M., Bi, F. T., Kamenan, A., Lobstein, A., … Anton, R. (2003). In vitro antiplasmodial activity and cytotoxicity of ethnobotanically selected Ivorian plants. *Journal of Ethnopharmacology, 87*(2–3), 221–225. Available from https://doi.org/10.1016/s0378-8741(03)00144-2.

WHO. (2008). World Health Statistics. <https://www.who.int/data/gho/publications/world-health-statistics> Accessed 15.05.21.

WHO. (2009). Methods for surveillance of antimalarial drug efficacy. <https://apps.who.int/iris/handle/10665/44048?locale-attribute = fr&>. Accessed 15.05.21.

WHO. (2013). WHO traditional medicine strategy: 2014–2023. <https://www.who.int/publications/i/item/9789241506096>. Accessed 15.05.21.

WHO. (2019). World malaria report 2019. <https://www.who.int/publications-detail-redirect/9789241565721> Accessed 15.05.20.

Weathers, P., Cornet-Vernet, L., Hassanali, A., & Schul, J. J. (2017). *Note to World Health Organisation on integration of Artmeisia annua into the strategy against malaria in Africa*.

Weniger, B., Lagnika, L., Vonthron-Sénécheau, C., Adjobimey, T., Gbenou, J., Moudachirou, M., … Sanni, A. (2004). Evaluation of ethnobotanically selected Benin medicinal plants for their in vitro antiplasmodial activity. *Journal of Ethnopharmacology, 90*(2–3), 279–284. Available from https://doi.org/10.1016/j.jep.2003.10.002.

Willcox, M. L. (1999). A clinical trial of "AM", a Ugandan herbal remedy for malaria. *Journal of Public Health Medicine, 21*(3), 318–324. Available from https://doi.org/10.1093/pubmed/21.3.318.

Willcox, M. L., Graz, B., Falquet, J., Sidibé, O., Forster, M., & Diallo, D. (2007). *Argemone mexicana* decoction for the treatment of uncomplicated *Falciparum malaria. Transactions of the Royal Society of Tropical Medicine and Hygiene, 101*(12), 1190–1198. Available from https://doi.org/10.1016/j.trstmh.2007.05.017.

Wykes, M. N., & Good, M. F. (2009). What have we learnt from mouse models for the study of malaria? *European Journal of Immunology, 39*, 2004–2007. Available from https://doi.org/10.1002/eji.200939552.

Zhang, Q. W., Lin, L. G., & Ye, W. C. (2018). Techniques for extraction and isolation of natural products: A comprehensive review. *Chinese Medicine (United Kingdom), 13*, 20. Available from https://doi.org/10.1186/s13020-018-0177-x.

Zirihi, G. N., Mambu, L., Guédé-Guina, F., Bodo, B., & Grellier, P. (2005). In vitro antiplasmodial activity and cytotoxicity of 33 West African plants used for treatment of malaria. *Journal of Ethnopharmacology, 98*(3), 281–285. Available from https://doi.org/10.1016/j.jep.2005.01.004.

PART II

Medicinal plants as anti-infectives: recent innovations and regulations

CHAPTER

9

Mycobacterial quorum quenching and biofilm inhibition potential of medicinal plants

Jonathan L. Seaman[1], *Carel B. Oosthuizen*[1], *Lydia Gibango*[1] *and Namrita Lall*[1,2,3]

[1]Department of Plant and Soil Sciences, Faculty of Natural and Agricultural Sciences, University of Pretoria, Pretoria, South Africa [2]School of Natural Resources, University of Missouri, Columbia, MO, United States [3]College of Pharmacy, JSS Academy of Higher Education and Research, Mysuru, Karnataka, India

Introduction

Tuberculosis (TB) caused by *Mycobacterium tuberculosis* (Mtb) is a complex communicable disease that has saturated the global disease burden and to this day remains the leading cause of death due to a single infectious organism. Remarkably, TB mortality rates have exceeded the rate of human immunodeficiency virus and acquired immunodeficiency syndrome, and thus global attention is required to address this pathogen and its biopsychosocial determinants. Approximately 1.7 billion people worldwide are infected with TB with an estimated 10 million new cases every year (WHO, 2019). Although the TB epidemic is of great concern to world health authorities, 58 million people worldwide have been cured throughout 2000–18 and thus the ability of effective diagnosis, treatment (6–9 months), and management in curbing the disease burden is entirely feasible. Ending the TB epidemic has achieved global recognition status and has been included in the health targets of the Sustainable Development Goals (SDG) as well as the assignment of a dedicated "End TB" strategy. As with any infectious disease, TB treatment requires a holistic medical approach that involves a complex interplay between the healthcare environment, healthcare practitioner, and the patient. The problem, however, lies in correcting

Medicinal Plants as Anti-infectives
DOI: https://doi.org/10.1016/B978-0-323-90999-0.00008-2

the inefficient healthcare practices of the past and striving toward innovative treatment strategies that are shorter with limited development of antimicrobial resistance. The highly versatile nature of Mtb, as a pathogen and its ability to be transferred via infectious aerosols, coupled with poor healthcare practices such as inaccurate identification and indiscriminate use of antibiotics, has laid the foundation for an extensive antimicrobial resistance capacity of Mtb (Sarathy et al., 2018). This has, in essence, crippled healthcare infrastructure and has forced scientists and healthcare workers alike to search for novel strategies to curb this disease. Although 1.7 billion people worldwide are infected with TB, it is essential to highlight the difference between latent and active infection.

Of the individuals infected with Mtb, only 5%–10% will develop active TB (WHO, 2019). Latent TB infections should not be ignored as these are a great contributor to bacterial persistence, which has several important implications for the development of post-treatment relapse (Zhang et al., 2014). Persisting bacterial cells are a population of dormant bacilli which continue to exist despite the presence of adverse environmental factors. Latent TB infection refers to the presence of infection without the development of observable disease symptoms. Active TB infection refers to the presence of infection with the development of disease symptoms which can be classified as pulmonary (affecting the lungs) or extrapulmonary TB (disseminated TB) (Feng et al., 2012). Reactivation of latent TB is characterized by the reemergence of disease symptoms in a latently infected individual. The progression of latent TB to active TB can occur when a latently infected individual is exposed to the infectious aerosols of an actively infected individual. This can lead to one of two scenarios; a progressive active infection or the reactivation of latent TB, which can be influenced by several risk factors including the immunocompromised state of the host and/or age (WHO, 2019). The dynamic nature of TB infections is represented in Fig. 9.1.

In order to evade host immune responses and antibiotic bombardment, TB bacilli have developed a strategy to encapsulate themselves within a protected matrix so that their susceptibility

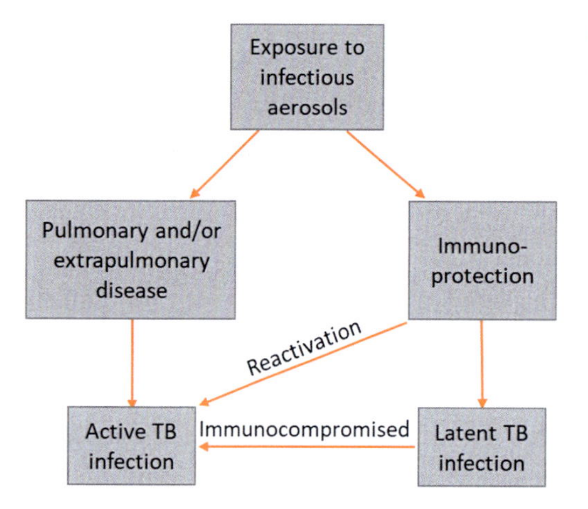

FIGURE 9.1 Flow diagram of the TB infection process.

and availability to antibiotics and host immune clearance are diminished. These dynamic structures are often in the form of bacterial biofilms and have been shown to lead to chronic infection and inflammation (Erhabor, Erhabor, & McGaw, 2019). Latent TB, and the persistence mechanisms, is thus an attractive target for further research as over 60% of infectious diseases treated in developed countries were due to the presence of biofilms (Chen & Wen, 2011).

Virulence factor production and biofilm formation are genetically regulated through the process of quorum sensing, whereby changes in cell density determine the expression of various pathogenic and nonpathogenic genes (Adonizio, Kong, & Mathee, 2008). In the case of Mtb, interference with the quorum sensing cascade, a process known as quorum quenching, is of beneficial importance as an adjunctive treatment modality for TB biofilm disruption. Selective pressure for bactericidal antibiotic resistance is minimal when it comes to quorum quenching strategies as these novel agents do not result in bactericidal effects, but rather act via the attenuation of virulence (Kordbacheh, Eftekhar, & Ebrahimi, 2017). Autoinducer molecules such as *N*-acyl-homoserine lactone (only Gram-negative bacteria), Autoinducer 2, and signaling polypeptides are the chemical signals used as the language of communication in cell-to-cell signaling (Wei et al., 2020). The targets for the inhibition of signaling include the inhibition of autoinducer synthesis, postsynthetic degradation as well as the disruption of signal reception and subsequent signal transduction (Wei et al., 2020).

Although synthetic agents aimed at targeting quorum sensing are currently being researched and developed, the role of natural products should not be ignored as the planet's rich biodiversity holds promise for the discovery of a biological gold mine of compounds that can be used to disease. Medicinal plants discovered via ethnobotanical methodologies are of great interest due to their abundance of bioactive secondary metabolites which function naturally as defense mechanisms; several of which have been shown to exhibit quorum sensing inhibition properties (Paul, Gopal, Kumar, & Manikandan, 2018). This chapter aims to summarize and highlight the potential of medicinal plants and natural products to act as quorum quenching antimycobacterial agents. In addition, the current state of research has been analyzed to indicate potential gaps within this interesting topic.

Significance of quorum quenching research

There has recently been a transition from The Millennium Development Goals of 2000–15 to the SDG of 2030. It has been realized that the attention is given to TB and the effect it has on quality of life has been inadequate and underestimated. This is not a lost battle, as worldwide attention is producing a myriad of ground-breaking initiatives and an array of global partnerships that can be used in the fight against the ever-evolving nature of TB. It is thus a global responsibility to ensure that ground-breaking treatment strategies are favored in comparison with conventional bactericidal pharmaceuticals as mechanisms that kill bacteria impose greater selective pressure than those that merely attenuate their virulence (Kordbacheh et al., 2017). Among the most notable global TB initiatives is the "End TB" strategy, as well as the first-ever United Nations (UN) ministerial high order meeting on TB, the outcome of which was a political declaration of commitment, dedication, and consistency in TB research and novel treatment strategies. The third SDG illustrates a desire for good health and well-being for all, and the SDG target 3.3 involves ending the TB epidemic by 2030. Three main areas of interest include reducing the TB incidence rate, reducing the mortality rate of TB, and finally reducing the catastrophic losses suffered by individuals and families affected by TB. Reductions of 80%, 90%, and 100%, respectively, are proposed for 2030 when compared to 2015 data. Formulated in 2014 by the World Health Assembly, the "End TB" strategy is to be used in conjunction with the SDG and other TB initiatives in the hopes of creating a planet free from TB and the socioeconomic downfall it entails.

TB has saturated the developing world with eight countries primarily classified as developing, contributing two-thirds of the world's global TB burden. These countries include India, South Africa, Philippines, and Nigeria (Phetlhu, Bimerew, Marie-Modeste, Naidoo, & Igumbor, 2018). There are several socioeconomic and environmental similarities shared amongst developing countries affected by TB, including the existence of exceptionally dense urban environments, extensive rural populations, and healthcare that is often lacking in availability and efficiency. Furthermore, education infrastructure is limited in these countries, which paves the way for poor compliance to treatment. The disease triangle involves a complex interplay between the host, environment, and pathogen. Each of these parameters needs to be addressed sufficiently to fully understand the versatile nature of opportunistic pathogens such as TB.

The significance of a TB adjuvant, focusing on quorum quenching, whose treatment regime is shorter, whose mechanism of action is selective for biofilm disruption, and whose availability to the public is improved, could not be more evident.

Current state of quorum quenching research

The raw bibliographic data collected were analyzed using VOSviewer version 1.6.15. This program allows for a comprehensive and interactive analysis of data acquired from scientific databases and subsequently generates a schematic representation of various trends, similarities and differences observed in the research. In the search for novel quorum quenching agents, such analytical interpretations are both relevant and required to produce high-quality research which is both innovative and justified. An illustration of a keyword-based search strategy and its subsequent analysis is shown in Fig. 9.2.

Bibliographic data was collected from a vast collection of journals using a host of scientific databases such as ScienceDirect, Web of Science, as well as Google Scholar. The search terms used included "mycobacterial quorum sensing," "mycobacterial quorum quenching," "mycobacterial biofilm formation," "mycobacterial virulence factor*," "medicinal plant* and quorum quenching," "medicinal plant* and quorum sensing," "*Mycobacterium tuberculosis*," "medicinal plant* and biofilm inhibition," "antimycobacterial medicinal plant*," and "phytochemical* and quorum quenching and medicinal plant* and mycobacterial quorum quenching." (* represents singular and/or plural.) The data was acquired and subsequently analyzed for important trends and research opportunities using a keyword map.

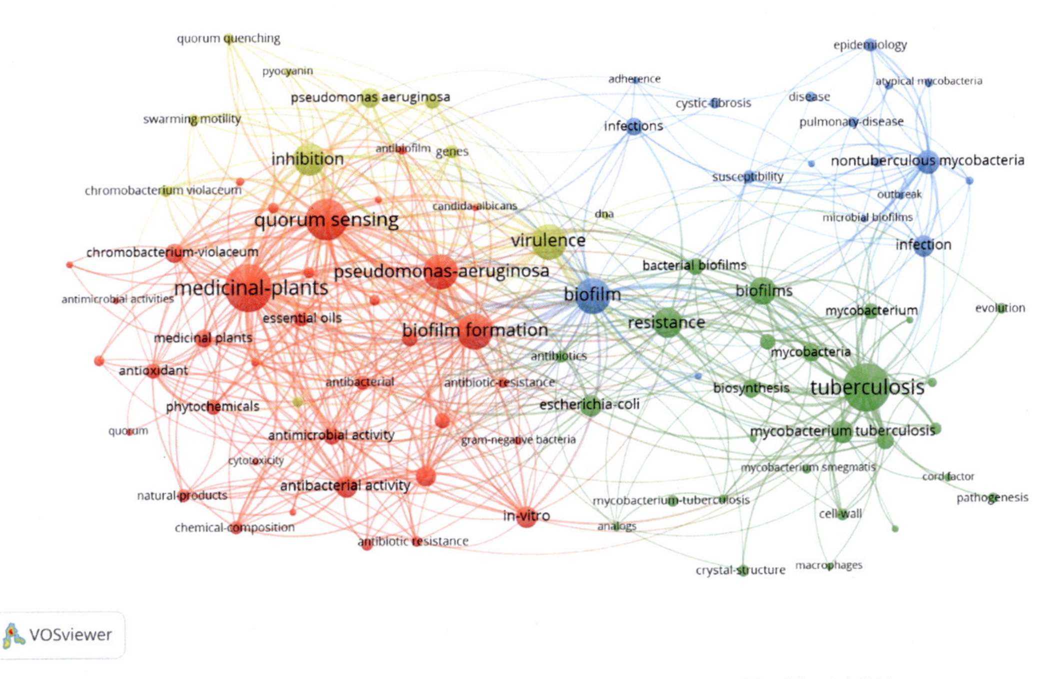

FIGURE 9.2 Spatial distribution of the current research in quorum sensing and biofilm inhibition.

Fig. 9.2 shows that extensive studies have been performed on medicinal plants and their respective phytoconstituents and that there is a significant link between these studies and quorum sensing studies. This is advantageous as preexisting research serves as a reference for information that can be used to optimize protocols and contextualize information. On the other hand, limitations regarding this observation relate to the test microorganisms used in these studies. *Pseudomonas aeruginosa* appears to be the most prevalent microorganism tested in quorum sensing studies along with *Chromobacterium violaceum*. Although this is beneficial for studies focusing on *P. aeruginosa* and *C. violaceum*, respectively, it limits the extrapolation of data to other species such as Mtb. This may, however, be surprisingly beneficial for novel quorum quenching research on Mtb., as few research provides greater opportunity for investigation into a previously underexplored topic. One of the most outstanding observations is the minimal degree of quorum quenching research in general and more specifically on mycobacteria and their associated mechanisms of virulence. As seen in Fig. 9.2, quorum quenching is a significant distance away from Mtb., with no observable link identified by current research initiatives. The fact that such little experimental evidence exists regarding mycobacterial quorum quenching using medicinal plants and natural products means that a significant gap in the research has been identified and that further investigation into this topic will likely yield innovative results which might be clinically relevant. The virulence of tuberculosis bacilli is associated with their ability to form biofilms. Although biofilm research has recently been quite concentrated, the way this mechanism can be inhibited in mycobacteria has drawn little attention.

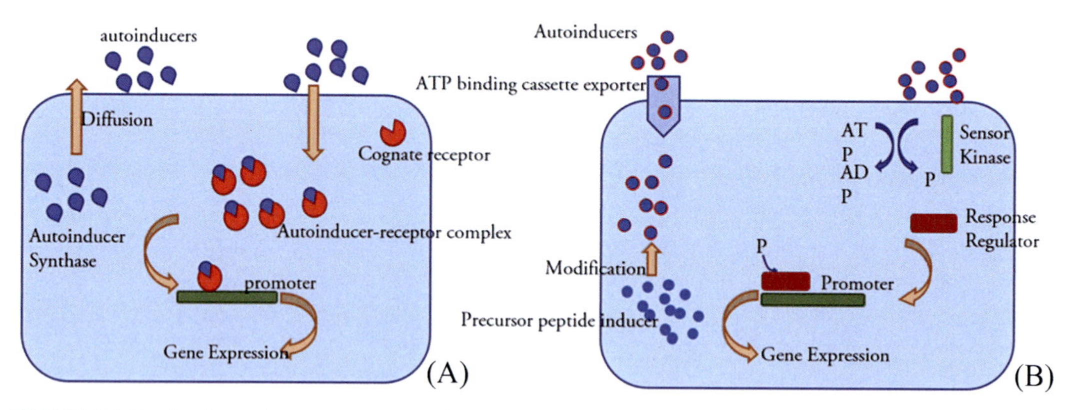

FIGURE 9.3 A schematic representation of quorum sensing in both Gram-negative (A) and Gram-positive bacteria (B). Autoinducer molecules are synthesized intracellularly and transported to the extracellular environment. When high enough concentrations of the autoinducer molecule are present in the extracellular environment, receptor binding and subsequent signal transduction occur. This results in the induction of gene expression cascades. Source: *From Paul, D., Gopal, J., Kumar, M., & Manikandan, M. (2018). Nature to the natural rescue: Silencing microbial chats.* Chemico-Biological Interactions, 280, *86–98. https://doi.org/10.1016/j.cbi.2017.12.018.*

Biofilms are essential virulence constituents of mycobacteria and thus there exists a need to develop a greater information base regarding biofilms and Mtb., as well as how biofilm formation and maturation can be inhibited using plant-based bioactives and natural products.

The process of quorum sensing, as illustrated in Fig. 9.3, relies on the presence and cooperation of three individual mechanisms; the production of autoinducer molecules, the subsequent recognition of autoinducer molecules via a membrane-bound or cytoplasmic receptor, as well as the effective gene response to autoinducer stimuli (Antonioli et al., 2019). Cell density thresholds are indicated by the concentration of the autoinducer molecule within the extracellular environment. Once the autoinducer molecule reaches a particular threshold concentration, coordinated behaviors are activated, resulting in population-dependent changes in gene expression, including biofilm formation and virulence factor production. Maintenance of the dynamic equilibrium in homogenous or heterogeneous bacterial populations relies on these intricate cell-to-cell communication strategies that ensure peaceful cohabitation of the ecological niche in question (Antonioli et al., 2019).

The petroleum ether fraction showed the best activity and was further fractionated and purified by using bioassay-guided isolation to yield the very potent compound 1 biofilm inhibitor (E)-2-(methyl (phenyl) amino) ethyl 2-(2-hydroxyundecanamido)-7, 11-dimethyl-3-oxotetradec-4-enoate, with an IC_{50} of 4-32 μg/mL (Jiang et al., 2019). Various biofilm parameters were investigated; namely, the ability of the extract and compound 1 to inhibit biofilm formation, disrupt mature biofilms, and disperse preformed mycobacterial biofilms, all of which are essential components of the dynamic nature of biofilms. At a concentration of 4 μg/mL, compound 1 was able to successfully inhibit the various mechanisms by which mycobacteria form biofilms thus decreasing the overall biofilm biomass (Jiang et al., 2019).

Quorum sensing versus quorum quenching

Bacterial cells are frequently known to recognize the cell density of their population and to regulate their gene expression accordingly. This process, known as quorum sensing, encapsulates the notion that when it comes to coordinated cellular strategies, it will be more fruitful to activate such mechanisms when the population is of a particular size or density threshold, resulting in a greater and more effective response (Antonioli, Blandizzi, Pacher, Guilliams, & Haskó, 2019). The process of quorum sensing, as illustrated in Figure 9.3, and relies on the presence and cooperation of three individual mechanisms; the production of autoinducer molecules, the subsequent recognition of autoinducer molecules via a membrane-bound or cytoplasmic receptor as well as the effective gene response to autoinducer stimuli (Antonioli et al., 2019). Cell density thresholds are indicated by the concentration of the autoinducer molecule within the extracellular environment. Once the autoinducer molecule reaches a particular threshold concentration, co-ordinated behaviours are activated, resulting in population-dependent changes in gene expression including biofilm formation and virulence factor production. Maintenance of the dynamic equilibrium in homogenous or heterogeneous bacterial populations relies on these intricate cell-to-cell communication strategies that ensure peaceful cohabitation of the ecological niche in question (Antonioli et al., 2019).

Novel treatment modalities that aim to interfere with the quorum sensing process have attained much attention as of late. The inhibitory process is known as quorum quenching; an attractive target for the inhibition of bacterial virulence (Paul et al., 2018). Many pathogenic organisms that infect humans, animals, and plants have been shown to regulate their virulence via effective quorum sensing strategies (Paul et al., 2018). Innovative quorum quenching agents should thus be able to selectively inhibit the process of quorum sensing, preventing the appearance of undesirable bacterial phenotypes. With further research and development, quorum quenching has the ability to hold great promise as a new avenue in treating bacterial and fungal infections and the various population-dependent responses that these cells exhibit (Tegos & Hamblin, 2013).

Biofilms

Background on biofilms

Although planktonic cells can move about freely, this level of exposure makes the bacilli susceptible to the environmental conditions in which they reside. To increase the likelihood of survival, bacteria form biofilms that reduce the level of exposure by creating a highly organized three-dimensional matrix consisting of sessile cells that are encapsulated within an extracellular polysaccharide structure. The formation of a biofilm relies on the random collision of planktonic cells and the subsequent attachment to a surface (biotic or abiotic) or to one another via a host of cellular appendages such as flagella and pili (Shirtliff, Mader, & Camper, 2002). Following attachment, the quorum sensing abilities of the bacteria are activated in an attempt to accelerate the maturation of the biofilm and to enhance cell to cell communication for coordinated responses (Shirtliff et al., 2002).

The maturation of the biofilm involves the secretion of greater volumes of the extracellular polymeric substance (EPS) and the generation of nutrient channels within the matrix. This complex configuration allows bacterial populations to maximize nutrient uptake and the subsequent retention of such nutrients within the biofilm (Shirtliff et al., 2002). Furthermore, biofilm formation often occurs in nutrient-dense locations thus making it a highly efficient and capable dynamic structure concerning nutrient acquisition. Another advantage of biofilm formation is that of resistance to physical detachment by shear stress. Due to the extensive adhesive interactions between bacteria and the matrix, bacterial cells are in essence anchored within the biofilm. Thirdly and most medically relevant is the inherent ability of bacterial cells in biofilms to resist penetration by antimicrobial substances as well as the ability to avoid consumption via phagocytic processes (Karami et al., 2020). This subsequently enables bacterial persistence and the generation of chronic infections (Zhang, 2014). A diagrammatic illustration of the process of biofilm formation and maturation is illustrated in Fig. 9.4.

The mechanisms by which bacterial cells within a biofilm acquire antibiotic resistance can be classified into two main categories, namely, the limited accessibility of cells in a biofilm to the environment and secondly the acquisition of genetic material conferring antibiotic resistance. Planktonic cells are naturally more exposed and are thus more susceptible to clearance via antimicrobials and immunological mechanisms. Due to their diminished exposure to the environment, bacterial cells within a biofilm resist penetration by antibiotics via the protective ability of the extracellular matrix and the presence of efflux pumps which increase the minimum inhibitory concentration (MIC) value and prevent drug accumulation within the biofilm and individual cells (Khaledi et al., 2016). Furthermore, in the case of TB, alveolar macrophages have limited access to bacterial cells within a biofilm thus hindering this immunological defense strategy. Although polysaccharides have been shown to form the core component of the extracellular matrix, proteins and exogenous DNA are also present in biologically significant concentrations (Khan, Jeong, Park, Kim, & Kim, 2019). This exogenous DNA exists in many forms such as transposons and plasmids; more importantly, however, is the potential presence of antibiotic

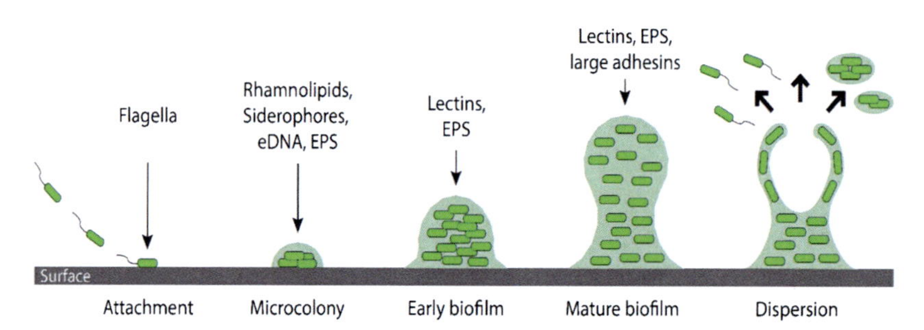

FIGURE 9.4 Schematic representation of the process of biofilm formation. Source: *From da Silva, D.P., Schofield, M.C., Parsek, M.R., & Tseng B.S. (2017). An Update on the Sociomicrobiology of Quorum Sensing in Gram-Negative Biofilm Development. Pathogens 2017, 6, 51. (CC BY-SA 4.0). Accessed January 2022. https://doi.org/10.3390/pathogens6040051.*

resistance genes contained within these mobile genetic elements. Uptake of this exogenous DNA is a possibility. However, bacteria rapidly regain their susceptibility to antibiotic treatment after exiting the biofilm and thus the transfer of antimicrobial resistance genes is less of a contributor to overall resistance than what was previously thought (Stewart & Costerton, 2001).

Furthermore, bacteria possess the ability to physiologically dissociate from the matrix in an attempt to develop additional colonies and biofilms within the host. This ability is critical in ensuring the sustainability of the species as biofilms are sessile structures that cannot evade changes in environmental conditions. Disassembly and dispersal of the biofilm are imperative in the innate ability of bacterial cells within a biofilm to relocate when local conditions deteriorate (Boles & Horswill, 2011).

Biofilms and *Mycobacterium tuberculosis*

The genus *Mycobacterium* includes both pathogenic species belonging to the Mtb complex, and environmental nontuberculous species. The genus *Mycobacterium* (family Mycobacteriaceae), currently includes more than 170 recognized species (Forbes, 2017). Mycobacteria have a common pathogenic factor being the formation of extracellular, polymeric matrices called biofilms. Biofilm formation is mainly influenced and regulated by the availability of nutrients, ions, and carbon sources. The first report of the concept of biofilms dates back to 1978 when initial observations were published (Costerton, Geesey, & Cheng, 1978; Esteban & García-Coca, 2018). Koch (1982) described the appearance of "cells which are pressed together and arranged in bundles." More scientific papers on the topic began to surface about a decade later, while mycobacterial cells forming aggregates or pellicles were described in earlier days of mycobacteriology (Calmette, 1936; Loöwenstein, 1920).

Mycobacterial biofilms are defined in the same way as any other biofilms. Some mycobacteria, however, have the ability to form biofilm structures on liquid–air media interfaces and exhibit sliding motility on agar surfaces (Ojha et al., 2008). The fast-growing nonpathogenic *M. smegmatis* is a model organism that has been extensively studied for mycobacterial biofilm formation (Chakraborty & Kumar, 2019; Danese, Pratt, & Kolter, 2000). Biofilm development starts with the bacterial adhesion and progresses through the different stages of surface attachment, sessile growth, biofilm maturation, and dispersal. Adhesins from the bacterial cell wall mediate the initial attachment of bacteria to the surfaces which is an important virulence trait of microbial pathogenesis. Following attachment to the surface, sessile bacteria initiate the synthesis of extracellular matrix which is composed of several glycopeptides, DNA, and other molecules. Mycobacteria lack surface fimbriae or pili and do not produce the usual exopolysaccharide components of extracellular matrix but can attach to different surfaces and form developed biofilms (Esteban & García-Coca, 2018).

Research studies have been conducted on biofilm formation of several mycobacterial species to better understand key components needed for its formation and how it is used for bacterial survival. A study conducted on different species of rapidly growing mycobacteria revealed that biofilm development follows a sigmoid growth kinetic. This study's findings were later confirmed in clinical strains. Nutrients, carbon sources (such as glucose and peptone), and ions (Ca^{2+}, Mg^{2+}, Zn^{2+}) are known to play an influential and regulatory role in bacterial behavior and biofilm formation. The study also indicated that tap water, as the nutrient source, can allow biofilms to form. This provides an explanation as to why mycobacterial biofilms can be found in water sources (Esteban & García-Coca, 2018; Esteban et al., 2008).

Different molecules in the formation of biofilms and their composition have been examined. In the well-studied mycobacterial model, *Mycobacterium smegmatis*, glycopeptidolipids are important for the initial surface attachment (Recht & Kolter, 2001). The same molecules have been shown to play a role in sliding motility, a property that several mycobacterial strains have that may be, although not always related, to biofilm spreading on surfaces. Shorter chain mycolic acids play a role in the structure of biofilms and are proposed to form a hydrophobic extracellular matrix. Mycolic acids are associated with higher resistance to disinfectants and antibiotics associated with these microorganisms. The mycolic acids found in the cell wall provide a permeability barrier. Other macromolecules such as GroEL1 chaperones play a role in biofilm development in *M. smegmatis*. The complexities of mycobacterial biofilm structure and development are continuously being investigated for a better understanding on its clinical impact and how to deal with its occurrence (Sharma, Misba, & Khan, 2019).

Mycobacterial biofilms harbor an extensive, drug-tolerant population of cells. The EPSs produced by mycobacteria disrupt the diffusion of antimicrobials throughout the matrix, thus protecting individual bacterial cells from exposure (Solokhina, Bonkat, Kulchavenya, & Braissant, 2018). This makes mycobacterial biofilms incredibly difficult to treat and enables chronic persistence within the human body (Ojha et al., 2008). Drug-tolerant bacteria are bacteria that exhibit a diminished response to the presence of high concentrations of antimicrobial agents and transiently survive in such environments (Crabbé, Jensen, Bjarnsholt, & Coenye, 2019). Drug-resistant bacteria are bacteria that possess the ability to significantly withstand the effects of a drug which is usually effective against them and generally increase the MIC value of the drug in question (Crabbé et al., 2019). The location in which mycobacterial biofilms form is of the utmost clinical relevance. Studies by Solokhina et al. (2018) have shown that necrotic granulomas and caseous foci are saturated with bacterial cells, and cavities are lined with mycobacterial biofilms despite their avascular nature. The presence of vasculature in infectious processes is important for the hematogenous dissemination of the infectious organism. Tuberculosis bacilli have been shown to possess the ability to overcome low oxygen tension environments as seen in avascular structures despite being a strict aerobe. Not only does this ensure their persistence, but it also ensures their inherent tolerance to antimicrobial compounds that are distributed via the vasculature. Remote lesions have also been shown to sequester an immunologically, chemically, and physically resistant population of cells which ensure recalcitrant infection processes (Solokhina et al., 2018).

During in vitro studies, disruption of mycobacterial biofilm formation is often achieved with the addition of a chemical detergent which prevents the aggregation and clumping of bacterial cells; a step critical in the formation of biofilms. Detergent-free media, on the other hand, enables the formation of mycobacterial biofilms on the air–media interface which is commonly known as a pellicle (Ojha et al., 2008). The question then arises, why not use detergents as biofilm disrupting agents or as adjuvants in TB treatments? The answer lies in both the effects that it has on the host as well as the effect it has on the bacterium. Detergents are generally classified as toxic to the human body and thus their usage and efficacy in vivo would be ethically questioned. Secondly, in terms of the bacterium, detergents have been shown to significantly alter several properties of the bacterial envelope, lipid structure, and function as well as the permeability of bacterial cell membranes. Therefore, the reliability of detergents to produce scientifically ethical and effective adjuvants is limited (Ojha et al., 2008).

Mycobacterial biofilms are unique concerning the content of the EPS they produce (Ben-Kahla & AL-Hajoj, 2016). Research has shown that novel lipids derived from mycolic acid as well as free mycolic acids are present in abundance in mycobacterial biofilms (Ojha et al., 2008). This has important implications regarding how Mtb form biofilms as well as how they ensure their subsequent maturation. During the late stages of biofilm maturation, mycolic acids in the EPS have been shown to increase in concentration with mutants defective in maturation potential such as the ΔgroEL1 mutant, lacking the ability to produce these lipids (Ojha et al., 2008). Not only do Mtb biofilms differ from other bacteria concerning the content of the EPS but they also exhibit extensive EPS variations between Mtb bacilli themselves (Ben-Kahla & AL-Hajoj, 2016). These variations are presumed to be due to the various evolutionary pressures that Mtb bacilli face when occupying a host of different ecological niches (Ojha et al., 2008). Despite playing a critical role in the cellular integrity of mycobacterial cells and the way in which Mtb interacts with the immune system, mycolic acid concentration is imperative in the phenotype of the biofilm produced with respect to its integrity and thickness (Ben-Kahla & AL-Hajoj, 2016). Mycolic acid synthesis is thus an important therapeutic target for the disruption of mycobacterial biofilms.

Conventional TB treatment is inefficient in the inhibition of mycobacterial quorum sensing and the subsequent clearance of biofilms, prompting the development of novel therapies aimed at combatting mycobacterial biofilm formation, maturation, and disassembly.

Virulence factors

Background on virulence factors

The pathogenicity of a bacterium, or the ability to cause disease within host cells, is dependent on an array of parameters that involve the complex interactions between pathogen and host. One of these essential parameters is the quantitative characteristic—virulence, which is defined as the ability to overcome several host defenses that may be initiated upon pathogen challenge and induce varying degrees of host-cell damage (Ufimtseva et al., 2018). The microbial virulence of a pathogen is a complex characteristic that involves the production and secretion of a host of virulence factors which are important regulators of pathogenicity. Virulence factor secretion is yet another trait regulated by quorum sensing, as the feasibility and success of infection is enhanced when bacteria achieve a certain threshold population density (Paul et al., 2018). Microorganisms produce virulence factors in varying quantities and in response to a myriad of different environmental stimuli.

Virulence factors and *Mycobacterium tuberculosis*

The way in which Mtb regulates its virulence is not via the secretion of a single virulence factor but rather via an extensive and complex combination of virulence responses that ensure its dynamic adaptation to host immunity and defenses (Madacki, Mas Fiol, & Brosch, 2019). Over the years, studies have shown that an increasing number of virulence genes are required for survival and replication of *M. tuberculosis* bacilli in vivo. An initial study by Sassetti and Rubin (2003) showed that over 196 genes associated with virulence were required for the survival of Mtb. Subsequent deep sequencing molecular advances have isolated an additional 400 genes required for the effective infection of host cells by Mtb (Zhang et al., 2013). Although a large proportion of these virulence genes encode functions in basic metabolism, the vast remainder of such gene products is on the front-line combatting host immune defenses. Examples of these virulence gene products include a host of ESX/type VII secretion systems as well as complex lipids of the cell envelope which ensure survival as well as a high degree of virulence in the host cell (Madacki et al., 2019). To examine the effects of these individual gene products, one needs to examine the host cell and its properties and the way in which host and pathogen interact on a physico-chemical and biological level.

The intracellular environment of host macrophages is a critical area of TB research as mycobacterial cells are ingested by these immunological cells upon entry into the pulmonary alveoli (Madacki et al., 2019). As part of the generic immunological response, macrophages form a phagosome post-bacilli ingestion. The phagosome fuses with a lysosome to produce a phagolysosome; a hostile acidic environment that is induced to clear the infected cell of infection. Tuberculosis bacilli have developed an array of microbial virulence strategies to prevent consumption by phagolysosomes and to ensure subsequent survival and replication success by delaying the formation or differentiation of such immunological structures. Studies have shown that escape from phagosomal confinement and the prevention of phagosomal acidification are regulated by the ESX-1 type VII secretion system as mutants deficient in this system were unable to escape from the phagosome via phagosomal rupture (Madacki et al., 2019). This secretion system is, therefore, a critical virulence factor during TB infection, as the ability to overcome phagosomal immune defenses is imperative for the survival of the species within the host.

Covalently linked to the outer mycobacterial cell wall layer, phenolic glycolipids are another example of a well-known mycobacterial virulence factor. A study by Reed et al. (2004) showed that the presence of these phenolic glycolipids resulted in a hypervirulent phenotype in a murine TB model and that all known pathogenic *Mycobacterium* strains contain these essential virulence factors (Reed et al., 2004). Furthermore, mutants deficient in this virulence factor were found to be attenuated in guinea pig TB models (Reed et al., 2004). From a host immunity perspective, phenolic glycolipids were also shown to inhibit the formation of pro-inflammatory cytokines which are critical in the formation of the inflammatory environment (Reed et al., 2004). If such an environment cannot be generated, TB bacilli have a head start in their ability to overcome host responses and ultimately secure a more efficient and long-lasting infection (Madacki et al., 2019).

Although the protein constituents of the mycomembrane are a vastly underexplored topic, recent studies have identified and characterized several integral proteins of the mycomembrane that contribute substantially to their virulence (Madacki et al., 2019). The channel-forming protein, CpnT, contains an N-terminal exotoxin domain which has been determined to be responsible for the ability of TB bacilli to induce necrosis in host cells (Danilchanka et al., 2014). Caseous necrosis is a debilitating complication of pulmonary mycobacterial infection and significantly alters the structural and functional integrity of the lung tissue. Following necrosis, CpnT limits cytokine production which allows for a silent escape and dissemination to both surrounding and distant tissues (Madacki et al., 2019). Due to the severity of disease that cooperating virulence factors can induce, it is necessary to develop novel pharmaceuticals that can be used to attenuate their toxic effects. As a quorum sensing-regulated process, virulence factor production can be addressed by innovative quorum quenching agents that aim to reduce their synthesis or inhibit their secretion. Using medicinal plants which are selected upon the basis of their ethnobotanical and ethnopharmacological uses, effective and ethically sound medicines can be discovered which target mycobacterial quorum sensing and subsequent virulence factor production.

Medicinal plants as quorum quenching agents

Nosocomial infections are infections that originate within the confines of the hospital environment and are those which were not incubating prior to hospital admission. *Pseudomonas aeruginosa* is an important cause of nosocomial respiratory infections especially in patients with preexisting pulmonary deficiencies (Vandeputte et al., 2010). The key determinant in its virulence lies in its ability to regulate communication and gene expression via quorum sensing. A study (Vandeputte et al., 2010) showed how leaf and bark extracts of *Combretum albiflorum* (Tul.) Jongkind were able to interfere with the transcription and regulation of quorum sensing genes in *P. aeruginosa* thereby inhibiting biofilm formation and maturation (Vandeputte et al., 2010). The inhibition of biofilm formation and maturation is critical in eradicating the chronicity of such respiratory conditions caused by *P. aeruginosa*. The bioactive compounds of *C. albiflorum* extract were fractionated and characterized, revealing primarily flavonoids with the particular abundance of the catechin, a flavan-3-ol derivative, which was found to be the active quorum sensing inhibitor (Vandeputte et al., 2010).

In another study by Vadakkan et al. (2018), violacein which is a purple pigment produced when bacterial populations reach a certain cell density was used as a reporter molecule for the visualization of quorum quenching activities of *Tribulus terrestris* L. root extract. *Chromobacterium violaceum* produces violacein during a natural quorum sensing response and is commonly used as a reporter strain due to its ability to visually illustrate the results of potential quorum quenching compounds. This study found that a three times daily exposure to 300 μg/mL of root extract was sufficient to significantly interfere with quorum sensing gene regulation by inhibiting the production of the reporter molecule, violacein (Vadakkan et al., 2018).

Although the ultimate goal of research and development into medicinal plants and their therapeutic actions is often to produce a pharmaceutical formulation, the role of the dietary ingestion of medicinal plants and their phytochemical constituents in quorum quenching strategies should not be set aside. Curcumin, the major phytochemical present in turmeric (*Curcuma longa* L.), has long been used in indigenous knowledge systems as a source of antimicrobial and antiinflammatory therapeutics (Packiavathy, Priya, Pandian, & Ravi, 2014). Scientific confirmation of anecdotal evidence was investigated in a study by Packiavathy et al. (2014) whereby curcumin was investigated for its quorum quenching potential in vitro against *Pseudomonas aeruginosa*, using a host of various quorum sensing processes as targets (Packiavathy et al., 2014). These targets included biofilm formation, swarming motility, and the production of various EPSs such as alginate, which are all regulated via quorum sensing. The study showed that curcumin had a substantial quorum quenching effect on biofilms via the disruption of biofilm architecture and the dislodging of biofilm biomass. At a concentration of 100 μg/mL, curcumin reduced biofilm biomass by 89% in *P. aeruginosa* (Packiavathy et al., 2014). Furthermore, alginate, a key component of the EPS in Pseudomonas biofilms, was inhibited by 63%. Finally, both swimming and swarming motility were inhibited at increasing concentrations of curcumin in a dose-dependent manner (Packiavathy et al., 2014). As swarming motility is a key regulator of biofilm formation, the inhibition of such a strategy would form a key component of novel quorum quenching agents. Even more promising was the fact that curcumin showed therapeutic synergism when combined with various antimicrobial drugs such as azithromycin, which holds great promise for its use in complementary medicine (Packiavathy et al., 2014).

Medicinal plants and mycobacterial quorum quenching

The chronicity of mycobacterial infections has long been associated with the presence of several quorum sensing strategies that mycobacteria employ, in an attempt to enhance the efficiency of infection, biofilm formation, and virulence factor production (Solokhina et al., 2018). Mycobacterial biofilms have been well described concerning their ability to harbor an extensive drug-tolerant population of cells which is known to complicate and extend the length of conventional treatment strategies (Jiang, Gan, An, & Yang, 2019). Novel agents that aim to induce a supportive, adjuvant, or synergistic effect have been vastly understudied and only a few known studies are currently in existence. Research has subsequently turned to the use of natural products in combatting the prolific rate of antimicrobial resistance and assisting in producing a therapeutic intervention that is efficient in mycobacterial quorum quenching and which is safe for human consumption.

Phytochemistry and ethnopharmacology research for new antimicrobial sources with novel modes of action has been of interest for many years. Plant extracts may be represented as potential superior sources of antimicrobial compounds than synthetic drugs (Romero et al., 2016). Aqueous extracts of *Vaccinium oxycoccos* L. (fresh fruit), *Azadirachta indica* A. Juss. (fresh leaves), *Hippophae rhamnoides* L. (dried fruit), *Juglans regia* L. (dried bark), and ground spices were evaluated for their efficacy for antibiofilm activity. *Azadirachta indica* was found to be the most effective in the study and had substantial biofilm reduction potential of *M. smegmatis*, followed by *Vaccinium oxycoccos* and spices with 28% and 26% biofilm reduction potential, respectively (Abidi et al., 2014). Leaves from *Parinari curatellifolia* Planch. ex Benth. have been studied for their efficacy against biofilm formation and the growth of *M. smegmatis*. The water, dichloromethanolic, and ethanolic leaf extracts were shown to inhibit biofilm formation. Phytochemical analysis of the plant revealed the presence of saponins, steroids, alkaloids, flavonoids, tannins, and cardiac glycosides (Bhunu, Mautsa, & Mukanganyama, 2017).

In addition to testing several plants, nature-inspired synthetic molecules have also been considered and heavily investigated. These naturally derived compounds have unlocked the development of a plethora of synthetic therapeutic agents and the discovery of many novel antibiofilm agents. Sesamol is an organic compound isolated from sesame seeds. Hans, Sharma, Hameed, and Fatima (2017) demonstrated that sesamol was a potent *M. smegmatis* biofilm inhibitor that reduced the metabolic activity and dry weight of biofilm biomass by 70% and 58%, respectively. Carvacrol [2-Methyl-5-(1-methylethyl) phenol] is a volatile monoterpene that is a major constituent of several essential oils of the Labiatae family. This compound has been approved as safe for usage in food products. Research has demonstrated its antioxidant, antitumor, antihepatotoxic, antibacterial, antiinflammatory, analgesic, and insecticidal biological activity (Hyldgaard, Mygind, & Meyer, 2012; Langeveld, Veldhuizen, & Burt, 2014; Magi, Marini, & Facinelli, 2015; Nostro & Papalia, 2012). Carvacrol has been reported to be effective against biofilm formations of rapidly growing mycobacteria, namely, *M. phlei*, *M. smegmatis*, and *M. fortuitum* (with antimycobacterial MIC values of 80–100 μg/mL). The study observed that concentrations above the MIC were able to cause a significant disaggregation effect on biofilm biomass and the metabolic activity of the cells protected by the biofilm matrix at a maturation phase. At concentrations below the MIC, the compound was able to disrupt biofilm formations (Marini et al., 2019). *Arisaema sinii* K. Krause, a Chinese medicinal herb, extract (80% ethanol in water) exhibited antibiofilm activity against *M. tuberculosis*. Historically speaking, this medicinal herb has been employed by local populations to treat pulmonary and lymphatic TB and thus its ethnobotanical and ethnopharmacological properties have directed further research into novel lead compounds isolated from *A. sinii*. The petroleum ether fraction showed the best activity and was further fractionated and purified by using bioassay-guided isolation to yield the very potent compound 1 (Fig. 9.5) biofilm inhibitor (E)-2-(methyl (phenyl) amino) ethyl 2-(2-hydroxyundecanamido)-7, 11-dimethyl-3-oxotetradec-4-enoate, with an IC_{50} of 4-32 μg/mL (Jiang et al., 2019). Various biofilm parameters were investigated; namely the ability of the extract and compound 1 to inhibit biofilm formation, disrupt mature biofilms and disperse pre-formed mycobacterial biofilms; all of which are essential components of the dynamic nature of biofilms. At a concentration of 4 μg/mL, compound 1 was able to successfully inhibit the

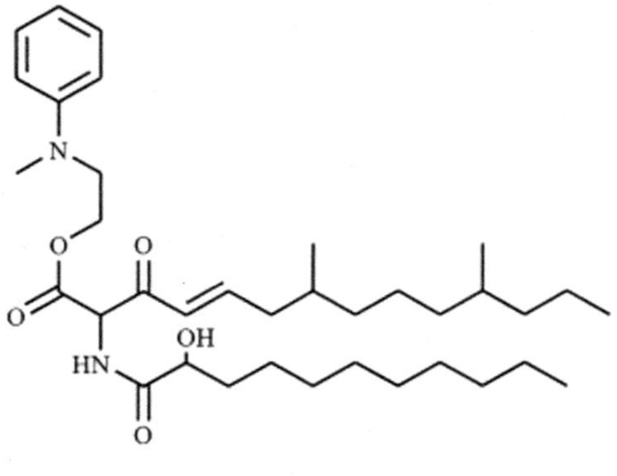

FIGURE 9.5 The chemical structure of (E)-2-(methyl (phenyl) amino) ethyl 2-(2-hydroxyundecanamido)-7,11-dimethyl-3-oxotetradec-4-enoate (compound 1). Source: *From Jiang, C.-H., Gan, M.-L., An, T.-T., & Yang, Z.-C. (2019). Bioassay-guided isolation of a* Mycobacterium tuberculosis *bioflim inhibitor from* Arisaema sinii *Krause.* Microbial Pathogenesis, *126, 351–356. https://doi.org/10.1016/j.micpath.2018.11.022.*

various mechanisms by which mycobacteria form biofilms thus decreasing the overall biofilm biomass (Jiang et al., 2019).

Although the inhibition of new biofilm formation is imperative, another important component of mycobacterial biofilm clearance is the ability to eliminate preexisting biofilms and biofilms that have matured prior to drug administration. The elimination of persistent bacterial populations is a known limitation of conventional antibiotic therapies (Jiang et al., 2019). Biofilm dispersal restores the susceptibility of TB bacilli to antibiotic treatment due to their resultant exposure to the unprotected environment outside the biofilm. In the study by Jiang et al. (2019), at a concentration of 8 μg/mL, compound 1 was able to disperse preformed biofilms (Jiang et al., 2019). The reason for the higher concentration of compound 1 required is due to the increased difficulty of dispersing biofilms when compared to inhibiting their formation (Jiang et al., 2019). Synergistic antibiotic activity and further disruption of mature biofilm architecture occurred using compound 1 at a concentration of 32 μg/mL. The disruption of the physically and chemically protective matrix allows conventional antibiotics to exert their full effect and results in the dramatic shortening of antibiotic therapy as TB bacilli are in an exposed planktonic state (Jiang et al., 2019). Mycobacterial biofilm disturbance is illustrated in Fig. 9.6, using no drug treatment as a negative control (A) and compound 1 as the test molecule (B) Jiang et al. (2019).

The limitations of using Mtb in an in vitro biofilm model are mainly related to the time taken for Mtb, biofilm formation. The above study highlighted that although scientifically significant results were generated, Mtb, biofilms took over a month to grow and testing could only begin 35 days postinoculation (Jiang et al., 2019). In an attempt to overcome this, other studies have focused on using *Mycobacterium smegmatis* as an alternative mycobacterial biofilm model due to its relative ease and rapidity of growth in vitro (Bonkat et al., 2012; Oosthuizen, Gasa, Hamilton, & Lall, 2019). In a study by Oosthuizen et al. (2019),

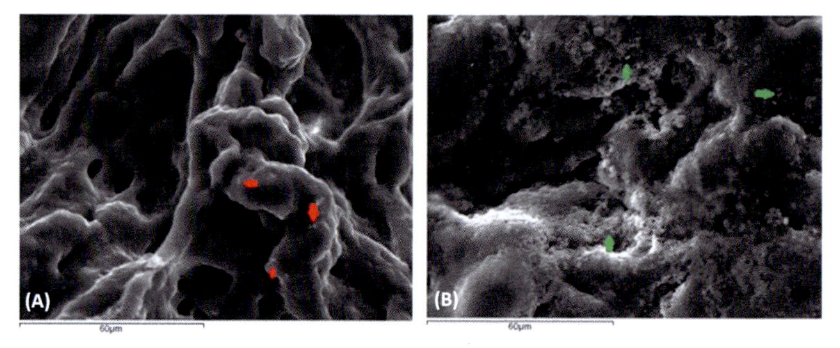

FIGURE 9.6 Scanning electron microscopy images of Mtb. (A) Scanning electron microscopy image of a mature Mtb, biofilm undisturbed with no drug treatment. Red arrows indicate TB bacilli physically embedded and protected within the biofilm matrix. (B) Scanning electron microscopy image of a disrupted Mtb, biofilm. Green arrows indicate exposed planktonic TB bacilli. Source: *From Jiang, C.-H., Gan, M.-L., An, T.-T., & Yang, Z.-C. (2019). Bioassay-guided isolation of a* Mycobacterium tuberculosis *bioflim inhibitor from* Arisaema sinii *Krause.* Microbial Pathogenesis, 126, *351–356. https://doi.org/10.1016/j.micpath.2018.11.022.*

ethnobotanically selected medicinal plants were evaluated for their antimycobacterial biofilm capabilities against *M. smegmatis* MC^2 155. *Leonotis leonurus* (L.) R. Br., *Sphedamnocarpus pruriens* (A. Juss.) Szyszył, and *Salvia africana-lutea* L. showed the most significant biofilm inhibitory capacity with EC_{50} values of 50.2, 62.2, and 95.8 μg/mL, respectively. Ciprofloxacin was the positive control utilized in this experiment. Although ciprofloxacin had a relatively low EC_{50} value of 1.98 μg/mL concerning biofilm formation inhibition, this value was still higher than the MIC value observed in the antimycobacterial assay conducted in the same study. This indicates that ciprofloxacin is more selective toward the inhibition of mycobacterial cell growth than mycobacterial biofilm formation inhibition (Oosthuizen et al., 2019). Other plants utilized in this study such as *Withania somnifera* (L.) Dunal showed far greater selectivity toward biofilm inhibition than antimycobacterial activity when compared to ciprofloxacin, with a selectivity index of 3.75.

Although it is often desirable to isolate phytochemicals with potent bioactivity, sometimes it may be necessary to enhance the bioactivity of these phytochemicals via derivatization. Not only can derivatives be chemically synthesized to enhance such activity, but they can also be designed to limit side effects that may be commonly experienced with the base compound (Junqueira et al., 2020). A study by Junqueira et al. (2020) serves as an example of such a technique whereby the researchers isolated licarin A, a neolignan with a dihydrobenzofuran structure from *Aristolochia taliscana* Hook. & Arn. and generated seven additional derivatives from the base compound. They tested the activity of these compounds, including licarin A, against mycobacterial biofilm formation, primarily using *Mycobacterium massiliense*, *Mycobacterium abscessus*, and *Mycobacterium fortuitum* as the test organisms (Junqueira et al., 2020). Licarin A is represented in Fig. 9.7.

The rationale behind the process of derivatization was to enhance the hydrophobicity of licarin A, since mycobacteria are characteristically well-known for their ability to produce an extensive concentration of mycolic acids, both in their cell wall as well as in the EPS of mycobacterial biofilms (Junqueira et al., 2020). By increasing the hydrophobicity of the base compound, the researchers postulated that this modification would have an advantageous effect with respect to the ability of the compound to penetrate the mycobacterial cell wall. This was, however, found not to be the case. Instead, the presence of a polar group

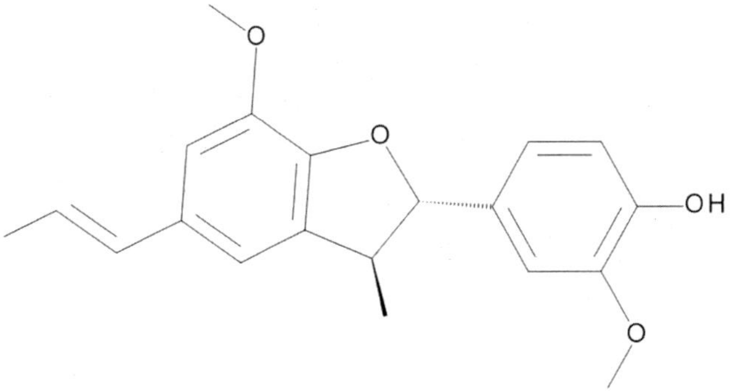

FIGURE 9.7 The chemical structure of licarin A.

such as a hydroxyl group was shown to greatly increase the susceptibility of the mycobacteria to drug treatment (Junqueira et al., 2020). The allylic alcohol, compound 9, was shown to have the greatest inhibitory potential on mycobacterial biofilm formation with minimum biofilm inhibitory concentration (MBIC) values of 2.44, 4.88, and 2.44 μg/mL when tested on *M. massiliense*, *M. fortuitum*, and *M. abscessus*, respectively (Junqueira et al., 2020). Licarin A was found to have MBIC values of 2.44, 9.76, and 9.76 μg/mL when tested against the same species, respectively (Junqueira et al., 2020). The limitations of this study were primarily related to the fact that although the above-mentioned compounds were capable of inhibiting biofilm formation, they were found to be relatively ineffective against preexisting biofilms which substantially increases the possibility that another drug may need to be added to achieve this effect. Nonetheless, the bioactive compound licarin A is widely distributed in the plant kingdom and is even found in nutmeg (*Myristica fragrans* Houtt.). Its presence in edible plant-based products is a significant advantage as individuals may receive a dual benefit of a food additive and an antimycobacterial compound in one.

Phytochemicals used in bacterial quorum quenching

The consumption of raw plant material is an impractical means of delivery of phyto-derived medications as it requires the consumption of a large amount of material that contains very little active constituent and is often unpalatable (Hoffman, 2003). There are thus several advantages in extracting bioactive metabolites from plants. First, the extraction of medicinal compounds from plant material is beneficial as it allows the formation of a more practical means of medication delivery. That being, a formulation with concentrated volumes of the active constituent, requiring the consumption of small volumes of plant-derived material which is in a far more palatable form (Hoffman, 2003).

In plant material, several constituents may interact with one another to produce physical, chemical, or therapeutic incompatibilities (Essien, Young, & Baroutian, 2020). Although this mainly applies to biochemically active compounds, several inert or unreactive compounds may result in unwanted therapeutic effects such as decreased efficiency of the therapeutic agent (Hoffman, 2003). It can therefore be beneficial to separate inert compounds from biologically active constituents through chromatography to allow the biologically active compounds to exert their full effect. The isolation and purification of pure compounds from vast amounts of raw material allow these compounds to be individually analyzed according to their physicochemical properties which subsequently determines their inclusion or exclusion in the final formulation to be administered (Essien et al., 2020).

Quorum sensing is an intricately regulated process that requires just as intricate quorum quenching molecules. The isolation and purification of these molecules favor selective interaction with the mechanism in question without the possibility of collateral cellular damage. Several examples of pure compounds isolated from medicinal plants with quorum quenching activity are illustrated in Table 9.1.

TABLE 9.1 Chemical structures of phytochemicals with potential usage in quorum quenching.

Phytochemical	Microorganism tested	Quorum quenching effect	Structure	Reference
β-sitosterol	*Listeria monocytogenes*	Disruption of cellular aggregation and subsequent biofilm formation	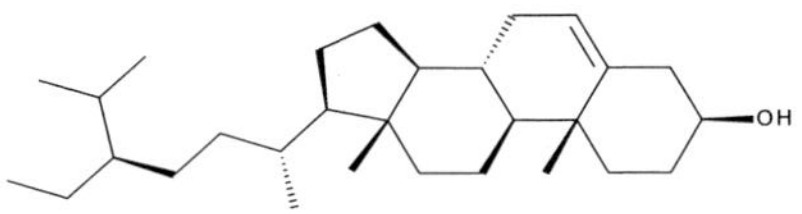 OH	Nyila, Leonard, Hussein, and Lall (2012)
Citral	*Staphylococcus aureus*	Inhibition of autoinducer 2 activity	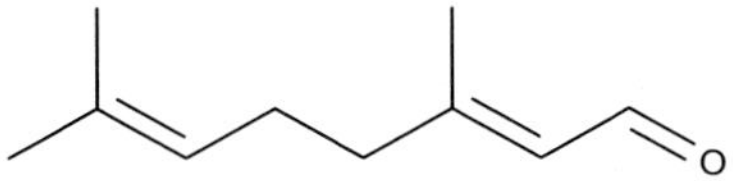 O	(Zhang et al., 2014)
Dihydroxybergamottin	*Escherichia coli*	Reduction in biofilm formation and autoinducer 2 signaling	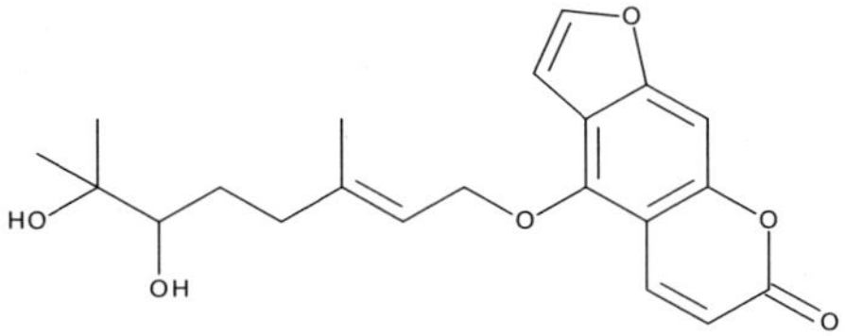 O, HO, OH, O, O, O	Cugini, Morales, and Hogan (2010)
Malic acid	*Escherichia coli* and *Salmonella typhimurium*	Autoinducer 2 inhibition	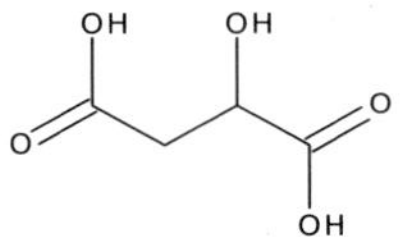 OH, OH, O, O, OH	(Almasoud et al., 2016)
Naringin	*Yersinia enterocolitica*	Inhibition of biofilm formation and acyl-homoserine lactone synthesis	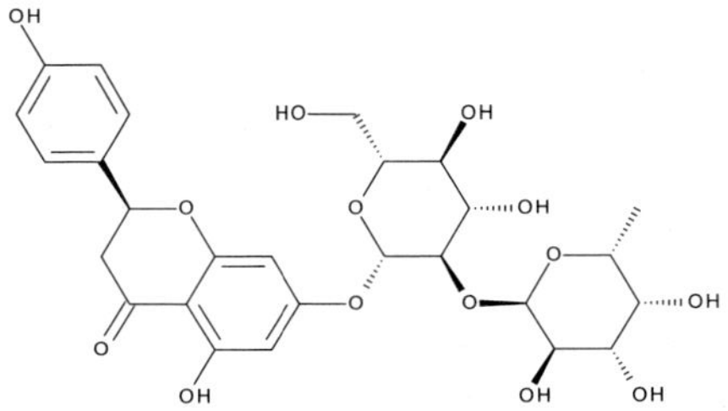 OH, HO, OH, O, OH, O, O, O, OH, O, O, OH, OH, OH	(Truchado et al., 2012)

(Continued)

TABLE 9.1 (Continued)

Phytochemical	Microorganism tested	Quorum quenching effect	Structure	Reference
Resveratrol	*Proteus mirabilis*	Reduction in swarming motility and flagellin production	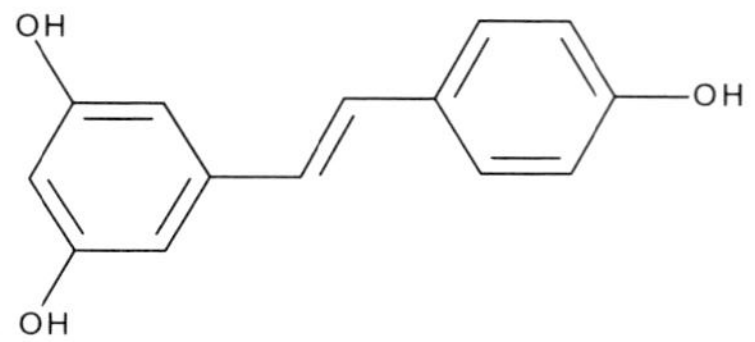	(Wang et al., 2006)
Salicylic acid	*Agrobacterium tumefaciens*	Modulation of 103 genes involved in bacterial virulence including acyl-homoserine lactone inhibition	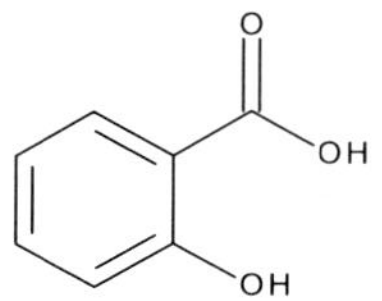	(Yuan et al., 2007)
Sesquiterpene lactones	*Pseudomonas aeruginosa*	Reduction in cell to cell communication via acyl-homoserine lactone inhibition	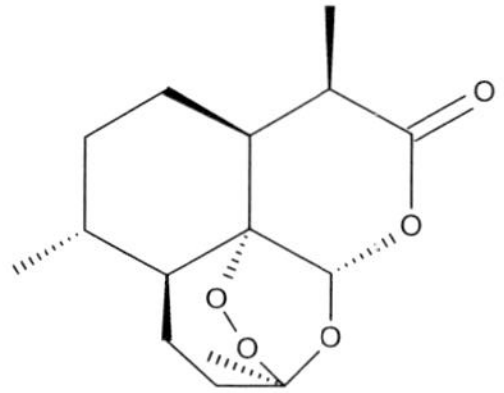	(Amaya et al., 2012)
Taxifolin	*Pseudomonas aeruginosa*	Reduction in quorum sensing regulated gene expression	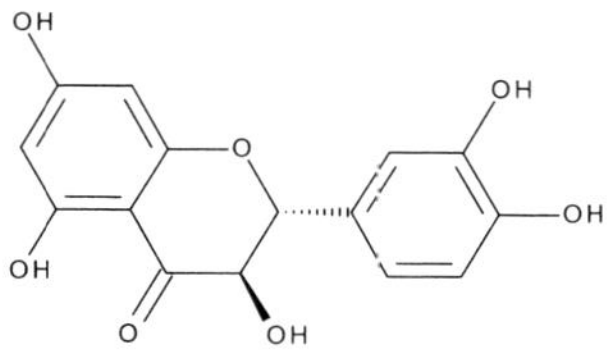	(Vandeputte et al., 2011)
Zingerone	*Chromobacterium violaceum*	Inhibition of violacein and pyocyanin production	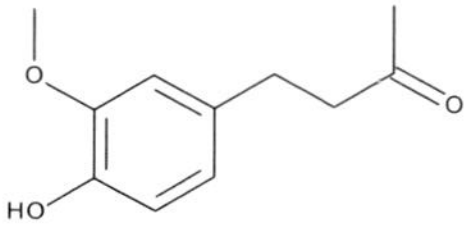	Vijendra Kumar, Murthy, Manjunatha, and Bettadaiah (2014)

Conclusion

TB, and the management thereof, remain a global health emergency considering over one-quarter of the world's population is infected with the dynamic pathogen, Mtb. Over the years unprecedented attempts to control this pathogen and how it influences human populations have been instituted; however, the efficiency of such treatment modalities has recently come under much scrutiny. As a means of sheltering themselves from the external environment and the various harmful substances that reside within, mycobacteria have devised a strategy to encapsulate themselves within a prolific and dynamic extracellular matrix, a structure known as a biofilm. This extracellular matrix not only serves as an effective mechanism to obtain and concentrate nutrients but furthermore, serves as an extensively resilient barrier to antimicrobial penetration. The haphazard use of antibiotics coupled with poor healthcare practices has led to a surge in antimicrobial resistance and has allowed the formation of resistant persisters which have been implicated as culprits in the chronicity of latent TB infections. The clinical implications of antimicrobial resistance are vast and devastating and have recently been shown to promote the emergence of multidrug resistant and extensively drug-resistant TB. Conventional treatment modalities and their mechanism of action primarily focus on their ability to inhibit the proliferation of planktonic cells with little scientific material available on the inhibition of populations of planktonic cells encapsulated within biofilms. The regulation of virulence factor production and biofilm formation and maturation lies in the quorum sensing cascade, an important target for quorum quenching compounds. Selective pressure for bactericidal antibiotic resistance is minimal when it comes to quorum quenching strategies as these novel agents do not result in bactericidal effects, but rather act via the attenuation of a host of mycobacterial virulence strategies. Throughout history, plants have been employed as medicinal alternatives in the treatment of infectious diseases. Several medicinal plants have been shown to possess highly bioactive secondary metabolites that function primarily as quorum sensing inhibitors. As an innovative treatment adjuvant, medicinal plants and their metabolic pools hold great promise in the treatment of infectious diseases such as TB. However, minimal research has been conducted in the field of medicinal plants as quorum quenching agents, and thus advances in this field will likely yield a myriad of novel compounds that can be used to combat the global TB epidemic and bring relief for individuals and populations suffering from this illness .

References

Abidi, S. H., Ahmed, K., Sherwani, S. K., Bibi, N., Kazmi, U., & Kazmi, S. (2014). Detection of *Mycobacterium smegmatis* biofilm and its control by natural agents. *International Journal of Current Microbiology and Applied Sciences*, *3*, 801–812.

Adonizio, A., Kong, K. F., & Mathee, K. (2008). Inhibition of quorum sensing-controlled virulence factor production in *Pseudomonas aeruginosa* by south Florida plant extracts. *Antimicrobial Agents and Chemotherapy*, *52*, 198–203. Available from https://doi.org/10.1128/AAC.00612-07.

Almasoud, A., Hettiarachchy, N., Rayaprolu, S., Babu, D., Kwon, Y. M., & Mauromoustakos, A. (2016). Inhibitory effects of lactic and malic organic acids on autoinducer type 2 (AI-2) quorum sensing of *Escherichia coli* O157: H7 and Salmonella Typhimurium. *LWT - Food Science and Technology*, *66*, 560–564. Available from https://doi.org/10.1016/j.lwt.2015.11.013.

Amaya, S., Pereira, J. A., Borkosky, S. A., Valdez, J. C., Bardón, A., & Arena, M. E. (2012). Inhibition of quorum sensing in *Pseudomonas aeruginosa* by sesquiterpene lactones. *Phytomedicine: International Journal of Phytotherapy and Phytopharmacology*, *19*, 1173–1177. Available from https://doi.org/10.1016/j.phymed.2012.07.003.

Antonioli, L., Blandizzi, C., Pacher, P., Guilliams, M., & Haskó, G. (2019). Rethinking communication in the immune system: The quorum sensing concept. *Trends in Immunology, 40*, 88–97. Available from https://doi.org/10.1016/j.it.2018.12.002.

Ben-Kahla, I., & AL-Hajoj, S. (2016). Drug-resistant tuberculosis viewed from bacterial and host genomes. *International Journal of Antimicrobial Agents, 48*, 353–360. Available from https://doi.org/10.1016/j.ijantimicag.2016.07.010.

Bhunu, B., Mautsa, R., & Mukanganyama, S. (2017). Inhibition of biofilm formation in *Mycobacterium smegmatis* by *Parinari curatellifolia* leaf extracts. *BMC Complementary and Alternative Medicine, 17*, 285. Available from https://doi.org/10.1186/s12906-017-1801-5.

Boles, B. R., & Horswill, A. R. (2011). Staphylococcal biofilm disassembly. *Trends in Microbiology, 19*, 449–455. Available from https://doi.org/10.1016/j.tim.2011.06.004.

Bonkat, G., Bachmann, A., Solokhina, A., Widmer, A. F., Frei, R., Gasser, T. C., & Braissant, O. (2012). Growth of mycobacteria in urine determined by isothermal microcalorimetry: Implications for urogenital tuberculosis and other mycobacterial infections. *Urology, 80*, 1163. Available from https://doi.org/10.1016/j.urology.2012.04.050, e12.

Calmette, A. (1936). Tuberculosis in man and animals.

Chakraborty, P., & Kumar, A. (2019). The extracellular matrix of mycobacterial biofilms: Could we shorten the treatment of mycobacterial infections? *Microbial Cell., 6*, 105–122. Available from https://doi.org/10.15698/mic2019.02.667.

Chen, L., & Wen, Y. (2011). The role of bacterial biofilm in persistent infections and control strategies. *International Journal of Oral Science, 3*, 66–73. Available from https://doi.org/10.4248/IJOS11022.

Costerton, J. W., Geesey, G. G., & Cheng, K. J. (1978). How bacteria stick. *Scientific American, 238*, 86–95. Available from https://doi.org/10.1038/scientificamerican0178-86.

Crabbé, A., Jensen, P. Ø., Bjarnsholt, T., & Coenye, T. (2019). Antimicrobial tolerance and metabolic adaptations in microbial biofilms. *Trends in Microbiology, 27*, 850–863. Available from https://doi.org/10.1016/j.tim.2019.05.003.

Cugini, C., Morales, D. K., & Hogan, D. A. (2010). *Candida albicans*-produced farnesol stimulates *Pseudomonas quinolone* signal production in LasR-defective *Pseudomonas aeruginosa* strains. *Microbiology (Reading, England), 156*, 3096–3107. Available from https://doi.org/10.1099/mic.0.037911-0.

da Silva, D. P., Schofield, M. C., Parsek, M. R., & Tseng, B. S. (2017). An Update on the Sociomicrobiology of Quorum Sensing in Gram-Negative Biofilm Development. *Pathogens* 2017, *6*, 51. (CC BY-SA 4.0). Accessed January 2022. Available form https://doi.org/10.3390/pathogens6040051.

Danese, P. N., Pratt, L. A., & Kolter, R. (2000). Exopolysaccharide production is required for development of *Escherichia coli* K-12 biofilm architecture. *Journal of Bacteriology, 182*, 3593–3596. Available from https://doi.org/10.1128/JB.182.12.3593-3596.2000.

Danilchanka, O., Sun, J., Pavlenok, M., Maueröder, C., Speer, A., Siroy, A., … Niederweis, M. (2014). An outer membrane channel protein of *Mycobacterium tuberculosis* with exotoxin activity. *Proceedings of the National Academy of Sciences of the United States of America, 111*, 6750–6755. Available from https://doi.org/10.1073/pnas.1400136111.

Erhabor, C. R., Erhabor, J. O., & McGaw, L. J. (2019). The potential of South African medicinal plants against microbial biofilm and quorum sensing of foodborne pathogens: A review. *South African Journal of Botany, 126*, 214–231. Available from https://doi.org/10.1016/j.sajb.2019.07.024.

Essien, S. O., Young, B., & Baroutian, S. (2020). Recent advances in subcritical water and supercritical carbon dioxide extraction of bioactive compounds from plant materials. *Trends in Food Science and Technology, 97*, 156–169. Available from https://doi.org/10.1016/j.tifs.2020.01.014.

Esteban, J., & García-Coca, M. (2018). Mycobacterium biofilms. *Frontiers in Microbiology, 8*, 2651. Available from https://doi.org/10.3389/fmicb.2017.02651.

Esteban, J., Martín-De-Hijas, N. Z., Kinnari, T. J., Ayala, G., Fernández-Roblas, R., & Gadea, I. (2008). Biofilm development by potentially pathogenic non-pigmented rapidly growing mycobacteria. *BMC Microbiology, 8*, 184. Available from https://doi.org/10.1186/1471-2180-8-184.

Feng, Y., Diao, N., Shao, L., Wu, J., Zhang, S., Jin, J., … Zhang, W. (2012). Interferon-gamma release assay performance in pulmonary and extrapulmonary tuberculosis. *PLoS One, 7*, e32652. Available from https://doi.org/10.1371/journal.pone.0032652.

Forbes, B. A. (2017). Mycobacterial taxonomy. *Journal of Clinical Microbiology, 55*, 380–383. Available from https://doi.org/10.1128/JCM.01287-16.

Hans, S., Sharma, S., Hameed, S., & Fatima, Z. (2017). Sesamol exhibits potent antimycobacterial activity: Underlying mechanisms and impact on virulence traits. *Journal of Global Antimicrobial Resistance, 10*, 228–237. Available from https://doi.org/10.1016/j.jgar.2017.06.007.

Hoffman, D. (2003). *Chapter 5: BOT 748. An introduction to phytochemistry in medical herbalism.*

Hyldgaard, M., Mygind, T., & Meyer, R. L. (2012). Essential oils in food preservation: Mode of action, synergies, and interactions with food matrix components. *Frontiers in Microbiology, 3*, 12. Available from https://doi.org/10.3389/fmicb.2012.00012.

Jiang, C.-H., Gan, M.-L., An, T.-T., & Yang, Z.-C. (2019). Bioassay-guided isolation of a *Mycobacterium tuberculosis* bioflim inhibitor from *Arisaema sinii* Krause. *Microbial Pathogenesis, 126*, 351–356. Available from https://doi.org/10.1016/j.micpath.2018.11.022.

Junqueira, A. D., Faria, M. L. M., Martins, O. L., de Oliveira, L. L. P. M., Anthony, H. J., Boas, R. B. V., ... Teixeira, C. D. (2020). Exploring how structural changes to new Licarin A derivatives effects their bioactive properties against rapid growing mycobacteria and biofilm formation. *Microbial Pathogenesis, 144*, 104203. Available from https://doi.org/10.1016/j.micpath.2020.104203.

Karami, P., Khaledi, A., Mashoof, R. Y., Yaghoobi, M. H., Karami, M., Dastan, D., & Alikhani, M. Y. (2020). The correlation between biofilm formation capability and antibiotic resistance pattern in *Pseudomonas aeruginosa*. *Gene Reports, 18*, 100561. Available from https://doi.org/10.1016/j.genrep.2019.100561.

Khaledi, A., Esmaeili, D., Jamehdar, S. A., Esmaeili, S. A., Neshani, A., & Bahador, A. (2016). Expression of MFS efflux pumps among multidrug resistant *Acinetobacter baumannii* clinical isolates. *Der Pharmacia Lettre, 8*, 262–267. Available from http://scholarsresearchlibrary.com/dpl-vol8-iss2/DPL-2016-8-2-262-267.pdf.

Khan, F., Jeong, M. C., Park, S. K., Kim, S. K., & Kim, Y. M. (2019). Contribution of chitooligosaccharides to biofilm formation, antibiotics resistance and disinfectants tolerance of Listeria monocytogenes. *Microbial Pathogenesis, 136*, 103673. Available from https://doi.org/10.1016/j.micpath.2019.103673.

Koch, R. (1982). The etiology of tuberculosis. *Reviews of Infectious Diseases, 4*, 1270–1274. Available from https://doi.org/10.1093/clinids/4.6.1270.

Kordbacheh, H., Eftekhar, F., & Ebrahimi, S. N. (2017). Anti-quorum sensing activity of *Pistacia atlantica* against *Pseudomonas aeruginosa* PAO1 and identification of its bioactive compounds. *Microbial Pathogenesis, 110*, 390–398. Available from https://doi.org/10.1016/j.micpath.2017.07.018.

Langeveld, W. T., Veldhuizen, E. J. A., & Burt, S. A. (2014). Synergy between essential oil components and antibiotics: A review. *Critical Reviews in Microbiology, 40*, 76–94. Available from https://doi.org/10.3109/1040841X.2013.763219.

Loöwenstein, E. (1920). Vorlesungen uöber Bakteriologie, Immunitaöt, spezifische Diagnostik und Therapie der Tuberkulose.

Madacki, J., Mas Fiol, G., & Brosch, R. (2019). Update on the virulence factors of the obligate pathogen *Mycobacterium tuberculosis* and related tuberculosis-causing mycobacteria. *Infection, Genetics and Evolution, 72*, 67–77. Available from https://doi.org/10.1016/j.meegid.2018.12.013.

Magi, G., Marini, E., & Facinelli, B. (2015). Antimicrobial activity of essential oils and carvacrol, and synergy of carvacrol and erythromycin, against clinical, erythromycin-resistant Group A *Streptococci*. *Frontiers in Microbiology, 6*, 165. Available from https://doi.org/10.3389/fmicb.2015.00165.

Marini, E., Di Giulio, M., Ginestra, G., Magi, G., Di Lodovico, S., Marino, A., ... Nostro, A. (2019). Efficacy of carvacrol against resistant rapidly growing mycobacteria in the planktonic and biofilm growth mode. *PLoS One, 14*, e0219038. Available from https://doi.org/10.1371/journal.pone.0219038.

Nostro, A., & Papalia, T. (2012). Antimicrobial activity of carvacrol: Current progress and future prospectives. *Recent Patents on Anti-Infective Drug Discovery, 7*, 28–35. Available from https://doi.org/10.2174/157489112799829684.

Nyila, M. A., Leonard, C. M., Hussein, A. A., & Lall, N. (2012). Activity of South African medicinal plants against Listeria monocytogenes biofilms, and isolation of active compounds from *Acacia karroo*. *South African Journal of Botany, 78*, 220–227. Available from https://doi.org/10.1016/j.sajb.2011.09.001.

Ojha, A. K., Baughn, A. D., Sambandan, D., Hsu, T., Trivelli, X., Guerardel, Y., ... Hatfull, G. F. (2008). Growth of *Mycobacterium tuberculosis* biofilms containing free mycolic acids and harbouring drug-tolerant bacteria. *Molecular Microbiology, 69*, 164–174. Available from https://doi.org/10.1111/j.1365-2958.2008.06274.x.

Oosthuizen, C. B., Gasa, N., Hamilton, C. J., & Lall, N. (2019). Inhibition of mycothione disulphide reductase and mycobacterial biofilm by selected South African plants. *South African Journal of Botany, 120*, 291–297. Available from https://doi.org/10.1016/j.sajb.2018.09.015.

Packiavathy, I. A. S. V., Priya, S., Pandian, S. K., & Ravi, A. V. (2014). Inhibition of biofilm development of uropathogens by curcumin - An anti-quorum sensing agent from *Curcuma longa*. *Food Chemistry*, *148*, 453–460. Available from https://doi.org/10.1016/j.foodchem.2012.08.002.

Paul, D., Gopal, J., Kumar, M., & Manikandan, M. (2018). Nature to the natural rescue: Silencing microbial chats. *Chemico-Biological Interactions*, *280*, 86–98. Available from https://doi.org/10.1016/j.cbi.2017.12.018.

Phetlhu, D. R., Bimerew, M., Marie-Modeste, R. R., Naidoo, M., & Igumbor, J. (2018). Nurses' knowledge of tuberculosis, HIV, and integrated HIV/TB care policies in rural Western Cape, South Africa. *Journal of the Association of Nurses in AIDS Care*, *29*, 876–886. Available from https://doi.org/10.1016/j.jana.2018.05.008.

Recht, J., & Kolter, R. (2001). Glycopeptidolipid acetylation affects sliding motility and biofilm formation in *Mycobacterium smegmatis*. *Journal of Bacteriology*, *183*, 5718–5724. Available from https://doi.org/10.1128/JB.183.19.5718-5724.2001.

Reed, M. B., Domenech, P., Manca, C., Su, H., Barczak, A. K., Kreiswirth, B. N., & Kaplan, G. (2004). A glycolipid of hypervirulent tuberculosis strains that inhibits the innate immune response. *Nature*, *431*, 84–87. Available from https://doi.org/10.1038/nature02837.

Romero, C. M., Vivacqua, C. G., Abdulhamid, M. B., Baigori, M. D., Slanis, A. C., de Allori, M. C. G., & Tereschuk, M. L. (2016). Biofilm inhibition activity of traditional medicinal plants from Northwestern Argentina against native pathogen and environmental microorganisms. *Revista da Sociedade Brasileira de Medicina Tropical*, *49*, 703–712. Available from https://doi.org/10.1590/0037-8682-0452-2016.

Sarathy, J. P., Via, L. E., Weiner, D., Blanc, L., Boshoff, H., Eugenin, E. A., … Dartois, V. A. (2018). Extreme drug tolerance of *Mycobacterium tuberculosis* in Caseum. *Antimicrobial Agents and Chemotherapy*, *62*. Available from https://doi.org/10.1128/AAC.02266-17, e02266-17.

Sassetti, C. M., & Rubin, E. J. (2003). Intracellular replication is essential for the virulence of *Salmonella typhimurium*. *Proceedings of the National Academy of Sciences of the United States of America.*, *88*, 11470–11474. Available from https://doi.org/10.1073/pnas.88.24.11470.

Sharma, D., Misba, L., & Khan, A. U. (2019). Antibiotics versus biofilm: An emerging battleground in microbial communities. *Antimicrobial Resistance and Infection Control*, *8*, 76. Available from https://doi.org/10.1186/s13756-019-0533-3.

Shirtliff, M. E., Mader, J. T., & Camper, A. K. (2002). Molecular interactions in biofilms. *Chemistry and Biology*, *9*, 859–871. Available from https://doi.org/10.1016/S1074-5521(02)00198-9.

Solokhina, A., Bonkat, G., Kulchavenya, E., & Braissant, O. (2018). Drug susceptibility testing of mature *Mycobacterium tuberculosis* H37Ra and *Mycobacterium smegmatis* biofilms with calorimetry and laser spectroscopy. *Tuberculosis*, *113*, 91–98. Available from https://doi.org/10.1016/j.tube.2018.09.010.

Stewart, P. S., & Costerton, J. W. (2001). Antibiotic resistance of bacteria in biofilms. *Lancet*, *358*, 135–138. Available from https://doi.org/10.1016/S0140-6736(01)05321-1.

Tegos, G. P., & Hamblin, M. R. (2013). Disruptive innovations: New anti-infectives in the age of resistance. *Current Opinion in Pharmacology*, *13*, 673–677. Available from https://doi.org/10.1016/j.coph.2013.08.012.

Truchado, P., Giménez-Bastida, J. A., Larrosa, M., Castro-Ibáñez, I., Espín, J. C., Tomás-Barberán, F. A., … Allende, A. (2012). Inhibition of quorum sensing (QS) in yersinia enterocolitica by an Orange extract rich in glycosylated flavanones. *Journal of Agricultural and Food Chemistry*, *60*, 8885–8894. Available from https://doi.org/10.1021/jf301365a.

Ufimtseva, E. G., Eremeeva, N. I., Petrunina, E. M., Umpeleva, T. V., Bayborodin, S. I., Vakhrusheva, D. V., & Skornyakov, S. N. (2018). *Mycobacterium tuberculosis* cording in alveolar macrophages of patients with pulmonary tuberculosis is likely associated with increased mycobacterial virulence. *Tuberculosis*, *112*, 1–10. Available from https://doi.org/10.1016/j.tube.2018.07.001.

Vadakkan, K., Gunasekaran, R., Choudhury, A. A., Ravi, A., Arumugham, S., Hemapriya, J., & Vijayanand, S. (2018). Response surface modelling through Box-Behnken approach to optimize bacterial quorum sensing inhibitory action of *Tribulus terrestris* root extract. *Rhizosphere*, *6*, 134–140. Available from https://doi.org/10.1016/j.rhisph.2018.06.005.

Vandeputte, O. M., Kiendrebeogo, M., Rajaonson, S., Diallo, B., Mol, A., Jaziri, M. E., & Baucher, M. (2010). Identification of catechin as one of the flavonoids from combretum albiflorum bark extract that reduces the production of quorum-sensing-controlled virulence factors in pseudomonas aeruginosa PAO1. *Applied and Environmental Microbiology*, *76*, 243–253. Available from https://doi.org/10.1128/AEM.01059-09.

Vandeputte, O. M., Kiendrebeogo, M., Rasamiravaka, T., Stévigny, C., Duez, P., Rajaonson, S., … el Jaziri, M. (2011). The flavanone naringenin reduces the production of quorum sensing-controlled virulence factors in pseudomonas aeruginosa PAO1. *Microbiology (Reading, England)*, *157*, 2120–2132. Available from https://doi.org/10.1099/mic.0.049338-0.

Vijendra Kumar, N., Murthy, P. S., Manjunatha, J. R., & Bettadaiah, B. K. (2014). Synthesis and quorum sensing inhibitory activity of key phenolic compounds of ginger and their derivatives. *Food Chemistry*, *159*, 451–457. Available from https://doi.org/10.1016/j.foodchem.2014.03.039.

Wang, W. B., Lai, H. C., Hsueh, P. R., Chiou, R. Y. Y., Lin, S. B., & Liaw, S. J. (2006). Inhibition of swarming and virulence factor expression in *Proteus mirabilis* by resveratrol. *Journal of Medical Microbiology*, *55*, 1313–1321. Available from https://doi.org/10.1099/jmm.0.46661-0.

Wei, Q., Bhasme, P., Wang, Z., Wang, L., Wang, S., Zeng, Y., ... Li, Y. (2020). Chinese medicinal herb extract inhibits PQS-mediated quorum sensing system in *Pseudomonas aeruginosa*. *Journal of Ethnopharmacology*, *248*, 112272. Available from https://doi.org/10.1016/j.jep.2019.112272.

WHO. (2019). Global tuberculosis report. Retrieved from https://www.who.int/teams/global-tuberculosis-programme/global-report-2019.

Yuan, Z. C., Edlind, M. P., Liu, P., Saenkham, P., Banta, L. M., Wise, A. A., ... Nester, E. W. (2007). The plant signal salicylic acid shuts down expression of the vir regulon and activates quormone-quenching genes in Agrobacterium. *Proceedings of the National Academy of Sciences of the United States of America*, *104*, 11790–11795. Available from https://doi.org/10.1073/pnas.0704866104.

Zhang, Y. (2014). Persisters, persistent infections and the Yin-Yang model. *Emerging Microbes and Infections*, *3*. Available from https://doi.org/10.1038/emi.2014.3.

Zhang, H., Zhou, W., Zhang, W., Yang, A., Liu, Y., Jiang, Y., ... Su, J. (2014). Inhibitory effects of citral, cinnamaldehyde, and tea polyphenols on mixed biofilm formation by foodborne *Staphylococcus aureus* and *Salmonella enteritidis*. *Journal of Food Protection*, *77*, 927–933. Available from https://doi.org/10.4315/0362-028X.JFP-13-497.

Zhang, Y. J., Reddy, M. C., Ioerger, T. R., Rothchild, A. C., Dartois, V., Schuster, B. M., ... Rubin, E. J. (2013). Tryptophan biosynthesis protects mycobacteria from CD4 T-Cell-mediated Killing. *Cell.*, *155*, 1296–1308. Available from https://doi.org/10.1016/j.cell.2013.10.045.

CHAPTER

10

Untargeted metabolomics for the study of antiinfective plants

Joshua J. Kellogg

Department of Veterinary and Biomedical Sciences, Pennsylvania State University, University Park, PA, United States

Introduction

Natural products represent a tremendously rich resource for drug discovery paradigms or as dietary supplements to improve human health. They arise from a variety of sources, including bacteria, fungi, marine organisms, and plants (Newman & Cragg, 2020). In their natural state, natural products are complex mixtures containing many structurally diverse components (Enke & Nagels, 2011). For the majority of natural products, the identities of potential bioactive constituents are often not known, and also likely to differ depending on the biological activity evaluated (Alvarez-Zapata et al., 2015; Simmler et al., 2016). A primary obstacle facing researchers is unraveling the intricacies of such complex mixtures and assigning structures and biological activities to the components. While traditional methods such as bioassay-guided fractionation (BGF) have had decades of success, newer methodologies based upon metabolomics are poised to delve deeper into the chemical make-up of natural products and are able to couple with biological activity evaluations in order to enhance bioactive compound discovery efforts. With this chapter, we review various strategies that can be employed and discuss their advantages and disadvantages in hunting for potential antiinfective compounds from plants.

Plants as sources of antiinfective agents

Natural products have long served as the foundation for antiinfective bioactive molecule discovery, though the majority of discoveries have arrived from microbial natural sources (Newman & Cragg, 2020). However, as analytical instrumentation has improved, along with the implementation of high-throughput techniques that can accommodate

Medicinal Plants as Anti-infectives
DOI: https://doi.org/10.1016/B978-0-323-90999-0.00017-3

more complex botanical samples (Harvey, Edrada-Ebel, & Quinn, 2015; Shen, 2015), plants have gained heightened attention for their potential bioactive secondary metabolites. Indeed, plants hold significant interest as potential sources of antiinfective agents due to their long history of biochemical defenses against microbial pathogens, (Adedeji & Babalola, 2020; Zaynab et al., 2018). Plants also possess a diverse array of structurally unique secondary metabolites from the estimated 374,000 species worldwide (Christenhusz & Byng, 2016), many of which are prominently featured in the traditional ecological knowledge and medicinal pharmacopeia of communities across the globe (Porras et al., 2020; Willis, 2017). However, with such diversity of both species and chemistry, accessing the relevant bioactive metabolites from a complex matrix can be challenging. Traditionally, iterative separative techniques have been employed to try and tease out the molecular underpinnings of bioactivity.

Bioassay-guided fractionation

Identifying novel chemical structures from natural sources can be a complex and lengthy process. The long-standing approach to isolation and identification of bioactive constituent(s) from plants (or other natural sources) is a robust repertoire of techniques broadly referred to as "bioassay-guided fractionation" (BGF) (Kinghorn et al., 1998). Bioassay-guided fractionation is an iterative process, with successive rounds of chromatographic separation and bioassay until a single bioactive entity (i.e., an isolated pure compound) is achieved (Fig. 10.1). This approach has remained the "gold standard" of bioactive discovery from natural resources for decades (Selander et al., 2015) and has resulted in the discovery of a host of critically important pharmaceutical agents, including camptothecin and taxol (paclitaxel) (Oberlies & Kroll, 2004; Wani, Taylor, Wall, Coggon, & McPhail, 1971), artemisinin (Tu, 2011), and vinblastine (Noble, 1990). However, a bioassay-guided fractionation approach possesses several limitations. It is possible that, at the end of multiple rounds of chromatographic separation on different resins and orthogonal techniques, isolation of a single bioactive agent fails due to chemical degradation on the chromatography matrices during separations, or "disappears" due to irreversible binding to the resin (Qiu et al., 2013). In addition, it is often possible that the observed activity evidenced by a complex sample is due to multiple compounds working in concert (e.g., synergy or additivity), and the biological effect is not observed once the purification separates them (Caesar & Cech, 2019). Because the process of refractionation and iterative bioassay is time-, labor-, and resource-consuming, some have begun to consider this approach too risky to pursue (Li & Vederas, 2009). There has thus been considerable interest in developing approaches that can help identify candidate metabolites earlier in the isolation workflow to prioritize and guide isolation efforts.

Metabolomics

The metabolome is defined as the complement of low-molecular-weight metabolites (ca. <1200 Da) present in a sample (e.g., environmental sample, biological fluid, cell, or organism) representing a snapshot of the system at a particular set of physiological

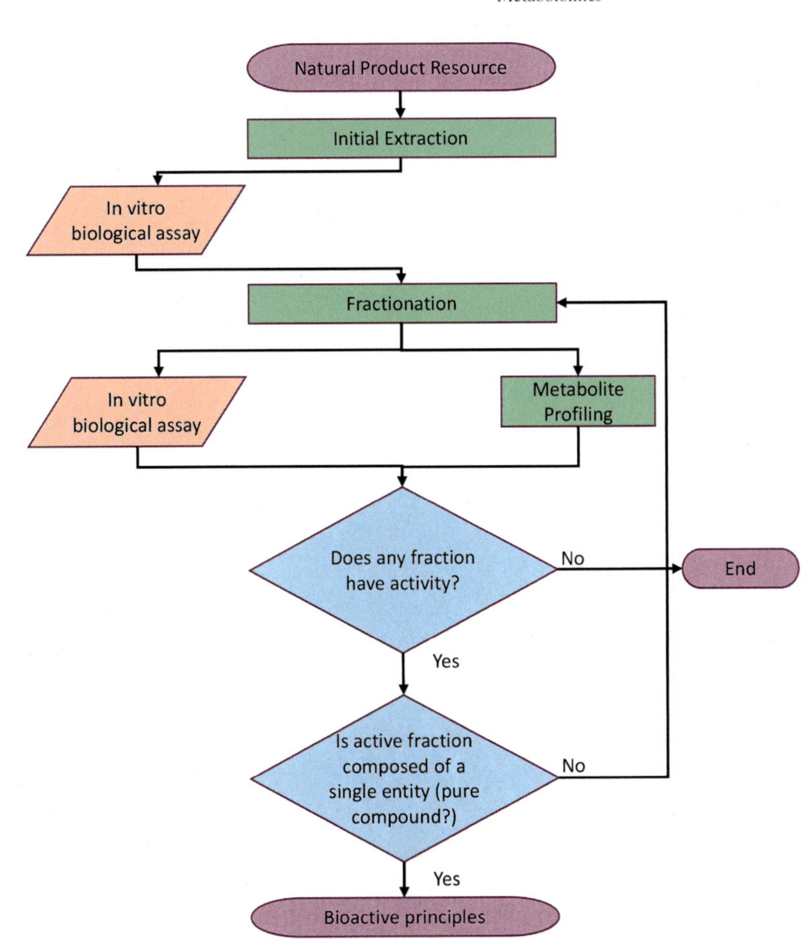

FIGURE 10.1 A generalized scheme for bioassay-guided fractionation and isolation of bioactive constituents from a complex mixture. The iterative process is generally: (i) extraction from the original resource (usually via solvent maceration); (ii) fractionation of the extract via chromatographic separation or partitioning; (iii) assaying the resulting fractions for activity; (iv) repeat (ii) and (iii) until a single molecule is isolated that demonstrates activity; and (v) structural identification of the molecule.

conditions. This could represent a multitude of different possible stimuli, including genetic variations, pathogenic infestation, or other abiotic or biotic stresses. Similarly, the goal of metabolomics is the chemical profiling of a phenotype through the qualitative and/or quantitative analysis of all measurable metabolites in a complex system and the measurement of the change in the metabolite profile due to some type of challenge or perturbation (Oliver, Winson, Kell, & Baganz, 1998). In ca. two decades since the term "metabolomics" was coined (Oliver et al., 1998), metabolomics has developed into an important analytical tool for a wide variety of applications, including studies on diseases (Gowda et al., 2008), toxicity (Ramirez et al., 2018), quality control (Shu et al., 2017), and natural products (Kellogg, Paine, McCune, Oberlies, & Cech, 2019).

Metabolomics strategies can be broadly divided into untargeted and targeted approaches, each with their own advantages and limitations. Untargeted metabolomics focuses on the analysis of all detectable metabolites within a sample, regardless of whether the chemical identity can be ascertained. However, the presence of such a large number of chemical constituents (especially unknown metabolites) makes quantification difficult (Cajka & Fiehn, 2016).

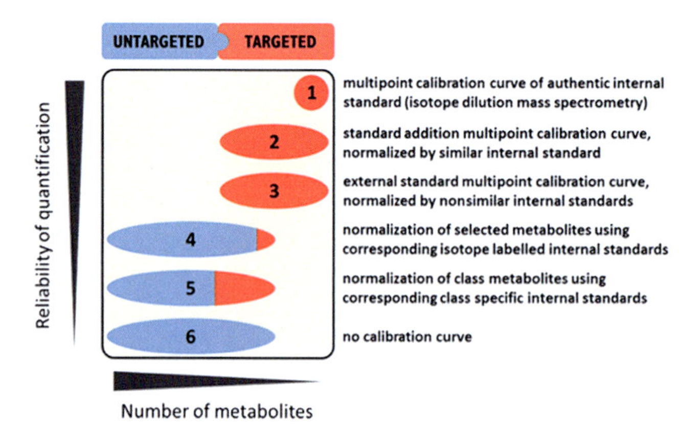

FIGURE 10.2 Untargeted and targeted metabolomics approaches. Untargeted and targeted metabolomics approaches differ in two critical areas: the number of metabolites detected (x-axis) and the ability to quantify the observed metabolites (y-axis). Source: *From Cajka, T., & Fiehn, O. (2016). Toward merging untargeted and targeted methods in mass spectrometry-based metabolomics and lipidomics.* Analytical Chemistry, *88, 524–545. https://doi.org/10.1021/acs.analchem.5b04491.*

By contrast, targeted metabolomics centers around the measurement of a defined group of metabolites, often with known chemical structures. While this does not provide as wide metabolite coverage as untargeted approaches, targeted methods have the potential to reliably quantify the targeted group of metabolites, via internal standards or multipoint calibration curves (Fig. 10.2). This allows for more quantitative measurements of the metabolome subset.

Methods of detection

There are a variety of analytical instruments currently employed to provide a chemical profile that forms the basis for metabolomics analyses. Studies have been carried out using several forms of spectroscopy, such as ultraviolet-visible spectroscopy (Anđelković et al., 2017) and Fourier-transformed infrared spectroscopy (Liu, Finley, Betz, & Brown, 2018). However, the two main analytical techniques currently employed for metabolomic studies are mass spectrometry (MS) and nuclear magnetic resonance (NMR). NMR-based metabolomic acquisition offers an unbiased assessment of a complex sample's composition, able to measure all metabolites therein, allowing for the simultaneous identification and quantification of diverse metabolites (Pauli et al., 2014). MS-based metabolomic methods have the advantages of greater sensitivity compared to NMR spectroscopy by several orders of magnitude and the ability to couple directly to separation methods such as gas chromatography (GC) or liquid chromatography (LC). However, a disadvantage of MS analysis is that ionization is required to detect sample components, yet not every molecule is ionized in a mass spectrometer (Cech & Yu, 2013), and those metabolites which are ionized are not universally uniformly ionized, which complicates quantification of metabolites in the sample. The advancements in these two analytical techniques have driven many developments in metabolomics profiling, and modern systems are able to provide exceptional sensitivity, signal-to-noise ratio, and levels of structural characterization. However, the purpose of this chapter is not to extensively review analytical techniques; the relative advantages of these approaches have been discussed elsewhere (Emwas et al., 2019; Jorge, Mata, & António, 2016; Markley et al., 2017; Moco et al., 2007).

Data analysis

Metabolomic analysis results in the generation of large datasets comprising both major and minor components from the sample, which requires statistical tools to parse the data and provide significant analysis. The most employed statistical tool for characterization and comparison of metabolomics data sets is the unsupervised method principal component analysis (Rajalahti & Kvalheim, 2011), while supervised methods for statistical analysis include soft independent modeling by class analogy (Wallace, Todd, Harnly, Cech, & Kellogg, 2020). However, it should be noted that these analyses consider only the chemical data from metabolomic profiling. Incorporating a dependent variable, such as bioactivity, requires different multivariate statistical analyses [e.g., partial least squares (PLS); see below for more details].

Many metabolomics studies focus on identifying potentially relevant features from a complex profile (or eliminating potentially irrelevant signals or noise). Advances in statistical analysis allow metabolomics studies to be utilized in more detailed or complex inquiries. One recent informatics tool that allows for these new inquiries is molecular networking, where molecules are clustered based upon their fragmentation MS/MS (aka MS^2, or MS^n) ion patterns (Watrous et al., 2012; Yang et al., 2013). The MS/MS fragmentation ions from each precursor feature are compared against each other and assigned a similarity score ("cosine score"); the underlying principle is that molecules with similar backbone structures contain comparable substructures upon fragmentation. The data are converted into a network diagram, with nodes representing MS/MS spectra of precursor ions at distinct *m/z* values, while the connections between nodes indicate a similarity in the fragmentation pattern (cosine score). Nodes are clustered based upon potential structural similarities as well as specific chemical properties (Watrous et al., 2012). These clusters can be supplemented with standards or database entries to facilitate annotation and structural identification (Guthals, Watrous, Dorrestein, & Bandeira, 2012).

Biochemometrics

The profiling characteristics of metabolomics studies possess several advantages when screening for bioactive metabolites, compared to the traditional bioassay-guided fraction approach (Calderon, 2017; Roberts et al., 2019). First, metabolomic profiling does not require multiple iterative separation and purification steps, and thus unstable compounds are more likely to be detected and measured. In addition, since there is less emphasis on chromatographic separation of the sample pre-analysis, additive or synergistic effects are more likely to be detected (Britton, Kellogg, Kvalheim, & Cech, 2018; Caesar & Cech, 2019). In addition, even in untargeted approaches, qualitative or relative quantitative differences in the metabolomic profiles of different samples can suggest potential chemical signals underlying the observed phenotypic changes (Prince & Pohnert, 2010). Thus, metabolomic fingerprinting has increasingly been integrated with bioactivity data to statistically model changing behavior as the chemical composition varies across different taxa, products, or fractions. These methods, collectively termed "biochemometrics" (Kellogg et al., 2016; Martens, Bruun, Adt, Sockalingum, & Kohler, 2006), have become a primary driver in improving the efficiency of bioactive molecule discovery from natural products and other resources (Wyss, Llivina, & Calderón, 2019). For the integration of these two disparate datasets into a single predictive model, multivariate statistical methods are required. Several different statistical approaches have been employed for this purpose, including Pearson correlations (Inui, Wang, Pro, Franzblau, & Pauli, 2012; Nothias et al., 2018; Richards et al., 2018), PLS (Britton et al., 2018; Kellogg et al., 2016; Kvalheim et al., 2011), PLS-discriminant analysis (PLS-DA, OPLS-DA) (Alvarez-Zapata et al., 2015; Chagas-Paula, Zhang, Da Costa, & Edrada-Ebel, 2015; Wen et al., 2018), and hierarchical cluster analysis (Patras et al., 2011). PLS has emerged as one of the foremost statistical modeling approaches for biochemometrics studies, and often generates multidimensional models, which can be difficult to deconstruct. To aid in the interpretation of the model, several visualization metrics and plots have been developed to interpret PLS models, with the variable importance in projection (VIP) method, the S-plot, and the selectivity ratio being the leading metrics (Farrés, Platikanov, Tsakovski, & Tauler, 2015; Kellogg et al., 2016; Rajalahti, Arneberg, Berven et al., 2009; Rajalahti, Arneberg, Kroksveen et al., 2009).

Metabolomics-driven antiinfective discovery from plants

The versatility and robustness of metabolomic and biochemometric analyses have improved the discovery of phytochemicals that target various infective agents without the intensive process of BGF. From the Pauli group, one study used Pearson correlation to identify constituents from the Alaskan ethnobotanical *Oplopanax horridus* (Sm.) Miq. (devil's club) possessing antituberculosis activity. Biochemometric analysis using orthogonal chromatographic methods (countercurrent and GC) coupled to MS identified the c. 100 most active constituents from the plant extract through Pearson correlations, which subsequently resulted in the annotation of 29 bioactive structures from three dominant structural classes which were present in the 19 initial fractions produced (Inui et al., 2012; Li et al., 2013; Qiu et al., 2013) (Fig. 10.3). Another respiratory-focused experiment investigated the antitussive and expectorant properties of *Tussilago farfara*, a traditional herbal medicine found in both European and Chinese pharmacopeia. This study examined multiple tissue types from *T. farfara*, including roots, leaves, and flower buds, and combined an in vivo expectorant test with 1H NMR metabolomic profiling and PCA analysis. It was revealed that the roots had no discernable antitussive or expectorant effects, while the leaves and flower buds both possessed efficacy. The biochemometric analysis from the untargeted metabolome indicated significantly different chemical profiles between the leaves, flower buds, and roots and highlighted three metabolites correlating with active constituents: chlorogenic acid, 3,5-dicaffeoylquinic acid, and rutin (Li et al., 2013).

Metabolomic profiling of essential oils has also revealed potential antibacterial compounds. Torch ginger (*Etlingera elatior* (Jack) R.M.Sm.) is an edible plant rich in

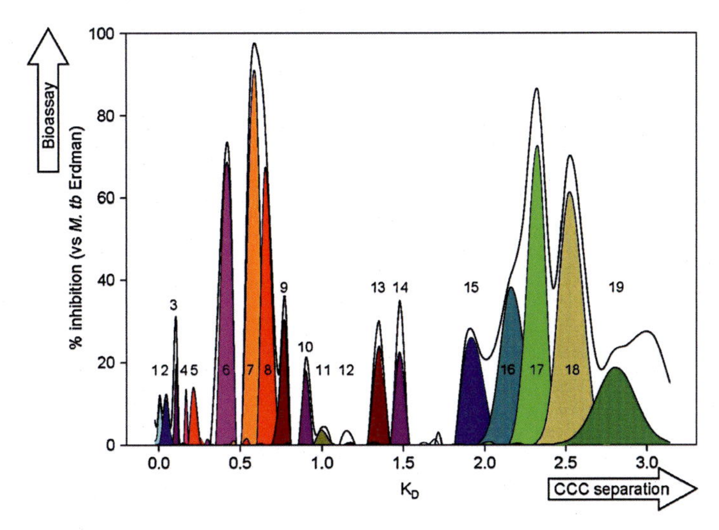

FIGURE 10.3 Biochromatogram of a fractionated *O. horridus* (Sm.) Miq. extract. The *x*-axis represents the countercurrent chromatographic fractions, while the *y*-axis indicates the antituberculosis activities of the fractions. The bioactivity observed at 50 μg/mL (black line) yielded 19 biopeaks (shaded), which represent the active components from devil's club and underwent further GC-MS analysis and biochemometrics using Pearson's correlation. Source: *From Inui, T., Wang, Y., Pro, S.M., Franzblau, S.G., & Pauli, G.F. (2012). Unbiased evaluation of bioactive secondary metabolites in complex matrices.* Fitoterapia, 83, *1218–1225.*

phytochemicals with well-documented pharmacological properties (Chan, Lim, & Wong, 2011). Torch ginger flower oil was subjected to metabolomics profiling using both gas chromatography-mass spectrometry (GC-MS) and ^{1}H-NMR approaches. Evaluating the antibacterial activity of the torch ginger flower oil via an agar diffusion assay demonstrated strong antibacterial activity against Gram-negative and Gram-positive bacterial strains *Salmonella typhimurium*, *Staphylococcus aureus*, and *Escherichia coli*, with sub-mg/mL MIC. The metabolomics analysis revealed 33 compounds using GC-MS, 15 of which were previously known for their antimicrobial activity. In addition, 16 metabolites were identified from the ^{1}H-NMR analysis and eight of those had antibacterial activity (Anzian et al., 2020). This study highlighted the potential benefits of using multiple analytical instruments to broaden coverage of the metabolome. *Cinnamomum camphora* (L.) J. Presl is one of the oldest herbal medicines used as a traditional medicine, and its essential oil contains a multitude of potentially bioactive compounds. The essential oil from cinnamon was profiled by GC-MS, and the methicillin-resistant *Staphylococcus aureus* (MRSA) metabolic profile in the presence of the essential oil was also analyzed with GC-MS-based metabolomics. Potential metabolites with antibacterial activity were believed to be linalool, eucalyptol, α-terpineol, isoborneol, β-phellandrene, and camphor (Chen et al., 2020). This study was also of note as it used metabolomics on both sides of the discovery process; profiling not only the botanical source of the antiinfective compound(s) but also looking at the response from the infectious agent, finding that 74 bacterial metabolites demonstrated significant differences (including 29 upregulated and 45 downregulated metabolites) across seven metabolic pathways (Chen et al., 2020).

Two recent studies looked at gastrointestinal nematodiasis, which can have substantial effects on ruminant health, and is experiencing a rise in resistance to commonly used antihelmenthics.

The first study used ^{1}H-NMR metabolomics to provide preliminary identification of antihelmintic compounds from *Lysiloma latisiliquum* (L.) Benth., an ethnoveterinary plant from Mexico (Hernández-Bolio, Ruiz-Vargas, & Peña-Rodríguez, 2019). A study by Hernández-Bolio, Kutzner, Eisenreich, de Jesús Torres-Acosta, and Peña-Rodríguez (2018) used OPLS-DA to identify the glycosylated compounds quercitrin and arbutin as possessing activity against the nematode *Haemonchus contortus* (Hernández-Bolio, Kutzner, Eisenreich, de Jesús Torres-Acosta, & Peña-Rodríguez, 2018). In addition, Borges et al. (2019) investigated the ovicidal activity of ethanol extracts from 17 plants collected from the Pantanal wetland in the state of Mato Grosso do Sul, Brazil. These plants were evaluated using the egg hatchability test of *Haemonchus placei*, and ethanol extracts were analyzed via HPLC-MS with Partial least squares regression discriminate analysis (PLS-DA) (Borges et al., 2019; Peña-Espinoza et al., 2020) (Fig. 10.4) and a univariate

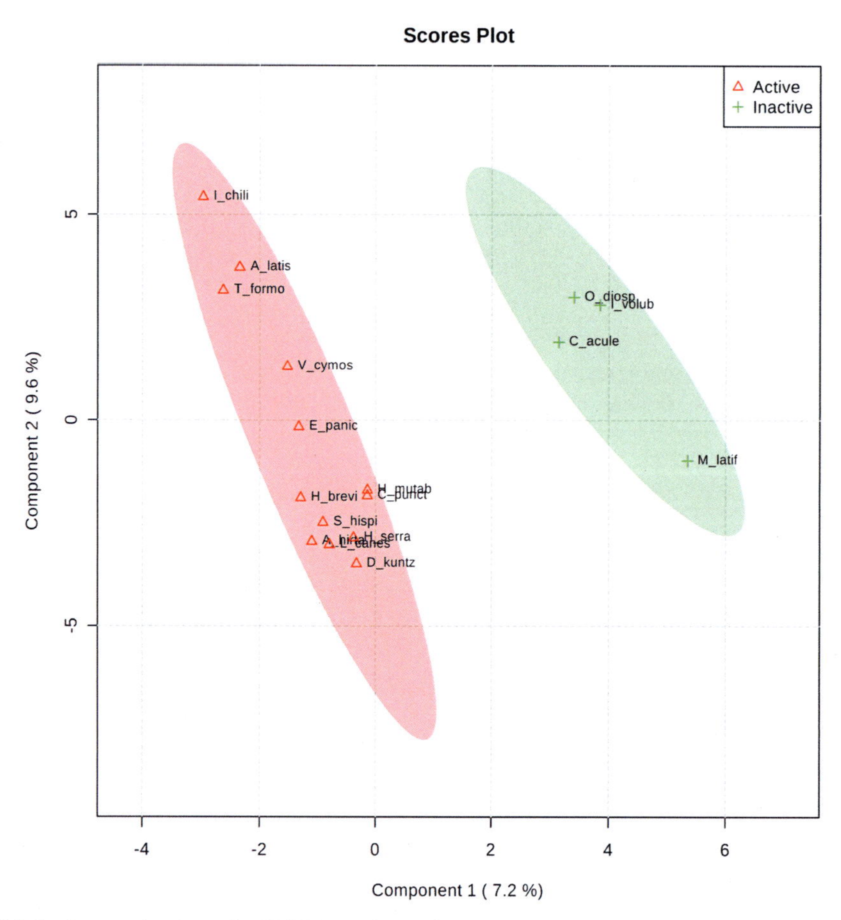

FIGURE 10.4 Scores plot from the PLS-DA analysis of 17 Brazilian plant species evaluated for ovicidal activity against *Haemonchus placei*. Active extracts are highlighted in red triangles, inactive are green crosses. *PLS-DA*, Partial least squares regression discriminate analysis. Source: *From Borges, D.G.L., Echeverria, J.T., de Oliveira, T.L., Heckler, R.P., de Freitas, M.G., Damasceno-Junior, G.A., Carollo, C.A., & de A. Borges, F. (2019). Discovery of potential ovicidal natural products using metabolomics.* PLoS One, 14, *e0211237.*

correlation mapping was used to detect compounds that positively correlated with ovicidal activity. Using multiple plant taxa enabled the discrimination of active plant compounds from the plant extracts. Ten metabolites were identified which had the strongest correlation with ovicidal activity, which spanned four different structural classes (phenylpropanoids, triterpene saponins, brevipolide, and flavonoid). And Peña-Espinoza et al. (2020) investigated the potential anthelmintic properties of chicory using untargeted LC–MS metabolomics, finding several sesquiterpene lactones that correlated with bioactivity. Notably, 11,13-dihydro-lactucopicrin was identified as the most correlative metabolite against the pig nematode *Ascaris suum* (Peña-Espinoza et al., 2020).

Metabolomics approaches have also been used to elucidate how external stimuli impact bioactive secondary metabolism in plants, both to understand biosynthetic mechanisms as well as potentially increase the supply of complex molecules for drug discovery research (Atanasov et al., 2015; Harvey et al., 2015). As an example, *Psiadia arguta* Voigt (Asteraceae) is a plant endemic to Mauritius, traditionally used to treat multiple ailments including as an expectorant or for the treatment of bronchitis and asthma. Preliminary biological screenings had suggested antimalarial (*Plasmodium falciparum*) activity from *P. arguta* leaves, and a phytochemical investigation of this plant led to the isolation and characterization of five antiplasmodial molecules (Mahadeo et al., 2019). However, due to its low occurrence naturally, and its protection from the collection due to its threatened status, Mahadeo et al. (2020) employed a ^{1}H NMR-based metabolomic approach to study the accumulation of antiplasmodial compounds during the growth of the plant. Young plants of *P. arguta* were cultured in vitro and then micropropagated plants at different stages of development were acclimatized in order to identify factors influencing the production of bioactive compounds. The metabolomics analysis revealed that four bioactive compounds (labdan-13(E)-en-8α-ol-15-yl acetate, labdan-8α-ol-15-yl acetate, labdan-13(E)-ene-8α-ol-15-diol, and (8R,13S)-labdan-8,15-diol) accumulated in the *P. arguta* leaves when the plants were subjected to biotic stress (Mahadeo et al., 2020).

The molecular networking approach has been adapted further to create a biochemometric modeling method, termed "bioactive molecular networking," where the chemical and bioactivity analyses are synchronized to visualize bioactive metabolites in the context of their structural relationship to one another. Nothias et al. developed a biochemometric workflow to overlay bioactivity results within a molecular network. They applied this approach to discover novel antiviral compounds from *Euphorbia dendroides* L. (Nothias et al., 2018). What was especially remarkable about this approach was that the bioactive molecular networks, built upon the untargeted metabolomics permitted the detection of bioactive molecules which were still unknown even after performing a classical BGF procedure.

The brown alga *Fucus vesiculosus* L. has demonstrated consistent antimicrobial activity of the extract against human pathogenic bacteria; at the same time, untargeted metabolomics analyses have suggested a variety of metabolites in the algal extract. Buedenbender, Astone, and Tasdemir (2020) also applied a "bioactive molecular networking" approach using the bioactive hexane and butanol partitions from an *F. vesiculosus* extract, and networked to identify the compounds responsible for antibacterial activity against MRSA. The first bioactive cluster identified by the bioactive molecular network consisted of galactolipids and allowed for subsequent targeted isolation efforts of six monogalactosyldiacylglycerol (MGDG) derivatives and one digalactosyldiacylglycerol (DGDG) (Buedenbender, Astone, & Tasdemir, 2020) (Fig. 10.5). Two of the MGDGs as well as the DGDG exhibited activity against MRSA. A second compound

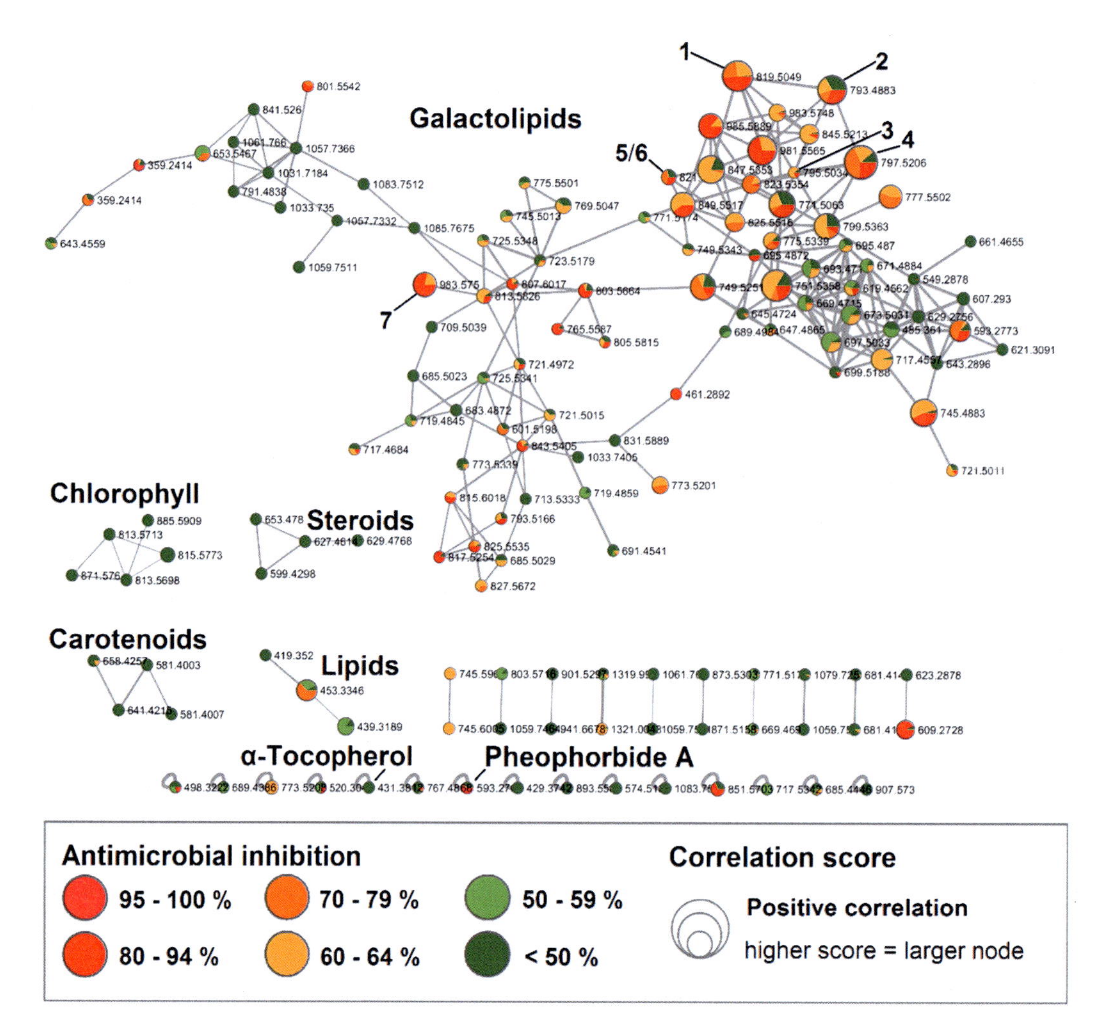

FIGURE 10.5 Biochemometric molecular network featuring fractions of the marine brown alga *Fucus vesiculosus* L. hexane extract fusing antimicrobial activity [MRSA inhibitory activity (%)]. MS data were recorded in positive ionization mode. Nodes are colored according to their bioactivity at 100 μg/mL, while the size of the node reflects the Pearson correlations of the ions. Compounds 1–7 were highlighted as promising leads and featured in the publication. *MRSA*, Methicillin-resistant *Staphylococcus aureus*. Source: *From Buedenbender, L., Astone, F.A., & Tasdemir, D. (2020). Bioactive molecular networking for mapping the antimicrobial constituents of the baltic brown alga* Fucus vesiculosus. Marine Drugs, 18, *311*.

structural class with enhanced bioactivity was phlorotannins, with phlorethol-type phlorotannins evidencing especially high correlations with antimicrobial activity, in which two active phlorotannins were isolated based on the bioactive molecular network approach (Buedenbender et al., 2020). This study further highlighted the analytical potential of combining molecular networking with bioactivity assaying as a complementary tool for the identification and targeted isolation of bioactive compounds from plant structures.

A study by Caesar, Kellogg, Kvalheim, Cech, & Cech (2018) investigated the antimicrobial properties of *Angelica keiskei* (Miq.) Koidz. (Apiaceae), or ashitaba, which is native to the southernmost islands of Japan and has been traditionally used to extend life expectancy, increase vitality, and treat a broad range of diseases and infections. Authors used a method integrating biochemometrics, and molecular networking was devised and applied to *Angelica keiskei* to comprehensively evaluate its antimicrobial activity against *Staphylococcus aureus*. This approach highlighted potential bioactive compounds and provided structural information on these structures. A set of chalcone analogs were prioritized for isolation, yielding 4-hydroxyderricin (MIC $\leq$ 4.6 μM, IC_{50} = 2.0 μM), xanthoangelol (MIC $\leq$ 4.0 μM, IC_{50} = 2.3 μM), and xanthoangelol K (IC_{50} = 168 μM), the latter of which had not been previously reported to possess antimicrobial activity (Caesar et al. 2018). The amalgamation of the two approaches, biochemometrics and molecular networking enabled a more complete understanding of the compounds responsible for *A. keiskei*'s antimicrobial activity.

Bioactive molecular networking also was employed to investigate chicory (*Cichorium intybus* L.), a plant rich in sesquiterpene lactones and demonstrated anthelmintic activity in livestock. From six different sources of chicory material, including industrial byproducts (fresh chicory root pulp, fresh leaves from chicory cv. Choice, and four samplings of fresh leaves from chicory cv. Spadona). The resulting extracts were tested for anthelmintic activity against the nematode model *Caenorhabditis elegans* as well as the pig nematode *Ascaris suum*. Untargeted metabolomics revealed that the chicory root pulp had a distinctly different chemical fingerprint compared to the fresh leaf chicory extracts. Molecular networking confirmed several sesquiterpene lactones and associated derivatives that may be responsible of its potent antihelminthic activity. Bioactivity-based molecular networking of chicory root pulp and the most potent forage chicory extracts used a Pearson correlation metric between a feature's relative abundance (LC-MS peak area) and the EC_{50} from the corresponding *A. suum* assays, and the subsequent analysis revealed a high predicted bioactivity for the guaianolide sesquiterpene 11,13-dihydro-lactucopicrin (Peña-Espinoza et al., 2020) (Fig. 10.6). Thus the study highlighted the potential of the agricultural or industrial by-product chicory root pulp as a livestock nutraceutical antihelminthic, as well as a source of new antiparasitic compounds (Peña-Espinoza et al., 2020)).

Challenges and future directions

Metabolomics has evolved significantly since its inception two decades ago, but still faces challenges. Three areas include improving the coverage of the metabolome, annotation, and identification of peaks in untargeted metabolomics, and questions surrounding multiple bioactive components working in combination.

Metabolome coverage

Given the incredible diversity in the structural chemistry and complexity, as well as the wide dynamic range of abundance of plant metabolites, the methods for detection are not as facile as those for genomic or transcriptomic analysis. Technological advancements continue to

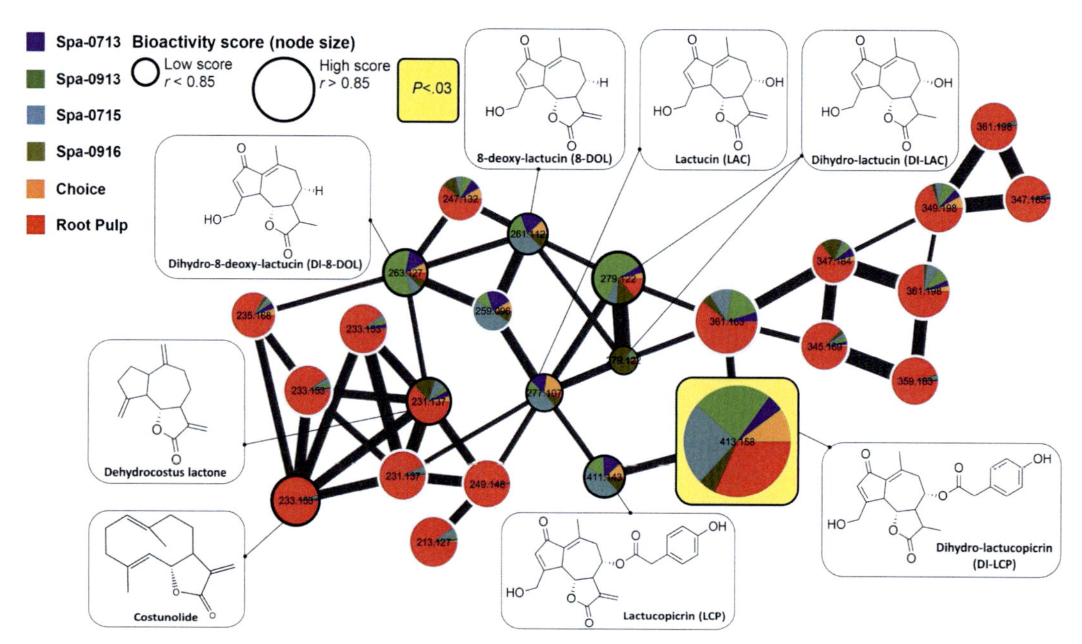

FIGURE 10.6 Bioactivity-based molecular networking of sesquiterpene lactones and derivatives in six chicory (*Cichorium intybus* L.) extracts based on their anthelmintic activity against the pig nematode *Ascaris suum*. Each nodes and edges represent one molecular ion and the pairwise spectral comparison between them, respectively. The node pie charts represent the relative quantitation (based on peak area) for the compound between the different chicory extracts, and node sizes represent the predicted bioactivity score of the feature, which was designated as the Pearson correlation coefficient (r) between the molecule relative abundance (peak area) and the EC_{50} value in the *A.* suum assays. The node highlighted in yellow represents dihydro-lactucopicrin, which evidenced a statistically significant high bioactivity score ($r < 0.85$ and a significance of $P < .03$). Compound annotations were determined by searching against the Global Natural Product Social Molecular Networking libraries. Source: *From Peña-Espinoza, M., Valente, A.H., Bornancin, L., Simonsen, H. T., Thamsborg, S.M., Williams, A.R., & López-Muñoz, R. (2020). Anthelmintic and metabolomic analyses of chicory* (Cichorium intybus) *identify an industrial by-product with potent in vitro antinematodal activity.* Veterinary Parasitology, 280, *109088.*

improve the sensitivity of analytical instruments, including the evolution of hyphenated methods for MS and NMR and could result in significant improvements in metabolome coverage. However, the most likely means to improve coverage is to utilize current techniques and instrumentation; however, this would require a considerable effort, by which multiple samples from a particular (or multiple) species would be evaluated by using different extraction techniques and conditions (Yanes, Tautenhahn, Patti, & Siuzdak, 2011; Yuliana, Khatib, Verpoorte, & Choi, 2011), separation technologies (Naser et al., 2018), MS ionization and detection methods (Nordström, Want, Northen, Lehtiö, & Siuzdak, 2008), and NMR experiments (Bingol, 2018; Emwas et al., 2019) to access the myriad different chemical structures. This would also be facilitated by the open exchange of metabolomics data and results across and between labs, enhancing cross-comparability of results; this has begun with open-access databases of metabolomics data, including the National Institute of Health's National Metabolomics Data Repository (https://www.metabolomicsworkbench.org/) and the MassIVE public repository in the Global Natural Product Social Molecular Networking (GNPS) system (https://gnps.ucsd.edu/).

There is one emergent strategy that warrants mention here: metabolite imaging. Metabolite imaging techniques provide a high spatial resolution of metabolite abundances. Spatial analysis of plant tissues has the potential to provide unique insights into the locations of biosynthesis, storage, and action of botanical natural products, including antiinfective agents, and produce insights into plant biology. In mass spectrometry imaging (MSI), the mass spec either samples small, discrete locations on the surface, analyzing the *m*/*z* values and then constructing a two-dimensional representation of the data, or larger sections that are analyzed multiple times and then stitched together as a single computational reconstruction of the sample (Boughton, Thinagaran, Sarabia, Bacic, & Roessner, 2016; Dreisewerd, 2003; Gonzalez et al., 2012; Kertesz & Van Berkel, 2010; Poulin & Pohnert, 2019; Stopka et al., 2019) (Fig. 10.7). Ionization of the samples can be achieved using a variety of methods, including desorption electrospray ionization (DESI), matrix-assisted laser desorption ionization (MALDI)-MSI,

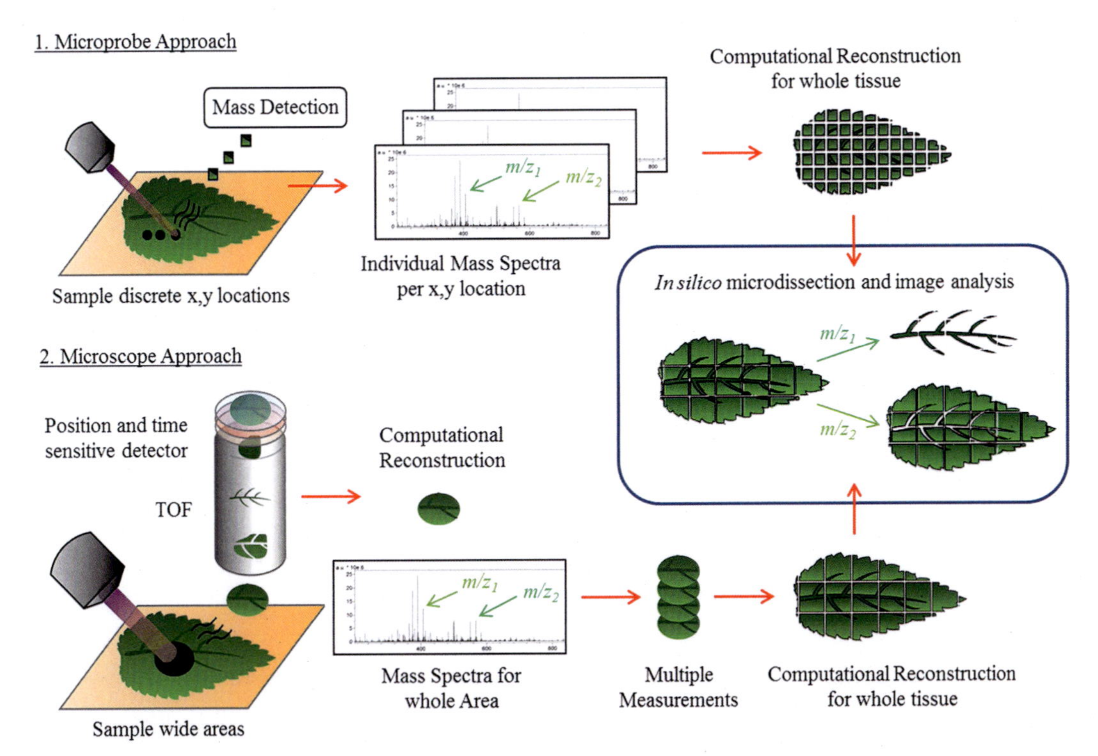

FIGURE 10.7 Scheme illustrating a mass spectrometry imaging workflow showing two main approaches, the microprobe (1) and microscope (2) technique. (1) In the microprobe approach, discrete locations on the sample surface are sampled and the *m*/*z* of the resulting ions is measured, after which the resulting mass spectra for each location (given as *x*,*y* coordinates) are reconstructed to form a final dataset. (2) With the microscope approach, a broadly focused laser permits sampling wide areas of tissue and the resulting ions are detected using both a position and time-sensitive time-of-flight (TOF) detector, which permits determination of the *m*/*z* and the spatial distribution of each ion within the sample area. Multiple samplings are required to cover very large surface areas, after which the data are computationally reconstructed to complete the dataset. Source: *From Boughton, B.A., Thinagaran, D., Sarabia, D., Bacic, A., & Roessner, U. (2016). Mass spectrometry imaging for plant biology: A review.* Phytochemistry Reviews, 15, *445–488.*

matrix-free laser desorption ionization MS (LDI–MS), secondary ion MS (SIMS), droplet-based liquid microjunction surface sampling probe (droplet-LMJ-SSP) (commonly known as the droplet probe), and laser-ablation electrospray ionization (LAESI)-MS (Gonzalez et al., 2012; Kertesz & Van Berkel, 2010; Poulin & Pohnert, 2019; Stopka et al., 2019). The most common mass spectrometry imaging ionization techniques are assisted laser desorption ionization, softer ionization methods that are used for direct measurement of molecular ions, some of which (e.g., MALDI and LAESI) are dependent upon a matrix layer deposited onto a sample surface to enable desorption of analytes from solid into the gas phase and to promote ionization (Dreisewerd, 2003).

As the spatial resolution of these imaging methods has advanced, along with the increased sensitivity and dynamic range of mass spectrometers, the size of samples has shrunk, enabling the investigation of metabolites or profiles in individual or small clusters of cells: the resolution of matrix-assisted laser desoprtion/ionization-mass spectrometry imaging (MALDI-MSI) has reached 5–20 μm (Dalisay et al., 2015; Soltwisch et al., 2015), while liquid metal ion guns for SIMS imaging have reduced the beam width to 200 nm (Fletcher & Vickerman, 2010). Imaging mass spec has enabled deeper insight into biosynthetic pathways as well as localization of metabolites within plant tissue. For example, the biosynthetic steps of the naphthodianthrone hypericin from *Hypericum perforatum* L. are still unestablished. Revuru, Bálintová, Henzelyová, Čellárová, and Kusari (2020) hypothesized that skyrin serves as a precursor in the biosynthesis of hypericin. They established the spatial distribution of skyrin and how it correlated to the distribution of hypericin using MALDI-high resolution mass spectrometry (MALDI-HRMS) imaging. The imaging across five different *Hypericum* species revealed a species-specific distribution and localization pattern of skyrin, which was similar to hypericin in the leaf tissue, and suggests an alternative biosynthetic pathway of hypericin and analogs (Revuru, Bálintová, Henzelyová, Čellárová, & Kusari, 2020).

However, many of these approaches have been limited to targeted metabolite analysis, rather than an untargeted metabolomics method. With advances in sampling methods and instrumentation, there has been a turn towards more metabolomics-based investigations of plant systems. Cahill, Riba, and Kertesz (2019) used a single-cell printer technology and liquid vortex capture-MS to investigate the lipid composition of the microalgae *Chlamydomonas reinhardtii* and *Euglena gracilis*. The imaging analysis discovered multiple diacylglyceryltrimethylhomo-Ser, phosphatidylcholine, MGDG, and DGDG lipids in single cells. The approach was also able to differentiate mixed cultures of the two algae by their lipidomic profile and specific levels of diacylglyceryltrimethylhomo-Ser and phosphatidylcholine lipids (Cahill, Riba, & Kertesz, 2019). To meet the challenges of limited sampling from small laser beam widths, Hansen and Lee (2018) developed a "multiplex MS imaging" technique to improve coverage and facilitate better annotation of the metabolites detected, while also developing new matrices capable of increasing the sensitivity from the small sampling sizes (Hansen & Lee, 2018). Feenstra, Hansen, and Lee (2015) sought to increase the diversity of chemical compounds that can be imaged and identified, using multiple ionization matrices to overcome the desorption/ionization bias for different metabolite classes and coupled it with dual polarity ionization and tandem MSI. They found that the use of multiple matrixes along with dual ionization polarities allowed the visualization of multiple compound classes, and the data-dependent MS^2

spectra permitted the identification of the compounds directly from the tissue. The team employed a test case of germinated corn seed, identifying 166 unique ions from the MS^2 spectra, 52 of which were identified as unique compounds. And, based upon the scans, the authors estimated over 500 metabolites could be potentially identified and visualized (Feenstra, Hansen, & Lee, 2015).

In addition, one in situ analysis technique incorporates chromatography, using a microextraction technique featuring a droplet-liquid microjunction-surface sampling probe (Kertesz & Van Berkel, 2013; Oberlies et al., 2019). This "droplet probe" has been used to perform microextraction on the surface of fungal (Knowles et al., 2019; Sica et al., 2015) and botanical samples (Kao, Henkin, Soejarto, Kinghorn, & Oberlies, 2018) for targeted as well as untargeted metabolomics. Kao et al. (2018) used the droplet probe to herbarium voucher specimens of *Garcinia mangostana* L. to profile multiple cytotoxic prenylated xanthones. The droplet probe has several advantages over other mass spec techniques, including the benefit of being a nondestructive imaging methodology (Oberlies et al., 2019). Put together, single-cell (or low-cell) level metabolomics will continue to advance and represent a potentially innovative way to explore how metabolite profiles vary in response to biotic and abiotic stressors with high spatial resolution, as well as impacting our understanding of the cooperative and antagonistic effects among metabolites.

Annotation/identification

Untargeted metabolomic approaches to bioactive discovery rely on the spectral data generated by the instrumentation for the independent variables. In MS-based metabolomics, these data are represented by the m/z value of the feature, which can be further specified by pairing it with retention time if a chromatography separation was used beforehand (e.g., LC-MS or GC-MS). For NMR metabolomics, the resulting peaks are then binned into narrow windows, which depends on the resolution of the instrument (Clark et al., 2016; Clendinen et al., 2015; Robinette, Brüschweiler, Schroeder, & Edison, 2012). However, neither m/z values nor binning provides any significant structural information about the detected features, especially if they are unknown. Because of this, and due to the fact that metabolomic profiling results in large datasets, annotating or identifying metabolites often faces challenges, and thus it is estimated that only 2%–10% of the features from untargeted metabolomics LC-MS studies can be annotated directly from the experiment (Aksenov, Da Silva, Knight, Lopes, & Dorrestein, 2017). However, annotation and structural information can be improved through the addition of other analyses. One of the most significant improvements in compound annotation and identification for Ms-based metabolomics analysis was improvements in mass accuracy (measured as parts per million, p.p.m.) and the corresponding resolving power ($m/\Delta m$) of the instruments, which can reach a resolution of more than 1,000,000 (at full-width half-maximum) (Kueger, Steinhauser, Willmitzer, & Giavalisco, 2012). With the advancements in relative isotopic abundance, in combination with sub-p.p.m. mass accuracies, these data help to distinguish between overlapping mass signals and aid in predicting the elemental composition of measured metabolic signals. In addition, one approach for accurate annotation of elemental compositions that is independent of mass spectrometer platform is the use of isotopically labeled compounds. Feeding a culture or organism with either a single labeled compound and follow its intercellular metabolism or isotopically label whole carbon, nitrogen, and/or sulfur input renders the whole metabolome isotopically labeled. These labeled metabolites are apparent through shifts in the composition-specific *m/z* and can be used to drastically improve precision in the elemental composition and reduce false positives, as well as differentiate between system compounds and contaminants and noise (Bueschl, Krska, Kluger, & Schuhmacher, 2013; Giavalisco, Köhl, Hummel, Seiwert, & Willmitzer, 2009).

As noted above, molecular networking clusters MS/MS data based upon similarity of fragmentation pattern and has become a dominant analytical method in metabolomics data analysis and annotation. The resulting molecular networks can be seeded with spectra of standards to provide enhanced annotation of structural features or the network can serve as a search against community-curated databases of MS/MS spectra, such as the GNPS platform (https://gnps.ucsd.edu/ProteoSAFe/static/gnps-splash.jsp) (Nguyen et al., 2013; Watrous et al., 2012; Yang et al., 2013). Other spectral databases, including MassBank (https://mona.fiehnlab.ucdavis.edu/, http://www.massbank.jp, https://massbank.eu/MassBank/) and the National Institute of Standards and Technology Mass Spectral Library (https://chemdata.nist.gov/), are additional reference sources. Beyond the ability to search through the publicly available libraries or seeded networks with reference MS/MS spectra, the molecular network can also provide an opportunity to extend an annotation to an unknown molecule (node) via an "analogues search" option, which expands and accelerates the capacity to ascertain similar structures (Quinn et al., 2017; Watrous et al., 2012; Yang et al., 2013). The annotation of the unknown can be propagated using differences in masses that correspond to known augmentations or features of a molecule; for example, a difference of 14 Da between two nodes would suggest a presumed methylene (CH_2) addition or deletion, a difference of 16 Da an oxygenation, and a difference of 34 Da could represent the substitution of proton by a chlorine atom. It is possible that, through thorough interpretation of these differences in masses coupled with diagnostic fragmentation patterns observed, one might be able to discern which part of the molecule is modified.

For unknown metabolites highlighted during an untargeted metabolomics study, with no commercially available standard, MS/MS databases or other mass spectral metrics only provide at best a tentative identification of the molecular structure. To definitively determine their structure, it is necessary to isolate these compounds for analysis. There are several approaches that can be employed to separate out individual compounds from a more complex extract (Agatonovic-Kustrin, Morton, & Yusof, 2015; Bucar, Wube, & Schmid, 2013; Kim et al., 2001; Michalkiewicz, Biesaga, & Pyrzynska, 2008; Pfoze, Myrboh, Kumar, & Rohman, 2014). Typically, compounds are isolated via a combination of solvent extraction, liquid–liquid partitioning, and column chromatography (Fig. 10.8). The extraction solvent, chromatographic phases (both stationary and mobile) employed for the various stages of fractionation and separation can be augmented to optimize the isolation of the bioactive compound(s) of interest. Once isolated (and purity is checked by a combination of spectroscopic means), structural elucidation can proceed via a combination of NMR and MS techniques; these have been extensively reviewed previously. (Bouslimani,

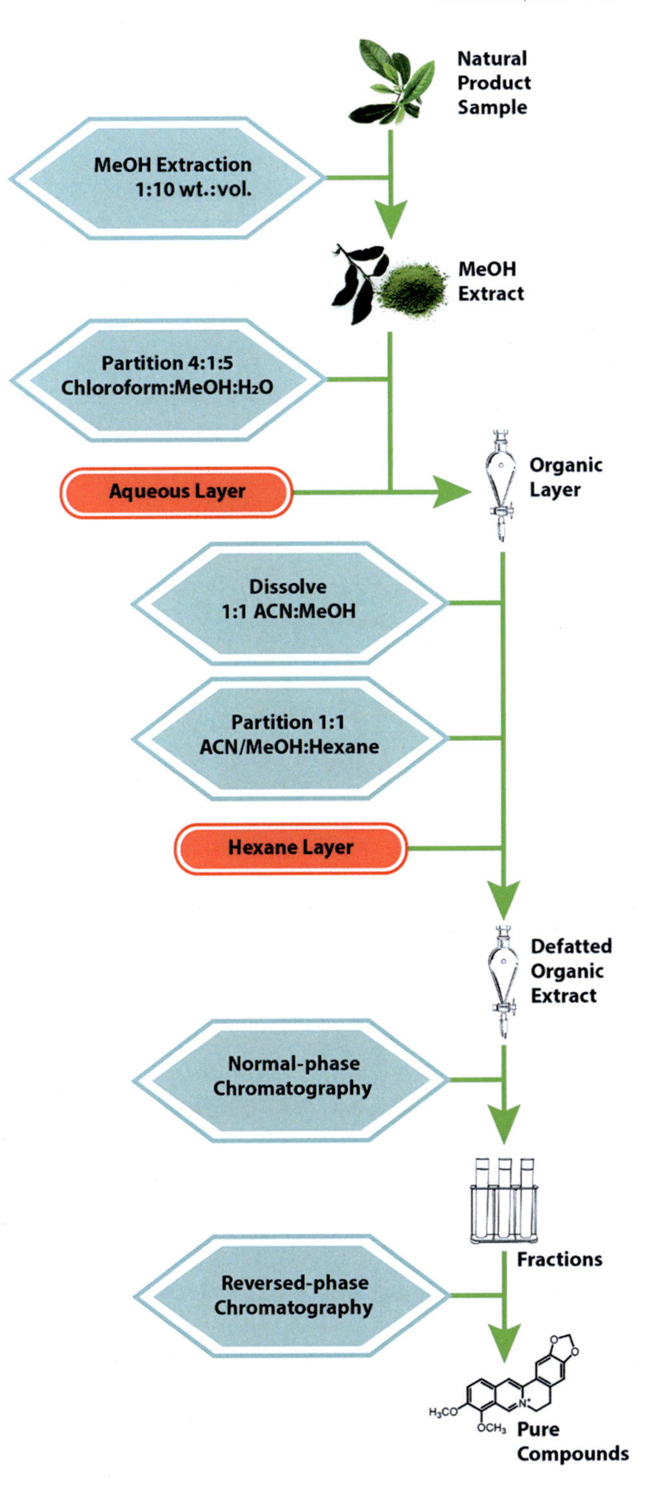

FIGURE 10.8 Generic schema for extraction and fractionation of a botanical natural product. Initial biomass is dried, ground, and mixed with an organic solvent, after which it is subjected to multiple liquid–liquid partitions to yield an organic extract. This extract is chromatographically separated using normal-phase and reverse-phase liquid chromatography systems to generate simpler fractions, and ultimately purified compounds. Source: *From Kellogg, J.J., Paine, M.F., McCune, J.S., Oberlies, N.H., & Cech, N.B. (2019). Selection and characterization of botanical natural products for research studies: A NaPDI center recommended approach.* Natural Product Reports, *36, 1196–1221.*

Sanchez, Garg, & Dorrestein, 2014; Bross-Walch, Kühn, Moskau, & Zerbe, 2005; Halabalaki, Vougogiannopoulou, Mikros, & Skaltsounis, 2014)

Synergy

Often in botanical samples, mixtures of compounds are more effective than their individual constituents in isolation due to additive or synergistic interactions among compounds. This is a potential advantage of plant natural products combatting infectious diseases (Patwardhan & Mashelkar, 2009), which may lead to increased efficacy and a diminished tendency for the evolution of resistance (Wagner & Ulrich-Merzenich, 2009). One recent study revealed that the antituberculosis effect of *Artemisia annua* L. was substantially stronger than equivalent concentrations of the putative bioactive artemisinin, suggesting that *A. annua* extracts exert their antituberculosis effects through a combination of additional compounds (Martini et al., 2020).

However, it remains difficult to ascertain whether compounds are in fact working in concert to produce a biological effect. Increasingly, metabolomics approaches have been suggested as a way to identify multiple components that could work in concert due to their de-emphasis of separation and isolation (Caesar & Cech, 2019). *Terminalia sericea* Burch. ex DC., a popular remedy in South Africa for the treatment of infectious diseases, was evaluated for antibacterial compounds. The resulting biochemometric analysis highlighted no consistent association between the phytochemical levels and the activity of the active or nonactive extracts. The authors deduced that multiple constituents of *T. sericea* root bark contributed to the observed activity; however, further investigation of the interactions of compounds present in the root bark was warranted (Anokwuru et al., 2020). Another study combined biochemometrics with synergy-directed fractionation to identify active compounds and/or synergists from the botanical *Hydrastis canadensis* L. (Britton et al., 2018). The analysis revealed a new synergistic flavonoid, 3,3′-dihydroxy-5,7,4′-trimethoxy-6,8-C-dimethylflavone which, when tested in combination with berberine, lowered berberine's IC_{50} from 132.2 ± 1.1 to 91.5 ± 1.1 μM, yet in isolation the flavonoid did not demonstrate antimicrobial activity (Britton et al., 2018). However, this approach suggested that the complexity of botanical extracts may necessitate several iterations of fractionation to simplify the mixture before biochemometric analysis yields useful results.

Conclusions

It should be noted that despite the great promise of untargeted metabolomics for bioactive molecule discovery, metabolomics is not a panacea for all botanical investigations. The data analysis of untargeted metabolomics datasets relies on statistical inference, and thus can only ascribe putative bioactivity to metabolites. Any such predictions require confirmation, which can be achieved via an isolation/structural isolation/bioassay approach (Fig. 10.8), genetic analysis of associated biosynthetic pathways followed by knockout assays resulting in defective mutants, and heterologous expression of the candidate biosynthetic pathway in a different organism to demonstrate a concomitant gain of antiinfective activity. Especially when considering the function and modes of action of bioactive metabolites, metabolomics alone is not sufficient for answering all of these queries, and it is here where other -omics tools and data-driven approaches should be employed to complement metabolomics analyses.

The developments over the last two decades have led metabolomics to emerge as a key means for analyzing complex mixtures of compounds, and within the last decade metabolomics has evolved into an instrument for the discovery of bioactive compounds. Given the vast untapped botanical diversity across the globe, there is tremendous potential for uncovering novel phytochemicals that could play crucial roles in combatting infectious diseases, and the metabolomics tools and approaches that have been described above, as well as the new tools that will continue to be developed, will increase the efficiency and robustness of the discovery effort. While this review noted the existing technical limitations that, for now, hamper certain metabolomics-driven inquiries, continued innovation in metabolomics data acquisition, data analysis, and statistical inference will propel the discipline forward. And, as public data repositories and molecular networking and chemical identification tools become more widespread and advanced, the challenges of annotating metabolomics datasets and structural elucidation of highlighted bioactive candidates will become more streamlined and robust. Furthermore, the evolution of new analytical technologies such as MSI, droplet probe, and single-cell or small culture analysis will help determine the spatial distribution of various antiinfective metabolites and provide clues to their biosynthesis and storage. Thus, as both the analytical and statistical capabilities of untargeted metabolomics continue to improve, these techniques will drive antiinfective natural product discovery from plants, annotating the responsible metabolites, and identifying biological roles of these small molecules.

References

Adedeji, A. A., & Babalola, O. O. (2020). Secondary metabolites as plant defensive strategy: A large role for small molecules in the near root region. *Planta*, *252*, 61. Available from https://doi.org/10.1007/s00425-020-03468-1.

Agatonovic-Kustrin, S., Morton, D. W., & Yusof, A. P. (2015). Thin-layer chromatography-bioassay as powerful tool for rapid identification of bioactive components in botanical extracts. *Modern Chemistry and Applications*, *3*, 120. Available from https://doi.org/10.4172/2329-6798.1000e120.

Aksenov, A. A., Da Silva, R., Knight, R., Lopes, N. P., & Dorrestein, P. C. (2017). Global chemical analysis of biology by mass spectrometry. *Nature Reviews Chemistry*, *1*, 0054.

Alvarez-Zapata, R., Sánchez-Medina, A., Chan-Bacab, M., García-Sosa, K., Escalante-Erosa, F., García-Rodríguez, R. V., … Peña-Rodríguez, L. M. (2015). Chemometrics-enhanced high performance liquid chromatography-ultraviolet detection of bioactive metabolites from phytochemically unknown plants. *Journal of Chromatography. A*, *1422*, 213–221. Available from https://doi.org/10.1016/j.chroma.2015.10.026.

Anokwuru, C., Tankeu, S., van Vuuren, S., Viljoen, A., Ramaite, I., Taglialatela-Scafati, O., … Combrinck, S. (2020). Unravelling the antibacterial activity of *Terminalia sericea* root bark through a metabolomic approach. *Molecules (Basel, Switzerland)*, *25*, 3683. Available from https://doi.org/10.3390/molecules25163683.

Anzian, A., Muhialdin, B. J., Mohammed, N. K., Kadum, H., Marzlan, A. A., Sukor, R., … Meor Hussin, A. S. (2020). Antibacterial activity and metabolomics profiling of torch ginger (*Etlingera elatior* Jack) flower oil extracted using subcritical carbon dioxide (CO_2). *Evidence-Based Complementary and Alternative Medicine*, *2020*, 4373401. Available from https://doi.org/10.1155/2020/4373401.

Anđelković, B., Vujisić, L., Vučković, I., Tešević, V., Vajs, V., & Gođevac, D. (2017). Metabolomics study of Populus type propolis. *Journal of Pharmaceutical and Biomedical Analysis*, *135*, 217–226. Available from https://doi.org/10.1016/j.jpba.2016.12.003.

Atanasov, A. G., Waltenberger, B., Pferschy-Wenzig, E. M., Linder, T., Wawrosch, C., Uhrin, P., ... Stuppner, H. (2015). Discovery and resupply of pharmacologically active plant-derived natural products: A review. *Biotechnology Advances*, *33*, 1582–1614. Available from https://doi.org/10.1016/j.biotechadv.2015.08.001.

Bingol, K. (2018). Recent advances in targeted and untargeted metabolomics by NMR and MS/NMR methods. *High-Throughput*, *7*, 9. Available from https://doi.org/10.3390/ht7020009.

Borges, D. G. L., Echeverria, J. T., de Oliveira, T. L., Heckler, R. P., de Freitas, M. G., Damasceno-Junior, G. A., ... Borges, A. (2019). Discovery of potential ovicidal natural products using metabolomics. *PLoS One*, *14*, e0211237. Available from https://doi.org/10.1371/journal.pone.0211237.

Boughton, B. A., Thinagaran, D., Sarabia, D., Bacic, A., & Roessner, U. (2016). Mass spectrometry imaging for plant biology: A review. *Phytochemistry Reviews*, *15*, 445–488. Available from https://doi.org/10.1007/s11101-015-9440-2.

Bouslimani, A., Sanchez, L. M., Garg, N., & Dorrestein, P. C. (2014). Mass spectrometry of natural products: Current, emerging and future technologies. *Natural Product Reports*, *31*, 718–729. Available from https://doi.org/10.1039/c4np00044g.

Britton, E. R., Kellogg, J. J., Kvalheim, O. M., & Cech, N. B. (2018). Biochemometrics to identify synergists and additives from botanical medicines: A case study with *Hydrastis canadensis* (Goldenseal). *Journal of Natural Products*, *81*, 484–493. Available from https://doi.org/10.1021/acs.jnatprod.7b00654.

Bross-Walch, N., Kühn, T., Moskau, D., & Zerbe, O. (2005). Strategies and tools for structure determination of natural products using modern methods of NMR spectroscopy. *Chemistry and Biodiversity*, *2*, 147–177. Available from https://doi.org/10.1002/cbdv.200590000.

Bucar, F., Wube, A., & Schmid, M. (2013). Natural product isolation-how to get from biological material to pure compounds. *Natural Product Reports*, *30*, 525–545. Available from https://doi.org/10.1039/c3np20106f.

Buedenbender, L., Astone, F. A., & Tasdemir, D. (2020). Bioactive molecular networking for mapping the antimicrobial constituents of the Baltic Brown Alga *Fucus vesiculosus*. *Marine Drugs*, *18*, 311. Available from https://doi.org/10.3390/md18060311.

Bueschl, C., Krska, R., Kluger, B., & Schuhmacher, R. (2013). Isotopic labeling-assisted metabolomics using LC-MS. *Analytical and Bioanalytical Chemistry*, *405*, 27–33. Available from https://doi.org/10.1007/s00216-012-6375-y.

Caesar, L. K., & Cech, N. B. (2016). A review of the medicinal uses and pharmacology of ashitaba. *Planta Medica*, *82*, 1236–1245. Available from https://doi.org/10.1055/s-0042-110496.

Caesar, L. K., & Cech, N. B. (2019). Synergy and antagonism in natural product extracts: When 1 + 1 does not equal 2. *Natural Product Reports*, *36*, 869–888. Available from https://doi.org/10.1039/c9np00011a.

Caesar, L. K., Kellogg, J. J., Kvalheim, O. M., Cech, R. A., & Cech, N. B. (2018). Integration of biochemometrics and molecular networking to identify antimicrobials in *Angelica keiskei*. *Planta Medica*, *84*, 721–728.

Cahill, J. F., Riba, J., & Kertesz, V. (2019). Rapid, untargeted chemical profiling of single cells in their native environment. *Analytical Chemistry*, *91*, 6118–6126. Available from https://doi.org/10.1021/acs.analchem.9b00680.

Cajka, T., & Fiehn, O. (2016). Toward merging untargeted and targeted methods in mass spectrometry-based metabolomics and lipidomics. *Analytical Chemistry*, *88*, 524–545. Available from https://doi.org/10.1021/acs.analchem.5b04491.

Calderon, A. I. (2017). Editorial: Combination of mass spectrometry and omics/chemometrics approaches to unravel bioactives in natural products mixtures. *Combinatorial Chemistry & High Throughput Screening*, *20*, 278. Available from https://doi.org/10.2174/138620732004170811113805.

Cech, N. B., & Yu, K. (2013). Mass spectrometry for natural products research: Challenges, pitfalls, and opportunities. *LCGC North America*, *31*, 938–947. Available from https://www.chromatographyonline.com/view/mass-spectrometry-natural-products-research-challenges-pitfalls-and-opportunities.

Chagas-Paula, D., Zhang, T., Da Costa, F., & Edrada-Ebel, R. (2015). A metabolomic approach to target compounds from the Asteraceae family for dual COX and LOX inhibition. *Metabolites*, *5*, 404–430. Available from https://doi.org/10.3390/metabo5030404.

Chan, E. W. C., Lim, Y. Y., & Wong, S. K. (2011). Phytochemistry and pharmacological properties of *Etlingera elatior*: A review. *Pharmacognosy Journal*, *3*, 6–10. Available from https://doi.org/10.5530/pj.2011.22.2.

Chen, J., Tang, C., Zhang, R., Ye, S., Zhao, Z., Huang, Y., ... Yang, D. (2020). Metabolomics analysis to evaluate the antibacterial activity of the essential oil from the leaves of *Cinnamomum camphora* (Linn.) Presl. *Journal of Ethnopharmacology*, *253*, 112652. Available from https://doi.org/10.1016/j.jep.2020.112652.

Christenhusz, M. J. M., & Byng, J. W. (2016). The number of known plants species in the world and its annual increase. *Phytotaxa*, *261*, 201–217. Available from https://doi.org/10.11646/phytotaxa.261.3.1.

Clark, T., Berrué, F., Calhoun, L., Boland, P., Kerr, R., Johnson, J., ... Gray, C. (2016). A comparison between the application of NMR and LC-HRMS based metabolomics on the discovery of natural products from endophytic fungi. *Planta Medica*, *81*, S1–S381. Available from https://doi.org/10.1055/s-0036-1596741.

Clendinen, C. S., Stupp, G. S., Ajredini, R., Lee-McMullen, B., Beecher, C., & Edison, A. S. (2015). An overview of methods using 13C for improved compound identification in metabolomics and natural products. *Frontiers in Plant Science*, *6*, 611. Available from https://doi.org/10.3389/fpls.2015.00611.

Dalisay, D. S., Kim, K. W., Lee, C., Yang, H., Rübel, O., Bowen, B. P., ... Lewis, N. G. (2015). Dirigent protein-mediated lignan and cyanogenic glucoside formation in flax seed: Integrated omics and MALDI mass spectrometry imaging. *Journal of Natural Products*, *78*, 1231–1242. Available from https://doi.org/10.1021/acs.jnatprod.5b00023.

Dreisewerd, K. (2003). The desorption process in MALDI. *Chemical Reviews*, *103*, 395–425. Available from https://doi.org/10.1021/cr010375i.

Emwas, A. H., Roy, R., McKay, R. T., Tenori, L., Saccenti, E., Nagana Gowda, G. A., ... Wishart, D. S. (2019). NMR spectroscopy for metabolomics research. *Metabolites*, *9*(7), 123. Available from https://doi.org/10.3390/metabo9070123.

Enke, C. G., & Nagels, L. J. (2011). Undetected components in natural mixtures: How many? What concentrations? Do they account for chemical noise? What is needed to detect them? *Analytical Chemistry*, *83*, 2539–2546. Available from https://doi.org/10.1021/ac102818a.

Farrés, M., Platikanov, S., Tsakovski, S., & Tauler, R. (2015). Comparison of the variable importance in projection (VIP) and of the selectivity ratio (SR) methods for variable selection and interpretation. *Journal of Chemometrics*, *29*, 528–536. Available from https://doi.org/10.1002/cem.2736.

Feenstra, A. D., Hansen, R. L., & Lee, Y. J. (2015). Multi-matrix, dual polarity, tandem mass spectrometry imaging strategy applied to a germinated maize seed: Toward mass spectrometry imaging of an untargeted metabolome. *Analyst*, *140*, 7293–7304. Available from https://doi.org/10.1039/c5an01079a.

Fletcher, J. S., & Vickerman, J. C. (2010). A new SIMS paradigm for 2D and 3D molecular imaging of bio-systems. *Analytical and Bioanalytical Chemistry*, *396*, 85–104. Available from https://doi.org/10.1007/s00216-009-2986-3.

Giavalisco, P., Köhl, K., Hummel, J., Seiwert, B., & Willmitzer, L. (2009). 13C isotope-labeled metabolomes allowing for improved compound annotation and relative quantification in liquid chromatography-mass spectrometry-based metabolomic research. *Analytical Chemistry*, *81*, 6546–6551. Available from https://doi.org/10.1021/ac900979e.

Gonzalez, D. J., Xu, Y., Yang, Y. L., Esquenazi, E., Liu, W. T., Edlund, A., ... Dorrestein, P. C. (2012). Observing the invisible through imaging mass spectrometry, a window into the metabolic exchange patterns of microbes. *Journal of Proteomics*, *75*, 5069–5076. Available from https://doi.org/10.1016/j.jprot.2012.05.036.

Gowda, G. N., Zhang, S., Gu, H., Asiago, V., Shanaiah, N., & Raftery, D. (2008). Metabolomics-based methods for early disease diagnostics. *Expert Review of Molecular Diagnostics*, *8*, 617–633. Available from https://doi.org/10.1586/14737159.8.5.617.

Guthals, A., Watrous, J. D., Dorrestein, P. C., & Bandeira, N. (2012). The spectral networks paradigm in high throughput mass spectrometry. *Molecular Biosystems*, *8*, 2535–2544. Available from https://doi.org/10.1039/c2mb25085c.

Halabalaki, M., Vougogiannopoulou, K., Mikros, E., & Skaltsounis, A. L. (2014). Recent advances and new strategies in the NMR-based identification of natural products. *Current Opinion in Biotechnology*, *25*, 1–7. Available from https://doi.org/10.1016/j.copbio.2013.08.005.

Hansen, R. L., & Lee, Y. J. (2018). High-spatial resolution mass spectrometry imaging: Toward single cell metabolomics in plant tissues. *The Chemical Record*, *18*, 65–77. Available from https://doi.org/10.1002/tcr.201700027.

Harvey, A. L., Edrada-Ebel, R., & Quinn, R. J. (2015). The re-emergence of natural products for drug discovery in the genomics era. *Nature Reviews Drug Discovery*, *14*, 111–129. Available from https://doi.org/10.1038/nrd4510.

Hernández-Bolio, G. I., Kutzner, E., Eisenreich, W., de Jesús Torres-Acosta, J. F., & Peña-Rodríguez, L. M. (2018). The use of 1 H-NMR metabolomics to optimise the extraction and preliminary identification of anthelmintic products from the leaves of *Lysiloma latisiliquum*. *Phytochemical Analysis*, *29*, 413–420. Available from https://doi.org/10.1002/pca.2724.

Hernández-Bolio, G. I., Ruiz-Vargas, J. A., & Peña-Rodríguez, L. M. (2019). Natural products from the Yucatecan flora: Structural diversity and biological activity. *Journal of Natural Products*, *82*(3), 647–656. Available from https://doi.org/10.1021/acs.jnatprod.8b00959.

Inui, T., Wang, Y., Pro, S. M., Franzblau, S. G., & Pauli, G. F. (2012). Unbiased evaluation of bioactive secondary metabolites in complex matrices. *Fitoterapia*, *83*, 1218–1225. Available from https://doi.org/10.1016/j.fitote.2012.06.012.

Jorge, T. F., Mata, A. T., & António, C. (2016). Mass spectrometry as a quantitative tool in plant metabolomics. *Philosophical Transactions of the Royal Society A: Mathematical, Physical and Engineering Sciences*, *374*(2079), 20150370. Available from https://doi.org/10.1098/rsta.2015.0370.

Kao, D., Henkin, J. M., Soejarto, D. D., Kinghorn, A. D., & Oberlies, N. H. (2018). Non-destructive chemical analysis of a *Garcinia mangostana* L. (Mangosteen) herbarium voucher specimen. *Phytochemistry Letters*, *28*, 124–129. Available from https://doi.org/10.1016/j.phytol.2018.10.001.

Kellogg, J. J., Todd, D. A., Egan, J. M., Raja, H. A., Oberlies, N. H., Kvalheim, O. M., ... Cech, N. B. (2016). Biochemometrics for natural products research: Comparison of data analysis approaches and application to identification of bioactive compounds. *Journal of Natural Products*, *79*, 376–386. Available from https://doi.org/10.1021/acs.jnatprod.5b01014.

Kellogg, J. J., Paine, M. F., McCune, J. S., Oberlies, N. H., & Cech, N. B. (2019). Selection and characterization of botanical natural products for research studies: A NaPDI center recommended approach. *Natural Product Reports*, *36*, 1196–1221. Available from https://doi.org/10.1039/c8np00065d.

Kertesz, V., & Van Berkel, G. J. (2010). Liquid microjunction surface sampling coupled with high-pressure liquid chromatography – electrospray ionization-mass spectrometry for analysis of drugs and metabolites in whole-body thin tissue sections. *Analytical Chemistry*, *82*, 5917–5921. Available from https://doi.org/10.1021/ac100954p.

Kertesz, V., & Van Berkel, G. J. (2013). Automated liquid microjunction surface sampling-HPLC-MS/MS analysis of drugs and metabolites in whole-body thin tissue sections. *Bioanalysis*, *5*, 819–826. Available from https://doi.org/10.4155/bio.13.42.

Kim, N. C., Oberlies, N. H., Brine, D. R., Handy, R. W., Wani, M. C., & Wall, M. E. (2001). Isolation of symlandine from the roots of common comfrey (*Symphytum officinale*) using countercurrent chromatography. *Journal of Natural Products*, *64*, 251–253. Available from https://doi.org/10.1021/np0004653.

Kinghorn, A. D., Fong, H. H. S., Farnsworth, N. R., Mehta, R. G., Moon, R. C., Moriarty, R. M., ... Pezzuto, J. M. (1998). Cancer chemoproventitive agents discovered by activity-guided fractionation: A review. *Current Organic Chemistry*, *2*, 597–612.

Knowles, S. L., Raja, H. A., Wright, A. J., Lee, A. M. L., Caesar, L. K., Cech, N. B., ... Oberlies, N. H. (2019). Mapping the fungal battlefield: Using in situ chemistry and deletion mutants to monitor interspecific chemical interactions between fungi. *Frontiers in Microbiology*, *10*, 285. Available from https://doi.org/10.3389/fmicb.2019.00285.

Kueger, S., Steinhauser, D., Willmitzer, L., & Giavalisco, P. (2012). High-resolution plant metabolomics: From mass spectral features to metabolites and from whole-cell analysis to subcellular metabolite distributions. *Plant Journal*, *70*, 39–50. Available from https://doi.org/10.1111/j.1365-313X.2012.04902.x.

Kvalheim, O. M., Yan Chan, H., Benzie, I. F. F., Tong Szeto, Y., Chung Tzang, A. H., Wah Mok, D. K., ... Tim Chau, F. (2011). Chromatographic profiling and multivariate analysis for screening and quantifying the contributions from individual components to the bioactive signature in natural products. *Chemometrics and Intelligent Laboratory Systems*, *107*, 98–105. Available from https://doi.org/10.1016/j.chemolab.2011.02.002.

Li, J. W. H., & Vederas, J. C. (2009). Drug discovery and natural products: End of an era or an endless frontier? *Science (New York, NY)*, *325*, 161–165. Available from https://doi.org/10.1126/science.1168243.

Li, Z. Y., Zhi, H. J., Zhang, F. S., Sun, H. F., Zhang, L. Z., Jia, J. P., ... Qin, X. M. (2013). Metabolomic profiling of the antitussive and expectorant plant *Tussilago farfara* L. by nuclear magnetic resonance spectroscopy and multivariate data analysis. *Journal of Pharmaceutical and Biomedical Analysis*, *75*, 158–164. Available from https://doi.org/10.1016/j.jpba.2012.11.023.

Liu, Y., Finley, J., Betz, J. M., & Brown, P. N. (2018). FT-NIR characterization with chemometric analyses to differentiate goldenseal from common adulterants. *Fitoterapia*, *127*, 81–88. Available from https://doi.org/10.1016/j.fitote.2018.02.006.

Mahadeo, K., Herbette, G., Grondin, I., Jansen, O., Kodja, H., Soulange, J., ... Frederich, M. (2019). Antiplasmodial diterpenoids from *Psiadia arguta*. *Journal of Natural Products*, *82*, 1361–1366. Available from https://doi.org/10.1021/acs.jnatprod.8b00698.

Mahadeo, K., Grondin, I., Herbette, G., Palama, T. L., Bouchemal, N., Soulange, J., ... Kodja, H. (2020). A 1H NMR-based metabolomic approach to study the production of antimalarial compounds from *Psiadia arguta* leaves (pers.) voigt. *Phytochemistry*, *176*, 112401. Available from https://doi.org/10.1016/j.phytochem.2020.112401.

Markley, J. L., Brüschweiler, R., Edison, A. S., Eghbalnia, H. R., Powers, R., Raftery, D., ... Wishart, D. S. (2017). The future of NMR-based metabolomics. *Current Opinion in Biotechnology, 43*, 34–40. Available from https://doi.org/10.1016/j.copbio.2016.08.001.

Martens, H., Bruun, S. W., Adt, I., Sockalingum, G. D., & Kohler, A. (2006). Pre-processing in biochemometrics: Correction for path-length and temperature effects of water in FTIR bio-spectroscopy by EMSC. *Journal of Chemometrics, 20*, 402–417. Available from https://doi.org/10.1002/cem.1015.

Martini, M. C., Zhang, T., Williams, J. T., Abramovitch, R. B., Weathers, P. J., & Shell, S. S. (2020). *Artemisia annua* and *Artemisia afra* extracts exhibit strong bactericidal activity against *Mycobacterium tuberculosis*. *Journal of Ethnopharmacology, 262*, 113191. Available from https://doi.org/10.1016/j.jep.2020.113191.

Michalkiewicz, A., Biesaga, M., & Pyrzynska, K. (2008). Solid-phase extraction procedure for determination of phenolic acids and some flavonols in honey. *Journal of Chromatography. A, 1187*, 18–24. Available from https://doi.org/10.1016/j.chroma.2008.02.001.

Moco, S., Vervoort, J., Moco, S., Bino, R. J., De Vos, R. C. H., & Bino, R. (2007). Metabolomics technologies and metabolite identification. *Trends in Analytical Chemistry, 26*, 855–866. Available from https://doi.org/10.1016/j.trac.2007.08.003.

Naser, F. J., Mahieu, N. G., Wang, L., Spalding, J. L., Johnson, S. L., & Patti, G. J. (2018). Two complementary reversed-phase separations for comprehensive coverage of the semipolar and nonpolar metabolome. *Analytical and Bioanalytical Chemistry, 410*, 1287–1297. Available from https://doi.org/10.1007/s00216-017-0768-x.

Newman, D. J., & Cragg, G. M. (2020). Natural products as sources of new drugs over the nearly four decades from 01/1981 to 09/2019. *Journal of Natural Products, 83*, 770–803. Available from https://doi.org/10.1021/acs.jnatprod.9b01285.

Nguyen, D. D., Wu, C. H., Moree, W. J., Lamsa, A., Medema, M. H., Zhao, X., ... Dorrestein, P. C. (2013). MS/MS networking guided analysis of molecule and gene cluster families. *Proceedings of the National Academy of Sciences of the United States of America, 110*, E2611–E2620. Available from https://doi.org/10.1073/pnas.1303471110.

Noble, R. L. (1990). The discovery of the vinca alkaloids - Chemotherapeutic agents against cancer. *Biochemistry and Cell Biology, 68*, 1344–1351. Available from https://doi.org/10.1139/o90-197.

Nordström, A., Want, E., Northen, T., Lehtiö, J., & Siuzdak, G. (2008). Multiple ionization mass spectrometry strategy used to reveal the complexity of metabolomics. *Analytical Chemistry, 80*, 421–429. Available from https://doi.org/10.1021/ac701982e.

Nothias, L. F., Nothias-Esposito, M., Da Silva, R., Wang, M., Protsyuk, I., Zhang, Z., ... Dorrestein, P. C. (2018). Bioactivity-based molecular networking for the discovery of drug leads in natural product bioassay-guided fractionation. *Journal of Natural Products, 81*, 758–767. Available from https://doi.org/10.1021/acs.jnatprod.7b00737.

Oberlies, N. H., & Kroll, D. J. (2004). Camptothecin and taxol: Historic achievements in natural products research. *Journal of Natural Products, 67*, 129–135. Available from https://doi.org/10.1021/np030498t.

Oberlies, N. H., Knowles, S. L., Amrine, C. S. M., Kao, D., Kertesz, V., & Raja, H. A. (2019). Droplet probe: Coupling chromatography to the in situ evaluation of the chemistry of nature. *Natural Product Reports, 36*, 944–959. Available from https://doi.org/10.1039/c9np00019d.

Oliver, S. G., Winson, M. K., Kell, D. B., & Baganz, F. (1998). Systematic functional analysis of the yeast genome. *Trends in Biotechnology, 16*, 373–378. Available from https://doi.org/10.1016/S0167-7799(98)01214-1.

Patras, A., Brunton, N. P., Downey, G., Rawson, A., Warriner, K., & Gernigon, G. (2011). Application of principal component and hierarchical cluster analysis to classify fruits and vegetables commonly consumed in Ireland based on in vitro antioxidant activity. *Journal of Food Composition and Analysis, 24*, 250–256. Available from https://doi.org/10.1016/j.jfca.2010.09.012.

Patwardhan, B., & Mashelkar, R. A. (2009). Traditional medicine-inspired approaches to drug discovery: Can Ayurveda show the way forward? *Drug Discovery Today, 14*, 804–811. Available from https://doi.org/10.1016/j.drudis.2009.05.009.

Pauli, G. F., Chen, S. N., Simmler, C., Lankin, D. C., Gödecke, T., Jaki, B. U., ... Napolitano, J. G. (2014). Importance of purity evaluation and the potential of quantitative 1H NMR as a purity assay. *Journal of Medicinal Chemistry, 57*, 9220–9231. Available from https://doi.org/10.1021/jm500734a.

Peña-Espinoza, M., Valente, A. H., Bornancin, L., Simonsen, H. T., Thamsborg, S. M., Williams, A. R., ... López-Muñoz, R. (2020). Anthelmintic and metabolomic analyses of chicory (*Cichorium intybus*) identify an industrial by-product with potent in vitro antinematodal activity. *Veterinary Parasitology, 280*, 109088. Available from https://doi.org/10.1016/j.vetpar.2020.109088.

Pfoze, N. L., Myrboh, B., Kumar, Y., & Rohman, M. R. (2014). Isolation of protoberberine alkaloids from stem bark of *Mahonia manipurensis* Takeda using RP-HPLC. *Journal of Medicinal Plants Studies, 2*, 48–57. Available from http://www.plantsjournal.com.

Porras, G., Chassagne, F., Lyles, J. T., Marquez, L., Dettweiler, M., Salam, A. M., … Quave, C. L. (2020). Ethnobotany and the role of plant natural products in antibiotic drug discovery. *Chemical Reviews, 121*(6), 3495–3560. Available from https://doi.org/10.1021/acs.chemrev.0c00922.

Poulin, R. X., & Pohnert, G. (2019). Simplifying the complex: Metabolomics approaches in chemical ecology. *Analytical and Bioanalytical Chemistry, 411*, 13–19. Available from https://doi.org/10.1007/s00216-018-1470-3.

Prince, E. K., & Pohnert, G. (2010). Searching for signals in the noise: Metabolomics in chemical ecology. *Analytical and Bioanalytical Chemistry, 396*, 193–197. Available from https://doi.org/10.1007/s00216-009-3162-5.

Qiu, F., Cai, G., Jaki, B. U., Lankin, D. C., Franzblau, S. G., & Pauli, G. F. (2013). Quantitative purity-activity relationships of natural products: The case of anti-tuberculosis active triterpenes from *Oplopanax horridus*. *Journal of Natural Products, 76*, 413–419. Available from https://doi.org/10.1021/np3007809.

Quinn, R. A., Nothias, L. F., Vining, O., Meehan, M., Esquenazi, E., & Dorrestein, P. C. (2017). Molecular networking as a drug discovery, drug metabolism, and precision medicine strategy. *Trends in Pharmacological Sciences, 38*, 143–154. Available from https://doi.org/10.1016/j.tips.2016.10.011.

Rajalahti, T., & Kvalheim, O. M. (2011). Multivariate data analysis in pharmaceutics: A tutorial review. *International Journal of Pharmaceutics, 417*, 280–290. Available from https://doi.org/10.1016/j.ijpharm.2011.02.019.

Rajalahti, T., Arneberg, R., Berven, F. S., Myhr, K. M., Ulvik, R. J., & Kvalheim, O. M. (2009). Biomarker discovery in mass spectral profiles by means of selectivity ratio plot. *Chemometrics and Intelligent Laboratory Systems, 95*, 35–48. Available from https://doi.org/10.1016/j.chemolab.2008.08.004.

Rajalahti, T., Arneberg, R., Kroksveen, A. C., Berle, M., Myhr, K. M., & Kvalheim, O. M. (2009). Discriminating variable test and selectivity ratio plot: Quantitative tools for interpretation and variable (biomarker) selection in complex spectral or chromatographic profiles. *Analytical Chemistry, 81*, 2581–2590. Available from https://doi.org/10.1021/ac802514y.

Ramirez, T., strigun, A., Verlohner, A., Huener, H. A., Peter, E., Herold, M., … van Ravenzwaay, B. (2018). Prediction of liver toxicity and mode of action using metabolomics in vitro in HepG2 cells. *Archives of Toxicology, 92*, 893–906. Available from https://doi.org/10.1007/s00204-017-2079-6.

Revuru, B., Bálintová, M., Henzelyová, J., Čellárová, E., & Kusari, S. (2020). MALDI-HRMS imaging maps the localization of Skyrin, the precursor of hypericin, and pathway intermediates in leaves of *Hypericum* species. *Molecules (Basel, Switzerland), 25*, 3964. Available from https://doi.org/10.3390/molecules25173964.

Richards, L. A., Oliveira, C., Dyer, L. A., Rumbaugh, A., Urbano-Muñoz, F., Wallace, I. S., … Jeffrey, C. S. (2018). Shedding light on chemically mediated tri-trophic interactions: A 1H-NMR network approach to identify compound structural features and associated biological activity. *Frontiers in Plant Science, 9*, 1155. Available from https://doi.org/10.3389/fpls.2018.01155.

Roberts, G. K., Gardner, D., Foster, P. M., Howard, P. C., Lui, E., Walker, L., … Rider, C. (2019). Finding the bad actor: Challenges in identifying toxic constituents in botanical dietary supplements. *Food and Chemical Toxicology, 124*, 431–438. Available from https://doi.org/10.1016/j.fct.2018.12.026.

Robinette, S. L., Brüschweiler, R., Schroeder, F. C., & Edison, A. S. (2012). NMR in metabolomics and natural products research: Two sides of the same coin. *Accounts of Chemical Research, 45*, 288–297. Available from https://doi.org/10.1021/ar2001606.

Selander, E., Kubanek, J., Hamberg, M., Andersson, M. X., Cervin, G., & Pavia, H. (2015). Predator lipids induce paralytic shellfish toxins in bloom-forming algae. *Proceedings of the National Academy of Sciences of the United States of America, 112*, 6395–6400. Available from https://doi.org/10.1073/pnas.1420154112.

Shen, B. (2015). A new golden age of natural products drug discovery. *Cell, 163*, 1297–1300. Available from https://doi.org/10.1016/j.cell.2015.11.031.

Shu, Y., Liu, Z., Zhao, S., Zong, S., He, D., Wang, M., … Liu, Y. (2017). Integrated and global pseudotargeted metabolomics strategy applied to screening for quality control markers of Citrus TCMs. *Analytical and Bioanalytical Chemistry, 409*, 4849–4865. Available from https://doi.org/10.1007/s00216-017-0428-1.

Sica, V. P., Raja, H. A., El-Elimat, T., Kertesz, V., Van Berkel, G. J., Pearce, C. J., … Oberlies, N. H. (2015). Dereplicating and spatial mapping of secondary metabolites from fungal cultures in situ. *Journal of Natural Products, 78*, 1926–1936. Available from https://doi.org/10.1021/acs.jnatprod.5b00268.

Simmler, C., Kulakowski, D., Lankin, D. C., McAlpine, J. B., Chen, S.-N., & Pauli, G. F. (2016). Holistic analysis enhances the description of metabolic complexity in dietary natural products. *Advances in Nutrition, 7*, 179–189. Available from https://doi.org/10.3945/an.115.009928.

Soltwisch, J., Kettling, H., Vens-Cappell, S., Wiegelmann, M., Müthing, J., & Dreisewerd, K. (2015). Mass spectrometry imaging with laser-induced postionization. *Science (New York, N.Y.), 348*, 211–215. Available from https://doi.org/10.1126/science.aaa1051.

Stopka, S. A., Samarah, L. Z., Shaw, J. B., Liyu, A. V., Veličković, D., Agtuca, B. J., ... Vertes, A. (2019). Ambient metabolic profiling and imaging of biological samples with ultrahigh molecular resolution using laser ablation electrospray ionization 21 Tesla FTICR mass spectrometry. *Analytical Chemistry, 91*, 5028–5035. Available from https://doi.org/10.1021/acs.analchem.8b05084.

Tu, Y. (2011). The discovery of artemisinin (qinghaosu) and gifts from Chinese medicine. *Nature Medicine, 17*, 1217–1220. Available from https://doi.org/10.1038/nm.2471.

Wagner, H., & Ulrich-Merzenich, G. (2009). Synergy research: Approaching a new generation of phytopharmaceuticals. *Phytomedicine: International Journal of Phytotherapy and Phytopharmacology, 16*, 97–110. Available from https://doi.org/10.1016/j.phymed.2008.12.018.

Wallace, E. D., Todd, D. A., Harnly, J. M., Cech, N. B., & Kellogg, J. J. (2020). Identification of adulteration in botanical samples with untargeted metabolomics. *Analytical and Bioanalytical Chemistry, 412*, 4273–4286. Available from https://doi.org/10.1007/s00216-020-02678-6.

Wani, M. C., Taylor, H. L., Wall, M. E., Coggon, P., & McPhail, A. T. (1971). Plant antitumor agents. VI. Isolation and structure of taxol, a novel antileukemic and antitumor agent from *Taxus brevifolia*. *Journal of the American Chemical Society, 93*, 2325–2327. Available from https://doi.org/10.1021/ja00738a045.

Watrous, J., Roach, P., Alexandrov, T., Heath, B. S., Yang, J. Y., Kersten, R. D., ... Dorrestein, P. C. (2012). Mass spectral molecular networking of living microbial colonies. *Proceedings of the National Academy of Sciences of the United States of America, 109*, E1743–E1752. Available from https://doi.org/10.1073/pnas.1203689109.

Wen, C., Wang, D., Li, X., Huang, T., Huang, C., & Hu, K. (2018). Targeted isolation and identification of bioactive compounds lowering cholesterol in the crude extracts of crabapples using UPLC-DAD-MS-SPE/NMR based on pharmacology-guided PLS-DA. *Journal of Pharmaceutical and Biomedical Analysis, 150*, 144–151. Available from https://doi.org/10.1016/j.jpba.2017.11.061.

Willis, K.J. (2017). State of the world's plants. Royal Botanic Gardens Kew. Richmond, Surrey, United Kingdom.

Wyss, K. M., Llivina, G. C., & Calderón, A. I. (2019). Biochemometrics and required tools in botanical natural products research: A review. *Combinatorial Chemistry & High Throughput Screening, 22*, 290–306. Available from https://doi.org/10.2174/1386207322666190704094003.

Yanes, O., Tautenhahn, R., Patti, G. J., & Siuzdak, G. (2011). Expanding coverage of the metabolome for global metabolite profiling. *Analytical Chemistry, 83*, 2152–2161. Available from https://doi.org/10.1021/ac102981k.

Yang, J. Y., Sanchez, L. M., Rath, C. M., Liu, X., Boudreau, P. D., Bruns, N., ... Dorrestein, P. C. (2013). Molecular networking as a dereplication strategy. *Journal of Natural Products, 76*, 1686–1699. Available from https://doi.org/10.1021/np400413s.

Yuliana, N. D., Khatib, A., Verpoorte, R., & Choi, Y. H. (2011). Comprehensive extraction method integrated with NMR metabolomics: A new bioactivity screening method for plants, adenosine a1 receptor binding compounds in orthosiphon stamineus benth. *Analytical Chemistry, 83*, 6902–6906. Available from https://doi.org/10.1021/ac201458n.

Zaynab, M., Fatima, M., Abbas, S., Sharif, Y., Umair, M., Zafar, M. H., ... Bahadar, K. (2018). Role of secondary metabolites in plant defense against pathogens. *Microbial Pathogenesis, 124*, 198–202. Available from https://doi.org/10.1016/j.micpath.2018.08.034.

CHAPTER

11

Value chains and DNA barcoding for the identification of antiinfective medicinal plants

Seethapathy G. Saroja[1], *Remya Unnikrishnan*[2], *Santhosh Kumar J. Urumarudappa*[3], *Xiaoyan Chen*[1] *and Jiangnan Peng*[1]

[1]Department of Chemistry, School of Computer, Mathematical and Natural Sciences, Morgan State University, Baltimore, MD, United States [2]Forest Genetics and Biotechnology Division, Kerala Forest Research Institute, Thrissur, India [3]Research Unit of DNA Barcoding of Thai Medicinal Plants, Department of Pharmacognosy and Pharmaceutical Botany, Faculty of Pharmaceutical Sciences, Chulalongkorn University, Bangkok, Thailand

Introduction

Taxonomy and DNA barcoding

Accurate species identification is fundamental to any branch of biological research. Carl Linnaeus (1707–78) gift to science was taxonomy: a classification system for the natural world to standardize the naming of species and order them according to their characteristics and relationships with one another. Linnaeus introduced a simple binomial system, based on the combination of two Latin names denoting genus and species. More than two centuries later, Linnaeus' work still plays a central role in biodiversity and biologists are still using Linnaeus' binomial system for the classification of life on Earth (Marta, 2007). However, traditional taxonomy consumes large amounts of time, and taxonomic expertise, which impedes the rapid assessment of biodiversity. One of the estimates suggests that our planet consists of 8.7 million eukaryotic species and merely 1.2 million (<15%) have been identified and classified so far. At this pace, it may take another 480 years to completely identify and classify the species on our planet (Lee, 2011).

Medicinal Plants as Anti-infectives
DOI: https://doi.org/10.1016/B978-0-323-90999-0.00009-4

DNA barcoding is one of the approaches proposed to speed up the task of classifying species. The term DNA barcoding was coined by Paul Hebert (Hebert, Alina, & deWaard, 2003) and can be defined as the use of short nuclear or organelle DNA sequences for the identification of organisms. Since it was first proposed, the technique has been found to be very useful in fingerprinting and identification of species to a remarkable 98%–100% percent accuracy in many organisms (mostly in Kingdom Animalia). In the last decade, DNA barcoding has been found to be a useful technique from fundamental research on biodiversity to enforcement of food laws, from teaching/educational tool to assess biodiversity in school campuses, from authentication of herbal products to assist in the quarantine process at border control and phytosanitary laws and protection of wildlife (Vijayan & Tsou, 2010). Apart from several applications of the DNA barcoding method, more specifically it has been proved to be a useful method in dealing with authenticity issues of medicinal plants and herbal products (Raclariu, Heinrich, Ichim, & de Boer, 2018).

Infectious diseases and antiinfective plants

Infectious diseases are caused by an organism that impairs the health of an individual and has the potential to spread from one person to another either directly or indirectly. The agents that cause infectious diseases fall into five groups: viruses, bacteria, fungi, protozoa, and helminths (worms) (Travers, Walport, & Shlomchik, 2001). Protozoa and worms are usually grouped together as parasites, whereas viruses, bacteria, and fungi are microbes (Travers et al., 2001). The modern health-care systems have used various measures such as antibiotics to effectively treat infectious diseases. However, these drugs may not be as effective as it was once owing to the microbial resistance to these drugs. Furthermore, the side effects of these drugs on human health have been a subject of concern and debate (Quinn, Banat, Abdelhameed, & Banat, 2020). The unaffordability of modern drugs, lack of accessibility to medical physicians in rural areas of developing countries leads people to rely on traditional medicine to treat infectious diseases (Willcox et al., 2007). One of the noteworthy recognitions of traditional medicine to treat infectious diseases is *Artemisia annua* L. Artemisinin was isolated in 1972 using information from the well-documented Chinese compendium of Materia Medica written by Shizhen Li (1518–93) and Ge Hong's "*handbook of prescriptions for emergencies*" from CE 340 to combat malaria as reviewed in Kong and Tan (2015) and Su and Miller (2015). The 2015 Nobel Prize in Physiology or Medicine was awarded to Youyou Tu for the discovery of artemisinin from *A. annua*. The well-known traditional medicines, like Ayurveda and traditional Chinese medicine, and ethnobotanical and/or ethnopharmacological studies have documented many medicinal plants and their formulations to treat many infectious diseases. For example, the flower buds (cloves) of *Syzygium aromaticum* (L.) Merr. & L.M. Perry (Myrtaceae) have been reported to be an antimicrobial agent against oral bacteria that are commonly associated with dental caries and periodontal diseases (Batiha et al., 2020). Likewise the *Embelia ribes* Burm.f. (Primulaceae) for its anthelmintics (Venkatasubramanian, Godbole, Vidyashankar, & Kuruvilla, 2013), *Azadirachta indica* A. Juss. (Meliaceae) for its antifungal (Suresh, Narasimhan, Masilamani, Partho, & Gopalakrishnan, 1997), and *Andrographis paniculata* (Burm.f.) Nees (Acanthaceae) for antiviral properties are a few of the well-documented and highly traded medicinal plants (Gupta, Mishra, & Ganju, 2017).

Herbal products, commercialization, and quality issues of antiinfective plants

Traditionally, assuring the quality and safety of traditional medicines was the responsibility of the traditional medicinal practitioner who collected and prepared the medicine in small amounts for curing diseases (Valiathan, 2006). In recent decades, traditionally used herbal medicines have continued to become mainstream commodities driven by the health industry from craft-based tradition to globalized industry. According to the CAMbrella consortium in Europe, over 100 million Europeans are traditional medicine or complementary and alternative medicine (CAM) users and the usage of herbal medicine is the most commonly reported CAM therapy in Europe (Fischer et al., 2014). Based on the National Health Interview Survey from 2002 to 2012, one-third of adults in the USA used some form of CAM (Clarke, Black, Stussman, Barnes, & Nahin, 2015). Herbal products as a commodity are not without quality concerns, and the growing commercial interest in herbal products increases the incentive for adulteration and substitution in the medicinal plants market (Raclariu et al., 2018). Adulteration could be deliberate due to economic profit or due to misidentification or substitution with allied congeneric species and geographically co-occurring species (Srirama et al., 2017). In addition, species adulteration might also arise due to the same vernacular name being applied to different species in various indigenous systems of medicine, or incorrect use of scientific generic names for the raw drugs (Srirama et al., 2017). The addition of synthetic substances to herbal products is also reported to be common under fraudulent adulteration (Calahan, Howard, Almalki, Gupta, & Calderón, 2016).

Advancements in quality control methods

The medicinal status and regulation of herbal medicines vary considerably in different parts of the world, and the authentication and identification methods for plant materials vary considerably in different parts of the world pharmacopeia (Sgamma et al., 2017). These include morphological characters, microscopy, and chemical fingerprinting (de Boer, Ichim, & Newmaster, 2015; Parveen, Gafner, Techen, Murch, & Khan, 2016) (Fig. 11.1). The inclusion of these methods was emphasized on the basis of single quick and thus cost-effective techniques for primary qualitative analysis or alternatively using hyphenated methods for quantification of chemical marker compounds (Raclariu et al., 2018). DNA barcoding method was originally developed for the identification of plant specimens and have been applied to the authentication of herbal drug materials for industrial quality assurance. This method is intended to be complementary to current morphological and chemical methods of identification (Pawar, Handy, Cheng, Shyong, & Grundel, 2017). Several reviews have provided a comprehensive overview of the background literature to DNA barcoding, DNA based authentication methods, and the detection of adulteration in commercial products (de Boer, Ichim, & Newmaster, 2015; Ganie, Upadhyay, Das, & Prasad Sharma, 2015; Parveen, Gafner, Techen, Murch, & Khan, 2016; Srirama et al., 2017).

European Medicines Agency, the US Food and Drug Administration, and the UK Medicines and Healthcare Products Regulatory Agency support the use of innovative analytical technologies such as DNA barcoding to complement the traditional identification methods in quality control of herbal raw materials and herbal products

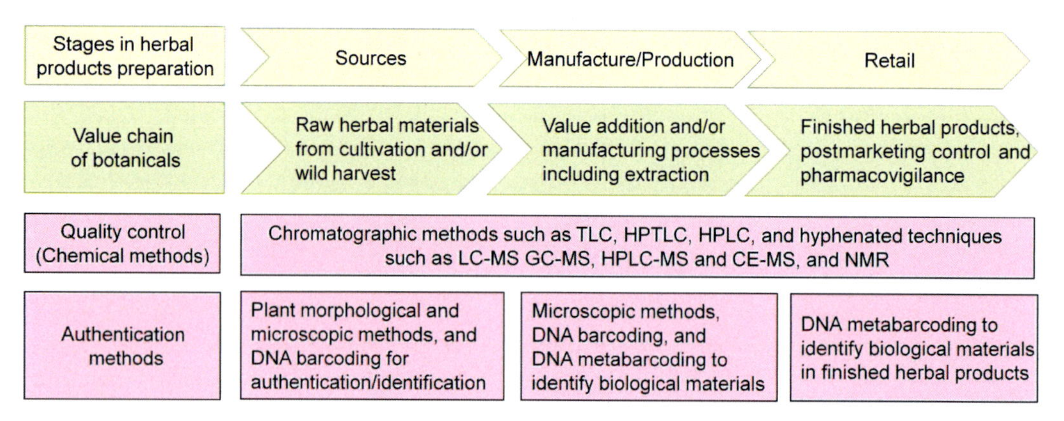

FIGURE 11.1 Different stages in herbal products production and the role of different methods in quality control and authentication. *Chemical methods are also commonly used for authentication. Source: *Adapted from Raclariu, A.C., Heinrich, M., Ichim, M.C., & de Boer, H. (2018). Benefits and limitations of DNA barcoding and metabarcoding in herbal product authentication.* Phytochemical Analysis, 29, *123–128.*

(Anantha Narayana & Johnson, 2019). In 2016 British Pharmacopoeia published "Deoxyribonucleic acid (DNA) based identification techniques for herbal drugs"—a new appendix method, with a focus on plant sampling, barcode regions, DNA extraction, purification, and amplification (B.P. Commission, 2017). In the United States, the sequences reference database United States Pharmacopoeia (USP) (Appendix 1, n.d.) monograph has got a section (DNA-based methods for authentication of articles of botanical origin), which describes DNA-based identification methods and the USP (Appendix 1, 2018) has provided detailed procedures for DNA-based methods and nucleic acid-based genotyping techniques. Chinese Pharmacopoeia Commission (CP Commission, 2017) has published guidelines on molecular DNA barcoding of Chinese Materia Medica. These pharmacopeias have been recognized and included in the general monographs with detailed guidance and test methods for using either DNA barcodes or testing for intact nucleic acid–base pairs. It was suggested that DNA-based testing can be done as an additional identity confirmation, particularly when other identity tests are inconclusive. However, none of these pharmacopeias have made DNA-based identification methods mandatory.

One of the major limitations of DNA barcoding is the inability of the DNA Sanger sequencing technique to detect DNA in herbal products containing more than one species such as polyherbal products (Ivanova, Kuzmina, Braukmann, Borisenko, & Zakharov, 2016). The synergistic action of polyherbal products is one of the fundamentals to the practices of traditional medicines and the use of such specific combinations of medicinal plants results in an enhanced outcome (Parasuraman, Thing, & Dhanaraj, 2014). To ensure the quality, safety, and efficacy of such polyherbal products each of the ingredients included in the product must be identified. Several comprehensive reviews on DNA-based authentication of botanicals have highlighted the merits and demerits of Sanger sequencing-based DNA barcoding (de Boer et al., 2015; Ganie et al., 2015). Nowadays, DNA metabarcoding, which is a combination of DNA barcoding and high-throughput DNA sequencing method, offers several key advantages over conventional DNA barcoding such as mass-amplification and sequencing of barcodes from a complex mixture of multiple species, analyzing samples with varying levels of DNA degradation, products containing fillers, or contaminants, and superior sensitivity of the method (de Boer et al., 2015; Taberlet, Coissac, Pompanon, Brochmann, & Willerslev, 2012).

In this chapter, we highlight the value chain of selected antiinfective plants, that is, how the plants are collected from the wild and traded in local markets (unregulated markets), and the possibilities of misidentification during collection and its consequences for drug safety. We also discuss the role of DNA barcoding in the identification of antiinfective plants/herbal products that are traded.

Materials and methods

A general list of antiinfective plants was derived from the Indian Medicinal Plant Database, National Medicinal Plants Board, Government of India http://www.medicinalplants.in/. Furthermore, using a set of criteria such as plants that have species complex, conservation status, highly traded, and are prone to adulteration, seven species have been shortlisted from 1178 species that are reported to be highly traded in India. Nomenclature follows The Plant List (The Plant List, 2013, http://www.theplantlist.org/) and Angiosperm Phylogeny Group IV (Chase et al., 2016).

Results and discussion

Embelia ribes—anthelmintic plant

Vidanga is an Ayurvedic herb, which is a first-line drug of choice for deworming (an anthelmintic drug) in Ayurvedic medicine. The Ayurvedic Pharmacopoeia of India correlates *Vidanga* to the fruits of *E. ribes* and *E. tsjeriam-cottam* (Roem. & Schult.) A. DC. as a substitute (Venkatasubramanian et al., 2013). The annual consumption of Vidanga during the year 2014–15 was estimated as 773 metric tons (dry weight) with an average price of ~7–10 USD/kg by Indian herbal industries (Goraya & Ved, 2017). Due to overexploitation, *E. ribes* has been listed in the priority species list' for cultivation by the National Medicinal Plant Board of India. Similarly, *E. tsjeriam-cottam* was also reported to be a species that need priority conservation management (Mhaskar et al., 2011). Owing to the high demand for *Vidanga* in herbal markets, it is reported that it is being adulterated with other closely related species such as *E. basaal* (Roem. & Schult.) A. DC., *Maesa indica* Roxb., and *Myrsine africana* L. (Venkatasubramanian et al., 2013; Devaiah & Venkatasubramanian, 2008).

Embelin (2,5-dihydroxy-3-undecyl-1,4-benzoquinone) is the major active constituent in the fruits of *E. ribes*. Fruits/berries of *E. ribes* and embelin have been extensively studied for their anthelmintic activity. Venkatasubramanian et al. (2013) comparatively studied the anthelmintic activity of *E. ribes*, *E. tsjeriam-cottam*, *Maesa indica*, and *Myrsine africana*, and reported that the bioactivity of *E. tsjeriam-cottam* was comparable to *E. ribes*, followed by *Myrsine africana* and *Maesa indica*. Furthermore, the embelin content in the fruits of *E. ribes*, *E. tsjeriam-cottam*, and *Myrsine africana* were 5.94%, 4.32%, and 1.85% (wt./wt.), respectively, whereas *Maesa indica* contains kiritiquinone at 4.4% (wt./wt.). Though *E. ribes* showed better bioactivity, the plant has been assessed as an endangered plant due to premature harvesting of fruits, habitat degradation, unsustainable and indiscriminate harvest. Therefore Venkatasubramanian et al. (2013) concluded *E. tsjeriam-cottam* can be used as a substitute anthelmintic drug for endangered *E. ribes* and suggested that it can be a good conservation strategy.

Peters, Balick, Kahn, and Anderson (1989) cited many examples of how harmful harvesting methods and excessive harvesting of various products from oligarchic forests of Amazonia have resulted in reduced fruit yields and abundance of the targeted species, as well as damage to forest structure. Similarly, Pandey and Shackleton (2012) reported the harvesting approaches for *E. tsjeriam-cottam* fruits in India. It was reported that generally, the harvesters target immature fruits (green) in October and November by cutting or breaking fruit-bearing branches. The branches were carried home and the fruits were plucked. The method of harvest leads to low-quality fruits because the bioactive compounds for the medicinal properties are reputed to accumulate with maturation. This harvesting method also removes the seed from the forest, thereby reducing the number of available seeds for natural regeneration. Furthermore, the concentration of embelin content increased as fruits ripened during December (3.58%–3.99%) and was at a minimum concentration in September when the fruits were immature (1.01%–1.45%) (Pandey & Shackleton, 2012).

Embelin is one of the major bioactive constituents and a marker compound in *E. ribes* fruits. A number of studies have utilized embelin as a marker compound for the standardization of *E. ribes*-based herbal drugs and for quality control. For example, Madhavan, Arimboor, and Arumughan (2011) developed reversed-phase high-performance liquid chromatography coupled with diode array detection for the quantification of embelin. It was reported that the solvent hexane showed a higher extractability of embelin than ethyl acetate, chloroform, and methanol. The linearity, limit of detection, limit of quantification, recovery, and precision for the developed method were reported to be 15–250, 3.97, 13.2 mg/mL, 99.4%–103.8%, and 1.43%–2.87%, respectively. Furthermore, analyzing the commercial phytopharmaceuticals labeled to contain *E. ribes* revealed the presence of the marker compound embelin in the phytopharmaceuticals (Madhavan et al., 2011). However, one of the disadvantages of the chemical methods is that the identity of species may not be ascertained due to the fact that the same marker compound can also be detected in similar congeneric species, such as in *E. tsjeriam-cottam* (Venkatasubramanian et al., 2013).

DNA markers do not correspond to the chemical profile; they are not tissue specific and thus can be detected at any stage of development, with a small amount of sample, in any physical form. Devaiah and Venkatasubramanian (2008) developed random amplified polymorphic DNA (RAPD)-based sequence-characterized amplified region (SCAR) for *E. ribes* that distinguishes from other traded plants *E. tsjeriam-cottam*, *Maesa indica*, and *Myrsine africana*. The developed RAPD-SCAR marker OPF 05 (Forward primer CTATCGCATGCCTCACAATATCATAAT; Reverse primer GAATCAAGTGGCTCTTGGGAGTAAGC) yields an amplicon of 594 bp only for *E. ribes*. Though RAPD-SCAR is still widely used, one of the disadvantages is the need for sequence data to design the SCAR primers (Ganie et al., 2015). Santhosh Kumar et al. (2018) utilized DNA barcoding method to authenticate the raw herbal drugs traded as *Vidanga* (eight samples) in India and revealed that only two samples were *E. ribes* and three samples each were *E. tsjeriam-cottam* and *Maesa indica*, respectively. The results indicated the endangered plant

E. ribes are still traded in India and the morphologically similar species *Maesa indica* in trade might reduce the safety and efficacy of anthelmintic drugs (Santhosh Kumar et al., 2018).

Swertia chirayita—antiviral plant

Swertia chirayita (Roxb.) Buch.-Ham. ex C.B. Clarke is also known as chirata; one of the highly traded perennial herb belonging to the family Gentianaceae spread across the high altitude Himalayas with a fragment distribution from India to Nepal and Bhutan (Scartezzini & Speroni, 2000). In India, 40 species of *Swertia* were recorded, of which, *S. chirayita* is considered the most important for its medicinal properties (Misra et al., 2010). *S. chirayita* is listed as one of the 32 high-priority medicinal plants by the National Medicinal Plant Board of India (Shukla, Dhakal, Uniyal, Paul, & Sahoo, 2017) and also included in the prioritized 30 medicinal plant species for economic development by the Government of Nepal. In 2014 the government of Nepal reported *S. chirayita* as one of the highest export revenue earning medicinal plants of the country (Cunningham, Brinckmann, Schippmann, & Pyakurel, 2018; Khanal, Shakya, Nepal, & Pant, 2014; Susanna & Kumar, 2011). However, only 5% of *S. chirayita* was utilized by Nepal and about 60% was exported to India and 35% to Tibet [about 675.6 of 711 metric tons (MT)] (Cunningham, Brinckmann, Schippmann, & Pyakurel, 2018; Susanna & Kumar, 2011). During the year 2014–15, 404.7 MT of *S. chirayita* were consumed by the herbal industries in India (Goraya & Ved, 2017). Overharvesting along with habitat destruction resulted in the drastic reduction of *S. chirayita* in the wild population, hence the wild harvest was prohibited and conservation management was prioritized by the Government of India (Goraya & Ved, 2017). In Nepal, about 90% of the medicinal plants traded are collected from the wild by local collectors and sold to the local traders. The trade of *S. chirayita* usually contains four levels of stakeholders: the local collectors, the local trader, road head trader, and the wholesaler. A comprehensive review on the trade route of *S. chirayita* is discussed by Cunningham, Brinckmann, Schippmann, and Pyakurel (2018). In India, the whole plant is collected during the flowering season of July and October. In Nepal, it is harvested between October and November with a seasonal prohibition on harvest and trade between May to September (Pyakurel & Baniya, 2011; Susanna & Kumar, 2011). The International Union of Conservation of Nature categorized *S. chirayita* as critically endangered.

Cultivation of *S. chirayita* has been also reported from the eastern parts of Nepal and India. In 2012–13, a total of 232 MT of cultivated *S. chirayita* was collected and traded from eastern Nepal, out of which 152 MT were exported to India and 80 MT were exported to Tibet (Cunningham, Brinckmann, Schippmann, & Pyakurel, 2018; Susanna & Kumar, 2011). Also, the German Federal Ministry for Economic Cooperation and Development (BMZ) and the German Federal Enterprise for International Cooperation are supporting companies in Nepal to uphold trade in the 30 "most exportable" medicinal plants, including *S. chirayita* (Cunningham, Brinckmann, Schippmann, & Pyakurel, 2018). Traditionally, decoctions of *S. chirayita* are used as anthelmintic, antimalarial, antifungal, antibacterial, antifatigue, antiinflammatory, antiaging, and antidiarrheal, and the extracts of *S. chirayita* showed antihepatitis B virus activities (Kumar & Staden, 2016; Zhou et al., 2015). This species was first documented in the Edinburgh Pharmacopeia in 1839 and is reported in Indian Ayurvedic Pharmacopeia, British, and American Pharmacopeia (Joshi & Dhawan, 2005). *S. chirayita* is found to have a number of bioactive chemical constituents and the first isolated dimeric xanthone was chiratanin present in different parts of *S. chirayita* and other major phytoconstituents including amarogentin, swertiamarin, mangiferin, swerchirin, sweroside, amaroswerin (Joshi & Dhawan, 2005; Kumar & Staden, 2016).

The trade and economic importance of chirata are not without adulteration/substitution concern. *S. chirayita* is often misidentified or substituted with allied congeneric species and geographically co-occurring species phenotypically very similar to *Swertia* species. The morphologically similar species that could easily be misidentified and mixed within herbal products are collected from the wild by local farmers or collectors who often rely only on their experience in identifying the species, and the services of specialists like taxonomists are rarely used for authentication (Susanna & Kumar, 2011). For example, other *Swertia* species such as *S. alata* C.B. Clarke, *S. angustifolia* Buch.-Ham. ex D. Don, *S. bimaculata* (Siebold & Zucc.) Hook.f. & Thomson ex C.B. Clarke, *S. ciliata* (D. Don ex G. Don) B.L. Burtt, *S. cordata* (Wall. ex G. Don) C.B. Clarke, *S. densifolia* (Griseb.) Kashyapa, *S. dilatata* C.B. Clarke, *S. elegans* Wall., *S. lawii* Burkill, *S. minor* T. Cooke, *S. paniculata* Wall., and *S. racemosa* (Wall. ex Griseb.) C.B. Clarke, considered to be inferior in medicinal quality, are mixed with *S. chirayita* (Barakoti, Chapagain, Thapa, & Bhusal, 1999; Khanal, Shakya, Thapa, & Pant, 2015). In Tibetan pharmacopeia, *Swertia mussotii* Franch., *Swertia ciliata*, and other species were recommended as a substitute of *S. chirayita* (Li et al., 2020). However, the chemical constituents differ in various species (e.g., iridoids, xanthones, and triterpenes) and have different therapeutic and pharmacological effects (Khanal et al., 2015; Kumar & Staden, 2016).

The adulteration in chirata is also often due to the same vernacular name being applied to different species in various indigenous systems of medicine or incorrect use of scientific generic names for the raw drugs. Species of other genera such as *Andrographis paniculata*, *Exacum tetragonum* Roxb., *E. pedunculatum* L., *Slevogtia orientalis* Griseb. are reported to be adulterants/substitutes due to similar vernacular names (Joshi & Dhawan, 2005). For example, in Unani medicine, the trade name of *A. paniculata* is "Chirayita Desi" (southern chirata), and both species share the name "Kiriyattu" in Malayalam language, "Kiratatikta" in Sanskrit, and "Nilavembu" in Tamil language (Goraya & Ved, 2017). Apparently, the herbs with identical vernacular names result in adulteration/substitution. However, the major chemical constituents of *A. paniculata* are kalmeghin, diterpenes: andrographolide, andrographiside, neoandrographolide as well as panicolide, caffeic acid, chlorogenic acid, and other polyphenolics (Li et al., 2007). These compounds are entirely different from that of *S. chirayita* and thus may result in different therapeutic and pharmacological effects.

Morphologically, the authentic *S. chirayita* can be distinguished from other substitutes and adulterants by its intense bitterness, brownish-purple stem (dark color), continuous yellowish pith, and petals with double nectaries (Joshi & Dhawan, 2005). Prasad (2010) has reported the diagnostic morphological characteristics that distinguish *S. chirayita* from other closely related species. Singh et al. (2019) developed an ultraperformance liquid chromatography method to distinguish *S. chirayita* from the adulterant species, viz., *S. bimaculata*, *S. cordata*, *S. ciliata*, *S. paniculata*, and *Halenia elliptica* D. Don. The marker compounds swertiamarin, mangiferin, gentiopicroside, and sweroside were evaluated, and the limit of detection and quantification of marker compounds were in the range of 1.40–2.06 and 4.57–6.27 g/mL respectively. The hierarchical clustering analysis and principal component analysis revealed that the samples are clustered into different groups and resulted in the discrimination of study species (Singh et al., 2019). Misra et al. (2010) developed amplified fragment length polymorphism-based DNA markers for six species, viz., *S. chirayita*, *S. angustifolia*, *S. bimaculata*, *S. ciliata*, *S. cordata*, and *S. alata*, and reported the species-specific polymorphic markers for the identification of the study species. Joshi and Li (2008) developed DNA barcodes for *S. chirayita* and several closely related species using a nuclear ribosomal internal transcribed spacer (nr-ITS) and chloroplast (trnL-F) regions, and revealed nr-ITS sequences are useful in differentiating Nepalese species commonly used in herbal medicine. Kshirsagar, Umdale, Chavan, and Gaikwad (2017) also indicate that nr-ITS sequences are more suitable DNA markers to distinguish *Swertia* species. In this study, six species, viz., *S. chirayita*, *S. densifolia*, *S. minor*, *S. lawii*, *S. corymbosa*, and *S. angustifolia* were used to evaluate four DNA barcode regions (nr-ITS, psbA-trnH, matK, and rbcL), and the results indicate that the highest interspecific divergence was in ITS (11.87%), followed by psbA-trnH (10.22%), matK (5.04%), and rbcL (0.99%). Furthermore, Stalin Nithaniyal used DNA barcoding and reported that *A. paniculata* is traded as *S. chirayita* in southern India (Nithaniyal et al., 2017). Li et al. (2020) have analyzed 36 commercial samples traded in Southwest China as dida (*S. chirayita*) using DNA barcoding and revealed that three samples (8.3%) were authenticated as *S. chirayita*, two samples (5.6%) as *S. mussotii*, three samples (8.3%) as *S. ciliata*, as recorded in the Tibetan Pharmacopeia. The other samples were authenticated as adulterants and all of them originated from common plants belonging to the genus *Saxifraga* and *Halenia*. These findings showed that DNA barcoding is an efficient tool for identification and authentication of *S. chirayita*.

Picrorhiza kurroa—antiviral plant

Picrorhiza kurroa Royle ex Benth., also known by its trade names Kutki or Kadu (dried roots), belongs to Plantaginaceae family and is distributed at an altitude of 3000–5200 m over the northwest of the central Himalayan region of the Indian subcontinent, China, Pakistan, Bhutan, and Nepal (Kapahi, Srivastava, & Sarin, 2008). Three *Picrorhiza* species are found in the Himalayan region, among which *P. kurroa* is used medicinally and highly traded. *P. kurroa* is listed as one of the high-priority medicinal plants by the National Medicinal Plant Board of India and is also listed as one of the 15 species with higher economic value (Alam & Belt, 2009). During the year 2014–15, 1000–2000 MT of *P. kurroa* has been utilized by the herbal industries in India with an average rate of 800–900 rupees/kg (Goraya & Ved, 2017). In India, *P. kurroa* is collected from the Himalayan forest by wild herb gatherers/local collectors. The collectors will gather together and proceed to high altitude and camp there for the collection (Alam & Belt, 2009). The whole plant is dug out, and the rootstock is separated from aerial parts and rootlets, washed and dried in the sun during the months of September and October (Alam & Belt, 2009; Kapahi et al., 2008).

The rhizomes of Kutki are widely used in Indian traditional medicine for its effectiveness as an antibiotic. It is also described by Ayurvedic literature as jvaraghna (antipyretic) and visaghna (detoxifying) (Kapahi et al., 2008). Other traditional uses of Kutki include the treatment of asthma, jaundice, fever, malaria, snakebite, and liver disorders (Kapahi et al., 2008). It is one of the major components of Arogyavardhini, an effective Ayurvedic preparation used to treat liver diseases (Upadhyay, Dash, Anandjiwala, & Nivsarkar, 2013). The bioactivity of *P. kurroa* is antimicrobial, antibacterial, antimutagenic, cardioprotective, hepatoprotective, antimalarial, antidiabetic, antiinflammatory, anticancer, antiulcer, and nephroprotective activities (Upadhyay et al., 2013). *P. kurroa* yields a crystalline product called "kutkin" which is a mixture of two major C9 iridoid glycosides such as picroside-I and -II and kutkoside, used in more than 2000 herbal formulations (Bhandari et al., 2009; Bhandari, Kumar, Singh, & Ahuja, 2010). A number of studies have reported the antiviral activity against SARS-CoV-2 (Maurya, Kumar, Bhatt, & Saxena, 2020), Chikungunya virus (Raghavendhar, Tripati, Ray, & Patel, 2019), and HIV (Win et al., 2019) of *P. kurroa* phytoconstituents.

Due to its overexploitation and collection from the wild, *P. kurroa* is categorized as an endangered plant and is in the need of conservation management. Around 300–400 plants are uprooted to get 1 kg of roots (Uniyal, Uniyal, & Rawat, 2011). Kutki is listed in CITES-Appendix II and the Indian Red List of endangered species which restricts the trade of noncultivated and nontraceable Kutki obtained from the wild (Alam & Belt, 2009). In order to meet market demands, *P. kurroa* is often adulterated with other species of *Picrorhiza* and *Lagotis cashmeriana* Rupr. (Plantaginaceae). *L. cashmeriana* is found growing with *P. kurroa*, at similar elevations and habitat of alpine Himalayas, between 3200 and 4500 m, and it is traded under the same name of Kutki (Kapahi et al., 2008). It is a small perennial herb with short rootstock and fleshy thick, root fibers. *P. kurroa* can easily be identified by its exerted stamens (Kapahi et al., 2008).

A number of chemical fingerprinting methods use the principle phytoconstituents: iridoid glycosides, picrosides I and II, and kutkoside as marker compounds. For example, Raj and Pal (2016) quantified the content of picroside-I and picroside-II in raw herbal drugs of Kutki using HPLC and found that the marker compounds decreased with the increase in storage duration (1–14 months) irrespective of storage condition. Furthermore, the authors concluded that at low temperature (4°C–6°C) the loss of picroside-I and picroside-II content in the drug Kutki is less during storage (Raj & Pal, 2016). Malik, Priya, and Babbar (2019) analyzed 22 raw herbal drugs collected from Indian herbal markets using DNA barcoding. Five DNA barcoding loci, viz., ITS, ITS2, matK, rbcL, and rpoC1 sequences were generated as reference sequences (from four voucher specimens) and revealed that only one sample collected as Kutki was authentic and other 21 samples were not *P. kurroa*, but rather matched with species such as *Berberis asiatica* Roxb. ex. DC., *Andrographis paniculata*, *Entada abyssinica* A. Rich., and *Erythrophleum ivorense* A. Chev.

Paris polyphylla—anthelmintic plant

Paris polyphylla Sm. (Melanthiaceae) is an important perennial medicinal plant of the Himalayas that is increasingly being used in traditional medicines and pharmaceutical industries (Kunwar et al., 2020). The genus *Paris* comprises 24 species which are distributed in Bhutan, China, north-eastern India, Laos, Myanmar, Nepal, Thailand, Vietnam, and one collection from Pakistan (Cunningham, Brinckmann, Bi et al., 2018; Liu & Ji, 2012). China has the highest number of species (22 species) with 12 endemic species (Cunningham, Brinckmann, Bi et al., 2018). In India, the genus is represented by two species, namely, *P. polyphylla* and *P. thibetica* Franch. with about six intraspecific taxa. In Vietnam, *P. polyphylla* is considered rare and listed as endangered and considered as vulnerable in Nepal, India, and China (Cunningham, Brinckmann, Bi et al., 2018). Unsustainable collection and harvesting practices along with other ecological factors have driven the species to be considered as vulnerable species. The rhizome is the main mode of regeneration though it regenerates from seeds. Overexploitation, indiscriminate harvesting of whole plant and collection of the species before its reproduction maturity affect the natural regeneration of *P. polyphylla* (Cunningham, Brinckmann, Bi et al., 2018; Kunwar et al., 2020).

P. polyphylla rhizomes are widely sold in traditional medicine markets in China and Nepal. China is one of the largest consumers of *P. polyphylla* where it is used as an ingredient in several Chinese herbal formulations (Kunwar et al., 2020). The rhizome of various species of genus *Paris* is used as a major source of raw material for "Yunnan Baiyao," a globally popular product used in Chinese medicine (Cunningham, Brinckmann, Bi et al., 2018; Kunwar et al., 2020). The trade of *P. polyphylla* rhizome from Nepal grew significantly after 2010 and the highest amounts, 58, 41, and 45 tons, were traded between 2011 and 2013 consecutively. The Ministry of Forest and Soil Conservation of Nepal reported that 47,753 kg was collected and exported to Tibet/China (76% of total) and 24% to India in the year 2014. The export to China often occurs through the Kathmandu airport or through border districts. Reliable export trade data for *P. polyphylla* are not available from Bhutan because the trade is mostly carried out informally. In India, all the harvested rhizomes of the *P. polyphylla* are traded to Myanmar and other southeast Asian countries illegally routed through Assam and Manipur. Illegal trading occurred either at local or directly to the regional level through middlemen and then outside of the country. It is also traded from Nepal to China and India. Illegal exporting of rhizomes of the *P. polyphylla* to Myanmar through Indo-Myanmar border by the local traders has also been reported. A comprehensive report on trade demand, trade route, and value chain is reviewed by Cunningham, Brinckmann, Bi et al. (2018) and Kunwar et al. (2020).

Paris polyphylla is known as kalchung (Tamil), paris root (English), satuwa (Nepali), Rhizoma Paridis (Chinese Pharmacopeia). It is used as anthelmintic, antispasmodic, digestive, and expectorant and to treat vermifuge problems, headache, and intestinal worms. Several other biological activities such as anticancer, antitumor, and cytotoxic, antimicrobial, antiangiogenic, immunostimulating, contractile, and hemostatic have also been reported (Negi, Bisht, Bhandari, Bhatt et al., 2014; Yang, Jin, Zhang, Zhang, & Wang, 2017). Secondary metabolites such as daucosterol, polyphyllin D, β-ecdysterone, Paris saponins I, II, V, VI, VII, H, dioscin, oligosaccharides, heptasaccharide, octasaccharide, trigofoenoside A, protogracillin, Paris yunnanosides G-J, padelaoside B, pinnatasterone, formosanin C, and 20-hydroxyecdyson saponins are the major chemical constituents identified in *P. polyphylla* (Negi, Bisht, Bhandari, Bhatt et al., 2014; Yang et al., 2017).

The lack of supply and high price resulted in adulteration of *P. polyphylla* with inferior herbs for monetary profit (deliberate adulteration) and misidentification leading to admixture with other species. More specifically, *P. polyphylla* is often adulterated with *P. thibetica*, *P. tengchongensis* Y.H. Ji, *P. forrestii* (Takht.) H. Li, *P. mairei* H. Lév., *Tupistra* spp., *Trillium tschonoskii* Maxim, *Trillium govanianum* Wall. ex D. Don, and *Polygonum paleaceum* Wall, and *Valeriana jatamansi* Jones (Duan et al., 2018). Liu and Ji (2012) developed polymerase chain reaction−restriction fragment length polymorphism (PCR-RFLP) method to distinguish *P. polyphylla* from 11 other congeneric species namely *P. dunniana* H. Lév., *P. daliensis*

H. Li & V.G. Soukup, *P. vietnamensis* (Takht.) H. Li, *P. mairei*, *P. cronquistii* (Takht.) H. Li, *P. delavayi var. delavayi* Franch., *P. thibetica*, *P. fargessi* Franch., *P. delavayi var. petiolata* (Baker ex C.H. Wright) H. Li, *P. marmomata* Stearn, and *P. axialis* H. Li. The restriction enzyme EaeI (C/GGCCA) has a restriction site specific to *P. polyphylla* nr-ITS sequence at position 200. The other 11 species did not have this restricted site. The specific sizes of the two digested products were 440 and 194 bp. The PCR-RFLP technique developed in this study can be used to discriminate between *P. polyphylla* and its related species (Liu & Ji, 2012). Yang, Zhai, Liu, Zhang, and Ji (2011) utilized the length variation in the chloroplast psbA-trnh loci to detect *V. jatamansi* in commercial medicinal *Paris* products and reported the presence of *V. jatamansi* tissue as adulterant in medicinal *Paris* products. Duan et al. (2018) reported DNA barcoding coupled with a high resolution melting method to distinguish *P. polyphylla* from *P. thibetica*, *P. tengchongensis*, *P. forrestii*, *P. mairei*, *Tupistra* spp., *T. tschonoskii*, and *P. paleaceum* using ITS2 sequences. The developed methods revealed the presence of adulteration in *P. polyphylla*. Out of 10 market samples, five species were identified as *P. polyphylla* var. *yunnanensis*, whereas the remaining five were adulterated with *P. tengchongensis* and *P. mairei*. Apart from DNA methods to distinguish *P. polyphylla* from its adulterants, Xue et al. (2009) developed microscopic methods and reported the diagnostic microscopical characteristics of 11 *Paris* species (Xue et al., 2009).

Saussurea costus—anthelminthic/antiparasitic plant

The genus *Saussurea* (Compositae) is a medicinally important genus consisting of 400 species, among which 62 species are reported from the Himalayan region (Butola & Samant, 2010). *Saussurea costus* (Falc.) Lipsch. is one such indigenous herb distributed at an altitude ranging from 2000 to 3500 m in the subalpine regions of the northwestern Himalaya (Pakistan, Jammu & Kashmir, Himachal Pradesh, and Uttaranchal). The species is listed as critically endangered in Jammu & Kashmir, endemic to Western Ghats, and included in Appendix I of the "Convention on International Trade in Endangered Species" (CITES) and the Wildlife (Protection) Act, 1972 that prohibits the export of the species (Inserted by Act 44 of 1991, w.e.f. 2-10-1991) (Kuniyal, Rawat, & Sundriyal, 2015; Rathore, Debnath, & Kumar, 2021).

S. costus is generally recognized as costus and by diverse vernacular names particularly in India like kuth, postkhai, kur, kustam, sepuddy, kut, koshta, kostum, kot, and kushta (Kuniyal et al., 2015). The declining wild populations of medicinal plant species due to harvesting pressure has prompted the government agencies like National Medicinal Plant Board in India and the domestic herbal industry to promote the cultivation of *S. costus* in India (Goraya & Ved, 2017). During 2014–15, around 164.65 MT of *S. costus* has been utilized by the herbal industries in India (Goraya & Ved, 2017). Even though the cultivation is reported, market samples of Kustha from six major herbal markets of India showed the presence of Ashwagandha [root of *Withania somnifera* (L.) Dunal] and Pushkarmool (roots of *Inula racemosa* Hook.f.) (Prasad & Subhaktha, 2002). The illegal trade of *S. costus* from India to Australia, Japan, Dubai, USA, Mauritius, and the Netherland was also reported by the Indian CITES Management Authority during the year 2004–05 (Kuniyal et al., 2015).

The major constituents of the plant are dehydrocostunolides, saussureal, saussureamines, lupeol palmitates, betulinic acid, flavone glycosides, and guaianolides (Nadda, Ali, Goyal, Khosla, & Goyal, 2020). The roots are traditionally used as an anthelmintic, antiepileptic, antiinflammatory, antilarvicidal, fumigant, and anticancer (Liu et al., 2012; Nadda et al., 2020; Negi, Bisht, Bhandari, Bhatt, Kuniyal et al., 2014). In addition to these, antifungal and antibacterial activities have been reported from the roots of *S. costus* (Alshubaily, 2019). The species has been used in traditional health-care systems and its medicinal properties are well documented in traditional Chinese medicine, the Tibetan system of medicine, and Indian system of medicine. *S. costus* is one of the main ingredients in about 175 formulations documented in *The Handbook of Traditional Tibetan Drugs* (Nadda et al., 2020).

To meet the demand, commercial cultivation is practiced at a higher scale now. It is cultivated in a forested area with similar conditions where it occurs naturally. China was the largest exporter of *S. costus*, it has exported 1024 tons from 1983 to 2009, and India was the second-largest exporter (Rathore et al., 2021). The earliest cultivation of *S. costus* was reported in the early 1940s in Himachal Pradesh, and during 2014–15, 250 ha has been utilized for the Kuth cultivation in India producing approximately 120 MT of root per year (Goraya & Ved, 2017). Chen et al. (2008) utilized nr-ITS sequences to distinguish *S. costus* from the reported adulterants namely, *Vladimiria berardioidea* (Franch.) Ling, *Vladimiria souliei* (Franch.) Ling, *Vladimiria souliei* (Franch.) Ling var. *mirabilis* Ling, *Inula helenium* L., *Inula racemosa* Hook.f., *Aristolochia debilis* Sieb. & Zucc. and *Aristolochia contorta* Bunge. These substitutes and adulterants bear the same common name Muxiang (Chinese) but with different chemical compositions. Sequencing results showed that the similarities of ITS1, ITS2, and 5S rRNA intergenic spacers among *S. costus* and related species were 56.3%–97.8%, 58.5%–97.0%, and 26.4%–77.9%, respectively, and the sequence variation may be used as differentiation markers (Chen et al., 2008).

Syzygium aromaticum—antimicrobial plant

Syzygium aromaticum (L.) Merr. & L.M. Perry (Myrtaceae), commonly known as clove (flower buds) which is the commercialized part of this tree, starts to produce flower buds after 4 years of plantation (Cortés-Rojas, de Souza, & Oliveira, 2014). Flower buds are collected in the maturation phase before flowering and the collection can be done manually or chemically mediated (Batiha et al., 2020; Cortés-Rojas et al., 2014). The largest producers of clove are Brazil, Indonesia, India, Malaysia, Sri Lanka, Madagascar, and Tanzania. In India, the annual trade of clove represents 500–1000 MT (Goraya & Ved, 2017), and in Brazil it is cultivated in approximately 8000 hectares producing near 2500 tons per year (Cortés-Rojas et al., 2014). Traditionally, cloves have been used to treat different microbial infections such as scabies, cholera, malaria, and tuberculosis. It was also used for inhibiting food-borne pathogens to treat viruses, worms, candida, and different bacterial and protozoan infections (Batiha et al., 2020; Deans, Noble, Hiltunen, Wuryani, & Pénzes, 1995). Clove essential oil is traditionally used as a pain reliever in dental care as well as for treating tooth infections and toothache. Moreover, eugenol has been widely used in dentistry because it can penetrate the dental pulp tissue and enter the bloodstream. Clove contains up to 18% of essential oil which is composed of roughly 89% of eugenol, 5%–15% of eugenol acetate and β-caryophyllene, and up to 2.1% of α-humulene. Other volatile compounds present in lower concentrations in clove essential oil are β-pinene, limonene, farnesol, benzaldehyde, 2-heptanone, and ethyl hexanoate (Batiha et al., 2020).

A number of chemical fingerprinting studies report essential oil of clove as being adulterated with inferior quality vegetables and mineral oils. For example, Bounaas et al. (2018) used Fourier-transform infrared spectroscopy method to analyze the quality of clove essential oil and revealed that the commercial samples were diluted with vegetables and mineral oils. Furthermore, Bruno et al. (2019) used the DNA metabarcoding method and analyzed food products that contain *S. aromaticum* (two products) as an ingredient and revealed that none of the products contained *S. aromaticum* as labeled.

Andrographis paniculata—antimicrobial plant

Due to its extremely bitter taste, *Andrographis paniculata* (Burm.f.) Nees (Acanthaceae) is often referred as "the king of bitters," and is used as a bitter tonic in Ayurvedic and other traditionally known health-care systems of India and many other Asian countries. The genus *Andrographis* comprises 40 species, and about 26 species of *Andrographis* are reported to occur in India (Neeraja et al., 2015). Traditionally, *A. paniculata* is used to treat stomachaches, inflammation, pyrexia, intermittent fevers, as an antidote for snakebite and poisonous stings of some insects, and to treat dyspepsia, influenza, dysentery, malaria, and respiratory infections (Hossain, Urbi, Sule, & Rahman, 2014; Okhuarobo et al., 2014). Andrographolide (structurally a labdane diterpenoid) is the major bitter-tasting secondary metabolite and bioactive constituent, reported to have a broad range of pharmacological effects including antidiarrheal, antihepatitis, anti-HIV, antimicrobial, antimalarial, cardiovascular, cytotoxic, hepatoprotective, and immunostimulatory activities, as well as the effect on sexual dysfunctions (Hossain et al., 2014; Okhuarobo et al., 2014).

The whole plant of *A. paniculata* is collected in preflowering stages and when dried, it constitutes the herbal drug. The trade name of *A. paniculata* in India is Kalmegh and Nilavembu, and the trade demand for Kalmegh for the year 2014–15 was approximately 2000 MT (Goraya & Ved, 2017). In the wild, the species has been assessed as vulnerable and also reported to be widely cultivated in India. Due to the species complexity, other species of *Andrographis* such as *A. alata* (Vahl) Nees, *A. lineata* Nees, *A. glandulosa* Nees, *A. echioides* (L.) Nees, *A. serpyllifolia* (Vahl) Wight, *A. macrobotrys* Nees, *A. neesiana* Wight, *A. elongata* (Vahl) T. Anderson, *A. nallamalayana* J.L. Ellis, and *A. wightiana* Arn. ex Nees are used as substitutes and adulterants and are being extensively used in folk medicine (Alagesa Boopathi & Andrographis, 2000; Arolla, Cherukupalli, Khareedu, & Vudem, 2015). Santhosh Kumar et al. (2018) utilized DNA barcoding method to authenticate the raw herbal drugs traded as Kalmegh (six samples) in India and reported no adulteration. Similarly, Osathanunkul et al. (2016) developed high resolution melting coupled with DNA barcoding method to identify *A. paniculata* herbal products available in markets in Thailand. In this study, rbcL DNA barcode loci were used to authenticate 10 commercial herbal products and it was found that all the tested herbal products melting curves profiles were similar to *A. paniculata* which indicated that all tested products contained the correct species as labeled (Osathanunkul et al., 2016). However, another study reported the presence of *Rhinacanthus nasutus* (L.) Kurzin three tested products out of 15 products (Osathanunkul, Madesis, & De Boer, 2015).

Future perspectives

Currently, the World Health Organization guidelines for assessing the quality of herbal medicines mainly include the methods that ensure the identity and safety of the raw plant materials by screening a specified chemical marker compound, and by checking the microbiological purity of the herbal products. Most herbal monographs specify the use of macroscopic and microscopic characterization, phytochemistry-based analysis of specific marker compounds, assays for toxic constituents such as heavy metals, and the use of different chromatographic approaches to detect adulteration. The appropriate utilization of these quality assurance methods, viz., morphological, microscopy, phytochemistry, and DNA-based methods for the identification of medicinal plants heavily depend on various stages of the value chain and are used on the basis of a case-by-case evaluation, starting from the plant material harvest, storage, and to the finished herbal products. Though British Pharmacopeia is one of the first to publish a specific methods section on DNA barcoding that creates a framework for compliance of DNA barcoding with regulatory requirements, DNA barcoding and metabarcoding are not yet widespread validated methods for use in the regulatory context of quality control. A number of studies support its usefulness for herbal product authentication and pharmacovigilance either as a standard method or as a complementary method.

In this chapter, we reviewed the development of DNA barcoding in a few selected anti-infective plants. As a widely accepted method, DNA barcoding has played an important role in the classification of medicinal plants, the identification of substitutes/adulterants, and the regulation of the herbal products market. The presence of adulterants and substitutes may be explained by various factors, including but not limited to the deliberate adulteration and unintentional substitution that may occur from the early stage of the supply chain of medicinal plants (i.e., cultivation, transport, and storage), to the manufacturing process and the commercialization of the final products. It was evident that the antiinfective plants categorized under any threat status were reported to be more adulterated and substituted. The existing technical means can effectively identify plant raw materials but for some processed products (e.g., tablets, pills, oral liquids, and injections, etc.), there is still a lack of effective, rapid, and standardized identification methods up to now, especially for the complex herbal products.

In the past, the development of DNA barcoding has been primarily focused on the selection of candidate DNA markers and the development of sequence databases. Although universal plant DNA barcodes have not yet been chosen, most studies utilize the four markers rbcL, matK, trnH-psbA, and nr-ITS, and the international DNA barcode database is maintained by BOLD in Canada (http://www.boldsystems.org/). Apart from the identification of specific DNA barcode markers for plants, the chloroplast genomes based identification of plants is being developed. Furthermore, the next and third-generation sequencing technologies have been sophisticated, and the research of chloroplast (or plastid) genomics of medical plants is underway. In short, DNA barcoding has a broad application in the field of pharmacovigilance and it will certainly help to assure the safety, quality, and efficacy of traditional medicines in countries around the globe.

References

Alagesa Boopathi, C. (2000). Andrographis SPP.: A source of bitter compounds for medicinal use. *Ancient Science of Life*, *19*, 164–168.

Alam, G., & Belt, J. (2009). *Developing a medicinal plant value chain: Lessons from an initiative to cultivate Kutki (Picrorhiza kurrooa) in Northern India*, KIT Work Pap Ser (WPS. 5).

Alshubaily, F. A. (2019). Enhanced antimycotic activity of nanoconjugates from fungal chitosan and *Saussurea costus* extract against resistant pathogenic *Candida strains*. *International Journal of Biological Macromolecules*, *141*, 499–503. Available from https://doi.org/10.1016/j.ijbiomac.2019.09.022.

Anantha Narayana, D. B., & Johnson, S. T. (2019). DNA barcoding in authentication of herbal raw materials, extracts and dietary supplements: A perspective. *Plant Biotechnology Reports*, *13*, 201–210. Available from https://doi.org/10.1007/s11816-019-00538-z.

Appendix 1. (n.d.). *United States pharmacopeial convention*.

Appendix 1. (2018). *United States Pharmacopeial Convention*. 5, 7357–7378.

Arolla, R. G., Cherukupalli, N., Khareedu, V. R., & Vudem, D. R. (2015). DNA barcoding and haplotyping in different species of Andrographis. *Biochemical Systematics and Ecology*, *62*, 91–97. Available from https://doi.org/10.1016/j.bse.2015.08.001.

Barakoti, T., Chapagain, T., Thapa, Y., & Bhusal, C. (1999). *Chiraito conservation and cultivation and cultivation workshop and achievement*. Kathmandu: Nepal Agricultural Research Council (NARC).

B. P. Commission. (2017). *Deoxyribonucleic acid (DNA) based identification techniques for herbal drugs, British Pharmacopoeia Appendix XI*.

Batiha, G. E. S., Alkazmi, L. M., Wasef, L. G., Beshbishy, A. M., Nadwa, E. H., & Rashwan, E. K. (2020). *Syzygium aromaticum* l. (myrtaceae): Traditional uses, bioactive chemical constituents, pharmacological and toxicological activities. *Biomolecules*, *10*, 202. Available from https://doi.org/10.3390/biom10020202.

Bhandari, P., Kumar, N., Singh, B., & Ahuja, P. S. (2010). Online HPLC-DPPH method for antioxidant activity of *Picrorhiza kurroa* Royle ex Benth. and characterization of kutkoside by ultra-performance LC-electrospray ionization quadrupole time-of-flight mass spectrometry. *Indian Journal of Experimental Biology*, *48*, 323–328. Available from http://nopr.niscair.res.in/bitstream/123456789/7410/1/IJEB%2048%283%29%20323-328.pdf.

Bhandari, P., Kumar, N., Singh, B., Gupta, A. P., Kaul, V. K., & Ahuja, P. S. (2009). Stability-indicating LC-PDA method for determination of picrosides in hepatoprotective Indian herbal preparations of picrorhiza kurroa. *Chromatographia*, *69*, 221–227. Available from https://doi.org/10.1365/s10337-008-0889-7.

Bounaas, K., Bouzidi, N., Daghbouche, Y., Garrigues, S., de la Guardia, M., & El Hattab, M. (2018). Essential oil counterfeit identification through middle infrared spectroscopy. *Microchemical Journal*, *139*, 347–356. Available from https://doi.org/10.1016/j.microc.2018.03.008.

Bruno, A., Sandionigi, A., Agostinetto, G., Bernabovi, L., Frigerio, J., Casiraghi, M., & Labra, M. (2019). Food tracking perspective: DNA metabarcoding to identify plant composition in complex and processed food products. *Genes*, *10*, 248. Available from https://doi.org/10.3390/genes10030248.

Butola, J. S., & Samant, S. S. (2010). *Saussurea* species in Indian Himalayan region: Diversity, distribution and indigenous uses. *International Journal of Plant Biology*, *1*, 43–51. Available from https://doi.org/10.4081/pb.2010.e9.

C. P. Commission. (2017). Guidelines for molecular DNA barcoding of Chinese materia medica. *Pharmacopoeia of the People's Republic of China*.

Calahan, J., Howard, D., Almalki, A. J., Gupta, M. P., & Calderón, A. I. (2016). Chemical adulterants in herbal medicinal products: A review. *Planta Medica*, *82*, 505–515. Available from https://doi.org/10.1055/s-0042-103495.

Chase, M. W., Christenhusz, M. J. M., Fay, M. F., Byng, J. W., Judd, W. S., Soltis, D. E., ... Weber, A. (2016). An update of the Angiosperm Phylogeny Group classification for the orders and families of flowering plants: APG IV. *Botanical Journal of the Linnean Society*, *181*, 1–20. Available from https://doi.org/10.1111/boj.12385.

Chen, F., Chan, H. Y. E., Wong, K. L., Wang, J., Yu, M. T., But, P. P. H., ... Shaw, P. C. (2008). Authentication of *Saussurea lappa*, an endangered medicinal material, by ITS DNA and 5S rRNA sequencing. *Planta Medica*, *74*, 889–892. Available from https://doi.org/10.1055/s-2008-1074551.

Clarke, T. C., Black, L. I., Stussman, B. J., Barnes, P. M., & Nahin, R. L. (2015). Trends in the use of complementary health approaches among adults: United States. *National Health Statistics Reports* (79), 1–16.

Cortés-Rojas, D. F., de Souza, C. R. F., & Oliveira, W. P. (2014). Clove (*Syzygium aromaticum*): A precious spice. *Asian Pacific Journal of Tropical Biomedicine*, *4*, 90–96. Available from https://doi.org/10.1016/S2221-1691(14)60215-X.

Cunningham, A. B., Brinckmann, J. A., Bi, Y. F., Pei, S. J., Schippmann, U., & Luo, P. (2018). Paris in the spring: A review of the trade, conservation and opportunities in the shift from wild harvest to cultivation of *Paris polyphylla* (Trilliaceae). *Journal of Ethnopharmacology*, *222*, 208–216. Available from https://doi.org/10.1016/j.jep.2018.04.048.

Cunningham, A. B., Brinckmann, J. A., Schippmann, U., & Pyakurel, D. (2018). Production from both wild harvest and cultivation: The cross-border *Swertia chirayita* (Gentianaceae) trade. *Journal of Ethnopharmacology*, *225*, 42–52. Available from https://doi.org/10.1016/j.jep.2018.06.033.

de Boer, H. J., Ichim, M. C., & Newmaster, S. G. (2015). DNA barcoding and pharmacovigilance of herbal medicines. *Drug Safety*, *38*, 611–620. Available from https://doi.org/10.1007/s40264-015-0306-8.

Deans, S. G., Noble, R. C., Hiltunen, R., Wuryani, W., & Pénzes, L. G. (1995). *Antimicrobial and antioxidant properties of* Syzygium aromaticum *(L.) Merr. & Perry: Impact upon bacteria, fungi and fatty acid levels in ageing mice. Flavour and Fragrance Journal* (10, pp. 323–328). Available from https://doi.org/10.1002/ffj.2730100507.

Devaiah, K. M., & Venkatasubramanian, P. (2008). Genetic characterization and authentication of *Embelia ribes* using RAPD-PCR and SCAR marker. *Planta Medica*, *74*, 194–196. Available from https://doi.org/10.1055/s-2008-1034279.

Duan, B. Z., Wang, Y. P., Fang, H. L., Xiong, C., Li, X. W., Wang, P., & Chen, S. L. (2018). Authenticity analyses of Rhizoma Paridis using barcoding coupled with high resolution melting (Bar-HRM) analysis to control its quality for medicinal plant product. *Chinese Medicine (United Kingdom)*, *13*, 8. Available from https://doi.org/10.1186/s13020-018-0162-4.

Fischer, F. H., Lewith, G., Witt, C. M., Linde, K., von Ammon, K., Cardini, F., . . . Brinkhaus, B. (2014). High prevalence but limited evidence in complementary and alternative medicine: Guidelines for future research. *BMC Complementary and Alternative Medicine*, *14*, 46. Available from https://doi.org/10.1186/1472-6882-14-46.

Ganie, S. H., Upadhyay, P., Das, S., & Prasad Sharma, M. (2015). Authentication of medicinal plants by DNA markers. *Plant Gene*, *4*, 83–99. Available from https://doi.org/10.1016/j.plgene.2015.10.002.

Goraya, G., & Ved, D. (2017). *Medicinal plants in India: An assessment of their demand and supply*. National Medicinal Plants Board, Ministry of AYUSH, Government of India.

Gupta, S., Mishra, K. P., & Ganju, L. (2017). Broad-spectrum antiviral properties of andrographolide. *Archives of Virology*, *162*, 611–623. Available from https://doi.org/10.1007/s00705-016-3166-3.

Hebert, P. D. N., Alina, C., & deWaard, B. S. L., Jr (2003). Biological identifications through DNA barcodes. *Proceedings of the Royal Society of London. Series B: Biological Sciences*, *270*(1512), 313–321. Available from https://doi.org/10.1098/rspb.2002.2218.

Hossain, M. S., Urbi, Z., Sule, A., & Rahman, K. M. H. (2014). *Andrographis paniculata* (Burm. f.) Wall. ex Nees: A review of ethnobotany, phytochemistry, and pharmacology. *Scientific World Journal*, *2014*, 274905. Available from https://doi.org/10.1155/2014/274905.

Ivanova, N. V., Kuzmina, M. L., Braukmann, T. W. A., Borisenko, A. V., & Zakharov, E. V. (2016). Authentication of herbal supplements using next-generation sequencing. *PLoS One*, *11*, e0168628. Available from https://doi.org/10.1371/journal.pone.0156426.

Joshi, K., & Li, J. (2008). *Phylogenetics of Swertia L.(gentinaceae–Swertiinae) and molecular differentiation of Swertia species in Nepalese medicinal herbs*.

Joshi, P., & Dhawan, V. (2005). Swertia chirayita - An overview. *Current Science*, *89*, 635–640. Available from http://www.ias.ac.in/currsci/aug252005/635.pdf.

Kapahi, B. K., Srivastava, T. N., & Sarin, Y. K. (2008). Description of *Picrorhiza kurroa*: A source of the ayurvedic drug kutaki. *International Journal of Pharmacognosy*, *31*(3), 217–222. Available from https://doi.org/10.3109/13880209309082945.

Khanal, S., Shakya, N., Nepal, N., & Pant, D. (2014). *Swertia chirayita*: The Himalayan herb. *International Journal of Applied Sciences and Biotechnology*, *2*, 389–392.

Khanal, S., Shakya, N., Thapa, K., & Pant, D. R. (2015). Phytochemical investigation of crude methanol extracts of different species of *Swertia* from Nepal. *BMC Research Notes*, *8*, 821. Available from https://doi.org/10.1186/s13104-015-1753-0.

Kong, L. Y., & Tan, R. X. (2015). Artemisinin, a miracle of traditional Chinese medicine. *Natural Product Reports*, *32*, 1617–1621. Available from https://doi.org/10.1039/c5np00133a.

Kshirsagar, P., Umdale, S., Chavan, J., & Gaikwad, N. (2017). Molecular authentication of medicinal plant, *Swertia chirayita* and its adulterant species. *Proceedings of the National Academy of Sciences India Section B - Biological Sciences*, *87*, 101–107. Available from https://doi.org/10.1007/s40011-015-0556-3.

Kumar, V., & Staden, J. V. (2016). A review of *Swertia chirayita* (Gentianaceae) as a traditional medicinal plant. *Frontiers in Pharmacology*, *6*, 308. Available from https://doi.org/10.3389/fphar.2015.00308.

Kuniyal, C. P., Rawat, D. S., & Sundriyal, R. C. (2015). Cultivation of *Saussurea costus* cannot be treated as "artificially propagated,". *Current Science*, *108*, 1587–1589. Available from http://www.currentscience.ac.in/Volumes/108/09/1587.pdf.

Kunwar, R. M., Adhikari, Y. P., Sharma, H. P., Rimal, B., Devkota, H. P., Charmakar, S., … Jentsch, A. (2020). Distribution, use, trade and conservation of *Paris polyphylla* Sm. in Nepal. *Global Ecology and Conservation*, *23*, e01081. Available from https://doi.org/10.1016/j.gecco.2020.e01081.

Lee, S. (2011). Number of species on Earth tagged at 8.7 million. *Nature*. Available from https://doi.org/10.1038/news.2011.498.

Li, R., Li, S., Yang, X., Guo, Y., Duan, B., Xu, L., & Xia, C. (2020). Identification of Tibetan medicine Dida based on DNA barcoding. *Mitochondrial DNA Part A: DNA Mapping, Sequencing, and Analysis*, *31*, 131–138. Available from https://doi.org/10.1080/24701394.2020.1741563.

Li, W., Xu, X., Zhang, H., Ma, C., Fong, H., Van Breemen, R., … Fitzloff, J. (2007). Secondary metabolites from *Andrographis paniculata*. *Chemical and Pharmaceutical Bulletin*, *55*, 455–458. Available from https://doi.org/10.1248/cpb.55.455.

Liu, T., & Ji, Y. (2012). Molecular authentication of the medicinal plant *Paris polyphylla* Smith var. yunnanensis (Melanthiaceae) and its related species by polymerase chain reaction restriction fragment length polymorphism (PCR-RFLP). *Journal of Medicinal Plants Research*, *6*, 1181–1186.

Liu, Z. L., He, Q., Chu, S. S., Wang, C. F., Du, S. S., & Deng, Z. W. (2012). Essential oil composition and larvicidal activity of *Saussurea lappa* roots against the mosquito *Aedes albopictus* (Diptera: Culicidae). *Parasitology Research*, *110*, 2125–2130. Available from https://doi.org/10.1007/s00436-011-2738-0.

Madhavan, S. N., Arimboor, R., & Arumughan, C. (2011). RP-HPLC-DAD method for the estimation of embelin as marker in Embelia ribes and its polyherbal formulations. *Biomedical Chromatography*, *25*, 600–605. Available from https://doi.org/10.1002/bmc.1489.

Malik, S., Priya, A., & Babbar, S. B. (2019). Employing barcoding markers to authenticate selected endangered medicinal plants traded in Indian markets. *Physiology and Molecular Biology of Plants*, *25*, 327–337. Available from https://doi.org/10.1007/s12298-018-0610-8.

Marta, P. (2007). There shall be order. *EMBO Reports*, *8*, 814–816. Available from https://doi.org/10.1038/sj.embor.7401061.

Maurya, V. K., Kumar, S., Bhatt, M. L. B., & Saxena, S. K. (2020). Antiviral activity of traditional medicinal plants from Ayurveda against SARS-CoV-2 infection. *Journal of Biomolecular Structure and Dynamics*. Available from https://doi.org/10.1080/07391102.2020.1832577.

Mhaskar, M., Joshi, S., Chavan, B., Joglekar, A., Barve, N., & Patwardhan, A. (2011). Status of *Embelia ribes* Burm f. (Vidanga), an important medicinal species of commerce from northern Western Ghats of India. *Current Science*, *100*, 547–552. Available from http://www.ias.ac.in/currsci/25feb2011/547.pdf.

Misra, A., Shasany, A. K., Shukla, A. K., Darokar, M. P., Singh, S. C., Sundaresan, V., … Khanuja, S. P. (2010). AFLP markers for identification of *Swertia* species (Gentianaceae). *Genetics and Molecular Research: GMR*, *9*, 1535–1544. Available from https://doi.org/10.4238/vol9-3gmr785.

Nadda, R. K., Ali, A., Goyal, R. C., Khosla, P. K., & Goyal, R. (2020). *Aucklandia costus* (Syn. *Saussurea costus*): Ethnopharmacology of an endangered medicinal plant of the Himalayan region. *Journal of Ethnopharmacology*, *263*, 113199. Available from https://doi.org/10.1016/j.jep.2020.113199.

Neeraja, C., Krishna, P. H., Reddy, C. S., Giri, C. C., Rao, K. V., & Reddy, V. D. (2015). Distribution of andrographis species in different districts of Andhra Pradesh. *Proceedings of the National Academy of Sciences India Section B - Biological Sciences*, *85*, 601–606. Available from https://doi.org/10.1007/s40011-014-0364-1.

Negi, J. S., Bisht, V. K., Bhandari, A. K., Bhatt, V. P., Singh, P., & Singh, N. (2014). *Paris polyphylla*: Chemical and biological prospectives. *Anti-cancer Agents in Medicinal Chemistry*, *14*, 833–839. Available from https://doi.org/10.2174/1871520614666140611101040.

Negi, J. S., Bisht, V. K., Bhandari, A. K., Kuniyal, C. P., Bhatt, V. P., & Bisht, R. (2014). Chemical fingerprinting and antibacterial activity of *Saussurea lappa* clarke. *Applied Biochemistry and Microbiology*, *50*, 588–593. Available from https://doi.org/10.1134/S0003683814060118.

Nithaniyal, S., Vassou, S. L., Poovitha, S., Raju, B., Parani, M., & Cristescu, M. E. (2017). Identification of species adulteration in traded medicinal plant raw drugs using DNA barcoding. *Genome/National Research Council Canada = Genome/Conseil National de Recherches Canada, 60*, 139–146. Available from https://doi.org/10.1139/gen-2015-0225.

Okhuarobo, A., Ehizogie Falodun, J., Erharuyi, O., Imieje, V., Falodun, A., & Langer, P. (2014). Harnessing the medicinal properties of *Andrographis paniculata* for diseases and beyond: A review of its phytochemistry and pharmacology. *Asian Pacific Journal of Tropical Disease, 4*, 213–222. Available from https://doi.org/10.1016/S2222-1808(14)60509-0.

Osathanunkul, M., Madesis, P., & De Boer, H. (2015). Bar-HRM for authentication of plant-based medicines: Evaluation of three medicinal products derived from *Acanthaceae* species. *PLoS One, 10*, e0128476. Available from https://doi.org/10.1371/journal.pone.0128476.

Osathanunkul, M., Suwannapoom, C., Khamyong, N., Pintakum, D., Lamphun, S. N., Triwitayakorn, K., ... Madesis, P. (2016). Hybrid analysis (barcode-high resolution melting) for authentication of Thai herbal products, *Andrographis paniculata* (Burm.f.) Wall.ex Nees. *Pharmacognosy Magazine, 12*, S71–S75. Available from https://doi.org/10.4103/0973-1296.176112.

Pandey, A. K., & Shackleton, C. M. (2012). The effect of harvesting approaches on fruit yield, embelin concentration and regrowth dynamics of the forest shrub, *Embelia tsjeriam-cottam*, in central India. *Forest Ecology and Management, 266*, 180–186. Available from https://doi.org/10.1016/j.foreco.2011.11.015.

Parasuraman, S., Thing, G. S., & Dhanaraj, S. A. (2014). Polyherbal formulation: Concept of ayurveda. *Pharmacognosy Reviews, 8*, 73–80. Available from https://doi.org/10.4103/0973-7847.134229.

Parveen, I., Gafner, S., Techen, N., Murch, S. J., & Khan, I. A. (2016). DNA barcoding for the identification of botanicals in herbal medicine and dietary supplements: Strengths and limitations. *Planta Medica, 82*, 1225–1235. Available from https://doi.org/10.1055/s-0042-111208.

Pawar, R. S., Handy, S. M., Cheng, R., Shyong, N., & Grundel, E. (2017). Assessment of the authenticity of herbal dietary supplements: Comparison of chemical and DNA barcoding methods. *Planta Medica, 83*, 921–936. Available from https://doi.org/10.1055/s-0043-107881.

Peters, C. M., Balick, M. J., Kahn, F., & Anderson, A. B. (1989). Oligarchic forests of economic plants in Amazonia: Utilization and conservation of an important tropical resource. *Conservation Biology, 3*, 341–349. Available from https://doi.org/10.1111/j.1523-1739.1989.tb00240.x.

Prasad, P. V., & Subhaktha, P. K. (2002). Medico-historical review of drug Kustha. *Bulletin of the Indian Institute of History of Medicine (Hyderabad), 32*, 79–92.

Prasad, R. D. (2010). Taxonomic study of some medicinally important species of *Swertia* L. (Gentianaceae) in Nepal. *Botanica Orientalis: Journal of Plant Science*, 18–24. Available from https://doi.org/10.3126/botor.v6i0.2906.

Pyakurel, D., & Baniya, A. (2011). *Impetus for conservation and livelihood support in Nepal: A reference book on ecology, conservation, product development and economic analysis of selected NTFPs of Langtang Area in the sacred Himalayan landscape* (pp. 35–41). Nepal: World Wildlife Fund (WWF).

Quinn, G. A., Banat, A. M., Abdelhameed, A. M., & Banat, I. M. (2020). Streptomyces from traditional medicine: Sources of new innovations in antibiotic discovery. *Journal of Medical Microbiology, 69*, 1040–1048. Available from https://doi.org/10.1099/jmm.0.001232.

Raclariu, A. C., Heinrich, M., Ichim, M. C., & de Boer, H. (2018). Benefits and limitations of DNA barcoding and metabarcoding in herbal product authentication. *Phytochemical Analysis, 29*, 123–128. Available from https://doi.org/10.1002/pca.2732.

Raghavendhar, S., Tripati, P. K., Ray, P., & Patel, A. K. (2019). Evaluation of medicinal herbs for Anti-CHIKV activity. *Virology, 533*, 45–49. Available from https://doi.org/10.1016/j.virol.2019.04.007.

Raj, T. P., & Pal, S. Y. (2016). Standardization of storage conditions and duration on picroside-I and picroside-II in raw material of drug kutki (*Picrorhiza kurroa* Royle ex Benth.). *Nepal Journal of Science and Technology, 17*, 23–26. Available from https://doi.org/10.3126/njst.v17i1.25059.

Rathore, S., Debnath, P., & Kumar, R. (2021). Kuth *Saussurea costus* (Falc.) Lipsch.: A critically endangered medicinal plant from Himalaya. *Journal of Applied Research on Medicinal and Aromatic Plants, 20*, 100277. Available from https://doi.org/10.1016/j.jarmap.2020.100277.

Santhosh Kumar, J. U., Krishna, V., Seethapathy, G. S., Ganesan, R., Ravikanth, G., & Shaanker, R. U. (2018). Assessment of adulteration in raw herbal trade of important medicinal plants of India using DNA barcoding. *3 Biotech, 8*, 135. Available from https://doi.org/10.1007/s13205-018-1169-3.

Scartezzini, P., & Speroni, E. (2000). Review on some plants of Indian traditional medicine with antioxidant activity. *Journal of Ethnopharmacology, 71*, 23–43. Available from https://doi.org/10.1016/S0378-8741(00)00213-0.

Sgamma, T., Lockie-Williams, C., Kreuzer, M., Williams, S., Scheyhing, U., Koch, E., ... Howard, C. (2017). DNA barcoding for industrial quality assurance. *Planta Medica, 83*, 1117–1129. Available from https://doi.org/10.1055/s-0043-113448.

Shukla, J. K., Dhakal, P., Uniyal, R. C., Paul, N., & Sahoo, D. (2017). Ex-situ cultivation at lower altitude and evaluation of *Swertia chirayita*, a critically endangered medicinal plant of Sikkim Himalayan region, India. *South African Journal of Botany, 109*, 138–145. Available from https://doi.org/10.1016/j.sajb.2017.01.001.

Singh, M., Mohan, R., Mishra, S., Goyal, N., Shanker, K., Gupta, N., ... Kumar, B. (2019). Ultra performance liquid chromatography coupled with principal component and cluster analysis of *Swertia chirayita* for adulteration check. *Journal of Pharmaceutical and Biomedical Analysis, 164*, 302–308. Available from https://doi.org/10.1016/j.jpba.2018.10.054.

Srirama, R., Santhosh Kumar, J. U., Seethapathy, G. S., Newmaster, S. G., Ragupathy, S., Ganeshaiah, K. N., ... Ravikanth, G. (2017). Species adulteration in the herbal trade: Causes, consequences and mitigation. *Drug Safety, 40*, 651–661. Available from https://doi.org/10.1007/s40264-017-0527-0.

Su, X. Z., & Miller, L. H. (2015). The discovery of artemisinin and the Nobel Prize in physiology or medicine. *Science China Life Sciences, 58*, 1175–1179. Available from https://doi.org/10.1007/s11427-015-4948-7.

Suresh, G., Narasimhan, N. S., Masilamani, S., Partho, P. D., & Gopalakrishnan, G. (1997). Antifungal fractions and compounds from uncrushed green leaves of *Azadirachta indica*. *Phytoparasitica, 25*, 33–39. Available from https://doi.org/10.1007/BF02981477.

Susanna, P., & Kumar, J. P. (2011). Trade and sustainable conservation of *Swertia chirayita* (Roxb. ex Fleming) h. Karst in Nepal. *Nepal Journal of Science and Technology, 11*, 125–132. Available from https://doi.org/10.3126/njst.v11i0.4134.

Taberlet, P., Coissac, E., Pompanon, F., Brochmann, C., & Willerslev, E. (2012). Towards next-generation biodiversity assessment using DNA metabarcoding. *Molecular Ecology, 21*, 2045–2050. Available from https://doi.org/10.1111/j.1365-294X.2012.05470.x.

Travers, P., Jr, Walport, M., & Shlomchik, M. J. (2001). *Infectious agents and how they cause disease. Immunobiology: The Immune System in Health and Disease*. New York: Garland Science.

Uniyal, A., Uniyal, S. K., & Rawat, G. S. (2011). Commercial extraction of *Picrorhiza kurrooa* royle ex benth. in the Western Himalaya. *Mountain Research and Development, 31*, 201–208. Available from https://doi.org/10.1659/MRD-JOURNAL-D-10-00125.1.

Upadhyay, D., Dash, R. P., Anandjiwala, S., & Nivsarkar, M. (2013). Comparative pharmacokinetic profiles of picrosides I and II from kutkin, *Picrorhiza kurroa* extract and its formulation in rats. *Fitoterapia, 85*, 76–83. Available from https://doi.org/10.1016/j.fitote.2013.01.004.

Valiathan, M. S. (2006). Ayurveda: Putting the house in order. *Current Science, 90*, 5–6. Available from http://www.ias.ac.in/currsci/jan102006/5.pdf.

Venkatasubramanian, P., Godbole, A., Vidyashankar, R., & Kuruvilla, G. R. (2013). Evaluation of traditional anthelmintic herbs as substitutes for the endangered *Embelia ribes*, using *Caenorhabditis elegans* model. *Current Science, 105*, 1593–1597. Available from http://www.currentscience.ac.in/Volumes/105/11/1593.pdf.

Vijayan, K., & Tsou, C. H. (2010). DNA barcoding in plants: Taxonomy in a new perspective. *Current Science, 99*, 1530–1541. Available from http://www.ias.ac.in/currsci/10dec2010/1530.pdf.

Willcox, M. L., Graz, B., Falquet, J., Sidibé, O., Forster, M., & Diallo, D. (2007). *Argemone mexicana* decoction for the treatment of uncomplicated falciparum malaria. *Transactions of the Royal Society of Tropical Medicine and Hygiene, 101*, 1190–1198. Available from https://doi.org/10.1016/j.trstmh.2007.05.017.

Win, N. N., Kodama, T., Lae, K. Z. W., Win, Y. Y., Ngwe, H., Abe, I., ... Morita, H. (2019). Bis-iridoid and iridoid glycosides: Viral protein R inhibitors from *Picrorhiza kurroa* collected in Myanmar. *Fitoterapia, 134*, 101–107. Available from https://doi.org/10.1016/j.fitote.2019.02.016.

Xue, D., Yin, H., Li, J., Liu, X., Zhang, H., & Peng, C. (2009). Application of microscopy in authentication and distinguishing of 11 Paris species in West Sichuan. *Microscopy Research and Technique, 72*, 744–754. Available from https://doi.org/10.1002/jemt.20726.

Yang, Y., Jin, H., Zhang, J., Zhang, J., & Wang, Y. (2017). Quantitative evaluation and discrimination of wild *Paris polyphylla* var. yunnanensis (Franch.) Hand.-Mazz from three regions of Yunnan Province using UHPLC–UV–MS and UV spectroscopy couple with partial least squares discriminant analysis. *Journal of Natural Medicines, 71*, 148–157. Available from https://doi.org/10.1007/s11418-016-1044-7.

Yang, Y., Zhai, Y., Liu, T., Zhang, F., & Ji, Y. (2011). Detection of *Valeriana jatamansi* as an adulterant of medicinal Paris by length variation of chloroplast psbA-trnH region. *Planta Medica, 77*, 87–91. Available from https://doi.org/10.1055/s-0030-1250072.

Zhou, N. J., Geng, C. A., Huang, X. Y., Ma, Y. B., Zhang, X. M., Wang, J. L., ... Chen, J. J. (2015). Anti-hepatitis B virus active constituents from *Swertia chirayita*. *Fitoterapia, 100*, 27–34. Available from https://doi.org/10.1016/j.fitote.2014.11.011.

CHAPTER

12

Fungal endophytes: a source of antibacterial and antiparasitic compounds

Romina Pacheco[1], *Sergio Ortiz*[1,2], *Mohamed Haddad*[1] *and Marieke Vansteelandt*[1]

[1]UMR 152 PharmaDev, Université de Toulouse, IRD, UPS, Toulouse, France [2]UMR 7200 Laboratoire d'Innovation Thérapeutique, Université de Strasbourg, CNRS, Strasbourg Drug Discovery and Development Institute (IMS), Illkirch-Graffenstaden, France

Introduction

Current situation of microbial infections

The golden era of the antibiotic occurred between the early 1930s and late 1960s, a period where most of the new classes of drugs were discovered. This period was followed by an important decline in drug discovery rate along with the increase and emergence of resistant microbes, including bacteria, viruses, fungi, and parasites (Aminov, 2010). In 2019 the WHO listed 32 antibiotics in clinical development that were addressed to the list of priority pathogens, of which only six were considered innovative. Antimicrobial resistance affects health care systems from all countries found in any level of development, generating a great economic burden while other medical procedures become more risky.

Protozoal diseases such as malaria (*Plasmodium* spp.), leishmaniasis (*Leishmania* spp.), and trypanosomiasis (*Trypanosoma* spp.) affect mainly tropical and subtropical regions with great health and economic impacts worldwide. These unicellular eukaryotic organisms are free-living organisms and their developmental stages include intracellular stages within the host cell or extracellular stages in body fluids, hollow organs, or between cells in the interstitial space. Additionally, they can possess dormant forms that help them to survive under extreme conditions over a long period of time. Their transmission is mainly vector-borne but also fecal–oral or predator–prey transmission.

Medicinal Plants as Anti-infectives
DOI: https://doi.org/10.1016/B978-0-323-90999-0.00006-9

According to the WHO, there are currently 20 diseases caused by bacteria, helminths, protozoa, or viruses considered as part of the neglected tropical diseases, including Chagas disease, human African trypanosomiasis, leishmaniasis, and many others. Individually, none of them represents a global priority, however, counted together, they affect approximately two billion people with a collective number of people affected equivalent to HIV/AIDS, tuberculosis, or malaria (Engels & Zhou, 2020).

In general, the arsenal of antimicrobial drugs is becoming very limited while the resistance development is rising against many of the main treatments. The problem of antibiotic resistance also involves natural environmental changes and proliferation of infections due to human activities. The absence of effective vaccines and the development of resistance to the current drugs situate these infections as major problems in endemic countries and highlight the need of novel molecules acting on new targets.

Microbial natural products as sources of new drugs

A significant number of compounds, including drugs and drug leads, are produced by microorganisms during the interaction with other microorganisms and with their environment. Almost 50% of the total small molecules approved as drugs between 1981 and 2019 are derived from natural products or from their synthetic variations (Newman & Cragg, 2020). In general, microorganisms respond to environmental signals which provide them better survival mechanisms and in this process, many secondary metabolites with different biological activities are produced. Most of the relevant classes that are currently clinically important are derived from fungi, actinomycetes, and other bacteria, besides the compounds obtained from medicinal plants (Hutchings, Truman, & Wilkinson, 2019). Environments with high biodiversity will potentially harbor new species or strains of microorganisms capable of producing novel compounds. These environments are considered as hot spots, such as the tropical forests, where plants represent excellent niches for interactions between microorganisms, including fungi and bacteria (Knight et al., 2003).

Species from fungal kingdom, including basidiomycetes and filamentous fungi, are responsible for 45% of the microbial metabolites production, mainly from species corresponding to *Penicillium*, *Aspergillus*, and *Trichoderma* genera (Bérdy, 2012). Massive whole-genome sequencing has shown the excellent capacity of fungi to produce a variety of secondary metabolites; however, the current challenge is to access these biosynthetic pathways that are usually silenced under standard culture conditions (Challinor & Bode, 2015). Secondary metabolites produced by fungi are chemically diverse including polyketides, aromatic compounds, alkaloids, or terpenoids, which are mainly synthesized by polyketide synthases and nonribosomal peptide synthetases. Additionally, there are other biosynthetic systems, including hybrid pathways that give rise to a greater chemical diversity (Keller, 2019). They exhibit a broad spectrum of biological activities such as antifungal, antibacterial, antiviral, antiparasitic, anticancer, immunosuppressant properties, and also possess high chemical originality, especially from fungal species obtained as endophytes or from marine ecosystems (Schueffler & Anke, 2014).

Endophytic fungi

Endophytes are microorganisms, mainly bacteria and fungi, which colonize internal plant tissues, including branches, leaves, seeds, roots, without generating visible harmful effects to the host plant. This successful endophyte colonization involves compatible interactions between the plant and its microbiome and the recognition of the endophyte during the invasion to the plant through the initiation of a cross-talk of signal molecules (Khare, Mishra, & Arora, 2018). It is important to consider that these interactions can be facultative or obligate which also depend on environmental conditions and on the physiological state of the plant, ranging from mutualism to commensalism and parasitism. Besides this, endophytic fungal communities can exhibit host plant species specificity and within the host plant, organ and tissue specificity (Aly, Debbab, & Proksch 2011).

It is considered that almost all plants on Earth are in constant association and interaction with microorganisms, especially fungal endophytes (Arnold & Lutzoni, 2007). This special relationship has evolved over long periods of time and approximately 18% of known plant metabolites can be produced by their associated fungi. The intrinsic relationships occur at different metabolic levels and have also been shown to improve plant growth and tolerance to biotic and abiotic stress factors including resistance to environment, herbivores attacks, via the production of antimicrobials, insecticides, and growth regulators while plants may provide them their nutrients and sugars. Specifically, a total of 46 families of medicinal plants distributed in tropical and subtropical or extreme environment, have been shown to exhibit a mutualistic relationship with their endophytes (Jia et al., 2016).

Taken together, the rationale for studying endophytic microbes as potential sources of new drugs is related to the fact that they represent a relatively unexplored area of biochemical diversity. In particular, the protection provided by endophytes to the host plant by various bioactive compounds increases the attractiveness of these compounds in medical applications. Furthermore, since toxicity to higher organisms is a concern of any potential drug, natural products isolated from endophytes may be an attractive alternative, since these compounds are produced in an eukaryotic system without apparent damage. Therefore, the host plant itself has naturally served as a selection system for microbes producing bioactive compounds with reduced toxicity to higher organisms.

Given the number of existing plant species on the planet, what are the justifications for selecting a plant to be studied for its endophytes? It is noteworthy that each of the approximately 300,000 plant species that exist on our planet is a host for one or more endophytes (Strobel & Daisy, 2003) and it is widely accepted that there are few plants without endophytes (Gennaro, Gonthier, & Nicolotti, 2003). It is estimated that there may be as many as one million different fungal species, only a few of which have been studied in detail (Ganley, Brunsfeld, & Newcombe, 2004). The presence of plant-associated microorganisms detected in fossilized stem and leaf tissues indicates that endophyte–host associations may have evolved since the appearance of higher plants on Earth (Andrzej, 2002; Strobel, 2003). Endophytes are detected in a wide variety of plant tissue types, such as seeds and ovules (Siegel, Latch, & Johnson, 1987), fruits (Schena, Nigro, Pentimone, Ligorio, & Ippolito, 2003), stems (Gutiérrez-Zamora & Martínez-Romero, 2001), roots (Germida, Siciliano, De Freitas, & Seib, 1998), leaves (Smith, Wingfield, & Petrini, 1996), tubers

(Sturz, Christie, & Matheson, 1998), buds (Ragazzi, Moricca, Capretti, & Dellavalle, 1999), xylem (Hoff et al., 2004), rachis (Rodrigues & Samuels, 1999), and bark (Raviraja, 2005).

There are different ways to select plants to study their endophytes. The selection of a plant will be guided by different criteria ranging from random selection of a number of different plants in an ecosystem to the search for a useful plant in the hope of finding a useful molecule (e.g., medicinal plants). This selection rationale may be based on a particular morphology of the plant, its age, its ecological niche, its environment, and/or its ethnobotanical history (medicinal uses) (Strobel & Daisy, 2003). For example, the research that led to the discovery of the paclitaxel-producing endophyte followed an ethnomedical rationale (isolated from a tree known to have medicinal uses, due in part to the presence of paclitaxel in its bark).

A total of 449 new secondary metabolites were isolated from endophytic fungi between 2017 and 2019, corresponding to terpenoids and polyketones, alkaloids, steroids, and other compounds (Zheng, Li, Zhang, & Zhao, 2021). Many of these compounds exhibited biological activities (Fig. 12.1) (Manganyi and Ateba, 2020) important for public health, agriculture and pharmaceutical industries (Gupta et al., 2020). For instance, the *in vivo* and *in vitro* screening of griseofulvin isolated from different endophytic fungi from the genus *Xylaria* have revealed its microbicidal activity against plant pathogens allowing its potential use in the control of crop diseases. Also, the compound lectin produced by endophytic fungus *Alternaria* sp. has shown its antidiabetic activity in rats and *in vitro* with no harmful effects compared to the side effects observed with some antidiabetic drugs (Adeleke and Babalola, 2021). In addition to this, 224 patents between 2001 and 2019 are related to

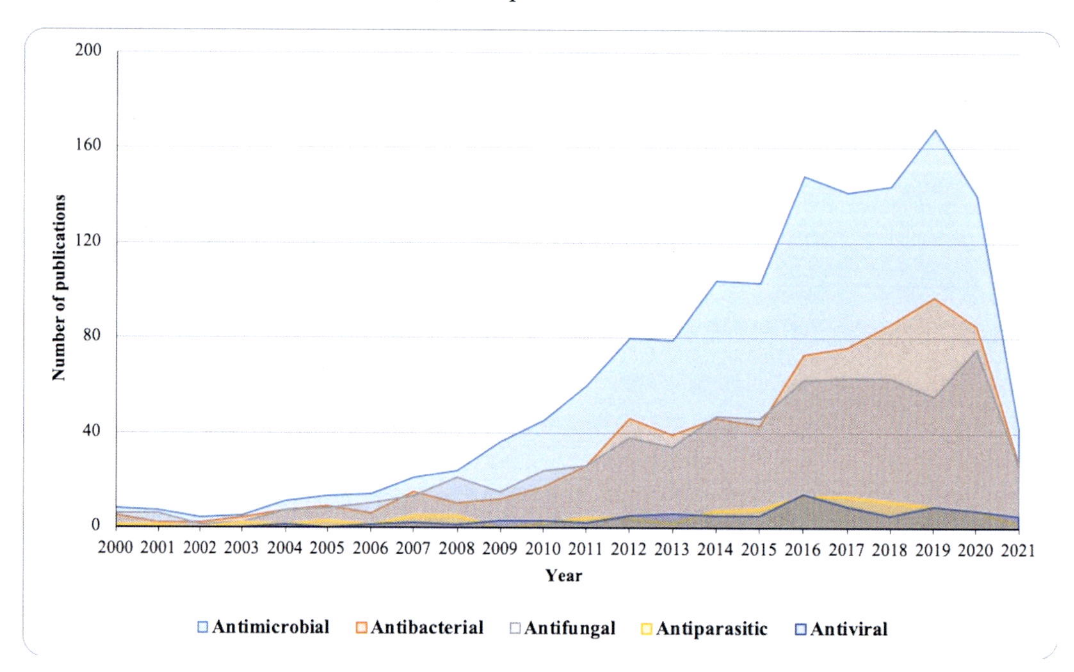

FIGURE 12.1 Number of publications using the Search query: ("antimicrobial" OR "antibacterial" OR "antifungal" OR "antiparasitic" OR "antiviral") AND ("endophytic fungi") in PubMed.

secondary metabolite production and biotechnological processes involving endophytic fungi from *Aspergillus*, *Fusarium*, *Penicillium*, *Trichoderma* and *Phomopsis* genera. For example, one submitted patent of *Trichoderma* acid is involved with the preparation of antifungal agents, another patent is about obtaining three isopiramane diterpenoids compounds from *Xylaria* sp. with antifungal activity and with potential application in agriculture and medicine. Moreover, another patent is related to the production of antifungal and immunosuppressive compounds by endophytic fungus *Colletotrichum* sp. (Ortega et al., 2020).

In the following section (Table 12.1), compounds with potent antibacterial activity isolated from endophytic fungi, obtained from mainly terrestrial but also some marine plants, have been listed. They exhibited activities with MIC values less than 20 μg/mL and/or with important inhibition zones (> 8 mm with at least 50 μg per disk).

Here, we focused on the compounds with IC_{50} values lower than 10 μM (Table 12.2) active against *Plasmodium* spp., *Leishmania* spp. and *Trypanosoma* spp. and we describe the culture methods and their potential cytotoxicity against different cancer cell lines. Interestingly, these compounds belong to several chemical classes, such as polyketides, terpenoids, alkaloids and polypeptides.

Fabaceae, Lamiaceae and Asteraceae are among the most represented families. These families are known to contain a large number of species producing bioactive compounds, as medicinal or toxic plants (Bessada, Barreira, & Oliveira, 2015; Moraes Neto et al., 2019; Setzer & Setzer, 2006; Sharma, Flores-Vallejo, Cardoso-Taketa, & Villarreal, 2017) and they may represent a great target for the researchers bioprospecting for endophytes. More generally, without focusing on antibacterial and antiparasitic properties, most of the bioactive endophytes studied between 2006 and 2016 were mainly isolated from plants corresponding to the families Fabaceae, Lamiaceae, Asteraceae and Araceae (Martinez-Klimova., 2017) while in a recent study, between 2016 and 2019, 254 plants families were studied for their endophytes, with Poaceae, Fabaceae, Pinaceae and Asteraceae as the most studied plant families (Harrison and Griffin, 2020).

The endophytic community present on the host plant is affected by different environmental factors such as the growth state and physiological conditions of the plant or the sampling season, while the culture conditions play an important role for their recovery (Gouda, Das, Sen, Shin, & Patra, 2016). For the isolation of endophytes, the selection of a healthy plant is mandatory to avoid the isolation of pathogenic fungi as well as the preservation of the tissue to be used within 24 h of harvesting to prevent death of the endophytic community. This process is followed by a series of sterilization steps to ensure the isolation of endophytes instead of epiphytes or root actinomycetes. Before surface sterilization, washing plant tissues under tap water can help to reduce the soil particles. Among the most used solutions for the surface sterilization are 2%–10% of hypochlorite solution, 3% of hydrogen peroxide, or 2% of potassium permanganate combined with 70%–96% of ethanol washing steps and rinses with sterile distilled water (Martinez-Klimova, Rodríguez-Peña, & Sánchez, 2017). Time in each bath depends on the plant material (leaves, roots, stems) especially their thickness. Sterilized tissues are cut into small fragments (3–5 mm) and dried under aseptic conditions to be later placed on solid agar media and incubated for 2–4 weeks. Usually, plates are discarded after this period of time to avoid the growth of contaminants. However, this can be corroborated with the control plates containing culture media with the sterile distilled water used for the surface sterilization. Some of the endophytes are not readily culturable or they are found to be underrepresented in the selected tissue, therefore some culture-independent techniques are used for studying or identifying the unculturable endophytes. For the identification of pure isolates, sequencing of 18S rDNA, internal transcribed spacer (ITS1 and ITS2), 5.8S rDNA, or 28S subunit of rDNA is usually done (Zhang, Song, & Tan, 2006).

Antimicrobial compounds from endophytic fungi

Endophytes are in close interaction with their host plant but also with the entire microbiome of the plant. The microbiome can also change over time, becoming a dynamic interaction and a constant microbial competition which at the same time stimulates metabolite production. This trend of life is a huge opportunity for researchers looking for new antiinfectives compounds such as antibacterial or antifungal metabolites. Endophytes are not naturally in interaction with pathogens of animals, such as *Plasmodium*, *Leishmania*, or other *Trypanosomatidae* parasites; however, their capacity to produce bioactive metabolites in answer to the presence of other microorganisms led the scientific community to screen fungal endophytes extracts for this kind of activity. Nowadays, there is an increasing interest for the research of antiparasitic compounds from fungal endophytes although these organisms do not share the same biotopes.

This work aims at reviewing the most active antibacterial and antiparasitic compounds isolated from endophytic fungi, between 2000 and 2019.

Antibacterial compounds

As briefly introduced before, drug and multidrug resistance in bacteria is a global health and development threat and is caused mainly by the large amounts of antibiotics used for human therapy, as well as for fish in aquaculture and for farm animals. The two principal mechanisms in the generation of drug resistance are the accumulation of resistance plasmids and the increase in the expression of the genes that code for drug efflux pumps. These mechanisms have produced bacterial strains resistant to methicillin, aminoglycosides, macrolides, tetracyclines, chloramphenicol, lincosamides, and/or penicillin agents. An important group of these drug/multidrug-resistant bacteria is the ESKAPE pathogens (*Enterococcus faecium*, *Staphylococcus aureus*, *Klebsiella pneumoniae*, *Acinetobacter baumannii*, *Pseudomonas aeruginosa*, and *Enterobacter* spp.) which are the leading cause of nosocomial infections throughout the world and one of the greatest challenges in clinical practice.

The urgency of finding new agents to fight these pathogens or new antibacterial mechanisms of actions has been one of the most important focuses of science in the last decades. A special focus on natural products has been applied in the research for new antibacterial compounds with marine, plant, or fungi origins. In the following section (Table 12.1), compounds with potent antibacterial activity isolated from endophytic fungi, obtained from mainly terrestrial but also some marine plants, have been listed. They exhibited activities with MIC values less than 20 μg/mL and/or with important inhibition zones (> 8 mm with at least 50 μg per disk).

In particular, the derivatives with MIC (minimal inhibitory concentration) values lower than 10 μg/mL are described and discussed. Compounds are organized according to families of

TABLE 12.1 Antibacterial compounds isolated from endophytic fungi (2000–19): activities with MIC values less than 20 μg/mL and/or with important inhibition zones (>8 mm with at least 50 μg per disk).

ID	Compound names	Chemical formula	Endophytic fungal strain	Host plant	Botanical family	Activity	References
Alkaloids							
1	(+)-flavipucine[a]	$C_{12}H_{15}NO_4$	*Phoma* sp. 7204	*Salsola oppositifolia* Desf.	Amaranthaceae	6, 17, and 11 mm zone inhibition against *B. subtilis*, *S. aureus*, and *E. coli*	Loesgen et al. (2011)
2	Brevianamide M	$C_{18}H_{15}N_3O_3$	*Aspergillus versicolor*	*Sargassum thunbergii* (Mertens ex Roth) Kuntze	Sargassaceae (Algae)	11–10 mm inhibition against *E. coli* and *S. aureus* at 30 μg/disk	Miao et al. (2012)
3	Cochliodinol	$C_{32}H_{30}N_2O_4$	*Chaetomium globosum* SNB-GTC2114	*Paspalum virgatum* L.	Poaceae	4 μg/mL MIC against *S. aureus*	Casella et al. (2013)
4	Fumigaclavine C	$C_{23}H_{30}N_2O_2$	*Aspergillus* sp. EJC08	*Bauhinia guianensis* Aubl.	Fabaceae	7.81 μg/mL MIC against *B. subtilis*	Pinheiro et al. (2013)
5	Fusapyridon A[a]	$C_{29}H_{43}NO_5$	*Fusarium* sp. YG-45	*Maackia chinensis* Takeda	Fabaceae	6.25 μg/mL MIC against *P. aeruginosa*	Tansuwan et al. (2007)
6	Neoaspergillic acid	$C_{12}H_{20}N_2O_2$	*Aspergillus* sp. FSY-01/*Aspergillus* sp. FSW-02 (coculture)	*Avicennia marina* (Forssk.) Vierh.	Acanthaceae	0.98–7.80 μg/mL MIC against *S. aureus*, *S. epidermidis*, *B. subtilis*, *B. dysenteriae*, *B. proteus*	Zhu, Wu, Chen, Lu, and Pan (2011)
7	Phomapyrrolidone B[a]	$C_{34}H_{41}NO_4$	*Phoma* sp. NRRL 46751	*Saurauia scaberrima* Lauterb.	Actinidiaceae	5.9 μM IC_{50} against *M. tuberculosis* H37Pv	Wijeratne et al. (2013)
8	Phomapyrrolidone C[a]	$C_{34}H_{41}NO_5$	*Phoma* sp. NRRL 46751	*Saurauia scaberrima* Lauterb.	Actinidiaceae	5.2 μM IC_{50} against *M. tuberculosis* H37Pv	Wijeratne et al. (2013)
9	Phomoenamide A[a]	$C_{14}H_{24}N_2O_4$	*Phomopsis* PSU-D15	*Garcinia dulcis* (Roxb.) Kurz	Clusiaceae	6.25 μg/mL MIC against *M. tuberculosis* H37Ra	Rukachaisirikul et al. (2008)
10	Piperine	$C_{17}H_{19}NO_3$	*Periconia* sp.	*Piper longum* L.	Piperaceae	2.62 and 1.74 MIC against *M. tuberculosis* and *M. smegmatis*	Verma, Lobkovsky, Gange, Singh, and Prakash (2011)
11	Pseurotin A	$C_{22}H_{25}NO_8$	*Aspergillus* sp. EJC08	*Bauhinia guianensis* Aubl.	Fabaceae	15.62 μg/mL MIC against *B. subtilis*	Pinheiro et al. (2013)
12	Pyrrocidine A	$C_{31}H_{37}NO_4$	*Acremonium zeae* NRRL 13540	*Zea mays* L.	Poaceae	0.25–2 μg/mL MIC against *S. aureus*, *S. haemolyticus*, *E. faecalis*, *E. faecium*	Wicklow and Poling (2009)

(*Continued*)

TABLE 12.1 (Continued)

ID	Compound names	Chemical formula	Endophytic fungal strain	Host plant	Botanical family	Activity	References
13	Pyrrocidine B	$C_{31}H_{39}NO_4$	*Acremonium zeae* NRRL 13540	*Zea mays* L.	Poaceae	4 μg/mL MIC against *S. aureus*, *S. haemolyticus*, *E. faecalis*, *E. faecium*	Wicklow and Poling (2009)
14	Pyrrocidine C[a]	$C_{34}H_{41}NO_4$	*Lewia infectoria* SNB-GTC2402	*Besleria insolita* C.V. Morton	Gesneriaceae	2 μg/mL MIC against *S. aureus*	Casella et al. (2013)
15	Semicochliodinol A	$C_{27}H_{22}N_2O_4$	*Chaetomium globosum* SNB-GTC2114	*Paspalum virgatum* L.	Poaceae	2 μg/mL MIC against *S. aureus*	Casella et al. (2013)
Fatty acids							
16	Cerebroside 1[a]	$C_{41}H_{77}NO_9$	*Fusarium* sp. IFB-121	*Quercus variabilis* Blume	Fagaceae	7.8–3.9 MIC against *B. subtilis*, *E. coli*, *P. fluorescens*	Shu et al. (2004)
17	Cerebroside 2	$C_{41}H_{75}NO_9$	*Fusarium* sp. IFB-122	*Quercus variabilis* Blume	Fagaceae	1.9–3.9 MIC against *B. subtilis*, *E. coli*, *P. fluorescens*	Shu et al. (2004)
Polyketides							
18	(12S)-12-hydroxymonocerin[a]	$C_{16}H_{22}O_7$	*Microdochium bolleyi*	*Fagonia cretica* L.	Zygophyllaceae	10 mm zone inhibition against *E. coli* at 50 μg/disk	Zhang et al. (2008)
19	(3R)-5-methylmellein	$C_{11}H_{12}O_3$	*Cytospora* sp.	*Ilex canariensis* Poir.	Aquifoliaceae	20 mm zone inhibition against *B. megaterium* at 50 μg/disk	Lu et al. (2011)
20	(4R,5R,6S)-6-acetoxy-4,5-dihydroxy-2-(hydroxymethyl) cyclohex-2-en-1-one	$C_9H_{12}O_6$	*Phoma* sp. 8889	*Salsola oppositifolia* Desf.	Amaranthaceae	10 mm zone inhibition 0.05 mg against *B. megaterium*	Qin et al. (2010)
21	(R)-5-((S)-acetate(phenyl)-methyl)dihydrofuran-2 (3H)-one[a]	$C_{13}H_{14}O_4$	*Cytospora* sp.	*Ilex canariensis* Poir.	Aquifoliaceae	20 mm zone inhibition against *B. megaterium* at 50 μg/disk	Lu et al. (2011)
22	(R)-5-((S)-hydroxy (phenyl)-methyl) dihydrofuran-2(3H)-one[a]	$C_{11}H_{12}O_3$	*Cytospora* sp.	*Ilex canariensis* Poir.	Aquifoliaceae	20 mm zone inhibition against *B. megaterium* at 50 μg/disk	Lu et al. (2011)
23	(S)-5-((S)-hydroxy (phenyl)-methyl) dihydrofuran-2(3H)-one	$C_{11}H_{12}O_3$	*Cytospora* sp.	*Ilex canariensis* Poir.	Aquifoliaceae	18 mm zone inhibition against *B. megaterium* at 50 μg/disk	Lu et al. (2011)

24	(S)-5-benzyl-dihydrofuran-2(3H)-one[a]	$C_{11}H_{12}O_2$	*Cytospora* sp.	*Ilex canariensis* Poir.	Aquifoliaceae	20 mm zone inhibition against *B. megaterium* at 50 μg/disk	Lu et al. (2011)
25	(S)-5-hydroxy-4-oxo-1,2,3,4-tetrahydronaphthalen-1-yl acetate[a]	$C_{12}H_{12}O_4$	*Cytospora* sp.	*Ilex canariensis* Poir.	Aquifoliaceae	15 mm zone inhibition against *B. megaterium* at 50 μg/disk	Lu et al. (2011)
26	2-hydroxy-3-(hydroxymethyl) anthraquinone[a]	$C_{15}H_{10}O_4$	*Coniothyrium* sp. Zw86	*Salsola oppositifolia* Desf.	Amaranthaceae	16 and 15 mm zone inhibition against *E. coli* and *B. megaterium* at 50 μg per disk	Sun et al. (2013)
27	2,3-didehydro-19a-hydroxy-14-epicochlioquinone B[a]	$C_{28}H_{38}O_7$	*Nigrospora* sp. MA75	*Pongamia pinnata* (L.) Pierre	Fabaceae	8–0.5 μg/mL MIC against MRSA, *E. coli, P. aeruginosa, P. fluorescens*	Shang, Li, Li, and Wang (2012)
28	3-(2,5-dihydro-4-hydroxy-5-oxo-3-phenyl-2-furyl) propionic acid	$C_{13}H_{12}O_5$	*Cytospora* sp.	*Ilex canariensis* Poir.	Aquifoliaceae	15 mm zone inhibition against *B. megaterium* at 50 μg/disk	Lu et al. (2011)
29	3-O-methylalaternin	$C_{16}H_{12}O_6$	*Ampelomyces* sp. EU143250	*Urospermum picroides* (L.) Scop. Ex F.W. Schmidt	Asteraceae	12.5 μg/mL MIC *S. epidermidis, E. faecalis, S. aureus*	
30	4-deoxytetrahydrobostrycin	$C_{16}H_{20}O_7$	*Nigrospora* sp. MA75	*Pongamia pinnata* (L.) Pierre	Fabaceae	4 μg/mL MIC against *E. coli*	Shang et al. (2012)
31	5-phenyl-4-oxopentanoic acid[a]	$C_{11}H_{12}O_3$	*Cytospora* sp.	*Ilex canariensis* Poir.	Aquifoliaceae	15 mm zone inhibition against *B. megaterium* at 50 μg/disk	Lu et al. (2011)
32	6-hydroxy-2-methyl-4-chromanone	$C_{10}H_{10}O_3$	*Periconia siamensis* CMUGE015	*Thysanolaena latifolia* (Roxb. Ex Hornem.) Honda	Poaceae	12.5–6.25 μg/mL MIC against *B. subtilis, L. monocytogenes, P. aeruginosa*	Bhilabutra et al. (2007)
33	6-hydroxyterrefuranone[a]	$C_{14}H_{20}O_4$	*Microdiplodia* sp. KS 75–1	*Pinus* sp.	Pinaceae	16 mm zone inhibition against *S. aureus* at 40 μg/disk	Shiono et al. (2012)
34	6(7)-dehydro-8-hydroxyterrefuranone[a]	$C_{14}H_{18}O_3$	*Microdiplodia* sp. KS 75–1	*Pinus* sp.	Pinaceae	15 mm zone inhibition against *S. aureus* at 40 μg/disk	Shiono et al. (2012)
35	7-amino-4-methylcoumarin[a]	$C_{10}H_9NO_2$	*Xylaria* sp. YX-28	*Ginkgo biloba* L.	Ginkgoaceae	20–4 μg/mL MIC against *S. aureus, E. coli, S. typhi, S. typhimurium, S. enteritidis, A. hydrophila, Shigella* sp., *V. parahaemolyticus*	Liu et al. (2008)
36	7,8-dihydonivefuranone A[a]	$C_{15}H_{20}O_4$	*Microdiplodia* sp. KS 75–1	*Pinus* sp.	Pinaceae	15 mm zone inhibition against *S. aureus* at 40 μg/disk	Shiono et al. (2012)

(*Continued*)

TABLE 12.1 (Continued)

ID	Compound names	Chemical formula	Endophytic fungal strain	Host plant	Botanical family	Activity	References
37	Altechromone A	$C_{11}H_{10}O_3$	*Alternaria brassicicola* ML-P08	*Malus halliana* Koehne	Rosaceae	3.9, 3.9, and 1.8 μg/mL MICs against *B. subtilis, E. coli and P. fluorescens*	Gu (2009)
38	Alterlactone	$C_{15}H_{12}O_6$	*Alternaria alternata* ZHJG5	*Cercis chinensis* Bunge	Fabaceae	16 μg/mL MIC value against *Xanthomonas oryzae pv. oryzae*	Zhao et al. (2020)
39	Alternariol	$C_{14}H_{10}O_5$	*Alternaria alternata* ZHJG5	*Cercis chinensis* Bunge	Fabaceae	4 μg/mL MIC value against *Xanthomonas oryzae pv. oryzae*	Zhao et al. (2020)
40	Altersolanol A	$C_{16}H_{16}O_8$	*Ampelomyces* sp. EU143251	*Urospermum picroides* (L.) Scop. Ex F.W. Schmidt	Asteraceae	12.5 μg/mL MIC *S. epidermidis, E. faecalis*	Aly et al. (2008)
41	Ambuic acid	$C_{19}H_{26}O_6$	Four strains of *Pestalotiopsis microspora, P. guepinii* MSU 214, *Monochaetia* sp. MSU-NC 202	*Taxus baccata* L., *Torreya taxifolia* Arn., *Taxodium distichum* (L.) Rich., *Dendrobium speciosum* Sm., *Wollemia nobilis* W.G. Jones, K.D. Hill & J.M. Allen, *Taxus wallichiana* Zucc.	Taxaceae, Taxaceae, Cupressaceae, Orchidaceae, Araucariaceae, Taxaceae	10 μM IC_{50} inhibited the production of gelatinase by *E. faecalis* OG1RF (anti-QS)	Nakayama et al. (2009), Li et al. (2001)
42	Cercosporamide	$C_{16}H_{13}NO_7$	*Phoma* sp. NG-25	*Saurauia scaberrima* Lauterb.	Actinidiaceae	2.0 μg/mL MIC against *S. aureus*	Hoffman et al. (2008)
43	Citreorosein	$C_{15}H_{10}O_6$	*Microsphaeropsis* sp. (strain 8875)	*Lycium intricatum* Boiss.	Solanaceae	9 mm zone inhibition against *E. coli*, 10 mm against *B. megaterium* at 50 μg/disk	Krohn et al. (2009)
44	Cytosporone D[a]	$C_{16}H_{22}O_5$	*Cytospora* sp. CR200 and *Diaporthe* sp. CR146	*Conocarpus erectus* L. and *Forsteronia spicata* (Jacq.) G. Mey.	Combretaceae and Apocynaceae	8 μg/mL MIC against *S. aureus* and *E. faecalis*	Brady et al. (2000)
45	Cytosporone E[a]	$C_{15}H_{20}O_5$	*Cytospora* sp. CR200 and *Diaporthe* sp. CR146	*Conocarpus erectus* L. and *Forsteronia spicata* (Jacq.) G. Mey.	Combretaceae and Apocynaceae	8 μg/mL MIC against *S. aureus* and *E. faecalis*	Brady et al. (2000)
46	Dicerandrol A[a]	$C_{34}H_{34}O_{14}$	*Phomopsis longicolla* MMW29	*Dicerandra frutescens* Shinners	Lamiaceae	11–10 mm inhibition against *B. subtilis* and *S. aureus*	Wagenaar et al. (2001)
47	Dicerandrol B[a]	$C_{36}H_{36}O_{15}$	*Phomopsis longicolla* MMW29	*Dicerandra frutescens* Shinners	Lamiaceae	9.5–8.5 mm inhibition against *B. subtilis* and *S. aureus*	Wagenaar et al. (2001)
48	Dicerandrol C[a]	$C_{38}H_{38}O_{16}$	*Phomopsis longicolla* MMW29	*Dicerandra frutescens* Shinners	Lamiaceae	8–7 mm inhibition against *B. subtilis* and *S. aureus*	Wagenaar et al. (2001)

49	Griseophenone C	$C_{16}H_{16}O_6$	*Nigrospora* sp. MA75	*Pongamia pinnata* (L.) Pierre	Fabaceae	2–0.5 μg/mL MIC against MRSA, *E. coli*, *P. aeruginosa*, *P. fluorescens*, *S. epidermidis*	Shang et al. (2012)
50	Guanacastepene A[a]	$C_{22}H_{30}O_5$	Strain CR115	*Daphnopsis americana* (Mill.) J.R. Johnst.	Thymelaeaceae	6–8 mm inhibition zone at 6.25 μg per spot against *S. aureus*, *E. faecium* and *E. coli*	Singh et al. (2000)
51	Integracins A	$C_{37}H_{56}O_8$	*Cytospora* sp.	*Ilex canariensis* Poir.	Aquifoliaceae	20 mm zone inhibition against *B. megaterium* at 50 μg/disk	Lu et al. (2011)
52	Integracins B	$C_{35}H_{54}O_7$	*Cytospora* sp.	*Ilex canariensis* Poir.	Aquifoliaceae	20 mm zone inhibition against *B. megaterium* at 50 μg/disk	Lu et al. (2011)
53	Isofusidienol A[a]	$C_{16}H_{12}O_6$	*Chalara* sp. (strain 6661)	*Artemisia vulgaris* L.	Asteraceae	23 mm zone inhibition 0.015 mg against *B. subtilis*	Lösgen et al. (2008)
54	Isofusidienol B[a]	$C_{16}H_{12}O_7$	*Chalara* sp. (strain 6661)	*Artemisia vulgaris* L.	Asteraceae	22 mm zone inhibition 0.015 mg against *B. subtilis*	Lösgen et al. (2008)
55	Isorhodoptilometrin	$C_{17}H_{14}O_6$	*Penicillium restrictum* G85ITS	*Silybum marianum* (L.) Gaertn.	Asteraceae	8.9 agr P3lux IC_{50} *S. aureus* MRSA (anti-QS)	Figueroa et al. (2014)
56	Javanicin	$C_{15}H_{14}O_6$	*Chloridium* sp.	*Azadirachta indica* A. Juss.	Meliaceae	10–2 μg/mL MIC against *P. fluorescens*, *P. aeruginosa*	Kharwar et al. (2009)
57	Lateropyrone	$C_{15}H_{10}O_8$	*Fusarium tricinctum*/ *Bacillus subtilis* 168 trpC2 (coculture)	*Aristolochia paucinervis* Pomel	Aristolochiaceae	8–2 μg/mL MIC against *B. subtilis*, *S. aureus*, *S. pneumoniae*, and *E. faecalis*, MRSA	Ola, Thomy, Lai, Brötz-Oesterhelt, and Proksch (2013)
58	Modiolide A	$C_{10}H_{14}O_4$	*Periconia siamensis* CMUGE015	*Thysanolaena latifolia* (Roxb. Ex Hornem.) Honda	Poaceae	12.5–3.12 μg/mL MIC against *B. subtilis*, *L. monocytogenes*, *P. aeruginosa*	Bhilabutra et al. (2007)
59	Monocerin	$C_{16}H_{20}O_6$	*Microdochium bolleyi*	*Fagonia cretica* L.	Zygophyllaceae	10 mm zone inhibition against *E. coli* at 50 μg/disk	Zhang et al. (2008)
60	Monomethylsulochrin	$C_{18}H_{18}O_7$	*Rhizoctonia* sp. Cy064	*Cynodon dactylon* (L.) Pers.	Poaceae	10 μg/mL MIC against *H. pylori* ATCC 43504	Ma et al. (2004)
61	Nivefuranone A	$C_{15}H_{18}O_4$	*Microdiplodia* sp. KS 75–1	*Pinus* sp.	Pinaceae	15 mm zone inhibition against *S. aureus* at 40 μg/disk	Shiono et al. (2012)
62	Norlichexanthone	$C_{14}H_{10}O_5$	*Nigrospora* sp. MA75	*Pongamia pinnata* (L.) Pierre	Fabaceae	0.5 μg/mL MIC against *S. epidermidis*	Shang et al. (2012)

(*Continued*)

TABLE 12.1 (Continued)

ID	Compound names	Chemical formula	Endophytic fungal strain	Host plant	Botanical family	Activity	References
63	Patulin	$C_7H_6O_4$	*Penicillium radicicola* (IBT 10696) and *Penicillium coprobium* (IBT 6895)	*Armoracia rusticana* P. Gaertn., B. Mey. & Scherb./-	Brassicaceae/-	Abolish of QS-controlled lasB induction at 80 μM (anti-QS)	Rasmussen et al. (2005)
64	Penicillic acid	$C_8H_{10}O_4$	*Penicillium radicicola* (IBT 10696) and *Penicillium coprobium* (IBT 6895)	*Armoracia rusticana* P. Gaertn., B. Mey. & Scherb./-	Brassicaceae/-	Abolish of QS-controlled lasB induction at 40 μM (anti-QS)	Rasmussen et al. (2005)
65	Phomodione[a]	$C_{20}H_{22}O_8$	*Phoma* sp. NG-25	*Saurauia scaberrima* Lauterb.	Actinidiaceae	MIC 1.6 μg/mL *S. aureus*	Hoffman et al. (2008)
66	Phomol[a]	$C_{22}H_{36}O_7$	*Phomopsis* sp. strain E02018	*Erythrina crista-galli* L.	Fabaceae	10 μg/mL MIC against *Corynebacterium insidiosum*	Weber et al. (2004)
67	Phomosine A	$C_{18}H_{18}O_7$	*Phomopsis* sp. 5686	*Ligustrum vulgare* L.	Oleaceae	11 mm zone inhibition against *B. megaterium* at 50 μg/disk	Krohn et al. (2011)
68	Phomosine B	$C_{19}H_{22}O_7$	*Phomopsis* sp. 5686	*Ligustrum vulgare* L.	Oleaceae	11 mm zone inhibition against *B. megaterium* at 50 μg/disk	Krohn et al. (1994)
69	Phomoxanthone A[a]	$C_{38}H_{38}O_{16}$	*Phomopsis* sp. BCC 1323	*Tectona grandis* L.f.	Lamiaceae	0.5 μg/mL against *M. tuberculosis* H37Ra	Isaka et al. (2001)
70	Phomoxanthone B[a]	$C_{38}H_{38}O_{16}$	*Phomopsis* sp. BCC 1323	*Tectona grandis* L.f.	Lamiaceae	0.5 μg/mL against *M. tuberculosis* H37Ra	Isaka et al. (2001)
71	Phyllostine	$C_7H_6O_4$	*Phomopsis* sp.	*Notobasis syriaca* (L.) Cass.	Asteraceae	10 mm zone inhibition against *E. coli* at 50 μg/disk	Hussain et al. (2011)
72	Primin	$C_{12}H_{16}O_3$	*Botryosphaeria mamane* PSU-M76	*Garcinia* x *mangostana* L.	Clusiaceae	8 μg/mL MIC against *S. aureus* and MRSA	Pongcharoen et al. (2007)
73	Pyrenocine L[a]	$C_{14}H_{20}O_6$	*Phomopsis* sp.	*Cistus salviifolius* L.	Cistaceae	10 mm zone inhibition against *E. coli* at 50 μg/disk	Hussain et al. (2012)
74	Secalonic acid B	$C_{32}H_{30}O_{14}$	*Blennoria* sp. 7064	*Carpobrotus edulis* (L.) N.E.Br.	Aizoaceae	15 mm zone inhibition against *B. megaterium* at 50 μg/disk	Zhang et al. (2008)
75	Spiropreussione A	$C_{19}H_{12}O_5$	*Preussia* sp. CGMCC 2022	*Aquilaria sinensis* (Lour.) Spreng.	Thymelaeaceae	16.4 mm zone inhibition against *S. aureus* at 5 μg/disk	Chen et al. (2009)

76	Tetrahydrobostrycin	$C_{16}H_{20}O_8$	*Nigrospora* sp. MA75	*Pongamia pinnata* (L.) Pierre	Fabaceae	2–0.5 μg/mL MIC against MRSA, *E. coli*	Shang et al. (2012)
77	Trypethelone	$C_{16}H_{16}O_4$	*Coniothyrium cereale*	*Enteromorpha* sp.	Ulvaceae (green Algae)	18, 14, and 12 mm, against *M. phlei*, *S. aureus* and *E. coli* at 20 μg/disk	Elsebai et al. (2011)
78	Usnic acid	$C_{18}H_{16}O_7$	*Phoma* sp. NG-25	*Saurauia scaberrima* Lauterb.	Actinidiaceae	MIC 2.0 μg/mL against *S. aureus*	Hoffman et al. (2008)
79	Xanthone 8-hydroxy-6-methyl-9-oxo-9H-xanthene-1-carboxylic acid methyl ester	$C_{16}H_{12}O_5$	*Microsphaeropsis* sp. (7177)	*Zygophyllum fontanesii* Webb & Berthel.	Zygophyllaceae	13 mm zone inhibition against *B. megaterium* at 50 μg/disk	Krohn et al. (2009)
80	Yicathin C[a]	$C_{16}H_{12}O_6$	*Aspergillus wentii* pt-1	*Gymnogongrus flabelliformis* Harvey	Phyllophoraceae (Algae)	12 mm inhibition against *E. coli* at 10 μg/disk	Sun et al. (2013)
81	γ-oxo-benzenepentanoic acid methyl ester[a]	$C_{12}H_{14}O_3$	*Cytospora* sp.	*Ilex canariensis* Poir.	Aquifoliaceae	25 mm zone inhibition against *B. megaterium* at 50 μg/disk	Lu et al. (2011)
82	ω-hydroxyemodin	$C_{15}H_{10}O_6$	*Penicillium restrictum* G85ITS	*Silybum marianum* (L.) Gaertn.	Asteraceae	8.1 agr P3lux IC_{50} *S. aureus* MRSA (anti-QS)	Figueroa et al. (2014)
Polyketides-fatty acids							
83	Biscogniazaphilone A[a]	$C_{24}H_{34}O_4$	*Biscogniauxia formosana* BCRC 33718	*Cinnamomum* sp.	Lauraceae	<5.12 μg/mL against *M. tuberculosis* H37RV	Cheng et al. (2012)
84	Biscogniazaphilone B[a]	$C_{25}H_{32}O_5$	*Biscogniauxia formosana* BCRC 33718	*Cinnamomum* sp.	Lauraceae	<2.52 μg/mL against *M. tuberculosis* H37RV	Cheng et al. (2012)
Polypeptides							
85	3,1′-didehydro-3[2″(3‴,3‴-dimethyl-prop-2-enyl)-3″-indolylmethylene]-6-methyl pipera-zine-2,5-dione[a]	$C_{14}H_{13}N_3O_2$	*Penicillium chrysogenum* MTCC 5108	*Porteresia coarctata* (Roxb.) Tateoka	Poaceae	14–16 mm zone inhibition against *V. cholerae* at 10 μg/disk	Devi et al. (2012)
86	Beauvericin	$C_{45}H_{57}N_3O_9$	*Fusarium redolens* Dzf2	*Dioscorea zingiberensis* C.H. Wright	Dioscoreaceae	26.6–18.4 μg/mL MIC against *A. tumefaciens*, *X. vesicatoria*, *B. subtilis*, *S. haemolyticus*	Xu et al. (2010)

(*Continued*)

TABLE 12.1 (Continued)

ID	Compound names	Chemical formula	Endophytic fungal strain	Host plant	Botanical family	Activity	References
87	Enniatin A1	$C_{35}H_{61}N_3O_9$	*Fusarium tricinctum*/ *Bacillus subtilis* 168 trpC2 (coculture)	*Aristolochia paucinervis* Pomel	Aristolochiaceae	8–2 μg/mL MIC against *B. subtilis*, *S. aureus*, *S. pneumoniae*, and *E. faecalis*, MRSA	Ola et al. (2013)
88	Enniatin B	$C_{33}H_{57}N_3O_9$	*Fusarium tricinctum* Salicorn 19	*Salicornia bigelovii* Torr.	Amaranthaceae	13–6 μM MIC against *B. subtilis*, *E. aerogenes*, *M. tetragenus*	Zhang et al. (2015)
89	Enniatin B1	$C_{34}H_{59}N_3O_9$	*Fusarium tricinctum*/ *Bacillus subtilis* 168 trpC2 (coculture)	*Aristolochia paucinervis* Pomel	Aristolochiaceae	16–4 μg/mL MIC against *B. subtilis*, *S. aureus*, *S. pneumoniae*, and *E. faecalis*, MRSA	Ola et al. (2013)
90	Halobacillin	$C_{53}H_{94}N_8O_{12}$	*Trichoderma asperellum*	*Panax notoginseng* (Burkill) F.H. Chen	Araliaceae	5.4 and 14.00 μg/mL IC_{50} against *E. faecium* and *S. aureus*	Ding et al. (2012)
91	PF1022F	$C_{40}H_{68}N_4O_{12}$	*Trichoderma asperellum*	*Panax notoginseng* (Burkill) F.H. Chen	Araliaceae	7.30 and 19.02 μg/mL IC_{50} against *E. faecium* and *S. aureus*	Ding et al. (2012)
92	Tardioxopiperazine A	$C_{24}H_{31}N_3O_2$	*Eurotium cristatum* EN-220	*Sargassum thunbergii* (Mertens ex Roth) Kuntze	Sargassaceae (Algae)	8 μg/mL MIC against *S. aureus* and *E. coli*	Du, Li, Li, Shang, and Wang (2012)
Steroids							
93	Helvolic acid	$C_{33}H_{44}O_8$	*Aspergillus* sp. CY725	*Cynodon dactylon* (L.) Pers.	Poaceae	5.0 μg/mL MIC against *H. pylori* ATCC 43504	Li et al. (2005)
Terpenoids							
94	1β-Hydroxy-α-cyperone	$C_{15}H_{22}O_2$	*Microsphaeropsis arundinis*	*Ulmus macrocarpa* Hance	Ulmaceae	11.4 μg/mL MIC against *S. aureus*	Luo et al. (2013)
95	Conidiogenol	$C_{20}H_{34}O_2$	*Penicillium chrysogenum* QEN-24S	*Laurencia* sp.	Rhodomelaceae (Algae)	16 μg/mL MIC against *P. fluorescens*, *S. epidermidis*	Gao et al. (2011)
96	Conidiogenone B	$C_{20}H_{30}O$	*Penicillium chrysogenum* QEN-24S	*Laurencia* sp.	Rhodomelaceae (Algae)	8 μg/mL MIC against MRSA, *P. aeruginosa*, *P. fluorescens*, *S. epidermidis*	Gao et al. (2011)
97	Fusarielin B	$C_{25}H_{40}O_5$	*Fusarium tricinctum* Salicorn 19	*Salicornia bigelovii* Torr.	Amaranthaceae	19–10 μM MIC against *M. smegmatis*, *B. subtilis*, *M. phlei*, *E. coli*	Zhang et al. (2006)

98	Fusariumin C[a]	$C_{21}H_{32}O_3$	*Fusarium oxysporum* ZZP-R1	*Rumex madaio* Makino	Polygonaceae	6.25 μg/mL MIC value against *S. aureus* ATCC 25923	Chen et al. (2019)
99	Fusartricin[a]	$C_{21}H_{40}O_8$	*Fusarium tricinctum* Salicorn 19	*Salicornia bigelovii* Torr.	Amaranthaceae	19 μM MIC against *E. aerogenes, M. tetragenus*	Zhang et al. (2006)
100	Periconicin A[a]	$C_{20}H_{28}O_3$	*Periconia* sp. OBW-15	*Taxus cuspidata* Siebold & Zucc.	Taxaceae	12.5–3.12 μg/mL MIC against *S. aureus, S. epidermidis, B. subtilis, K. pneumoniae, S. typhimurium*	Kim et al. (2004)
Other							
101	3-nitropropionic acid	$C_3H_5NO_4$	*Phomopsis longicolla*	*Trichilia elegans* A. Juss.	Meliaceae	2.3–1.1 mm zone inhibition against *Xanthomonas axonopodis, Micrococcus luteus, S. typhi*	Flores et al. (2013)

[a]*Compounds isolated for the first time; IC50: half maximal inhibitory concentration;* MIC, *minimal inhibitory concentration; -: not identified or not specified;* agr, *accessory gene regulator quorum-sensing system;* MRSA, *multi-resistant* Staphylococcus aureus; *Bacterial strains used for bioassay:* Aeromonas hydrophila, Agrobacterium tumefaciens, Bacillus dysenteriae, Bacillus megaterium, Bacillus proteus, Bacillus subtilis, Corynebacterium insidiosum, Enterococcus faecalis, Enterococcus faecium, Escherichia coli, Helicobacter pylori, Klebsiella pneumoniae, Listeria monocytogenes, Micrococcus luteus, Micrococcus tetragenus, Mycobacterium phlei, Mycobacterium smegmatis, Mycobacterium tuberculosis, Pseudomonas aeruginosa, Pseudomonas fluorescens, Salmonella enteritidis, Salmonella typhimurium, Shigella *sp.*, Staphylococcus aureus, Staphylococcus epidermidis, Staphylococcus haemolyticus, Streptococcus pneumoniae, Vibrio cholerae, Vibrio parahaemolyticus, Xanthomonas axonopodis, Xanthomonas oryzae, Xanthomonas vesicatoria.

metabolites and the possible mechanisms of action proposed in the literature will be mentioned. Finally, we describe some cases of antivirulence compounds against different pathogens of human concern in order to evaluate new strategies to fight bacterial infections.

Alkaloids

Pyrazin-2-one

Neoaspergillic acid (**6**), an alkaloid pyrazinoic derivative, was isolated from a coculture of two *Aspergillus* strains (A. FSY-01 and A. FSW-02), which were isolated from the mangrove tree *Avicennia marina* (Acanthaceae) and showed antibacterial activity in a serial dilution assay with MIC values of 0.98 and 0.49 μg/mL against *Staphylococcus aureus* and *S. epidermidis* bacteria species (Zhu et al., 2011). In another study, neoaspergillic acid significantly inhibited the cell proliferation of SGC-7901 and K562 cancer cell lines with IC_{50} values of 8.24 and 7.99 μg/mL respectively, and moderate cytotoxicity to SPA-A-1 and BEL-7402 cell lines with IC_{50} values of 22.19 and 24.19 μg/mL, respectively (Fig. 12.2).

Piperine

Piperine (**10**), a well-known plant-produced alkaloid, was found to be produced by an endophytic strain of *Periconia* sp. isolated from *Piper longum* L (Piperaceae). Additionally, piperine was reported to possess an MIC value of 1.74 μg/mL against a multidrug resistance

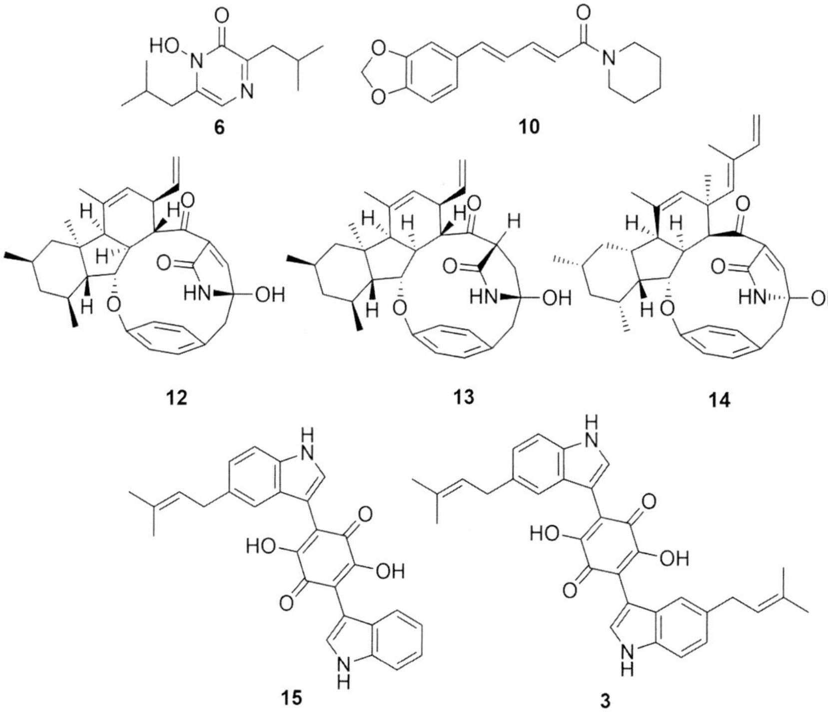

FIGURE 12.2 Chemical structures of neoaspergillic acid (**6**), piperine (**10**), pyrrocidine A (**12**), pyrrocidine B (**13**), semicochliodinol A (**15**), and cochliodinol (**3**).

strain of *M. tuberculosis* (Verma et al., 2011). There are several reports in the literature that confirm the activity of piperine against *Mycobacterium* spp. and its role in the inhibition of rv1258c, a putative multidrug efflux pump of *M. tuberculosis* (Sharma et al., 2010).

Pyrrocidines

The 13-member macrocycle alkaloids, pyrrocidine A (**12**) and B (**13**) were isolated for the first time from the endophytic fungi *Acremonium zeae* NRRL 13540, which was isolated from maize, and they showed MICs values of 0.25–2 and 4–8 μg/mL, respectively, against four strains of *Staphylococcus aureus* (including two piperacillin-resistant strains); 0.25 and 8 μg/mL, respectively, against *S. haemolyticus* GC4546; 0.5 and 4–8 μg/mL against three strains of *Enterococcus faecalis* and 0.5–1 and 4–8 μg/mL, respectively, against three strains (including two vancomycin-resistant strains) of *Enterococcus faecium* (He et al., 2002; Wicklow & Poling, 2009). In addition, pyrrocidine A also exerted moderate antifungal activity against a strain of *Candida albicans* with an MIC value of 8 μg/mL. Another member of this family, pyrrocidine C (**14**) was isolated from the endophytic fungi *Lewia infectoria* SNB-GTC2402, which was obtained from the amazonian tree *Besleria insolita* (Gesneriaceae), and showed antibacterial activity against *S. aureus* ATCC 29213 with an MIC value of 2 μg/mL (Casella et al., 2013). In addition, pyrrocidine C showed cytotoxicity activity against the cervical uterine cell line KB with an IC_{50} value of 10 μM, however, it was inactive against MDA-MB-435 and MRC-5 cell lines.

Bisindoles

From the cultures of the same strain of *L. infectoria* described above, two bisindole alkaloids identified as semicochliodinol A (**15**) and cochliodinol (**3**) were also isolated and showed MICs values of 2 and 4 μg/mL, respectively, against *Staphylococcus aureus* ATCC 29213 strain. In addition, both derivatives showed cytotoxic activity against the KB cancer cell line with IC_{50} values of 0.31 and 0.53 μM, respectively. Cochliodinol also showed antifungal activity against *Candida albicans* ATCC 10213 strain with an MIC value of 2 μg/mL. Semicochliodinol A has also been reported with anti-HIV activity by inhibiting the HIV-1 protease (IC_{50} of 0.37 μM) and as an anticancer agent by inhibiting the epidermal growth factor receptor protein tyrosine kinase (IC_{50} of 20 μM) (Casella et al., 2013; Fredenhagen et al., 1997).

Peptides

Dipeptides

The 2,5-diketopiperazine derivative, tardioxopiperazine A (**92**) (Fig. 12.3), was isolated from culture extract of *Eurotium cristatum* EN-220, an endophytic fungus isolated from the marine alga *Sargassum thunbergii* (Du et al., 2012). This compound exerted antibacterial activity in a well-diffusion method against a strain of *Staphylococcus aureus*, with an MIC value of 8 μg/mL. In addition, tardioxopiperazine A showed no cytotoxic activity against P388, HL-60, BEL-7402, and A549 cell lines in vitro ($IC_{50} > 50$ μM) (Wang et al., 2007).

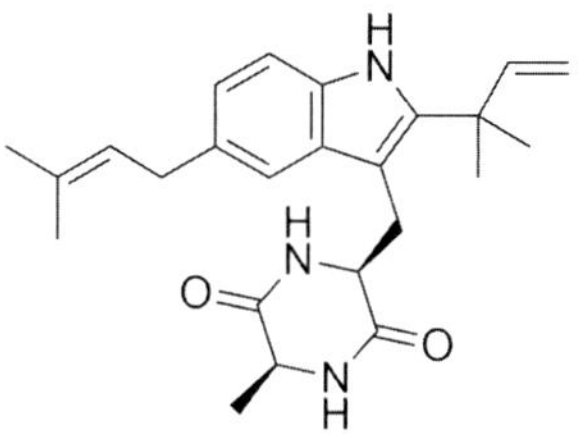

FIGURE 12.3 Chemical structures of tardioxopiperazine A (**92**), PF1022F (**91**), halobacillin (**90**), and enniatin A1 (**87**).

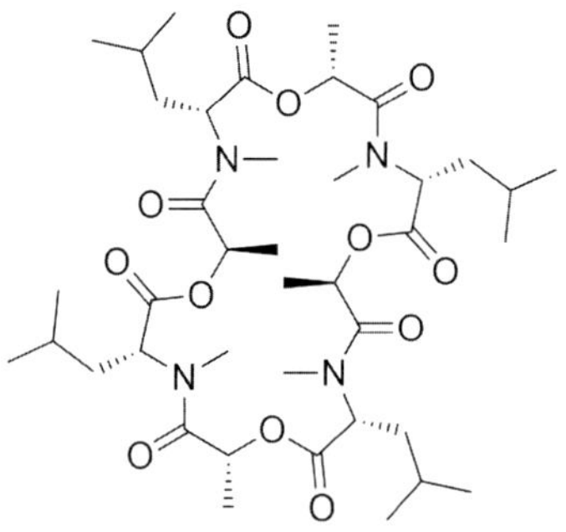

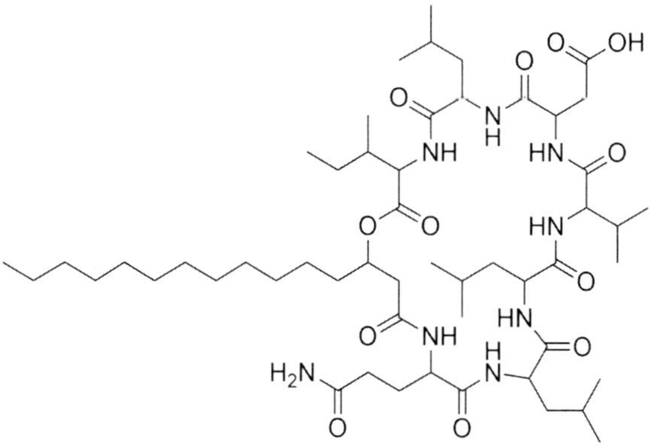

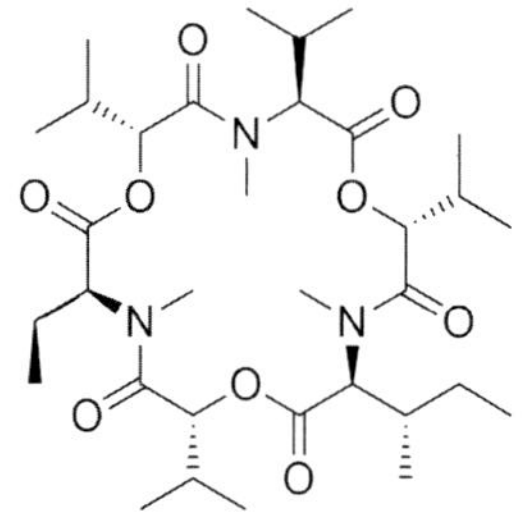

Polypeptides

Two cyclopeptides PF1022F (**91**) and halobacillin (**90**) have been isolated from the endophytic fungus *Trichoderma asperellum*, residing in the traditional Chinese medicinal plant *Panax notoginseng*. Both compounds showed antibacterial activity in vitro in a serial dilution model against *Enterococcus faecium* CGMCC 1.2025 strain with MICs values of 7.3 and 5.4 μg/mL, respectively (Ding et al., 2012). Halobacillin, previously isolated from cultures of *Bacillus* sp. (CND-914) obtained from a deep-sea sediment core, showed also strong cytotoxic activity, inhibiting the growth of human colon tumor cells (HCT-116) with an IC_{50} of 0.98 μg/mL (Trischman, Jensen, & Fenical, 1994).

In 2013 the cyclic depsipeptide enniatin A1 (**87**), a cyclic hexadepsipeptide of alternating D-α-hydroxyisovaleric acids and N-methyl-L-amino acids, was isolated from the fungal endophyte *Fusarium tricinctum* with antibacterial activity showing MIC value of 2 μg/mL against *Staphylococcus pneumoniae* ATCC 49619. Remarkably, the coculturing of *F. tricinctum* with the bacterium *Bacillus subtilis* 168 trpC2 resulted in an up to 78-fold increase in the production of enniatin A1 (Ola et al., 2013). Enniatin A1 also possesses anticarcinogenic properties by induction of apoptosis of HepG2 liver cancer cell line and disruption of extracellular signal-regulated kinases signaling pathway. In addition, this compound has been probed as an enzyme inhibitor of acyl-CoA:cholesterol acyltransferase with an IC_{50} of 49 μM in rat liver microsomes (Juan-Garcia, Ruiz, Font, & Manyes, 2015; Tomoda et al., 1992).

Polyketides

Chromones

The chromone derivative altechromone A (**37**) (Fig. 12.4), first isolated from a strain of *Alternaria* sp. in 1992, which was obtained from an ear of oat, was reported as a plant growth promoter. Also isolated from several fungi including *Hypoxylon truncatum*, *Ascomycota* spp., and from *Alternaria brassicicola*, altechromone A has shown antibacterial activity with an MIC value of 3.9, 3.9, 1.8 and 7.8 μg/mL against *Bacillus subtilis* CGMCC1.1162, *Escherichia coli* CGMCC1.1571, *Pseudomonas fluorescens* CGMCC1.1828, and *Staphylococcus aureus* CGMCC1.1361 strains, respectively (Gu, 2009; Kimura, Mizuno, Nakajima, & Hamasaki, 1992).

Alternariol (**39**), a chromone derivative isolated from the crude extract of *Alternaria alternata* ZHJG5 (an endophytic fungus isolated from healthy leaves of *Cercis chinensis*), showed antibacterial activity against the phytopathogen *Xanthomonas oryzae* pv. oryzae

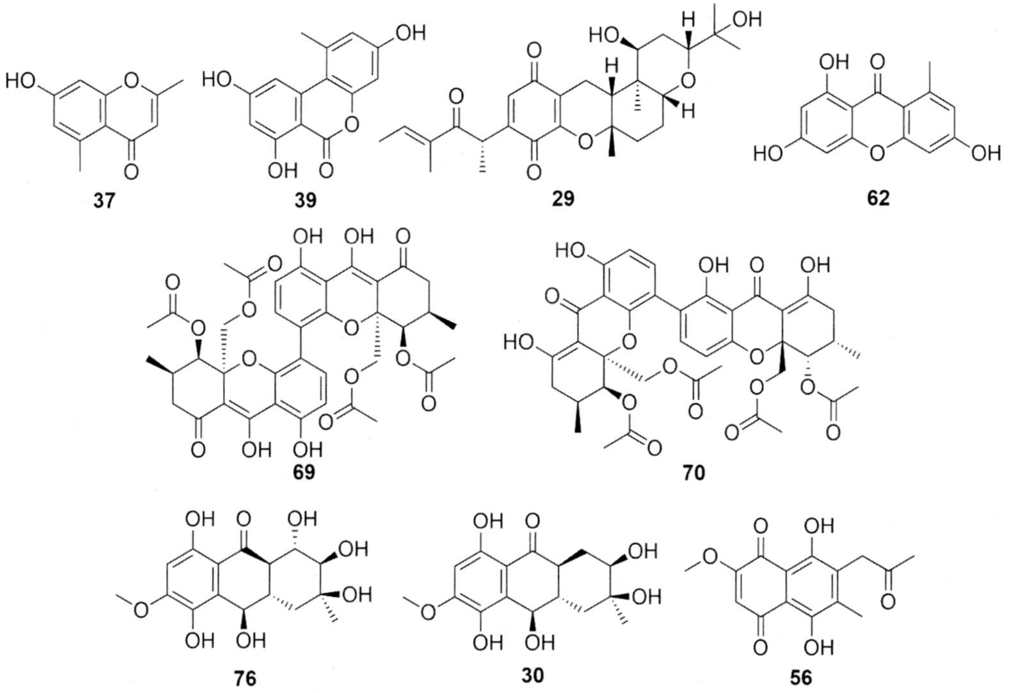

FIGURE 12.4 Chemical structures of altechromone A (**37**), alternariol (**39**), 2,3-didehydro-19a-hydroxy-14-epi-cochlioquinone B (**29**), norlichexanthone (**62**), phomoxanthone A (**69**), phomoxanthone B (**70**), tetrahydrobostrycin (**76**), 4-deoxytetrahydrobostrycin (**30**), and javanicin (**56**).

KACC10331 with an MIC value of 4 μg/mL. This activity was further evaluated in a protective rice-infected model, showing protection efficacy of 52.8% and 66.2% at 100 and 200 μg/mL, respectively, against the same pathogen. Alternariol was found to inhibit the β-ketoacyl-acyl carrier protein synthase III, an essential protein for bacterial fatty acid biosynthesis (Zhao et al., 2020).

Quinones

A new quinone derivative, the 2,3-didehydro-19a-hydroxy-14-epicochlioquinone B (**29**) was isolated from the endophytic fungus *Nigrospora* sp. MA75 obtained from the marine semimangrove plant *Pongamia pinnata* (Fabaceae), and showed antibacterial activity in a serial dilution assay with MICs values of 8.0, 4.0, 4.0, 0.5 and 0.5 μg/mL against a methicillin-resistant *Staphylococcus aureus*, *Escherichia coli*, *Pseudomonas aeruginosa*, *P. fluorescens*, and *S. epidermidis*, respectively. In addition, this compound showed cytotoxic effect in vitro against several tumor cell lines with IC_{50} values of 4, 5, and 7 μg/mL for MCF-7, SW1990, and SMMC7721 strains, respectively (Shang et al., 2012).

From the endophyte fungus *Nigrospora* sp. MA75, described previously, two anthraquinones derivatives, tetrahydrobostrycin (76) and 4-deoxytetrahydrobostrycin (30), were isolated with antibacterial activity with MICs values of 0.5 and 4.0 μg/mL against a strain of *Escherichia coli*, respectively. Tetrahydrobostrycin was active also against a methicillin-resistant *Staphylococcus aureus* with an MIC value of 2.0 μg/mL (Shang et al., 2012). In the same study, only deoxytetrahydrobostrycin showed moderate cytotoxic activity against HeLa cell line with an IC_{50} of 22.0 μg/mL.

Javanicin (56), a highly functionalized naphthoquinone, was isolated from an endophytic fungus, *Chloridum* sp. (J. F. H. Beyma), which was obtained from the roots of *Azadirachta indica* (Meliaceae). This compound showed antibacterial activity in a standard plate bioassay test method against *Pseudomonas* spp., *P. aeruginosa*, and *P. fluorescens* with an MIC value of 2.0 μg/mL against both species. In addition, javanicin has shown antifungal activity against the fungal plant pathogen species *Cercospora arachidicola*, with an MIC value of 5.0 μg/mL (Kharwar et al., 2009).

Xanthones

From the same endophytic fungus of *Nigrospora* sp. MA75 described previously, a xanthone derivative, norlichexanthone (**62**) was isolated and showed antibacterial activity against *Staphylococcus epidermidis* with an MIC value of 0.5 μg/mL. This compound showed cytotoxic effects against HepG2 cell line with an IC_{50} of 15 μg/mL (Shang et al., 2012). In addition, norlichexanthone has been identified as a selective ligand of estrogen receptor-alpha (ERα) by activating the Gal4/DBD-ERα/LDB fusion protein dose-dependently with an EC_{50} of 38.8 nM. The potential efficacy to prevent osteoporosis of norlichexanthone was evaluated in an ovariectomized mouse model, preventing bone loss at the dosage of 1 mg/kg (Wang et al., 2021).

Two novel xanthone dimers derivatives, phomoxanthone A (**69**) and B (**70**), were isolated from culture of the endophytic fungi *Phomopsis* sp. BSS 1323, an endophytic fungus isolated from *Tectona grandis* (Lamiaceae) and showed antitubercular activity against the strain *Mycobacterium tuberculosis* H37Ra with MIC values of 0.5 and 6.25 μg/mL, respectively (Isaka et al., 2001). In addition, these two xanthone dimers have been shown to inhibit the proliferation of cancer cell lines (IC_{50} of 0.9 and 4.1 μg/mL against KB cells, 0.51 and 0.70 μg/mL against BC-1 cells, and 1.4 and 1.8 μg/mL against vero cells, respectively) and inhibit the tyrosine phosphatase SHP1 in MCF-7 cells leading to the upregulation of inflammatory factors (Yang et al., 2020).

Benzofurans

Usnic acid (**78**) and two of its derivatives, cercosporamide (**42**) and phomodione (**65**) (Fig. 12.5), were isolated from the endophytic fungus *Phoma* sp. NG-25, which was isolated from *Saurauia scaberrima* (Actinidiaceae) and they showed antibacterial activity with MIC values of 2.0 μg per disk with an halo of inhibition of 0.5 mm against the strain *Staphylococcus aureus* ATCC 25923 (Hoffman et al., 2008). Previously, isolated from a lichen, usnic acid showed antibacterial activity against *Clostridium perfringens* ATCC 13124, *Propionibacterium acnes*, and 29 clinical isolated of *Enterococcus faecalis* and eight clinically isolated strains of methicillin-resistant *S. aureus* with MICs values of 4.0, 2.0, 8.0, and 8.0 μg/mL, respectively (Lauterwein, Oethinger, Belsner, Peters, & Marre, 1995). Usnic acid showed also antibiofilm activity, inhibiting the formation of biofilm by loaded *S. aureus* cells and altering the biofilm morphology formed by *Pseudomonas aeruginosa* cells, indicating the possibility of interference in signaling pathways (Francolini, Norris, Piozzi, Donelli, & Stoodley, 2004). Recently, the strong inhibition of RNA and DNA synthesis in *Bacillus subtilis* and *S. aureus*, by usnic acid has been demonstrated as a possible antibacterial mechanism of action. Interestingly, DNA synthesis was halted rapidly, suggesting interference of usnic acid with the elongation of DNA replication. It was also observed a slight inhibition of RNA synthesis in a Gram-negative bacterium, *Vibrio harveyi* (Maciag-Dorszynska, Wegrzyn, & Guzow-Krzeminska, 2014).

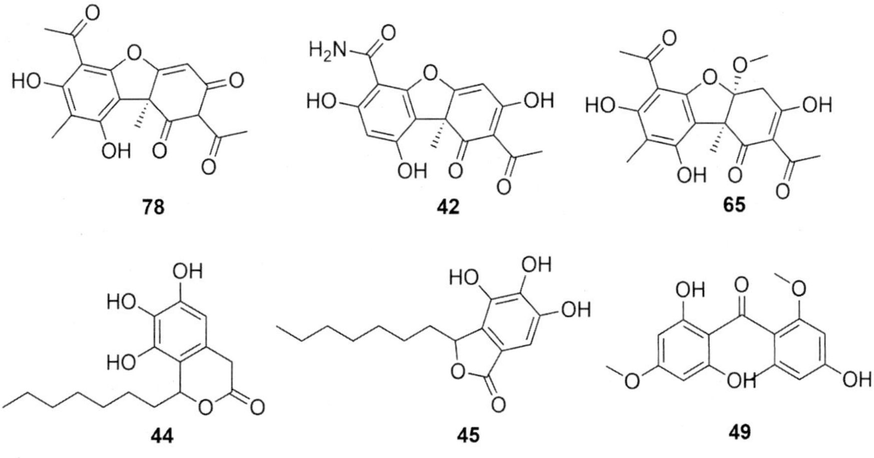

FIGURE 12.5 Chemical structures of usnic acid (**78**), cercosporamide (**42**), phomodione (**65**), cytosporone D (**44**), cytosporone E (**45**), and griseophenone C (**49**).

Octaketides

Two new octaketides, cytosporone D (**44**) and E (**45**), were isolated from extracts obtained from the endophytic fungus *Cytospora* sp. CR200, isolated from the tissue of the mangrove shrub *Conocarpus erecta* (Combretaceae). These compounds showed antibacterial activity against representative strains of *Staphylococcus aureus* and *Enterococcus faecalis* with MICs values of 8 μg/mL. In addition, cytosporone D also showed antifungal activity with an MIC value of 4 μg/mL against the fungus *Candida albicans*.

Benzophenones

From the endophyte fungus *Nigrospora* sp. MA75, described previously, a benzophenone identified as griseophenone C (**49**) was also isolated. This compound showed MIC value of 0.5 μg/mL against *Pseudomonas aeruginosa*, *P. fluorescens*, and a methicillin-resistant *Staphylococcus aureus* strain.

Terpenoids

A meroterpene with a cyclohexane acid moiety, named fusariumin C (**98**) (Fig. 12.6), was isolated from a crude extract of an endophytic fungus *Fusarium oxysporum* ZZP-R1, obtained from the roots of coastal plant *Rumex madaio* (Polygonaceae). This terpenoid possesses significant antibacterial activity against *Staphylococcus aureus* ATCC 25923 strain with an MIC value of 6.25 μg/mL (Chen et al., 2019).

Antivirulence compounds

In the last years, and in order to find new strategies to fight some persistent and multidrug pathogens, the expression, production, secretion, or activity of different virulence factors have been targeted. Some of these factors are essential in the invasion, colonization, and pathogenicity of bacterial infections such as the production of gelatinase, which is inhibited by ambuic acid (**41**) isolated from different strains of *Pestalotiopsis* sp. and *Monochaetia* sp. (Li et al., 2001; Nakayama et al., 2009). In the exploration of new bioactive compounds, scientifics have been screening isolated compounds from endophytic fungi.

During a screening of 100 extracts from 50 *Penicillium* species, 33 were found to produce QS inhibitory compounds. From the extracts of the endophytic fungus of *Penicillium radicicola* (IBT 10696) and *Penicillium coprobium* (IBT 6895) bioguided isolation led to the identification of penicillic acid (**64**) and patulin (**63**) (Fig. 12.7), respectively, as potent quorum system inhibitors of a *Pseudomonas aeruginosa* PA01 strain. These compounds were found

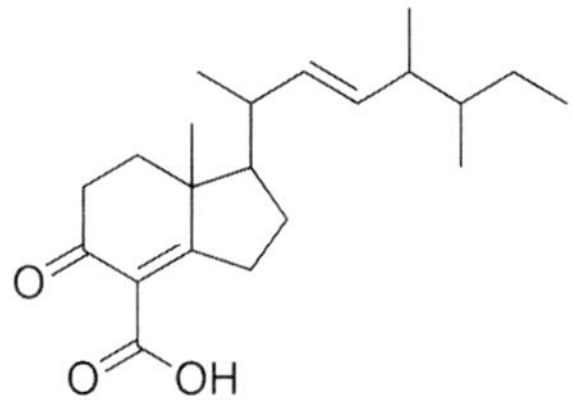

O OH O
98

FIGURE 12.6 Chemical structure of fusariumin C (**98**).

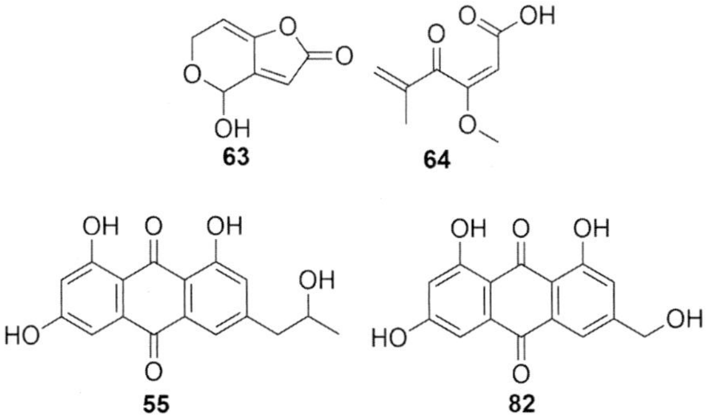

FIGURE 12.7 Chemical structures of patulin (**63**), penicillic acid (**64**), ω-hydroxyemodin (**82**), and isorhodoptilometrin (**55**).

to target the transcription of different genes encoding the expression of RhlR and LasR proteins at 80–40 μM (Rasmussen et al., 2005).

From the endophytic fungi *Penicillium restrictum* G85ITS, obtained from healthy parts of *Silybum marianum* (Asteraceae), a series of antiquorum sensing polyhydroxyanthraquinones were isolated. The most active compounds were identified as ω-hydroxyemodin (**82**) and isorhodoptilometrin (**55**) with IC_{50} values of 8.1 and 8.4, respectively, against the expression of a functional accessory gene regulator, which is required for the infection of clinical strains of methicillin-resistant *Staphylococcus aureus* (Figueroa et al., 2014).

Antiparasitic compounds

Among all the works carried out on the search for bioactive compounds from endophytic fungi, many studies have focused on their activity against parasitic infections (Hzounda Fokou et al., 2021; Ibrahim, Mohamed, Al Haidari, El-Kholy, & Zayed, 2018; Toghueo, 2019). The reported antiparasitic activities in the literature can vary up to $IC_{50} > 290$ μM against *P. falciparum* such as the lactones isolated from *Xylaria* sp. isolated from the leaves of *Siparuna* sp. (Siparunaceae) (Jiménez-Romero, Ortega-Barría, Arnold, & Cubilla-Rios, 2008) or up to $IC_{50} > 169$ μM against *L. donovani* such as the polyketides isolated from *Mycoleptodiscus indicus* isolated from the plant *Borreria verticillata* (Rubiaceae) (Andrioli et al., 2014). Here, we focused on the compounds with IC_{50} values lower than 10 μM (Table 12.2) active against *Plasmodium* spp., *Leishmania* spp. and *Trypanosoma* spp. and we describe the culture methods and their potential cytotoxicity against different cancer cell lines. Interestingly, these compounds belong to several chemical classes, such as polyketides, terpenoids, alkaloids and polypeptides.

Antileishmanial compounds

Among the most active compounds isolated from endophytic fungi active against *Leishmania donovani* with an IC_{50} below 1 μM, we can mention preussomerin EG1 (IC_{50} 0.12 μM) followed by FD-838 (IC_{50} 0.2 μM), fumiquinone B and pseurotin D (IC_{50} 0.5 μM), palmarumycin CP18 (IC_{50} 0.62 μM), and purpureone (IC_{50} 0.87 μM). The compounds exhibiting a moderate activity against *L. donovani* with an IC_{50} between 1 and 5 μM are palmarumycin CP17 (IC_{50} 1.34 μM) followed by integracide J (IC_{50} 3.29 μM), palmarumycin CP2

TABLE 12.2 Antiparasitic compounds isolated from endophytic fungi (2000–19), with IC50 < 10 μM against *Plasmodium* spp., *Leishmania* spp., and *Trypanosoma* spp.

ID	Compound names	Chemical formula	Endophytic fungal strain	Host plant	Botanical family	Parasite strain	Activity	Reference
Alkaloids								
102	18-Deoxy-19,20-epoxycytochalasin C	$C_{30}H_{37}NO_6$	*Nemania* sp. UM10M	*Torreya taxifolia* Arn.	Taxaceae	*P. falciparum* (D6 and W2)	IC_{50} 0.56 μM (280 ng/mL) in D6; 0.19 μM (100 ng/mL) in W2	Kumarihamy et al. (2019)
103	19,20-Epoxycytochalasin C	$C_{30}H_{37}NO_7$	*Nemania* sp. UM10M	*Torreya taxifolia* Arn.	Taxaceae	*P. falciparum* (D6 and W2)	IC_{50} 0.07 μM (37 ng/mL) in D6; 0.05 μM (28 ng/mL) in W2	Kumarihamy et al. (2019)
104	19,20-Epoxycytochalasin D	$C_{30}H_{37}NO_7$	*Nemania* sp. UM10M	*Torreya taxifolia* Arn.	Taxaceae	*P. falciparum* (D6 and W2)	IC_{50} 0.04 μM (22 ng/mL) in D6; 0.04 μM (20 ng/mL) in W2	Kumarihamy et al. (2019)
			Diaporthe sp.	*Guapira standleyana* Woodson	Nyctaginaceae	*P. falciparum* 3D7	IC_{50} 136 nM (0.136 μM)	Calcul et al. (2013)
105	2,5-Dihydroxy-1-(hydroxymethyl)pyridin-4-one[a]		Strain BB4	*Tinospora crispa* (L.) Hook. f. & Thomson	Menispermaceae	*P. falciparum* 3D7	IC_{50} 0.02 μg/mL (0.127 μM)	Elfita, Muharni, Munawar, Legasari, and Darwati (2011)
106	7-Hydroxy-3,4,5-trimethyl-6-on 2,3,4,6-tetrahydroisoquinoline-8-carboxylic acid[a]	$C_{13}H_{15}O_3N$	Strain BB4	*Tinospora crispa* (L.) Hook. f. & Thomson	Menispermaceae	*P. falciparum* 3D7	IC_{50} 0.03 μg/mL (0.129 μM)	Elfita et al. (2011)
107	Cissetin	$C_{22}H_{31}NO_5$	*Preussia* sp.	*Enantia chlorantha* Oliv.	Annonaceae	*P. falciparum* NF54	IC_{50} 10.3 μM	Talontsi, Lamshöft, Douanla-Meli, Kouam, and Spiteller (2014)
108	Cytochalasin D	$C_{30}H_{37}NO_6$	*Diaporthe* sp., *Xylaria* sp.	*Kandelia obovata* Sheue, H.Y. Liu & J. Yong, *Avicennia marina* (Forssk.) Vierh., *Lumnitzera racemosa* Willd.	Rhizophoraceae, Acanthaceae, Combretaceae	*P. falciparum* 3D7	IC_{50} 25.8 nM (0.0258 μM)	Calcul et al. (2013)
109	Cytochalasin H	$C_{30}H_{39}NO_5$	*Diaporthe* sp.	*Kandelia obovata* Sheue, H.Y. Liu & J. Yong, *Avicennia marina* (Forssk.) Vierh., *Lumnitzera racemosa* Willd.	Rhizophoraceae, Acanthaceae, Combretaceae	*P. falciparum* 3D7	$IC_{50} < 20$ nM (<0.02 μM)	Calcul et al. (2013)

No.	Compound	Formula	Endophyte	Host plant	Family	Target	Activity	Reference
110	Cytochalasin J	$C_{28}H_{37}NO_4$	*Diaporthe* sp.	*Kandelia obovata* Sheue, H.Y. Liu & J. Yong, *Avicennia marina* (Forssk.) Vierh., *Lumnitzera racemosa* Willd.	Rhizophoraceae, Acanthaceae, Combretaceae	*P. falciparum* 3D7	$IC_{50} < 20$ nM (<0.02 μM)	Calcul et al. (2013)
111	Cytochalasin O	$C_{28}H_{37}NO_4$	*Verticillium* sp.	*Kandelia obovata* Sheue, H.Y. Liu & J. Yong, *Avicennia marina* (Forssk.) Vierh., *Lumnitzera racemosa* Willd.	Rhizophoraceae, Acanthaceae, Combretaceae	*P. falciparum* 3D7	$IC_{50} < 20$ nM (<0.02 μM)	Calcul et al. (2013)
112	Epoxycytochalasin H	$C_{30}H_{39}NO_5$	*Diaporthe miriciae* UFMGCB 9720	*Vellozia gigantea* N. L. Menezes & Mello-Silva	Velloziaceae	*P. falciparum* (D6 and W2)	IC_{50} 51.70 ng/mL in D6 (0.1 μM) and 39.40 ng/mL in W2 (0.07 μM)	Ferreira et al. (2017)
Polyketides								
113	11-hydroxymonocerin[a]	$C_{16}H_{20}O_7$	*Exserohilum rostratum*	*Stemona* sp.	Stemonaceae	*P. falciparum* K1	IC_{50} 7.70 μM	Sappapan et al. (2008)
114	2-chloro-5-methoxy-3-methylcyclohexa-2,5-diene-1,4-dione[a]	$C_8H_7ClO_3$	*Xylaria* sp. PB-30	*Sandoricum koetjape* (Burm.f.) Merr.	Meliaceae	*P. falciparum* K1	IC_{50} 1.84 μM	Tansuwan et al. (2007)
115	7-butyl-6,8-dihydroxy-3(R)-pentylisochroman-1-one[a]	$C_{18}H_{26}O_4$	*Geotrichum* sp.	*Crassocephalum crepidioides* (Benth.) S. Moore	Asteraceae	*P. falciparum* K1	IC_{50} 2.6 μg/mL (8.48 μM)	Kongsaeree, Prabpai, Sriubolmas, Vongvein, and Wiyakrutta (2003)
116	Altenusin	$C_{15}H_{14}O_6$	*Alternaria* sp. UFMGCB55	*Trixis vauthieri* DC.	Asteraceae	Inhibits trypanothione reductase	IC_{50} 4.3–0.3 μM	Cota et al. (2008)
117	Asterric acid	$C_{17}H_{16}O_8$	*Preussia* sp.	*Enantia chlorantha* Oliv.	Annonaceae	*P. falciparum* NF54	IC_{50} 8.67 μM	Talontsi et al. (2014)
118	Cercosporin	$C_{29}H_{26}O_{10}$	*Mycosphaerella* sp. nov. strain F2140	*Psychotria horizontalis* Spreng. ex DC.	Rubiaceae	*L. donovani*, *P. falciparum*, *T. cruzi*	IC_{50} 0.46 μM, 1.03 μM, 1.08 μM	Moreno et al. (2011)
119	Cercosporin derivative tetraacetylated[a] (#)	$C_{37}H_{34}O_{14}$	*Mycosphaerella* sp. nov. strain F2140	*Psychotria horizontalis* Spreng. ex DC.	Rubiaceae	*L. donovani*, *P. falciparum*, *T. cruzi*	IC_{50} 0.64 μM, 2.99 μM, 0.78 μM	Moreno et al. (2011)
120	Chaetoxanthone B[a]	$C_{20}H_{18}O_6$	*Chaetomium* sp.	Algal species	–	*P. falciparum* K1	IC_{50} 0.5 μg/mL (1.41 μM)	Pontius, Krick, Kehraus, Brun, and König (2008)

(*Continued*)

TABLE 12.2 (Continued)

ID	Compound names	Chemical formula	Endophytic fungal strain	Host plant	Botanical family	Parasite strain	Activity	Reference
121	Chaetoxanthone C[a]	$C_{20}H_{19}ClO_6$	*Chaetomium* sp.	Algal species	–	*T. cruzi* C4	IC_{50} 1.5 μg/mL (3.83 μM)	Pontius et al. (2008)
122	CJ-12,371	$C_{20}H_{16}O_4$	*Edenia* sp.	*Petrea volubilis* L.	Verbenaceae	*L. donovani* amastigotes	IC_{50} 8.40 μM	Martínez-Luis et al. (2008)
123	Cochlioquinone A	$C_{30}H_{44}O_8$	*Cochliobolus* sp.	*Piptadenia adiantoides* (Spreng.) J.F. Macbr.	Fabaceae	*L. amazonensis* promastigotes	EC_{50} 1.7 μM	Campos et al. (2008)
124	Dicerandrol D[a]	$C_{36}H_{37}O_{15}$	*Diaporthe* sp.	*Kandelia obovata* Sheue, H.Y. Liu & J. Yong, *Avicennia marina* (Forssk.) Vierh., *Lumnitzera racemosa* Willd.	Rhizophoraceae, Acanthaceae, Combretaceae	*P. falciparum* 3D7	IC_{50} 600 nM (0.6 μM)	Calcul et al. (2013)
125	Fumiquinone B	$C_8H_8O_5$	*Aspergillus* sp. F1544	*Guapira standleyana* Woodson	Nyctaginaceae	*L. donovani; P. falciparum*	IC_{50} 0.5 μM, 5.4 μM	Martińez-Luis et al. (2012)
126	Isocochlioquinone A	$C_{30}H_{44}O_8$	*Cochliobolus* sp.	*Piptadenia adiantoides* (Spreng.) J.F. Macbr.	Fabaceae	*L. amazonensis* promastigotes	EC_{50} 4.1 μM	Campos et al. (2008)
127	KS-501a	$C_{33}H_{48}O_{10}$	*Acremonium* sp. BCC 14080	Palm tree	Arecaceae	*P. falciparum* K1	IC_{50} 9.9 μM	Bunyapaiboonsri, Yoiprommarat, Khonsanit, and Komwijit (2008)
128	Mollicellin E	$C_{22}H_{19}ClO_8$	*Chaetomium brasiliense*	Hala-Bala evergreen forest	–	*P. falciparum* K1	IC_{50} 3.2 μg/mL (7.16 μM)	Khumkomkhet, Kanokmedhakul, Kanokmedhakul, Hahnvajanawong, and Soytong (2009)
129	Mollicellin K[a]	$C_{21}H_{18}O_7$	*Chaetomium brasiliense*	Hala-Bala evergreen forest	–	*P. falciparum* K1	IC_{50} 1.2 μg/mL (3.13 μM)	Khumkomkhet et al. (2009)
130	Mollicellin L[a]	$C_{22}H_{20}O_7$	*Chaetomium brasiliense*	Hala-Bala evergreen forest	–	*P. falciparum* K1	IC_{50} 3.4 μg/mL (8.6 μM)	Khumkomkhet et al. (2009)
131	Mollicellin M[a]	$C_{21}H_{17}ClO_7$	*Chaetomium brasiliense*	Hala-Bala evergreen forest	–	*P. falciparum* K1	IC_{50} 2.9 μg/mL (6.95 μM)	Khumkomkhet et al. (2009)
132	Monocerin	$C_{16}H_{20}O_6$	*Exserohilum rostratum*	*Stemona* sp.	Stemonaceae	*P. falciparum* K1	IC_{50} 0.68 μM	Sappapan et al. (2008)
133	Palmarumycin CP17[a]	$C_{20}H_{14}O_5$	*Edenia* sp.	*Petrea volubilis* L.	Verbenaceae	*L. donovani* amastigotes	IC_{50} 1.34 μM	Martínez-Luis et al. (2008)

No.	Compound	Formula	Endophyte	Host plant	Family	Target	Activity	Reference
134	Palmarumycin CP18[a]	$C_{20}H_{14}O_5$	*Edenia* sp.	*Petrea volubilis* L.	Verbenaceae	*L. donovani* amastigotes	IC_{50} 0.62 μM	Martínez-Luis et al. (2008)
135	Palmarumycin CP2	$C_{20}H_{14}O_4$	*Edenia* sp.	*Petrea volubilis* L.	Verbenaceae	*L. donovani* amastigotes	IC_{50} 3.93 μM	Martínez-Luis et al. (2008)
136	Phomoxanthone A[a]	$C_{38}H_{38}O_{16}$	*Phomopsis* sp. BCC 1323	*Tectona grandis* L.f.	Lamiaceae	*P. falciparum* K1	IC_{50} 0.11 μg/mL (0.15 μM)	Isaka et al. (2001)
137	Phomoxanthone B[a]	$C_{38}H_{38}O_{16}$	*Phomopsis* sp. BCC 1323	*Tectona grandis* L.f.	Lamiaceae	*P. falciparum* K1	IC_{50} 0.33 μg/mL (0.44 μM)	Isaka et al. (2001)
138	Preussiafuran A[a]	$C_{18}H_{16}O_7$	*Preussia* sp.	*Enantia chlorantha* Oliv.	Annonaceae	*P. falciparum* NF54	IC_{50} 8.76 μM	Talontsi et al. (2014)
139	Preussomerin EG1	$C_{20}H_{12}O_6$	*Edenia* sp.	*Petrea volubilis* L.	Verbenaceae	*L. donovani* amastigotes	IC_{50} 0.12 μM	Martínez-Luis et al. (2008)
140	Purpureone[a]	$C_{32}H_{30}O_{14}$	*Purpureocillium lilacinum*	*Rauvolfia macrophylla* Ruiz & Pav.	Apocynaceae	*L. donovani* amastigotes	MIC 0.63 μg/mL (0.87 μM)	Lenta et al. (2016)
141	Xylariaquinone A[a]	$C_{15}H_{12}O_5$	*Xylaria* sp. PB-30	*Sandoricum koetjape* (Burm.f.) Merr.	Meliaceae	*P. falciparum* K1	IC_{50} 6.68 μM	Tansuwan et al. (2007)
Polyketides-alkaloids								
142	14-Norpseurotin A	$C_{21}H_{23}NO_8$	*Aspergillus* sp. F1544	*Guapira standleyana* Woodson	Nyctaginaceae	*L. donovani*	IC_{50} 4.4 μM	Martińez-Luis et al. (2012)
143	Codinaeopsin[a]	$C_{32}H_{40}N_2O_3$	*Codinaeopsis gonytrichoides* CR127A	*Vochysia guatemalensis* Donn. Sm.	Vochysiaceae	*P. falciparum* 3D7	IC_{50} 2.3 μg/mL (4.7 μM)	Kontnik and Clardy (2008)
144	FD-838	$C_{22}H_{21}NO_7$	*Aspergillus* sp. F1544	*Guapira standleyana* Woodson	Nyctaginaceae	*L. donovani*	IC_{50} 0.2 μM	Martińez-Luis et al. (2012)
145	Pseurotin A	$C_{22}H_{25}NO_8$	*Aspergillus* sp. F1544	*Guapira standleyana* Woodson	Nyctaginaceae	*L. donovani*	IC_{50} 5.8 μM	Martińez-Luis et al. (2012)
146	Pseurotin D	$C_{22}H_{25}NO_8$	*Aspergillus* sp. F1544	*Guapira standleyana* Woodson	Nyctaginaceae	*L. donovani*	IC_{50} 0.5 μM	Martińez-Luis et al. (2012)
Polypeptides								
147	12,12a-dihydroantibiotic PI 016[a]	$C_{31}H_{36}N_2O_9S_2$	*Menisporopsis theobromae* BCC 3975	Seed plant	–	*P. falciparum* K1	IC_{50} 2.95 μM	Chinworrungsee, Kittakoop, Saenboonrueng, Kongsaeree, and Thebtaranonth (2006)
148	Apicidin B[a]	$C_{33}H_{47}N_5O_6$	*Fusarium pallidoroseum*	*Acacia* sp.	Fabaceae	*P. falciparum*	MIC 189 nM (0.189 μM)	Singh et al. (2001)

(*Continued*)

TABLE 12.2 (Continued)

ID	Compound names	Chemical formula	Endophytic fungal strain	Host plant	Botanical family	Parasite strain	Activity	Reference
149	Apicidin C[a]	$C_{33}H_{47}N_5O_6$	*Fusarium pallidoroseum*	*Acacia* sp.	Fabaceae	*P. falciparum*	MIC 69 nM (0.069 μM)	Singh et al. (2001)
150	Beauvericin	$C_{45}H_{57}N_3O_9$	*Fusarium* sp. [KF611679]	*Caesalpinia echinata* Lam.	Fabaceae	*T. cruzi*	IC_{50} 2.43 μM	Campos et al. (2015)
151	Fusaripeptide A[a]	$C_{46}H_{75}N_7O_{11}$	*Fusarium* sp.	*Mentha longifolia* (L.) L.	Lamiaceae	*P. falciparum* (DC6)	IC_{50} 0.34 μM	Ibrahim et al. (2018)
152	Pullularin A[a]	$C_{42}H_{57}N_5O_9$	*Pullularia* sp. BCC 8613	*Calophyllum* sp.	Clusiaceae	*P. falciparum* K1	IC_{50} 3.6 μg/mL (4.64 μM)	Isaka, Berkaew, Intereya, Komwijit, and Sathitkunanon (2007)
153	Pullularin B[a]	$C_{43}H_{59}N_5O_9$	*Pullularia* sp. BCC 8613	*Calophyllum* sp.	Clusiaceae	*P. falciparum* K1	IC_{50} 3.3 μg/mL (4.18 μM)	Isaka et al. (2007)
Terpenoids								
154	12,13-Deoxyroridin E	$C_{29}H_{38}O_7$	CY-3923 (not identified)	*Kandelia obovata* Sheue, H.Y. Liu & J. Yong, *Avicennia marina* (Forssk.) Vierh., *Lumnitzera racemosa* Willd.	Rhizophoraceae, Acanthaceae, Combretaceae	*P. falciparum* 3D7	IC_{50} <20 nM (<0.02 μM)	Calcul et al. (2013)
155	7α,10α-Dihydroxy-1βmethoxyeremophil-11 (13)-en-12,8β olide[a]	$C_{16}H_{24}O_5$	*Xylaria* sp. BCC 21097	*Licuala spinosa* Wurmb	Arecaceae	*P. falciparum* K1	IC_{50} 8.1 μM	Isaka, Chinthanom, Boonruangprapa, Rungjindamai, and Pinruan (2010)
156	Integracide H[a]	$C_{36}H_{54}O_7$	*Fusarium* sp.	*Mentha longifolia* (L.) L.	Lamiaceae	*L. donovani* amastigotes	IC_{50} 4.75 μM	Ibrahim, Abdallah, Mohamed, and Ross (2016)
157	Integracide J[a]	$C_{39}H_{54}O_7$	*Fusarium* sp.	*Mentha longifolia* (L.) L.	Lamiaceae	*L. donovani* amastigotes	IC_{50} 3.29 μM	Ibrahim et al. (2016)
158	Phomoarcherin B[a]	$C_{23}H_{28}O_5$	*Phomopsis archeri*	*Vanilla albidia* Blume	Orchidaceae	*P. falciparum* K1	IC_{50} 0.79 μg/mL (2.05 μM)	Hemtasin et al. (2011)
159	Roridin E	$C_{29}H_{38}O_8$	CY-3923 (not identified)	*Kandelia obovata* Sheue, H.Y. Liu & J. Yong, *Avicennia marina* (Forssk.) Vierh., *Lumnitzera racemosa* Willd.	Rhizophoraceae, Acanthaceae, Combretaceae	*P. falciparum* 3D7	IC_{50} <20 nM (<0.02 μM)	Calcul et al. (2013)

[a]*Compounds isolated for the first time.* IC50, *Half maximal inhibitory concentration;* MIC, *minimal inhibitory concentration;* EC50, *half maximal effective concentration; -: not identified or not specified; #: acetylated derivative of a natural compound; Parasitic species used for bioassay:* Leishmania amazonensis, Leishmania donovani, Plasmodium falciparum, Trypanosoma cruzi.

(IC_{50} 3.93 μM), 14-norpseurotin A (IC_{50} 4.4 μM), and integracide H (IC_{50} 4.75 μM). Similarly, cochlioquinone A and isocochlioquinone A exhibited moderate activities with EC_{50} values of 1.7 and 4.1 μM, respectively, and were tested against *Leishmania amazonensis* promastigotes (Campos et al., 2008). Finally, the compounds presenting a weaker activity against *L. donovani* with an IC_{50} between 5 and less than 10 μM are pseurotin A (IC_{50} 5.8 μM) and CJ-12,371 (IC_{50} 8.4 μM). Additionally, amphotericin B is one of the reference drugs to be used in biological assays against *Leishmania donovani* exhibiting an IC_{50} value of 0.07 μM, besides being part of one of the most important antileishmanial current treatments.

Polyketides

Phytotoxic compounds, isocochlioquinone A (**126**) and cochlioquinone A (**123**) (Fig. 12.8), were isolated for the first time from the fungal pathogen *Bipolaris bicolor* (*Cochliobolus* sp.) from a lesioned leaf of *Eleusine coracana* (Poaceae) (Miyagawa et al., 2014). Isocochlioquinone A and cochlioquinone A were later isolated from a different strain of *Cochliobolus* sp. from the plant *Piptadenia adiantoides* (Fabaceae). These compounds were tested against *L. amazonensis* amastigotes, exhibiting an EC_{50} value (effective concentration to kill 50% of the parasites) of 4.1 and 1.7 μM, respectively. Additionally, they are considered to possess some degree of selectivity since no cytocidal activity was detected against three human cancer cell lines [UACC-62 (melanoma), MCF-7 (breast), and TK-10 (renal)] (Campos et al., 2008). Amphotericin B was used at 0.2 μg/mL as a control drug. In

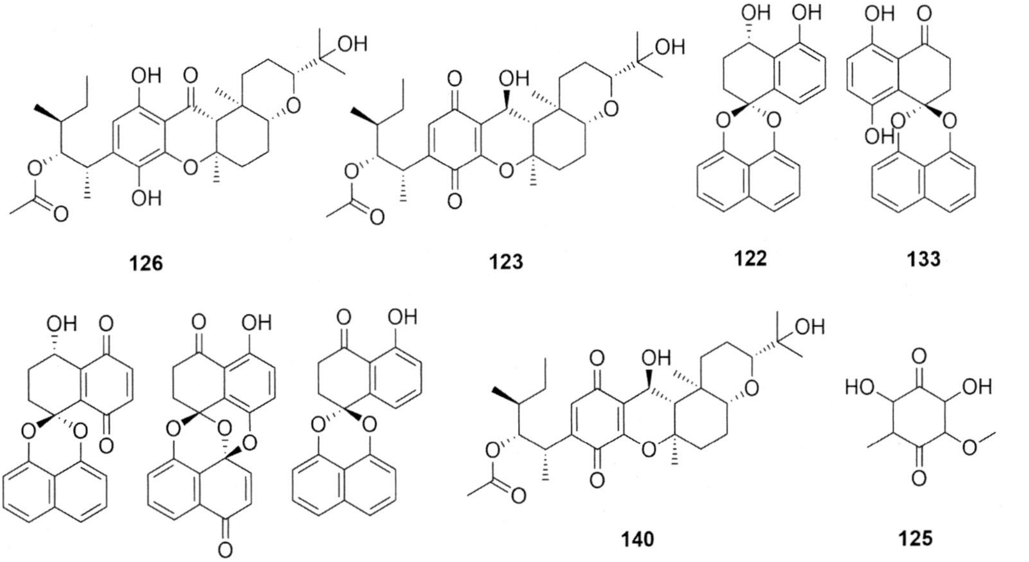

126 123 122 133

134 139 135 140 125

FIGURE 12.8 Chemical structures of isocochlioquinone A (**126**), cochlioquinone A (**123**), CJ-12,371 (**122**), palmarumycin CP17 (**133**), palmarumycin CP18 (**134**), preussomerin EG1 (**139**), palmarumycin CP2 (**135**), purpureone (**140**), and fumiquinone B (**125**).

another study, the same compounds were isolated but not separated and therefore their activity was tested as a mixture against *L. amazonensis* promastigotes, exhibiting an IC_{50} value of 10.16 μM (do Nascimento et al., 2015).

CJ-12,371—This compound with DNA-gyrasie activity was previously isolated from a nonidentified fungus (N983–46) (Sakemi et al., 1995) and recently isolated from *Edenia* sp. (Martínez-Luis et al., 2008). In this last study, CJ-12,371 (**122**) exhibited a weak activity with an IC_{50} value of 8.40 μM against the amastigotes of *Leishmania donovani*. Amphotericin B was used as the positive control with an IC_{50} value between 0.07 and 0.12 μM.

Palmarumycin CP17, palmarumycin CP18 (news), preussomerin EG1, palmarumycin CP2—Two new members of the palmarumycin family with potent antileishmanial activity were isolated from an endophytic strain of *Edenia* sp. from a mature leaf of *Petrea volubilis* (Verbenaceae). This fungus was cultivated in malt extract media for 15 days before the mycelium extraction. Palmarumycin CP17 (**133**) and CP18 (**134**) exhibited an IC_{50} of 1.34 and 0.62 μM, respectively, against the amastigotes of *Leishmania donovani* (Martínez-Luis et al., 2008). In addition, two already known compounds isolated from *Edenia* sp. exhibited potent antileishmanial activity, being preussomerin EG1 (**139**) the most active (IC_{50} = 0.12 μM), followed by palmarumycin CP2 (**135**, IC_{50} = 3.93 μM). Amphotericin B was used as the positive control with an IC_{50} value between 0.07 and 0.12 μM. Preussomerin EG1 was first described in *Edenia gomezpompae*, an endophytic fungus isolated from *Callicarpa acuminata* (Verbenaceae) (Macías-Rubalcava et al., 2008) while palmarumycin CP2 was first obtained from *Coniothyrium* sp. isolated from the forest soil in West Borneo (Indonesia), being also the first described member of the palmarumycin family (Krohn et al., 1994).

Purpureone—The new ergochrome compound was isolated for the first time from the endophytic fungus *Purpureocillium lilacinum* from the roots of the medicinal plant *Rauvolfia macrophylla* (Apocynaceae). This endophyte was cultivated in a solid rice medium for 3 weeks at room temperature. Purpureone (**140**) possesses a potent activity against the amastigotes of *Leishmania donovani* with an IC_{50} value of 0.87 μM, showing also a good selectivity index (SI = 49.5) while control drug miltefosine showed an IC_{50} value of 0.35 μM. Additionally, it exhibited good antimicrobial activity, signaling the promising future investigations for drug development (Lenta et al., 2016).

Fumiquinone B—This compound was first isolated from the culture of a soil strain of *Aspergillus fumigatus* and showed a nematicidal activity against *Bursaphelenchus xylophilus* (pine wood nematode) (Hayashi et al., 2007). Later, fumiquinone B (**125**) was also isolated from *Aspergillus* sp. F1544, a fungal endophyte isolated from a mature leaf of *Guapira standleyana* (Nyctaginaceae) in Panama. *Aspergillus* sp. F1544 was cultured in potato dextrose agar and in Czapek Dox medium at 30°C for 15 days at 150 rpm for whole culture extraction. Fumiquinone B exhibited a strong activity against the amastigotes of *Leishmania donovani* with an IC_{50} value of 0.5 μM and a moderate activity against *Plasmodium falciparum* with an IC_{50} value of 5.4 μM. Amphotericin B was used as the positive control with an IC_{50} value between 0.07 and 0.12 μM. Additionally, it presented a moderate cytotoxicity against the human breast cancer cell MCF-7 with an IC_{50} value of 23.9 μM (Martiñez-Luis et al., 2012).

Polyketide-alkaloids

14-norpseurotin A, FD-838, pseurotin A, pseurotin D—Along with the isolation of fumiquinone B from endophytic fungus *Aspergillus* sp. F1544, four other known compounds

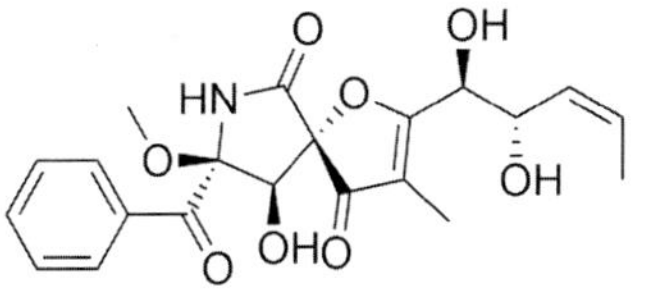

FIGURE 12.9 Chemical structures of 14-norpseurotin A (**142**), FD-838 (**144**), pseurotin A (**145**), and pseurotin D (**146**).

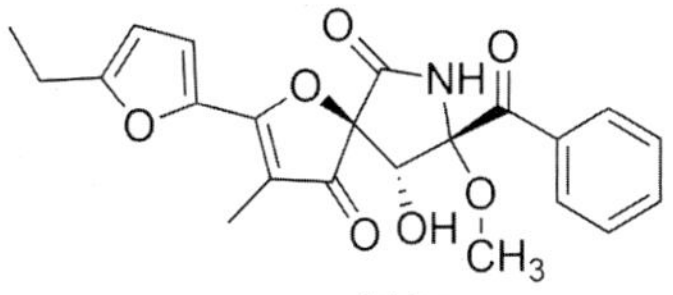

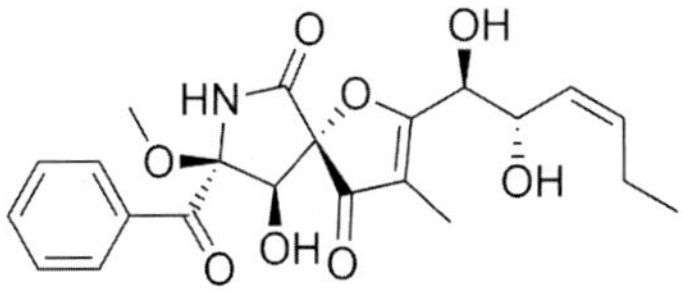

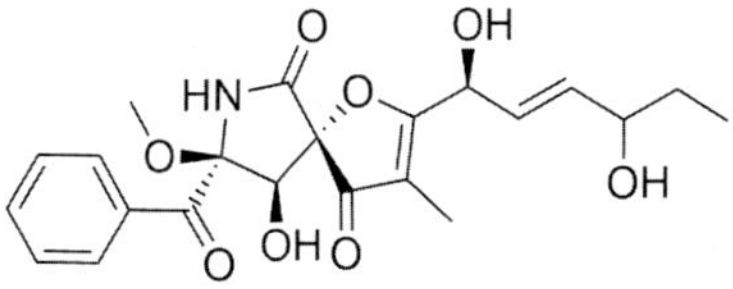

were also isolated, 14-norpseurotin A (**142**), FD-838 (**144**), pseurotin A (**145**), and pseurotin D (**146**) (Martińez-Luis et al., 2012) (Fig. 12.9). 14-norpseurotin was first isolated from a marine-derived strain of *Aspergillus sydowii* as a new oxaspirol[4,4]lactam exhibiting a significant antimicrobial activity (Zhang et al., 2008). Pseurotin A was first isolated from *Pseudeurotium ovalis* as a minor metabolite from the culture of this fungus (Bloch, Tamm, Bollinger, Petcher, & Weber, 1976) while pseurotin D was isolated 5 years later from the same strain along with other compounds from the same family (pseurotin B, C, D, and E) which possess a rare substituted skeleton found in natural product chemistry (Breitenstein, Chexal, Mohr, & Tamm, 1981). When 14-norpseurotin A, pseurotin A, and pseurotin D were tested against the amastigotes of *Leishmania donovani*, they exhibited an IC_{50} value of 4.4, 5.8, and 0.5 μM, respectively. However, compound FD-838 (patent application EP0216607) exhibited the highest antileishmanial activity with an IC_{50} value of 0.2 μM (Martińez-Luis et al., 2012). Amphotericin B was the positive control of the assay exhibiting IC_{50} values between 0.07 and 0.12 μM.

Terpenoids

Integracide H, integracide J—Two new tetracyclic triterpenoids with antileishmanial activity were isolated from *Fusarium* sp. isolated from the roots of *Mentha longifolia* (Lamiaceae). The fungal endophyte was cultivated in a rice solid medium for 30 days at 27°C before extraction. Integracides H (**156**) and J (**157**) exhibited moderate activity against the amastigotes of *Leishmania donovani* with an IC_{50} value of 4.75 and 3.29 μM, respectively (Ibrahim et al., 2016) (Fig. 12.10). Pentamidine was used as the positive control with an IC_{50} value of 6.35 μM. Moreover, integracide H presented a high cytotoxicity against the human cancer cell lines BT-549, SKOV, and KB with IC_{50} values of 1.82, 1.32, and 0.18 μM, respectively; while integracide H exhibited IC_{50} values of 2.46, 3.01, and 2.54 μM, respectively (Ibrahim et al., 2016).

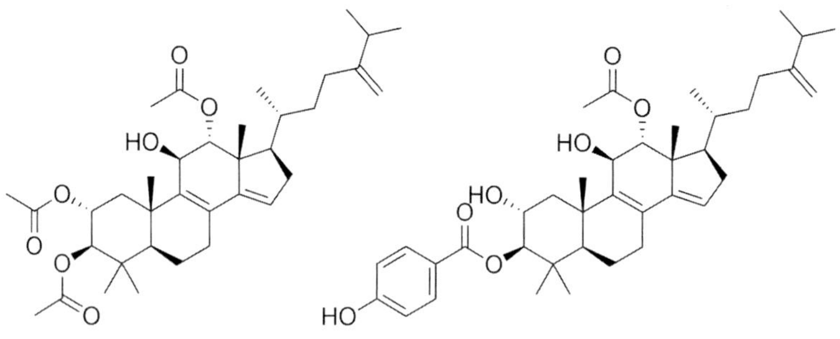

FIGURE 12.10 Chemical structures of integracide H (**156**) and integracide J (**157**).

Antiplasmodial

The following compounds isolated from endophytic fungi have been tested against different strains of *Plasmodium falciparum*: 3D7 (drug sensitive), D6 (chloroquine sensitive), W2 (chloroquine resistant), NF54 (chloroquine resistant), and K1 (multidrug resistant).

Among the most active compounds against *P. falciparum* 3D7 are cytochalasins H, J and O, 12,13-deoxyroridin E and roridin E with IC_{50} values <0.02 μM, followed by cytochalasin D (IC_{50} 0.0258 μM), 2,5-dihydroxy-1-(hydroxymethyl)pyridin-4-one (IC_{50} 0.127 μM), 7-hydroxy-3,4,5-trimethyl-6-on 2,3,4,6-tetrahydroisoquinoline-8-carboxylic acid (IC_{50} 0.129 μM), 19,20-epoxycytochalasin D (IC_{50} 0.136 μM), and dicerandrol D (IC_{50} 0.6 μM). Compounds with potent activity against *P. falciparum* D6 and W2 are 19,20-epoxycytochalasin D with an IC_{50} value of 0.04 μM followed by 19,20-epoxycytochalasin C (IC_{50} 0.05–0.07 μM), epoxycytochalasin H (IC_{50} 0.07–0.1 μM) and 18-deoxy-19,20-epoxycytochalasin C (IC_{50} 0.19–0.56 μM) while fusaripeptide A was tested only against *P. falciparum* D6 and it exhibited an IC_{50} value of 0.34 μM. Three other compounds presented potent activity against *P. falciparum* K1 being phomoxanthone A with an IC_{50} value of 0.15 μM followed by phomoxanthone B (IC_{50} 0.44 μM) and monocerin (IC_{50} 0.68 μM). Additionally, apicidins B and C were tested against *P. falciparum* (strain not specified) exhibiting IC_{50} values of 0.189 and 0.069 μM, respectively.

Compounds presenting a moderate activity in terms of IC_{50} were tested against *P. falciparum* K1, a multidrug-resistant strain. Chaetoxanthone B was the most active, with an IC_{50} value of 1.41 μM followed by 2-chloro-5-methoxy-3-methylcyclohexa-2,5-diene-1,4-dione (IC_{50} 1.84 μM), phomoarcherin B (IC_{50} 2.05 μM), 12,12a-dihydro antibiotic PI 016 (IC_{50} 2.95 μM), mollicellin K (IC_{50} 1.84 μM), pullularin B (IC_{50} 4.18 μM), and pullularin A (IC_{50} 4.64 μM). Codinaeopsin was tested against *P. falciparum* 3D7 and exhibited a moderate activity with an IC_{50} value of 4.7 μM.

Finally, those exhibiting a weaker activity in terms of IC_{50} were also mainly tested against *P. falciparum* K1. Xylariaquinone A presented an IC_{50} value of 6.68 μM followed by mollicellin M (IC_{50} 6.95 μM), mollicellin E (IC_{50} 7.16 μM), 11-hydroxymonocerin (IC_{50}

7.7 μM), 7α,10α-dihydroxy-1βmethoxyeremophil-11(13)-en-12,8β olide (IC_{50} 8.1 μM), 7-butyl-6,8-dihydroxy-3(R)-pentylisochroman-1-one (IC_{50} 8.48 μM), mollicellin L (IC_{50} 8.6 μM), and KS-501a (IC_{50} 9.9 μM). Asterric acid, preussiafuran A, and cissetin were tested against *P. falciparum* NF54 exhibiting weak activities with IC_{50} values of 8.67, 8.76, and 10.3 μM, respectively.

Alkaloids

18-deoxy-19,20-epoxycytochalasin C; 19,20-epoxycytochalasin C; 19,20-epoxycytochalasin D—Three known cytochalasins were obtained from the endophytic fungus *Nemania* sp. isolated from the diseased leaves of *Torreya taxifolia* Arnott (Taxaceae) (Kumarihamy et al., 2019). Compounds 19,20-epoxycytochalasin C (**103**) and D (**104**) were first isolated from the fungus *Xylaria hypoxylon* and by this time, there were many biological activities, such as antibiotic and antitumor activities, which were already reported for this family of compounds (Espada, Rivera-Sagredo, de la Fuente, Hueso-Rodríguez, & Elson, 1997) (Fig. 12.11). Moreover, 18-deoxy-19,20-epoxycytochalasin C was first isolated from a nonidentified fungus (KL-1.1) isolated from the leaves of the medicinal plant *Psidium guajava* (Myrtaceae) (Okoye, Nworu, Debbab, Esimone, & Proksch, 2015).

For the isolation of secondary metabolites, *Nemania* sp. was cultivated in potato dextrose broth at 27°C for 30 days at 100 rpm. These three known cytochalasins exhibited potent in vitro activity against two strains of *Plasmodium falciparum* (D6 and W2). Compound 19,20-epoxycytochalasin C (**103**) was isolated as the major compound and it exhibited IC_{50} values of 0.07 and 0.05 μM against the D6 and W2 strains, respectively. Compound 18-deoxy-19,20-epoxycytochalasin C (**102**) exhibited IC_{50} values of 0.56 and

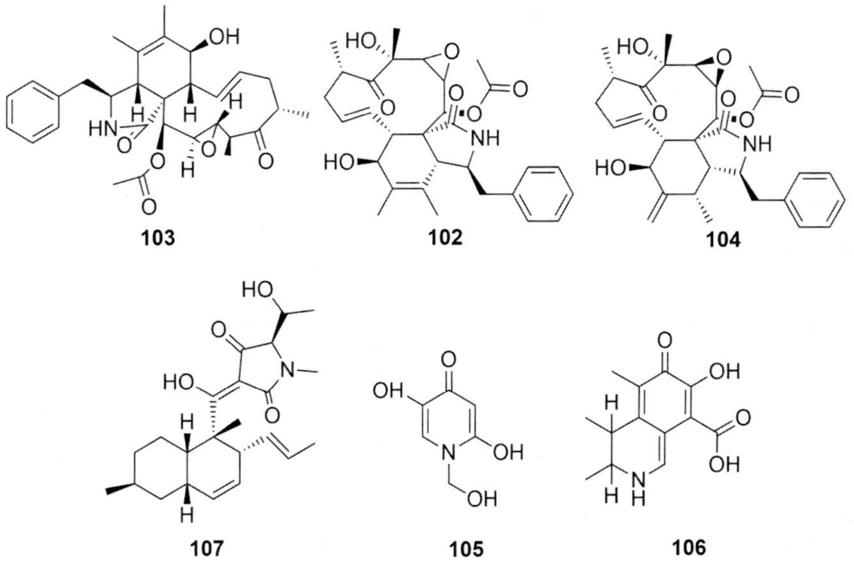

FIGURE 12.11 Chemical structures of 18-deoxy-19,20-epoxycytochalasin C (**102**), 19,20-epoxycytochalasin C (**103**), 19,20-epoxycytochalasin D (**104**), cissetin (**107**), 2,5-dihydroxy-1-(hydroxymethyl)pyridin-4-one (105), and 7-hydroxy-3,4,5-trimethyl-6-on-2,3,4,6-tetrahydroisoquinoline-8-carboxylic acid (**106**).

0.19 μM against the D6 and W2 strains, respectively, while compound 19,20-epoxycytochalasin D (**104**) exhibited an IC_{50} value of 0.04 μM for both *P. falciparum* strains (Kumarihamy et al., 2019). Positive standard controls used were chloroquine (IC_{50} 0.03 μM) and artemisinin (IC_{50} 0.02 μM). Similarly, when the compound 19,20-epoxycytochalasin D was isolated from endophytic fungus *Diaporthe* sp., it exhibited an IC_{50} value of 0.136 μM against *P. falciparum* 3D7 (Calcul et al., 2013).

Additionally, all three compounds presented moderate toxicity against different tumor cell lines, 19,20-epoxycytochalasin C exhibited an IC_{50} value of 8.02 μM against SK-MEL (malignant melanoma); 19,20-epoxycytochalasin D exhibited IC_{50} values of 7.84 and 8.4 μM against BT-549 (breast ductal carcinoma) and LLC-PK_{11} (kidney epithelial cells), respectively. However, 18-Deoxy-19,20-epoxycytochalasin C exhibited an IC_{50} value of 6.89 μM against BT-549. None of these compounds was toxic against Vero Cells (kidney fibroblast), therefore showing their selectivity to *Plasmodium falciparum* strains (Kumarihamy et al., 2019).

Cissetin—Cissetin (**107**) was first isolated from a nonidentified fungus (OSI 50185) isolated from a decomposed plant in Peru, showing antibacterial activity against penicillin-resistant *Streptococcus pneumoniae* (MIC 2 μg/mL) (Boros, Dix, Katz, Vasina, & Pearce, 2003). This compound was later isolated from endophytic fungus *Preussia* sp. isolated from the leaves and barks of the plant *Enantia chlorantha* (Annonaceae) (Talontsi et al., 2014). For the isolation of compounds, *Preussia* sp. was cultivated in a rice medium at 25°C for 30 days under static conditions. Cissetin exhibited a weak activity against *Plasmodium falciparum* (NF54) with an IC_{50} value of 10.3 μM and a weak cytotoxicity against L6 cells (rat skeletal myoblast) with an IC_{50} value of 42.1 μM.

2,5-dihydroxy-1-(hydroxymethyl)pyridin-4-one; 7-hydroxy-3,4,5-trimethyl-6-on 2,3,4,6-tetrahydroisoquinoline-8-carboxylic acid—Two new alkaloids were isolated from a nonidentified endophytic fungus (strain BB4) isolated from the stem and leaves of *Tinospora crispa* (Menispermaceae), a plant traditionally used for malaria treatment. This endophytic fungal strain was cultivated in potato dextrose broth for 3 weeks for further extraction. Compounds 2,5-dihydroxy-1-(hydroxymethyl)pyridin-4-one (**105**) and 7-hydroxy-3,4,5-trimethyl-6-on-2,3,4,6-tetrahydroisoquinoline-8-carboxylic acid (**106**) exhibited potent activity against *Plasmodium falciparum* (3D7) with IC_{50} values of 0.127 and 0.129 μM, respectively (Elfita et al., 2011). Positive standard chloroquine presented an IC_{50} value of 0.02 μM.

Cytochalasin D, cytochalasin H, cytochalasin J, cytochalasin O—Cytochalasin D (**108**) was first isolated from endophytic fungus, *Xylaria* sp. from the plant *Palicourea marcgravii* (Rubiaceae), showing antifungal activity (Cafêu et al., 2005). This compound was able to reduce the attachment of the promastigotes of *Leishmania* by the destabilization of the actin cytoskeleton of macrophages, accompanied by a reduction of the intracellular amastigote load (Roy, Kumar, Jafurulla, Mandal, & Chattopadhyay, 1838). Cytochalasin H (**109**) and J (**110**) were previously isolated from *Phomopsis* sp. (Izawa, Hirose, Shimizunée Tomioka, Koyama, & Natori, 1989; Wells, Cutler, & Cole, 1976) while cytochalasin O (**111**) was isolated from the fungus *Hypoxylon terricola* (Edwards, Maitland, & Whalley, 1989) (Fig. 12.12). As part of a bigger study, endophytic fungi *Diaporthe* sp., *Xylaria* sp., and *Verticillium* sp. were isolated from the barks and leaves of three different plants: *Kandelia obovata* (Rhizophoraceae), *Avicennia marina* (Acanthaceae), and *Lumnitzera racemosa* (Combretaceae). These fungi were cultivated in a liquid medium

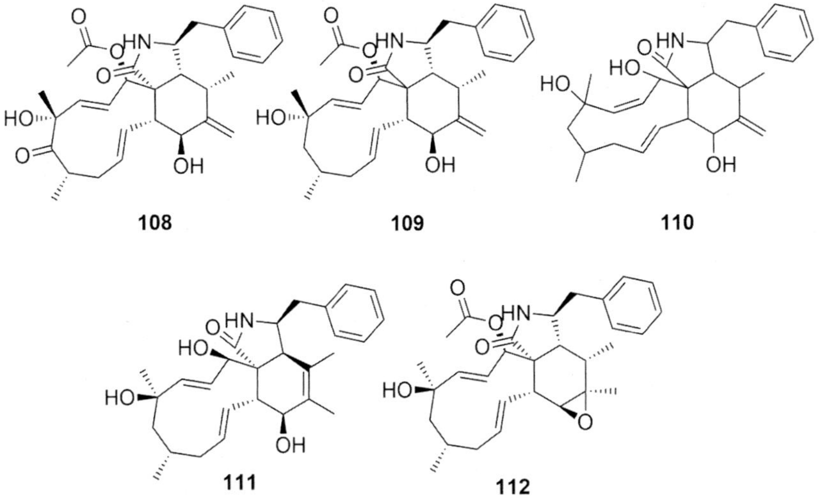

FIGURE 12.12 Chemical structures of cytochalasin D (**108**), cytochalasin H (**109**), cytochalasin J (**110**), cytochalasin O (**111**), and epoxycytochalasin H (**112**).

containing 1% wt./vol. glucose, 0.1% wt./vol. yeast extract, and 0.2% wt./vol. peptone during 3 weeks for isolation of secondary metabolites. Cytochalasin D exhibited potent activity against *Plasmodium falciparum* (3D7) with an IC_{50} value of 0.0258 μM while cytochalasins H, J, and O presented IC_{50} values of <0.02 μM (Calcul et al., 2013). In this study, the positive standard used was chloroquine presenting an IC_{50} value of 0.0045 μM and dihydroartemisinin with an IC_{50} value of 0.0003 μM.

Epoxycytochalasin H—This compound (**112**) was first isolated from *Phomopsis sojae* isolated from soybean seeds (Cole et al., 1982), and later it was also obtained from the endophytic fungus *Diaporthe miriciae* isolated from the plant *Vellozia gigantea* (Velloziaceae) showing potent antiplasmodial activity. Endophytic fungus *Diaporthe miriciae* was cultured in potato dextrose agar at 25°C for 15 days for compound extraction. Epoxycytochalasin H exhibited an IC_{50} value of 0.1 and 0.07 μM against D6 and W2 strains of *Plasmodium falciparum*, respectively, while no toxicity was detected against Vero cells (mammalian kidney) (Ferreira et al., 2017). Chloroquine and artemisinin were used as positive standards for both plasmodial strains.

Polyketides

11-hydroxymonocerin, monocerin—Monocerin (**132**) and a new analog of monocerin (**113**) were isolated from endophytic fungus *Exserohilum rostratum* isolated from the leaves and roots of *Stemona* sp. (Stemonaceae) (Sappapan et al., 2008) (Fig. 12.13). Monocerin was first isolated from *Helminthosporium monoceras* along with other related benzopyrans (Aldridge & Turner, 1970). Fungus *Exserohilum rostratum* was cultivated in yeast sucrose extract at 31°C during 21 days under static conditions for supernatant extraction. Compound 11-hydroxymonocerin was weakly active against *Plasmodium falciparum* K1 with an IC_{50} value of 7.7 μM while monocerin exhibited a potent antiplasmodial activity with an IC_{50} value of 0.68 μM. According to the

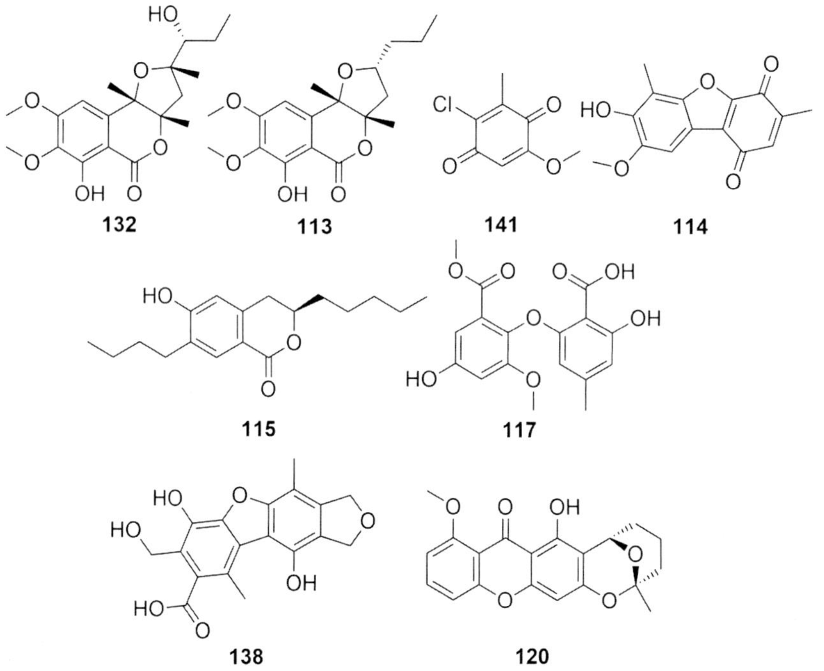

FIGURE 12.13 Chemical structures of 11-hydroxymonocerin (**113**), monocerin (**132**), 2-chloro-5-methoxy-3-methylcyclohexa-2,5-diene-1,4-dione (**114**), xylariaquinone A (**141**), dihydroisocoumarin (**115**), asterric acid (**117**), preussiafuran A (**138**), and chaetoxanthone B (**120**).

authors, the presence of the additional −OH group in 11-hydroxymonocerin caused the lower activity compared to monocerin. Dihydroartemisinin was used as the positive standard with an IC_{50} value of 0.004 μM. Additionally, the compounds showed no cytotoxicity against five human tumor cell lines BT474 (breast carcinoma), CHAGO (lung carcinoma), Hepg2 (hepatocarcinoma), KATO-3 (gastric carcinoma), and SW-620 (colon carcinoma) (Sappapan et al., 2008).

2-chloro-5-methoxy-3-methylcyclohexa-2,5-diene-1,4-dione, xylariaquinone—A two novel benzoquinones with antiplasmodial activity were isolated from fungal endophyte *Xylaria* sp. isolated from the leaves of *Sandoricum koetjape* (Meliaceae) (Tansuwan et al., 2007). Fungus was cultivated in malt extract broth at 30°C for 5 weeks under static conditions for supernatant extraction. Xylariaquinone A (**141**) was weakly active against *Plasmodium falciparum* K1 with an IC_{50} value of 6.68 μM while 2-chloro-5-methoxy-3-methylcyclohexa-2,5-diene-1,4-dione (**114**) exhibited a moderate antiplasmodial activity with an IC_{50} value of 1.84 μM and cytotoxic activity against Vero cells with an IC_{50} value of 1.35 μM. Positive standard dihydroartemisinin exhibited an IC_{50} value of 0.0033 μM.

7-butyl-6,8-dihydroxy-3(R)-pentylisochroman-1-one—A novel dihydroisocoumarin (**115**) was isolated from fungal endophyte *Geotrichum* sp. isolated from the healthy stems of *Crassocephalum crepidioides* (Asteraceae). The fungus was cultivated in Czapek broth at 25°C for 21 days for supernatant extraction. The 7-butyl-6,8-dihydroxy-3(R)-pentylisochroman-1-one was weakly active against *Plasmodium falciparum* K1 with an IC_{50} value of 8.48 μM

(Kongsaeree et al., 2003). The standard compound chloroquine presented an IC_{50} value of 0.31 μM.

Asterric acid, preussiafuran A Asterric acid (**117**) and a novel dibenzofuran (**138**) were isolated from fungal endophyte *Preussia* sp. isolated from the healthy leaves and barks of *Enantia chlorantha* (Annonaceae) (Talontsi et al., 2014). Asterric acid was first isolated from *Aspergillus terreus* (Curtis, Hassall, Jones, & Williams, 1960). The endophytic strain of *Preussia* sp. was cultivated in a rice medium at 25°C for 30 days under static conditions. Asterric acid and preussiafuran A were weakly active against *Plasmodium falciparum* NF54 and they exhibited IC_{50} values of 8.67 and 8.76 μM, respectively. Chloroquine was used as the positive standard. Moreover, asterric acid and preussiafuran A presented moderate cytotoxicity against L6 cell lines (rat skeletal myoblasts) with IC_{50} values of 14.8 and 36.7 μM, respectively.

Chaetoxanthone B—A new compound (**120**) was isolated from the marine-derived fungus *Chaetomium* sp. isolated from an algal species (taxonomy not determined) (Pontius et al., 2008). This fungus was cultivated in malt extract yeast agar at room temperature for 15 days for compound extraction. Chaetoxanthone B exhibited moderate activity against *Plasmodium falciparum* K1 with an IC_{50} value of 1.41 μM while it showed no cytotoxicity against L6 cells (rat skeletal myoblasts). Chloroquine was used as the positive standard.

Phomoxanthone A, phomoxanthone B—Two novel xanthone dimers were isolated from the fungal endophyte *Phomopsis* sp. isolated from the leaves of *Tectona crispa* (Lamiaceae). This fungus was cultivated in bacto-malt extract broth at 22°C for 20 days for mycelium extraction. Phomoxanthone A (**136**) and B (**137**) exhibited potent activity against *Plasmodium falciparum* K1 with IC_{50} values of 0.15 and 0.44 μM, respectively. However, they presented elevated cytotoxicity against three cancer cell lines, KB cells, BC-1 cells, and Vero cells (Isaka et al., 2001). Chloroquine diphosphate and artemisinin were used as positive standards.

Dicerandrol D—A new dimeric tetrahydroxanthone was isolated from the fungal endophyte *Diaporthe* sp. along with other known cytochalasins previously mentioned (Calcul et al., 2013). Dicerandrol D (**124**) exhibited a potent activity against *Plasmodium falciparum* 3D7 with an IC_{50} value of 0.6 μM and a low cytotoxicity against A549 cells (adenocarcinomic human alveolar epithelial cells) with an IC_{50} value of 7.8 μM (Fig. 12.14).

KS-501a—KS-501a (**127**) was first isolated as a potent inhibitor of Ca^{2+} from *Sporothrix* sp. from a fallen leaf in Japan (Nakanishi, Ando, Kawamoto, & Kase, 1989). Later, it was also obtained from the culture of *Acremonium* sp. isolated from a palm leaf in Thailand. Here, *Acremonium* sp. was cultivated in bacto-malt extract broth at 25°C for 32 days at 200 rpm for the isolation of compounds from the supernatant. This compound exhibited a weak activity against *Plasmodium falciparum* K1 with an IC_{50} value of 9.9 μM while its cytotoxic activity possessed an IC_{50} value of 8.8 μM against BC (human breast cancer) cell lines (Bunyapaiboonsri et al., 2008). Dihydroartemisinin was used as the positive standard with an IC_{50} value of 0.004 μM.

Mollicellin E, mollicellin K, mollicellin L, mollicellin M—Three novel depsidones (mollicellin K, L, and M, **129**, **130**, **131**) and one known (mollicellin E, **128**) were isolated from endophytic fungus *Chaetomium brasiliense*, isolated from a leaf collected in the Hala-Bala evergreen forest (Thailand) (Khumkomkhet et al., 2009) (Fig. 12.15). Mollicellin E was first isolated from *Chaetomium mollicellum* and its mutagenic and bactericidal activity against *Salmonella typhimurium* was reported (Stark et al., 1978). The strain of *C. brasiliense* was cultured in potato dextrose broth between 25°C and 28°C for 4 weeks under static conditions

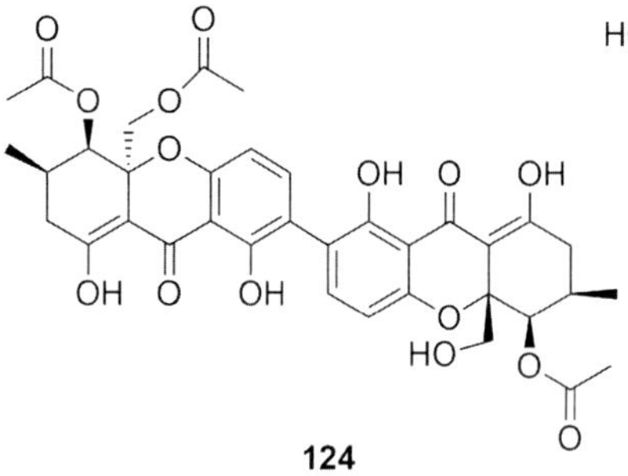

FIGURE 12.14 Chemical structures of dicerandrol D (**124**) and KS-501a (**127**).

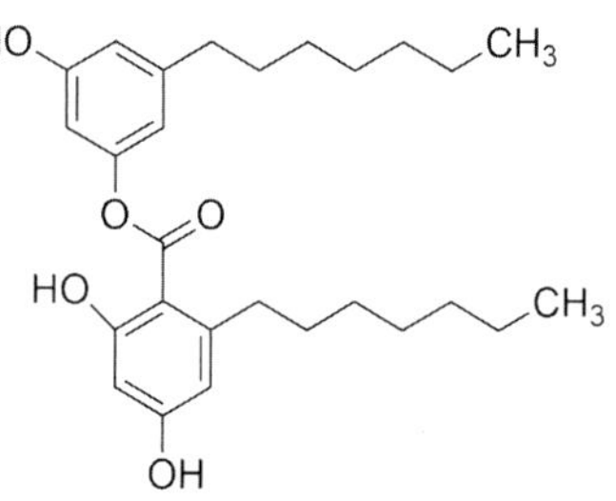

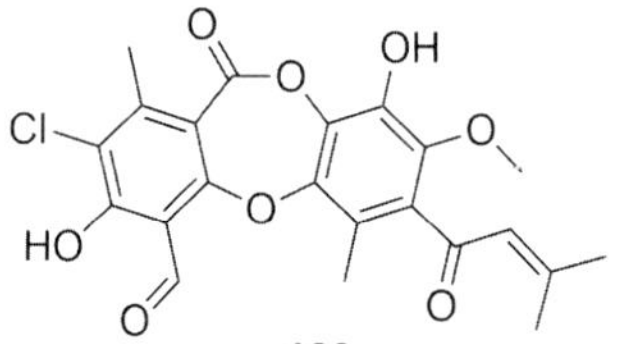

FIGURE 12.15 Chemical structures of mollicellin E (**128**), mollicellin K (**129**), mollicellin L (130), and mollicellin M (**131**).

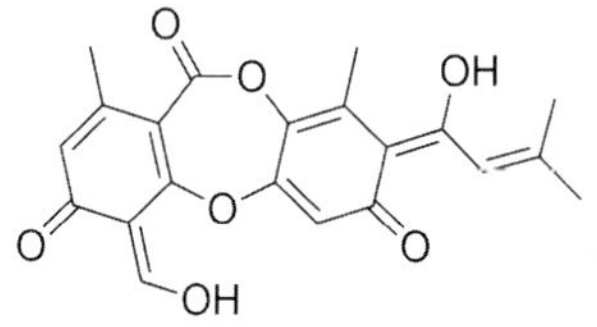

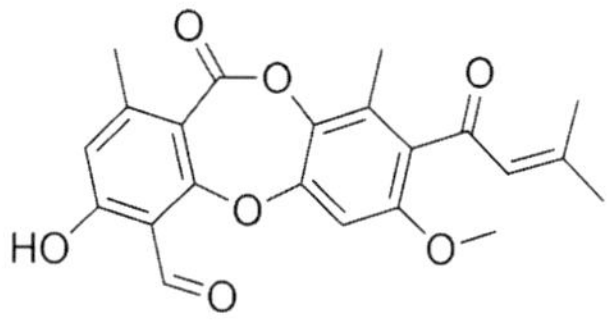

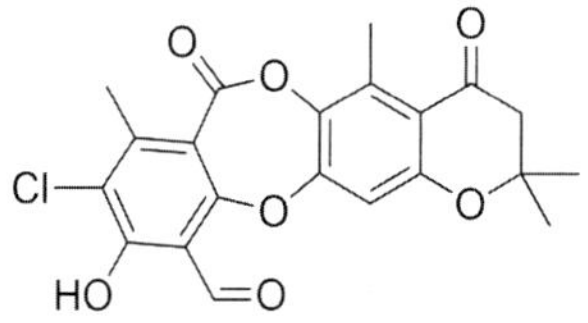

for mycelium extraction. Mollicellin E, L, and M presented a weak activity against *Plasmodium falciparum* K1, exhibiting IC_{50} values of 7.16, 8.6, and 6.95 μM, respectively. Mollicellin K presented a moderate antiplasmodial activity with an IC_{50} value of 3.13 μM and cytotoxicity against KB cells (human epidermoid carcinoma of the mouth) and NCI-H187 (human small cell lung cancer) with IC_{50} values of 1.9 and 0.35 μM, respectively (Khumkomkhet et al., 2009). The standard compound used was artemisinin. Additionally, the four compounds exhibited significant cytotoxicity against five cholangiocarcinoma cell lines (KKU-100, KKU-M139, KKU-M156, KKU-M213, and KKUM214).

Polyketide-alkaloid

Codinaeopsin—A novel tryptophan-polyketide hybrid (**143**) was isolated from fungal endophyte *Codinaeopsis gonytrichoidesi* isolated from the tree *Vochysia guatemalensis* (Vochysiaceae). This fungus was cultivated in a rich seed medium at 25°C for 21 days for compound extraction. Codinaeopsin exhibited moderate activity against *Plasmodium falciparum* 3D7 with an IC_{50} value of 4.7 μM (Kontnik & Clardy, 2008) (Fig. 12.16).

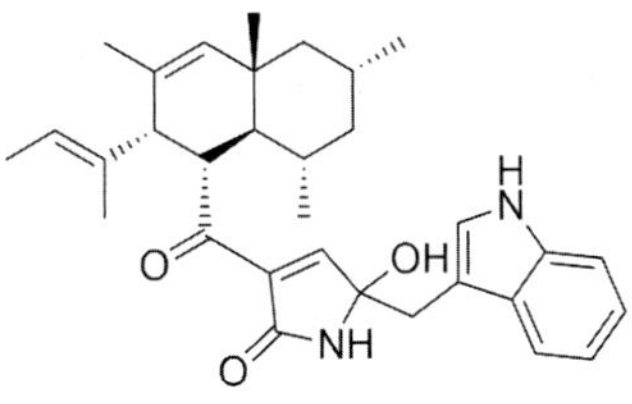

FIGURE 12.16 Chemical structure of codinaeopsin (**143**).

Polypeptides

12,12a-dihydroantibiotic PI 016—A novel compound (**147**) with antimalarial activity was isolated from fungal endophyte *Menisporopsis theobromae* isolated from the seeds of a plant (not specified) in Thailand (Fig. 12.17). This fungus was grown in potato dextrose broth at 25°C for 24 days at 200 rpm for whole culture extraction. The compound 12,12a-dihydroantibiotic PI 016 exhibited a moderate activity against *Plasmodium falciparum* K1 with an IC_{50} value of 2.95 μM (Chinworrungsee et al., 2006). Moreover, this compound presented cytotoxicity against NCI-H187 cell lines (small cell lung cancer) with an IC_{50} value of 20.3 μM. Dihydroartemisinin was used as the positive standard with an IC_{50} value between 0.004 and 0.014 μM.

Apicidin B, apicidin C—Two novel cyclic tetrapeptides with antiplasmodial activity were isolated from fungal endophyte *Fusarium pallidoroseum* isolated from the branches of *Acacia* sp. (Fabaceae). Apicidin B (**148**) and C (**149**) exhibited significant activity against *Plasmodium falciparum* with MIC values of 0.189 and 0.069 μM, respectively (Singh et al., 2001).

Fusaripeptide A—A novel cyclodepsipeptide was isolated from endophytic fungus *Fusarium* sp. isolated from the roots of *Mentha longifolia* (Lamiaceae). Fungus was cultivated in rice solid medium at room temperature for 30 days. Fusaripeptide A (**151**) exhibited a potent activity against *Plasmodium falciparum* (D6) with an IC_{50} value of 0.34 μM (Ibrahim, Abdallah, et al., 2018). Additionally, this compound presented significant cytotoxicity against L5178Y (mouse lymphoma) and PC12 (rat brain cancer) cell lines with IC_{50} values of 5.71 and 9.55 μM, respectively. Fusaripeptide A also presented high antifungal activity against three different species of *Candida* and one strain of *Aspergillus fumigatus*. Positive control artemisinin presented an IC_{50} value of 0.57 μM.

Pullularin A, pullularin B—Two new cyclohexadepsipeptides with antiplasmodial activity were isolated from endophytic fungus *Pullularia* sp. (yeast-like) isolated from the leaves of *Calophyllum* sp. (Clusiaceae). This endophyte was cultivated in potato dextrose broth at 25°C for 19 days at static conditions for supernatant extraction. Pullularins A (**152**) and B (**153**) exhibited moderate activity against *Plasmodium falciparum* K1 with IC_{50} values of 4.64 and 4.18 μM, respectively (Isaka et al., 2007). Pullularin A presented cytotoxic activity against Vero cells with an IC_{50} value of 36 μM. Dihydroartemisinin was used as the positive standard.

Terpenoids

7α,10α-Dihydroxy-1βmethoxyeremophil-11(13)-en-12,8β olide—A novel eremophilanolide sesquiterpenoid was isolated from *Xylaria* sp. isolated from the palm *Licuala spinosa* (Arecaceae). This fungus was cultivated in potato dextrose broth at 25°C for 30 days at 200 rpm. The compound (**155**) exhibited weak activity against *Plasmodium falciparum* K1

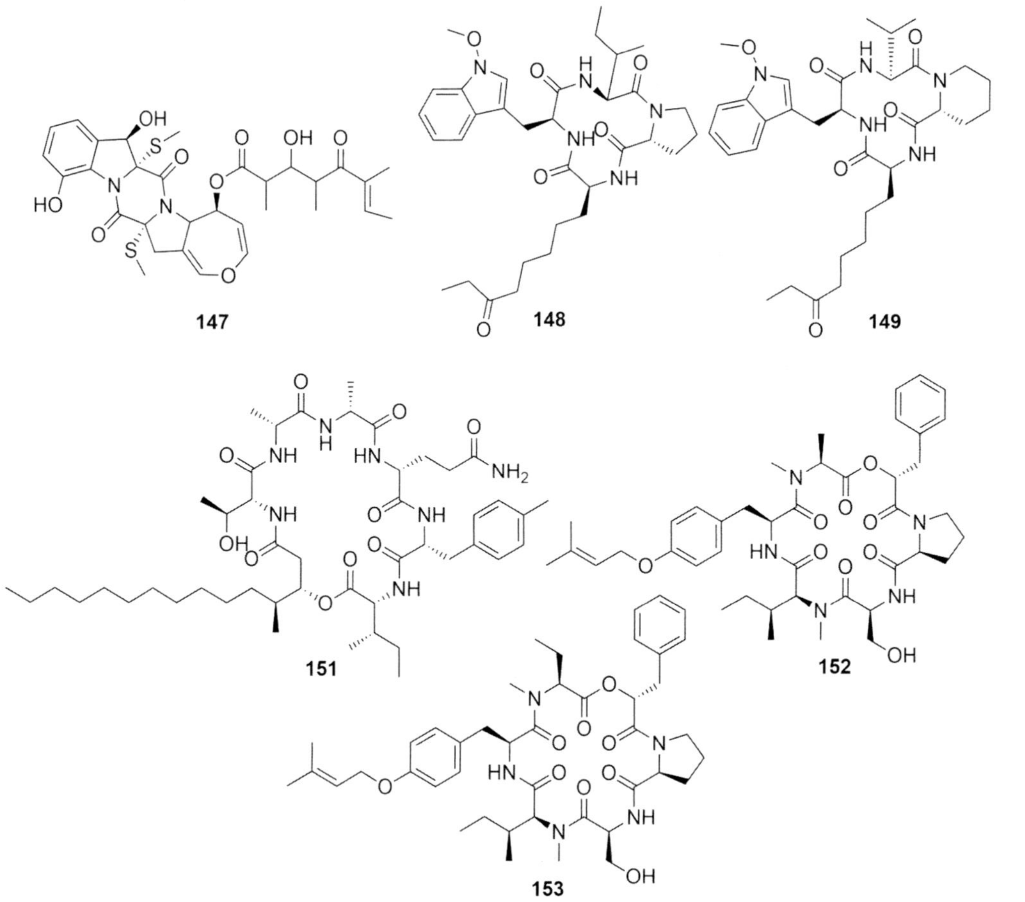

FIGURE 12.17 Chemical structures of 12,12a-dihydroantibiotic PI 016 (**147**), apicidin B (**148**), apicidin C (**149**), fusaripeptide A (**151**), pullularin A (**152**), and pullularin B (**153**).

with an IC_{50} value of 8.1 μM (Isaka et al., 2010) (Fig. 12.18). Moreover, it presented cytotoxic activity against KB (oral human epidermoid carcinoma), MCF-7 (human breast cancer), NCI-H187 (human small cell lung cancer), and Vero cells (African green monkey kidney fibroblasts) with IC_{50} values of 21, 15, 7.2, and 8.5 μM, respectively. Dihydroartemisinin was used as the positive standard with an IC_{50} value of 0.004 μM.

Phomoarcherin B—A new sesquiterpene was isolated from fungal endophyte *Phomopsis archeri* isolated from the cortex stem of *Vanilla albida* (Orchidaceae). The fungus was cultivated in potato dextrose broth at 25°C–28°C for 4 weeks for mycelium extraction. Phomoarcherin B (**158**) exhibited moderate activity against *Plasmodium falciparum* K1 with an IC_{50} value of 2.05 μM (Hemtasin et al., 2011). Additionally, it presented low cytotoxicity against the KB cell line with an IC_{50} value of 9.4 μM. However, it possessed significant toxicity against KKU-M139 and KKU-M156 (cholangiocarcinoma cell lines) with IC_{50} values

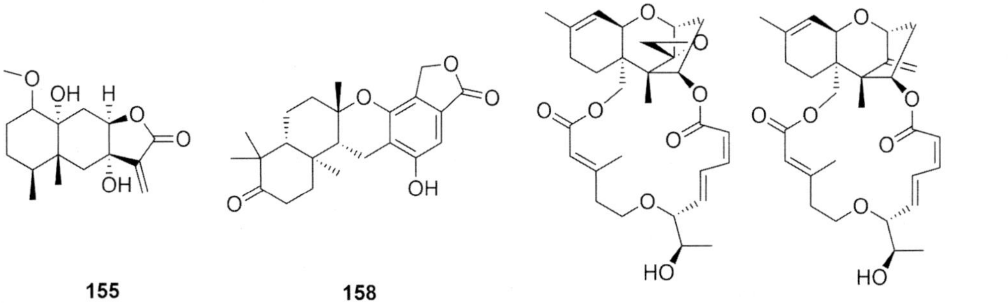

FIGURE 12.18 Chemical structures of 7α,10α-dihydroxy-1βmethoxyeremophil-11(13)-en-12,8β olide (**155**), phomoarcherin B (**158**), 12,13-deoxyroridin E (**154**), and roridin E (**159**).

of 0.1 and 2 μM, respectively. Dihydroartemisinin was used as the positive standard with an IC_{50} value of 0.015 μM.

12,13-deoxyroridin E, roridin E—Along with the cytochalasins D, H, J, and O and dicerandrol D previously mentioned, two already known terpenoids with potent antiplasmodial activity were isolated from a nonidentified strain (CY-3923) isolated from the barks and leaves of three different plants; *Kandelia obovata* (Rhizophoraceae), *Avicennia marina* (Acanthaceae), and *Lumnitzera racemosa* (Combretaceae) (Calcul et al., 2013). The compounds 12,13-deoxyroridin E (**154**) and roridin E (**159**) exhibited activity against *Plasmodium falciparum* 3D7, both with an IC_{50} value <0.02 μM. Moreover, they showed no cytotoxicity against A549 cell lines (adenocarcinomic human alveolar epithelial cells).

Antitrypanosomal/antiplasmodial/antileishmanial compounds

The most active compounds against *Trypanosoma cruzi* are cercosporin, exhibiting an IC_{50} value of 1.08 μM while its semisynthetic tetra-acetylated derivative presented an IC_{50} value of 0.78 μM. Cercosporin and its tetra-acetylated derivative also exhibited activity against *Leishmania donovani* with IC_{50} values of 0.46 and 0.64 μM, respectively, and activity against *Plasmodium falciparum* with IC_{50} values of 1.03 and 2.99 μM, respectively. Moreover, compound beauvericin exhibited a moderate activity against *T. cruzi* with an IC_{50} value of 2.43 μM followed by chaetoxanthone C with an IC_{50} value of 3.83 μM. Altenusin was tested in an assay with the recombinant enzyme trypanothione reductase (TR), considered as a drug target (Cota et al., 2008) and it exhibited an IC_{50} value of 4.3 μM .

Polyketides

Altenusin—Altenusin (**116**) was first isolated from *Alternaria tenuis* (Thomas, 1961) (Fig. 12.19). It was also obtained from endophytic fungus *Alternaria* sp. isolated from the leaves of *Trixis vauthieri* DC (Asteraceae). *Alternaria* sp. was cultivated in malt extract agar at 28°C for 9 days at 150 rpm. Altenusin exhibited a moderate activity against the recombinant enzyme TR from *Trypanosoma cruzi* with an IC_{50} value of 4.3 μM but this activity was not observed for the amastigotes of *Leishmania amazonensis* (Cota et al., 2008). It is suggested that altenusin might not be able to reach the intracellular compartments in

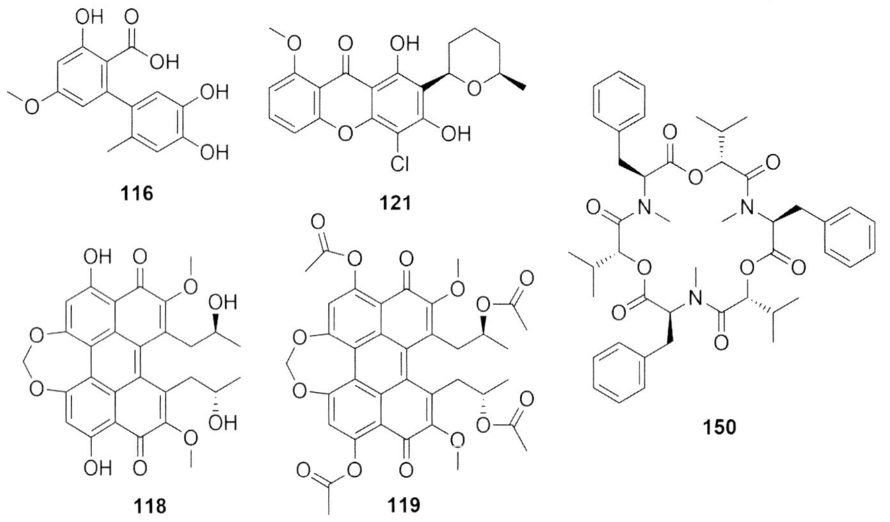

FIGURE 12.19 Chemical structures of altenusin (**116**), chaetoxanthione C (**121**), cercosporin (**118**), tetra-acetyl cercosporin derivative (**119**), and beauvericin (**150**).

Leishmania parasites in sufficient quantities to inhibit parasite survival. In this assay, clomipramine was used as a standard TR inhibitor at a concentration of 6 μg/mL.

Chaetoxanthone C—Chaetoxanthone C (**121**) was isolated from endophytic fungus *Chaetomium* sp. along with chaetoxanthone B, which exhibited a moderate antiplasmodial activity as it was previously mentioned (Pontius et al., 2008). Chaetoxanthone C also exhibited a moderate activity against *Trypanosoma cruzi* Tulahuen C4 with an IC_{50} value of 3.83 μM and a low cytotoxicity against L6 cells (rat skeletal myoblasts) with an IC_{50} value of 46.7 μM. Benznidazole was used as a standard compound showing an IC_{50} value of 1.15 μM.

Cercosporin, cercosporin derivative tetra-acetylated—Cercosporin (**118**) was first isolated as a pigment from the cultured mycelia of *Cercospora kikuchii* (Kuyama & Tamura, 1957). Later, it was also obtained from the endophytic fungus *Mycosphaerella* sp. isolated from *Psychotria horizontalis* (Rubiaceae). The fungus was cultivated in potato dextrose at 28°C for 15 days at 200 rpm for mycelium extraction. In this study, cercosporin was acetylated to obtain a tetra-acetylated derivative (**119**), both with potent antiparasitic activities. Cercosporin and its tetra-acetylated derivative exhibited activity against *Leishmania donovani* with IC_{50} values of 0.46 and 0.64 μM, respectively, against *Plasmodium falciparum* with IC_{50} values of 1.03 and 2.99 μM, respectively and activity against *Trypanosoma cruzi* with IC_{50} values of 1.08 and 0.78 μM, respectively (Moreno et al., 2011). The drugs used as positive controls were amphotericin (IC_{50} 0.08–0.13 μM) for *L. donovani*, chloroquine (IC_{50} 0.08–0.1 μM) for *P. falciparum*, and nifurtimox (IC_{50} 10–16 μM) for *T. cruzi*. Additionally, cercosporin and its tetra-acetylated derivative presented cytotoxic activities against MCF-7 (human breast cancer) with IC_{50} values of 4.68 and 3.56 μM, respectively, and against Vero cells (African green monkey kidney fibroblasts) with IC_{50} values of 1.54 and 1.24 μM, respectively.

Polypeptides

Beauvericin—The depsipeptide beauvericin (**150**) was first isolated from the fungus *Beauveria bassiana*, known to be a potent insect pathogen (Hamill, Higgens, Boaz, & Gorman, 1969). This compound was also isolated from the endophytic fungus *Fusarium* sp. isolated from the stems and barks of the endangered species *Caesalpinia echinata* Lam (Fabaceae) (Campos et al., 2015). This fungus was cultivated in potato dextrose agar at 28°C for 14 days for compound extraction. Beauvericin exhibited a moderate activity against *Trypanosoma cruzi* with an IC_{50} value of 2.43 μM. Benznidazole was used as a standard compound showing an IC_{50} value of 3.8 μM. Beauvericin showed cytotoxic activity against L929 (mouse fibroblasts) with an IC_{50} value of 6.38 μM.

Discussion and conclusion

This work listed 101 antibacterial and 58 antiparasitic compounds isolated from endophytic fungi over the last two decades (2000–19). These endophytic fungi were obtained from host plants of different botanical families and as shown in Fig. 12.20, Fabaceae, Lamiaceae and Asteraceae are among the most represented families. These families are known to contain a large number of species producing bioactive compounds, as medicinal or toxic plants (Bessada, Barreira, & Oliveira, 2015; Moraes Neto et al., 2019; Setzer & Setzer, 2006; Sharma, Flores-Vallejo, Cardoso-Taketa, & Villarreal, 2017) and they may represent a great target for the researchers bioprospecting for endophytes. More generally, without focusing on antibacterial and antiparasitic properties, most of the bioactive endophytes studied between 2006 and 2016 were mainly isolated from plants corresponding to the families Fabaceae, Lamiaceae, Asteraceae and Araceae (Martinez-Klimova et al., 2017) while in a recent study, between 2016 and 2019, 254 plants families were studied for their endophytes, with Poaceae, Fabaceae, Pinaceae and Asteraceae as the most studied plant families (Harrison and Griffin, 2020).

In this review, endophytes were mostly isolated from leaves tissues, which represent one of the largest microbial habitats on the planet (Oono, Rasmussen, & Lefèvre, 2017). Most of the endophytic fungal strains belong to the Pezizomycotina subdivision of Ascomycetes: 53% of the compounds were isolated from Sordariomycetes (85 compounds from 48 fungal strains), 28% from Dothideomycetes (45 compounds from 27 fungal strains), and 12% from Eurotiomycetes (19 compounds from 12 fungal strains). According to Arnold (Arnold & Lutzoni, 2007; Arnold, 2007), Dothideomycetes, Sordariomycetes, and Eurotiomycetes are the three most common classes of foliar endophytes found from Canadian arctic to the lowland tropical forest of central Panama. Only one Basidiomycetes, a strain belonging to the genus *Rhizoctonia*, has been listed here, as a producer of an antibacterial compound, monomethylsulochrin, a polyketide active against *Helicobacter pylori* (MIC = 10 μg/mL).

Almost 60% of the compounds described herein are polyketides (94 over the 159), 16% are alkaloids (26/159), 9% are polypeptides (15/159), and 8% belong to the terpenoids (13/159). Polyketides are a large class of secondary metabolites with diverse bioactivities, including immunosuppressive, anticancer, and antibacterial properties (Yin & Dickschat, 2021).

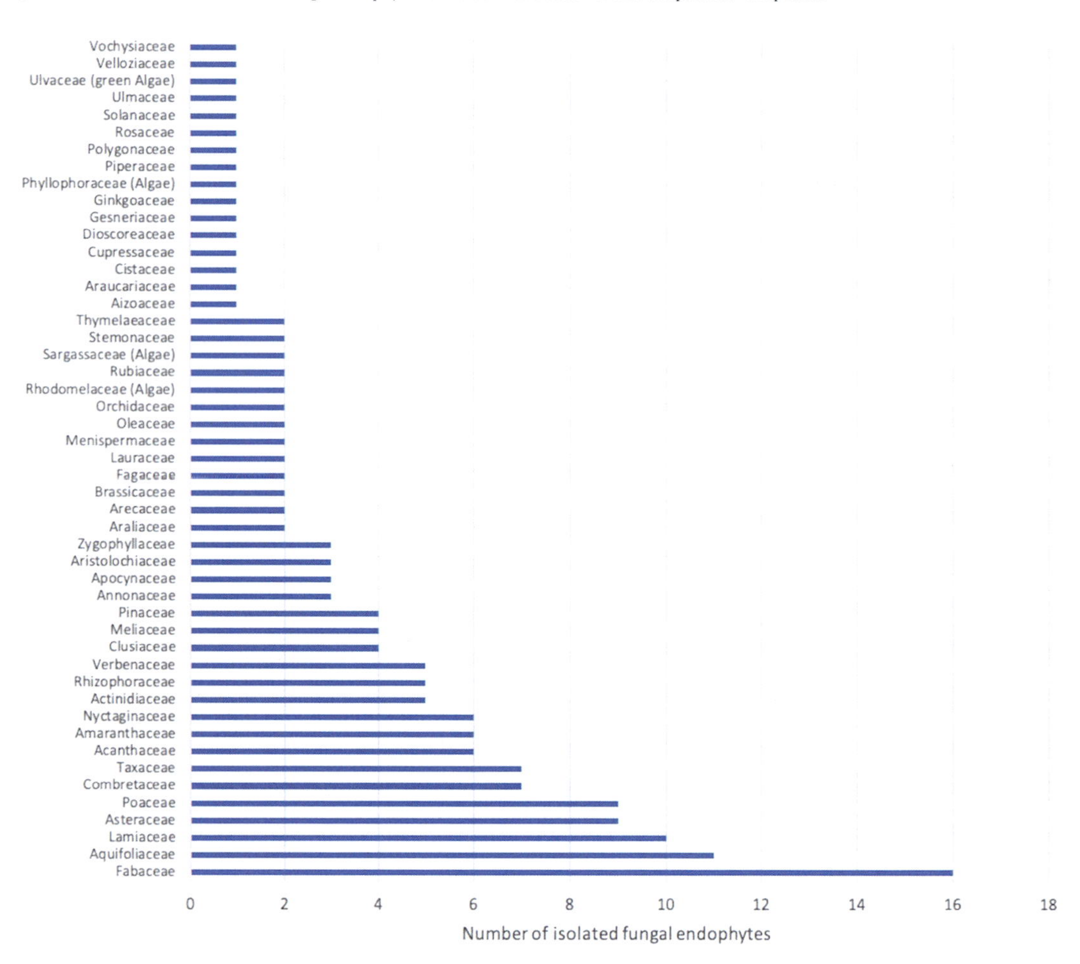

FIGURE 12.20 Botanical families of host plants reported in this work.

More than half of the 58 antiparasitic compounds (30 over the 58) and almost 40% (40 over the 101) of the antibacterial compounds listed here were described for the first time, highlighting the potential of novelty of endophytic fungal metabolites. Almost all the listed compounds were isolated after traditional axenic cultures. Two hypotheses could be made: due to their specific environment these endophytic fungal strains present specific metabolism leading to novel compounds; indeed, interactions between endophytes and their host plant could enhance the biosynthesis of bioactive specialized metabolites. The second hypothesis is that fungal endophytes listed here may belong to species chemically understudied. Only four antibacterial compounds, neoaspergillic acid, lateropyrone, enniatin A1, and enniatin B1, were isolated from extracts of cocultures with other fungi or bacteria. Surprisingly, all of them were already known, and no new antibacterial or antiparasitic compounds seemed to be de novo induced by cocultures. Nevertheless, interestingly their production rate could be increased by these coculture conditions, such as enniatin A1.

Nowadays, the development of new methods is commonly used in order to reveal some biosynthetic pathway and to induce the production of original specialized metabolites.

Despite the great number of gene clusters found in fungi thanks to molecular techniques, the conditions for gene expression are not fully understood and the presence of silent or cryptic genes and lowly expressed genes represent one of the main challenges for the isolation of novel secondary metabolites (Brakhage & Schroeckh, 2011). For the activation of cryptic genes, many approaches are proposed, broadly classified in culture-independent and culture-dependent techniques. Regarding the culture-based techniques, the modification of cultivation conditions focuses on providing a better environment to trigger the expression of cryptic gene clusters.

The set of variations of culture conditions is known as the OSMAC approach (One strain—Many compounds) which is based on the potential of a single strain to expand its chemical diversity under small changes (Bode, Bethe, Höfs, & Zeeck, 2002). These small changes involve different carbon or nitrogen sources, solid or liquid media, temperature variations, aeration or shaking conditions, salinity or pH levels, light conditions, and time of development in culture (Romano, Jackson, Patry, & Dobson, 2018). Under this approach, many endophytic fungi have already been shown to expand their chemical diversity, being able to produce novel secondary metabolites. For example, the endophytic fungus *Trichocladium* sp. was cultured under the OSMAC approach, considering variations of culture media, rice medium, and peas medium, allowing the isolation of a new amidepsine derivative and a new reduced spiro azaphilone derivative (Tran-Cong et al., 2019).

In addition to this, chemical elicitation during fungal culture such as treatment with epigenetic modifiers has also been shown to elicit secondary metabolism. Biosynthetic gene clusters in fungi are transcriptionally regulated by epigenetic regulators, mainly by DNA methyltransferases (DNMT) and histone deacetylases (HDAC), which change the condensation state of the chromosomes (Brakhage, 2013). The use of inhibitors of these enzymes induces the transcription and expression of silenced genes. For instance, the treatment of the endophytic fungus *Dimorphoporicola tragani* with 5-azacytidine (DNMT inhibitor) and valproic acid (HDAC inhibitor) elicited the production of three different mycotoxins dendrodolides (González-Menéndez et al., 2019).

Similarly, the addition of biosynthetic precursors such as the use of amino acids has been used for the induction of cryptic biosynthetic pathways in fungi. The culture of the marine-derived fungus *Dichotomomyces cejpii* was supplemented with the amino acids L-tryptophan and L-phenylalanine allowing the production of two new polyketides, dichocetides B and C and two new alkaloids, dichotomocejs E and F, and the induction of other known compounds (Wu et al., 2018).

Moreover, considering the natural conditions under which endophytic fungi are found, constant interactions with other microorganisms are occurring, situations that induce the production of different secondary metabolites useful for competition, communication, or defense. For this reason, the coculture of endophytic fungi with another bacterium or fungus might mimic microbial interactions, triggering different signaling pathways and responses. The coculture of the endophytic fungus *Bionectria* sp. with *Bacillus subtilis* or *Streptomyces lividans* triggered the production of the new molecules, bionectriamines A and B, and two other known compounds (Kamdem et al., 2018). To conclude, this review shows the great potential of fungal endophytes as renewable sources of antiparasitic and antibacterial compounds. With the advent of original cultures techniques, fungal endophytes are far from having revealed all their secrets.

References

Adeleke, B., & Babalola, O. (2021). Pharmacological potential of fungal endophytes associated with medicinal plants: A review. *Journal of Fungi*, *7*, 147. Available from https://doi.org/10.3390/jof7020147.

Aldridge, D. C., & Turner, W. B. (1970). Metabolites of *Helminthosporium monoceras*: Structures of monocerin and related benzopyrans. *Journal of the Chemical Society C: Organic*, 2598. Available from https://doi.org/10.1039/j39700002598.

Aly, A. H., Debbab, A., & Proksch, P. (2011). Fungal endophytes: Unique plant inhabitants with great promises. *Applied Microbiology and Biotechnology*, *90*, 1829–1845. Available from https://doi.org/10.1007/s00253-011-3270-y.

Aly, A. H., Edrada-Ebel, R. A., Wray, V., Müller, W. E. G., Kozytska, S., Hentschel, U., ... Ebel, R. (2008). Bioactive metabolites from the endophytic fungus *Ampelomyces* sp. isolated from the medicinal plant *Urospermum picroides*. *Phytochemistry*, *69*(8), 1716–1725. Available from https://doi.org/10.1016/j.phytochem.2008.02.013.

Aminov, R. I. (2010). A brief history of the antibiotic era: Lessons learned and challenges for the future. *Frontiers in Microbiology*, *1*, 134. Available from https://doi.org/10.3389/fmicb.2010.00134.

Andrioli, W. J., Conti, R., Araújo, M. J., Zanasi, R., Cavalcanti, B. C., Manfrim, V., ... Bastos, J. K. (2014). Mycoleptones A-C and polyketides from the endophyte mycoleptodiscus indicus. *Journal of Natural Products*, *77*, 70–78. Available from https://doi.org/10.1021/np4006822.

Andrzej, C. (2002). Coexistence and coevolution: Various levels of interactions I. Internal fungi (endophytes) and their biological significance in coevolution. *Wiadomości Botaniczne*, *46*, 35–44.

Arnold, A. E. (2007). Understanding the diversity of foliar endophytic fungi: Progress, challenges, and frontiers. *Fungal Biology Reviews*, *21*, 51–66. Available from https://doi.org/10.1016/j.fbr.2007.05.003.

Arnold, A. E., & Lutzoni, F. (2007). Diversity and host range of foliar fungal endophytes: Are tropical leaves biodiversity hotspots? *Ecology*, *88*, 541–549. Available from https://doi.org/10.1890/05-1459.

Bérdy, J. (2012). Thoughts and facts about antibiotics: Where we are now and where we are heading. *Journal of Antibiotics*, *65*, 385–395. Available from https://doi.org/10.1038/ja.2012.27.

Bessada, S. M. F., Barreira, J. C. M., & Oliveira, M. B. P. P. (2015). Asteraceae species with most prominent bioactivity and their potential applications: A review. *Industrial Crops and Products*, *76*, 604–615. Available from https://doi.org/10.1016/j.indcrop.2015.07.073.

Bhilabutra W., Techowisan, T., Peberdy, J. F., & Lumyong, S. (2007). Antimicrobial activity of bioactive compounds from Periconia siamensis CMUGE015. *Research Journal of Microbiology*, *2(10)*, 749–755. Available from https://doi.org/10.3923/jm.2007.749.755

Bloch, P., Tamm, C., Bollinger, P., Petcher, T. J., & Weber, H. P. (1976). Pseurotin, a new metabolite of *Pseudeurotium ovalis* STOLK having an unusual hetero-spirocyclic system (Preliminary Communication). *Helvetica Chimica Acta*, *59*, 133–137. Available from https://doi.org/10.1002/hlca.19760590114.

Bode, H. B., Bethe, B., Höfs, R., & Zeeck, A. (2002). Big effects from small changes: Possible ways to explore nature's chemical diversity. *ChemBioChem: A European Journal of Chemical Biology*, *3*, 619–627. Available from https://doi.org/10.1002/1439-7633(20020703)3:7 < 619::AID-CBIC619 > 3.0.CO;2-9.

Boros, C., Dix, A., Katz, B., Vasina, Y., & Pearce, C. (2003). Isolation and identification of cissetin – A setin-like antibiotic with a novel cis-octalin ring fusion. *Journal of Antibiotics*, *56*, 862–865. Available from https://doi.org/10.7164/antibiotics.56.862.

Brady, S. F., Wagenaar, M. M., Singh, M. P., Janso, J. E., & Clardy, J. (2000). The cytosporones, new octaketide antibiotics isolated from an endophytic fungus. *Organic Letters*, 2(25), 4043–4046. Available from https://doi.org/10.1021/ol006680s.

Brakhage, A. A. (2013). Regulation of fungal secondary metabolism. *Nature Reviews Microbiology*, *11*, 21–32. Available from https://doi.org/10.1038/nrmicro2916.

Brakhage, A. A., & Schroeckh, V. (2011). Fungal secondary metabolites – Strategies to activate silent gene clusters. *Fungal Genetics and Biology*, *48*, 15–22. Available from https://doi.org/10.1016/j.fgb.2010.04.004.

Breitenstein, W., Chexal, K. K., Mohr, P., & Tamm, C. (1981). Pseurotin B, C, D, and E. Further new metabolites of *Pseudeurotium ovalis* STOLK. *Helvetica Chimica Acta, 64*, 379–388. Available from https://doi.org/10.1002/hlca.19810640203.

Bunyapaiboonsri, T., Yoiprommarat, S., Khonsanit, A., & Komwijit, S. (2008). Phenolic glycosides from the filamentous fungus *Acremonium* sp. BCC 14080. *Journal of Natural Products, 71*, 891–894. Available from https://doi.org/10.1021/np070689m.

Cafêu, M. C., Silva, G. H., Teles, H. L., Bolzani, V. D. S., Araújo, A. R., Young, M. C. M., & Pfenning, L. H. (2005). Substâncias antifúngicas de Xylaria sp., um fungo endofítico isolado de palicourea marcgravii (Rubiaceae). *Quimica Nova, 28*, 991–995. Available from https://doi.org/10.1590/S0100-40422005000600011.

Calcul, L., Waterman, C., Ma, W. S., Lebar, M. D., Harter, C., Mutka, T., ... Baker, B. J. (2013). Screening mangrove endophytic fungi for antimalarial natural products. *Marine Drugs, 11*, 5036–5050. Available from https://doi.org/10.3390/md11125036.

Campos, F. F., Rosa, L. H., Cota, B. B., Caligiorne, R. B., Teles Rabello, A. L., Alves, T. M. A., ... Zani, C. L. (2008). Leishmanicidal metabolites from *Cochliobolus* sp., an endophytic fungus isolated from Piptadenia adiantoides (Fabaceae). *PLoS Neglected Tropical Diseases, 2*. Available from https://doi.org/10.1371/journal.pntd.0000348.

Campos, F. F., Sales Junior, P. A., Romanha, A. J., Araújo, M. S. S., Siqueira, E. P., Resende, J. M., Alves, T. M. A., Martins-Filho, O. A., Dos Santos, V. L., Rosa, C. A., Zani, C. L., & Cota, B. B. (2015). Bioactive endophytic fungi isolated from *Caesalpinia echinata* Lam. (Brazilwood) and identification of beauvericin as a trypanocidal metabolite from *Fusarium* sp. *Memorias do Instituto Oswaldo Cruz, 110*, 65–74. Available from https://doi.org/10.1590/0074-02760140243.

Casella, T. M., Eparvier, V., Mandavid, H., Bendelac, A., Odonne, G., Dayan, L., ... Stien, D. (2013). Antimicrobial and cytotoxic secondary metabolites from tropical leaf endophytes: Isolation of antibacterial agent pyrrocidine C from Lewia infectoria SNB-GTC2402. *Phytochemistry, 96*, 370–377. Available from https://doi.org/10.1016/j.phytochem.2013.10.004.

Challinor, V. L., & Bode, H. B. (2015). Bioactive natural products from novel microbial sources. *Annals of the New York Academy of Sciences, 1354*, 82–97. Available from https://doi.org/10.1111/nyas.12954.

Chen, J., Bai, X., Hua, Y., Zhang, H., & Wang, H. (2019). Fusariumins C and D, two novel antimicrobial agents from *Fusarium oxysporum* ZZP-R1 symbiotic on Rumex madaio Makino. *Fitoterapia, 134*, 1–4. Available from https://doi.org/10.1016/j.fitote.2019.01.016.

Chen, X., Shi, Q., Lin, G., Guo, S., & Yang, J. (2009). Spirobisnaphthalene analogues from the endophytic fungus *Preussia* sp. *Journal of Natural Products, 72*(9), 1712–1715. Available from https://doi.org/10.1021/np900302w.

Cheng, M. J., Wu, M. Der, Yanai, H., Su, Y. S., Chen, I. S., Yuan, G. F., Hsieh, S. Y., & Chen, J. J. (2012). Secondary metabolites from the endophytic fungus *Biscogniauxia formosana* and their antimycobacterial activity. *Phytochemistry Letters, 5(3)*, 467–472. Available from https://doi.org/10.1016/j.phytol.2012.04.007

Chinworrungsee, M., Kittakoop, P., Saenboonrueng, J., Kongsaeree, P., & Thebtaranonth, Y. (2006). Bioactive compounds from the seed fungus Menisporopsis theobromae BCC 3975. *Journal of Natural Products, 69*, 1404–1410. Available from https://doi.org/10.1021/np0601197.

Cole, R. J., Wilson, D. M., Harper, J. L., Cox, R. H., Cochran, T. W., Cutler, H. G., & Bell, D. K. (1982). Isolation and identification of two new [11]cytochalasins from *Phomopsis sojae*. *Journal of Agricultural and Food Chemistry, 30*, 301–304. Available from https://doi.org/10.1021/jf00110a021.

Cota, B. B., Rosa, L. H., Caligiorne, R. B., Rabello, A. L. T., Almeida Alves, T. M., Rosa, C. A., & Zani, C. L. (2008). Altenusin, a biphenyl isolated from the endophytic fungus *Alternaria* sp., inhibits trypanothione reductase from *Trypanosoma cruzi*. *FEMS Microbiology Letters, 285*, 177–182. Available from https://doi.org/10.1111/j.1574-6968.2008.01221.x.

Curtis, R. F., Hassall, C. H., Jones, D. W., & Williams, T. W. (1960). 940. The biosynthesis of phenols. Part II. Asterric acid, a metabolic product of aspergillus terreus thom. *Journal of the Chemical Society (Resumed)*, 4838. Available from https://doi.org/10.1039/jr9600004838.

Devi, P., Rodrigues, C., Naik, C. G., & D'Souza, L. (2012). Isolation and characterization of antibacterial compound from a mangrove-endophytic fungus, *Penicillium chrysogenum* MTCC 5108. *Indian Journal of Microbiology, 52*(4), 617–623. Available from https://doi.org/10.1007/s12088-012-0277-8.

Ding, G., Chen, A. J., Lan, J., Zhang, H., Chen, X., Liu, X., & Zou, Z. (2012). Sesquiterpenes and cyclopeptides from the endophytic fungus *Trichoderma asperellum* Samuels, Lieckf. & Nirenberg. *Chemistry & Biodiversity, 9*, 1205–1212. Available from https://doi.org/10.1002/cbdv.201100185.

do Nascimento, A. M., Soares, M. G., da Silva Torchelsen, F. K. V., de Araujo, J. A. V., Lage, P. S., Duarte, M. C., … do Nascimento, A. M. (2015). Antileishmanial activity of compounds produced by endophytic fungi derived from medicinal plant Vernonia polyanthes and their potential as source of bioactive substances. *World Journal of Microbiology and Biotechnology*, *31*, 1793–1800. Available from https://doi.org/10.1007/s11274-015-1932-0.

Du, F. Y., Li, X. M., Li, C. S., Shang, Z., & Wang, B. G. (2012). Cristatumins A-D, new indole alkaloids from the marine-derived endophytic fungus *Eurotium cristatum* EN-220. *Bioorganic and Medicinal Chemistry Letters*, *22*, 4650–4653. Available from https://doi.org/10.1016/j.bmcl.2012.05.088.

Edwards, R. L., Maitland, D. J., & Whalley, A. J. S. (1989). Metabolites of the higher fungi. Part 24. Cytochalasin N, O, P, Q, and R. New cytochalasins from the fungus *Hypoxylon terricola* mill. *Journal of the Chemical Society, Perkin Transactions 1*, 57–65. Available from https://doi.org/10.1039/p19890000057.

Elfita, E., Muharni, M., Munawar, M., Legasari, L., & Darwati, D. (2011). Antimalarial compounds from endophytic fungi of brotowall (*Tinaspora crispa* L). *Indonesian Journal of Chemistry*, *11*(1), 53–58. Available from https://doi.org/10.22146/ijc.21420.

Elsebai, M. F., Natesan, L., Kehraus, S., Mohamed, I. E., Schnakenburg, G., Sasse, F., Shaaban, S., Gütschow, M., & König, G. M. (2011). HLE-inhibitory alkaloids with a polyketide skeleton from the marine-derived fungus *Coniothyrium cereale*. *Journal of Natural Products*, *74*(10), 2282–2285. Available from https://doi.org/10.1021/np2004227.

Engels, D., & Zhou, X. N. (2020). Neglected tropical diseases: An effective global response to local poverty-related disease priorities. *Infectious Diseases of Poverty*, *9*, 10. Available from https://doi.org/10.1186/s40249-020-0630-9.

Espada, A., Rivera-Sagredo, A., de la Fuente, J. M., Hueso-Rodríguez, J. A., & Elson, S. W. (1997). New cytochalasins from the fungus *Xylaria hypoxylon*. *Tetrahedron*, *53*, 6485–6492. Available from https://doi.org/10.1016/s0040-4020(97)00305-0.

Ferreira, M. C., Cantrell, C. L., Wedge, D. E., Gonçalves, V. N., Jacob, M. R., Khan, S., … Rosa, L. H. (2017). Antimycobacterial and antimalarial activities of endophytic fungi associated with the ancient and narrowly endemic neotropical plant *Vellozia gigantea* from Brazil. *Memorias do Instituto Oswaldo Cruz*, *112*, 692–697. Available from https://doi.org/10.1590/0074-02760170144.

Figueroa, M., Jarmusch, A. K., Raja, H. A., El-Elimat, T., Kavanaugh, J. S., Horswill, A. R., … Oberlies, N. H. (2014). Polyhydroxyanthraquinones as quorum sensing inhibitors from the guttates of *Penicillium restrictum* and their analysis by desorption electrospray ionization mass spectrometry. *Journal of Natural Products*, *77*, 1351–1358. Available from https://doi.org/10.1021/np5000704.

Flores, A. C., Pamphile, J. A., Sarragiotto, M. H., & Clemente, E. (2013). Production of 3-nitropropionic acid by endophytic fungus *Phomopsis longicolla* isolated from *Trichilia elegans* A. JUSS ssp. elegans and evaluation of biological activity. *World Journal of Microbiology and Biotechnology*, *29*(5), 923–932. Available from https://doi.org/10.1007/S11274-013-1251-2/FIGURES/4.

Francolini, I., Norris, P., Piozzi, A., Donelli, G., & Stoodley, P. (2004). Usnic acid, a natural antimicrobial agent able to inhibit bacterial biofilm formation on polymer surfaces. *Antimicrobial Agents and Chemotherapy*, *48*, 4360–4365. Available from https://doi.org/10.1128/AAC.48.11.4360-4365.2004.

Fredenhagen, A., Petersen, F., Tintelnot-Blomley, M., Rösel, J., Mett, H., & Hug, P. (1997). Semicochliodinol A and B: Inhibitors of HIV-1 protease and EGF-R protein tyrosine kinase related to asterriquinones produced by the fungus *Chrysosporium merdarium*. *The Journal of Antibiotics*, *50*, 395–401. Available from https://doi.org/10.7164/antibiotics.50.395.

Gao, S.-S., Li, X.-M., Zhang, Y., Li, C.-S., & Wang, B.-G. (2011). Conidiogenones H and I, two new diterpenes of cyclopiane class from a marine-derived endophytic fungus *Penicillium chrysogenum* QEN-24S. *Chemistry & Biodiversity*, *8*(9), 1748–1753. Available from https://doi.org/10.1002/cbdv.201000378.

Ganley, R. J., Brunsfeld, S. J., & Newcombe, G. (2004). A community of unknown, endophytic fungi in western white pine. *Proceedings of the National Academy of Sciences of the United States of America*, *101*, 10107–10112. Available from https://doi.org/10.1073/pnas.0401513101.

Gennaro, M., Gonthier, P., & Nicolotti, G. (2003). Fungal endophytic communities in healthy and declining *Quercus robur* L. and *Q. cerris* L. trees in Northern Italy. *Journal of Phytopathology*, *151*, 529–534. Available from https://doi.org/10.1046/j.1439-0434.2003.00763.x.

Germida, J. J., Siciliano, S. D., De Freitas, J. R., & Seib, A. M. (1998). Diversity of root-associated bacteria associated with field-grown canola (*Brassica napus* L.) and wheat (*Triticum aestivum* L.). *FEMS Microbiology Ecology*, *26*, 43–50. Available from https://doi.org/10.1016/S0168-6496(98)00020-8.

González-Menéndez, V., Crespo, G., Toro, C., Martín, J., De Pedro, N., Tormo, J. R., & Genilloud, O. (2019). Extending the metabolite diversity of the endophyte *Dimorphosporicola tragani*. *Metabolites*, *9*, 197. Available from https://doi.org/10.3390/metabo9100197.

Gouda, S., Das, G., Sen, S. K., Shin, H. S., & Patra, J. K. (2016). Endophytes: A treasure house of bioactive compounds of medicinal importance. *Frontiers in Microbiology*, *7*, 1538. Available from https://doi.org/10.3389/fmicb.2016.01538.

Gu, W. (2009). Bioactive metabolites from *Alternaria brassicicola* ML-P08, an endophytic fungus residing in malus halliana. *World Journal of Microbiology and Biotechnology*, *25*, 1677–1683. Available from https://doi.org/10.1007/s11274-009-0062-y.

Gupta, S., Chaturvedi, P., Kulkarni, M. G., & Van Staden, J. (2020). A critical review on exploiting the pharmaceutical potential of plant endophytic fungi. *Biotechnology Advances*, *39*, 107462. Available from https://doi.org/10.1016/j.biotechadv.2019.107462.

Gutiérrez-Zamora, M. L., & Martínez-Romero, E. (2001). Natural endophytic association between *Rhizobium etli* and maize (*Zea mays* L.). *Journal of Biotechnology*, *91*, 117–126. Available from https://doi.org/10.1016/s0168-1656(01)00332-7.

Hamill, R. L., Higgens, C. E., Boaz, H. E., & Gorman, M. (1969). The structure of beauvericin, a new depsipeptide antibiotic toxic to *Artemia salina*. *Tetrahedron Letters*, *10*, 4255. Available from https://doi.org/10.1016/S0040-4039(01)88668-8.

Harrison, J. G., & Griffin, E. A. (2020). The diversity and distribution of endophytes across biomes, plant phylogeny and host tissues: How far have we come and where do we go from here? *Environmental Microbiology*, *22*, 2107–2123. Available from https://doi.org/10.1111/1462-2920.14968.

Hayashi, A., Fujioka, S., Nukina, M., Kawano, T., Shimada, A., & Kimura, Y. (2007). Fumiquinones A and B, nematicidal quinones produced by *Aspergillus fumigatus*. *Bioscience, Biotechnology, and Biochemistry*, *71*, 1697–1702. Available from https://doi.org/10.1271/bbb.70110.

He, H., Yang, H. Y., Bigelis, R., Solum, E. H., Greenstein, M., & Carter, G. T. (2002). Pyrrocidines A and B, new antibiotics produced by a filamentous fungus. *Tetrahedron Letters*, *43*, 1633–1636. Available from https://doi.org/10.1016/S0040-4039(02)00099-0.

Hemtasin, C., Kanokmedhakul, S., Kanokmedhakul, K., Hahnvajanawong, C., Soytong, K., Prabpai, S., & Kongsaeree, P. (2011). Cytotoxic pentacyclic and tetracyclic aromatic sesquiterpenes from phomopsis archeri. *Journal of Natural Products*, *74*, 609–613. Available from https://doi.org/10.1021/np100632g.

Hoff, J. A., Klopfenstein, N. B., Mcdonald, G. I., Tonn, J. R., Kim, M. S., Zambino, P. J., … Carris, L. M. (2004). Fungal endophytes in woody roots of Douglas-fir (*Pseudotsuga menziesii*) and ponderosa pine (*Pinus ponderosa*). *Forest Pathology*, *34*, 255–271. Available from https://doi.org/10.1111/j.1439-0329.2004.00367.x.

Hoffman, A. M., Mayer, S. G., Strobel, G. A., Hess, W. M., Sovocool, G. W., Grange, A. H., … Kelley-Swift, E. G. (2008). Purification, identification and activity of phomodione, a furandione from an endophytic Phoma species. *Phytochemistry*, *69*, 1049–1056. Available from https://doi.org/10.1016/j.phytochem.2007.10.031.

Hutchings M., A. Truman, B. Wilkinson, Antibiotics: Past, present and future, Current Opinion in Microbiology. 51 (2019) 72–80. https://doi.org/10.1016/j.mib.2019.10.008.

Hussain, H., Ahmed, I., Schulz, B., Draeger, S., & Krohn, K. (2012). Pyrenocines J-M: Four new pyrenocines from the endophytic fungus, *Phomopsis* sp. *Fitoterapia*, *83(3)*, 523–526. Available from https://doi.org/10.1016/j.fitote.2011.12.017

Hussain, H., Tchimene, M. K., Ahmed, I., Meier, K., Steinert, M., Draeger, S., Schulz, B., & Krohn, K. (2011). Antimicrobial chemical constituents from the endophytic fungus *Phomopsis* sp. from *Notobasis syriaca*. *Natural Product Communications*, *6(12)*, 1934578X1100601. Available from https://doi.org/10.1177/1934578X1100601228.

Hzounda Fokou, J. B., Dize, D., Etame Loe, G. M., Nko'o, M. H. J., Ngene, J. P., Ngoule, C. C., & Boyom, F. F. (2021). Anti-leishmanial and anti-trypanosomal natural products from endophytes. *Parasitology Research*, *120*, 785–796. Available from https://doi.org/10.1007/s00436-020-07035-1.

Ibrahim, S. R. M., Mohamed, G. A., Al Haidari, R. A., El-Kholy, A. A., & Zayed, M. F. (2018). Potential antimalarial agents from endophytic fungi: A review. *Mini Reviews in Medicinal Chemistry*, *18*, 1110–1132. Available from https://doi.org/10.2174/1389557518666180305163151.

Ibrahim, S. R. M., Abdallah, H. M., Elkhayat, E. S., Al Musayeib, N. M., Asfour, H. Z., Zayed, M. F., & Mohamed, G. A. (2018). Fusaripeptide A: New antifungal and anti-malarial cyclodepsipeptide from the endophytic fungus *Fusarium* sp. *Journal of Asian Natural Products Research*, *20*, 75–85. Available from https://doi.org/10.1080/10286020.2017.1320989.

Ibrahim, S. R. M., Abdallah, H. M., Mohamed, G. A., & Ross, S. A. (2016). Integracides H-J: New tetracyclic triterpenoids from the endophytic fungus *Fusarium* sp. *Fitoterapia, 112*, 161–167. Available from https://doi.org/10.1016/j.fitote.2016.06.002.

Isaka, M., Jaturapat, A., Rukseree, K., Danwisetkanjana, K., Tanticharoen, M., & Thebtaranonth, Y. (2001). Phomoxanthones A and B, novel xanthone dimers from the endophytic fungus *Phomopsis* species. *Journal of Natural Products, 64*, 1015–1018. Available from https://doi.org/10.1021/np010006h.

Isaka, M., Berkaew, P., Intereya, K., Komwijit, S., & Sathitkunanon, T. (2007). Antiplasmodial and antiviral cyclohexadepsipeptides from the endophytic fungus *Pullularia* sp. BCC 8613. *Tetrahedron, 63*, 6855–6860. Available from https://doi.org/10.1016/j.tet.2007.04.062.

Isaka, M., Chinthanom, P., Boonruangprapa, T., Rungjindamai, N., & Pinruan, U. (2010). Eremophilane-type sesquiterpenes from the fungus *Xylaria* sp. BCC 21097. *Journal of Natural Products, 73*, 683–687. Available from https://doi.org/10.1021/np100030x.

Izawa, Y., Hirose, T., Shimizunée Tomioka, T., Koyama, K., & Natori, S. (1989). Six new 10-pheynl-[11]cytochalasans, cytochalasins N - S from phomopsis SP. *Tetrahedron, 45*, 2323–2335. Available from https://doi.org/10.1016/S0040-4020(01)83434-7.

Jia, M., Chen, L., Xin, H. L., Zheng, C. J., Rahman, K., Han, T., & Qin, L. P. (2016). A friendly relationship between endophytic fungi and medicinal plants: A systematic review. *Frontiers in Microbiology, 7*, 906. Available from https://doi.org/10.3389/fmicb.2016.00906.

Jiménez-Romero, C., Ortega-Barría, E., Arnold, A. E., & Cubilla-Rios, L. (2008). Activity against *Plasmodium falciparum* of lactones isolated from the endophytic fungus *Xylaria* sp. *Pharmaceutical Biology, 46*, 700–703. Available from https://doi.org/10.1080/13880200802215859.

Juan-Garcia, A., Ruiz, M. J., Font, G., & Manyes, L. (2015). Enniatin A1, enniatin B1 and beauvericin on HepG2: Evaluation of toxic effects. *Food and Chemical Toxicology: An International Journal Published for the British Industrial Biological Research Association, 84*, 196. Available from https://doi.org/10.1016/j.fct.2015.08.030.

Kamdem, R. S. T., Wang, H., Wafo, P., Ebrahim, W., Özkaya, F. C., Makhloufi, G., Janiak, C., Sureechatchaiyan, P., Kassack, M. U., Lin, W., Liu, Z., & Proksch, P. (2018). Induction of new metabolites from the endophytic fungus *Bionectria* sp. through bacterial co-culture. *Fitoterapia, 124*, 132–136. Available from https://doi.org/10.1016/j.fitote.2017.10.021.

Keller, N. P. (2019). Fungal secondary metabolism: Regulation, function and drug discovery. *Nature Reviews Microbiology, 17*, 167–180. Available from https://doi.org/10.1038/s41579-018-0121-1.

Khare, E., Mishra, J., & Arora, N. K. (2018). Multifaceted interactions between endophytes and plant: Developments and prospects. *Frontiers in Microbiology, 9*, 2732. Available from https://doi.org/10.3389/fmicb.2018.02732.

Kharwar, R. N., Verma, V. C., Kumar, A., Gond, S. K., Harper, J. K., Hess, W. M., ... Strobel, G. A. (2009). Javanicin, an antibacterial naphthaquinone from an endophytic fungus of neem, *Chloridium* sp. *Current Microbiology, 58*, 233–238. Available from https://doi.org/10.1007/s00284-008-9313-7.

Khumkomkhet, P., Kanokmedhakul, S., Kanokmedhakul, K., Hahnvajanawong, C., & Soytong, K. (2009). Antimalarial and cytotoxic depsidones from the fungus *Chaetomium brasiliense*. *Journal of Natural Products, 72*, 1487–1491. Available from https://doi.org/10.1021/np9003189.

Kimura, Y., Mizuno, T., Nakajima, H., & Hamasaki, T. (1992). Altechromones A and B, new plant growth regulators produced by the fungus, *Alternaria* sp. *Bioscience, Biotechnology, and Biochemistry, 56*, 1664–1665. Available from https://doi.org/10.1271/bbb.56.1664.

Kim, S., Shin, D. S., Lee, T., & Oh, K. B. (2004a). Periconicins, two new fusicoccane diterpenes produced by an endophytic fungus *Periconia* sp. with antibacterial activity. *Journal of Natural Products, 67*(3), 448–450. Available from https://doi.org/10.1021/np030384h.

Knight, V., Sanglier, J. J., DiTullio, D., Braccili, S., Bonner, P., Waters, J., ... Zhang, L. (2003). Diversifying microbial natural products for drug discovery. *Applied Microbiology and Biotechnology, 62*, 446–458. Available from https://doi.org/10.1007/s00253-003-1381-9.

Kongsaeree, P., Prabpai, S., Sriubolmas, N., Vongvein, C., & Wiyakrutta, S. (2003). Antimalarial dihydroisocoumarins produced by *Geotrichum* sp., an endophytic fungus of *Crassocephalum crepidioides*. *Journal of Natural Products, 66*, 709–711. Available from https://doi.org/10.1021/np0205598.

Kontnik, R., & Clardy, J. (2008). Codinaeopsin, an antimalarial fungal polyketide. *Organic Letters, 10*, 4149–4151. Available from https://doi.org/10.1021/ol801726k.

Krohn, K., Kouam, S. F., Kuigoua, G. M., Hussain, H., Cludius-Brandt, S., Flörke, U., ... Schulz, B. (2009). Xanthones and oxepino[2,3-b]chromones from three endophytic fungi. *Chemistry - A European Journal, 15*(44), 12121–12132. Available from https://doi.org/10.1002/chem.200900749.

Krohn, K., Michel, A., Flörke, U., Aust, H., Draeger, S., & Schulz, B. (1994). Biologically active metabolites from fungi, 5. Palmarumycins C1–C16 from *Coniothyrium* sp.: Isolation, structure elucidation, and biological activity. *Liebigs Annalen Der Chemie, 1994*, 1099–1108. Available from https://doi.org/10.1002/jlac.199419941108.

Kumarihamy, M., Ferreira, D., Croom, E. M., Sahu, R., Tekwani, B. L., Duke, S. O., ... Dhammika Nanayakkara, N. P. (2019). Antiplasmodial and cytotoxic cytochalasins from an endophytic fungus, *Nemania* sp. UM10M, isolated from a diseased *Torreya taxifolia* leaf. *Molecules (Basel, Switzerland), 24*, 777. Available from https://doi.org/10.3390/molecules24040777.

Kuyama, S., & Tamura, T. (1957). Cercosporin. A pigment of Cercosporina Kikuchii Matsumoto et Tomoyasu. I. Cultivation of fungus, isolation and purification of pigment. *Journal of the American Chemical Society, 79*, 5725–5726. Available from https://doi.org/10.1021/ja01578a038.

Lauterwein, M., Oethinger, M., Belsner, K., Peters, T., & Marre, R. (1995). In vitro activities of the lichen secondary metabolites vulpinic acid, (+)-usnic acid, and (-)-usnic acid against aerobic and anaerobic microorganisms. *Antimicrobial Agents and Chemotherapy, 39*. Available from https://doi.org/10.1128/AAC.39.11.2541.

Lenta, B. N., Ngatchou, J., Frese, M., Ladoh-Yemeda, F., Voundi, S., Nardella, F., ... Sewald, N. (2016). Purpureone, an antileishmanial ergochrome from the endophytic fungus *Purpureocillium lilacinum*. *Zeitschrift Für Naturforschung B, 71*, 1159–1167. Available from https://doi.org/10.1515/znb-2016-0128.

Li, J. Y., Harper, J. K., Grant, D. M., Tombe, B. O., Bashyal, B., Hess, W. M., & Strobel, G. A. (2001). Ambuic acid, a highly functionalized cyclohexenone with antifungal activity from *Pestalotiopsis* spp. and *Monochaetia* sp. *Phytochemistry, 56*, 463–468. Available from https://doi.org/10.1016/S0031-9422(00)00408-8.

Li, Y., Song, Y. C., Liu, J. Y., Ma, Y. M., & Tan, R. X. (2005). Anti-helicobacter pylori substances from endophytic fungal cultures. *World Journal of Microbiology and Biotechnology, 21*(4), 553–558. Available from https://doi.org/10.1007/s11274-004-3273-2.

Liu, X., Dong, M., Chen, X., Jiang, M., Lv, X., & Zhou, J. (2008). Antimicrobial activity of an endophytic *Xylaria* sp. YX-28 and identification of its antimicrobial compound 7-amino-4-methylcoumarin. *Applied Microbiology and Biotechnology, 78*(2), 241–247. Available from https://doi.org/10.1007/s00253-007-1305-1.

Loesgen, S., Bruhn, T., Meindl, K., Dix, I., Schulz, B., Zeeck, A., & Bringmann, G. (2011). (+)-Flavipucine, the missing member of the pyridione epoxide family of fungal antibiotics. *European Journal of Organic Chemistry, 2011*(26), 5156–5162. Available from https://doi.org/10.1002/ejoc.201100284.

Lösgen, S., Magull, J., Schulz, B., Draeger, S., & Zeeck, A. (2008). Isofusidienols: novel chromone-3-oxepines produced by the endophytic fungus *Chalara* sp. *European Journal of Organic Chemistry, 2008*(4), 698–703. Available from https://doi.org/10.1002/ejoc.200700839.

Lu, S., Draeger, S., Schulz, B., Krohn, K., Ahmed, I., Hussain, H., Yi, Y., Li, N., & Zhang, W. (2011). Bioactive Aromatic derivatives from endophytic fungus, *Cytospora* sp. *Natural Product Communications, 6*(5), 661–666. Available from https://doi.org/10.1177/1934578x1100600518.

Luo, J., Liu, X., Li, E., Guo, L., & Che, Y. (2013). Arundinols A-C and arundinones A and B from the plant endophytic fungus *Microsphaeropsis arundinis*. *Journal of Natural Products, 76*(1), 107–112. Available from https://doi.org/10.1021/np300806a.

Ma, Y. M., Li, Y., Liu, J. Y., Song, Y. C., & Tan, R. X. (2004). Anti-Helicobacter pylori metabolites from *Rhizoctonia* sp. Cy064, an endophytic fungus in *Cynodon dactylon*. *Fitoterapia, 75*(5), 451–456. Available from https://doi.org/10.1016/j.fitote.2004.03.007.

Maciag-Dorszynska, M., Wegrzyn, G., & Guzow-Krzeminska, B. (2014). Antibacterial activity of lichen secondary metabolite usnic acid is primarily caused by inhibition of RNA and DNA synthesis. *FEMS Microbiology Letters, 353*, 57–62. Available from https://doi.org/10.1111/1574-6968.12409.

Macías-Rubalcava, M. L., Hernández-Bautista, B. E., Jiménez-Estrada, M., González, M. C., Glenn, A. E., Hanlin, R. T., ... Anaya, A. L. (2008). Naphthoquinone spiroketal with allelochemical activity from the newly discovered endophytic fungus Edenia gomezpompae. *Phytochemistry, 69*, 1185–1196. Available from https://doi.org/10.1016/j.phytochem.2007.12.006.

Manganyi, M. C., & Ateba, C. N. (2020). Untapped potentials of endophytic fungi: A review of novel bioactive compounds with biological applications. *Microorganisms, 8*, 1–25. Available from https://doi.org/10.3390/microorganisms8121934.

Martinez-Klimova, E., Rodríguez-Peña, K., & Sánchez, S. (2017). Endophytes as sources of antibiotics. *Biochemical Pharmacology*, *134*, 1–17. Available from https://doi.org/10.1016/j.bcp.2016.10.010.

Martínez-Luis, S., Della-Togna, G., Coley, P. D., Kursar, T. A., Gerwick, W. H., & Cubilla-Rios, L. (2008). Antileishmanial constituents of the panamanian endophytic fungus *Edenia* sp. *Journal of Natural Products*, *71*, 2011–2014. Available from https://doi.org/10.1021/np800472q.

Martińez-Luis, S., Cherigo, L., Arnold, E., Spadafora, C., Gerwick, W. H., & Cubilla-Rios, L. (2012). Antiparasitic and anticancer constituents of the endophytic fungus *Aspergillus* sp. strain F1544. *Natural Product Communications*, *7*, 165–168. Available from https://doi.org/10.1177/1934578x1200700207.

Miao, F.-P., Li, X.-D., Liu, X.-H., Cichewicz, R. H., & Ji, N.-Y. (2012). Secondary metabolites from an algicolous *Aspergillus versicolor* strain. *Marine Drugs*, *10*(12), 131–139. Available from https://doi.org/10.3390/md10010131.

Miyagawa, H., Nagai, S., Tsurushima, T., Sato, M., Ueno, T., & Fukami, H. (2014). Phytotoxins produced by the plant pathogenic fungus bipolar is bicolor El-1. *Bioscience, Biotechnology, and Biochemistry*, *58*, 1143–1145. Available from https://doi.org/10.1271/bbb.58.1143.

Moraes Neto, R. N., Setúbal, R. F. B., Higino, T. M. M., Brelaz-de-Castro, M. C. A., da Silva, L. C. N., dos, A. S., & Aliança, S. (2019). Asteraceae plants as sources of compounds against Leishmaniasis and Chagas disease. *Frontiers in Pharmacology*, *10*, 477. Available from https://doi.org/10.3389/fphar.2019.00477.

Moreno, E., Varughese, T., Spadafora, C., Arnold, A. E., Coley, P. D., Kursar, T. A., Gerwick, W. H., & Cubilla-Rios, L. (2011). Chemical constituents of the new endophytic fungus *Mycosphaerella* sp. nov. and their anti-parasitic activity. *Natural Product Communications*, *6*, 835–840. Available from https://doi.org/10.1177/1934578x1100600620.

Nakanishi, S., Ando, K., Kawamoto, I., & Kase, H. (1989). KS-501 and KS-502, new inhibitors of Ca^{2+} and calmodulin-dependent cyclic-nucleotide phosphodiesterase from *Sporothrix* sp. *The Journal of Antibiotics*, *42*, 1049–1055. Available from https://doi.org/10.7164/antibiotics.42.1049.

Nakayama, J., Uemura, Y., Nishiguchi, K., Yoshimura, N., Igarashi, Y., & Sonomoto, K. (2009). Ambuic acid inhibits the biosynthesis of cyclic peptide quormones in Gram-positive bacteria. *Antimicrobial Agents and Chemotherapy*, *53*, 580–586. Available from https://doi.org/10.1128/AAC.00995-08.

Newman, D. J., & Cragg, G. M. (2020). Natural products as sources of new drugs over the nearly four decades from 01/1981 to 09/2019. *Journal of Natural Products*, *83*, 770–803. Available from https://doi.org/10.1021/acs.jnatprod.9b01285.

Okoye, F. B. C., Nworu, C. S., Debbab, A., Esimone, C. O., & Proksch, P. (2015). Two new cytochalasins from an endophytic fungus, KL-1.1 isolated from *Psidium guajava* leaves. *Phytochemistry Letters*, *14*, 51–55. Available from https://doi.org/10.1016/j.phytol.2015.09.004.

Ola, A. R. B., Thomy, D., Lai, D., Brötz-Oesterhelt, H., & Proksch, P. (2013). Inducing secondary metabolite production by the endophytic fungus *Fusarium tricinctum* through coculture with *Bacillus subtilis*. *Journal of Natural Products*, *76*, 2094–2099. Available from https://doi.org/10.1021/np400589h.

Oono, R., Rasmussen, A., & Lefèvre, E. (2017). Distance decay relationships in foliar fungal endophytes are driven by rare taxa. *Environmental Microbiology*, *19*, 2794–2805. Available from https://doi.org/10.1111/1462-2920.13799.

Ortega, H. E., Torres-Mendoza, D., & Cubilla-Rios, L. (2020). Patents on endophytic fungi for agriculture and bio-and phytoremediation applications. *Microorganisms*, *8*, 1–26. Available from https://doi.org/10.3390/microorganisms8081237.

Pinheiro, E. A. A., Carvalho, J. M., Dos Santos, D. C. P., De Oliveira Feitosa, A., Marinho, P. S. B., Guilhon, G. M. S. P., De Souza, A. D. L., Da Silva, F. M. A., & Andrey, A. M. (2013). Antibacterial activity of alkaloids produced by endophytic fungus *Aspergillus* sp. EJC08 isolated from medicinal plant *Bauhinia guianensis*. *Natural Product Research*, *27*(18), 1633–1638. Available from https://doi.org/10.1080/14786419.2012.750316.

Pongcharoen, W., Rukachaisirikul, V., Phongpaichit, S., & Sakayaroj, J. (2007). A New Dihydrobenzofuran Derivative from the Endophytic Fungus *Botryosphaeria mamane* PSU-M76. *Chemical &Pharmaceutical Bulletin*, *55* (9), 1404–1405. Available from https://doi.org/10.1248/cpb.55.1404.

Pontius, A., Krick, A., Kehraus, S., Brun, R., & König, G. M. (2008). Antiprotozoal activities of heterocyclic-substituted xanthones from the marine-derived fungus *Chaetomium* sp. *Journal of Natural Products*, *71*, 1579–1584. Available from https://doi.org/10.1021/np800294q.

Qin, S., Hussain, H., Schulz, B., Draeger, S., & Krohn, K. (2010). Two new metabolites, epoxydine A and B, from *Phoma* sp. *Helvetica Chimica Acta*, *93*(1), 169–174. Available from https://doi.org/10.1002/hlca.200900199.

Ragazzi, A., Moricca, S., Capretti, P., & Dellavalle, I. (1999). Endophytic presence of Discula quercina on declining *Quercus cerris*. *Journal of Phytopathology*, *147*, 437–440. Available from https://doi.org/10.1046/j.1439-0434.1999.00408.x.

Rasmussen, T. B., Skindersoe, M. E., Bjarnsholt, T., Phipps, R. K., Christensen, K. B., Jensen, P. O., . . . Givskov, M. (2005). Identity and effects of quorum-sensing inhibitors produced by *Penicillium* species. *Microbiology (Reading, England)*, *151*, 1325–1340. Available from https://doi.org/10.1099/mic.0.27715-0.

Raviraja, N. S. (2005). Fungal endophytes in five medicinal plant species from Kudremukh Range, Western Ghats of India. *Journal of Basic Microbiology*, *45*, 230–235. Available from https://doi.org/10.1002/jobm.200410514.

Rodrigues, K. F., & Samuels, G. J. (1999). Fungal endophytes of *Spondias mombin* leaves in Brazil. *Journal of Basic Microbiology*, *39*, 131–135. Available from https://doi.org/10.1002/(SICI)1521-4028(199905)39:2 < 131::AID-JOBM131 > 3.0.CO;2-9.

Romano, S., Jackson, S. A., Patry, S., & Dobson, A. D. W. (2018). Extending the "one strain many compounds" (OSMAC) principle to marine microorganisms. *Marine Drugs*, *16*, 244. Available from https://doi.org/10.3390/md16070244.

Roy, S., Kumar, G. A., Jafurulla, M., Mandal, C., & Chattopadhyay, A. (1838). Integrity of the actin cytoskeleton of host macrophages is essential for *Leishmania donovani* infection. *Biochimica et Biophysica Acta - Biomembranes*, *2014*, 2011–2018. Available from https://doi.org/10.1016/j.bbamem.2014.04.017.

Rukachaisirikul, V., Sommart, U., Phongpaichit, S., Sakayaroj, J., & Kirtikara, K. (2008). Metabolites from the endophytic fungus *Phomopsis* sp. PSU-D15. *Phytochemistry*, *69*(3), 783–787. Available from https://doi.org/10.1016/j.phytochem.2007.09.006.

Sakemi, S., Inagaki, T., Kaneda, K., Hirai, H., Iwata, E., Sakakibara, T., . . . Kojima, N. (1995). CJ-12,371 and CJ-12,372, two novel DNA gyrase inhibitors. Fermentation, isolation, structural elucidation and biological activities. *The Journal of Antibiotics*, *48*, 134–142. Available from https://doi.org/10.7164/antibiotics.48.134.

Sappapan, R., Sommit, D., Ngamrojanavanich, N., Pengpreecha, S., Wiyakrutta, S., Sriubolmas, N., & Pudhom, K. (2008). 11-Hydroxymonocerin from the plant endophytic fungus *Exserohilum rostratum*. *Journal of Natural Products*, *71*, 2080. Available from https://doi.org/10.1021/np8006167.

Schena, L., Nigro, F., Pentimone, I., Ligorio, A., & Ippolito, A. (2003). Control of postharvest rots of sweet cherries and table grapes with endophytic isolates of *Aureobasidium pullulans*. *Postharvest Biology and Technology*, *30*, 209–220. Available from https://doi.org/10.1016/S0925-5214(03)00111-X.

Schueffler, A., & Anke, T. (2014). Fungal natural products in research and development. *Natural Product Reports*, *31*, 1425–1448. Available from https://doi.org/10.1039/C4NP00060A.

Setzer, W. N., & Setzer, M. C. (2006). *Antitrypanosomal agents from higher plants. Biologically active natural products for the 21st century* (pp. 47–95). Trivandrum: Research Signpost.

Shang, Z., Li, X. M., Li, C. S., & Wang, B. G. (2012). Diverse secondary metabolites produced by marine-derived fungus *Nigrospora* sp. MA75 on various culture media. *Chemistry and Biodiversity*, *9*, 1338–1348. Available from https://doi.org/10.1002/cbdv.201100216.

Sharma, A., Flores-Vallejo, Rd. C., Cardoso-Taketa, A., & Villarreal, M. L. (2017). Antibacterial activities of medicinal plants used in Mexican traditional medicine. *Journal of Ethnopharmacology*, *208*, 264–329. Available from https://doi.org/10.1016/j.jep.2016.04.045.

Sharma, S., Kumar, M., Sharma, S., Nargotra, A., Koul, S., & Khan, I. A. (2010). Piperine as an inhibitor of Rv1258c, a putative multidrug efflux pump of *Mycobacterium tuberculosis*. *Journal of Antimicrobial Chemotherapy*, *65*, 1694–1701. Available from https://doi.org/10.1093/jac/dkq186.

Shiono, Y., Hatakeyama, T., Murayama, T., & Koseki, T. (2012). Polyketide metabolites from the endophytic fungus Microdiplodia sp. KS 75-1. *Natural Product Communications*, *7*(8), 1934578X1200700. Available from https://doi.org/10.1177/1934578X1200700825.

Shu, R. G., Wang, F. W., Yang, Y. M., Liu, Y. X., & Tan, R. X. (2004). Antibacterial and xanthine oxidase inhibitory cerebrosides from *Fusarium* sp. IFB-121, an endophytic fungus in *Quercus variabilis*. *Lipids*, *39*(7), 667–673. Available from https://doi.org/10.1007/s11745-004-1280-9.

Siegel, M. R., Latch, G. C. M., & Johnson, M. C. (1987). Fungal endophytes of grasses. *Annual Review of Phytopathology*, *25*, 293–315. Available from https://doi.org/10.1146/annurev.py.25.090187.001453.

Singh, M. P., Janso, J. E., Luckman, S. W., Brady, S. F., Clardy, J., Greenstein, M., & Maiese, W. M. (2000). Biological activity of guanacastepene, a novel diterpenoid antibiotic produced by an unidentified fungus CR115. *Journal of Antibiotics*, *53*(3), 256–261. Available from https://doi.org/10.7164/antibiotics.53.256.

Singh, S. B., Zink, D. L., Liesch, J. M., Dombrowski, A. W., Darkin-Rattray, S. J., Schmatz, D. M., & Goetz, M. A. (2001). Structure, histone deacetylase, and antiprotozoal activities of apicidins B and C, congeners of apicidin with proline and valine substitutions. *Organic Letters*, *3*, 2815–2818. Available from https://doi.org/10.1021/ol016240g.

Smith, H., Wingfield, M. J., & Petrini, O. (1996). *Botryosphaeria dothidea* endophytic in *Eucalyptus grandis* and *Eucalyptus nitens* in South Africa. *Forest Ecology and Management, 89*, 189–195. Available from https://doi.org/10.1016/S0378-1127(96)03847-9.

Stark, A. A., Kobbe, B., Matsuo, K., Büchi, G., Wogan, G. N., & Demain, A. L. (1978). Mollicellins: Mutagenic and antibacterial mycotoxins. *Applied and Environmental Microbiology, 36*, 412–420. Available from https://doi.org/10.1128/aem.36.3.412-420.1978.

Strobel, G., & Daisy, B. (2003). Bioprospecting for microbial endophytes and their natural products. *Microbiology and Molecular Biology Reviews, 67*, 491–502. Available from https://doi.org/10.1128/MMBR.67.4.491-502.2003.

Strobel, G. A. (2003). Endophytes as sources of bioactive products. *Microbes and Infection, 5*, 535–544. Available from https://doi.org/10.1016/S1286-4579(03)00073-X.

Sturz, A. V., Christie, B. R., & Matheson, B. G. (1998). Associations of bacterial endophyte populations from red clover and potato crops with potential for beneficial allelopathy. *Canadian Journal of Microbiology, 44*, 162–167. Available from https://doi.org/10.1139/cjm-44-2-162.

Sun, P., Huo, J., Kurtán, T., Mándi, A., Antus, S., Tang, H., Draeger, S., Schulz, B., Hussain, H., Krohn, K., Pan, W., Yi, Y., & Zhang, W. (2013). Structural and stereochemical studies of hydroxyanthraquinone derivatives from the endophytic fungus *Coniothyrium* sp. *Chirality, 25*(2), 141–148. Available from https://doi.org/10.1002/chir.22128.

Sun, R.-R., Miao, F.-P., Zhang, J., Wang, G., Yin, X.-L., & Ji, N.-Y. (2013). Three new xanthone derivatives from an algicolous isolate of *Aspergillus wentii*. *Magnetic Resonance in Chemistry, 51*(1), 65–68. Available from https://doi.org/10.1002/mrc.3903.

Talontsi, F. M., Lamshöft, M., Douanla-Meli, C., Kouam, S. F., & Spiteller, M. (2014). Antiplasmodial and cytotoxic dibenzofurans from *Preussia* sp. harboured in *Enantia chlorantha* Oliv. *Fitoterapia, 93*, 233–238. Available from https://doi.org/10.1016/j.fitote.2014.01.003.

Tansuwan, S., Pornpakakul, S., Roengsumran, S., Petsom, A., Muangsin, N., Sihanonta, P., & Chaichit, N. (2007). Antimalarial benzoquinones from an endophytic fungus, *Xylaria* sp. *Journal of Natural Products, 70*, 1620–1623. Available from https://doi.org/10.1021/np0701069.

Thomas, R. (1961). Studies in the biosynthesis of fungal metabolites. Alternariol monomethyl ether and its relation to other phenolic metabolites of *Alternaria tenuis*. *Biochemical Journal, 80*, 234–240. Available from https://doi.org/10.1042/bj0800234.

Toghueo, R. M. K. (2019). Anti-leishmanial and anti-inflammatory agents from endophytes: A review. *Natural Products and Bioprospecting, 9*, 311–328. Available from https://doi.org/10.1007/s13659-019-00220-5.

Tomoda, H., Nlshida, H., Huang, X. H., Masuma, R., Kim, Y. K., & Omura, S. (1992). New cyclodepsipeptides, enniatins D, E and F produced by *Fusarium* sp. Fo-1305. *Journal of Antibiotics (Tokyo), 45*, 1207–1215. Available from https://doi.org/10.7164/antibiotics.45.1207.

Tran-Cong, N. M., Mándi, A., Kurtán, T., Müller, W. E. G., Kalscheuer, R., Lin, W., Liu, Z., & Proksch, P. (2019). Induction of cryptic metabolites of the endophytic fungus: *Trichocladium* sp. through OSMAC and co-cultivation. *RSC Advances, 9*, 27279–27288. Available from https://doi.org/10.1039/c9ra05469c.

Trischman, J. A., Jensen, P. R., & Fenical, W. (1994). Halobacillin: A cytotoxic cyclic acylpeptide of the iturin class produced by a marine *Bacillus*. *Tetrahedron Letters, 35*, 5571–5574. Available from https://doi.org/10.1016/S0040-4039(00)77249-2.

Verma, V. C., Lobkovsky, E., Gange, A. C., Singh, S. K., & Prakash, S. (2011). Piperine production by endophytic fungus *Periconia* sp. isolated from *Piper longum* L. *Journal of Antibiotics, 64*, 427–431. Available from https://doi.org/10.1038/ja.2011.27.

Wang, K., Chen, Y., Gao, S., Wang, M., Ge, M., Yang, Q., ... Zhou, H. (2021). Norlichexanthone purified from plant endophyte prevents postmenopausal osteoporosis by targeting ERα to inhibit RANKL signaling. *Acta Pharmaceutica Sinica B, 11*, 442–455. Available from https://doi.org/10.1016/j.apsb.2020.09.012.

Wang, W. L., Lu, Z. Y., Tao, H. W., Zhu, T. J., Fang, Y. C., Gu, Q. Q., & Zhu, W. M. (2007). Isoechinulin-type alkaloids, variecolorins A-L, from halotolerant *Aspergillus variecolor*. *Journal of Natural Products, 70*, 1558–1564. Available from https://doi.org/10.1021/np070208z.

Wagenaar, M. M., & Clardy, J. (2001). Dicerandrols, new antibiotic and cytotoxic dimers produced by the fungus *Phomopsis longicolla* isolated from an endangered mint. *Journal of Natural Products, 64*(8), 1006–1009. Available from https://doi.org/10.1021/np010020u.

Weber, D., Sterner, O., Anke, T., Gorzalczancy, S., Martino, V., & Acevedo, C. (2004). Phomol, a new antiinflammatory metabolite from an endophyte of the medicinal plant Erythrina crista-galli. *Journal of Antibiotics, 57*(9), 559–563. Available from https://doi.org/10.7164/antibiotics.57.559.

Wells, J. M., Cutler, H. G., & Cole, R. J. (1976). Toxicity and plant growth regulator effects of cytochalasin H isolated from *Phomopsis* sp. *Canadian Journal of Microbiology, 22*, 1137–1143. Available from https://doi.org/10.1139/m76-165.

Wicklow, D. T., & Poling, S. M. (2009). Antimicrobial activity of pyrrocidines from *Acremonium zeae* against endophytes and pathogens of maize. *Phytopathology, 99*, 109–115. Available from https://doi.org/10.1094/PHYTO-99-1-0109.

Wijeratne, E. M. K., He, H., Franzblau, S. G., Hoffman, A. M., & Gunatilaka, A. A. L. (2013). Phomapyrrolidones A-C, antitubercular alkaloids from the endophytic fungus *Phoma* sp. NRRL 46751. *Journal of Natural Products, 76*(10), 1860–1865. Available from https://doi.org/10.1021/np400391p.

Wu, D. L., Li, H. J., Smith, D. R., Jaratsittisin, J., Xia-Ke-Er, X. F. K. T., Ma, W. Z., Guo, Y. W., Dong, J., Shen, J., Yang, D. P., & Lan, W. J. (2018). Polyketides and alkaloids from the marine-derived fungus *Dichotomomyces cejpii* F31-1 and the antiviral activity of scequinadoline A against dengue virus. *Marine Drugs, 16*, 229. Available from https://doi.org/10.3390/md16070229.

Xu, L., Wang, J., Zhao, J., Li, P., Shan, T., Wang, J., Li, X., & Zhou, L. (2010). Beauvericin from the endophytic fungus, Fusarium redolens, isolated from Dioscorea zingiberensis and its antibacterial activity. *Natural Product Communications, 5*(5), 1934578X1000500. Available from https://doi.org/10.1177/1934578X1000500527.

Yang, R., Dong, Q., Xu, H., Gao, X. H., Zhao, Z., Qin, J., … Luo, D. (2020). Identification of phomoxanthone A and B as protein tyrosine phosphatase inhibitors. *ACS Omega, 5*, 25927–25935. Available from https://doi.org/10.1021/acsomega.0c03315.

Yin, Z., & Dickschat, J. S. (2021). Cis double bond formation in polyketide biosynthesis. *Natural Product Reports, 38*, 1445–1468. Available from https://doi.org/10.1039/d0np00091d.

Zhang, H. W., Song, Y. C., & Tan, R. X. (2006). Biology and chemistry of endophytes. *Natural Product Reports, 23*, 753–771. Available from https://doi.org/10.1039/b609472b.

Zhang, J., Liu, D., Wang, H., Liu, T., & Xin, Z. (2015). Fusartricin, a sesquiterpenoid ether produced by an endophytic fungus Fusarium tricinctum Salicorn 19. *European Food Research and Technology, 240*(4), 805–814. Available from https://doi.org/10.1007/s00217-014-2386-6.

Zhang, M., Wang, W. L., Fang, Y. C., Zhu, T. J., Gu, Q. Q., & Zhu, W. M. (2008). Cytotoxic alkaloids and antibiotic nordammarane triterpenoids from the marine-derived fungus *Aspergillus sydowi*. *Journal of Natural Products, 71*, 985–989. Available from https://doi.org/10.1021/np700737g.

Zhao, S., Xiao, C., Wang, J., Tian, K., Ji, W., Yang, T., … Ye, Y. (2020). Discovery of natural FabH inhibitors using an immobilized enzyme column and their antibacterial activity against *Xanthomonas oryzae* pv. oryzae. *Journal of Agricultural and Food Chemistry, 68*, 14204–14211. Available from https://doi.org/10.1021/acs.jafc.0c06363.

Zheng, R., Li, S., Zhang, X., & Zhao, C. (2021). Biological activities of some new secondary metabolites isolated from endophytic fungi: A review study. *International Journal of Molecular Sciences, 22*, 1–75. Available from https://doi.org/10.3390/ijms22020959.

Zhu, F., Wu, J., Chen, G., Lu, W., & Pan, J. (2011). Biosynthesis, characterization and biological evaluation of Fe (III) and Cu(II) complexes of neoaspergillic acid, a hydroxamate siderophore produced by co-cultures of two marine-derived mangrove epiphytic fungi. *Natural Product Communications, 6*, 1137–1140. Available from https://doi.org/10.1177/1934578x1100600824.

CHAPTER

13

Antiviral potential of medicinal plants: a case study with guava tree against dengue virus using a metabolomic approach

Thomas Vial[1,2], *Chiobouaphong Phakeovilay*[1,3,*], *Satoru Watanabe*[2], *Kitti Wing Ki Chan*[2], *Minhua Peng*[2], *Eric Deharo*[1,4], *François Chassagne*[1], *Subhash G. Vasudevan*[2] and *Guillaume Marti*[1,3]

[1]UMR 152 PharmaDev, IRD, UPS, Université de Toulouse, Toulouse, France [2]Programme in Emerging Infectious Diseases, Duke-NUS Medical School, Singapore [3]Laboratoire de Recherche en Sciences Végétales and Metatoul-AgromiX Platform, MetaHUB, National Infrastructure for Metabolomics and Fluxomics, LRSV, Université de Toulouse, CNRS, UPS, Toulouse, France [4]Institut de Recherche pour le Développement, Vientiane, Lao PDR

Introduction

Dengue disease

Dengue is an arthropod-borne viral disease (arbovirus) that currently infects about 400 million people every year throughout the tropical and subtropical world (Bhatt et al., 2013). Dengue virus (DENV) is the most widespread arbovirus. It emerged in the second part of the 20th century (Gubler, 2002), and ever since its incidence has increased 30-fold,

* Contributed equally.

Medicinal Plants as Anti-infectives
DOI: https://doi.org/10.1016/B978-0-323-90999-0.00010-0

according to World Health Organization (WHO). DENV is transmitted by the bite of an infected female mosquito of the genus *Aedes*. Due to the *Aedes* geographical distribution, more than 40% of the global population, living in more than 120 countries, is at risk to contract dengue (Kraemer et al., 2018; Simmons, Farrar, Van Vinh Chau, & Wills, 2012). *Aedes* distribution which now extends to North America and Europe (Kraemer et al., 2018), in addition to the subtropical regions, where it is normally prevalent, has significant implications for increasing the burden of dengue (Fig. 13.1). DENV belongs to genus *Flavivirus*. There are four DENV serotypes: DENV-1, DENV-2, DENV-3, and DENV-4 (Simmons et al., 2012). Dengue is classified as a neglected tropical disease and development of many vaccines and novel antidengue drugs as well as novel vector control measures have been investigated in the last decade (Horstick, Tozan, & Wilder-Smith, 2015). However, although research activity has increased substantially in the past decade, findings for antiviral therapy have so far been unsatisfactory.

About 75% of people infected with DENV remain asymptomatic (not registered in hospital). The other 25% represents around 100 million patients that experience a range of different symptoms from mild flu-like illness to more severe dengue characterized by vascular leakage, hemorrhages, organ failure, and shock (Fig. 13.2) (Bhatt et al., 2013). WHO Dengue Guideline described the clinical manifestations of dengue fever with an incubation period of 3–7 days after viral infection by the mosquito. Dengue has a broad range of clinical symptoms and in some cases no symptoms appear up to 14 days. The most common symptom is the immediate onset of fever together with headaches, myalgia, arthralgia, pain behind the eyes, abdominal pain, nausea, and flushing of the face. A rash is often seen and can be maculopapular, morbilliform, macular, and scarlatiniform. WHO classified dengue fever manifestation into dengue

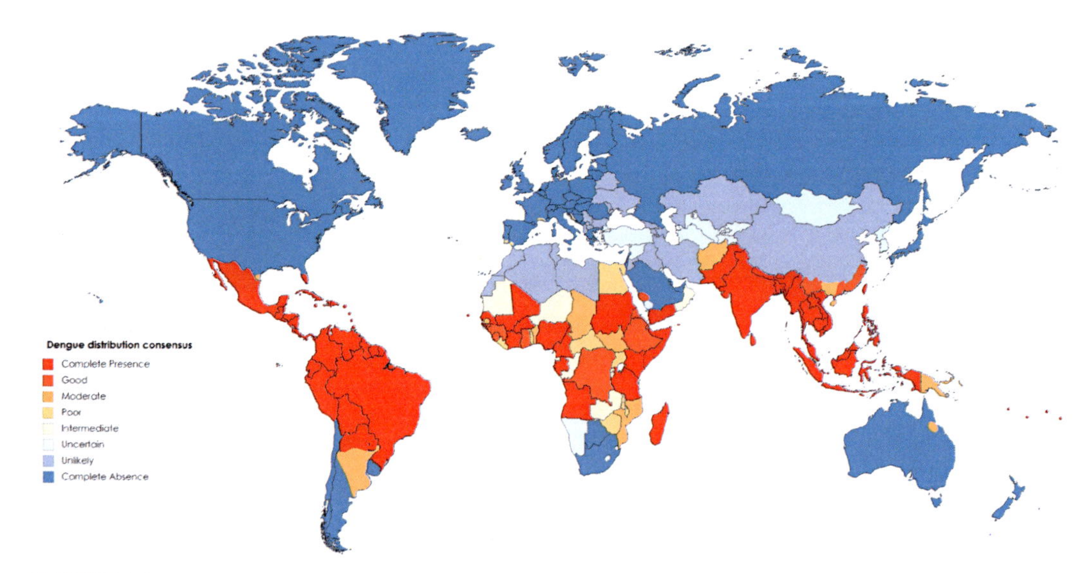

FIGURE 13.1 The global distribution of dengue. Source: *Adapted from CDC Dengue heatmap (http://www.healthmap.org/dengue), and data from the European Centre for Disease Prevention and Control (ECDC) (http://www.ecdc.europa.eu/en/dengue). National and local consensus of complete presence (red) or absence (blue) reported by local transmission. This map was created with mapchart.net and does not take into account imported dengue cases by travelers.*

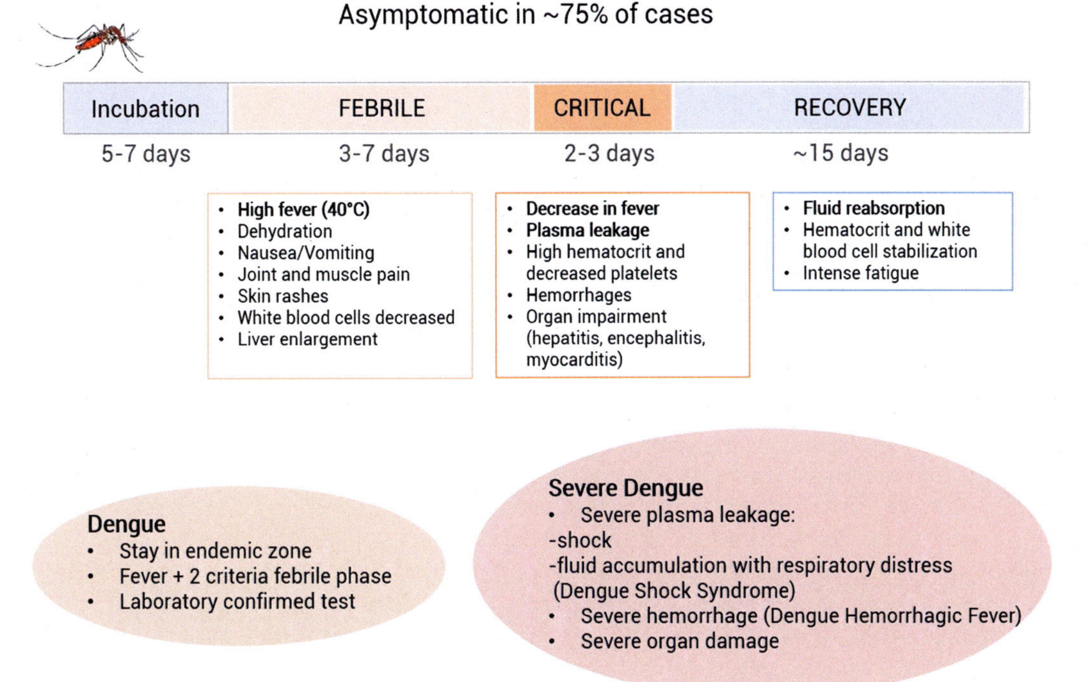

FIGURE 13.2 Dengue clinical phases and classification. Source: *Adapted from WHO Dengue Guidelines 2009. https://apps.who.int/iris/handle/10665/44188.*

and severe dengue. Dengue is characterized by abdominal pain, vomiting, clinical fluid accumulation, mucosal bleeding, lethargy, liver enlargement, and increase in hematocrit concurrent with a rapid decrease in platelet number. Severe dengue signs comprise severe plasma leakage leading to shock, severe bleeding, and severe organ impairment. Secondary infection from other serotypes increases the risk of developing severe dengue (World Health Organization, 2012).

Conventional treatment

No antiviral against DENV is currently available and only symptomatic care is provided to patients (Malina, Boon, Aurapa, & Azmath, 2019). It is recommended to stay hydrated and avoid anticoagulant such as aspirin-containing drugs. For severe dengue patients with shock syndrome, intravenous fluid supplementation is essential and prophylactic platelet transfusion is performed, although the latter does not prevent bleeding (Lye et al., 2017).

Multiple drugs with in vitro antiviral activity against DENV have been tested in clinical trials (chloroquine, balapiravir, celgosivir, lovastatin, prednisolone, ribavirin, zinc bisglycinate, vitamin E, and UV-4B), without success in preventing disease or lowering viremia (Wilder-Smith, Ooi, Horstick, & Wills, 2019). Several clinically approved drugs for other diseases have been evaluated for antidengue activity (Dighe et al., 2019). For example, ivermectin went in clinical trial in phase II/III in children and adult patients

(ClinicalTrials.gov number NCT02045069) and shown also a 50% decrease in infection rate of DENV-2-infected *Aedes albopictus* mosquito treated with ivermectin and almost complete clearance of DENV RNA (Xu et al., 2018). Ivermectin also inhibits in vitro replication of flaviviruses, mainly YFV and DENV with a lower effect for the latter, by targeting viral nonstructural (NS) protein NS3 helicase activity (Mastrangelo et al., 2012). Other drugs that target NS have been studied, but none reached clinical trials (Hernandez-Morales et al., 2017; Luo, Vasudevan, & Lescar, 2015). Neutralizing monoclonal antibodies are also candidates for dengue treatment (Sun, Chen, & Lai, 2018). However, none of them have been approved as anti-dengue drugs (Beesetti, Khanna, & Swaminathan, 2016; Dighe et al., 2019).

Medicinal plants

Given the lack of efficacy of repurposed drugs against the DENV and the difficulty in developing a universal vaccine with no contraindications, it is necessary to find new directions in the search for antivirals. The medicinal plants are an important source of active compounds for various diseases in the world, also for viral infections. The medicinal plant extracts and their derivatives have been suggested by WHO in the fight against dengue disease as they are comparatively less harmful, and cheaper than synthetic drugs. Currently, the identification of the active compounds against DENV from medicinal plants has attracted the researchers' interests. The use of medicinal plants in the management of dengue is widely reported, aiming to demonstrate antiviral effects of plant extracts or isolated compounds (Abd Kadir, Yaakob, & Mohamed, 2013; Ali, Chorsiya, Anjum, Khasimbi, & Ali, 2020; Ferreira et al., 2017; Qadir et al., 2015; Tang, Ling, Koh, Chye, & Voon, 2012). Although several studies have methodological variations in the experimental design, the biological model, or limited to an in silico model, medicinal plants products have the potential for anti-DENV development (Ferreira et al., 2017). An overview of the main natural extracts and their respective biosources depicted as active against dengue fever are presented in Table 13.1 (Kaushik, Kaushik, Sharma, & Yadav, 2018).

Case study: metabolomics reveal antidengue compounds isolated from *Psidium guajava*

Introduction

Psidium guajava: a potential antidengue medicinal plant

Psidium guajava Linn. or guava has been reported for its potential antidengue activity as mentioned in our literature review. In this context, studying the ethanolic leaf extract of guava should be of interest to find new active compounds against DENV. Guava belongs to the *Psidium* genus and Myrtaceae family. Guava is native to Mexico (Gutiérrez, Mitchell, & Solis, 2008), and distributed throughout the tropical and subtropical areas. Moreover, guava has been reported in traditional medicine in many regions, and its pharmacological activities have been widely studied for various diseases. Finally, some bioactive compounds were also reported from this plant species (Gutiérrez et al., 2008).

TABLE 13.1 Summary of reported antidengue medicinal plants.

Plant species	Family	Active form	Activity
Andrographis paniculata (Burm.f.) Nees	Acanthaceae	Methanolic extract	Inhibited the activity of DENV-1 in in vitro assays (doi:10.1186/1472-6882-12-3)
Azadirachta indica A. Juss.	Meliaceae	Leaves aqueous extract	Inhibited the DENV-2 replication by the absence of Dengue-related clinical symptoms in suckling mice confirmation (doi:10.1016/S0378-8741(01)00395-6)
Boesenbergia rotunda (L.) Mansf.	Zingiberaceae	4-hydroxypanduratin A and panduratin A	Inhibited the DENV-2 NS3 protease (doi:10.1016/j.bmcl.2005.12.075)
Carica papaya L.	Caricaceae	Leaves aqueous extracts	Exhibited potential activity against dengue fever in 45-year-old patient by increasing the platelets count, WBC, and neutrophils (doi:10.1016/S2221-1691(11)60055-5)
Castanospermum australe A. Cunn. & C. Fraser	Fabaceae	Castanospermine	Castanospermine inhibited at the level of secretion and infective of virus particles (doi:10.1128/JVI.79.14.8698-8706.2005)
Cissampelos pareira L.	Menispermaceae	Ethanolic extracts	Showed antiviral activity against all types of dengue virus and reduced the production of TNF-α in Wistar rats-induced severe dengue disease (doi:10.1371/journal.pntd.0004255)
Cryptonemia crenulata (J. Agardh) J. Agardh	Halymeniaceae	Carrageenan G3d and DL-galactan hybrid C2S-3	Active against DENV-2 at the early infection stage means these compounds target virus absorption and internalization (doi:10.1016/j.antiviral.2005.02.001)
Euphorbia hirta L.	Euphorbiaceae	Leaf extract	Inhibited the viral serotype 1 and increased platelet count in the animal study (doi:10.1155/2018/2048530)
Gastrodia elata Blume	Orchidaceae	D-glucans sulfated derivatives	Strongly interfering with the DENV-2 infections mainly in virus adsorption, in a very early stage of the virus cycle (doi:10.1016/j.carres.2007.06.021)
Gymnogongrus griffithsiae (Turner) C. Martius	Phyllophoraceae	Carrageenan G3d and DL-galactan hybrid C2S-3	Active against DENV-2 at the early infection stage means these compounds target virus absorption and internalization (doi:10.1016/j.antiviral.2005.02.001)
Gymnogongrus torulosus (J.D. Hooker & Harvey) F. Schmitz	Phyllophoraceae	DL-galactan hybrid	Inhibited the DENV-2 serotype in Vero cells (doi:10.1177/095632020201300202)
Houttuynia cordata Thunb.	Saururaceae	Aqueous extract (hyperoside)	Significantly reduce intracellular DENV-2 RNA production in HepG2 cells (doi:10.1111/j.1745-4514.2010.00514.x)
Hippophae rhamnoides L.	Elaeagnaceae	Leaf extract	Be able to maintain cell viability of dengue infected cells, decreases in TNF-α, and increases IFN-γ (doi:10.1016/j.phymed.2008.04.017)
Lippia alba (Mill.) N.E.Br. ex Britton & P. Wilson	Verbenaceae	Essential oil	Inactivated virus before adsorption on host cells (doi:10.1590/S0074-02762010000300010)

(*Continued*)

TABLE 13.1 (Continued)

Plant species	Family	Active form	Activity
Leucaena leucocephala (Lam.) de Wit	Fabaceae	Seeds extract	Demonstrated activity against DENV-1 in vitro and in vivo level (doi:10.1016/S0166-3542(03)00175-X)
Meristiella gelidium (J. Agardh) D.P. Cheney & P.W. Gabrielson	Solieriaceae	Extract and carrageenan derivatives	Antiviral activity against DENV-2 (doi:10.1016/j.carbpol.2005.09.020)
Mimosa scabrella Benth.	Fabaceae	Seeds extract	Demonstrated activity against DENV-1 in vitro and in vivo level (doi:10.1016/S0166-3542(03)00175-X)
Myrtopsis corymbosa (Labill.) Guillaumin	Rutaceae	Myrsellinol, ramosin, and myrsellin	Inhibited about 87% against DENV polymerase (doi:10.4103/phrev.phrev_2_18)
Ocimum sanctum L.	Labiatae	Leaves extract	Inhibited the DENV-1 serotype in cell lines (doi:10.4103/phrev.phrev_2_18)
	Phyllanthus urinaria L.		Phyllanthaceae Aqueous and methanolic extract Showed strongest inhibitory activity against DENV-2 with more than 90% of virus reduction (doi:10.1186/1472-6882-13-192)
Psidium guajava L.	Mrytaceae	Bark extract	Demonstrated good activity on DENV-2 serotype (doi:10.1186/s12906-019-2695-1)
	Tephrosia madrensis Seem.		Fabaceae Flavonoids isolated glabranine and 7-O-methyl-glabranine Strongly inhibited dengue virus replication in Rhesus monkey epithelial cells (doi:10.4103/phrev.phrev_2_18)
Uncaria tomentosa (Willd. ex Schult.)DC.	Rutaceae	Alkaloidal fraction	Reduced monocyte infection rates and cytokine levels (doi:10.1016/j.intimp.2007.11.010)
Zostera marina L.	Zosteraceae	p-sulfoxycinnamic acid	Antidengue activity against DENV-2 serotype in LLC-MK2 cell lines (doi:10.1016/j.antiviral.2008.05.007)

The ethanolic leaf extract of guava was reported to have a good activity against the DENV-2 with an IC_{50} of 7.2 μg/mL (Saptawati et al., 2017). Two years later, Correa et al. reported that the ethanolic extracts and fractions of guava bark demonstrated a good activity on DENV-2, along with five identified compounds (gallic acid, naringin, quercetin, catechin, and hesperidin) from the most active fractions with $EC_{50} = 17.7$ μg/mL and SI = 35.4. Catechin was the most active with more than 90% inhibition. From the in silico molecular docking with virus protein NS5 and E protein, naringin and hesperidin had a better interaction score than the theoretical threshold that showed higher affinity to NS5 protein (Trujillo-Correa et al., 2019).

A metabolomic approach in antiviral compound identification

Metabolomics is an emerging "omics" that identifies and quantifies metabolites from cells, biofluids, tissues, and organisms (Schrimpe-Rutledge, Codreanu, Sherrod, & McLean, 2016). Furthermore, metabolomic analysis is a powerful method to identify active compounds. Metabolomics was previously described to decipher the most active redox compounds in crude extracts of *Viola alba* subsp. *dehnhardii*, Violaceae (Chervin et al., 2017) and to identify antihepatocarcinogenic compounds from plants used to treat liver cancer in Cambodia (Chassagne et al., 2018). Metabolomic approaches can help to identify compounds having potential antiviral activity, as in the case of herpes virus simplex (Haggag et al., 2019; Prinsloo & Vervoort, 2018).

Objectives

The aim of this study is to find new antidengue compounds from leaf extracts of *P. guajava* L. using a metabolomic approach and correlation analysis. Our approach aims to link the chemical profile variability using a fast crude fractionation method to bioassay results by means of multivariate data analysis (MVA). This holistic method has been implemented to rapidly understand the diversity and role of chemical components involved in bioactivity and is in contrast to the reductionist approach that involves successive fractionation steps for the purification of compounds responsible for the activity (Ayouni et al., 2016). The use of partial least square (PLS) regression models will provide a ranking of putative active features detected from liquid chromatography—mass spectrometry (LC—MS) profiles. The annotation of peaks of interests was based on high-resolution mass spectrometry (HRMS) and tandem mass spectrometry (MS/MS) patterns mirrored to in silico fragmentation and confirmed by purification or commercial authentic standards (Tsugawa et al., 2016).

Results

UHPLC-HRMS-based metabolomics approach

We followed the same steps to validate our approach as previously described (Chervin et al., 2017). Briefly, nine extracts collected in Lao PDR (three crude extracts from the Champasak province, three crude extracts from the Savannakhet province, and three crude extracts from the Vientiane province) were pooled according to their collection area and submitted to a rapid solid-phase extraction (SPE) step. This fractionation provided eight fractions of increasing polarity for each pooled extract (a total of 24 fractions). The ultra-high-performance liquid chromatography–high resolution mass spectrometry (UHPLC-HRMS) profiles displayed 448 features (*m/z*-RT pairs) in negative ionization (NI) mode and 163 features (*m/z*-RT pairs) in positive ionization (PI) mode after data processing using MS-CleanR workflow (Fraisier-Vannier et al., 2020).

As an initial step, principal component analysis (PCA) was applied to obtain an unsupervised overview of the LC-MS fingerprints of 24 SPE fractions. PCA gathers all independent biological replicates from the same fraction. The corresponding results are presented in Fig. 13.3. Crude extracts [denoted quality control (QC)] were plotted near the center of the PCA score plot while each independent fraction clustered together according to their respective polarity. The first principal component axis separated polar and apolar fractions. These results indicated that SPE fractionation and LC-MS workflows highlighted variability in the data set and were reproducible. Moreover, it demonstrated a stable chemical composition within extracts of guava collected in the three distinct areas.

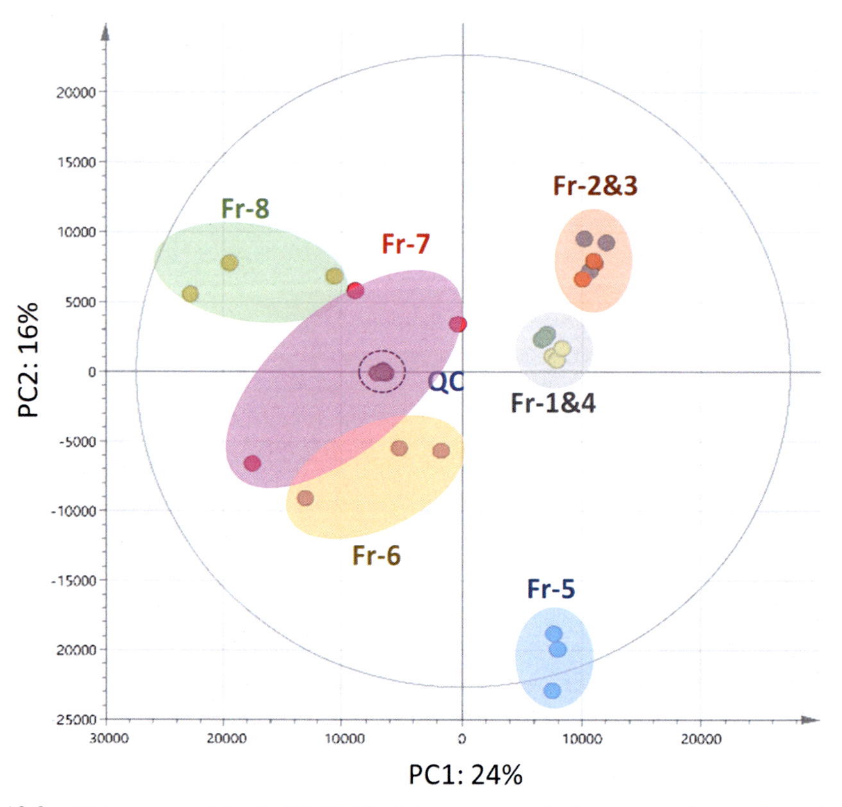

FIGURE 13.3 PCA score plot of ESI-NI/PI dataset grouped between polar and apolar fractions. *PCA*, Principal component analysis; *NI*, negative ionization; *PI*, positive ionization.

Antidengue activity

The crude extracts and fractions were evaluated for their antiviral activity against DENV-2 serotype in a cell-based infection assay as previously described (Cannalire et al., 2020; Watanabe et al., 2016). The inhibition percentages were evaluated at three concentrations (200, 100, and 10 μg/mL) by both cotreatment and posttreatment. Under the cotreatment condition, Huh-7 cells were supplemented by crude extracts or fractions and DENV2 at the same time. All crude extracts of guava showed 100% DENV2 inhibition at 100 and 200 μg/mL and over 70% DENV-2 inhibition at 10 μg/mL. Under posttreatment condition, Huh-7 cells were infected by DENV-2 and then treated with the crude extracts or fractions. For the 24 SPE fractions tested in cotreatment or posttreatment, the most active were the apolar fractions (fractions 7 and 8) that showed more than 90% DENV-2 inhibition activity at 100 and 200 μg/mL and over 70% DENV-2 inhibition at 10 μg/mL. Guava crude extracts and fractions displayed low cytotoxicity at 10 μg/mL (Fig. 13.4).

An orthogonal projection to latent structures (OPLS) regression model was built to get a ranking of the loadings (*m/z*-RT pairs) toward the antidengue inhibition percentage (input Y). This supervised technique provided a classification of potentially antidengue compounds according to their regression coefficient values. The classification is presented in Fig. 13.5A: positive coefficients were related to the high correlation with antidengue activity, whereas negative coefficients were related to low correlation with antidengue activity. The coefficient plots of the four top-ranked features are presented in Fig. 13.5B. From the list of first features (Fig. 13.5C), we implemented the annotation procedure based on their respective accurate masse and Ms/MS spectra.

Identification of putative antidengue compounds

The first four ranked compounds from the OPLS regression model were putatively annotated by matching experimental mass spectra to several databases [Universal Natural

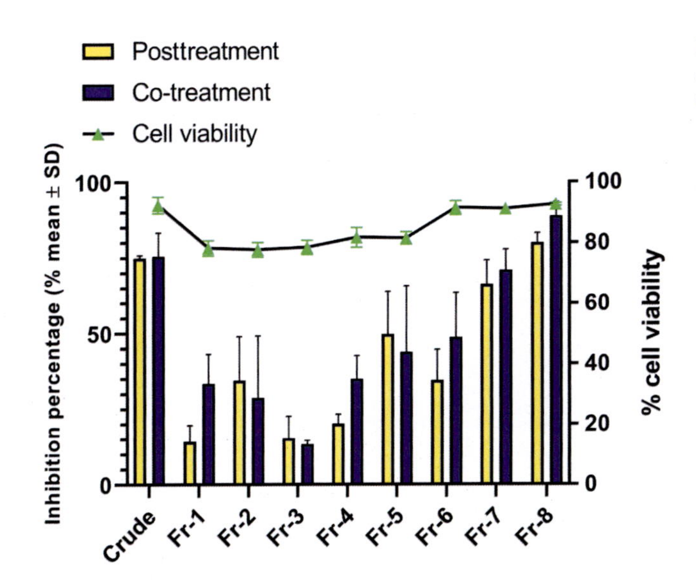

FIGURE 13.4 DENV-2 inhibition percentage of each fraction (Fr) obtained by C18-SPE at 10 μg.

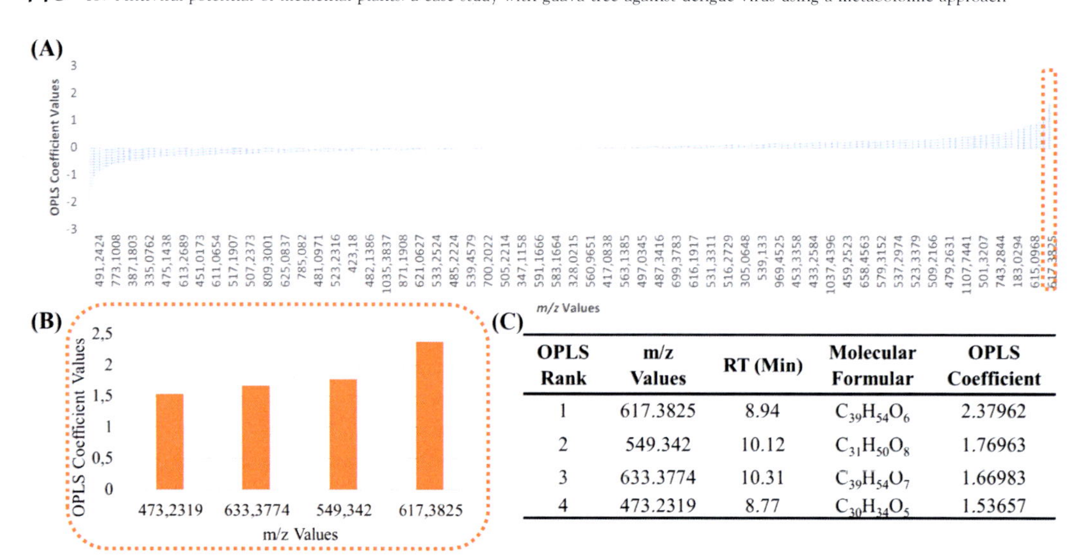

OPLS Rank	m/z Values	RT (Min)	Molecular Formular	OPLS Coefficient
1	617.3825	8.94	$C_{39}H_{54}O_6$	2.37962
2	549.342	10.12	$C_{31}H_{50}O_8$	1.76963
3	633.3774	10.31	$C_{39}H_{54}O_7$	1.66983
4	473.2319	8.77	$C_{30}H_{34}O_5$	1.53657

FIGURE 13.5 OPLS regression analysis and classification of putative antidengue. (A) Coefficient plot obtained by OPLS regression; (B) emphasis on the first loadings; (C) First ranked feature details. *OPLS*, Orthogonal projection to latent structures.

Products Database (UNPD), KNApSAc, PlantCyc, Dictionary of Natural Products (DNP, CRC Press, v25.2) and CheBI]. Compounds belonging to the Myrtaceae family and the *Psidium* genus were prioritized to narrow down the possibilities. For each compound, some candidates were ranked and proposed based on their similarity score according to the comparison between experimental Ms/MS fragmentation and the in silico spectra of candidates using MS-FINDER (Tsugawa et al., 2016). This process resulted in the annotation of four candidates belonging to the *Psidium* genus, and all were classified as triterpenoid compounds. Annotation results were supported by UV spectra of each peak (Table 13.2; column "UV"). The UV spectra of two triterpenoids (compounds 1 and 3) showed an absorption band at 310 nm highlighting the presence of a conjugated radical. Four compounds were putatively identified, jacoumaric acid (JA), guavacoumaric acid, and guajadial B which were already discovered in the *Psidium* genus, and terminolic acid first isolated from *Terminalia* genus (King, King, & Ross, 1955) (Fig. 13.6). Interestingly, three of them have a ursane triterpenoid backbone, two of them (guavacoumaric acid and JA) are position isomer of coumaric acid moiety. For three compounds, JA, terminolic acid, and guajadial B, commercial authentic standards were used to confirm feature annotation and biological activity against DENV.

Antidengue assay of pure authentic standards

Antidengue assays on DENV serotypes DENV-1, DENV-2, DENV-3, and DENV-4 were carried out for the first three compounds dereplicated by using chemical standards. The results, presented in Table 13.3, showed that the first ranked compound, JA, is more active than the other two authentic standards (terminolic acid and guajadial B) with an EC_{50} of 9.62, 4.62, 4.31, and 2.52 μg/mL, respectively. Furthermore, JA also

TABLE 13.2 Summary of all the compounds identified or dereplicated for antidengue activity.

OPLS rank[a]	*m/z* values	RT	Adduct type	MF	Δ *m/z* (mDa)	Main MS/Ms fragments	UV (nm)	Putative annotation	Score	Chemical class	Coefficient value	Biology source
1	617.3825	10.31	[M-H]-	$C_{39}H_{54}O_6$	1.563	601.1249 497.3494 373.2319	310	Jacoumaric acid[b,c]	6.2645	Ursane triterpenoid	2.37962	*Jacaranda cauranda*
2	549.342	7.1	[M + FA-H]-	$C_{30}H_{48}O_6$	1.2918	529.1945 435.2639 282.9619 161.1879	ND	Terminolic acid[b,c]	6.1864	Oleanane triterpenoid	1.76963	*Terminalia glaucescens*
3	633.3774	8.94	[M-H]-	$C_{39}H_{54}O_7$	1.0776	612.5383 513.4674 380.8951	310	Guavacoumaric acid[a]	5.24	Ursane triterpenoid	1.66983	*Plumeria obtusa*
4	473.2319	12.46	[M-H]-	$C_{30}H_{34}O_5$	1.0477	445.3014 356.8398 270.1221	ND	Guajadial B[b,c]	5.6465	Meroterpenoid	1.53657	*Psidium guajava*

[a]*Ranking based on OPLS regression coefficients, served as compound number.*
[b]*Determined by in silico Ms/MS fragmentation with Ms-FINDER.*
[c]*Confirmed by commercial authentic standard compounds.*
ND, Not detected; *MF*, molecular formula; *Ms/MS*, tandem mass spectrometry; *OPLS*, orthogonal partial least squares/projections to latent structures; *UV*, ultraviolet.

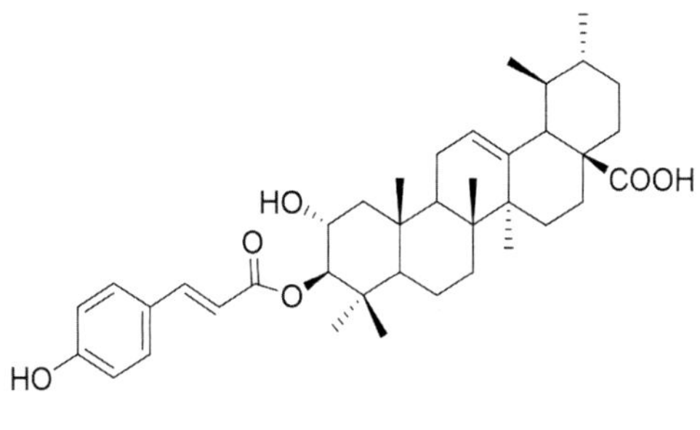

jacoumaric acid (**1**)

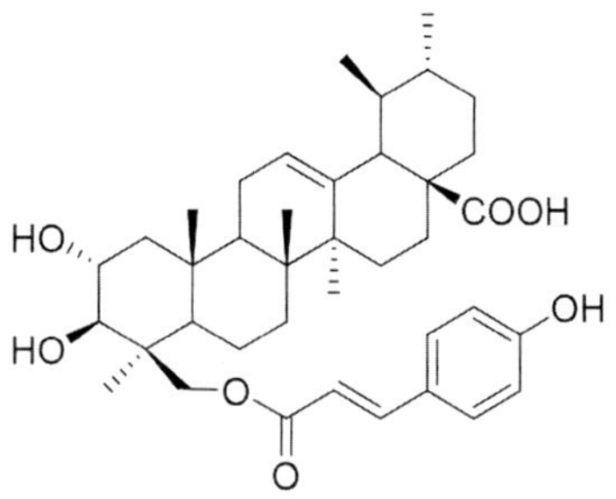

guavacoumaric acid (**3**)

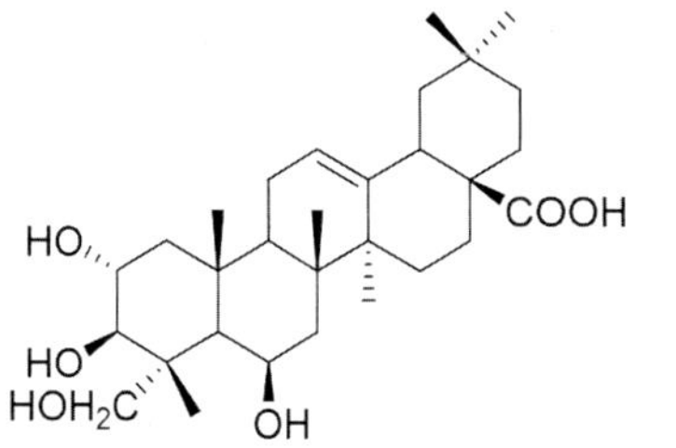

terminolic acid (**2**)

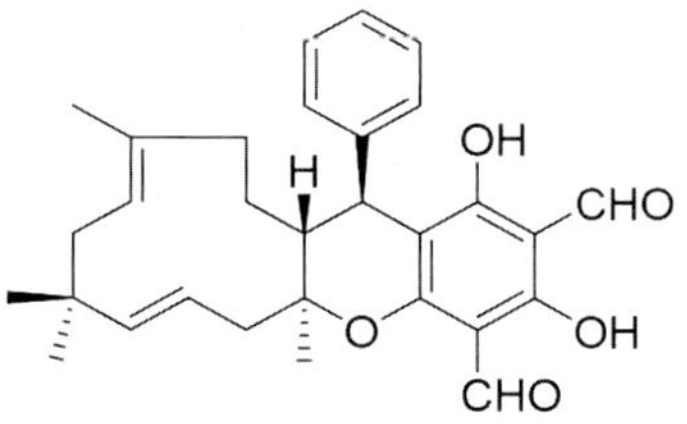

guajadial B (**4**)

FIGURE 13.6 Chemical structure of annotated compounds (see Table 13.3 for detail).

TABLE 13.3 Confirmation of the antidengue activity of top-ranked compounds by the OPLS regression model.

Compound	Virus	EC_{50} (μg/mL)	CC_{50} (μg/mL)	SI
Jacoumaric acid	DENV-1	9.62	30	3.12
	DENV-2	4.62		6.49
	DENV-3	4.31		6.96
	DENV-4	2.52		11.9
Terminolic acid	DENV-1	–	>50	–
	DENV-2			
	DENV-3			
	DENV-4			
Guajadial B	DENV-1	–	>50	–
	DENV-2			
	DENV-3			
	DENV-4			

OPLS, Orthogonal projection to latent structures.

showed a good selectivity towards the DENV serotypes 24 with a selective index value over 5 for DENV 2 to 4.

Discussion

The aim of this study was to identify antidengue compounds from the ethanolic leaf extract of *P. guajava*. According to the inhibition percentage at 10 μg/mL, crude extract of *P. guajava* could inhibit more than 70% of DENV-2 serotype, which is more than the ethyl acetate fraction of *Euphorbia hirta* L. (Tayone, Tayone, & Hashimoto, 2014). Other studies carried out on eight Indonesian plant extracts also confirmed that *P. guajava* and *Carica papaya* leaf extracts were active against DENV-2 serotype at an EC_{50} value of 7.2 and 6.57 μg/mL, respectively (Saptawati et al., 2017), but no information about compounds supporting this activity has been provided so far. Our metabolomics approach fills this knowledge gap by using a fast fractionation step before hyphenation of LC-Ms profiles to biological assays results. The apolar fractions obtained from SPE fractionation showed over 60% inhibition against DENV-2. To rank statistically related compounds to the biological activity against DENV, we applied an OPLS regression model. These analyses allowed the identification of jacoumaric aid, terminolic acid, guavacoumaric acid, and guajadial B as the putative most active compounds of apolar fractions.

Guava is widely used throughout the world for food and traditional medicine. The *Psidium guajava* leaf has been used as an infusion or a decoction. Guava is mainly used as an antidiarrheal. Also, guava was reported to be used as an antiinflammatory, for diabetes, hypertension, wound, pain relief, and fever (Gutiérrez et al., 2008). All parts of guava are used, especially its leaves (Dakappa, Adhikari, Timilsina, & Sajjekhan, 2013; Elbert, 1964; Gutiérrez et al., 2008; Heinrich, Ankli, Frei, Weimann, & Sticher, 1998). Previous studies reported various biological activities of guava including antioxidant, antiinflammatory, antibacterial, lipid-lowering agent, anticough, antidiarrheal, antidiabetic, cardioprotective, antimutagenic, hepatoprotective, and larvicidal effects (Ngbolua et al., 2018). Various types of phytochemicals were reported from the *P. guajava* including primary metabolites; mineral, enzymes, proteins, and secondary metabolites; triterpenoids, sesquiterpenoids, alkaloids, glycosides, flavonoids (guajaverin and quercetin), tannins, and saponins (oleic acid, morin-3-*O*-α-L-lyxopyranoside, and morin-3-*O*-α-L-arabopyranoside) (Arima & Danno, 2002; Dakappa et al., 2013). Moreover, the leaves of guava contain tannins, triterpenes, flavonoids, alkaloids, phytosterols, and essential oil rich in cineol (Manikandan, Anand, & Muthumani, 2013). Many compounds related to the triterpenoid class were isolated from guava leaves, some of these compounds were already known and others were newly identified (Begum, Hassan, & Siddiqui, 2002; Begum, Hassan, Siddiqui, Shaheen, et al., 2002; Shao et al., 2012).

A previous study identified different antidengue compounds obtained from *Psidium guajava* extract from the bark, belonging to phenolic acid and flavonoid, such as gallic acid, quercetin, and catechin (Trujillo-Correa et al., 2019). Catechin had the highest antiviral activity. The difference observed with our study, where triterpenoids were identified with high antiviral affinity from the leaf, could be explained by the part of the plant used. Other triterpenoids purified from the fungi *Ganoderma lucidum* have been identified as a potential compound with antidengue activity, such as ganodermanontriol and celastrol (Bharadwaj et al., 2019; Yu et al., 2017). Furthermore, celastrol has the ability to inhibit replication of all four DENV serotypes, associated with innate immune stimulation by the induction of the interferon expression and antiviral response in vitro. Another widespread flavonoid, luteolin, which can be isolated from traditional Chinese medicinal plant, can inhibit the replication of DENV1-4 serotypes by reducing infectious virus particle formation. Antiviral activity of luteolin was also demonstrated in vivo in DENV-infected mice (Peng et al., 2017). Luteolin acts on a host protease and counteracts the maturation process of newly produced viruses (Peng et al., 2018). This illustrates both the diversity of compound classes, such as flavonoids and triterpenoids, that may have antidengue activity, and the diverse activities on the viral infectious cycle of these compounds isolated from medicinal plants.

We reported here the antidengue activity of three of the annotated compounds (JA, terminolic acid, and guajadial B), confirmed by measuring the activity of their authentic standards. The tests showed that JA was the most active against DENV14 with EC_{50} values of 9.62, 4.62, 4.31, and 2.52 μg/mL, respectively. Based on the EC_{50} value, jacoumaric acid is more active than previously reported compounds from *P. guajava* bark, that is, gallic acid (25.8 μg/mL), naringin (47.9 μg/mL), quercetin (19.2 μg/mL), catechin (33.7 μg/mL), and hesperidin (225.8 μg/mL) against DENV-2 serotype (Trujillo-Correa et al., 2019). One of the major advantages of JA is its activity against all four serotypes of DENV. JA was identified as potential histone deacetylase inhibitor by in silico methods, which could play a role in epigenetic modifications and cell proliferation (Adewole & Ishola, 2020). JA could have antimicrobial activity, as we previously described against leishmania using the same metabolomic approach (Phakeovilay et al., 2019). Although JA was identified as potential HIV-1 protease inhibitor by in silico virtual screening (Yanuar, Suhartanto, Mun'im, Anugraha, & Syahdi, 2014), to our knowledge, no further studies have shown any activity against HIV.

Based on the results of our study, a metabolomic workflow has been successfully applied to rapidly identify the antidengue compounds from the *P. guajava* leaf extract. Our dereplication approach led to identify JA as the most active compounds against DENV1–4 serotypes in vitro. This is the first report of antidengue activity of JA. Further studies would be required to understand the mechanism of action of JA against the DENV both in vitro and in vivo.

Materials and methods

Plant collection

Nine samples of *Psidium guajava* L. (Pg) leaves were collected from middle and south part of Lao PDR (Champasak: Pg1 to Pg 3, Savannakhet: Pg4 to Pg6 and Vientiane province: Pg7 to Pg9, Fig. 13.7). Samples were washed and dried before being grounded into powder to obtain 1 kg of each. A voucher specimen of each sample was collected and deposited at the herbarium of the Institute of traditional medicine of Lao, Vientiane, Lao PDR. Permission for collection has been obtained from the competent authorities.

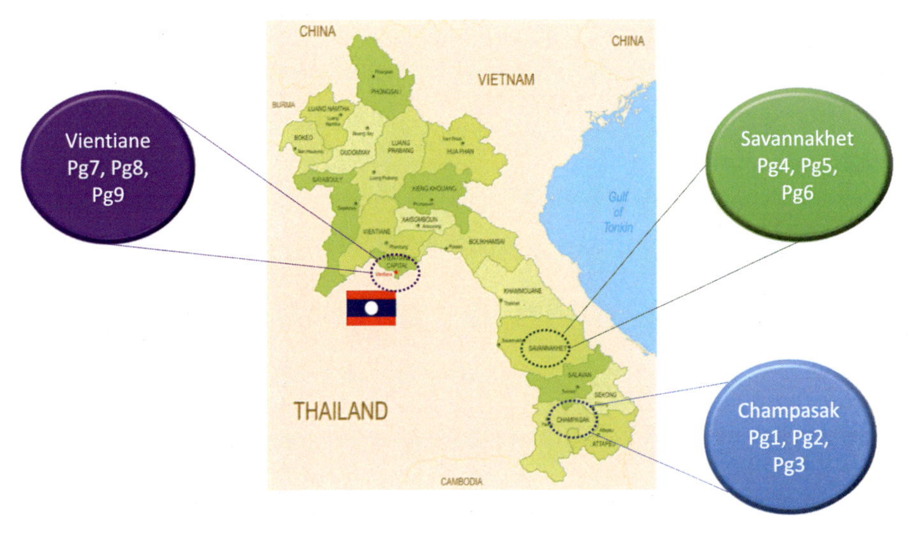

FIGURE 13.7 Collection areas of *Psidium guajava* L.

Leaf extraction

For each accession area, one sample of *P. guajava* leaves (250 g) was extracted by 2.5 L of 80% ethanol (EtOH) under agitation at room temperature for 24 h. The filtrated solution was evaporated under reduced pressure (Buchi rotavapor R-114). Then, 100 mg of each crude extract was dissolved in 1 mL of water and fractionated by SPE (1 g Sep-Pak C18 cartridge, Waters, Milford, MA, USA). Each extract was separated using eight aqueous methanolic solutions of decreasing polarity (H_2O/MeOH; 100/0, 90/10, 80/20, 70/30, 40/60, 30/70, 20/80, 0/100). Eight fractions were obtained: fraction 1 (25 mg), fraction 2 (5 mg), fraction 3 (10 mg), fraction 4 (8 mg), fraction 5 (6 mg), fraction 6 (8.5 mg), fraction 7 (4 mg), and fraction 8 (11 mg). All fractions were separated into two parts, first part was used for UHPLC-HRMS profiling at 1 mg/mL in 80% of MeOH, and second part was sent to Singapore laboratory for antidengue assay.

Cells and virus

Huh-7 [hepatocellular carcinoma cells, *Aedes albopictus* cell line (ATCC)] cells were cultured in a DMEM medium containing 10% FBS, 1% penicillin/streptomycin (P/S) at 37°C in 5% CO_2. BHK-21 (baby hamster kidney fibroblast cells, ATCC) cells were cultured in RPMI1640 medium containing 10% FBS, 1% P/S, at 37°C in 5% CO_2. C6/36, an ATCC, was maintained in RPMI1640 medium containing 25 mM HEPES, 10% FBS, and 1% P/S, at 28°C in the absence of CO_2.

DENV-2 EDEN 3295 (GenBank accession EU081177) was obtained from the Early Dengue infection and outcome (EDEN) study in Singapore (Low et al., 2006). Virus was grown in C6/36 cells and the supernatants were stored at −80°C. Virus titer was determined by plaque assay on BHK-21 cells.

Extracts preparation

The three crude extracts and 24 fractions were sent to the laboratory of Duke-NUS Medical School to evaluate their antidengue activity and were prepared by dissolution in DMSO to obtain solutions at 250 mg/mL.

Cell viability assay

For measurement of compound cytotoxicity, Huh-7 cells were seeded at 2×10^4 cells per well with 10% FBS medium in 96-well white flat-bottom plate. Cells were incubated for 48 h with various concentrations of compounds tested. Cell viability was measured using the CellTiterGlo Luminescent cell viability Assay (Promega) kit according to the manufacturer's instructions. Luminescence was measured on a microplate reader (Tecan Infinite 200 PRO) with a 100 Ms integration time. Cell viability is expressed as the percentage of luminescence derived from treated samples relative to that of the untreated control.

Virus infection

Huh-7 cells were seeded in a 24-well plate at 1×10^5 cells per well. Three compound concentrations (200, 100, and 10 μg/mL) were used. For cotreatment, cells were infected with DENV-2 at a multiplicity of infection (MOI) of 0.3 in the presence of compounds for 1 h. Virus/drug inoculums were removed and a fresh medium containing the indicated concentrations of compounds was added. For posttreatment, cells were infected with DENV-2 at a MOI of 0.3 for 1 h. Virus inoculums were removed and a fresh medium containing the indicated concentrations of compounds was added. Cells were incubated for additional 48 h at 37°C and the supernatants were collected. Virus titers in the supernatants were determined by plaque assay using BHK-21 cells. The molecule NITD008 was used as a positive control for DENV infection (Yin et al., 2009).

UHPLC-HRMS profiling

All extracts were profiled using a UHPLC-DAD-LTQ Orbitrap XL instrument (Ultimate 3000, Thermo Fisher Scientific, Hemel Hempstead, UK). The UV detection was performed by a diode array detector (DAD) from 210 to 400 nm. Mass detection was performed using an electrospray source in positive (PI) and NI modes at 15,000 resolving power [full width at half maximum (FWHM) at 400 *m/z*]. The mass scanning range was *m/z* 100–1500 Da. The capillary temperature was 300°C and voltage was fixed at 4.2 kV (positive mode) and 3.0 kV (negative mode). Mass measurement was externally calibrated before starting the acquisition. Each full Ms scan was followed by data-dependent Ms/MS on the four most intense peaks using collision-induced dissociation (35% normalized collision energy, isolation width 2 Da, activation *Q* of 0.250). The LC–MS system was run in binary gradient mode using a BEH C18 Acquity column (100 × 2.1 mm i.d., 1.7 μm, Waters, MA, USA) equipped with a guard column. Mobile phase A (MPA) was 0.1% formic acid (FA) in water and mobile phase B was 0.1% FA in acetonitrile. Gradient conditions were: 0 min, 95% MPA; 0.5 min 95% MPA; 12 min, 5% MPA; 15 min, 5% MPA; 15.5 min, 95% MPA; 19 min, 95% MPA. The flow rate was 0.3 mL/min, column temperature 40°C, and injection volume 2 μL.

Data processing

The UHPLC-HRMS raw data were converted to abf files (Reifycs Abf Converter) and processed with Ms-DIAL version 2.56 (Tsugawa et al., 2015) for mass signal extraction between 100 and 1500 Da from 0 to 15 min. Respective Ms1 and Ms2 tolerance were set to 0.01 and 0.2 Da in centroid mode. The optimized detection threshold was set to 2.5×10^4 for Ms1 and 5 for Ms2. Adducts and complexes were identified to exclude them from the final peak list. Finally, the peaks were aligned on a QC reference file with a retention time tolerance of 0.1 min and a mass tolerance of 0.025 Da. The resulting peak list was treated with MS-CleanR (Fraisier-Vannier et al., 2020) and then exported to comma-separated value format prior to MVA using SIMCA-P + (version 14.0, Umerics, Umea, Sweden).

Statistical analysis

For multivariate data analysis, all data were log-transformed and Pareto scaled. The OPLS regression analysis was done with antidengue inhibition values as Y input. Coefficient scores were used to rank variables according to their DENV-inhibition potential. For each model, a leave-one-subject-out cross-validation was performed to assess the model fit. The validity of the discriminant model was verified using permutation tests (Y-scrambling).

Identification of significant features

Molecular formulae of significant features were calculated with Ms-FINDER 2.12 (Tsugawa et al., 2016). Various parameters were used in order to reduce the number of potential candidates, such as the element selection exclusively including C, H, O; mass tolerance fixed to Ms1:0.01 Da and Ms2:0.2 Da, and the isotopic ratio tolerance set to 20%. Only natural product databases focused on plants were selected from UNPD, KNApSAc, PlantCyc, DNP (CRC Press, v25:2), and CheBI. Compounds from *Psidium* genus or Myrtaceae family were prioritized. The results were presented as a list of compounds sorted according to the score value of the match. This value encompassed uncertainty on accurate mass, the isotopic pattern score, and the experimental MS/MS fragmentation mirrored to in silico matches. Only chemical identities were retained with a final score above 5.

Acknowledgments

Chiobouaphong Phakeovilay was supported by Pierre Fabre Foundation and Thomas Vial by IRD. We also thank Julien Pompon for his collaborative role in the intermediation of the different teams.

References

Abd Kadir, S. L., Yaakob, H., & Mohamed Zulkifli, R. (2013). Potential anti-dengue medicinal plants: A review. *Journal of Natural Medicines*, *67*, 677–689. Available from https://doi.org/10.1007/s11418-013-0767-y.

Adewole, K. E., & Ishola, A. A. (2020). A computational approach to investigate the HDAC6 and HDAC10 binding propensity of *Psidium guajava*-derived compounds as potential anticancer agents. *Current Cancer Drug Targets*, *18*(3), 423–436. Available from https://doi.org/10.2174/1568009620666200502013657.

Ali, F., Chorsiya, A., Anjum, V., Khasimbi, S., & Ali, A. (2020). A systematic review on phytochemicals for the treatment of dengue. *Phytotherapy Research*, *35*(4), 1782–1816. Available from https://doi.org/10.1002/ptr.6917.

Arima, H., & Danno, G. I. (2002). Isolation of antimicrobial compounds from guava (*Psidium guajava* L.) and their structural elucidation. *Bioscience, Biotechnology, and Biochemistry, 66*, 1727–1730. Available from https://doi.org/10.1271/bbb.66.1727.

Ayouni, K., Berboucha-Rahmani, M., Kim, H. K., Atmani, D., Verpoorte, R., & Choi, Y. H. (2016). Metabolomic tool to identify antioxidant compounds of *Fraxinus angustifolia* leaf and stem bark extracts. *Industrial Crops and Products, 88*, 65–77. Available from https://doi.org/10.1016/j.indcrop.2016.01.001.

Beesetti, H., Khanna, N., & Swaminathan, S. (2016). Investigational drugs in early development for treating dengue infection. *Expert Opinion on Investigational Drugs, 25*, 1059–1069. Available from https://doi.org/10.1080/13543784.2016.1201063.

Begum, S., Hassan, S. I., Siddiqui, B. S., Shaheen, F., Nabeel Ghayur, M., & Gilani, A. H. (2002). Triterpenoids from the leaves of *Psidium guajava*. *Phytochemistry, 61*, 399–403. Available from https://doi.org/10.1016/S0031-9422(02)00190-5.

Begum, S., Hassan, S. I., & Siddiqui, B. S. (2002). Two new triterpenoids from the fresh leaves of *Psidium guajava*. *Planta Medica, 68*, 1149–1152. Available from https://doi.org/10.1055/s-2002-36353.

Bharadwaj, S., Lee, K. E., Dwivedi, V. D., Yadava, U., Panwar, A., Lucas, S. J., … Kang, S. G. (2019). Discovery of *Ganoderma lucidum* triterpenoids as potential inhibitors against Dengue virus NS2B-NS3 protease. *Scientific Reports, 9*, 19059. Available from https://doi.org/10.1038/s41598-019-55723-5.

Bhatt, S., Gething, P. W., Brady, O. J., Messina, J. P., Farlow, A. W., Moyes, C. L., … Hay, S. I. (2013). The global distribution and burden of dengue. *Nature, 496*, 504–507. Available from https://doi.org/10.1038/nature12060.

Cannalire, R., Ki Chan, K. W., Burali, M. S., Gwee, C. P., Wang, S., Astolfi, A., … Manfroni, G. (2020). Pyridobenzothiazolones exert potent anti-dengue activity by hampering multiple functions of NS5 polymerase. *ACS Medicinal Chemistry Letters, 11*, 773–782. Available from https://doi.org/10.1021/acsmedchemlett.9b00619.

Chassagne, F., Haddad, M., Amiel, A., Phakeovilay, C., Manithip, C., Bourdy, G., … Marti, G. (2018). A metabolomic approach to identify anti-hepatocarcinogenic compounds from plants used traditionally in the treatment of liver diseases. *Fitoterapia, 127*, 226–236. Available from https://doi.org/10.1016/j.fitote.2018.02.021.

Chervin, J., Perio, P., Martins-Froment, N., Pharkeovilay, C., Reybier, K., Nepveu, F., … Marti, G. (2017). Dereplication of natural products from complex extracts by regression analysis and molecular networking: Case study of redox-active compounds from *Viola alba* subsp. dehnhardtii. *Metabolomics, 13*, 96. Available from https://doi.org/10.1007/s11306-017-1227-6.

Dakappa, S. S., Adhikari, R., Timilsina, S. S., & Sajjekhan, S. (2013). A review on the medicinal plant *Psidium guajava* Linn. (Myrtaceae). *Journal of Drug Delivery and Therapeutics, 3*, 162–168. Available from https://doi.org/10.22270/jddt.v3i2.404.

Dighe, S. N., Ekwudu, O., Dua, K., Chellappan, D. K., Katavic, P. L., & Collet, T. A. (2019). Recent update on anti-dengue drug discovery. *European Journal of Medicinal Chemistry, 176*, 431–455. Available from https://doi.org/10.1016/j.ejmech.2019.05.010.

Frederico, É. H. F. F., Cardoso, A. L. B. D., Moreira-Marconi, E., de Sá-Caputo, D. d. C., Guimarães, C. A. S., Dionello, C. d. F., … Filho, M. (2017). Anti-viral effects of medicinal plants in the management of dengue: A systematic review. *African Journal of Traditional, Complementary and Alternative Medicines, 14*, 33–40. Available from https://doi.org/10.21010/ajtcam.v14i4S.5.

Fraisier-Vannier, O., Chervin, J., Cabanac, G., Puech, V., Fournier, S., Durand, V., … Marti, G. (2020). MS-CleanR: A feature-filtering workflow for untargeted LC-MS based metabolomics. *Analytical Chemistry, 92*, 9971–9981. Available from https://doi.org/10.1021/acs.analchem.0c01594.

Gubler, D. J. (2002). The global emergence/resurgence of arboviral diseases as public health problems. *Archives of Medical Research, 33*, 330–342. Available from https://doi.org/10.1016/S0188-4409(02)00378-8.

Gutiérrez, R. M. P., Mitchell, S., & Solis, R. V. (2008). *Psidium guajava*: A review of its traditional uses, phytochemistry and pharmacology. *Journal of Ethnopharmacology, 117*, 1–27. Available from https://doi.org/10.1016/j.jep.2008.01.025.

Haggag, E. G., Elshamy, A. M., Rabeh, M. A., Gabr, N. M., Salem, M., Youssif, K. A., … Abdelmohsen, U. R. (2019). Antiviral potential of green synthesized silver nanoparticles of *Lampranthus coccineus* and *Malephora lutea*. *International Journal of Nanomedicine, 14*, 6217–6229. Available from https://doi.org/10.2147/IJN.S214171.

Heinrich, M., Ankli, A., Frei, B., Weimann, C., & Sticher, O. (1998). Medicinal plants in Mexico: Healers' consensus and cultural importance. *Social Science & Medicine, 47*, 1859–1871. Available from https://doi.org/10.1016/s0277-9536(98)00181-6.

Hernandez-Morales, I., Geluykens, P., Clynhens, M., Strijbos, R., Goethals, O., Megens, S., ... Van Loock, M. (2017). Characterization of a dengue NS4B inhibitor originating from an HCV small molecule library. *Antiviral Research, 147*, 149–158. Available from https://doi.org/10.1016/j.antiviral.2017.10.011.

Horstick, O., Tozan, Y., & Wilder-Smith, A. (2015). Reviewing dengue: Still a neglected tropical disease? *PLoS Neglected Tropical Diseases, 9*, e0003632. Available from https://doi.org/10.1371/journal.pntd.0003632.

Kaushik, S., Kaushik, S., Sharma, V., & Yadav, J. P. (2018). Antiviral and therapeutic uses of medicinal plants and their derivatives against dengue viruses. *Pharmacognosy Reviews, 12*, 177–185. Available from https://doi.org/10.4103/phrev.phrev_2_18.

King, F. E., King, T. J., & Ross, J. M. (1955). The chemistry of extractives from hardwoods. Part XXIII. The isolation of a new triterpene (terminolic acid) from *Terminalia ivorensis*. *Journal of the Chemical Society (Resumed)*, 1333. Available from https://doi.org/10.1039/jr9550001333.

Kraemer, M. U., Sinka, M. E., Duda, K. A., Mylne, A. Q., Shearer, F. M., Barker, C. M., Moore, C. G., Carvalho, R. G., Coelho, G. E., Bortel, W. V., et al. (2018). *The global distribution of the arbovirus vectors Aedes aegypti and Ae albopictus, 4*, e08347. Available from https://doi.org/10.7554/eLife.08347.002.

Little, E. L., & Wadsworth, F. H. (1964). *Common tree of Puerto Rico and the Virgin Islands*. U.S. Department of Agriculture, Forest Service, Agriculture Handbook 249.

Low, J. G. H., Ooi, E. E., Tolfvenstam, T., Leo, Y. S., Hibberd, M. L., Ng, L. C., ... Ong, A. (2006). Early dengue infection and outcome study (EDEN) - Study design and preliminary findings. *Annals of the Academy of Medicine, Singapore, 35*, 783–789.

Luo, D., Vasudevan, S. G., & Lescar, J. (2015). The flavivirus NS2B-NS3 protease-helicase as a target for antiviral drug development. *Antiviral Research, 118*, 148–158. Available from https://doi.org/10.1016/j.antiviral.2015.03.014.

Lye, D. C., Archuleta, S., Syed-Omar, S. F., Low, J. G., Oh, H. M., Wei, Y., ... Leo, Y. S. (2017). Prophylactic platelet transfusion plus supportive care vs supportive care alone in adults with dengue and thrombocytopenia: A multicentre, open-label, randomised, superiority trial. *The Lancet, 389*, 1611–1618. Available from https://doi.org/10.1016/S0140-6736(17)30269-6.

Malina, J., Boon, Y. W., Aurapa, S., & Azmath, J. (2019). Current prevention and potential treatment options for dengue infection. *Journal of Pharmacy & Pharmaceutical Sciences, 22*, 440–456. Available from https://doi.org/10.18433/jpps30216.

Manikandan, R., Anand, A. V., & Muthumani, G. D. (2013). Phytochemical and in vitro anti-diabetic activity of methanolic extract of *Psidium guajava* leaves. *International Journal of Current Microbiology and Applied Sciences, 2*, 15–19.

Mastrangelo, E., Pezzullo, M., De burghgraeve, T., Kaptein, S., Pastorino, B., Dallmeier, K., De lamballerie, X., Neyts, J., Hanson, A. M., Frick, D. N., Bolognesi, M., & Milani, M. (2012). Ivermectin is a potent inhibitor of flavivirus replication specifically targeting NS3 helicase activity: New prospects for an old drug. *Journal of Antimicrobial Chemotherapy, 67*, 1884–1894. Available from https://doi.org/10.1093/jac/dks147.

Ngbolua, K.-N. J.-P., Lufuluabo, L. G., Moke, L. E., Bongo, G. N., Liyongo, C. I., Ashande, C. M., ... Zoawe, B. G. (2018). A review on the phytochemistry and pharmacology of *Psidium guajava* L. (Myrtaceae) and future direction. *Discovery Phytomedicine, 5*, 7–13. Available from https://doi.org/10.15562/phytomedicine.2018.58.

Peng, M., Swarbrick, C. M. D., Chan, K. W. K., Luo, D., Zhang, W., Lai, X., ... Vasudevan, S. G. (2018). Luteolin escape mutants of dengue virus map to prM and NS2B and reveal viral plasticity during maturation. *Antiviral Research, 154*, 87–96. Available from https://doi.org/10.1016/j.antiviral.2018.04.013.

Peng, M., Watanabe, S., Chan, K. W. K., He, Q., Zhao, Y., Zhang, Z., ... Li, G. (2017). Luteolin restricts dengue virus replication through inhibition of the proprotein convertase furin. *Antiviral Research, 143*, 176–185. Available from https://doi.org/10.1016/j.antiviral.2017.03.026.

Phakeovilay, C., Bourgeade-Delmas, S., Perio, P., Valentin, A., Chassagne, F., Deharo, E., ... Marti, G. (2019). Antileishmanial compounds isolated from *Psidium guajava* L. using a metabolomic approach. *Molecules (Basel, Switzerland), 24*, 4536. Available from https://doi.org/10.3390/molecules24244536.

Prinsloo, G., & Vervoort, J. (2018). Identifying anti-HSV compounds from unrelated plants using NMR and LC–MS metabolomic analysis. *Metabolomics: Official Journal of the Metabolomic Society, 14*, 134. Available from https://doi.org/10.1007/s11306-018-1432-y.

Qadir, M. I., Abbas, K., Tahir, M., Irfan, M., Bukhari, S. F. R., Ahmed, B., ... Ali, M. (2015). Review: Dengue fever: Natural management. *Pakistan Journal of Pharmaceutical Sciences, 28*, 647–655. Available from http://www.pjps.pk/wp-content/uploads/pdfs/28/2/Paper-36.pdf.

Saptawati, L., Febrinasari, R. P., Yudhani, R. D., Faza, A. G., Ummiyati, H. S., Sudiro, T. M., & Dewi, B. E. (2017). In vitro study of eight Indonesian plants extracts as anti Dengue virus. *Health Science Journal of Indonesia, 8*, 12–18.

Schrimpe-Rutledge, A. C., Codreanu, S. G., Sherrod, S. D., & McLean, J. A. (2016). Untargeted metabolomics strategies—Challenges and emerging directions. *Journal of the American Society for Mass Spectrometry*, *27*, 1897–1905. Available from https://doi.org/10.1007/s13361-016-1469-y.

Shao, M., Wang, Y., Huang, X. J., Fan, C. L., Zhang, Q. W., Zhang, X. Q., & Ye, W. C. (2012). Four new triterpenoids from the leaves of *Psidium guajava*. *Journal of Asian Natural Products Research*, *14*, 348–354. Available from https://doi.org/10.1080/10286020.2011.653964.

Simmons, C. P., Farrar, J. J., Van Vinh Chau, N., & Wills, B. (2012). Current concepts: Dengue. *New England Journal of Medicine*, *366*, 1423–1432. Available from https://doi.org/10.1056/NEJMra1110265.

Sun, H., Chen, Q., & Lai, H. (2018). Development of antibody therapeutics against flaviviruses. *International Journal of Molecular Sciences*, *19*, 54. Available from https://doi.org/10.3390/ijms19010054.

Tang, L. I. C., Ling, A. P. K., Koh, R. Y., Chye, S. M., & Voon, K. G. L. (2012). Screening of anti-dengue activity in methanolic extracts of medicinal plants. *BMC Complementary and Alternative Medicine*, *12*, 3. Available from https://doi.org/10.1186/1472-6882-12-3.

Tayone, W. C., Tayone, J. C., & Hashimoto, M. (2014). Isolation and structure elucidation of potential anti-dengue metabolites from tawa-tawa (*Euphorbia hirta* Linn.). *Walailak Journal of Science and Technology*, *11*, 825–832.

Trujillo-Correa, A. I., Quintero-Gil, D. C., Diaz-Castillo, F., Quiñones, W., Robledo, S. M., & Martinez-Gutierrez, M. (2019). In vitro and in silico anti-dengue activity of compounds obtained from *Psidium guajava* through bioprospecting. *BMC Complementary and Alternative Medicine*, *19*, 298. Available from https://doi.org/10.1186/s12906-019-2695-1.

Tsugawa, H., Cajka, T., Kind, T., Ma, Y., Higgins, B., Ikeda, K., ... Arita, M. (2015). MS-DIAL: Data-independent MS/MS deconvolution for comprehensive metabolome analysis. *Nature Methods*, *12*, 523–526. Available from https://doi.org/10.1038/nmeth.3393.

Tsugawa, H., Kind, T., Nakabayashi, R., Yukihira, D., Tanaka, W., Cajka, T., ... Arita, M. (2016). Hydrogen rearrangement rules: Computational MS/MS fragmentation and structure elucidation using MS-FINDER software. *Analytical Chemistry*, *88*, 7946–7958. Available from https://doi.org/10.1021/acs.analchem.6b00770.

Watanabe, S., Chan, K. W. K., Dow, G., Ooi, E. E., Low, J. G., & Vasudevan, S. G. (2016). Optimizing celgosivir therapy in mouse models of dengue virus infection of serotypes 1 and 2: The search for a window for potential therapeutic efficacy. *Antiviral Research*, *127*, 10–19. Available from https://doi.org/10.1016/j.antiviral.2015.12.008.

Wilder-Smith, A., Ooi, E.-E., Horstick, O., & Wills, B. (2019). Dengue. *The Lancet*, *393*, 32560–32561. Available from https://doi.org/10.1016/S0140-6736.

World Health Organization. (2012). *Global strategy for dengue prevention and control, 2012–2020*. Geneva: WHO Press.

Xu, T. L., Han, Y., Liu, W., Pang, X. Y., Zheng, B., Zhang, Y., & Zhou, X. N. (2018). Antivirus effectiveness of ivermectin on dengue virus type 2 in *Aedes albopictus*. *PLoS Neglected Tropical Diseases*, *12*, e0006934. Available from https://doi.org/10.1371/journal.pntd.0006934.

Yanuar, A., Suhartanto, H., Mun'im, A., Anugraha, B. H., & Syahdi, R. R. (2014). Virtual screening of Indonesian herbal database as HIV-1 protease inhibitor. *Bioinformation*, *10*, 52–55. Available from https://doi.org/10.6026/97320630010052.

Yin, Z., Chen, Y. L., Schul, W., Wang, Q. Y., Gu, F., Duraiswamy, J., ... Shi, P. Y. (2009). An adenosine nucleoside inhibitor of dengue virus. *Proceedings of the National Academy of Sciences of the United States of America*, *106*, 20435–20439. Available from https://doi.org/10.1073/pnas.0907010106.

Yu, J. S., Tseng, C. K., Lin, C. K., Hsu, Y. C., Wu, Y. H., Hsieh, C. L., & Lee, J. C. (2017). Celastrol inhibits dengue virus replication via up-regulating type I interferon and downstream interferon-stimulated responses. *Antiviral Research*, *137*, 49–57. Available from https://doi.org/10.1016/j.antiviral.2016.11.010.

CHAPTER

14

How history can help present research of new antimicrobial strategies: the case of cutaneous infections' remedies containing metals from the Middle Age Arabic pharmacopeia

Véronique Pitchon[1], *Elora Aubert*[2], *Catherine Vonthron*[2] *and Pierre Fechter*[3]

[1]CNRS, UMR 7044, Archaeology and Ancient History: Mediterranean - Europe, MISHA, Strasbourg University, Strasbourg, France [2]CNRS, UMR 7200, Laboratory of Therapeutic Innovation, Medalis LabEx, Faculty of Pharmacy, Strasbourg University, Strasbourg, France [3]CNRS, UMR 7242, Biotechnology and Cell Signaling, Strasbourg University, Illkirch-Graffenstaden, France

Introduction

Plants are a valuable source of a wide range of secondary metabolites, which are used as drugs with different pharmacological activities. Herbal medicine was used extensively in dermatology for the treatment of acne, wound and burn, viral, fungal, bacterial, parasitic infections, dermatitis, psoriasis, vitiligo, alopecia, skin cancer, and other skin complaints. The medical treatment of skin diseases has the advantage of being able to use so-called topical drugs, which is to say by direct application to the skin. This allows the use of products that could prove to be toxic upon ingestion but which by surface application allows the treatment of infections without necessarily affecting the cells in depth. This is the case with metals that are widely used in ancient dermatology, although they are rarely ingested due to their well-known toxicity. This is why we have chosen to do here a review

Medicinal Plants as Anti-infectives
DOI: https://doi.org/10.1016/B978-0-323-90999-0.00016-1

TABLE 14.1 List of commented and discussed sources.

Title	Author	Document
Therapeutic properties of medicinal plants: a review of their antibacterial activity (part 1)	Al-Snafi A.E.	Review & Research, 6(3), (2015), pp. 137–1258
Therapeutic properties of medicinal plants: a review of their dermatological effects (part 1)	Al-Snafi A.E.	International Journal of Pharmacy Review & Research, 5 (4), (2015), pp. 328–337
The contributions of Arabs physicians in dermatology	Bachour, H.T.	JISHIM, 2, (2002), pp. 43–45
On the transmission of Indian medical texts to the Arabs in the early Middle Ages	Kahl O.	Arabica, 66, no. 1 et 2, (2019), pp. 82–97
Arabian contributors to dermatology	Marquis, L.	International Journal of Dermatology 24, (1985), pp. 60–64
Dermatologie infectieuse	Mokni, M., Dupin N., del Giuduce P.	Dir. Dan Michael Lipsker - Collection Dermatologie, Elsevier Masson (2014)
Traditional Arabic medicine in dermatology	Oumeish Youssef	Clinics in Dermatology, 17(1), (1999), pp. 13–20
Medicinal plants for skin and hair care	Sharma Laxmikant, Agarwal Gaurav, Kumar Ashwani	Indian Journal of Traditional Knowledge, 2 (1), (2003), pp. 62–68
La Parasitologie et la Zoologie dans l'œuvre d'Avenzoar	Théodoridès J.	Revue d'histoire des Sciences, 8(2), (1955), pp. 137–145

of drugs used for the skin, whether they are herbal but contain one or more metals. We will see that this involves specificities, especially in the preparation because the metal must be homogeneously dispersed in the medium that receives it and that it does not affect the active principle of the plants to which it is added. It comments and discusses different surveys of these pharmacopeias, described in Table 14.1.

Brief history of Arabic medicine

One of the sciences which experienced a peak during the medieval period is that of pharmacology, which was one of the last to be supplanted by modern science. The use of this pharmacopeia is still ongoing in the Middle East and India. In the realm of science, the Arabs are the creators of the real development of pharmacology (Saad & Said, 2011) (Fig. 14.1).

At first, Arab scientists focused on translating Greek medical knowledge. This flourishing civilization in the 8th century CE integrated medicine the humoral theories of Hippocrates then was interested in the practical aspects of medicine by integrating the texts of Galen, in particular, the texts of dietetics and, as regards with pharmacology, enriched the arsenal of Greek plants by integrating the knowledge of Ayurveda and the

FIGURE 14.1 14th century Arabic pharmacy.

Indo-Persian therapeutic arsenal. The 11th century CE, with the Arabic translation of *De materia medica* by Dioscorides, marked a real turning point and the work of the Greek botanist has known an unprecedented impact among Arab scientists. It inspired many medieval Arabic works in botany and allowed the radical evolution of the Arabic scientific lexicon. To get around the difficulty of translating Greek botanical terms into Arabic, the translators will go through Syriac to develop a whole new nomenclature in Arabic.

Principles of Arab medicine: theoretical aspects

Upstream of medical practice through drug or diet treatment, there was a common theoretical framework stipulating that the universe is made up of four elements: fire, air, water, and earth. In this vision, each element has its own quality, of which it is both holder and producer; humidity for water, heat for fire, dryness for air, and cold for the earth. In this cosmological conception, the Greek influence appears very clear. One easily finds the resonance and the influence of the philosophical system of Empedocles, of the theory of Aristotle, of the treatises of Hippocrates, and of the writings of Galen.

The numerous treatises of pharmacology, drugs, food, and medicinal plants which multiplied in parallel with the development of Arab-Muslim medicine emphasized the properties of drugs and their qualities. There are thus descriptions of preparations of medicinal products based on plants, animal, or mineral matter incorporating data concerning the qualities of each component with the mention of mixing them, as appropriate, with fresh, cold, lukewarm water, or rainwater.

The disturbance of the balance of the humors and the return to the state of equilibrium meaning healing was treated by the addition of heat during an overproduction of cold or by a humidifying recipe when it comes to balance an excess of drought. The treatments also took into account the influence of the climate, the seasons, the age, the nature of the organ to be treated, the disease, and the patient's temperament. These four Aristotelian categories (quality, quantity, time, and manner) formed the framework of this medical discourse.

Arab sources of pharmacology: the aqrābādhīn, a constituted literature

Muslims were excellent organizers of knowledge and this ability to organize led to the production of forms of texts specially dedicated to pharmacology. These forms of text can be placed in more or less precise categories which generally indicate the main directions of research and thinking and to place our reflection over a long period of time, we have studied a wide range of sources written between the 9th and 13th centuries.

There are many sources relating to pharmacology because indeed every medieval Arab doctor left a trace not only of the theoretical aspects of medicine but of its practical aspects by writing pharmacopeias that list several thousand drugs.

The Arab-Muslims developed texts on the basis of the Greek classification but also generated major new types of pharmacological literary models. Their approach was not only enumerative but much more flexible. In the pharmacological literature a number of subjects flourished exploring new horizons of knowledge, resulting in new understanding and new areas to invest.

The texts of Arabic pharmacology can be divided into two categories, that relating directly to the subject and that having a more or less distant relation with this same subject. For example, for this second category, we find works on how to cure during travel or on the *hisba*, which is the common law regulating more particularly medicine and pharmacy, medical biographies as well as works on botany, zoology, which can also shed light on pharmacology; one finds in these texts scattered but nevertheless, useful information. However, the knowledge produced existed in the form of purely pharmacological texts and meticulously followed all possible educational directions. As a result, these texts belong to perfectly delimited groups which can be placed in the following categories:

- medical formularies (*aqrābādhīn*)
- books on poisons
- lists of synonyms in alphabetical order
- tables, synoptic treatises
- alphabetical lists of plants, minerals (book of simples)
- chapters of medical treatises
- methods to check purity of drugs
- special books on particular diseases (eyes, sexual organs, pulse, etc.)

In general, these works contain more or less descriptive information on the diseases, symptoms, drugs and procedure, and sometimes dosage and form of the drug. There is no systematization of the written form but in general, the works are divided into chapters, relating either to the nature of the diseases affecting an organ or to the forms taken by the drug (pill, sirup, ointment, cataplasm). While some authors have opted for a classification in tables with an alphabetical arrangement, others give a set of generalities on the nature of the diseases and their links with the theory of humors before proposing a set of medical preparations useful for the care of a disease or a type of diseases (Table 14.2).

TABLE 14.2 List of sources used (9–13th centuries CE).

Title	Author	Période	Document	Geographical zone
Kitāb jawāhit al-ṭib al mufrada	Yuḥanā ibn Māsawaih	9th CE	Books of simple	Baghdad
Al-aqrābādhīn	Sābur ibn Sahl	9th CE	Medical formularies	Baghdad
Al-aqrābādhīn	Al-Kindī	9th CE	Medical formularies	Baghdad
Kitāb al-sumūm	Ibn Wahshīya	10th CE	Books on poisons	Kufa
Qānūn fī al-ṭibb	Ibn Sīna	XI s.	Medical treatise	Persia
Al-aqrābādhīn	Ibn al-Tilmīdh	12th CE	Medical formularies	Baghdad
*Taysīr f ī'l-mudāwāt wa'l-*tadbīr	Ibn Zuhr	12th CE	Medical treatise	Al-Andalus
Kitab al-Musta'ni	Ibn Biklārish	12th CE	Synoptic table	Al-Andalus
Shar' asmā' al-'uqqār	Maïmonides	12th CE	Glossary	Al-Andalus/ Cairo
Kitāb al-qarābādhīn ala tartīb al-ilal	al-Samarqandī	13th CE	Medical formularies	Baghdad
Kitāb al-ghāmi li mufradāt al adawiga wa al aġḍiya	Ibn Baytar	13th CE	Book of simples	Al-Andalus/ Damas
Minhāj al-dukkān wa-dustur al-a'yan fī a'māl wa-tarākīb al-adwiya al-nāfi'a lil-insān	al-Kūhīn al-'Aṭṭār al-Isrā'īlī	13th CE	Manual of pharmacy	Cairo

Cutaneous infections and medications

The specificity of skin and eye diseases in the pharmacopeias and the nature of the diseases treated

Dermatological diseases have been particularly well understood by Arab doctors because the skin constitutes a particular element of the human body. Visible to everyone, human skin, the outer covering of the body, is the largest organ in the body. It is also its first line of defense. The skin is the most superficial and most visible organ. Very early on, it offered to doctors the possibility of observations, some of which are still relevant.

The skin is a complex organ of the human body. Its surface extends for approximately 2 m^2; it weighs approximately 3 kg for an adult with a total weight of 70 kg: it already forms the most extensive and heaviest apparatus in the body as a whole. One square centimeter contains three blood vessels, 10 hairs, 12 nerves, 15 sebaceous glands, 100 sweat, and three million cells. The medical formularies books reveal the methods of Arab physicians adopted in the field of dermatology. This branch of learning falls into three categories. The first includes skin diseases which are a branch of internal diseases, such as tumors, ulcers, pustules, leprosy, and others. The second covers pure skin diseases such as vitiligo and others. The third category covers diseases related to the skin appendages such as hair and nails, classified according to a special order called (al-Zina), that is, cosmetology.

Arab physicians investigated the nature of these diseases, their causes and effects, and alluded to the fact that illnesses as such were affected by what was called the four complexions or (humors) (*al-akhlat*). This theory was a common feature of the age.

The number of skin diseases listed in medical forms is large and probably covers all skin conditions as they are described today (ibn, al-Qaršī, Ibn al-Nafīs, Chadli, & Hemrit, 2005; Bouamrane, 2010) (Table 14.3).

The number of different internal and dermal diseases is large and reveals a particular interest of physicians in skincare. They studied ulcers and diseases of the head, ears, nose, mouth, lips, teeth, eyes, neck, lungs, heart, as well as on types of fevers, and epidemics.

Skin conditions can be classified into different categories: rashes, viral infections, bacterial infections, fungal infections, parasitic infections, pigmentation disorders, tumors and cancers, trauma, and other conditions such as wrinkles, rosacea, spider veins, and varicose veins. Several herbal remedies possessed a wide range of dermatological effects, including antibacterial, antifungal, antiviral, antiparasitic, anticancer, stimulating hair growth, healing wounds, and burns, for the treatment of eczema, acne, cancer, vitiligo, and psoriasis, as a skin lightener, as a skin protection therapy, and to slow skin aging.

Parasitic diseases are one example that deserves attention. Ibn Zuhr as early as the 12th century described hitherto unidentified skin diseases. In the first part of his work (Bouamrane, 2010), it is a question of ringworm, the parasitic etiology of which was of course unknown at that time, but whose lesions were known since antiquity. Ibn Zuhr also mentions in this same book tonsuring herpes (ophiasis), head lice, and their nits. He also devotes an entire treatise to ocular affections (Treatise No. 8), it is very interesting to see the phthiriasis of the eyelids cited, an infrequent localization of the pubic louse or "crab louse" (*Phiirius inguinalis*).

Much work is devoted to parasitology. Chapters XXXII and XXXIII relate, respectively, to animals that can enter the ear and leeches. In Treaty VII of Book II, Chapter XVIII is devoted to a parasitic animalcule "assoab." This is surely the scabies mite.

Plants and metals useful for skin diseases

Let us now turn to medicinal preparations and to plants and metals more specifically used to treat skin diseases. In the Middle Eastern region, there are more than 2600 known plant species; about 200–250 of them are still in use for the treatment and prevention of various diseases (Snafi, 2018).

Some are specific to skin diseases, but what interests us here is the metal–plant combinations particularly used for certain ailments that we will identify.

TABLE 14.3 Table of diseases listed in two medical treatises written by Ibn al-Nafis and by Ibn Zuhr.

Name	Description
Acne	
Erythrose facial	
Ringworm	
Ophiase	Variety of alopecia areata in which the alopecic plaques form sinuosities and form a bare crown around the skull
Baldness	
Pustulose	Skin disease characterized by the presence of pustules
Canities	Physiological or pathological bleaching of hair
Capillary asymmetry	
Pityriasis capitis	Dandruff
Phthiriasis	Lice
Nits	
Accidental wounds	
Tumors	Swelling, growth of tissue
Ear ulcers	
Nose ulcers	
Polyps of the nose	
Tongue affections	Tumors, ulcers
Oral granulations	
Lip cracks	
Canker sores	
Tongue cracks	
Scabies, warts, scrofula	
Chalazion	
Blepharitis	Trachoma, a nonspecific, contagious bacterial eye infection caused by chlamydia trachomatis
Trichiasis	Deviation of the implantation of the eyelashes toward the inside of the eye
Falling eyelashes	
Eye's phthiriasis	Lice
Stye	
Dacryocystitis	Tumor between the inner corner of the eye and the nose
Tear duct disorders	Enlarged or atrophy of the lacrimal caruncle

(*Continued*)

TABLE 14.3 (Continued)

Name	Description
Conjunctival pustules and ulcers	
Pupil disorders	Deformation, dilatation
Conjunctivitis	
Corneal pannus	
Boils	
Pruritic	
Vitiligo	
Leucodermia	Refers to several skin diseases characterized by a lasting decrease or loss of pigmentation
Nevi	Congenital malformation of the skin visible as a colored patch or small tumor
Warts	
Lichen	
Peeling of the skin	Exfoliation
Leprosy	
Medina worm	Parasite, worm
Cysticercosis	Parasite, worm
Cracked feet and hands	
Nail cracking	
Pyoderma	Purulent skin lesion resulting from infection
Pannus	
Gale	
Eczema	
Papules et pustules	
Abscess	
Urticaria	
Bruises	

Therapeutics was classically subdivided into three parts, namely, pharmacology, dietetics, and surgery. The first area was arguably of the greatest practical importance in medieval medicine. Pharmacology is further subdivided into two distinctly different parts: the use of simple remedies on the one hand, and compounds on the other. These treatises on pharmacology reveal the methods of Arab physicians adopted in the field of dermatology. This branch of learning falls into three categories. Reading the pharmacy treatises, we see that the use of metals as ingredients of drugs used to treat diseases is not uncommon. This is due to the very early antibacterial properties observed when a metal is introduced into a topical preparation such as cream, ointment, poultice, and so on. Minerals, for example, those containing antimony, arsenic, copper, gold, iron, lead, mercury, sulfur, or zinc, were used frequently.

One area of medicine where metals have been used frequently in the therapeutic composition is that of the treatment of diseases of the eye. There is a treatise on the eye written by Ḥunayn ibn Isḥāq in the 9th CE. This pupil of Yuḥanā ibn Māsawaih, originally from al-Hira in Iraq, is best known for his translations of Greek works, especially medical texts by Galen, towards Syriac, the language of culture of his religious community, and Arabic, was a renowned doctor who left a hundred works including a set of 10 ophthalmology treatises in which he describes in detail the ailments of the eye and the various ways of treating them. The use of plant–metal combinations is frequent there. Among the drugs of mineral origin G. A. Anawati has identified 16 which are listed below (Anawati, Médicaments, & L'Oeil, 1905).

- al-shāḍana, hematite
- al-milḥ, salt
- al-nushāḍir, ammoniac
- al-zarnīkh, arsenic
- al-zinjār, rust
- al-iqlīmiyya, cadmie (calamine)
- al-bawraq, saltpeter
- al-zāj, white vitriol
- al-raṣaṣ, lead
- al-itmid, antimony
- al-qalqant, blue vitriol
- al-qalqadīs, red vitriol
- al-nuḥās, copper
- al-isfīdāj, lead carbonate
- al-tūtya, zinc oxide
- tūbāl al-ḥadīd, iron scoria

Medicines are of several kinds; in the first place the topicals directly applied to the skin, they come in the form of balms, oils, ointments, and creams. They correspond to mixtures with different concentrations of different natural compounds which can be either crushed or ground. Several recipes for different diseases have been proposed. For example, in the treatment of eczema, Ibn Sinā suggests the following formulation (ibn, al-Qaršī, Ibn al-Nafīs, Chadli, & Hemrit, 2005):

- *Cover the lesion with a mixture of lentils, pomegranate peels, barley flour, planta, all finely pounded. For miliary eczema, add a purgative and a little turbith and small moments. Fresh milk is very effective, pomegranate peels, Armenian clay in vinegar, and rose water.*

For pruritus and scabies, Ibn Sinā recommends a diet based on chicory, chard, purslane, and spinach cooked with kid meat and pomegranate grains. For the topical to be applied regularly in addition to the diet, here is the composition:

- *You need sulfur, lily, soap, gum ammonia, gum ammonia. One of these products is mixed with litharge or white lead and an equivalent amount of white salt and pomegranate seeds are added. Grill everything and add rose oil, violet oil, rose water, green cilantro water, and vinegar. Camphor can be used.*

Of course, due to their antibacterial and antiinfectious properties, metals are often found in preparations that are used to treat skin diseases where there is presence of pus such as abscesses and boils. We have noted a few examples of Al-Kindī (ibn, al-Qaršī, Ibn al-Nafīs, Chadli, & Hemrit, 2005), a doctor who practiced in Baghdad during the 9th CE.

For a nasal ulcer:

- *Two parts of scoria of gold, one part of Lycium juice, and one part of saffron.*
- *It is sieved with a silk cloth and taken nasally with dripping juice of garden cress. After this, it is taken nasally (aspired) with good violet. It is very useful, with the leave of God.*

For boils:

- *It is tested and wonderful, one part of each.*
- *Long aristolochia.*
- *Electuary of plantain.*
- *Litharge (white lead).*
- *Scoria of silver.*
- *Aloe vera.*
- *Myrrh.*
- *They are pulverized with vinegar and rose oil. The boil is daubed with this mixture which removes it in two or three applications, with God's leave.*

Drug for an abscess:

- *It is powdered on; then natural cotton is placed on it. It clears away the defects, dries the flesh, and heals the abscess. The mixture should be warm for effectiveness and contains ½ part of verdigris and one part of each: Aloe, myrrh, gum ammoniac, frankincense, wild pomegranate flower, and minium (red lead).*

Toxicity of metals

Metals have often been associated with plants to combat symptoms reminiscent of bacterial infections. They have been abandoned because of their toxicity, which is poorly known and poorly controlled. In addition, certain unfortunate and inappropriate use of metals has largely contributed to the current mistrust concerning their use in medicine (Bonah & du Stalinon, 2007; Mikulski et al., 2017; Parascandola, 2009; Riva, Lafranconi, D'Orso, & Cesana, 2012). Furthermore, the presence of toxic metals in different traditional preparations, mainly due to the contamination of soils by these metals, and that lead to intoxications enhances even more the suspicion regarding metals (i.e., Chen, Zou, Sun, Qin, & Yang, 2021; Ernst, 2002; Saper et al., 2008; Shaw, Leon, Kolev, & Murray, 1997). Promoting their use from traditional medicines will thus require addressing the toxicity issues of these metals/remedies.

The great difficulty in studying the toxicity of metals is that it is linked to a large number of mechanisms, based in particular on the competition that there may be between the metals themselves for binding to proteins (many metals share similar structural and functional properties), and to the generation of oxidative stress (Lemire, Harrison, & Turner, 2013; Squibb & Fowler, 1981; Xianglin, Castranova, Vallyathan, & Perry, 2001). This can lead to the dysfunction of many enzymes, the generation of ROS, or a depletion of the antioxidative potential, genotoxicity, odd membrane function, interference with the assimilation of nutrients (Egorova & Ananikov, 2017; Schatzschneider, 2018) (Fig. 14.2).

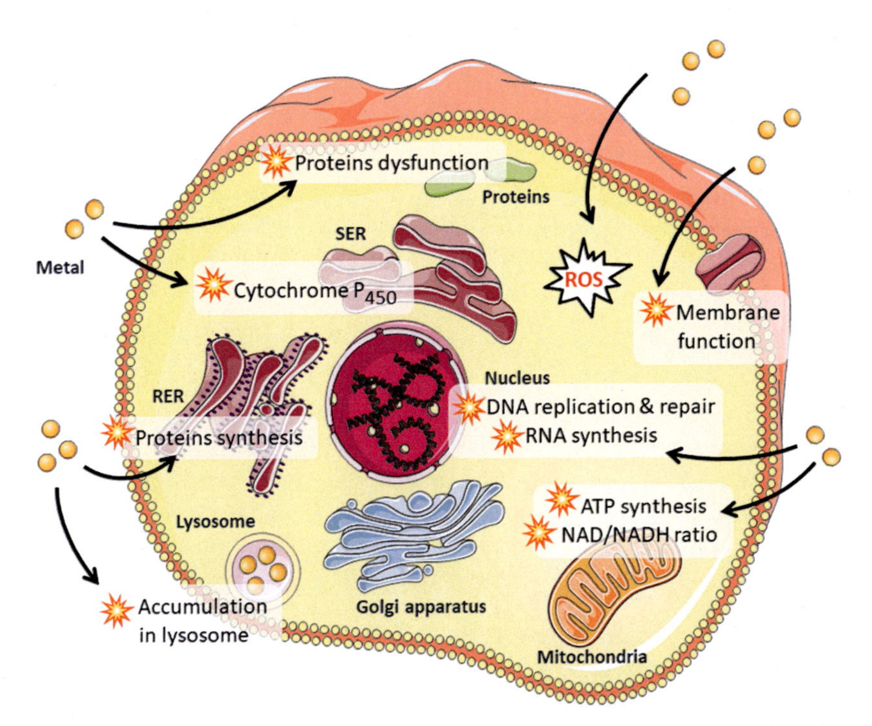

FIGURE 14.2 Metal toxicity on eukaryote cells. Over the past decades, many researches have been conducted to elucidate the mechanisms of metal toxicity. While this toxicity remains metal-specific, some general trends are observed, some of which are represented here. Metals can bind and so disturb the function of many proteins and enzymes, they can affect membranes permeability/nutrient import, induce the formation of ROS, which in turn will lead to molecule oxidation (proteins/nucleic acids/lipids). Metal can therefore affect major cellular function: they can inhibit DNA/RNA synthesis and repair, reduce the level of ATP, affect the ratios NAD/NADH in the mitochondria, disrupt proteins synthesis in the rough endoplasmic reticulum, decrease the level of cytochromes P450 (hemoproteins involved in most of oxidoreduction reactions) in the smooth endoplasmic reticulum, compromise detoxification mechanisms by accumulation in lysosome. See text for details. Source: *From Lemire, Harrison, & Turner, 2013; Squibb & Fowler, 1981; Xianglin, Castranova, Vallyathan, & Perry, 200; Egorova & Ananikov, 2017; Schatzschneider, 2018. Servier Medical Art by Servier. (n.d.). Retrieved from https://smart.servier.com/. Figure modified from Servier Medical Art by Servier, licensed under a Creative Commons Attribution 3.0 Unported License.*

Renewed interest in metal–organic molecule combinations

For the past 20 years, linked to the antibiotic crisis, but also because the mechanisms of action and toxicity of metals have become better known, metals have enjoyed a revival of interest (Lemire et al., 2013), which in turn has led to a renewed focus on these ancient remedies combining plants and metals. In order to study and understand this past use, we felt necessary to take a look at the current use of metals, and their formulation. With regard to the fight against bacterial infections, metals can be used as drug adjuvants, incorporated into organometallic molecules, or metal nanoparticles.

Metal salts are used as an adjuvant in some drugs. For example, Pylera, a drug prescribed to combat *Helicobacter pylori* infections, is a combination of four antibiotics and a metal, bismuth (Hu, Zhu, & Lu, 2017). Another way of using metals in the form of elementary particles is to use a so-called Trojan horse strategy, whereby the cell incorporates a toxic metal in place of a biological metal; for example, some strategies aim at replacing iron, a metal that is essential for the catalytic activity of many enzymes, with gallium, a toxic metal that is structurally close to iron (Abd-El-Aziz, Agatemor, & Etkin, 2017; Albada & Metzler-Nolte, 2017; Banin et al., 2008; Chitambar, 2016; Golonka, Yeoh, & Vijay-Kumar, 2019; Greenberg, Banin, Banin, Berenshtein, & Chevion, 2007; Gunawan, Teoh, Marquis, & Amal, 2011; Mislin & Schalk, 2014; Nairz et al., 2018; Ong & Gasser, 2019; Patra, Gasser, & Metzler-Nolte, 2012; Ruparelia, Chatterjee, Duttagupta, & Mukherji, 2008; Sierra, Casarrubios, & de la Torre, 2019; Sotiriou & Pratsinis, 2010) (Fig. 14.3).

Recent studies have shown that such molecules (elemental metal; organometallic molecules; metal nanoparticles) were already used in the past, exploiting the available wealth of the environment. Through a few examples from the Indo-Arabic tradition, we will show that ancient scientists already possessed relatively precise notions of the toxicity of metals and that through well-mastered protocols, they were able to transform these metals to give them different formulations, functions, and coatings, capable of profoundly changing the activity and toxicity of metals. The arsenal that we are beginning to develop was already in place centuries ago (i.e., Barillo & Marx, 2014; Dolwett & Sorensen, 1985; Lev, 2002), and can reveal quite original and unexpected combinations, but their characterization will require important investigations.

Elementary metal particle

In past Arab or Middle Eastern tradition, we find many metal salts associated with treatments against symptoms reminiscent of bacterial infections, but more generally for external care (eye care, skincare, etc.) and less integrated into ingested remedies; as metal salts are not very soluble at physiological pH, there was a high risk that they would precipitate and accumulate in digestive organs without reaching their target. Different salts could be used for a single metal. Chemists recognized the different salts, assigned them different functions, and used them for different treatments. Different salts of copper were commonly used to treat different conditions (Le Bonniec & Histoire naturelle, 1953). For example, black copper oxide was given with honey to remove intestinal worms; misy is found in eye drops (Dolwett and Sorensen, 1985). In addition, practitioners were aware of the toxic effects of metals, thus although Pb has been shown to be highly toxic to humans,

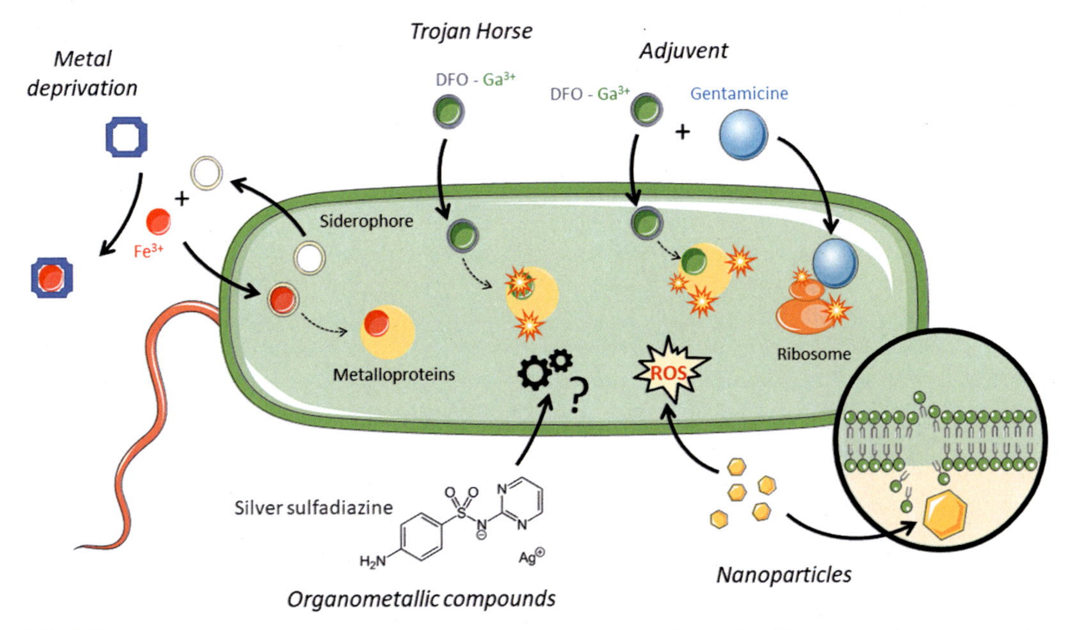

FIGURE 14.3 Metals in antibacterial strategies. In this scheme are depicted different strategies using metals to inhibit bacterial growth. Trojan horse: For their iron supply, bacteria produce siderophore molecules (small iron chelators) that are exported outside the cell where they bind iron before being imported back into the bacteria. One Trojan Horse strategy consists in fooling the bacteria by giving siderophore molecules bound to a toxic metal, which will be imported as natural iron–siderophore complexes. For example, gallium is imported through the siderophore desferrioxamine (DFO) by the pathogen Pseudomonas aeruginosa. In the cell, gallium takes the place of iron in different iron-dependent molecules and inactivates them. Metal deprivation: Other iron chelators can be used to deprive the cell from this metal, and thus inhibit bacterial growth. Adjuvant: metals can also be used in combination with antibiotics to improve their efficacy, like gentamycin which antibacterial activity is improved when combined to DFO–gallium. Nanoparticles: among their different effect on bacteria, they accumulate on the membrane, disrupt it, and generate ROS. Organometallic compounds: many different organometallic compounds have been designed, like silver sulfadiazine, to inhibit bacterial growth, although their mechanism of action is not always well understood. See text for details. Source: *From Abd-El-Aziz, Agatemor, & Etkin, 2017; Albada & Metzler-Nolte, 2017; Banin et al., 2008; Chitambar, 2016; Golonka, Yeoh, & Vijay-Kumar, 2019; Greenberg, Banin, Banin, Berenshtein, & Chevion, 2007; Gunawan, Teoh, Marquis, & Amal, 2011; Mislin & Schalk, 2014; Nairz et al., 2018; Ong & Gasser, 2019; Patra, Gasser, & Metzler-Nolte, 2012; Ruparelia, Chatterjee, Duttagupta, & Mukherji, 2008; Sierra, Casarrubios, & de la Torre, 2019; Sotiriou & Pratsinis, 2010.* Servier Medical Art by Servier. *(n.d.). Retrieved from https://smart.servier.com/. Figure modified from Servier Medical Art by Servier, licensed under a Creative Commons Attribution 3.0 Unported License.*

it was widely used in the past, notably as an antiinfective (either in make-up to promote protective immunity or in the remedies themselves), which is still debatable and would call into question the reasoned use of metals at that time. Nevertheless, the toxicity of Pb was known even in antiquity. Its therapeutic use was documented and regulated as reported by Lanoë (2002). As an antiinfective, Pb was mainly used externally, which limited the diffusion of this toxic metal in the body, while concentrating its action on the site of the infection. Knowledge of metal toxicity was not limited to Pb. For example, Hg (mercury) formulations were also used (Vaidya & Mehendale, 2014). Its toxicity was also known since ancient times, as reported by Pliny the Elder (Le Bonniec & Histoire naturelle, 1953), and chemists in the past knew that different formulations of Hg resulted in different toxicity. Hg was therefore preferably used in the form of cinnabar (sulfur salt, HgS), whose toxicity has recently been estimated to be >3 logs lower than that of the well-known and toxic mercury salts [Me-Hg, $HgCl_2$; (Wu et al., 2011)]. It is therefore possible that these ancient and prudent precautions have altered the benefit/risk balance of the use of these metals. In the past, medical doctors were therefore aware of the toxicity of metals, and their use could be carefully controlled.

Recent studies have further shown that past chemists mastered the transformation of metals and that various metal fabrications have evolved to improve the therapeutic potentials of metals. Two studies have revealed this mastery of the transformation of these metal salts. Litharge (PbO) compounds have been widely used since antiquity either in the formulation of make-up materials or in remedies. A study on the chemical analysis of make-up samples found in Egyptian tombs, confirmed by the reconstruction of ancient recipes, showed that two unnatural Pb chlorides (laurionite Pb(OH)Cl and phosgenite $Pb_2Cl_2CO_3$) were purposely synthesized from litharge and triggered pro-inflammatory activity (Tapsoba, Arbault, Walter, & Amatore, 2010) to prevent infections. We recently showed that, through a different transformation process, physicians could obtain Pb acetate with the opposite antiinflammatory properties, which promoted wound healing (Abdallah et al.; manuscript in revision). It seems that, even without knowing the structure of metal salts, these practitioners knew that different formulations could have different properties, they knew how to process the metal to obtain these formulations, and used these salts appropriately.

Metals could also have another role in treatment, which can be found in eye care. The warm, dry air of Middle Eastern countries dried out the eyes, making them more susceptible to infection. As metals precipitate in physiological fluids, the irritation from these metal particles caused tears and thus contributed to better eye protection.

Organometallic molecule

Metals can also be found in a complex form with organic molecules. Many polyphenolic derivatives from plants show the ability to bind metals. This metal chelating property has been widely characterized through the study of siderophores, small molecules produced by bacteria to import iron. These molecules are produced by the bacteria, exported to the outside environment to chelate iron, and then reimported to ensure iron supply (Miethke & Marahiel, 2007; Schalk & Cunrath, 2016). These molecules generally belong to four large families, carboxylates, phenolates, catecholates, or hydroxamates (Wilson, Bogdan, Miyazawa, Hashimoto, & Tsuji, 2016). Other molecules with similar structures have since shown chelating properties for iron or other metals. Plants in particular produce numerous phenolic derivatives, some examples of which are shown in Fig. 14.4.

These phenolic compounds have generated interest due to their different properties, including their antimicrobial activity (Alu'datt et al., 2017; Sakihama, Cohen, Grace, & Yamasaki, 2002). They have been recognized largely as beneficial antioxidants that can scavenge harmful active oxygen species including $O^{2-\cdot}$, H_2O_2, OH. In contrast to their antioxidant activity, phytophenolics also have the potential to act as prooxidants under

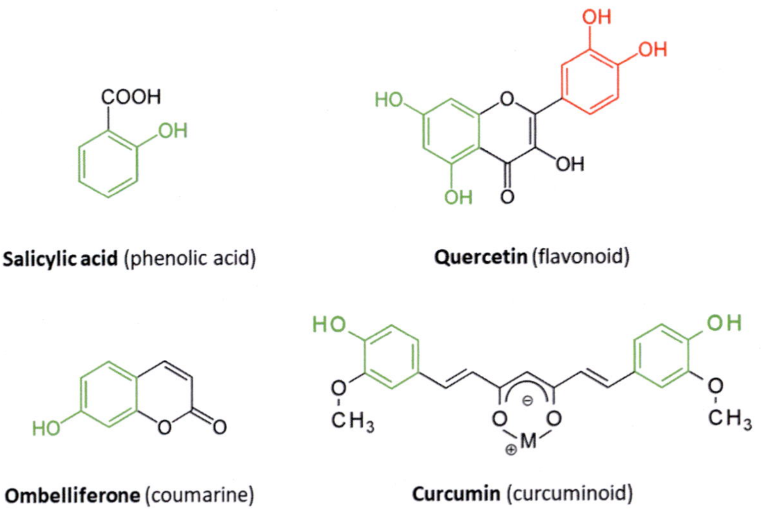

Salicylic acid (phenolic acid)

Quercetin (flavonoid)

Ombelliferone (coumarine)

Curcumin (curcuminoid)

FIGURE 14.4 Examples of active phytophenolic compounds. Phenolic compounds are prevalent in plants and are known to have, among others capacities, antioxidant, antibacterial, antiinflammatory, and antitumoral properties. Besides, some of them may bind metals thanks to the presence of chemical groups also present in siderophores like phenolate (green) and catecholate (red). The effect of this metal chelation potential is still not known, except for some examples like curcumin.

certain conditions. For example, flavonoids and dihydroxycinnamic acids can nick DNA via the production of radicals in the presence of Cu and O^2. Phenoxyl radicals can also initiate lipid peroxidation. Recently, Al, Zn, Ca, Mg, and Cd have been found to stimulate phenoxyl radical-induced lipid peroxidation (Sakihama et al., 2002). Their ability to chelate iron, already compared in rat with a current iron-chelation therapy (El-Sheikh, Ameen, & AbdEl-Fatah, 2018) can also deprive bacteria of this precious metal and stop its growth. All these activities related to their ability to bind metals may explain their antibacterial activity, although the precise mechanisms have not been studied, nor how these compounds are imported into the cell, which may be favored by the presence of metals: some of these compounds, chelated with metals, could indeed be recognized by the cell as metal-siderophore derivatives and thus be imported into the cell, as it has been shown for human catecholamine (Perraud et al., 2020).

An example well studied is from turmeric (Bagchi, Mukherjee, Bhowmick, & Raha, 2015; Hatamie et al., 2012; Hatcher, Planalp, Cho, Torti, & Torti, 2008; Refat, 2013; Song et al., 2009; Wanninger, Lorenz, Subhan, & Edelmann, 2015).

Metal nanoparticles

Another form in which metals can be found is nanoparticles. Simple or composite metal nanoparticles are composed of clusters of atoms and have a size range of 1–100 nm. Nanoparticles have been fabricated using various metals combined with organic and inorganic moieties and metallic nanomaterials (in particular those made of Cu and Ag) and can have strong antibacterial properties (Benoit, Sims, & Fraser, 2019; Hemeg, 2017; Lemire et al., 2013; Vimbela, Ngo, Fraze, Yang, & Stout, 2017). They could further help combine organic and inorganic functions, but it is important to notice that their functional properties, their solubility, their 'transit', their toxicity differ widely from the elementary particles. A striking capability that has been reported for many nanoparticles is their ability to physically interact with the cell surfaces of some bacteria. In principle, the nature of these interactions could be exploited to target metal nanomaterials to specific bacterial cells [(Lemire et al., 2013) and ref. therein]. Nanoparticle toxicity could be due to several attributes, including traits that are particle specific (such as size, shape, or surface charge) and traits that control the release of metal ions. The toxic mode of action of nanoparticles has also been associated with ROS generation and membrane disruption. A growing number of reports indicate that the release of ions is a driving force behind the antimicrobial properties of nanoparticles.

The formation of metal nanoparticles has been demonstrated in remedies from traditional Indian pharmacopeias. These pharmacopeias make extensive use of metals in remedies whose preparation is complex and involves various successive heating/calcination steps, in which various plants are gradually added. A few pioneering studies have shown that these steps lead to the formation of metal nanoparticles (i.e., Kannan, Balaji, & Kumar, 2017; Ruddaraju, Pammi, Guntuku, Padavala, & Kolapalli, 2020; Sharma, Sharma, Kumar, & Yadav, 2015; Singh et al., 2019; Wijenayake, Abayasekara, Pitawala, & Bandara, 2016). These nanoparticles will have very different properties from plants/metals alone, and may therefore lead to different prescriptions, different modes of administration properties that remain to be studied.

Although various approaches have been developed to synthesize nanoparticles, recent approaches have enabled us to synthesize them in a simple and environmentally friendly way (Ruddaraju et al., 2020). Many plants produce antioxidants, which are essential for the formation of nanoparticles, and it turns out that certain metals spontaneously form nanoparticles in a plant extract. Various authors have therefore used this ability to synthesize nanoparticles from therapeutic plant extracts, in order to obtain effective nanoparticles, the characteristics of which would be reproducible, allowing more rigorous clinical analysis (i.e., Chinnasamy, Chandrasekharan, & Bhatnagar, 2019; Khoobchandani et al., 2020; Tiloke, Anand, Gengan, & Chuturgoon, 2018). This would also make it possible to combine traditional medicines and modern medical (nano)technologies. Various gold, silver, or copper nanoparticles have been obtained on this basis, some with antibacterial properties (Devanesan, Ponmurugan, Alsalhi, & Al-Dhabi, 2020; Tripathi, Modi, Narayan, & Rai, 2019). It is not impossible that traditional remedies, combining plants and metals, could lead to the formation of such nanoparticles, without necessarily requiring thermal transformation steps (heating, calcination, etc.)

Characterization of plant–metal combinations

Although the prospects may be interesting, the study of such combinations from traditional remedies is still in its infancy. Many obstacles still stand in the way. The processing of these remedies is based on well-defined preparation protocols, which must be followed faithfully. However, finding and analyzing these protocols from the historical literature do not always allow for a faithful reconstruction. The resources are not always exactly defined, their names may have changed, finding their current equivalent is not guaranteed. In addition, the protocols are only partial. The heating temperature, for example, is often not well explained, certain 'shortcuts' used in the past are not easy to decipher, the translation of the texts sometimes leaves some uncertainties.

Structurally and functionally characterizing these organometallic formulations will also require to work on the whole remedy and not on each individual ingredient. The formation of organometallic molecules can only be obtained after scrupulously following the manufacturing protocol. It is therefore also necessary to review the protocols for the identification and isolation of fractions/active molecules, which were often based on extractions from the various dried plants alone, which were then fractionated by organic solvents. The characterization of these molecules/organometallic particles will require the adaptation of purification and characterization methods, which are often more technology-intensive. It would also not be surprising if the minerals could be found in different forms in the remedy (elemental, particulate) and that all these forms contribute to the properties of the remedy. Some properties will also be difficult to reproduce in vitro (e.g., the irritant effect) and will necessarily require an animal model.

Finally, reusing these remedies will require a great effort in terms of reproducibility/standardization. Indeed, very fine variations in the protocol can have important consequences on the formation of the different active molecules. Moreover, as these minerals could be found in different forms, the toxicity of the remedy could reflect the balance between the different forms, which would be more difficult to control.

Conclusion

There is no original origin for pharmacy. It was born with man, who has always sought to relieve his physical suffering by using the remedies available to him in nature. Ancient and medieval pharmacopeias contain thus vast information on the use of medical material, whether vegetable, mineral, or animal. It is not unrealistic to think that these 1000-year-old pharmacopeias still hold secrets that could have a significant impact on the composition of today's medicines (Harrison & Connely, 2019). Their study could thus provide more than only understanding the medieval mind.

References

Abd-El-Aziz, A. S., Agatemor, C., & Etkin, N. (2017). Antimicrobial resistance challenged with metal-based antimicrobial macromolecules. *Biomaterials, 118*, 27–50. Available from https://doi.org/10.1016/j.biomaterials.2016.12.002.

Albada, B., & Metzler-Nolte, N. (2017). Highly potent antibacterial organometallic peptide conjugates. *Accounts of Chemical Research, 50*, 2510–2518. Available from https://doi.org/10.1021/acs.accounts.7b00282.

Alu'datt, M. H., Rababah, T., Alhamad, M. N., Al-Mahasneh, M. A., Almajwal, A., Gammoh, S., ... Alli, I. (2017). A review of phenolic compounds in oil-bearing plants: Distribution, identification and occurrence of phenolic compounds. *Food Chemistry, 218*, 99–106. Available from https://doi.org/10.1016/j.foodchem.2016.09.057.

Anawati, G. C., Médicaments, L., & L'Oeil, D. (1905). Chez Hunayn Ibn Ishãq. *Arabica, 21*.

Bagchi, A., Mukherjee, P., Bhowmick, S., & Raha, A. (2015). Synthesis, characterization and antibacterial activity of a novel curcumin metal complex. *International Journal of Drug Development and Research, 7*, 11–14. Available from http://ijddr.in/drug-development/synthesis-characterization-and-antibacterial-activity-of-a-novel-curcuminmetal-complex.pdf.

Banin, E., Lozinski, A., Brady, K. M., Berenshtein, E., Butterfield, P. W., Moshe, M., ... Banin, E. (2008). The potential of desferrioxamine-gallium as an anti-Pseudomonas therapeutic agent. *Proceedings of the National Academy of Sciences of the United States of America, 105*, 16761–16766. Available from https://doi.org/10.1073/pnas.0808608105.

Barillo, D. J., & Marx, D. E. (2014). Silver in medicine: A brief history BC 335 to present. *Burns: Journal of the International Society for Burn Injuries*, *40*, S3–S8. Available from https://doi.org/10.1016/j.burns.2014.09.009.

Benoit, D. S. W., Sims, K. R., & Fraser, D. (2019). Nanoparticles for oral biofilm treatments. *ACS Nano*, *13*, 4869–4875. Available from https://doi.org/10.1021/acsnano.9b02816.

Bonah, C., & du Stalinon, L. 'affaire (2007). Et ses conséquences réglementaires, 1954–1959, LA. *Revue du Praticien*, *57*, 1501–1505.

Bouamrane, F. (2010). *Le traité médical (Kitâb al-Taysir) de Ibn Zuhr de Séville, vrin.*

Chen, Y., Zou, J., Sun, H., Qin, J., & Yang, J. (2021). Metals in traditional Chinese medicinal materials (TCMM): A systematic review. *Ecotoxicology and Environmental Safety*, *207*, 111311. Available from https://doi.org/10.1016/j.ecoenv.2020.111311.

Chinnasamy, G., Chandrasekharan, S., & Bhatnagar, S. (2019). Biosynthesis of silver nanoparticles from Melia azedarach: Enhancement of antibacterial, wound healing, antidiabetic and antioxidant activities. *International Journal of Nanomedicine*, *14*, 9823–9836. Available from https://doi.org/10.2147/IJN.S231340.

Chitambar, C. R. (2016). Gallium and its competing roles with iron in biological systems. *Biochimica et Biophysica Acta - Molecular Cell Research*, *1863*, 2044–2053. Available from https://doi.org/10.1016/j.bbamcr.2016.04.027.

Devanesan, S., Ponmurugan, K., Alsalhi, M. S., & Al-Dhabi, N. A. (2020). Cytotoxic and antimicrobial efficacy of silver nanoparticles synthesized using a traditional phytoproduct, asafoetida gum. *International Journal of Nanomedicine*, *15*, 4351–4362. Available from https://doi.org/10.2147/IJN.S258319.

Dolwett, H. H. A., & Sorensen, J. R. J. (1985). Historic uses of copper compounds in medicine. *Trace Elements in Medicine*, *2*, 80–87.

Egorova, K. S., & Ananikov, V. P. (2017). Toxicity of metal compounds: Knowledge and myths. *Organometallics*, *36*, 4071–4090. Available from https://doi.org/10.1021/acs.organomet.7b00605.

El-Sheikh, A. A., Ameen, S. H., & AbdEl-Fatah, S. S. (2018). Ameliorating iron overload in intestinal tissue of adult male rats: Quercetin vs deferoxamine. *Journal of Toxicology*, *2018*, 8023840. Available from https://doi.org/10.1155/2018/8023840.

Ernst, E. (2002). Heavy metals in traditional Indian remedies. *European Journal of Clinical Pharmacology*, *57*, 891–896. Available from https://doi.org/10.1007/s00228-001-0400-y.

Golonka, R., Yeoh, B. S., & Vijay-Kumar, M. (2019). The iron tug-of-war between bacterial siderophores and innate immunity. *Journal of Innate Immunity*, *11*, 249–262. Available from https://doi.org/10.1159/000494627.

Greenberg, E. P., Banin, E., Banin, E., Berenshtein, E., & Chevion, M. (2007). *Metallo-desferrioxamine complexes and their use in the treatment of bacterial infections*. Hadasit Medical Research Services and Development Co.

Gunawan, C., Teoh, W. Y., Marquis, C. P., & Amal, R. (2011). Cytotoxic origin of copper(II) oxide nanoparticles: Comparative studies with micron-sized particles, leachate, and metal salts. *ACS Nano.*, *5*, 7214–7225. Available from https://doi.org/10.1021/nn2020248.

Harrison, F., & Connely, E. (2019). Could medieval medicine help the fight against antimicrobial resistance? *Making the medieval relevant*. De Gruyter.

Hatamie, S., Nouri, M., Karandikar, S. K., Kulkarni, A., Dhole, S. D., Phase, D. M., & Kale, S. N. (2012). Complexes of cobalt nanoparticles and polyfunctional curcumin as antimicrobial agents. *Materials Science and Engineering C*, *32*, 92–97. Available from https://doi.org/10.1016/j.msec.2011.10.002.

Hatcher, H., Planalp, R., Cho, J., Torti, F. M., & Torti, S. V. (2008). Curcumin: From ancient medicine to current clinical trials. *Cellular and Molecular Life Sciences*, *65*, 1631–1652. Available from https://doi.org/10.1007/s00018-008-7452-4.

Hemeg, H. A. (2017). Nanomaterials for alternative antibacterial therapy. *International Journal of Nanomedicine*, *12*, 8211–8225. Available from https://doi.org/10.2147/IJN.S132163.

Hu, Y., Zhu, Y., & Lu, N. H. (2017). Novel and effective therapeutic regimens for *Helicobacter pylori* in an era of increasing antibiotic resistance. *Frontiers in Cellular and Infection Microbiology*, *7*, 168. Available from https://doi.org/10.3389/fcimb.2017.00168.

Kannan, N., Balaji, S., & Kumar, N. V. A. (2017). Structural and elemental characterization of traditional Indian Siddha formulation: Thalagak karuppu. *Journal of Ayurveda and Integrative Medicine*, *8*, 184–189. Available from https://doi.org/10.1016/j.jaim.2016.11.005.

Khoobchandani, M., Katti, K. K., Karikachery, A. R., Thipe, V. C., Srisrimal, D., Mohandoss, D. K. D., ... Katti, K. V. (2020). New approaches in breast cancer therapy through green nanotechnology and nano-ayurvedic medicine – pre-clinical and pilot human clinical investigations. *International Journal of Nanomedicine, 15*, 181–197. Available from https://doi.org/10.2147/IJN.S219042.

Lanoë, C. (2002). La céruse dans la fabrication des cosmétiques sous l'Ancien Régime (XVIe-XVIIIe siècles). In: *Techniques & Culture. Revue Semestrielle d'anthropologie Des Techniques*. Available from https://doi.org/10.4000/tc.224.

Le Bonniec, H., & Histoire naturelle, P l'A. (1953). *Livre XXXIV (Des Métaux et de la sculpture)*, Les Belles Lettres.

Lemire, J. A., Harrison, J. J., & Turner, R. J. (2013). Antimicrobial activity of metals: Mechanisms, molecular targets and applications. *Nature Reviews Microbiology, 11*, 371–384. Available from https://doi.org/10.1038/nrmicro3028.

Lev, E. (2002). Medicinal exploitation of inorganic substances in the Levant in the Medieval and early Ottoman periods. *Adler Museum Bulletin, 28*, 11–16.

Miethke, M., & Marahiel, M. A. (2007). Siderophore-based iron acquisition and pathogen control. *Microbiology and Molecular Biology Reviews, 71*, 413–451. Available from https://doi.org/10.1128/MMBR.00012-07.

Mikulski, M. A., Wichman, M. D., Simmons, D. L., Pham, A. N., Clottey, V., & Fuortes, L. J. (2017). Toxic metals in ayurvedic preparations from a public health lead poisoning cluster investigation. *International Journal of Occupational and Environmental Health, 23*, 187–192. Available from https://doi.org/10.1080/10773525.2018.1447880.

Mislin, G. L. A., & Schalk, I. J. (2014). Siderophore-dependent iron uptake systems as gates for antibiotic Trojan horse strategies against *Pseudomonas aeruginosa*. *Metallomics, 6*, 408–420. Available from https://doi.org/10.1039/c3mt00359k.

Nairz, M., Dichtl, S., Schroll, A., Haschka, D., Tymoszuk, P., Theurl, I., & Weiss, G. (2018). Iron and innate antimicrobial immunity—Depriving the pathogen, defending the host. *Journal of Trace Elements in Medicine and Biology, 48*, 118–133. Available from https://doi.org/10.1016/j.jtemb.2018.03.007.

Ong, Y. C., & Gasser, G. (2019). Organometallic compounds in drug discovery: Past, present and future. *Drug Discovery Today: Technologies*. Available from https://doi.org/10.1016/j.ddtec.2019.06.001.

Parascandola, J. (2009). From mercury to miracle drugs: Syphilis therapy over the centuries. *Pharmacy in History, 51*, 14–23.

Patra, M., Gasser, G., & Metzler-Nolte, N. (2012). Small organometallic compounds as antibacterial agents. *Dalton Transactions, 41*, 6350–6358. Available from https://doi.org/10.1039/c2dt12460b.

Perraud, Q., Kuhn, L., Fritsch, S., Graulier, G., Gasser, V., Normant, V., ... Schalk, I. J. (2020). Opportunistic use of catecholamine neurotransmitters as siderophores to access iron by *Pseudomonas aeruginosa*. *Environmental Microbiology*. Available from https://doi.org/10.1111/1462-2920.15372.

Refat, M. S. (2013). Synthesis and characterization of ligational behavior of curcumin drug towards some transition metal ions: Chelation effect on their thermal stability and biological activity. *Spectrochimica Acta - Part A: Molecular and Biomolecular Spectroscopy, 105*, 326–337. Available from https://doi.org/10.1016/j.saa.2012.12.041.

Riva, M. A., Lafranconi, A., D'Orso, M. I., & Cesana, G. (2012). Lead poisoning: Historical aspects of a paradigmatic \occupational and environmental disease\. *Safety and Health at Work, 3*, 11–16. Available from https://doi.org/10.5491/SHAW.2012.3.1.11.

Ruddaraju, L. K., Pammi, S. V. N., Guntuku, G. S., Padavala, V. S., & Kolapalli, V. R. M. (2020). A review on antibacterials to combat resistance: From ancient era of plants and metals to present and future perspectives of green nano technological combinations. *Asian Journal of Pharmaceutical Sciences, 15*, 42–59. Available from https://doi.org/10.1016/j.ajps.2019.03.002.

Ruparelia, J. P., Chatterjee, A. K., Duttagupta, S. P., & Mukherji, S. (2008). Strain specificity in antimicrobial activity of silver and copper nanoparticles. *Acta Biomaterialia, 4*, 707–716. Available from https://doi.org/10.1016/j.actbio.2007.11.006.

Saad, B., & Said, O. (2011). *Greco-Arab and Islamic herbal medicine: Traditional system, ethics, safety, efficacy, and regulatory Issues*. Wiley.

Sakihama, Y., Cohen, M. F., Grace, S. C., & Yamasaki, H. (2002). Plant phenolic antioxidant and prooxidant activities: Phenolics-induced oxidative damage mediated by metals in plants. *Toxicology, 177*, 67–80. Available from https://doi.org/10.1016/S0300-483X(02)00196-8.

Saper, R. B., Phillips, R. S., Sehgal, A., Khouri, N., Davis, R. B., Paquin, J., ... Kales, S. N. (2008). Lead, mercury, and arsenic in US- and Indian-manufactured Ayurvedic medicines sold via the internet. *Journal of the American Medical Association, 300*, 915–923. Available from https://doi.org/10.1001/jama.300.8.915.

Schalk, I. J., & Cunrath, O. (2016). An overview of the biological metal uptake pathways in *Pseudomonas aeruginosa*. *Environmental Microbiology, 18*, 3227–3246. Available from https://doi.org/10.1111/1462-2920.13525.

Schatzschneider, U. (2018). Antimicrobial activity of organometal compounds: Past, present, and future prospects. *Advances in bioorganometallic chemistry* (pp. 173–192). Germany: Elsevier. Available from https://doi.org/10.1016/B978-0-12-814197-7.00009-1.

Sharma, K. C., Sharma, U., Kumar, V., & Yadav, Y. (2015). Some observations of the zinc-metals based preparations and its properties in ayurveda with regard to stability of medicines in the nano state. *International Journal of Research in Ayurveda & Pharmacy*, *6*(1), 30–34. Available from https://doi.org/10.7897/2277-4343.0618.

Shaw, D., Leon, C., Kolev, S., & Murray, V. (1997). Traditional remedies and food supplements. A 5-year toxicological study (1991-1995). *Drug Safety*, *17*, 342–356. Available from https://doi.org/10.2165/00002018-199717050-00006.

Sierra, M. A., Casarrubios, L., & de la Torre, M. C. (2019). Bio-organometallic derivatives of antibacterial drugs. *Chemistry - A European Journal*, *25*, 7232–7242. Available from https://doi.org/10.1002/chem.201805985.

Singh, R. K., Kumar, S., Aman, A. K., Karim, S. M., Kumar, S., & Kar, M. (2019). Study on physical properties of ayurvedic nanocrystalline Tamra Bhasma by employing modern scientific tools. *Journal of Ayurveda and Integrative Medicine*, *10*, 88–93. Available from https://doi.org/10.1016/j.jaim.2017.06.012.

Snafi, A. E. A. (2018). Arabian medicinal plants with dermatological effects - Plant based review (part 1). *IOSR Journal of Pharmacy (IOSRPHR)*, *8*, 44–73.

Song, Y. M., Xu, J. P., Ding, L., Hou, Q., Liu, J. W., & Zhu, Z. L. (2009). Syntheses, characterization and biological activities of rare earth metal complexes with curcumin and 1,10-phenanthroline-5,6-dione. *Journal of Inorganic Biochemistry*, *103*, 396–400. Available from https://doi.org/10.1016/j.jinorgbio.2008.12.001.

Sotiriou, G. A., & Pratsinis, S. E. (2010). Antibacterial activity of nanosilver ions and particles. *Environmental Science and Technology*, *44*, 5649–5654. Available from https://doi.org/10.1021/es101072s.

Squibb, K. S., & Fowler, B. A. (1981). Relationship between metal toxicity to subcellular systems and the carcinogenic response. *Environmental Health Perspectives*, *40*, 181–188.

Tapsoba, I., Arbault, S., Walter, P., & Amatore, C. (2010). Finding out Egyptian Gods' secret using analytical chemistry: Biomedical properties of Egyptian black makeup revealed by amperometry at single cells. *Analytical Chemistry*, *82*, 457–460. Available from https://doi.org/10.1021/ac902348g.

Tiloke, C., Anand, K., Gengan, R. M., & Chuturgoon, A. A. (2018). *Moringa oleifera* and their phytonanoparticles: Potential antiproliferative agents against cancer. *Biomedicine and Pharmacotherapy*, *108*, 457–466. Available from https://doi.org/10.1016/j.biopha.2018.09.060.

Tripathi, D., Modi, A., Narayan, G., & Rai, S. P. (2019). Green and cost effective synthesis of silver nanoparticles from endangered medicinal plant *Withania coagulans* and their potential biomedical properties. *Materials Science and Engineering C*, *100*, 152–164. Available from https://doi.org/10.1016/j.msec.2019.02.113.

Vaidya, V. S., & Mehendale, H. M. (2014). Mercuric chloride ($HgCl_2$). *Encyclopedia of toxicology* (Third Edition, pp. 203–206). United States: Elsevier. Available from https://doi.org/10.1016/B978-0-12-386454-3.00330-4.

Vimbela, G. V., Ngo, S. M., Fraze, C., Yang, L., & Stout, D. A. (2017). Antibacterial properties and toxicity from metallic nanomaterials. *International Journal of Nanomedicine*, *12*, 3941–3965. Available from https://doi.org/10.2147/IJN.S134526.

Wanninger, S., Lorenz, V., Subhan, A., & Edelmann, F. T. (2015). Metal complexes of curcumin – Synthetic strategies, structures and medicinal applications. *Chemical Society Reviews*, 4986–5002. Available from https://doi.org/10.1039/C5CS00088B.

Wijenayake, A. U., Abayasekara, C. L., Pitawala, H. M. T. G. A., & Bandara, B. M. R. (2016). Antimicrobial potential of two traditional herbometallic drugs against certain pathogenic microbial species. *BMC Complementary and Alternative Medicine*, *16*, 365. Available from https://doi.org/10.1186/s12906-016-1336-1.

Wilson, B. R., Bogdan, A. R., Miyazawa, M., Hashimoto, K., & Tsuji, Y. (2016). Siderophores in iron metabolism: From mechanism to therapy potential. *Trends in Molecular Medicine*, *22*, 1077–1090. Available from https://doi.org/10.1016/j.molmed.2016.10.005.

Wu, Q., Lu, Y. F., Shi, J. Z., Liang, S. X., Shi, J. S., & Liu, J. (2011). Chemical form of metals in traditional medicines underlines potential toxicity in cell cultures. *Journal of Ethnopharmacology*, *134*, 839–843. Available from https://doi.org/10.1016/j.jep.2011.01.031.

Xianglin, S., Castranova, V., Vallyathan, V., & Perry, W. G. (2001). *Molecular mechanisms of metal toxicity and carcinogenesis*. Springer.

ibn Ā al-Ḥazm al-Qaršī, Ā., al-Dn Ibn al-Nafīs, Q., Chadli, A., & Hemrit, A. (2005). *Abrégé du Canon d'Avicenne*, Simpact.

CHAPTER

15

Improved traditional medicine for infectious disorders in Mali

Rokia Sanogo[1,2], *Mahamane Haïdara*[2] *and Adama Dénou*[2]

[1]Department of Traditional Medicine, Bamako, Mali [2]Faculty of Pharmacy, University of Sciences, Techniques and Technologies of Bamako, Mali

Introduction

African traditional medicine (TM) is a cultural heritage and a reservoir of knowledge still largely untapped. It offers possibilities of effective, available, accessible, acceptable, and affordable treatments for the common diseases in African rural communities. The World Health Organization (WHO) estimates that the vast majority of rural populations in developing countries use the resources of TM for their health needs (WHO, 2001). The use of plant-based systems continues to play an essential role in health care. It has been estimated that approximately 80% of the population in developing countries depend on TM for their primary health care (World Health Organization, 2002). In Ghana, Mali, Nigeria, and Zambia, the first line of treatment for 60% of the children with high fevers, resulting from malaria, is the use of herbal medicines at home (WHO, 2004). Since the declaration of Alma Ata in 1978, the WHO has recommended the inclusion of TM and the pharmacopeia in primary health care (WHO, 1978).

In 2000 WHO Africa reaffirmed the usefulness and desirability of the inclusion in the armamentarium of traditional remedies, which gave evidence of safety, efficacy, and quality (WHO, Fiftieth session of the WHO regional committee for Africa, 2002). In 2001 the Abuja Declaration of Heads of State and Government of the African Union focused on the research about the resources of TM for supporting Malaria, HIV AIDS, tuberculosis, and other priority diseases treatments. The same year, the WHO Africa launched the African Decade of TM (2001–10). In 2008 30 years after Alma Ata, the WHO reaffirmed the need of the revival of primary health care (WHO, Ouagadougou Declaration on Primary Health Care, & Health Systems in Africa: Achieving Better Health for Africa in the New Millennium, 2008). The fifth session of the Conference of Ministers of Health in Windhoek, Namibia, in April 2011, approved the report of the end of decade review (African Union,

Medicinal Plants as Anti-infectives
DOI: https://doi.org/10.1016/B978-0-323-90999-0.00004-5

Fifth Ordinary Session of the African Union Conference of Ministers of Health, 2011) and recommended the renewal of the decade and the development of the plan of action for implementation of the second Decade of African TM (2011–20). An interesting point of the plan of action of this new decade is to continue the research and the development of TM and medicinal plants to produce evidence on the safety, efficacy, and quality of new phytodrugs.

In Mali, the sanitary situation is characterized by a predominance of many endemic and epidemic diseases, with a lack of qualified health workers, medicines, and equipment. To improve their state of health, people are using both conventional medicine and TM. The TM, being a significant element in the cultural patrimony, still remains the main recourse for a large majority of people for treating health problems. A high level of Malian Government commitment supported the research and development on TM, with the creation of the first institute for the study on medicinal plants in 1968. Today, this institute is the Department of Traditional Medicine (DMT as abbreviation of its French name) within the National Institute for Research on Public Health (INRSP as abbreviation of its French name), technical institution of the Ministry of Health and Ministry of Scientific Research of Mali, burdened for the policy of TM resources valorization (Practitioners–Practics–Products). This institute is a collaborating center of the WHO in the research on TM. The main objectives of the DMT are to assure the collaboration between traditional and conventional medicines and to produce medicines from local resources, in particular medicinal plants (Ms/INRSP/DTM, 2005).

The DMT has been a collaborating center of the WHO and is today a center of excellence of the West African Health Organization (WAHO) of the ECOWAS region in terms of the valorization of TM resources. The DMT has a strategic partnership with the holders of traditional knowledge and know-how in the field of health, mainly through the Malian Federation of Associations of Therapists and Herbalists.

The main activity of the DMT is the research and development of medicinal products from medicinal plants, called improved traditional medicines (ITMs). These ITMs make a specific contribution to people's access to quality health care for the management, among other things, of infectious diseases.

This article will review ITMs that are indicated for the management of infectious diseases.

General information on improved traditional medicines

Definition

ITMs are medicines derived from the local traditional pharmacopeia, with determined toxicity limits, pharmacological activity confirmed by scientific research, quantified dosage, and quality controlled when they are marketed (Ms/INRSP/DTM, 2005).

Regulatory framework

Decree No. 04–557/P-RM of December 01, 2004 establishing marketing authorizations (MA) for medicinal products for human and veterinary use.

Interministerial Order No. 05–2203/Ms-MEP-SG of September 20, 2005 determining the terms and conditions for applying for MA for medicinal products for human and veterinary use.

Interministerial Order No. 05–2440/Ms-MEF-MEP-SG of October 12, 2005 setting the rate and terms of the fixed duty relating to MAs for medicinal products for human and veterinary use. The MA application files for pharmaceutical specialties for human or veterinary use concern (1) Generic essential drugs in International Nonproprietary Name from the National list of essential drugs and (2) Medicinal plant-based drugs (drugs from the traditional pharmacopeia).

Categories of improved traditional medicines

The categories are defined in Article 10:

Article 10: The term "drug derived from the traditional category 1 pharmacopeia" is understood to mean any drug prepared by the traditional health practitioner for a patient, extemporaneously with fresh or dry raw materials generally of short duration.

The term "drug derived from the traditional category 2 pharmacopeia" means any drug derived from the traditional pharmacopeia commonly used in the community, prepared in advance and whose active components of which it is composed are crude raw materials.

The term "medicinal product derived from the traditional category 3 pharmacopeia" means any medicinal product resulting from scientific research and whose active components are standardized extracts, prepared in advance.

By "drug derived from the traditional pharmacopeia of category 4" is meant any drug resulting from scientific research and whose active components are purified molecules.

Marketing authorization files for ITMs in Mali

In Mali, the MA application files for pharmaceutical specialties for human or veterinary use, in particular, (1) Generic essential drugs in International Nonproprietary Name from the National list of essential drugs: (2) Medicines based on medicinal plants (medicines derived from the traditional pharmacopeia) classified according to the categories mentioned above.

Category 1 drugs are not subject to approval and MA.

Categories 2, 3, and 4 must be the object of obtaining the MA and in Mali the MA request is addressed to the Directorate of Pharmacy and Medicines. The composition of the file depends on the category. MA is granted by decision of the competent authority after technical advice from the National Marketing Authorization Commission, made up of experts from different fields (Arrêté Ministériel No. 05–2203/Ms-MEP-SG du 20 Septembre 2005, 2005):

According to Article 11, the MA application file for traditional herbal medicines written in French must include:

For traditional drugs of category 2, the file is said to be of the light type, and comprises three subfiles which are:

Administrative subfile: it includes:

- A letter of motivation addressed to the Ministry of Health
- A presentation of the production institution
- A copy of the deed authorizing the creation of the production institution
- Samples (10) of the sale model of the product
- Proof of payment of registration fees
- A wholesale price proposal excluding tax (PGHT from the French spelling).

Pharmaceutical subfile:

- Complete monographs of plants used as raw material
- The methods and stages of preparation and production
- A report on good manufacturing practices (GMP)
- An analytical expert report specifying
- The method of quality control of raw materials
- The results of stability and quality control tests of raw materials and excipients
- The method and results of control of products during manufacture
- The results of the quality control of the finished product
- The results of the stability tests of the finished product.

Toxico-clinical subfile:

- An expert report showing a long experience of using the drug in its current form or in its traditional form (at least 20 years)
- Known toxicological risks must be presented in detail (risks of dose-dependent and/or independent toxicity)
- The risks associated with the misuse of the drug as well as the possibilities of physical or psychological dependence must also be indicated.

For traditional drugs of category 3 and 4, the file includes the following subfiles: Administrative subfile: it includes

- A letter of motivation addressed to the Ministry of Health
- A presentation of the production institution
- A copy of the deed authorizing the creation of the production institution
- Samples (10) of the sale model of the product
- Proof of payment of registration fees
- The memorandum of understanding between the manufacturer and the research institute
- A wholesale price proposal excluding tax (PGHT from the French spelling).

Pharmaceutical subfile (category 3):

- Complete monographs of plants used as raw material
- The methods and stages of preparation and production
- A report on GMP
- An analytical expert report
- The method of quality control of raw materials

- The results of stability and quality control tests of raw materials and excipients
- The method and results of control of products during manufacture
- The results of the quality control of the finished product
- The results of the stability tests of the finished product.

Pharmacological and toxicological subfile:

- Pharmacodynamic data
- Review of the literature on pharmacology and toxicology
- The results of acute and subchronic toxicity tests
- An expert report on the tests carried out.

Clinical subfile:

- An authorization for clinical trials, issued by the competent national authority (recognized body or ethics committee)
- A clinical trial protocol according to standard methods (Phase I and II)
- The results of clinical trials
- An expert report on the clinical trials carried out.

Constitution of safety, efficiency, and quality data:

- The safety, efficacy, and quality data obtained are used to constitute the MA files.
- For the development of ITMs, DMT is carrying out research activities using a methodology that goes through a long process.
- The different stages consist of collecting information on traditional recipes for the management of infectious diseases, identifying medicinal plants that are used to prepare recipes, harvesting plant raw materials, phytochemical investigations, preclinical studies (toxicological, pharmacodynamic), drug formulation tests, ethnomedical evidence studies and clinical trials (Sanogo, 2014).
- Collection of information: Ethnobotanical surveys are carried out by a multidisciplinary team in order to collect information on traditional recipes and medicinal plants used in the traditional treatment of infectious diseases. The approach consists in selecting remedies and medicinal plants according to the indication of traditional healers in certain contexts of use. The indication is generally related to the symptoms and signs of the diseases. Other data, such as preparation of remedies, claims for use, including dosage, are also collected.
- Identification of medicinal plants and bibliographical review: This involves conducting research to review scientific literature to collect preclinical and clinical data on plants primarily used for the management of the same disease.
- Preclinical and clinical studies: Additional studies are carried out on selected plants, including characterization of chemical constituents of aqueous and organic extracts, toxicity estimation. The results of the preclinical study allow, in addition to confirming the safety and efficacy of the extracts, the determination of the doses, dosages, and form of use. The proposed formulation is subject to observational studies and if necessary for a clinical trial carried out on the basis of a protocol approved by a scientific and ethics committee.

Historical development of ITMs in Mali

Since 1979, the success of research at DMT has allowed the development of about 20 ITMs. Since 1990, seven ITMs have obtained a MA and are on the list of Essential Medicines of the National Therapeutic Form of Mali in the same way as conventional medicines: (1) BALEMBO© adult syrup and children's syrup (Antitussive), (2) DYSENTERAL© packets (Antiamoebic), (3) GASTROSEDAL© packet (Antiulcer), (4) HEPATISANE© packets (Choleretic), (5) LAXA CASSIA© packets (Laxative), (6) MALARIAL 5© packets (Antimalarial), (7) PSOROSPERMINE© ointment (Antieczematous) (National Therapeutic Form of Mali) (Fig. 15.1).

These ITMs are produced by DMT and distributed in pharmacies and community health centers. These ITMs constitute an alternative safe, efficient, and financially accessible to the populations compared to imported drugs.

The safety, efficacy, and quality data of around 10 new ITMs are available for the constitution of technical files for the MA application.

ITMs in the management of infectious diseases

The main ITMs are indicated for the treatment of diseases caused among others by parasites (malaria and amoebic dysentery), viruses (hepatitis and HIV), bacteria (gastric ulcer associated with *Helicobacter pylori*), and fungi (dermatoses).

For each disease treated, the ITM will be presented as well as the data on the medicinal plants that are used to prepare the ITM.

FIGURE 15.1 ITMs with marketing authorization. *ITMs*, improved traditional medicines.

ITMs for the management of malaria

Malarial 5

Presentation of ITM Malarial 5

Phytomedicine registered on the list of essential drugs in Mali (AMM No.: 290/00–1983), it is an MA for an ITM of category 2.

This is how the first research into DMT led to the marketing of an ITM, Malarial 5, which is a mixture of drug powders from three plants in the following proportions: *Senna occidentalis* L. (Caesalpinaceae) 62%; *Lippia chevalieri* Moldenke (Verbenaceae) 32%; *Acmella oleracea* (L.) R.K. Jansen (Asteraceae) 6%.

- Therapeutic class: Antimalarials
- Properties: Schizonticides and febrifuges
- Indications: Febrile states linked to malaria; influenza, and para-influenza syndromes
- Contraindications: Children under 5 years old
- Pregnancy and breast-feeding: nothing to report to date
- Side effects: Nothing to report to date
- Interaction: Nothing to report to date
- Galenical forms: Package of 11 sachets of 10 g of powder of *Senna occidentalis, Lippia chevalieri*, and *Acmella oleracea* to be used as a decoction
- Precautions for use: The drug should be taken after meals
- Storage: cool and dry places.

Plant data

MALARIAL is an ITM based on the leaves of *Lippia chevalieri* (Verbenaceae) (32%), the leaves of *Senna occidentalis* (Leguminosae) (62%), and the flower heads of *Acmella oleracea* [synonyms: *Spilanthes oleracea* L. (Jansen)] from the Asteraceae family (6%).

Traditional uses: *Acmella oleracea* is an annual herb plant native to the tropics of Africa and America with a yellow flower head. It is used to treat toothaches, sore throats, stomatitis, and malaria (Cheng et al., 2015). All parts of the plant are pungent, but the flower heads are by far the pungentest, exhibiting an inherent characteristic taste, causing itching and salivation. In TM, they are usually chewed to relieve toothaches, throat, and gum conditions, as well as to paralyze the tongue. It is also used for the acute or long-term treatment of microbial infections, in particular pathogenic oral microorganisms, dental caries, periodontosis, gum disease, bleeding gums and/or reduction of dental plaque. An infusion of the flowers is used in the treatment of urinary stones and as a diuretic (Dias et al., 2012).

Lippia chevalieri is an aromatic herb widely used in West Africa. It is used not only as an analgesic in TM to treat pathologies linked to malaria but also in the treatment of respiratory diseases. The leaves are used in the treatment of diarrhea and high blood pressure. It is traditionally used as an antimalarial, as a sedative, and also for the treatment of respiratory diseases (Jean et al., 2012). An infusion of the leaves is used as an antiinfluenza, stimulant, sedative, and relaxant (Pousset, 1989).

Senna occidentalis is a pantropical herb widely used for the treatment of fever, typhoid fever, malaria, and hepatitis (Muhammad, 2020). The roots, leaves, flowers, and seeds are used as laxatives and purgatives. It is used as a febrifuge, dewormer, and anticonvulsant (Saganuwan & Gulumbe, 2006). The leaves are particularly recommended for childbirth by pregnant women because they are oxytocic. The leaves and roots are used in sterility and impotence. The leaves are also used as depuratives (Pousset, 1989).

Phytochemical data: The leaves of *Lippia chevalieri* are very rich in essential oils (Jean et al., 2012). They contain other metabolites such as flavonoids, tannins, sterols, and triterpenes and saponosides (Diarra et al., 2019).

The leaves of *Senna occidentalis* contain alkaloids, anthracenosides, flavonoids, and saponins (Muhammad, 2020; Saganuwan & Gulumbe, 2006). The molecules isolated belong to the family of anthracenosides and flavonoids (Yadav et al., 2010).

Alkylamides, 3-acetylaleuritolic acid, β-sitostenone, scopoletin, vanillic acid, trans-ferulic acid, and trans-isoferulic acid are the major constituents isolated from the flowers of *Acmella oleracea* (Cheng et al., 2015). These alkylamides are: (2E,5Z)-N-isobutylundeca-2,5-diene-8,10-diynamide, (2E)-N-isobutyl-2-undecene-8,10-diynamide, (2E)-N-(2-methylbutyl)-2-undecene-8,10-diynamide, spilanthol, et (2E,6Z,8E)-N-(2-methylbutyl)-2,6,8-decatrienamide. Dias et al. (2012) also identified the presence of spilanthol in the flowers of *Acmella oleracea*.

Efficiency and safety data: The experimental studies carried out by DMT to determine the effectiveness of Malarial 5 have mainly focused, on the one hand, on clinical trials aimed at comparing the effect of Malarial 5 with that of chloroquine and, on the other hand, on in vitro and in vivo studies intended to measure the effectiveness of Malarial 5 and the plants that compose it. These studies have made it possible to demonstrate the beneficial effects of Malarial 5 in the treatment of malaria and to prove that the in vitro antiparasitic activity of Malarial 5 is mainly due to *Lippia chevalieri* and *Acmella oleracea*. The results demonstrated a very weak schizonticidal action of Malarial 5, without any comparison, however, with the action of chloroquine. The effectiveness of Malarial 5 is more evident with regard to clinical signs, patients feel relieved from the 48th hour of treatment (Gasquet et al., 1993; Guindo, 1988; Keita et al., 1990; Koita, 1989, 1991).

Sumafura Tiemoko Bengaly

Presentation of ITM Sumafura Tiemoko Bengaly

This is a new ITM, in the form of *Argemone mexicana* L. leaf powder in a sachet of 30 g to boil with 500 mL of water for 30 min (as directed by the Traditional Health Practitioner). The efficacy, safety, and quality data made it possible to propose an ITM of category 2. A monograph of the plant exists in the West African Pharmacopeia published in 2013. The data exist to make the request for MA.

- Therapeutic class: Antimalarials
- Therapeutic indications: Simple malaria and dracunculiasis
- Precaution for use: Do not use it for more than a week
- Contraindications: Children and pregnant women

- Galenical forms: Package of 21 single-dose sachets of 30 g of powder of *Argemone mexicana* to be used as a decoction
- Storage: Cool and dry places, protected from light.

Plant data

Traditional uses: SUMAFOURA TIEMOGO BENGALY is an ITM based on the seedless aerial part of *Argemone mexicana* (Papaveraceae). *Argemone mexicana* is a plant widely distributed in Mali. The leaves are traditionally used in the treatment of uncomplicated and severe malaria (Diallo et al., 2007). The leaves are useful in coughs, wounds, ulcers, warts, cold sores, skin conditions, skin diseases, itching, and so on (Saranya, Arun, & Iyappan, 2012).

Phytochemical data: The leaves of *Argemone mexicana* contain alkaloids, anthracenosides, flavonoids, carbohydrates, tannins, saponins, and terpenoids (Apu et al., 2012; Khan & Bhadauria, 2019; Saranya et al., 2012; Sarkar, Mitra, Acharyya, & Sadhu, 2019).

Efficiency and safety data: The aqueous and methanolic extracts of the leaves demonstrated antiplasmodial activity in vitro on strains of *Plasmodium falciparum* K1 with inhibitory concentrations 50 (IC_{50}) lower than 10 μg/mL. The best activity was obtained with the methanolic extract with an IC_{50} of 1 μg/mL (Diallo et al., 2007).

An observational clinical study confirmed the ethnomedical use of the plant decoction in the treatment of uncomplicated malaria in patients over 5 years of age, with 89% adequate clinical response (Willcox et al., 2007). In a randomized, controlled trial, the herbal decoction demonstrated clinical efficacy in the treatment of uncomplicated malaria, with similar results compared to a combination therapy based on artemisinin. In both cases, progression to severe malaria remained below 5% (Graz et al., 2010).

Leaf decoction at single doses of 1000, 2000, and 3000 mg/kg produced no signs of toxicity or mortality in mice during the 14 days observation period. The LD_{50} is therefore greater than 3000 mg/kg orally in mice (Sanogo, Maiga, Djimdé, Doumbia, & Guirou, 2008). Repeated administration of 300 mg/kg aqueous extract (per os) 5 days a week for 4 weeks did not affect the biochemical parameters of the blood, liver, and kidneys in rats (Sanogo et al., 2008).

The 80% hydroethanolic extract of the aerial part administered at a dose of 4000 mg/kg per os in mice did not cause mortality and signs of toxicity after 14 days of observation. The LD_{50} of this extract is therefore greater than 4000 mg/kg (Sarkar et al., 2019).

Wolotisane

Presentation of ITM Wolotisane

It is a new ITM, in powder form dosed at 10 g of *Terminalia macroptera* Guill. & Perr. to be used as a decoction.

Efficacy, safety, and quality data made it possible to propose an ITM of category 2. The data exist to apply for MA.

- Therapeutic class: Antimalarials
- Therapeutic actions: Antiplasmodial, analgesic, antiinflammatory, hepatoprotective
- Therapeutic indications: Simple malaria and hepatitis
- Precaution for use: Respect the preparation method
- Contraindications: Not reported to date

- Galenical forms: Package of 14 sachets of powder dosed at 10 g of *Terminalia macroptera* to be used as a decoction
- Storage: cool and dry places.

Plant data

Traditional uses: WOLOTISANE is an ITM made from the leaves or roots of *Terminalia macroptera*, a plant widely distributed in West Africa. *Terminalia macroptera* is used in the treatment of hepatic disorders (jaundice, hepatic syndromes), malaria, diarrhea, as a healing agent in the case of wounds, in conjunctivitis, and vulvovaginitis. Different parts of the plant can be used but the leaf, bark, and root are the most frequently used. Decoction and maceration are the main forms of use and administration is by mouth or topically (Haidara, 2018).

Phytochemical data: Tannins, flavonoids, saponins, anthracene derivatives, oses and holosides, mucilages, sterols, and triterpenes have been detected in the powder of leaves and roots (Haidara, 2018; Yakubu, Adoum, Wudil, & Ladan, 2015). The qualitative analysis by LC-CAD-Ms of the ethanolic extract of the roots and leaves made it possible to identify compounds belonging mainly to the tannins' family and triterpenoids (Haïdara et al., 2020; Haidara, 2018).

Efficiency and safety data: The aqueous extract of the root bark demonstrated antiplasmodial activity in vitro on strains of *Plasmodium falciparum* W2 with an IC_{50} of 1 μg/mL (Sanon et al., 2003). The 70% ethanolic extract of the root bark demonstrated antiplasmodial activity in vitro on strains of *Plasmodium falciparum* K1 with an IC_{50} of 6.8 μg/mL (Traore et al., 2014). The 90% ethanolic extracts of *Terminalia macroptera* leaves and roots are active in vitro on chloroquine-resistant strains of *Plasmodium falciparum* FcB1 with an IC_{50} of 1.2 and 1.6 μg/mL, respectively (Haidara, 2018).

Ethanolic extracts from the leaves and roots of *Terminalia macroptera* (100 mg/kg) administered orally reduced parasitaemia (37.2% and 46.4%, respectively) in mice infected with *Plasmodium chabaudi chabaudi* on the seventh day of infection compared to noninfected mice treated (Haidara, 2018). The same treatments also improved the survival of mice infected with *P. berghei ANKA* with 50%–66.6% survival, respectively, on day 9 post-infection with leaves and roots (Haidara, 2018).

Ethanolic extracts from the leaves and roots of *Terminalia macroptera* have also demonstrated analgesic, antipyretic, antiinflammatory, and hepatoprotective properties in vivo on rodents which may be beneficial in the symptomatic management of malaria (Haidara, 2018).

Ethanolic extracts from leaves and roots showed low cytotoxicity on VERO cells with a cytotoxic concentration (CC) CC50 of 118.2 and 128 μg/mL, respectively. The selectivity index against *Plasmodium falciparum* FcB1 was 98.5 and 80, respectively (Haidara, 2018).

Ethanolic extracts of leaves and roots, administered by gavage at a dose of 2000 mg/kg in albino mice did not cause mortality. The oral lethal dose 50 (LD_{50}) for these extracts is therefore greater than 2000 mg/kg in Swiss albino mice (Haidara, 2018).

Yakubu et al. (2015) evaluated the acute and subchronic toxicity of ethanolic root extract in rats. The results of the study showed that the LD_{50} by the intraperitoneal route was greater than 5000 mg/kg. Repeated administration of the extracts at doses of 500, 1000, and 1500 mg/kg did not cause toxic effects.

ITM for the management of dysentery

Dysenteral

Presentation of ITM Dysenteral

Phytomedicine is registered on the list of essential drugs in Mali (AMM No.: 290/00−1983), it is an MA for an ITM of category 2. A monograph of the plant exists in the West African Pharmacopeia published in 2013.

- Therapeutic actions: Antiamoebic agent on *Entamoeba histolytica* and on *Giardia intestinalis*
- Therapeutic indications: Amoebic dysentery, diarrhea
- Precaution for use: Respect the preparation method
- Contraindications: Not reported to date
- Galenical forms: Package of nine sachets of powder dosed at 10 g of *Euphorbia hirta* L. to be used in the form of a decoction
- Storage: cool and dry places.

Plant data

Traditional uses: DYSENTERAL is an ITM made from the whole plant of *Euphorbia hirta* (Euphorbiaceae). *Euphorbia hirta* is a small herbaceous plant, widely distributed throughout the tropics and subtropics of the world. It is widely used in TM for a wide variety of therapeutic indications such as gastrointestinal infections (diarrhea, dysentery, intestinal parasitosis, etc.), liver, and bronchial and respiratory diseases (asthma, bronchitis, hay fever, etc.) (Galvez et al., 1993; Lanhers, Nicolas, Fleurentin, & Weniger, 2005). It is also used for its diuretic, galactagogue, and sedative effects (Lanhers et al., 2005). It is often used traditionally to treat female disorders, respiratory ailments (cough, coryza, bronchitis, asthma), infestations in children, dysentery, jaundice, pimples, gonorrhea, digestive problems, and tumors (Kumar, Malhotra, & Kumar, 2010).

It is widely used as a decoction or infusion to treat various ailments such as intestinal parasites, diarrhea, heartburn, vomiting, amoebic dysentery, asthma, bronchitis, hay fever, kidney stones, menstrual problems, infertility, and venereal disease. Additionally, the plant is also used to treat ailments of the skin and mucous membranes including warts, scabies, ringworm, thrush, canker sores, fungal diseases, measles, guinea worm, and as an antiseptic to treat wounds and conjunctivitis (Basma, Zakaria, Latha, & Sasidharan, 2011).

Phytochemical data: Phytochemical screening of *Euphorbia hirta* leaf extract revealed the presence of reducing sugars, terpenoids, alkaloids, steroids, tannins, flavonoids, and phenolic compounds (Ahmad, Singh, & Kumar, 2017; Basma et al., 2011; Chichioco-Hernandez & Paguigan, 2010; Ogbulie, Ogueke, Okoli, & Anyanwu, 2007). Thirty-one compounds, including 14 triterpenoids, seven coumarins, four lignans, and six diterpenes have been isolated from the aerial part of *Euphorbia hirta* (Li et al., 2015).

Efficiency and safety data: Numerous studies have shown antimicrobial activity of *Euphorbia hirta* extracts on strains of bacteria that can cause diarrhea (Abubakar, 2009; Ahmad et al., 2017; Ogbulie et al., 2007; Suresh, 2008). Whole plant decoction and quercitrin isolated from whole plant demonstrated antidiarrhoeal activity in several experimental models of diarrhea in rats (Galvez et al., 1993). Numerous studies have shown good activity of the aqueous extracts of the whole plant of *Euphorbia hirta* on *Entamoeba histolytica* (Duez, Livaditis, Guissou, Sawadogo, & Hanocq, 1991; Pechangou, Moundipa, & Sehgal, 2014; Tona et al., 1999, 2000; Tsheko, Bualufu, & Kazembe, 2019).

Two clinical trials have shown that the hydroalcoholic extract in liquid form or in tablet form is effective in treating amoebic dysentery (Martin et al., 1964; Ridet & Chartol, 1964).

An unpublished clinical trial was conducted in Mali on the "DYSENTERAL" treatment (10 g to boil in 500 mL of water for 3 days). The results of the study showed that the trophozoites of *E. histolytica* disappeared from the stool after 2 days, equivalent to treatment with metronidazole (Keita, 1994; Willcox et al., 2012).

The oral LD_{50} of aqueous leaf extract in rats is >5000 mg/kg; however, there was evidence of CNS depression. Histopathological examination showed multifocal submucosal hemorrhages and foci of perivascular hemorrhages and neuronal necrosis, indicating, respectively, acute hemorrhagic enteritis and cerebral hemorrhage (Capule, 2013).

The methanolic extract of whole plant at a single dose of 5000 mg/kg in rats produced no signs of toxicity or mortality during the 14 days observation period. Therefore, the LD_{50} of this plant has been estimated to be greater than 5000 mg/kg (Yuet Ping, Darah, Chen, Sreeramanan, & Sasidharan, 2013). In the 90-day repeated dose oral toxicity study, administration of 50, 250, and 1000 mg/kg per day of *Euphorbia hirta* extract revealed no significant difference in consumption of food and water, weight change, hematological and biochemical parameters, relative organ weights compare to control group. Macropathological and histopathological examinations of all organs, including the liver, did not reveal any morphological alteration (Yuet Ping et al., 2013).

ITM for the management of viral hepatitis

Samanere

Presentation of ITM Samanere

It is a new ITM, in powdered form from the roots of *Entada africana* Guill. & Perr. (Leguminosae) to be used as an infusion.

The efficacy, safety, and quality data made it possible to propose an ITM of category 2. The data exist to make the request for MA.

- Therapeutic actions: Hepatoprotective, antiviral
- Therapeutic indications: Viral hepatitis and jaundice syndrome
- Precaution for use: Respect the preparation method
- Contraindications: Pregnant women and children under 5 years old
- Galenical forms: Package of 14 sachets of powder dosed at 10 g of *Entada africana* to be used in the form of an infusion
- Storage: Cool and dry places.

Plant data

Traditional uses: SAMANERE is an ITM made from the roots of *Entada africana* (Leguminosae). *Entada africana* is used in the treatment of many diseases such as infectious disorders (cough, cold, tuberculosis), hepatitis, fever, and malaria. Maceration of stem bark is used to treat bronchitis and coughs (Yusuf & Abdullahi, 2019). The stem bark and the roots are used to treat diseases of the respiratory tract (Occhiuto et al., 1999).

Phytochemical data: The main chemical constituents of the root of *Entada africana* are flavonoids, tannins, saponins, and sugars (Bako, Bakfur, John, & Bala, 2005; Diarra, 2011; Doumbia, 2011; Olajide, Fadimu, Osaguona, & Saliman, 2013; Tibiri et al., 2010; Yusuf & Abdullahi, 2019). Six triterpene saponins have been isolated from the root (Cioffi et al., 2006). The root also contains arabinogalactan protein type II polysaccharides (Diallo, Paulsen, Liljebäck, & Michaelsen, 2001).

Efficiency and safety data: Extracts from the roots have demonstrated antiviral properties against hepatitis A virus in vitro (Keïta, Renaudet, Girond, Grance, & Deloince, 1994). Extracts from the roots have also demonstrated hepatoprotective properties in vivo (Sanogo, Germanò, D'Angelo, Guglielmo, & De Pasquale, 1998) and immunomodulatory properties in vitro (Diallo et al., 2001) which may be beneficial in the management of hepatitis.

A clinical study carried out on patients with hepatitis B in Mali showed that after a month and a half of treatment, jaundice disappeared in 93.33% of cases with normalization to 100% of SGOT and SGPT transaminases (Douare, 1991).

Many studies have shown the safety of extracts from the roots of *Entada africana*. The lethal dose 50 (LD_{50}) of the decoction is greater than 2000 mg/kg in mice (Doumbia, 2011). Daily administration of the root decoction of *Entada africana* to rats for 3 months did not cause toxicity (Diarra, 2011).

ITM for the management of gastric ulcer associated with *Helicobacter pylori*

Calmogastryl

Presentation of ITM Calmogastryl

The efficacy, safety, and quality data made it possible to propose an ITM of category 2. The data exist to make the request for MA.

- Therapeutic class: Gastric ulcer associated with *Helicobacter pylori*
- Therapeutic actions: Anti-*Helicobacter pylori*
- Therapeutic indications: Peptic ulcer and gastritis
- Precaution for use: Respect the preparation method.
- Contraindications: Not reported to date
- Galenical forms: Packet of 225 g of powder of *Pteleopsis suberosa* Engl. & Diels to use (1 tablespoon or about 15 g) in the form of a decoction.

Plant data

Traditional uses: CALMOGASTRYL is an ITM made from the stem bark of *Pteleopsis suberosa* (Combretaceae). Decoction from the stem bark is used in the treatment of gastritis, peptic ulcer disease (Germanò et al., 2008), digestive mycoses (Gbogbo et al., 2013), and dysentery (Lagnika, Fantodji, & Sanni, 2012).

Phytochemical data: The stem barks of *Pteleopsis suberosa* are rich in tannins (De Pasquale, Germanò, Keita, Sanogo, & Iauk, 1995; Gbogbo et al., 2013; Lagnika et al., 2012; Sanogo, 2010). It also contains triterpenoid and steroidal glucosides and saponins (De Pasquale et al., 1995; Sanogo, 2010).

Thirteen oleanane saponins have been identified in the stem bark (Leo et al., 2006).

Sanni and Omotoyinbo (2016) identified 12 compounds in the methanol−chloroform (2:1) extract of stem bark. The majority of compounds were: friedelan-3-ol followed by di-n-octyl phthalate, 9-octadecenoic acid (Z), diethyl hexadecanedioate, glycerol-1,2-dipalmitate, 9,12-octadecadienoic acid (Z,Z)-, and n-hexadecanoic acid.

Efficiency and safety data: The decoction and the methanolic extract of the stem bark demonstrated antibacterial activity on clinical strains of *Helicobacter pylori* with minimum inhibitory concentrations (MIC) of between 62.5 and 500 μg/mL for the decoction and 31.25−250 μg/mL for the methanolic extract (Germanò et al., 1998).

The decoction (1000 mg/kg per os) demonstrated antiulcer activity by reducing the number and severity of the ethanol-induced ulcer in rats and by increasing the secretion of gastric mucus (Germanò et al., 1998).

Recent studies have further confirmed the gastroprotective properties of the butanol fraction (50−100 and 200 mg/kg per os) of the methanolic extract of the stem bark of *P. suberosa* (Germanò et al., 2008). Arjunglucoside I (saponin oleanane) demonstrated anti-*Helicobacter pylori* activity against three strains resistant to metronidazole with an MIC of between 1.9 and 7.8 μg/mL (Leo et al., 2006).

The acute toxicity of stem bark decoction was evaluated in rats. The results showed that the administration of the doses of 1, 2, and 4 g/kg per os does not cause mortality and side effects. The lethal dose 50 (LD_{50}) of the decoction by the oral route is therefore greater than 4000 mg/kg per os in rats (Sanogo, 2014).

ITM for the management of dermatosis

Mitradermine

Presentation of ITM Mitradermine

This is a new ITM, the efficacy, safety, and quality data will make it possible to constitute the MA file for a category 3 ITM.

- Therapeutic actions: antifungal, antibacterial
- Therapeutic indications: Dermatoses
- Precaution for use: Avoid in children under 1 year old
- Contraindications: Not reported to date
- Galenical forms: Can of 30 g from a 5% ethanolic extract of powdered roots of *Mitracarpus scaber* Zucc. ex Schult. & Schult.f
- Storage: Dry and cool place.

Plant data

Traditional uses: MITRADERMINE is an ITM based on the aerial part of *Mitracarpus scaber* (Rubiaceae). *Mitracarpus scaber* is widely used in TM in West Africa for headaches, toothaches, amenorrhea, dyspepsia, liver disease, venereal disease, and leprosy. Among the folkloric uses, the juice of the plant is applied topically for the treatment of skin diseases (infectious dermatitis, eczema, and scabies). In Malian folk medicine, the aerial parts of *Mitracarpus scaber* are used in the treatment of skin diseases such as superficial fungal diseases, in the form of lotion and skin ointment (Bisignano et al., 2000; Sanogo, Germanò, De Pasquale, Keita, & Bisignano, 1996).

Phytochemical data: The leaves of *Mitracarpus scaber* contain alkaloids, tannins, saponins (Abere, Onwukaeme, & Eboka, 2007; Kporou, Ibourahema, Ackah, & Ouattara, 2017), flavonoids, terpenoids, and phenols (Kporou et al., 2017; Sani, Bello, & Abdul-Kadir, 2014). Bisignano et al. (2000) isolated gallic acid, 4-methoxyacetophenone, 3,4,5-trimethoxyacetophenone, 3,4,5-trimethoxybenzoic acid, kaempferol-3-O-rutinoside, rutin, and psoralen in the methanol extract of the aerial part (Bisignano et al., 2000). Azaanthraquinone has been isolated from the ethanolic extract of the aerial part (Okunade, Clark, Hufford, & Oguntimein, 1999). Oleanolic acid and ursolic acid have been isolated from the ethanolic extract of the aerial part (Gbaguidi, Accrombessi, Moudachirou, & Quetin-Leclercq, 2005).

Efficiency and safety data: The extracts of the aerial part have demonstrated antifungal activity on strains of *Candida* and antibacterial activity on strains of *Staphylococcus*. Diethyl ether extract has fairly good antifungal activity against standard strains and clinical isolates of *Candida* sp. and antibacterial on standard strains and clinical isolates of *Staphylococcus* (Sanogo et al., 1996). The methanolic extract of *Mitracarpus scaber* and the isolated molecules possess both antibacterial and antimycotic activities (Bisignano et al., 2000).

Numerous other studies have demonstrated the antibacterial and antifungal activities of extracts from the leaves or from the aerial part (Ali-Emmanuel, Moudachirou, Akakpo, & Leclercq, 2002; Ekpendu, Akah, Adesomoju, & Okogun, 1994; Kporou, Koffi, Ouattara, & Guede-Guina, 2010; Kporou et al., 2017; Ouadja, Anani, Djeri, Ameyapoh, & Karou, 2018).

The 2-azaanthraquinone isolated from the aerial part of *Mitracarpus scaber* demonstrated antimycotic activity on *Dermatophilus congolensis* with an MIC of 7.5 μg/mL (Gbaguidi, Muccioli, Accrombessi, Moudachirou, & Quetin-Leclercq, 2005). It has also demonstrated antimicrobial activity on several strains of bacteria and fungi (Okunade et al., 1999).

Oleanolic acid and ursolic acid demonstrated antimycotic activity against *Dermatophilus congolensis* with an MIC of 15 μg/mL (Gbaguidi et al., 2005).

In Mali, the antimicrobial and antifungal activities of "MITRADERMINE OINTMENT" (5 g of hydroalcoholic extract of *M. scaber* in 100 g of shea butter) were tested in vivo by local application on the lesions three times per day (Sanogo et al., 1996).

A 20% ointment based on ethanolic extract of the leaves and shea butter promoted the healing of wounds infected with *P. aeruginosa* in rats (Tettegah et al., 2009).

The LD_{50} by intraperitoneal route in rats was 980 ± 14 and 885 ± 23 mg/kg for petroleum ether extract and methanolic extract, respectively (Ekpendu et al., 1994).

Animal test results have shown that the plant extracts (leaves) are safe for topical application to the skin (Shinkafi, 2014).

Conclusion and perspectives

The Malian flora is very rich in medicinal plants, traditionally used in the treatment of frequent pathologies. Safety, efficacy, and quality data will contribute to the constitution of MA dossiers and increase the number of ITMs with MA to be used in the fight against infectious diseases.

In the context of the coronavirus pandemic, with the absence of specific treatment, traditional preparations, in particular, based on medicinal plants, and certain ITMs of DMT can constitute an opportunity and contribute to the care of affected people. It is about using the African pharmacopeia to help care for people with COVID-19.

In this perspective, we have made a collection of ITMs and medicinal plants with therapeutic potential for antiviral, antibacterial, cough suppressant, bronchodilator, antiinflammatory, antioxidant, and immune system strengthening properties. We can offer herbal teas for integrated and adapted care for people with COVID-19, in collaboration with hospitals based on validated scientific protocols. We are also carrying out in vitro studies of medicinal plants extracts for their antiviral activity on SARS-CoV-2, in collaboration with the research team on malaria which has just succeeded in sequencing the virus circulating in Mali.

Index of ITMs

- CALMOGASTRYL 16
- DYSENTERAL 13
- MALARIAL 8
- MITRADERMINE POMMADE 17
- SAMANERE 15
- SUMAFURA 10
- WOLOTISANE 11

References

Abere, A. T., Onwukaeme, D. N., & Eboka, C. J. (2007). Pharmacognostic evaluation of the leaves of *Mitracarpus scaber* Zucc (Rubiaceae). *Tropical Journal of Pharmaceutical Research*, *6*, 849–853. Available from https://doi.org/10.4314/tjpr.v6i4.14669.

Abubakar, E. M. M. (2009). Antibacterial activity of crude extracts of *Euphorbia hirta* against some bacteria associated with enteric infections. *Journal of Medicinal Plants Research*, *3*, 498–505. Available from http://www.academicjournals.org/JMPR/PDF/pdf2009/July/El-Mahmood.pdf.

African Union, Fifth Ordinary Session of the African Union Conference of Ministers of Health. (2011). Windhoek, Namibia, 17–21 April 2011.

Ahmad, W., Singh, S., & Kumar, S. (2017). Phytochemical screening and antimicrobial study of *Euphorbia hirta* extracts. *Journal of Medicinal Plants Studies*, *5*, 183–186.

Ali-Emmanuel, N., Moudachirou, M., Akakpo, A.J., & Leclercq, J.Q. (2002). Activités antibactériennes in vitro de Cassia alata, Lantana camara et Mitracarpus scaber sur Dermatophilus congolensis isolé au Bénin. Revue d'élevage et de médecine vétérinaire des pays tropicaux.

Arrêté Ministériel No. 05–2203/MS-MEP-SG du 20 Septembre 2005. (2005). Déterminant les modalités de demande des autorisations de mise sur le marché (AMM) des médicaments à usage humain et vétérinaire, Arrêté Interministériel No. 05–2203/MS-MEP-SG Du 20 Septembre 2005.

Apu, A. S., Al-Baizyd, A. H., Ara, F., Bhuyan, S. H., Matin, M., & Hossain, M. F. (2012). Phytochemical analysis and bioactivities of *Argemone mexicana* Linn. leaves. *Pharmcologyonline*, *3*, 16–23.

Bako, S. P., Bakfur, M. J., John, I., & Bala, E. I. (2005). Ethnomedicinal and phytochemical profile of some Savanna plant species in Nigeria. *International Journal of Botany*, *1*, 147–150. Available from https://doi.org/10.3923/ijb.2005.147.150.

Basma, A. A., Zakaria, Z., Latha, L. Y., & Sasidharan, S. (2011). Antioxidant activity and phytochemical screening of the methanol extracts of *Euphorbia hirta* L. *Asian Pacific Journal of Tropical Medicine*, *4*, 386–390. Available from https://doi.org/10.1016/S1995-7645(11)60109-0.

Bisignano, G., Sanogo, R., Marino, A., Aquino, R., D'angelo, V., Germano, M. P., ... Pizza, C. (2000). Antimicrobial activity of *Mitracarpus scaber* extract and isolated constituents. *Letters in Applied Microbiology, 30*, 105–108. Available from https://doi.org/10.1046/j.1472-765x.2000.00692.x.

Capule, F. R. (2013). Acute oral toxicity of the crude aqueous extract of the whole plant of *Euphorbia hirta* L. (Family euphorbiaceae). *Acta Medica Philippina, 47*, 23–29. Available from https://www.actamedicaphilippina.com.ph/.

Cheng, Y. B., Liu, R. H., Ho, M. C., Wu, T. Y., Chen, C. Y., Lo, I. W., ... Chang, F. R. (2015). Alkylamides of *Acmella oleracea*. *Molecules (Basel, Switzerland), 20*, 6970–6977. Available from https://doi.org/10.3390/molecules20046970.

Chichioco-Hernandez, C. L., & Paguigan, N. D. (2010). Phytochemical profile of selected Philippine plants used to treat asthma. *Pharmacognosy Journal, 2*, 198–202. Available from https://doi.org/10.1016/S0975-3575(10)80092-6.

Cioffi, G., Dal Piaz, F., De Caprariis, P., Sanogo, R., Marzocco, S., Autore, G., & De Tommasi, N. (2006). Antiproliferative triterpene saponins from *Entada africana*. *Journal of Natural Products, 69*, 1323–1329. Available from https://doi.org/10.1021/np060257w.

De Pasquale, R., Germanò, M. P., Keita, A., Sanogo, R., & Iauk, L. (1995). Antiulcer activity of *Pteleopsis suberosa*. *Journal of Ethnopharmacology, 47*(1), 55–58. Available from https://doi.org/10.1016/0378-8741(95)01256-d.

Diallo, D., Paulsen, B. S., Liljebäck, T. H. A., & Michaelsen, T. E. (2001). Polysaccharides from the roots of *Entada africana* Guill. et Perr., Mimosaceae, with complement fixing activity. *Journal of Ethnopharmacology, 74*, 159–171. Available from https://doi.org/10.1016/S0378-8741(00)00361-5.

Diallo, D., Diakité, C., Mounkoro, P. P., Sangaré, D., Graz, B., Falquet, J., & Giani, S. (2007). La prise en charge du paludisme par les therapeutes traditionnels dans les aires de sante de Kendie (Bandiagara) et de Finkolo (Sikasso) au Mali. *Le Mali Médical, 22*, 1–8.

Diarra, B. (2011). Effets de l'administration répétée du décocté des racines de Entada africana Guill et Perr (Mimosaceae) sur certains paramètres biologiques chez les rats. Thèse de doctorat en Pharmacie, Faculté de Médecine, de Pharmacie et d'Odonto- Stomatologie, Université de Bamako, Mali.

Diarra, M. L., Haïdara, M., Dénou, A., Doumbia, S., Dembélé, D., Diallo, D., & Noba, K. (2019). Contrôle Physicochimique des Feuilles de Lippia chevalieri Moldenke Cultivé. *European Scientific Journal, 15*, 1857–7881.

Dias, A. M. A., Santos, P., Seabra, I. J., Júnior, R. N. C., Braga, M. E. M., & De Sousa, H. C. (2012). Spilanthol from *Spilanthes acmella* flowers, leaves and stems obtained by selective supercritical carbon dioxide extraction. *Journal of Supercritical Fluids, 61*, 62–70. Available from https://doi.org/10.1016/j.supflu.2011.09.020.

Douare, I. (1991). Contribution à l'étude d'une préparation traditionnelle utilisant les racines de Entada africana SAMANERE pour le traitement de l'hépatite virale. Thèse de Doctorat en médecine, Ecole Nationale de Médecine et de Pharmacie, Mali.

Doumbia, S. (2011). Etude de la phytochimie et des activités biologiques des feuilles, des écorces de tronc et des racines de Entada africana Guill et Perr (Mimosaceae). Thèse de doctorat en Pharmacie, Faculté de Médecine, de Pharmacie et d'Odonto-Stomatologie, Université de Bamako, Mali.

Duez, P., Livaditis, A., Guissou, P. I., Sawadogo, M., & Hanocq, M. (1991). Use of an *Amoeba proteus* model for in vitro cytotoxicity testing in phytochemical research. Application to Euphorbia hirta extracts. *Journal of Ethnopharmacology, 34*, 235–246. Available from https://doi.org/10.1016/0378-8741(91)90042-C.

Ekpendu, T. O., Akah, P. A., Adesomoju, A. A., & Okogun, J. I. (1994). Antiinflammatory and antimicrobial activities of *Mitracarpus scaber* extracts. *International Journal of Pharmacognosy, 32*, 191–196.

Galvez, J., Zarzuelo, A., Crespo, M. E., Lorente, M. D., Ocete, M. A., & Jimenez, J. (1993). Antidiarrhoeic activity of *Euphorbia hirta* extract and isolation of an active flavonoid constituent. *Planta Medica, 59*, 333–336. Available from https://doi.org/10.1055/s-2006-959694.

Gasquet, M., Delmas, F., Timon-David, P., Keita, A., Guindo, M., Kiota, N., ... Doumbo, O. (1993). Evaluation in vitro and in vivo of a traditional antimalarial, "Malarial 5". *Fitoterapia, 64*, 423–426.

Gbaguidi, F., Accrombessi, G., Moudachirou, M., & Quetin-Leclercq, J. (2005). HPLC quantification of two isomeric triterpenic acids isolated from *Mitracarpus scaber* and antimicrobial activity on *Dermatophilus congolensis*. *Journal of Pharmaceutical and Biomedical Analysis, 39*, 990–995. Available from https://doi.org/10.1016/j.jpba.2005.05.030.

Gbaguidi, F., Muccioli, G., Accrombessi, G., Moudachirou, M., & Quetin-Leclercq, J. (2005). Densitometric HPTLC quantification of 2-azaanthraquinone isolated from *Mitracarpus scaber* and antimocribial activity against *Dermatophilus congolensis*. *Journal of Planar Chromatography - Modern TLC, 18*, 377–379. Available from https://doi.org/10.1556/JPC.18.2005.5.8.

Gbogbo, K. A., Agban, A., Woegan, Y. A., Amana, E. K., Hoekou, P. Y., Batawila, K., & Akpagana, K. (2013). Evaluation de l'activité antimicrobienne de *Momordica charantia* (cucurbitaceae), *Psidium guajava* (myrtaceae) et *Pteleopsis suberosa* (combretaceae). *European Scientific Journal., 9*, 411–421.

Germanò, M. P., Sanogo, R., Guglielmo, M., De Pasquale, R., Crisafi, G., & Bisignano, G. (1998). Effects of *Pteleopsis suberosa* extracts on experimental gastric ulcers and *Helicobacter pylori* growth. *Journal of Ethnopharmacology, 59*, 167–172. Available from https://doi.org/10.1016/s0378-8741(97)00109-8.

Germanò, M. P., D'Angelo, V., Biasini, T., Miano, T. C., Braca, A., De Leo, M., … Sanogo, R. (2008). Anti-ulcer, anti-inflammatory and antioxidant activities of the n-butanol fraction from *Pteleopsis suberosa* stem bark. *Journal of Ethnopharmacology, 115*, 271–275. Available from https://doi.org/10.1016/j.jep.2007.10.001.

Graz, B., Willcox, M. L., Diakite, C., Falquet, J., Dackuo, F., Sidibe, O., … Diallo, D. (2010). *Argemone mexicana* decoction vs artesunate-amodiaquine for the management of malaria in Mali: Policy and public-health implications. *Transactions of the Royal Society of Tropical Medicine and Hygiene, 104*, 33–41. Available from https://doi.org/10.1016/j.trstmh.2009.07.005.

Guindo, M. (1988). Contribution à l'étude du traitement traditionnel du «suma» (paludisme). Thése de Pharmacie. Ecole Nationale de Médecine et de Pharmacie du Mali.

Haidara, M. (2018). Contribution à l'étude de l'activité pharmacologique de Terminalia macroptera Guill. et Perr. (Combretaceae) dans le but de l'élaboration d'un médicament traditionnel amélioré au Mali. Thèse de doctorat, Université Paul Sabatier, Toulouse, France.

Haïdara, M., Dénou, A., Haddad, M., Camara, A., Traoré, K., Aubouy, A., … Sanogo, R. (2020). Evaluation of Anti-inflammatory. *Anti-pyretic, Analgesic, and, Planta Medica International Open, 07*, e58–e67.

Jean, B. M., Abarca, N. A., Nâg-Tiero, M. R., Mouhibatou, Y. Z., Jeanne, M. R., & Germaine, N. O. (2012). Lippia chevalieri Moldenke a brief review of traditional uses, phytochemistry and pharmacology. *International Journal of Drug Delivery, 4*. Available from https://doi.org/10.5138/ijdd.v4i3.795.

Keita, A. (1994). Activities of the traditional medicine department in Mali. International workshop by the GIFTS of Health, Mbarara, Uganda.

Keita, A., Doumbo, O., Koita, N., Diallo, D., Guindo, M., & Traore, A. K. (1990). Recherche expérimentale sur un antimalarique traditionnel. *Bulletin de Médecine Traditionnelle et Pharmacopée, 4*, 139–146.

Keïta, A., Renaudet, J., Girond, S., Grance, J.M., & Deloince, R. (1994). Effet antiviral de deux plantes de la pharmacopée malienne sur la multiplication du virus de l'hépatite A (VHA) in vitro. Phyllanthus amarus et Entada africana. Médecine et pharmacopée africaine.

Khan, A. M., & Bhadauria, S. (2019). Analysis of medicinally important phytocompounds from Argemone mexicana. *Journal of King Saud University - Science, 31*, 1020–1026. Available from https://doi.org/10.1016/j.jksus.2018.05.009.

Koita, N. (1989). *A comparative study of the traditional remedy "Suma–Kala" and chloroquine as teeatment for malaria in rural area of Mali* (M.Sc. Dissertation). London: London School of Hygiene and Tropical Medicine, University of London.

Koita, N. (1991). A comparative study of the traditional remedy "Suma –Kala" and chloroquine as treatment for malaria in rural area of Mali. In Proceedings of an International Conference of Experts from Developing Country on Traditional Medicinal Plants.

Kporou, E. K., Koffi, M., Ouattara, S., & Guede-Guina, F. (2010). Evaluation de l'activité antifongique de *Mitracarpus scaber*, une rubiaceae codifiée MISCA sur *Candida glabrata. Therapie, 65*, 271–274. Available from https://doi.org/10.2515/therapie/2010013.

Kporou, E. K., Ibourahema, C., Ackah, A. J., & Ouattara, S. (2017). Antifungal potential of improved crude extracts of *Mitracarpus scaber* (Zucc) against *Candida guilliermondii* and *Candida parapsilosis. Pharmaceutical and Biological Evaluations, 4*, 90–96. Available from https://doi.org/10.26510/2394-0859.pbe.2017.14.

Kumar, S., Malhotra, R., & Kumar, D. (2010). Euphorbia hirta: Its chemistry, traditional and medicinal uses, and pharmacological activities. *Pharmacognosy Reviews, 4*, 58–61. Available from https://doi.org/10.4103/0973-7847.65327.

Lagnika, L., Fantodji, M. H., & Sanni, A. (2012). Phytochemical study and antibacterial, antifungal and antioxidant properties of *Bridelia ferruginea* and *Pteleopsis suberosa. International Journal of Pharmaceutical Sciences and Research, 3*, 2130–2136.

Lanhers, M. C., Nicolas, J. P., Fleurentin, J., & Weniger, B. (2005). *Euphorbia hirta* L. *Ethnopharmacologia, 36*, 9–23.

Leo, M. D., Tommasi, N. D., Sanogo, R., D'Angelo, V., Germanò, M. P., Bisignano, G., & Braca, A. (2006). Triterpenoid saponins from *Pteleopsis suberosa* stem bark. *Phytochemistry, 67*, 2623–2629. Available from https://doi.org/10.1016/j.phytochem.2006.07.017.

Li, E. T., Liu, K. H., Zang, M. H., Zhang, X. L., Jiang, H. Q., Zhou, H. L., … Wu, Y. (2015). Chemical constituents from *Euphorbia hirta. Biochemical Systematics and Ecology, 62*, 204–207. Available from https://doi.org/10.1016/j.bse.2015.09.007.

Martin, M., Chartol, A., Porte, L., Ridet, J., Biot, J., & Bezon, A. (1964). Action thérapeutique de l'extrait d'Euphorbia hirta dans l'amibiase intestinale. A propos de 150 observations. *Médecine Tropicale: Revue du Corps de Santé Colonial, 24*, 250–261.

MS/INRSP/DTM. (2005). Politique Nationale de Médecine Traditionnelle du Mali.

Muhammad, D. A. (2020). Ethnobotanical survey and in vivo assessment of the antimalarial activities of a locally used medicinal plant (*Senna occidentalis*) for "Malaria Suspected" fever in potiskum and nangere local government areas of Yobe state. *Asian Journal of Research in Infectious Diseases, 4*(4), 35–43. Available from https://doi.org/10.9734/ajrid/2020/v4i430155.

Occhiuto, F., Sanogo, R., Germano, M. P., Keita, A., D'Angelo, V., & De Pasquale, R. (1999). Effects of some Malian medicinal plants on the respiratory tract of guinea-pigs. *Journal of Pharmacy and Pharmacology, 51*, 1299–1303. Available from https://doi.org/10.1211/0022357991776877.

Ogbulie, J. N., Ogueke, C. C., Okoli, I. C., & Anyanwu, B. N. (2007). Antibacterial activities and toxicological potentials of crude ethanolic extracts of *Euphorbia hirta*. *African Journal of Biotechnology, 6*, 1544–1548. Available from http://www.academicjournals.org/AJB/PDF/pdf2007/4Jul/Ogbulie%20et%20al.pdf.

Okunade, A. L., Clark, A. M., Hufford, C. D., & Oguntimein, B. O. (1999). Azaanthraquinone: An antimicrobial alkaloid from Mitracarpus scaber. *Planta Medica, 65*, 447–448. Available from https://doi.org/10.1055/s-2006-960807.

Olajide, O. B., Fadimu, O. Y., Osaguona, P. O., & Saliman, M. I. (2013). Ethnobotanical and phytochemical studies of some selected species of leguminoseae of Northern Nigeria: A study of Borgu Local Government Area. *International Journal of Science and Nature, 4*, 546–551.

Ouadja, B., Anani, K., Djeri, B., Ameyapoh, Y. O., & Karou, D. S. (2018). Evaluation of the phytochemical composition, antimicrobial and anti-radical activities of *Mitracarpus scaber* (Rubiaceae). *Journal of Medicinal Plants Research, 12*, 493–499. Available from https://doi.org/10.5897/jmpr2018.6631.

Pechangou, S. N., Moundipa, P. F., & Sehgal, R. (2014). In vitro susceptibilities of the clinical isolate of *Entamoeba histolytica* to *Euphorbia hirta* (Euphorbiaceae) aqueous extract and fractions. *African Journal of Microbiology Research, 8*, 3354–3361.

Pousset, J. L. (1989). Plantes médicinales Africaines, Ellipses, Agence de Coopération Culturelle et Technique, Paris.

Ridet, J., & Chartol, A. (1964). Les Propriétés antidysentériques de l'Euphorbia hirta. *Médecine Tropicale: Revue du Corps de Santé Colonial, 24*, 119–143.

Saganuwan, A. S., & Gulumbe, M. L. (2006). Evaluation of in-vitro antimicrobial activities and phytochemical constituents of *Cassia occidentalis*. *Animal Research International, 3*, 566–569. Available from https://doi.org/10.4314/ari.v3i3.40793.

Sani, I., Bello, F., & Abdul-Kadir, D. (2014). Phytochemical screening and antibacterial activity of *Allium sativum, Calotropis procera, Acacia nilotica* and *Mitracarpus scaber* mixed hexane extracts. *World Journal of Pharmaceutical Research, 3*, 142–149.

Sanni, D. M., & Omotoyinbo, O. V. (2016). GC-MS analysis of *Pteleopsis suberosa* stem bark methanol-chloroform extract. *Journal of Plant Sciences, 4*, 37–40.

Sanogo, B. (2010). *Etude des activités antioxydantes et anticonvulsivantes de deux plantes médicinales du Mali* (Thèse de doctorat en Pharmacie). Mali: Faculté de Médecine, de Pharmacie et d'Odonto-Stomatologie, Université de Bamako.

Sanogo, R. (2014). *Pteolopsis suberosa* Engl. et Diels (Combretaceae): une plante à activité antiulcère et anti *Helicobacter pylori*. *Hegel, 4*, 148–153.

Sanogo, R. (2014). Development of phytodrugs from indigenous plants: The Mali experience. *Novel plant bioresources: Applications in food, medicine and cosmetics* (pp. 191–203). Mali: Wiley Blackwell. Available from https://doi.org/10.1002/9781118460566.ch15.

Sanogo, R., Germanò, M. P., De Pasquale, R., Keita, A., & Bisignano, G. (1996). Selective antimicrobial activities of *Mitracarpus scaber* Zucc. against *Candida* and *Staphylococcus* sp. *Phytomedicine: International Journal of Phytotherapy and Phytopharmacology, 2*, 265–268. Available from https://doi.org/10.1016/s0944-7113(96)80053-5.

Sanogo, R., Germanò, M. P., D'Angelo, V., Guglielmo, M., & De Pasquale, R. (1998). Antihepatotoxic properties of *Entada africana* (Mimosaceae). *Phytotherapy Research, 12*, S157–S159. Available from https://doi.org/10.1002/(SICI)1099-1573(1998)12:1 + S157:AID-PTR2823.0.CO;2-H.

Sanogo, R., Maiga, A., Djimdé, A., Doumbia, L., Guirou, C., et al. (2008). Etude de la toxicité subchronique du décocté de Argemone Mexicana. *Médecine et Pharmacopée Traditionnelles, 15*, 26–31.

Sanon, S., Ollivier, E., Azas, N., Mahiou, V., Gasquet, M., Ouattara, C. T., ... Fumoux, F. (2003). Ethnobotanical survey and in vitro antiplasmodial activity of plants used in traditional medicine in Burkina Faso. *Journal of Ethnopharmacology*, *86*, 143–147. Available from https://doi.org/10.1016/S0378-8741(02)00381-1.

Saranya, M. S., Arun, T., & Iyappan, P. (2012). In vitro antibacterial activity and preliminary phytochemical analysis of leaf extracts of Argemone Mexicana Linn–A medicinal plant. *International Journal of Current Pharmaceutical Research*, *4*, 85–87.

Sarkar, K. K., Mitra, T., Acharyya, R. N., & Sadhu, S. K. (2019). Phytochemical screening and evaluation of the pharmacological activities of ethanolic extract of *Argemone mexicana* Linn. aerial parts. *Oriental Pharmacy and Experimental Medicine*, *19*, 91–106. Available from https://doi.org/10.1007/s13596-018-0357-3.

Shinkafi, S. A. (2014). Antidermatophytic activities of column chromatographic fractions and toxicity studies of *Pergularia tomentosa* L. and *Mitracarpus scaber* Zucc used in the treatment of dermatophytoses. *Advancement in Medicinal Plant Research*, *2*, 7–15.

Suresh, K. (2008). Antimicrobial and phytochemical investigation of the leaves of *Carica papaya* L., *Cynodon dactylon* (L.) Pers., *Euphorbia hirta* L., *Melia azedarach* L. and *Psidium guajava* L, ethnobotanical leaflets (p. 157).

Tettegah, M., Eklu-Kadegbeku, K., Aklikokou, A. K., Agbonon, A., De Souza, C., & Gbeassor, M. (2009). Infected wound healing and antimicrobial effects of *Chenopodium ambrosioides* and *Mitracarpus scaber*. *International Journal of Biological and Chemical Sciences*, *3*, 623–627. Available from https://doi.org/10.4314/ijbcs.v3i3.45329.

Tibiri, A., Sawadogo, R. W., Ouedraogo, N., Banzouzi, J. T., Guissou, I. P., & Nacoulma, G. O. (2010). Evaluation of antioxidant activity, total phenolic and flavonoid contents of *Entada africana* guill. et Perr. (Mimosaceae) organ extracts. *Research Journal of Medical Sciences*, *4*, 81–87. Available from https://doi.org/10.3923/rjmsci.2010.81.87.

Tona, L., Kambu, K., Mesia, K., Cimanga, K., Apers, S., De Bruyne, T., ... Vlietinck, A. J. (1999). Biological screening of traditional preparations from some medicinal plants used as antidiarrhoeal in Kinshasa, Congo. *Phytomedicine: International Journal of Phytotherapy and Phytopharmacology*, *6*, 59–66. Available from https://doi.org/10.1016/S0944-7113(99)80036-1.

Tona, L., Kambu, K., Ngimbi, N., Mesia, K., Penge, O., Lusakibanza, M., ... Vlietinck, A. J. (2000). Antiamoebic and spasmolytic activities of extracts from some antidiarrhoeal traditional preparations used in Kinshasa, Congo. *Phytomedicine: International Journal of Phytotherapy and Phytopharmacology*, *7*, 31–38. Available from https://doi.org/10.1016/S0944-7113(00)80019-7.

Traore, M. S., Diane, S., Diallo, M. S. T., Balde, E. S., Balde, M. A., Camara, A., ... Balde, A. M. (2014). In vitro antiprotozoal and cytotoxic activity of ethnopharmacologically selected Guinean plants. *Planta Medica*, *80*, 1340–1344. Available from https://doi.org/10.1055/s-0034-1383047.

Tsheko, M., Bualufu, N. B., & Kazembe, M. (2019). Étude de l'activité antiamibienne des quelques plantes médicinales utilisées par les tradipraticiens de la ville de Lubumbashi, R.D.Congo. *Journal of Applied Biosciences*, *137*, 13953–13960. Available from https://doi.org/10.4314/jab.v137i1.3.

WHO. (2001). Promotion du rôle de la médecine traditionnelle dans le systéme de santé: Stratégie de la région africaine. AFR/RC50/09.

World Health Organization. (2002). *Programme on traditional medicine, Stratégie de l' OMS pour la médecine traditionnelle pour 2002–2005*. Retrieved from https://apps.who.int/iris/handle/10665/67313.

WHO. (2004). *Tools for institutionalizing traditional medicine in health systems of countries in the WHO African Region, AFR/TRM/04.3*. Retrieved from https://www.afro.who.int/sites/default/files/2017-06/tools-for-institutionalizing-traditional-medicine-in-health-systems-%28final-version%29.pdf. (Accessed 11 January 2021).

WHO. (1978). Primary health care: report of the International Conference on primary health care, Alma-Ata, USSR, 6–12 September 1978.

WHO, Fiftieth session of the WHO regional committee for Africa (2002). Retrieved from https://www.afro.who.int/sites/default/files/sessions/final-reports/AFR-RC50-17%20Final%20Report.pdf.

WHO, Ouagadougou Declaration on Primary Health Care and Health Systems in Africa: Achieving Better Health for Africa in the New Millennium. (2008). Retrieved from https://www.afro.who.int/publications/ouagadougou-declaration-primary-health-care-and-health-systems-africa.

Willcox, M., Sanogo, R., Diakite, C., Giani, S., Paulsen, B. S., & Diallo, D. (2012). Improved traditional medicines in Mali. *Journal of Alternative and Complementary Medicine*, *18*, 212–220. Available from https://doi.org/10.1089/acm.2011.0640.

Willcox, M. L., Graz, B., Falquet, J., Sidibé, O., Forster, M., & Diallo, D. (2007). Argemone mexicana decoction for the treatment of uncomplicated falciparum malaria. *Transactions of the Royal Society of Tropical Medicine and Hygiene*, *101*, 1190–1198. Available from https://doi.org/10.1016/j.trstmh.2007.05.017.

Yadav, J. P., Arya, V., Yadav, S., Panghal, M., Kumar, S., Dhankhar, S., & Cassia occidentalis, L. (2010). A review on its ethnobotany, phytochemical and pharmacological profile. *Fitoterapia*, *81*, 223–230. Available from https://doi.org/10.1016/j.fitote.2009.09.008.

Yakubu, Y., Adoum, O. A., Wudil, A. M., & Ladan, Z. (2015). Toxicity study of ethanol root extract of *Terminalia macroptera* Guill. Perr. (Combretaceae) and assessment of some heavy metals. *African Journal of Pure and Applied Chemistry*, *9*(9), 193–196. Available from https://doi.org/10.5897/AJPAC2015.0647.

Yuet Ping, K., Darah, I., Chen, Y., Sreeramanan, S., & Sasidharan, S. (2013). Acute and subchronic toxicity study of *Euphorbia hirta* L. methanol extract in rats. *BioMed Research International*, *2013*, 182064. Available from https://doi.org/10.1155/2013/182064.

Yusuf, A. J., & Abdullahi, M. I. (2019). The phytochemical and pharmacological actions of *Entada africana* Guill, & Perr. *Heliyon.*, *5*(9), e02332.

CHAPTER

16

Selecting the most promising local treatments: retrospective treatment-outcome surveys and reverse pharmacology

Joëlle Houriet[1,2], *Jean-Luc Wolfender*[1,2] *and Bertrand Graz*[3]

[1]School of Pharmaceutical Sciences, University of Geneva, CMU, Geneva, Switzerland [2]Institute of Pharmaceutical Sciences of Western Switzerland, University of Geneva, CMU, Geneva, Switzerland [3]Antenna Foundation, Geneva, Switzerland

Introduction

Plants have always accompanied humanity as a source of food, clothes, shelter, and medicine. Some of the oldest references of medicinal herbs go back to Mesopotamian and Egyptian civilizations, with, for example, the use of the *Papaver somniferum* L., which is still an essential plant in our therapeutic arsenal as the source of morphine (Presley & Lindsley, 2018). The plants used in ancient times are not only a source of monosubstance (i.e., pure active compound) for drug discovery purposes but also represent complex mixtures of bioactive ingredients that can be used together as herbal drug preparations.

In the 2013 *Traditional Medicine Strategy of the WHO* (WHO traditional medicine strategy, 2013), it is highlighted that people increasingly include traditional and complementary medicines (T&CMs) in their health care choices. This trend has increased the economic importance of T&CM, making it more important that governments integrate them in their national policies. Thus, based on international recommendations, national authorities asked that treatments in T&CMs get evaluated in terms of quality, efficacy, and safety with scientific standards. However, the field of T&CMs—and in particular herbal medicines—was considered as the "stepchild of politics" in the European context (Wiesner & Knöss, 2014). Indeed, despite the interest of the general public, public authorities rarely

DOI: https://doi.org/10.1016/B978-0-323-90999-0.00003-3

allocate funding to this field of research. Moreover, the low level of reimbursement of herbal preparations by public insurance at the European level does not encourage private companies to invest in this field (Wiesner & Knöss, 2014).

Overall, the situation is paradoxical: on the one hand, the lack of financial incentives does not encourage the establishment of scientific research dedicated to T&CM, whether in the academic or industrial context. On the other hand, the general public in the West consumes a lot of herbal preparations in the form of food supplements. This interest enhances substantially the economic importance of the food supplement market, which is the main source of herbal preparation on the market. Companies performing market analysis summarized the trends in the so-called "global herbal supplement market" and described it as an "enormously aggressive business environment" which is expected to reach more than USD 86 billion by 2022 (Report by Zion Market Research ZMR on Herbal Supplement Market, 2018). Paradoxically, in response to the growing demand for traditional treatments a huge market has emerged in the midst of a legislative gray zone. There are some rare pharmaceutical companies that have developed state-of-the-art herbal drugs, but the majority of the herbal market is linked to the food supplement market (Abdel-Tawab, 2018).

This context makes the risks associated with the use of plants go beyond the toxicity of the plant alone (side effects or unwanted treatment interactions). These risks include, at a minimum, the use of poor quality, adulterated or counterfeit herbal products (Abdel-Tawab, 2018; Skalicka-Woźniak, Georgiev, & Orhan, 2017; Wang et al., 2018), exposure to misleading or unreliable information (Bilia, 2015), unqualified practitioners, and/or misdiagnosis, delayed diagnosis or failure to use effective conventional treatments (WHO Traditional Medicine Strategy, 2013).

Traditional plant extracts are often referred to as herbal preparations (Pferschy-Wenzig & Bauer, 2015) and "botanicals" (Butterweck & Nahrstedt, 2012). The regulatory framework distinguishes two main categories: pharmaceutical herbal products (herbal drugs, herbal drug preparations, botanical drugs) and food supplements. Nutraceuticals and functional foods are two additional types that are currently included in the food supplements category. The legislation and the nomenclature often change from country to country. Table 16.1 summarizes the various terms used to designate plant extracts in the diverse available forms.

Ideally, an integrated approach that combines clinical testing with pharmacological and phytochemical investigations should be conducted to document proofs based on evidence-based medicine (EBM) for using a given herbal medicine. Such an approach would allow standardization (possibly on active principles) and rational quality control of herbal preparations that show proper clinical and pharmacological activities, with demonstrated safety. In the investigations of plants recognized for their traditional indications, this approach can directly start with clinical studies (retrospective/observational studies followed by prospective/interventional trials).

Then, through different laboratory studies, the identification of active principles with defined modes of action can be made to tentatively understand the clinical effect at the molecular level. This type of investigation is the reverse of the classical development of drugs in the West, for which a thorough investigation of the activity and toxicity must be performed first, and that is subsequently followed by clinical studies after many steps (classical drug discovery). This chapter begins by recalling the definition of clinical efficacy and its role in the development of any drug, then continues with the different steps of the

TABLE 16.1 Terms used for defining preparations in which medicinal plants can be included.

Designation	Definition
Herbal drugs, herbal substances	Whole, fragmented, or broken plants, parts of plants, fungi, or lichen, in an unprocessed state usually in dried form but sometimes fresh [PhEur 9: 1433 (Directive.2004/24/EC, 2004)]
Herbal teas	Consist exclusively of one or more herbal drugs intended for oral aqueous preparations by means of decoction, infusion, or maceration, immediately before use (PhEur 9: 1435)
Herbal drug preparations, herbal preparations	Homogeneous products obtained by subjecting herbal drug to treatments such as extraction, distillation, expression, fractionation, purification, concentration, fermentation {extracts, essential oils, expressed juice, processed exudates, and herbal drugs submitted to size reduction for herbal teas or powdered for encapsulation [PhEur 9: 1434 (Directive.2004/24/EC, 2004)]}
Botanical drugs	Consist of vegetable materials, which may include plant materials, algae, macroscopic fungi, or combination thereof. May be available as (but not limited to) a solution (e.g., tea), powder, tablet, capsule, elixir, topical, or injection[a]
Herbal drug extracts	Liquid (liquid extraction preparations), semisolid (soft extracts and oleoresins), or solid preparations from herbal drugs using suitable solvents (dry extracts) (PhEur 9: 0765)
Food supplement	Foodstuffs the purpose of which is to supplement the normal diet and which are concentrated sources of nutrients or other substances with a nutritional or physiological effect, alone or in combination, marketed in dose form (Directives.2002/46/EC, 2002)
Nutraceuticals	Concentrated and administrated in the proper pharmaceutical form and they are capable of providing beneficial health effects, including the prevention and/or the treatment of a disease[b]
Functional food	As any food or ingredient that has a positive impact on an individual's health, physical performance, or state of mind, in addition to its nutritive value[c]

[a] *According to Food and Drug Administration (2019).*
[b] *According to Santini et al. (2018).*
[c] *According to Hardy (2000).*
Medicinal plants are referred to in many ways in different contexts, whether legislative or research. This table presents the main definitions.

proposed reverse approach, named *reverse pharmacology*. This reverse approach is illustrated by different examples of plants for which clinical evidence has been documented. Finally, we present promising approaches aiming to decipher the pharmacological activities of herbal preparations, and we discuss their potential benefits and limitations.

Clinical efficacy

In addition to the safety and toxicity that must be verified, the question of efficacy is essential to develop an herbal drug preparation in an evidence-based manner. To do this, the correlation between traditional usage of given herbs and/or the consumption habits of given botanical food supplements and the improvement of patient's health has to be made with scientific criteria. Since herbal medicines are already on the market and consumed, a way to prove the efficacy is to conduct clinical trials.

What is clinical efficacy? It is related to the clinical evaluation of a drug in a well-designed clinical study (Misra, 2012). The randomized, double-blind, controlled, clinical trial is the gold standard and is considered to have the highest level of medical evidence. Randomized controlled studies are studies that aim at evaluating the effect of a clinical intervention (drug ingestion) on an outcome (Misra, 2012). The outcome is defined as the patient's progress after a treatment (Graz, Willcox, & Elisabetsky, 2015). Comparison of the chosen intervention with a placebo, that is an inert substance, allows the evaluation of the specific effects of the chosen intervention (Graz, 2013). Indeed, on the one hand, spontaneous healing is expected for many ailments, and on the other hand, interactions between the patient and medical staff can influence outcomes. The design of double-blind studies makes it possible to estimate the specific effect, by subtracting the placebo effect in the strictest sense, that is the part of nonspecific effects expected when one of the parties, patients or caregivers, knows the nature of the intervention (Misra, 2012). In place of placebo, the control group can receive the reference treatment, then we have a "pragmatic" trial (Pawson et al., 2019).

In the case of phytotherapy, unfortunately, such well-defined clinical trials are rare (Fürst & Zündorf, 2015). Indeed, clinical trials often involve too few patients or are of too short duration. There are many reasons for this, but the most important is the high cost of clinical trials (Fürst & Zündorf, 2015), and the lack of financial incentives, as described above (Wiesner & Knöss, 2014).

Efficacy is closely related to the concept of EBM, which is defined as the application of the best available research to clinical care, which requires the integration of evidence with clinical expertise and patient values. The cornerstone of EBM is the integration of scientific evidence from clinical research to the individual needs of patients. To integrate traditional practices into EBM, evaluation of clinical efficacy is a necessary step. The reverse pharmacology workflow proposes one approach to conduct these clinical studies by adapting to the particular context surrounding T&CM.

Reverse pharmacology approach

In the conventional process of discovering and developing a new drug, clinical studies to evaluate efficacy occur after several years of preclinical research. For plants used in T&CMs, population uses can be studied to accelerate the evaluation of clinical efficacy. The approach called *reverse pharmacology* is dedicated to this study (Willcox et al., 2011). Reverse pharmacology can be defined as an adaptation of the drug discovery and development process to traditional herbal preparations. Its purpose is not to deny the importance of the conventional process, but to adapt it to the inherent specificity and complexity of the traditional herbal preparations.

The term *reverse pharmacology* was first used in the context of Ayurveda to define a new process of drug discovery and development process which should start by the clinical documentation of Ayurvedic drugs (Patwardhan, Vaidya, Chorghade, & Joshi, 2008). Reverse was used to emphasize that the routine "laboratory to clinic" path of drug discovery pipeline should be reversed to a "clinic to laboratory" path (Patwardhan et al., 2008), that is a "bedside-to-bench" workflow. It is important to underline that the same term "reverse pharmacology" was previously used to define the search for a new pharmacological target from a given molecule (Harrigan, Brackett, & Boros, 2005), that is not the way the term is being used in our current context.

The bedside-to-bench workflow was conceptually modified around a decade ago, by proposing a step before the clinical evaluation, called the retrospective treatment-outcome (RTO) study. The reverse pharmacology approach was thus defined as a four steps workflow (Fig. 16.1), starting with this special type of ethnopharmacological survey, the RTO study. The second step is dedicated to assessing the best dose-response directly in humans. The third step consists of evaluating the efficacy in the frame of clinical trials. Finally, the fourth step is defined as the laboratory studies, which in this case, and contrary to drug discovery, are thus postclinical (Willcox et al., 2011).

Step 1: Retrospective treatment outcome study

The first step in a reverse pharmacology approach is a "retrospective treatment outcome" study (RTO). Such clinical observational study is of the case–control type and is

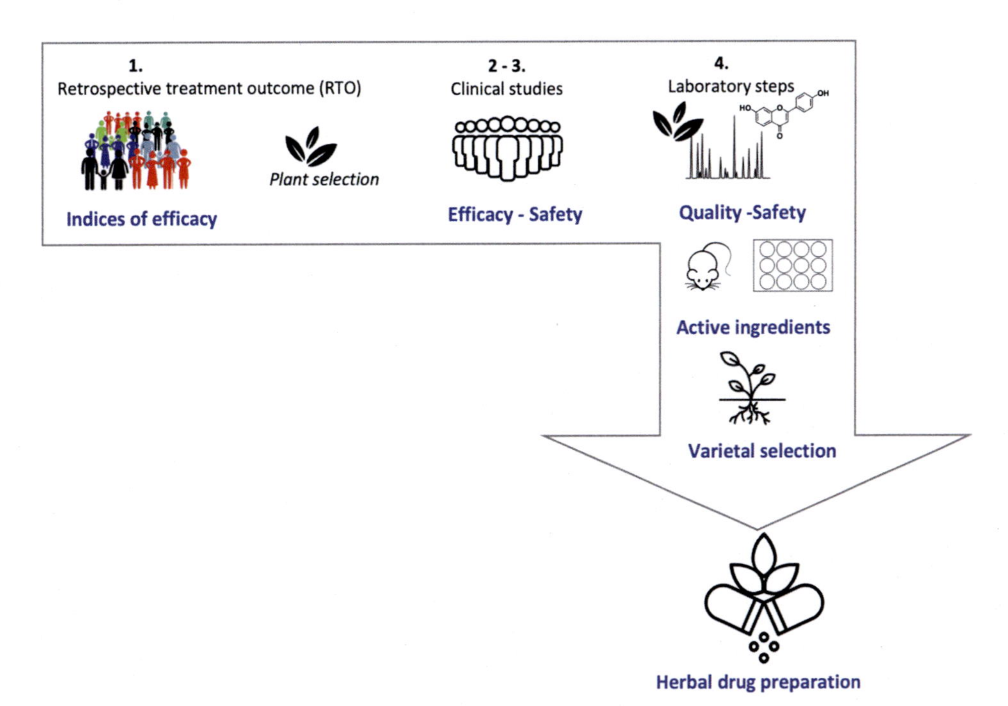

FIGURE 16.1 Schematic representation of the four steps of the reverse pharmacology approach. Source: *From Graz, B. (2013). What is "clinical data"? Why and how can they be collected during field surveys on medicinal plants?* Journal of Ethnopharmacology, *150, 775–779.*

performed retrospectively on a large number of patients. This first step aims to collect valuable information on plant usages to select herbal preparations correlated to the best outcomes (Willcox et al., 2011). The RTO methodology is the result of a deep thinking about the design of research on T&CM. To choose an appropriate research design, the research question and objectives must be very clear and explicit (Graz, Elisabetsky, & Falquet, 2007; Graz, Falquet, & Elisabetsky, 2010). The RTO study modified the methodology of the ethnomedical survey to obtain indices of efficacy and safety of a traditional preparation as perceived and reported by users (Graz et al., 2015). It adds two essential elements to the classical ethnomedical research: (1) clinical information and (2) statistical analysis (Willcox et al., 2011). Clinical information is collected retrospectively on the presentation and progress of a defined and recent disease episode. Treatments and outcomes are analyzed to elicit statistically significant correlations between them. Such an approach requires a large sample if there are many different treatments in order to have enough users of the same treatment and thus be able to compare reported outcomes in groups of users of different treatments. Therefore, this method makes it possible to identify the remedy with the highest statistical correlation with a particular reported clinical outcome. The hypothesis is that an observed correlation between a treatment used and a favorable outcome is an index of efficacy, which can then be further tested in a prospective comparative clinical trial (Willcox et al., 2011). Conversely, if treatment is often associated with failure, this is considered a marker of lack of clinical effect.

When preparing an RTO study, one needs to understand local concepts and terms used for the disease in focus. The objective is for respondents to give information about the disease that the researchers are studying—and not about another condition with similar symptoms. For example, if obesity is being studied, there will be little risk of confusion if the respondents have measured their weight; however, if the study focuses on Covid-19 infection, many other acute respiratory diseases can be responsible for similar symptoms, so it will be challenging to know if respondents really had Covid-19 unless they had a specific test.

The first RTO study was conducted in Mali on malaria and resulted in a database of treatments used in 952 households. The study inquired about recent episodes of "uncomplicated cases" of the disease and whether the patients had healed or not (Diallo et al., 2006, 2007). The analysis was an iterative process starting with a test of correlation between reported clinical outcome and the herbal preparations used. Since some recipes contained more than one herb, a second step was to adjust for this in the analysis (grouping all mixtures with a given herb) in an attempt to determine whether individual components were associated with clinical outcomes. From the 66 plants used, alone or in various combinations, the one associated with the best outcome was a decoction of the aerial parts of *Argemone mexicana* L. (Papaveraceae) (*A. mexicana*), which was reported to provide a complete cure with very few side effects (Table 16.2) (Diallo et al., 2007). The clinical outcomes were the same when this herb was used alone or in combination with other plants. This herb was therefore assessed alone in its traditional mode of preparation in further studies.

Surprisingly, this first RTO also revealed that most traditional medicines were not obtained from specialists (traditional practitioners) but from family members (parents usually provided these substances for their children, see Fig. 16.2).

TABLE 16.2 Correlation between plant used and reported outcome in a study on traditional treatments for malaria in Mali.

Herbs	# Cases reporting use	# Cases reporting clinical recovery	# Treatment failure	Correlation with clinical recovery[a]	P (Fisher)
Argemone mexicana	30	30	0	100% (88–100)	Reference
Carica papaya	33	28	5	85% (68–95)	.05
Anogeissus leiocarpus	33	27	6	82% (64–93)	.03

[a]*95% Confidence Interval.*
From Willcox, M.L., Graz, B., Falquet, J., Diakite, C., Giani, S., & Diallo, D. (2011). A reverse pharmacology approach for developing an anti-malarial phytomedicine. Malaria Journal, 10, *S8.*

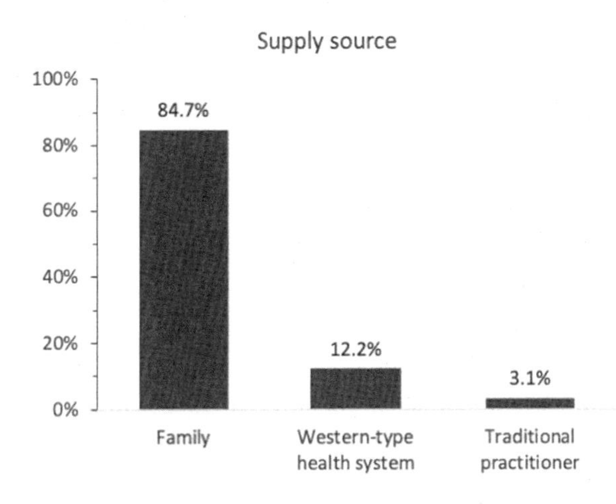

FIGURE 16.2 Source of supply to obtain a first treatment for uncomplicated malaria in Finkolo district in Mali. The number of informants was $n = 469$, and the recall period was of 15 days. Source: *From Willcox, M.L., Graz, B., Falquet, J., Diakite, C., Giani, S., & Diallo, D. (2011). A reverse pharmacology approach for developing an anti-malarial phytomedicine.* Malaria Journal, 10, *S8.*

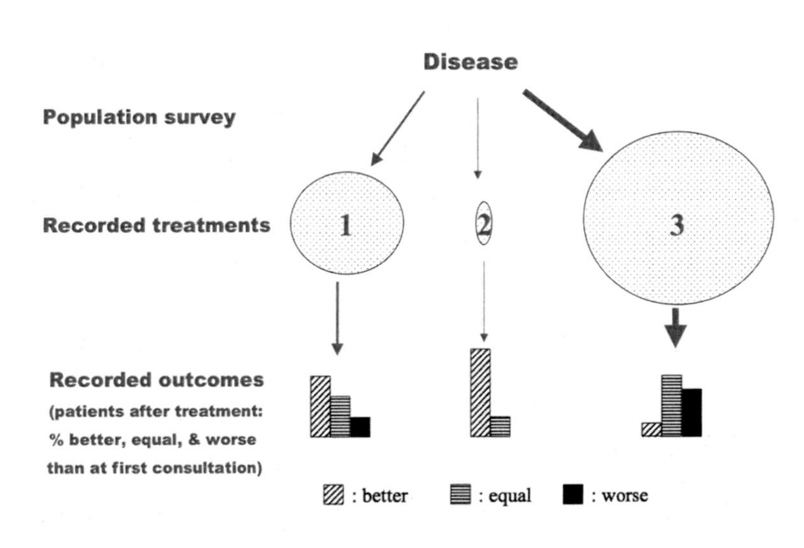

FIGURE 16.3 The treatments with the best-associated outcome are not the most frequent. Treatment number (2) was not used by many respondents but was found to be associated with the best outcome (patient condition described as "better" in the survey).

In addition, the treatments with the best-associated outcome were not the most frequent, a phenomenon that has been observed in another RTO study (Fig. 16.3; Graz, Kitalong, & Yano, 2015). A proposed explanation was that differences between various treatment outcomes are not apparent without the power of statistical analysis of a large number of reported cases. As a result, individual users and traditional healers do not have a reliable picture of the relative efficacy of different treatments used in their area. So, a very effective treatment is sometimes used only in a single village, even if its users are not trying to be secretive about it.

RTO was applied to different situations in addition to the project presented above, such as malaria in Ethiopia (Gurmu et al., 2018), premenstrual symptoms in Switzerland (Graz, Savoy, Buclin, & Bonvin, 2014) as well as noncommunicable diseases (NCDs) in the Republic of Palau (Graz et al., 2015). An RTO can also compile all treatments in the context of a selected disease and not only the traditional ones. For example, the study on premenstrual symptoms showed that the first alternative treatment was the hot water bottle, which was associated with a better outcome than herbal preparations (Graz et al., 2014). In this case, a prospective randomized study had already looked at the use of heat to relieve pain and had demonstrated that local heat was as effective as ibuprofen and provided faster relief (Akin et al., 2001). However, the results of the study had not been widely translated into official treatment recommendations by gynecologists.

Several other RTO studies have been undertaken. One was conducted in China in 2019 and focused on respiratory diseases. Some 20,000 participants responded to the online questionnaire, and the results are currently being prepared for publication. Similarly, in the context of the Covid-19 pandemic, a large study is also underway involving 15 countries (Covid-19 Survey: Our Selfcare & Its Effects, 2020). The online questionnaire was launched in the summer of 2020 and has so far (November 2021) been completed by over 70,000 participants. The questionnaire included a wide range of treatments potentially used by the participants, from conventional medications to complementary medicine treatments, but also for example physical activities and diets. Data analysis is underway, and results will begin to be published in the course of 2022.

Steps 2 and 3: clinical evaluations

Following the RTO study, a traditional treatment is selected for the next steps (Fig. 16.1). The two following steps of reverse pharmacology are dedicated to prospective clinical evaluations. The second step evaluates the best dose–response directly in humans (Willcox et al., 2007). Starting directly with humans is justified because these substances have been used traditionally and can thus be considered a priori safe. Indeed, it seems safe to make the assumption that these substances which have been used for centuries do not present any toxicity, and this can be declared with even more certainty when the selected product is also used as a common food (Graz et al., 2015). The WHO has assessed that regulatory restrictions and the obligation of conducting toxicology studies should not be applied in the case of plants with proven traditional use (World Health Organization, 2000). The RTO study is essential here, as it distinguishes between preparations that are actually used by patients from the so-called traditional preparations mentioned in books, in course materials, by oral transmission, which are said to be useful, but which no one ever sees used.

For the project in Mali which selected the decoction of *A. mexicana*, a dose-escalating prospective study showed a dose–response phenomenon (Willcox et al., 2011). After this, a prospective, randomized, controlled trial was organized in a remote village: the "control" treatment was the standard artemisinin combination therapy (ACT) artesunate-amodiaquine, and the aim was to test the hypothesis that *A. mexicana* would be equivalent to ACT. The clinical study proved that the hypothesis was true and thus demonstrated the efficacy of the decoction in nonsevere cases of malaria (Graz et al., 2010; Willcox et al., 2007). Deterioration to severe malaria was 1.9% in both groups in children aged ≤ 5 years (with 0% coma/convulsions), and no deterioration in patients aged >5 years were observed. A 3-month follow-up study confirmed that, even when all parasites were not eliminated, the rate of severe malaria and anemia remained low. Total parasite elimination did not make a difference in this hyperendemic context, where reinfection would occur anyway. And remaining parasites after *A. mexicana* treatment did not appear to become resistant to the plant, as *A. mexicana* has been in use for at least four generations in the study setting. Resistance might be slower to appear when a plant product with several active ingredients is used. The same principle is at the basis of the combination treatment ACT: having several active substances together lowers the chances that parasites will become resistant.

Today, *A. mexicana* decoctions are proposed as a possible alternative to standard modern drugs in high-transmission areas to reduce the development of resistance to ACT. It is recognized as an "improved traditional medicine" (*médicament traditionnel amélioré*) and constitutes a first-aid treatment when access to other antimalarial treatment is delayed (Graz et al., 2010; Willcox et al., 2011).

Another reverse pharmacology project reached the clinical evaluation stage. It was launched in the Republic of Palau (Pacific Ocean) in the context of NCD (Graz et al., 2015). The plant selected by the RTO, *Phaleria nisidai* Kaneh., was clinically evaluated as an adjuvant in patients with poorly controlled diabetes under conventional treatment (Kitalong et al., 2017).

Step 4: Laboratory stage

The fourth and final step of reverse pharmacology is defined as the laboratory stage (Fig. 16.1) (Willcox et al., 2011). The selected herbal preparation, which is clinically validated at this stage, is brought into the laboratory for what can be defined as *postclinical* studies in opposition to a *preclinical* step in classical drug discovery. The goals of the laboratory stages are diverse. First of all, assessing the chemical composition of the selected herbal preparation is used for quality control purposes (Willcox et al., 2011). The next aim is to understand the pharmacological effects of the herbal preparation and try to decipher how each individual constituent participates in the observed clinical efficacy. This step should ideally explain the mechanism of action (Willcox et al., 2011). This understanding could predict possible side effects when used by a larger number of people, as well as possible interactions with other drugs. Once the active constituents are discovered, they can be used for a meaningful standardization. In addition, if laboratory studies are successful in identifying active constituents, agronomic research can be carried out to select varieties that are enriched in active constituents and suitable for large-scale cultivation (Willcox et al., 2011). Besides, identifying active constituents with well-defined pharmacological properties can also serve classical drug discovery and be a valuable source of new molecules for the development of drugs. The fact that a molecule comes from a clinically evaluated plant is perceived as an advantage. Indeed, prior clinical evaluation increases the chances of identifying a molecule with drug-likeness quality (Willcox et al., 2011).

In the *A. mexicana* project, the fourth step of reverse pharmacology was conducted after the RTO and clinical investigations (Simoes-Pires et al., 2014). A conventional method of phytochemical bioguided fractionation was first applied. The freeze-dried decoction of *A. mexicana* was fractionated and tested in a common antiplasmodial assay (chloroquine-resistant strains of *Plasmodium falciparum*). This process identified three alkaloids (berberine, allocryptopine, and protopine) as active ingredients in vitro. Quantitative analyses then verified that the dose of decoction ingested during the clinical studies contained a relevant concentration of these three alkaloids (Simoes-Pires et al., 2014). Next, a pilot in vivo experiment with the freeze-dried *A. mexicana* decoction and berberine sulfate were conducted on mice infected with *P. berghei* but no reduction of parasitemia could be observed. Thus, to date, the mechanism of action of the decoction could not be deciphered. If the clinical study had not been conducted before the laboratory research, the in vitro data would have led to proclaim the discovery of active substances, while the in vivo data would have caused to abandon the research on this plant (Simoes-Pires et al., 2014).

The example of the decoction of *A. mexicana* illustrated that the pharmacological investigations are far from being straightforward. It should be noted that so far, the ideal fourth step of reverse pharmacology has not been too detailed since the workflow has been more discussed from a medical point of view. Nevertheless, the question of evidence is essential. Classically, the level of medical evidence for a drug is represented by a pyramid, the highest level of which is the randomized controlled double-blind clinical study (Straus et al, 2011). By reversing the research process and starting by studying clinical efficacy, the most significant medical evidence is reached earlier. Thus, laboratory studies are initiated on a solid clinical basis to provide pharmacological evidence. In the context of herbal preparations, pharmacological evidence encompasses the study of the chemical composition of the herbal preparations, whether to develop quality control methods or to elucidate the mechanism of action of the various constituents. This pharmacological evidence may then enhance the potential applications of the results of clinical studies. In the following sections, promising approaches to achieve the goals of the laboratory stage are presented.

New and promising approaches for the laboratory stage in a reverse pharmacology approach

The objectives of the laboratory stage range from establishing quality markers to understanding pharmacological mechanism at the molecular level. This stage is challenging because of the chemical complexity of herbal preparations. The following sections present new areas of research and their methodologies that can provide scientific evidence of the detailed pharmacological mechanisms of a given herbal preparation in humans at the molecular level. The paradigms underlying these new areas of pharmaceutical research are first presented.

Changes of paradigms

Over the last century, herbs have been studied as a source of potentially interesting chemical entities in the frame of the modern approach of drug discovery and development (Newman & Cragg, 2016). In this context, the term *natural products* (NPs) emerged to refer to a molecule from a natural source, and the concept of "pharmacognosy" was defined as the science of drugs prepared from natural sources, including preparations from plants, animals, and other organisms (Medical Subject Headings: Pharmacognosy, 2019). This approach fed the pipeline of drug discovery and met numerous successes in terms of drugs to cure important diseases (Cragg & Newman, 2013; Newman & Cragg, 2016).

In the frame of the conventional drug discovery approach, the herbs selected through an ethnopharmacological survey or other methods are tested on a battery of tests in vitro and fractionated until the isolation of one or several active(s) principle(s). These isolated bioactive NPs can then be used as such or structurally modified to improve their bioactivity or bioavailability. However, the field of NPs encountered a high number of failures, in such extent that authors defined the concept of *background noise data* to refer to irrelevant studies, mainly in vitro (Butterweck & Nahrstedt, 2012; Gertsch, 2009). The irrelevance lies both in methodological issues and in extrapolations in terms of biological activity when the tests are performed in vitro. For instance, one famous irrelevance concerns the shockingly high extract concentration of NP used in vitro, leading to the now classic nonsense that 627 bottles of red wine should be daily drunk to reach the antiaging effect provided by resveratrol (Butterweck & Nahrstedt, 2012). This case highlights the importance of the claims related to studies in vitro, particularly when these claims are used outside of the scientific context to advocate the use of a given plant (Newman & Cragg, 2016). A clear distinction should be made between the in vitro activity of isolated substances and the potential activity of herbal preparations in humans.

Thus, the field related to herbal medicine needed to question itself and to find more relevant approaches and new paradigms (Cordell, 2017). By definition, *pharmacology* is the study of the mechanism of action of drugs, including the actions of a drug on an organism (pharmacodynamic) and the fate of the drug in the organism (pharmacokinetic) (Liu, Wu, Jiang, Yang, & Guo, 2013). What underpinned all pharmacological research in the 20th century is the paradigm of the key and the lock, also called the *magic bullet* (Williams, 2009). This paradigm models the disease by a single therapeutic target, which can be reached by a single and specific molecule. What is also called the "reductionist approach" is still an important paradigm of the pharmacological research. The apparition of "systems biology" enabled a shift of paradigm toward a more holistic approach (Fig. 16.4) (Verpoorte, Choi, & Kim, 2005; Van Der Greef, 2011). Systems biology was recently defined as "an integrative discipline connecting the molecular components within a single biological scale and also among different scales (e.g., cells, tissues and organ systems) to physiological functions and organismal phenotypes through quantitative reasoning, computational models, and high-throughput experimental technologies" (Tavassoly, Goldfarb, & Iyengar, 2018). The key point of this definition is that the physiological function and organismal phenotypes seek to be linked to molecular components at different levels, from intracellular machinery to the whole organism. Two approaches are distinguished, the bottom-up approach, which starts from small functional units to build a

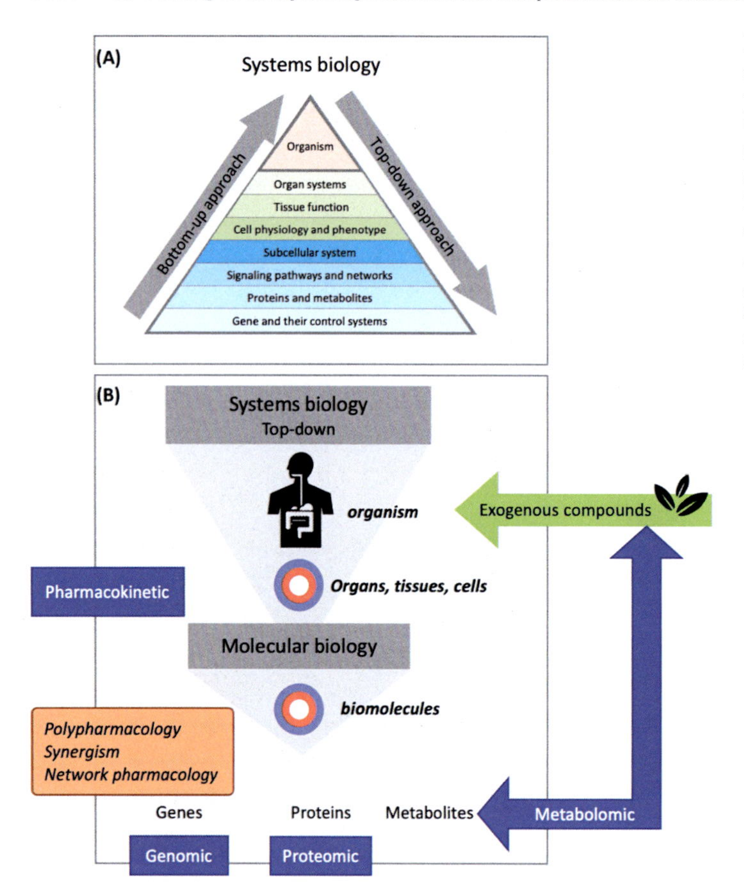

FIGURE 16.4 Systems biology, a holistic approach to studying the pharmacology of herbal preparations. (A) The different levels of systems biology, (B) new tools (*blue*) and paradigms (*orange*) for studying herbal preparations in a top-down systems biology approach. Source: *Adapted from Tavassoly, I., Goldfarb, J., & Iyengar, R. (2018). Systems biology primer: The basic methods and approaches.* Essays in Biochemistry, *62, 487–500.*

system, and the top-down approach which starts from the global overview of the system (Fig. 16.4A) (Tavassoly et al., 2018). The emergence of omics techniques (mainly genomics, proteomics, and metabolomics) comes from systems biology (Fig. 16.4B). Metabolomics aims at identifying and quantifying all metabolites in a biological system (Tavassoly et al., 2018) (see chapter 10 from Prof. Joshua J. Kellogg). Large data sets are acquired from biological samples mainly by liquid chromatography-mass spectrometry and nuclear magnetic resonance. Multivariate data analyses are then used to retrieve meaningful information from these large data sets (Eriksson, Johansson, Trygg, & Vikström, 2013).

In a top-down approach, systems biology begins with clinical or in vivo studies, that is on complete organisms (Fig. 16.4B). Biofluid analyses are the preferred gateway to understanding the state of an organism, whether in response to intervention (ingestion of an herbal preparation in our case) or by comparing a group of diseased organisms to a healthy group. Hypotheses made at the organism level can then be studied at different levels (e.g., tissue or cell level) (Fig. 16.4B). At the organism level, one of the goals of the systems biology is to define biomarkers that can reflect the states of physiological pathways. From a pharmacological perspective, the goal would be to observe whether a pharmaceutical intervention influences these biomarkers. Systems biology, with its techniques, was perceived as an approach capable of studying the biological activity of herbal preparations in a way that is both more complete and representative of their complexity (Verpoorte et al., 2005). What is difficult to grasp by a reductionist approach is the phenomenon of polypharmacology, synergy, and prodrugs. By using a holistic approach, these phenomena could be better assessed (Verpoorte et al., 2005). Polypharmacology and synergism were proposed as the new paradigms of the field of NPs (David, Wolfender, & Dias, 2015; Gertsch, 2011). Polypharmacology is defined as the ability of a molecule to interact with several therapeutic targets (Gertsch, 2011). In the case of a typical herbal preparation and provided that several constituents are bioactive, the situation is even more complex, potentially resulting from multiple compounds acting in a polypharmacological manner. Indeed, interactions between compounds can also be antagonistic or additive (Caesar & Cech, 2019; Wagner & Ulrich-Merzenich, 2009).

To encompass multiple molecule–target interaction, *network pharmacology* has also emerged as an additional paradigm. It results from the observation of the robustness of a phenotype. The robustness is explained by the redundancy of functions in an organism and the pathways that compensate each other (Hopkins, 2008; Kitano, 2007). Robustness was considered as the central framework for drug design from a system-based perspective in particular for complex diseases such as diabetes and cancers (Kitano, 2007). From this perspective, two main drug development strategies can be distinguished. The first is to reach several redundant targets with a single molecule, which is defined by polypharmacology. The second is the multidrug combination, which is already proposed for antiretroviral HIV therapy, cancer treatments, and tuberculosis treatment (Hopkins, 2008).

Thus, considering the complexity of organisms and the chemical composition of herbal preparations, the paradigm of the single therapeutic target "targeted" by a specific molecule has to be fully revised. To return to the representation of the previous paradigm, the image of the key and the lock extends and becomes a castle (organism) with multiple floors (pathways) accessible by many access roads (targets and network pharmacology) that several tools (molecules) can open in several ways (polypharmacology), including by combining their strength (synergism). This rich image does not consider the means deployed by the castle to eliminate exogenous molecules, that is the pharmacokinetic aspects. Indeed, the bioavailability of the components of herbal preparations is an essential aspect to be considered in understanding their mechanism of action (Butterweck & Nahrstedt, 2012). Understanding the fate of the numerous components of herbal preparations is one of the main challenges of the field, especially because herbal preparations are not formulated and undergo heavy metabolism.

Pharmacokinetics

Pharmacokinetics (PK) studies the fate of exogenous substances in contact with a living organism and how that organism can transform the substance, which includes, according to Nordberg, Duffus, and Templeton (2004), "the process of substances uptake (absorption), the biotransformation they undergo, the distribution of the substances and their metabolites in the tissues and their elimination," often referred as Absorption, Distribution, Metabolism, and Elimination. Metabolism (or "biotransformation") designates "the chemical conversions of a substance mediated by a living organism" and concerns various conversions such as the first-pass effect occurring in the liver, the phase 1 and 2 enzymatic reactions, as well as the metabolic activation of a substance to a more biologically active metabolite (prodrug) (Nordberg et al., 2004).

Absorption is an important step in the pharmacokinetics of herbal preparations, because they are mainly ingested. Absorption studies have so far observed that multiple constituents influence each other. Indeed, the absorption of some constituents has been increased in the presence of other constituents of the source plant (Li et al., 2012, 2013). Similar phenomena have been observed between constituents of different plants combined in a formula (Liu et al., 2013). This type of interaction is called pharmacokinetic synergies. The opposite phenomenon was also observed, demonstrating that some toxic constituents, such as aristolochic acid, were less absorbed in the presence of other substances from other source plant and therefore less likely to be harmful when their source plant is used in combination with other plants (Liu et al., 2013). This interaction can be described as an antagonist interaction with positive effect.

Biotransformation occurs in different organs and involves several types of reactions. Phase 1 reactions are mediated by enzymes (mainly cytochromes P450s) that modify substances mainly by oxidation. Phase 2 reactions are conjugation biotransformation, which bind exogenous substances with endogenous substances. Both types of reactions make exogenous substances more polar, which facilitates their elimination by the kidneys (Nordberg et al., 2004; Sjögren et al., 2014). The liver and small intestine are the two main organs responsible for this biotransformation (Fritz et al., 2019).

The biotransformation that occurs in the colon is also important. Indeed, the bacterial flora present in the colon has many chemical transformation capacities. This is currently a hot topic, as studies have demonstrated its influence on several diseases (Gilbert et al., 2018), and large-scale research projects have been set up this topic. For example, the *American Gut Project* is a crowed–sourced citizen science research project that has resulted in the collection of more than 10,000 human samples (McDonald et al., 2018). For herbal preparations, microbiome-mediated biotransformation is of interest for understanding the fate of plant constituents, but the role of herbal preparations constituents on the microbiome is also interesting (Pferschy-Wenzig, Koskinen, Moissl-Eichinger, & Bauer, 2017). Microflora-mediated biotransformations are mainly reductive and hydrolytic reactions, generating rather apolar and low-molecular-weight metabolites (Sousa et al., 2008). These biotransformations are therefore distinct from oxidation and conjugation reactions in the liver and small intestine, which produce more polar metabolites with a higher molecular weight (Sousa et al., 2008). Since herbal preparations contain many glycosides and polyphenols, microbiome-mediated biotransformations transform them into more apolar metabolites, which can then be absorbed better than the source compound (Pferschy-Wenzig et al., 2017). A recent review summarized the many phenomena described so far in the context of traditional Chinese medicine (TCM) (Feng, Ao, Peng, & Yan, 2019). Concerning the metabolites resulting from microbiome-mediated biotransformation, these phenomena include the increase or decrease in bioactivity, but also in toxicity and bioavailability.

Models to study absorption and biotransformation of natural products and herbal preparations

Important steps in PK are absorption and biotransformation, which occur mainly in the intestines and the liver. Understanding these steps at an early stage of laboratory research can be used to guide pharmacodynamic testing. These studies may include PK steps in different organs, particularly in the intestines and liver. To move from the level of the organism to lower levels (organs, tissue, cells, etc.), the choice of models is important. Regarding absorption, different models coexist, which range from in vivo, in situ, ex vivo, in vitro and in silico (Liu et al., 2013). Among them, the simplest is the parallel artificial membrane permeability assay (PAMPA). It permits to evaluate the passive absorption of NPs alone or included in herbal preparations and has the advantage to be a high-throughput assay (Petit et al., 2016). The Caco-2 cell line model is more complex to use than PAMPA but has the advantage to show active transport mechanisms (Hubatsch, Ragnarsson, & Artursson, 2007). The main limitations of Caco-2 cell line are the absence of the different characteristics of each intestinal segment (duodenum, jejunum, and ileum), the low level of the main enzyme of phase 1 metabolism (CYP3A4), and the overexpression of the efflux transporter P-glycoprotein (Pgp) (Sjöberg et al., 2013). Caco-2 cell line has been used to evaluate the absorption of NPs alone (e.g., in Fang, Cao, Xia, Pan, & Xu, 2017; Walgren, Walle, & Walle, 1998) or from their source plants, including mixtures of plants (Du et al., 2018; Wang et al., 2017; Zheng et al., 2018).

A third model uses *ex vivo* viable intestinal tissues inserted into an Ussing chamber system (Sjöberg et al., 2013; Ussing & Zerahn, 1951). This system can monitor the transepithelial resistance of intestinal membrane, which informs about the viability of the tissue. The Ussing chamber contains a donor and an acceptor compartments. The intestinal membrane is inserted between these two compartments. In addition to the transepithelial resistance, this system heats the compartments to physiological temperature. It also allows the supply of oxygen to ensure the viability of the membrane and the circulation of the solution in the compartments. Since the human intestine is difficult to access, rat tissues were most frequently used (Lennernäs, 2007). Porcine tissues are known to have a higher similarity with the human intestine than that of rats (Patterson, Lei, & Miller, 2008). Both rat (Matsumoto, Matsukawa, Mineo, Chiji, & Hara, 2004; Wolffram, Blöck, & Ader, 2002) and porcine tissue (Deußer et al., 2013; Erk et al., 2014; Vermaak, Viljoen, Chen, & Hamman, 2011; Houriet et al., 2021) were used in a few studies involving pure NPs and plant extracts.

The liver is an important organ for phase 1 and 2 biotransformations. To study hepatic biotransformations, the model mainly used for NPs consists of using microsomes, that is an enzyme pool containing cytochromes, or unique cytochromes (supersomes) (Asha & Vidyavathi, 2010). These studies focus on biotransformations (e.g., in Tolleson, Doerge, Churchwell, Marques, & Roberts, 2002), but also on the possible inhibition of cytochromes by NPs (e.g., in Kopečná-Zapletalová, Krasulová, Anzenbacher, Hodek, & Anzenbacherová, 2017). While several other models in vitro exist for the liver (Asha & Vidyavathi, 2010; Soldatow, Lecluyse, Griffith, & Rusyn, 2013), few studies on NPs have been conducted. For example, rat liver tissue slices were used to study the biotransformation of curcumin (Hoehle, Pfeiffer, Sólyom, & Metzler, 2006). To our knowledge, no studies on herbal preparations were conducted. It should be noted that the potential hepatotoxicity of herbal preparations is a concern (Brewer & Chen, 2017; Licata, Macaluso, & Craxì, 2013) and that in vitro models for routine evaluation would be required. Indeed, the herbal hepatotoxicity has been described as a hidden epidemic (Licata et al., 2013) and remains little understood. It concerns certain classes of chemical compounds, such as pyrrolizidinic alkaloids, whose hepatotoxic mechanism has been described. The use of weight loss products is particularly problematic, as it is common for patients to use high doses. High-dose toxicities, as well as cases of interaction with other hepatotoxic drugs, have been documented (Licata et al., 2013). Clinical studies of low statistical values combined with the use of chemically poorly defined herbal preparations are incriminated to explain this situation. In addition, the absence of a well-established phytovigilance system has the potential consequence that cases of hepatotoxicity are underreported (Licata et al., 2013).

Biotransformations mediated by the colon microbiome have attracted the attention of researchers (Feng et al., 2019). One of the studies on herbal preparations concerned white willow bark extract (*Salix alba* L.) (Pferschy-Wenzig et al., 2017). The experiments consisted in incubating the extract with feces from a healthy volunteer. A follow-up of the bacteria populations was done in parallel with the follow-up of the extract. The monitoring of the extract was performed by ultrahigh performance liquid chromatography-high-resolution mass spectrometry (UHPLC-HRMS) in an untargeted metabolomic approach. As a result of this monitoring, 58 compounds linked to the willow bark extract were annotated, and metabolism pathways for glycosylated flavonoids and salicylic alcohols were proposed (Pferschy-Wenzig et al., 2017).

Another study focused on an herbal preparation in a metabolomic approach (Zhou et al., 2016). This study takes an interesting look at polysaccharides, which are still largely unknown in the field of preparations and highlights their potential role on the microbiome (Koropatkin, Cameron, & Martens, 2012). This in vivo study with rats focused on ginseng decoction, with an attention on the potential role of polysaccharides present in this decoction (Zhou et al., 2016). The effect of ginseng polysaccharides was evaluated in vivo on an over-fatigue and acute cold stress model generated by forced swimming events. Four groups were formed, one blank group and three groups undergoing forced swimming events. Of these three groups, one model group received a saline solution, one positive control group received oligofructose, and one test group received polysaccharides from ginseng decoction. Two weeks of forced swimming events were performed before each group was given a solution of ginsenosides. First, the feces were analyzed to monitor the microbiome population. Then, feces, plasma, and urine were monitored in a targeted and nontargeted manner by UHPLC-MS. All the analyses showed that polysaccharides were able to restore the microbiome destabilized by the forced swimming events. However, absorption and biotransformation of certain ginsenosides were improved in the polysaccharide group and the positive group (Zhou et al., 2016).

Deciphering the mode of action of herbal preparations: successes and limitations

The new research areas presented above have emerged recently and have generated great enthusiasm especially in the research areas of TCM (Wang et al., 2011). Nevertheless, when traditional knowledge is considered at the beginning of a laboratory research process, two categories of herbs can be distinguished with respect to their pharmacological activities. The first category is made of herbs with one active molecule that accounts for most of the bioactivity and from which monosubstance drugs can be obtained, such as the morphine of the sleeping poppy (*Papaver somniferum* L.) (Presley & Lindsley, 2018) or the artemisinin of *Artemisia annua* L. (Ma, Zhang, Liao, Jiang, & Tu, 2020). The second category contains herbs for which the bioactivity is not carried by a single molecule but by many constituents of the plant extract. These herbs are by far the most challenging to study. To our knowledge, among the most studied plants such as *Salvia miltiorrhiza* (more than 3000 publications in English) (Jung, Kim, Moon, Lee, & Kim, 2020), whose effect does not rely on a very active molecule, we are not aware of cases of clinical efficacy that were fully deciphered, considering polypharmacological and pharmacokinetics mechanisms. However, each study contributing to decipher a part of the mechanism improves our understanding of the mode of action. In TCM, the focus is currently on improving clinical research (Dai et al., 2019) and verifying the quality of herbal preparations (Li, Shen, Yao, & Guo, 2020). Regarding quality control, the most recent methods are no longer based on the monitoring of a single constituent, but on metabolite profiling methods, by checking a set of representative constituents of the plant (Hou et al., 2011). State-of-the-art techniques allow the acquisition of the necessary data to document extract composition in detail at the molecular level (Wolfender, Nuzillard, Van Der Hooft, Renault, & Bertrand, 2019) even if the interpretation of these data remains challenging. The monograph of *Uncaria stem with hooks* [*U. rhynchophylla* (Miq.) Miq ex Havil] is an example of a new method of quality control incorporated in the European Pharmacopoeia (Ph.Eur. 9.0).

Examples of application of metabolomic studies in human

Metabolomic approaches are widely employed to compare the chemical composition of herbal preparations and for biochemometric studies to identify active compounds on in vitro/in vivo models (see chapter 10 from Prof. Joshua J. Kellogg). Metabolomics studies considering the ingestion of herbal preparations in humans are still rare. We present these studies in humans in this chapter because they represent potentially interesting novel approaches in reverse pharmacology. Metabolomic studies in humans mainly focused on the monitoring of endogenous metabolites following an intervention. These interventions consisted of ingesting herbal preparations for therapeutic or food-type purposes. These studies have in common a small number of people enrolled (ranging from eight to less than 50 people) and can be considered preliminary in terms of interpretation. Nevertheless, these metabolomic analyses all revealed changes in the fingerprinting of endogenous metabolites related to ingestions of herbal preparations and hypotheses on the mechanisms leading to these changes could be proposed.

The first study using a metabolomic approach and an herbal preparation followed the urine of 14 healthy volunteers consuming chamomile tea (*Matricaria chamomilla* L.) (Wang et al., 2005). The fingerprints were acquired by ^{1}H-NMR. During this study, the considerable interindividual variation was highlighted in the urine analyses. However, endogenous differential substances were identified, including hippurate, known to be a product of renal and hepatic syntheses of glycine and benzoic acid. These two substances are in turn known to be degradation products of polyphenols. The authors thus hypothesized that hippurate could be a sign of the degradation of polyphenols present in chamomile tea (Wang et al., 2005), and also be considered an exogenous substance.

Similar studies performed on urine samples concerned the intake of *Origanum dictamnus* L. (Takis, Oraiopoulou, Konidaris, & Troganis, 2016), as well as lingonberries (*Vaccinium vitis-idaea* L.) (Lehtonen et al., 2013), red grape, and dry red wine (Van Dorsten et al., 2010). These studies have successfully observed differences between the different groups and have in common the observation of the alteration of the levels of hippuric acid derivatives in the urine.

Another study differs from the previous ones by analyzing plasma rather than urine and this by UHPLC-HRMS. In a study involving 30 healthy volunteers, a garlic intake as a dietary supplement was evaluated, and its impact on phospholipid metabolism was mainly observed (Fernández-Ochoa et al., 2018).

All the above studies focused on the level of endogenous metabolites modulated by the herbal preparation intake and on identifying metabolic pathways influenced by this intervention. Other studies, we shall look at now, have attempted to more accurately link the fate of exogenous constituents present in the herbal preparation once in contact with the human body.

The first study of this type focused on black tea polyphenols, monitoring urines by ^{1}H-NMR, in a randomized, placebo-controlled, double-blind, full crossover study design in which 20 healthy volunteers were enrolled (Van Velzen et al., 2009). This study was also designed to measure the elimination parameters of given metabolites by collecting urines for 48 h after ingestion. This part was defined as nutrikinetic by the authors. After metabolite discrimination by metabolomics analysis, three metabolites from the biotransformation of polyphenols by colon microflora were monitored in the nutrikinetic study. The elimination parameters permitted the authors to infer that the limiting step in their elimination was due to microbial biotransformation and not to conjugation biotransformation (phase 2) due to their slow elimination rates. In addition, the nutrikinetic responses of these three metabolites were observed as relatively distinct from one person to another, whereas they were more consistent for a given individual.

Coffee intake was investigated in another study with a randomized, double-blind, placebo-controlled, crossover design, in which 10 healthy volunteers participated (Madrid-Gambin et al., 2016). In this study, urines, as well as the coffee beverage were monitored by ^{1}H-RMN. As some coffee constituents had previously been the subject of pharmacokinetic studies, the authors were able to relate their observations to these data, without conducting any pharmacokinetic/nutrikinetic studies themselves. Potential urinary biomarkers directly related to coffee ingestion have been proposed: trigonelline, an unchanged coffee constituent, and 2-fluroylglycine, a product resulting from the biotransformation of two constituents. In addition, the final products of the degradation of chlorogenic acids by intestinal microflora were observed. Finally, the increase in some endogenous metabolites highlighted the impact of coffee on central energy metabolism.

In the TCM context, a few studies published in English have applied a metabolomic approach. These studies have in common a design that first looked for biomarkers of a selected pathology by comparing a diseased group with a healthy group. Then, an intervention in the form of a TCM formula was performed, where biomarkers identified in the first part were followed. These studies involved patients with primary dysmenorrhea (Su et al., 2013), depression (Tian et al., 2014, 2016), and psoriasis (Lu, Deng, Li, Wang, & Li, 2014), as well as patients with a syndrome called "suboptimal health status" whose study will be detailed here (Tian et al., 2016). Indeed, this study aimed to define biomarkers that could reflect the suboptimal health status (Tian et al., 2016). This suboptimal health condition results in fatigue, lack of energy, or disturbed sleep among other nonspecific states, which is therefore difficult to diagnose. This study compared a healthy group (23 volunteers) to an affected group (22 patients) by analyzing the plasma of participants with ^{1}H-NMR, which allowed them to highlight potential biomarkers. To verify the relevance of the selected biomarkers, the affected patients then received a 4-week treatment consisting of a TCM formula of two plants [lily bulb (*Lilium lancifolium* Thunb) and radix rehmanniae (*Rehmannia glutinosa* Libosch) (synonyms of *Rehmannia glutinosa* (Gaertn.) DC.)]. The level of the proposed endogenous biomarkers was checked after the TCM intervention, showing that some were within normal values while others were at least significantly modified. This preliminary study highlighted that a series of nine markers distinguished the unaffected people from the affected people. This first study was completed by a second clinical study involving a similar number of patients and focused more on the mechanism of action of the formula (Tian et al., 2019). Plasma analyses were performed by UHPLC-HRMS and gas chromatography-mass spectrometry, in order to investigate in-depth the mechanisms related to the TCM formula intervention. In this second study, 22 potential biomarkers of the disease were identified, 15 of which were regulated by the TCM intervention. A pharmacological network was then established with the pathways related to the highlighted biomarkers. It also incorporated proteins and genes known to be the targets of the compounds in the formula. Through this approach, three targets were identified on two metabolic pathways affected by the syndrome. Overall, this study provided hypotheses about both the disease and the mechanism of action of the formula. It was based on preexisting activity data of the constituents of the formula and did not address the pharmacokinetic aspect.

Such metabolomic studies do not, of course, provide proof of a product's efficacy from the perspective of EBM and should not be used in this sense. Instead, their interest lies elsewhere: they generate fruitful hypotheses, both on pathogenic and therapeutic mechanisms.

Case study: perspectives on the *Phaleria nisidai* decoction study

Within the reverse pharmacology project conducted in the Republic of Palau, a local plant was selected through an RTO study, *Phaleria nisidai* (*P. nisidai*), which was then evaluated in a clinical study (Graz et al., 2015; Kitalong et al., 2017). *P. nisidai* is a tree endemic to certain Pacific islands and grows mainly in places related to current or past human activity (Kitalong, 2014). Local populations use the leaves of these trees growing in their gardens or near their villages (Kulakowski et al., 2015). *P. nisidai* belongs to the family of Thymelaeaceae, which is known to contain toxic compounds (Borris, Blaskó, & Cordell, 1988). Mimicking the traditional preparation was an essential point to avoid extracting toxic substances (Kulakowski et al., 2015). The tragic cases caused by the nontraditional use of *Piper methysticum* (kava kava) serve as an example in this context (Whitton, Lau, Salisbury, Whitehouse, & Evans, 2003). Indeed, standardized extracts of *Piper methysticum* were reported to cause severe hepatic damage on humans, while the traditional preparation has been used for many years by the population of the South Pacific islands without this adverse effect. Concerning *P. nisidai*, traditional knowledge in Palau reported that fresh leaves could be used as abortifacient. It can be assumed that the fresh leaves contain abortifacient substances that the decoction eliminates when the cooking time is sufficient, especially if a lid is not placed on the pot so that the toxic compounds evaporate (Kitalong, 2014). Daphnane-type diterpenes esters described previously in the methanolic extract of *P. nisidai* (Kulakowski et al., 2015) have also been described as abortive (Borris et al., 1988). These findings supported the choice to conduct laboratories studies on the traditional extract of *P. nisidai* leaves instead of using exhaustive extracts with organic solvents.

The case of *P. nisidai* can be used to make a link with important recommendations made by ecopharmacognosy. Proposed as both a philosophy and a way of practicing the development of natural resources, ecopharmacognosy is defined as "the study of sustainable, biologically active natural resources" (Cordell, 2017). Sustainability is a critical aspect when considering traditional medicines and it is important to consider the impact that scientific research can have on a plant. Indeed, in the context of a potentially medical use of a given extract, the supply of dry leaves must be considered. *P. nisidai* is a tree that is endemic in a limited part of the Pacific islands. Tissue culture projects are underway, but this is made difficult by the fact that *P. nisidai* grows slowly. In the reverse pharmacology approach, the aim was first to propose an effective and culturally accepted antidiabetic intervention on the Island of Palau (Graz et al., 2015). Since the results of the clinical studies were published (Kitalong et al., 2017), the decoction of *P. nisidai* is produced on a regular basis for type-2 diabetic patients who collect it from an outpatient clinic at Palau; the treatment is reimbursed by the local health insurance. At a minimum, the current study encourages local use of *P. nisidai* in a population severely affected by Type 2 diabetes, as well as the planning of larger-scale clinical studies. This traditional preparation is currently in stage 4 laboratory studies to tentatively identify possible active principles and decipher their mode of action.

Conclusion and perspectives

To provide evidence for a rational use of herbal preparations and effective quality control, the reverse pharmacology workflow is an approach not only from *bedside to bench* but also from *population to bench*. The ideal project is initiated by a survey on traditional uses and associated outcomes as reported by users, in order to provide indices of efficacy retrospectively (Fig. 16.5A). The plants with the highest indices of efficacy are then evaluated under different aspects (Fig. 16.5B): The clinical trials follow a prospective randomized design—if possible double-blind and in some cases crossover design (Fig. 16.5C). Crossover designs can be practical when studying chronic conditions, as a way of considering a high interindividual variability while keeping the sample rather small. During clinical studies, if biofluids are collected, monitoring metabolites from herbal preparations may be performed (Fig. 16.5D). To decipher the molecular mechanisms of herbal preparation constituents, the metabolites circulating in plasma serve as starting points for direct in vitro tests on relevant models, depending on the pathology studied (Fig. 16.5E). To better understand bioactivity, animal models are used as long as the herbal preparation is used in its complete form or as complex subfractions whose pharmacokinetics are not

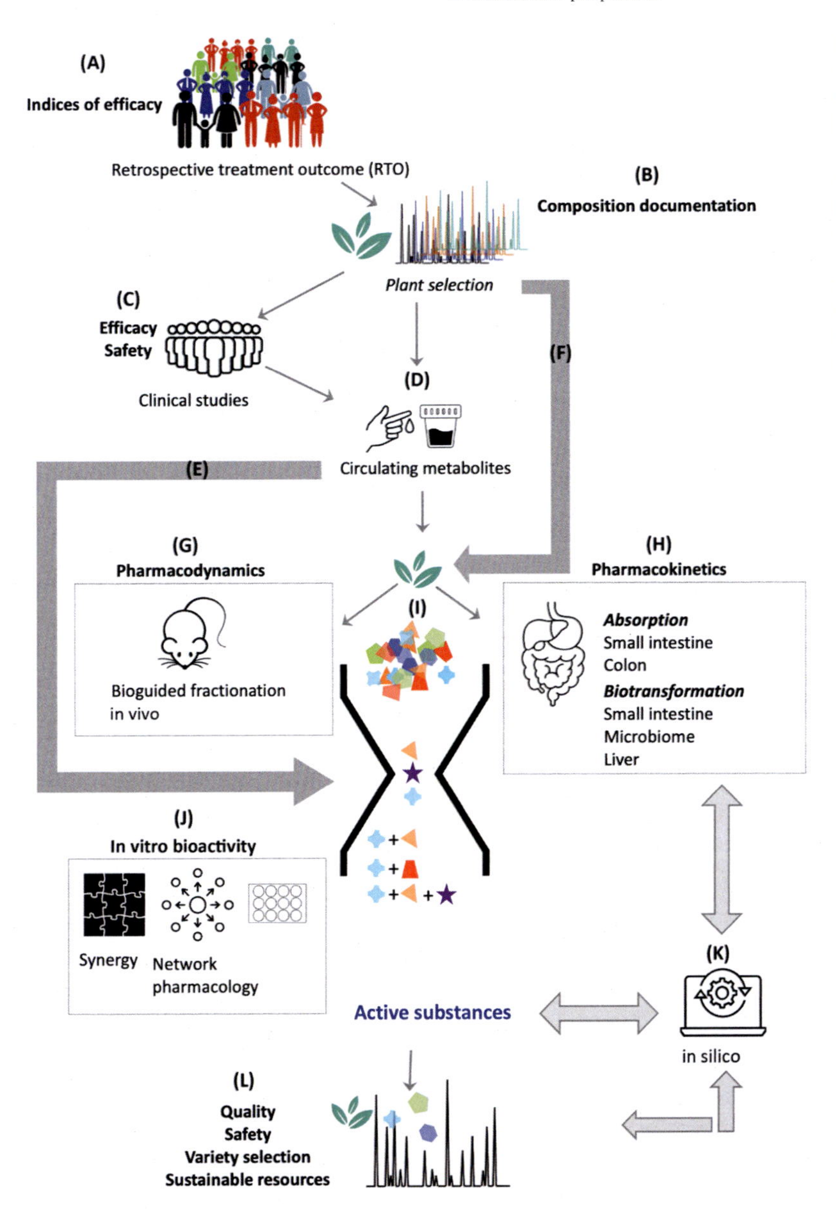

FIGURE 16.5 Ideal workflow for deciphering the pharmacological activity of herbal preparations. This figure proposes an ideal workflow to study a plant-based preparation in a reverse pharmacology approach, combining laboratory research after selecting a plant preparation by a Reverse Treatment Outcome survey.

deciphered (Fig. 16.5G). Models of intestinal permeation, as well as intestinal, colonic, and hepatic biotransformation, are used to verify and understand plasma observations (Fig. 16.5H). Considering that direct links "between clinic and laboratory" are difficult to establish in the current context, we can suggest an ideal workflow based only on laboratory studies performed in a postclinical approach (Fig. 16.5F).

All observations made for a given substance, whether on its PK fate or its action on a given target, can be integrated into the databases used for systemic analyses, in a loop aimed at increasing knowledge levels (Fig. 16.5K). All the knowledge gathered on the bioactivity of herbal preparation is finally integrated into the quality control method, in the form used by patients (Fig. 16.5L). The sources of supply are considered as part of a sustainable development approach, agronomic research can be performed for the selection of varieties and their cultivation. The whole process aims to understand the action of the herbal preparation constituents at the level of the organism by highlighting the metabolic pathways impacted. This allows us to consider the interactions that can occur during associated treatment or when other pathologies occur.

This ideal workflow is a complex process: the activity of multicomponent preparations on whole organisms is difficult to explain in a simple way, and holistic methods are necessary. This observation does not preclude the use of plants for which relevant clinical studies have been conducted, provided that quality control methods are used to verify the preparation, and dosages are correct.

While many studies focus on documenting clinical effects on the one hand and chemical composition on the other, efforts must be made to try to build a bridge between these different levels. The integration of systems biology approaches into this framework would be appropriate and could potentially be used to explain at the molecular level the mode of action of a given herbal preparation for which all clinical efficacy data have been acquired. With the extremely rapid progress of omics methods and bioinformatics, there is no doubt that, in the incoming era of *big data*, the mode of action of herbal preparations will become more decipherable.

From a medical perspective, the reverse pharmacology approach was born in a public health context and has sought to evaluate inexpensive practices that are already accepted and used by the population. The RTO and clinical steps provide the evidence for health care and a solid foundation for the laboratory research process. From a pharmacological perspective, the study of herbal preparations is challenging and requires the development of new methods. These new methods are still being developed and it is essential they be based on the highest level of medical evidence necessary to evaluate all medical treatments.

Acknowledgments

The authors acknowledge Dr. Elisabeth Rivera-Minten and Dr. Justin Veuthey for checking the English, and the creators of the icons from Nounproject (https://thenounproject.com/) used in the Figs. 16.1 and 16.5: Marie Van den Broeck (family), Oksana Latysheva (clinical study), Ker'is (leaves), Alina Oleynik (mouse), Julie Ko (in vitro), Iyikon (plant with roots), Mambu (open pill with leaf), Lafs (blood sample), Pham Thank Lôc (urine sample), Olena Panasovska (digestive system), Davo Sime (puzzle), Kokota (network), and Creative Stall (in silico).

References

Abdel-Tawab, M. (2018). Do we need plant food supplements? A critical examination of quality, safety, efficacy, and necessity for a new regulatory framework. *Planta Medica*, *84*, 372–393. Available from https://doi.org/10.1055/s-0043-123764.

Akin, M. D., Weingand, K. W., Hengehold, D. A., Goodale, M. B., Hinkle, R. T., & Smith, R. P. (2001). Continuous low-level topical heat in the treatment of dysmenorrhea. *Obstetrics and Gynecology*, *97*, 343–349. Available from https://doi.org/10.1016/S0029-7844(00)01163-7.

Asha, S., & Vidyavathi, M. (2010). Role of human liver microsomes in in vitro metabolism of drugs - A review. *Applied Biochemistry and Biotechnology, 160*, 1699–1722. Available from https://doi.org/10.1007/s12010-009-8689-6.

Bilia, A. R. (2015). Herbal medicinal products vs botanical-food supplements in the european market: State of art and perspectives. *Natural Product Communications, 10*, 125–131. Available from https://doi.org/10.1177/1934578x1501000130.

Borris, R. P., Blaskó, G., & Cordell, G. A. (1988). Ethnopharmacologic and phytochemical studies of the Thymelaeaceae. *Journal of Ethnopharmacology, 24*, 41–91. Available from https://doi.org/10.1016/0378-8741(88)90138-9.

Brewer, C. T., & Chen, T. (2017). Hepatotoxicity of herbal supplements mediated by modulation of cytochrome P450. *International Journal of Molecular Sciences, 18*, 2353. Available from https://doi.org/10.3390/ijms18112353.

Butterweck, V., & Nahrstedt, A. (2012). What is the best strategy for preclinical testing of botanicals? A critical perspective. *Planta Medica, 78*, 747–754. Available from https://doi.org/10.1055/s-0031-1298434.

Caesar, L. K., & Cech, N. B. (2019). Synergy and antagonism in natural product extracts: When 1 + 1 does not equal 2. *Natural Product Reports, 36*, 869–888. Available from https://doi.org/10.1039/c9np00011a.

Cordell, G. A. (2017). Sixty challenges - A 2030 perspective on natural products and medicines security. *Natural Product Communications, 12*, 1371–1379. Available from https://doi.org/10.1177/1934578x1701200849.

Covid-19 Survey: Our Selfcare and Its Effects. (2020). Retrieved from https://www.rtocovid19.com/. (Accessed 15 June 2021).

Cragg, G. M., & Newman, D. J. (2013). Natural products: A continuing source of novel drug leads. *Biochimica et Biophysica Acta - General Subjects, 1830*, 3670–3695. Available from https://doi.org/10.1016/j.bbagen.2013.02.008.

Dai, L., Cheng, C. W., Tian, R., Zhong, L. L., Li, Y. P., Lyu, A. P., ... Bian, Z. X. (2019). Standard protocol items for clinical trials with traditional Chinese medicine 2018: Recommendations, explanation and elaboration (SPIRIT-TCM Extension 2018). *Chinese Journal of Integrative Medicine, 25*, 71–79. Available from https://doi.org/10.1007/s11655-018-2999-x.

David, B., Wolfender, J. L., & Dias, D. A. (2015). The pharmaceutical industry and natural products: Historical status and new trends. *Phytochemistry Reviews, 14*, 299–315. Available from https://doi.org/10.1007/s11101-014-9367-z.

Deußer, H., Rogoll, D., Scheppach, W., Volk, A., Melcher, R., & Richling, E. (2013). Gastrointestinal absorption and metabolism of apple polyphenols ex vivo by the pig intestinal mucosa in the Ussing chamber. *Biotechnology Journal, 8*, 363–370. Available from https://doi.org/10.1002/biot.201200303.

Diallo, D., Graz, B., Falquet, J., Traoré, A. K., Giani, S., Mounkoro, P. P., ... Diakité, C. (2006). Malaria treatment in remote areas of Mali: Use of modern and traditional medicines, patient outcome. *Transactions of the Royal Society of Tropical Medicine and Hygiene, 100*, 515–520. Available from https://doi.org/10.1016/j.trstmh.2005.08.003.

Diallo, D., Diakité, C., Mounkoro, P. P., Sangaré, D., Graz, B., Falquet, J., & Giani, S. (2007). La prise en charge du paludisme par les therapeutes traditionnels dans les aires de sante de Kendie (Bandiagara) et de Finkolo (Sikasso) au Mali. *Le Mali médical, 22*, 1–8.

Du, T., Zeng, M., Chen, L., Cao, Z., Cai, H., & Yang, G. (2018). Chemical and absorption signatures of Xiao Chai Hu Tang. *Rapid Communications in Mass Spectrometry, 32*, 1107–1125. Available from https://doi.org/10.1002/rcm.8114.

Eriksson, L., Johansson, B. T., Trygg, E., & Vikström, J. (2013). *Multi- and megavariate data analysis*. Malmö, Sweden: MKS Umetrics AB.

Erk, T., Hauser, J., Williamson, G., Renouf, M., Steiling, H., Dionisi, F., & Richling, E. (2014). Structure- and dose-absorption relationships of coffee polyphenols. *Biofactors (Oxford, England), 40*, 103–112. Available from https://doi.org/10.1002/biof.1101.

Fang, Y., Cao, W., Xia, M., Pan, S., & Xu, X. (2017). Study of structure and permeability relationship of flavonoids in CaCo-2 cells. *Nutrients, 9*, 1301. Available from https://doi.org/10.3390/nu9121301.

Feng, W., Ao, H., Peng, C., & Yan, D. (2019). Gut microbiota, a new frontier to understand traditional Chinese medicines. *Pharmacological Research, 142*, 176–191. Available from https://doi.org/10.1016/j.phrs.2019.02.024.

Fernández-Ochoa, Á., Borrás-Linares, I., Baños, A., García-López, J. D., Guillamón, E., Nuñez-Lechado, C., ... Segura-Carretero, A. (2018). A fingerprinting metabolomic approach reveals deregulation of endogenous metabolites after the intake of a bioactive garlic supplement. *Journal of Functional Foods, 49*, 137–145. Available from https://doi.org/10.1016/j.jff.2018.08.003.

Food & Drug Administration. (2019). What is a botanical drug. Retrieved from https://www.fda.gov/about-fda/center-drug-evaluation-and-research-cder/what-botanical-drug. (Accessed 15 July 2021).

Fritz, A., Busch, D., Lapczuk, J., Ostrowski, M., Drozdzik, M., & Oswald, S. (2019). Expression of clinically relevant drug-metabolizing enzymes along the human intestine and their correlation to drug transporters and nuclear receptors: An intra-subject analysis. *Basic and Clinical Pharmacology and Toxicology, 124*, 245–255. Available from https://doi.org/10.1111/bcpt.13137.

Fürst, R., & Zündorf, I. (2015). Evidence-based phytotherapy in Europe: Where do we stand? *Planta Medica, 81*, 962–967. Available from https://doi.org/10.1055/s-0035-1545948.

Gertsch, J. (2009). How scientific is the science in ethnopharmacology? Historical perspectives and epistemological problems. *Journal of Ethnopharmacology, 122*, 177–183. Available from https://doi.org/10.1016/j.jep.2009.01.010.

Gertsch, J. (2011). Botanical drugs, synergy, and network pharmacology: Forth and back to intelligent mixtures. *Planta Medica, 77*, 1086–1098. Available from https://doi.org/10.1055/s-0030-1270904.

Gilbert, J. A., Blaser, M. J., Caporaso, J. G., Jansson, J. K., Lynch, S. V., & Knight, R. (2018). Current understanding of the human microbiome. *Nature Medicine, 24*, 392–400. Available from https://doi.org/10.1038/nm.4517.

Graz, B. (2013). What is \clinical data\? Why and how can they be collected during field surveys on medicinal plants? *Journal of Ethnopharmacology, 150*, 775–779. Available from https://doi.org/10.1016/j.jep.2013.08.036.

Graz, B., Elisabetsky, E., & Falquet, J. (2007). Beyond the myth of expensive clinical study: Assessment of traditional medicines. *Journal of Ethnopharmacology, 113*, 382–386. Available from https://doi.org/10.1016/j.jep.2007.07.012.

Graz, B., Willcox, M. L., Diakite, C., Falquet, J., Dackuo, F., Sidibe, O., ... Diallo, D. (2010). Argemone mexicana decoction vs artesunate-amodiaquine for the management of malaria in Mali: Policy and public-health implications. *Transactions of the Royal Society of Tropical Medicine and Hygiene, 104*, 33–41. Available from https://doi.org/10.1016/j.trstmh.2009.07.005.

Graz, B., Falquet, J., & Elisabetsky, E. (2010). Ethnopharmacology, sustainable development and cooperation: The importance of gathering clinical data during field surveys. *Journal of Ethnopharmacology, 130*, 635–638. Available from https://doi.org/10.1016/j.jep.2010.04.044.

Graz, B., Savoy, M., Buclin, T., & Bonvin, E. (2014). Dysménorrhée: Patience, pilules ou bouillotte? *Revue Medicale Suisse, 10*, 2285–2288. Available from http://rms.medhyg.ch/load_pdf.php?ID_ARTICLE = RMS_452_2285.

Graz, B., Willcox, M., & Elisabetsky, E. (2015). Retrospective treatment-outcome as a method of collecting clinical data in ethnopharmacological surveys. *Ethnopharmacology* (pp. 251–261). Switzerland: Wiley. Available from https://doi.org/10.1002/9781118930717.ch22.

Graz, B., Kitalong, C., & Yano, V. (2015). Traditional local medicines in the republic of Palau and non-communicable diseases (NCD), signs of effectiveness. *Journal of Ethnopharmacology, 161*, 233–237. Available from https://doi.org/10.1016/j.jep.2014.11.047.

Gurmu, A. E., Kisi, T., Shibru, H., Graz, B., & Willcox, M. (2018). Treatments used for malaria in young Ethiopian children: a retrospective study. *Malaria Journal, 17*(1), 451. Available from https://doi.org/10.1186/s12936-018-2605-x.

Hardy, G. (2000). Nutraceuticals and functional foods: Introduction and meaning. *Nutrition (Burbank, Los Angeles County, Calif.), 16*, 688–689. Available from https://doi.org/10.1016/S0899-9007(00)00332-4.

Harrigan, G. G., Brackett, D. J., & Boros, L. G. (2005). Medicinal chemistry, metabolic profiling and drug target discovery: A role for metabolic profiling in reverse pharmacology and chemical genetics. *Mini-Reviews in Medicinal Chemistry, 5*, 13–20. Available from https://doi.org/10.2174/1389557053402800.

Hoehle, S. I., Pfeiffer, E., Sólyom, A. M., & Metzler, M. (2006). Metabolism of curcuminoids in tissue slices and subcellular fractions from rat liver. *Journal of Agricultural and Food Chemistry, 54*, 756–764. Available from https://doi.org/10.1021/jf058146a.

Hopkins, A. L. (2008). Network pharmacology: The next paradigm in drug discovery. *Nature Chemical Biology, 4*, 682–690. Available from https://doi.org/10.1038/nchembio.118.

Hou, J. J., Wu, W. Y., Da, J., Yao, S., Long, H. L., Yang, Z., ... Guo, D. A. (2011). Ruggedness and robustness of conversion factors in method of simultaneous determination of multi-components with single reference standard. *Journal of Chromatography. A, 1218*, 5618–5627. Available from https://doi.org/10.1016/j.chroma.2011.06.058.

Houriet, J., Arnold, Y. E., Pellissier, L., Kalia, Y. N., & Wolfender, J. L. (2021). Using Porcine Jejunum Ex Vivo to Study Absorption and Biotransformation of Natural Products in Plant Extracts: Pueraria lobata as a Case Study. *Metabolites, 11*(8). Available from https://doi.org/10.3390/metabo11080541.

Hubatsch, I., Ragnarsson, E. G. E., & Artursson, P. (2007). Determination of drug permeability and prediction of drug absorption in CaCo-2 monolayers. *Nature Protocols, 2*, 2111–2119. Available from https://doi.org/10.1038/nprot.2007.303.

Jung, I., Kim, H., Moon, S., Lee, H., & Kim, B. (2020). Overview of *Salvia miltiorrhiza* as a potential therapeutic agent for various diseases: An update on efficacy and mechanisms of action. *Antioxidants*, *9*, 1–40. Available from https://doi.org/10.3390/antiox9090857.

Kitalong, C. (2014). *Ethnomedical, ecological and phytochemical studies of the Palauan Flora* (Dissertation thesis). CUNY Academic Works.

Kitalong, C., Nogueira, R. C., Benichou, J., Yano, V., Espangel, V., Houriet, J., ... Graz, B. (2017). "DAK," a traditional decoction in Palau, as adjuvant for patients with insufficient control of diabetes mellitus type II. *Journal of Ethnopharmacology*, *205*, 116–122. Available from https://doi.org/10.1016/j.jep.2017.05.003.

Kitano, H. (2007). A robustness-based approach to systems-oriented drug design. *Nature Reviews Drug Discovery*, *6*, 202–210. Available from https://doi.org/10.1038/nrd2195.

Kopečná-Zapletalová, M., Krasulová, K., Anzenbacher, P., Hodek, P., & Anzenbacherová, E. (2017). Interaction of isoflavonoids with human liver microsomal cytochromes P450: Inhibition of CYP enzyme activities. *Xenobiotica*, *47*, 324–331. Available from https://doi.org/10.1080/00498254.2016.1195028.

Koropatkin, N. M., Cameron, E. A., & Martens, E. C. (2012). How glycan metabolism shapes the human gut microbiota. *Nature Reviews Microbiology*, *10*, 323–335. Available from https://doi.org/10.1038/nrmicro2746.

Kulakowski, D., Kitalong, C., Negrin, A., Tadao, V.-R., Balick, M. J., & Kennelly, E. J. (2015). Traditional preparation of *Phaleria nisidai*, a Palauan tea, reduces exposure to toxic daphnane-type diterpene esters while maintaining immunomodulatory activity. *Journal of Ethnopharmacology*, *173*, 273–279. Available from https://doi.org/10.1016/j.jep.2015.06.023.

Lehtonen, H. M., Lindstedt, A., Järvinen, R., Sinkkonen, J., Graça, G., Viitanen, M., ... Gil, A. M. (2013). 1H NMR-based metabolic fingerprinting of urine metabolites after consumption of lingonberries (Vaccinium vitis-idaea) with a high-fat meal. *Food Chemistry*, *138*, 982–990. Available from https://doi.org/10.1016/j.foodchem.2012.10.081.

Lennernäs, H. (2007). Animal data: The contributions of the Ussing Chamber and perfusion systems to predicting human oral drug delivery in vivo. *Advanced Drug Delivery Reviews*, *59*, 1103–1120. Available from https://doi.org/10.1016/j.addr.2007.06.016.

Li, C. R., Zhang, L., Wo, S. K., Zhou, L. M., Lin, G., & Zuo, Z. (2012). Pharmacokinetic interactions among major bioactive components in Radix Scutellariae via metabolic competition. *Biopharmaceutics and Drug Disposition*, *33*, 487–500. Available from https://doi.org/10.1002/bdd.1815.

Li, Y., Shen, Y., Yao, C. L., & Guo, D. A. (2020). Quality assessment of herbal medicines based on chemical fingerprints combined with chemometrics approach: A review. *Journal of Pharmaceutical and Biomedical Analysis*, *185*, 113215. Available from https://doi.org/10.1016/j.jpba.2020.113215.

Licata, A., Macaluso, F. S., & Craxì, A. (2013). Herbal hepatotoxicity: A hidden epidemic. *Internal and Emergency Medicine*, *8*, 13–22. Available from https://doi.org/10.1007/s11739-012-0777-x.

Liu, J. Y., Lee, K. F., Sze, C. W., Tong, Y., Tang, S. C. W., Ng, T. B., & Zhang, Y. B. (2013). Intestinal absorption and bioavailability of traditional Chinese medicines: A review of recent experimental progress and implication for quality control. *Journal of Pharmacy and Pharmacology*, *65*, 621–633. Available from https://doi.org/10.1111/j.2042-7158.2012.01608.x.

Liu, X., Wu, W. Y., Jiang, B. H., Yang, M., & Guo, D. A. (2013). Pharmacological tools for the development of traditional Chinese medicine. *Trends in Pharmacological Sciences*, *34*, 620–628. Available from https://doi.org/10.1016/j.tips.2013.09.004.

Lu, C., Deng, J., Li, L., Wang, D., & Li, G. (2014). Application of metabolomics on diagnosis and treatment of patients with psoriasis in traditional Chinese medicine. *Biochimica et Biophysica Acta - Proteins and Proteomics*, *1844*, 280–288. Available from https://doi.org/10.1016/j.bbapap.2013.05.019.

Ma, N., Zhang, Z., Liao, F., Jiang, T., & Tu, Y. (2020). The birth of artemisinin. *Pharmacology & Therapeutics*, *216*, 107658. Available from https://doi.org/10.1016/j.pharmthera.2020.107658.

Madrid-Gambin, F., Garcia-Aloy, M., Vázquez-Fresno, R., Vegas-Lozano, E., de Villa Jubany, M. C. R., Misawa, K., ... Andres-Lacueva, C. (2016). Impact of chlorogenic acids from coffee on urine metabolome in healthy human subjects. *Food Research International*, *89*, 1064–1070. Available from https://doi.org/10.1016/j.foodres.2016.03.038.

Matsumoto, M., Matsukawa, N., Mineo, H., Chiji, H., & Hara, H. (2004). A soluble flavonoid-glycoside, αG-rutin, is absorbed as glycosides in the isolated gastric and intestinal mucosa. *Bioscience, Biotechnology, and Biochemistry*, *68*, 1929–1934. Available from https://doi.org/10.1271/bbb.68.1929.

McDonald, D., Hyde, E., Debelius, J. W., Morton, J. T., Gonzalez, A., Ackermann, G., ... Gunderson, B. (2018). American gut: An open platform for citizen science microbiome research. *MSystems*, *3*(3). Available from https://doi.org/10.1128/mSystems.00031-18.

Medical Subject Headings: Pharmacognosy. (2019). Retrieved from https://www.ncbi.nlm.nih.gov/mesh/?term = pharmacognosy. (Accessed 15 July 2021).

Misra, S. (2012). Randomized double blind placebo control studies, the \gold Standard\ in intervention based studies. *Indian Journal of Sexually Transmitted Diseases*, *33*, 131–134. Available from https://doi.org/10.4103/0253-7184.102130.

Newman, D. J., & Cragg, G. M. (2016). Natural products as sources of new drugs from 1981 to 2014. *Journal of Natural Products*, *79*, 629–661. Available from https://doi.org/10.1021/acs.jnatprod.5b01055.

Nordberg, M., Duffus, J., & Templeton, D. M. (2004). Glossary of terms used in toxicokinetics (IUPAC Recommendations 2003). *Pure and Applied Chemistry*, *76*, 1033–1082. Available from https://doi.org/10.1351/pac200476051033.

Patterson, J. K., Lei, X. G., & Miller, D. D. (2008). The pig as an experimental model for elucidating the mechanisms governing dietary influence on mineral absorption. *Experimental Biology and Medicine*, *233*, 651–664. Available from https://doi.org/10.3181/0709-MR-262.

Patwardhan, B., Vaidya, A. D. B., Chorghade, M., & Joshi, S. P. (2008). Reverse pharmacology and systems approaches for drug discovery and development. *Current Bioactive Compounds*, *4*, 201–212. Available from https://doi.org/10.2174/157340708786847870.

Pawson, R. (2019). Pragmatic trials and implementation science: grounds for divorce? *BMC Medical Research Methodology*, *19*, 176. Available from https://doi.org/10.1186/s12874-019-0814-9.

Petit, C., Bujard, A., Skalicka-Woźniak, K., Cretton, S., Houriet, J., Christen, P., ... Wolfender, J. L. (2016). Prediction of the passive intestinal absorption of medicinal plant extract constituents with the parallel artificial membrane permeability assay (PAMPA). *Planta Medica*, *82*, 424–431. Available from https://doi.org/10.1055/s-0042-101247.

Pferschy-Wenzig, E.-M., & Bauer, R. (2015). The relevance of pharmacognosy in pharmacological research on herbal medicinal products. *Epilepsy & Behavior*, *52*, 344–362. Available from https://doi.org/10.1016/j.yebeh.2015.05.037.

Pferschy-Wenzig, E. M., Koskinen, K., Moissl-Eichinger, C., & Bauer, R. (2017). A combined LC-MS metabolomics- and 16S rRNA sequencing platform to assess interactions between herbal medicinal products and human gut bacteria in vitro: A pilot study on Willow bark extract. *Frontiers in Pharmacology*, *8*, 893. Available from https://doi.org/10.3389/fphar.2017.00893.

Presley, C. C., & Lindsley, C. W. (2018). DARK classics in chemical neuroscience: Opium, a historical perspective. *ACS Chemical Neuroscience*, *9*, 2503–2518. Available from https://doi.org/10.1021/acschemneuro.8b00459.

Report by Zion Market Research (ZMR) on Herbal Supplement Market. (2018). Retrieved from https://www.zionmarketresearch.com/news/herbal-supplement-market. (Accessed 15 July 2021).

Santini, A., Cammarata, S. M., Capone, G., Ianaro, A., Tenore, G. C., Pani, L., & Novellino, E. (2018). Nutraceuticals: Opening the debate for a regulatory framework. *British Journal of Clinical Pharmacology*, *84*, 659–672. Available from https://doi.org/10.1111/bcp.13496.

Simoes-Pires, C., Hostettmann, K., Haouala, A., Cuendet, M., Falquet, J., Graz, B., & Christen, P. (2014). Reverse pharmacology for developing an anti-malarial phytomedicine. The example of *Argemone mexicana*. *International Journal for Parasitology: Drugs and Drug Resistance*, *4*, 338–346. Available from https://doi.org/10.1016/j.ijpddr.2014.07.001.

Sjöberg, A., Lutz, M., Tannergren, C., Wingolf, C., Borde, A., & Ungell, A. L. (2013). Comprehensive study on regional human intestinal permeability and prediction of fraction absorbed of drugs using the Ussing chamber technique. *European Journal of Pharmaceutical Sciences*, *48*, 166–180. Available from https://doi.org/10.1016/j.ejps.2012.10.007.

Sjögren, E., Abrahamsson, B., Augustijns, P., Becker, D., Bolger, M. B., Brewster, M., ... Langguth, P. (2014). In vivo methods for drug absorption - Comparative physiologies, model selection, correlations with in vitro methods (IVIVC), and applications for formulation/API/excipient characterization including food effects. *European Journal of Pharmaceutical Sciences*, *57*, 99–151. Available from https://doi.org/10.1016/j.ejps.2014.02.010.

Skalicka-Woźniak, K., Georgiev, M. I., & Orhan, I. E. (2017). Adulteration of herbal sexual enhancers and slimmers: The wish for better sexual well-being and perfect body can be risky. *Food and Chemical Toxicology*, *108*, 355–364. Available from https://doi.org/10.1016/j.fct.2016.06.018.

Soldatow, V. Y., Lecluyse, E. L., Griffith, L. G., & Rusyn, I. (2013). In vitro models for liver toxicity testing. *Toxicology Research*, *2*, 23–39. Available from https://doi.org/10.1039/c2tx20051a.

Sousa, T., Paterson, R., Moore, V., Carlsson, A., Abrahamsson, B., & Basit, A. W. (2008). The gastrointestinal microbiota as a site for the biotransformation of drugs. *International Journal of Pharmaceutics*, *363*, 1–25. Available from https://doi.org/10.1016/j.ijpharm.2008.07.009.

Straus, S. E., Glasziou, P., Richardson, W. S., & Haynes, R. B. (2011). *Evidence-based medicine: How to practice and teach it* (4th ed., p. 2011)Edinburgh: Churchill Livingstone, Elsevier.

Su, S., Duan, J., Wang, P., Liu, P., Guo, J., Shang, E., ... Tang, Z. (2013). Metabolomic study of biochemical changes in the plasma and urine of primary dysmenorrhea patients using UPLC–MS coupled with a pattern recognition approach. *Journal of Proteome Research*, *12*, 852–865. Available from https://doi.org/10.1021/pr300935x.

Takis, P. G., Oraiopoulou, M. E., Konidaris, C., & Troganis, A. N. (2016). 1H-NMR based metabolomics study for the detection of the human urine metabolic profile effects of *Origanum dictamnus* tea ingestion. *Food and Function*, *7*, 4104–4115. Available from https://doi.org/10.1039/c6fo00560h.

Tavassoly, I., Goldfarb, J., & Iyengar, R. (2018). Systems biology primer: The basic methods and approaches. *Essays in Biochemistry*, *62*, 487–500. Available from https://doi.org/10.1042/ebc20180003.

Tian, J., Xia, X., Wu, Y., Zhao, L., Xiang, H., Du, G., ... Qin, X. (2016). Discovery, screening and evaluation of a plasma biomarker panel for subjects with psychological suboptimal health state using 1H-NMR-based metabolomics profiles. *Scientific Reports*, *6*, 33820. Available from https://doi.org/10.1038/srep33820.

Tian, J. S., Peng, G. J., Gao, X. X., Zhou, Y. Z., Xing, J., Qin, X. M., & Du, G. H. (2014). Dynamic analysis of the endogenous metabolites in depressed patients treated with TCM formula Xiaoyaosan using urinary 1H NMR-based metabolomics. *Journal of Ethnopharmacology*, *158*, 1–10. Available from https://doi.org/10.1016/j.jep.2014.10.005.

Tian, J. S., Peng, G. J., Wu, Y. F., Zhou, J. J., Xiang, H., Gao, X. X., ... Du, G. H. (2016). A GC–MS urinary quantitative metabolomics analysis in depressed patients treated with TCM formula of Xiaoyaosan. *Journal of Chromatography B: Analytical Technologies in the Biomedical and Life Sciences*, *1026*, 227–235. Available from https://doi.org/10.1016/j.jchromb.2015.12.026.

Tian, J. S., Meng, Y., Wu, Y. F., Zhao, L., Xiang, H., Jia, J. P., & Qin, X. M. (2019). A novel insight into the underlying mechanism of Baihe Dihuang Tang improving the state of psychological suboptimal health subjects obtained from plasma metabolic profiles and network analysis. *Journal of Pharmaceutical and Biomedical Analysis*, *169*, 99–110. Available from https://doi.org/10.1016/j.jpba.2019.02.041.

Tolleson, W. H., Doerge, D. R., Churchwell, M. I., Marques, M. M., & Roberts, D. W. (2002). Metabolism of biochanin A and formononetin by human liver microsomes in vitro. *Journal of Agricultural and Food Chemistry*, *50*, 4783–4790. Available from https://doi.org/10.1021/jf025549r.

Ussing, H. H., & Zerahn, K. (1951). Active transport of sodium as the source of electric current in the short-circuited isolated frog skin. *Acta Physiologica Scandinavica*, *23*, 110–127. Available from https://doi.org/10.1111/j.1748-1716.1951.tb00800.x.

Van Der Greef, J. (2011). Perspective: All systems go. *Nature*, *480*, S87. Available from https://doi.org/10.1038/480S87a.

Van Dorsten, F. A., Grün, C. H., Van Velzen, E. J. J., Jacobs, D. M., Draijer, R., & Van Duynhoven, J. P. M. (2010). The metabolic fate of red wine and grape juice polyphenols in humans assessed by metabolomics. *Molecular Nutrition & Food Research*, *54*, 897–908. Available from https://doi.org/10.1002/mnfr.200900212.

Van Velzen, E. J. J., Westerhuis, J. A., Van Duynhoven, J. P. M., Van Dorsten, F. A., Grün, C. H., Jacobs, D. M., ... Smilde, A. K. (2009). Phenotyping tea consumers by nutrikinetic analysis of polyphenolic end-metabolites. *Journal of Proteome Research*, *8*, 3317–3330. Available from https://doi.org/10.1021/pr801071p.

Vermaak, I., Viljoen, A. M., Chen, W., & Hamman, J. H. (2011). In vitro transport of the steroidal glycoside P57 from *Hoodia gordonii* across excised porcine intestinal and buccal tissue. *Phytomedicine*, *18*, 783–787. Available from https://doi.org/10.1016/j.phymed.2011.01.017.

Verpoorte, R., Choi, Y. H., & Kim, H. K. (2005). Ethnopharmacology and systems biology: A perfect holistic match. *Journal of Ethnopharmacology*, *100*, 53–56. Available from https://doi.org/10.1016/j.jep.2005.05.033.

Wagner, H., & Ulrich-Merzenich, G. (2009). Synergy research: Approaching a new generation of phytopharmaceuticals. *Phytomedicine*, *16*, 97–110. Available from https://doi.org/10.1016/j.phymed.2008.12.018.

Walgren, R. A., Walle, U. K., & Walle, T. (1998). Transport of quercetin and its glucosides across human intestinal epithelial CaCo-2 cells. *Biochemical Pharmacology*, *55*, 1721–1727. Available from https://doi.org/10.1016/S0006-2952(98)00048-3.

Wang, Q., Kuang, Y., Song, W., Qian, Y., Qiao, X., Guo, D. A., & Ye, M. (2017). Permeability through the CaCo-2 cell monolayer of 42 bioactive compounds in the TCM formula Gegen-Qinlian Decoction by liquid chromatography tandem mass spectrometry analysis. *Journal of Pharmaceutical and Biomedical Analysis, 146*, 206–213. Available from https://doi.org/10.1016/j.jpba.2017.08.042.

Wang, X., Sun, H., Zhang, A., Sun, W., Wang, P., & Wang, Z. (2011). Potential role of metabolomics apporoaches in the area of traditional Chinese medicine: As pillars of the bridge between Chinese and Western medicine. *Journal of Pharmaceutical and Biomedical Analysis, 55*, 859–868. Available from https://doi.org/10.1016/j.jpba.2011.01.042.

Wang, X. B., Zheng, J., Li, J. J., Yu, H. Y., Li, Q. Y., Xu, L. H., ... Liu, B. J. (2018). Simultaneous analysis of 23 illegal adulterated aphrodisiac chemical ingredients in health foods and Chinese traditional patent medicines by ultrahigh performance liquid chromatography coupled with quadrupole time-of-flight mass spectrometry. *Journal of Food and Drug Analysis, 26*, 1138–1153. Available from https://doi.org/10.1016/j.jfda.2018.02.003.

Wang, Y., Tang, H., Nicholson, J. K., Hylands, P. J., Sampson, J., & Holmes, E. (2005). A metabonomic strategy for the detection of the metabolic effects of chamomile (*Matricaria recutita* L.) ingestion. *Journal of Agricultural and Food Chemistry, 53*, 191–196. Available from https://doi.org/10.1021/jf0403282.

Whitton, P. A., Lau, A., Salisbury, A., Whitehouse, J., & Evans, C. S. (2003). Kava lactones and the kava-kava controversy. *Phytochemistry, 64*, 673–679. Available from https://doi.org/10.1016/S0031-9422(03)00381-9.

WHO Traditional Medicine Strategy. (2013). World Health Organization, 2014–2023.

Wiesner, J., & Knöss, W. (2014). Future visions for traditional and herbal medicinal products - A global practice for evaluation and regulation? *Journal of Ethnopharmacology, 158*, 516–518. Available from https://doi.org/10.1016/j.jep.2014.08.015.

Willcox, M. L., Graz, B., Falquet, J., Sidibé, O., Forster, M., & Diallo, D. (2007). Argemone mexicana decoction for the treatment of uncomplicated falciparum malaria. *Transactions of the Royal Society of Tropical Medicine and Hygiene, 101*, 1190–1198. Available from https://doi.org/10.1016/j.trstmh.2007.05.017.

Willcox, M. L., Graz, B., Falquet, J., Diakite, C., Giani, S., & Diallo, D. (2011). A reverse pharmacology approach for developing an anti-malarial phytomedicine. *Malaria Journal, 10*, S8. Available from https://doi.org/10.1186/1475-2875-10-S1-S8.

Williams, K. J. (2009). The introduction of "chemotherapy" using arsphenamine - The first magic bullet. *Journal of the Royal Society of Medicine, 102*, 343–348. Available from https://doi.org/10.1258/jrsm.2009.09k036.

Wolfender, J. L., Nuzillard, J. M., Van Der Hooft, J. J. J., Renault, J. H., & Bertrand, S. (2019). Accelerating metabolite identification in natural product research: Toward an ideal combination of liquid chromatography-high-resolution tandem mass spectrometry and NMR profiling, in silico databases, and chemometrics. *Analytical Chemistry, 91*, 704–742. Available from https://doi.org/10.1021/acs.analchem.8b05112.

Wolffram, S., Blöck, M., & Ader, P. (2002). Quercetin-3-glucoside is transported by the glucose carrier SGLT1 across the brush border membrane of rat small intestine. *Journal of Nutrition, 132*, 630–635. Available from https://doi.org/10.1093/jn/132.4.630.

World Health Organization. (2000). General guidelines for methodologies on research and evaluation of traditional medicine.

Zheng, Y., Feng, G., Sun, Y., Liu, S., Pi, Z., Song, F., & Liu, Z. (2018). Study on the compatibility interactions of formula Ding-Zhi-Xiao-Wan based on their main components transport characteristics across CaCo-2 monolayers model. *Journal of Pharmaceutical and Biomedical Analysis, 159*, 179–185. Available from https://doi.org/10.1016/j.jpba.2018.06.067.

Zhou, S. S., Xu, J., Zhu, H., Wu, J., Xu, J. D., Yan, R., ... Li, S. L. (2016). Gut microbiota-involved mechanisms in enhancing systemic exposure of ginsenosides by coexisting polysaccharides in ginseng decoction. *Scientific Reports, 6*, 22474. Available from https://doi.org/10.1038/srep22474.

CHAPTER

17

Nagoya Protocol and access to genetic resources

Bruno David

Green Mission Pierre Fabre, Pierre Fabre Research Institute, Toulouse, France

Introduction

Since the dawn of time, plants and other biological resources considered as common goods of humanity have been circulating freely through migrations, explorations, and botanical exchanges. Then in 1992, in the context of environmental concern, the Convention on Biological Diversity (CBD) put an end to the historically free and open access by placing animals, plants, fungi, and microorganisms under the sovereignty of the source countries. Countries then have the right to request benefit sharing in return for access, this is Access and Benefit Sharing (ABS). When genetic resources (GR) fell within the scope of the CBD, this quickly created legal uncertainty because most economic exchanges concern nongenetic resources. This point was resolved by the Nagoya Protocol (NP), which extended the scope of ABS regulations beyond GR by formally including biological resources that do not contain functional units of heredity. By ratifying the CBD and/or the NP, nations commit themselves to implement regulations on access to biodiversity, the modalities of which they can freely determine. The history and evolution of these legislations are the subject of this chapter.

History and evolution of concepts

Development of environmental awareness

Environmental issues are very old, having already been expressed by Plato in ancient Greece in his dialogue Critias. In June 1864, President Abraham Lincoln signed the Yosemite Valley Grant Act to preserve this natural heritage and the Mariposa giant sequoia grove (US Congress, 1864). In the middle of the 20th century, the urge for environmental protection became more

Medicinal Plants as Anti-infectives
DOI: https://doi.org/10.1016/B978-0-323-90999-0.00011-2

global and led to the creation of the IUCN (International Union for Conservation of Nature) in October 1948 in Fontainebleau, France. IUCN was the first global ecological union created and it has played a fundamental role in the creation of key international conventions such as the environmental CBD (IUCN, 2021). In the 1960s, Rachel Carson, with her book "Silent Spring" (Carson, 1962) warned about the environmental and human danger of the indiscriminate use of pesticides. She succeeded in banning the environmentally toxic dichloro-diphenyl-trichloroethane and initiated modern ecological thinking. She warned of the environmental impact of human activities and recommended preserving the quality of life for future generations. Then the pictures of the earth from the lunar surface taken by William Anders during the Apollo 8 mission on December 24, 1968, have changed the vision that humanity has of its planet. These photographs became the environmentalism movement's icon (Overbye, 2018). They have strengthened environmental awareness by reminding us that our planet is limited, small, and fragile (IUCN, 2020). The message "S.O.S. environment" signed by 2,200 environmental scientists from 23 countries, including four Nobel Prize winners was handed to the United Nations Secretary-General in a ceremony in New York on May 11, 1971, and published in The Unesco Courier (Collective, 1971). This manifesto originally drafted in the French town of Menton and known as the "Menton message" warned humanity of the unprecedented dangers posed to the human race by environmental destruction. Then in January 1972, the magazine «The Ecologist» asked for an urgent need to preserve the environment in its famous issue "A Blueprint for survival" (Kozlov, 2021). In April of the same year, the Meadows report, commissioned by the Club of Rome, stimulated considerable public attention by denouncing the limits of growth of the industrial consumerist lifestyle, the exhaustion of natural resources, and by asking for sustainable development (Meadows, Meadows, Randers, & Behrens, 1972). These three influential calls to "respect ecosystems essential to our lives" were published in advance of the world's first United Nations Environment summit that was going to be held in June 1972 in Stockholm (United Nations, 1972).

Concept of biodiversity

The word "Biodiversity" appeared in the form "Biological Diversity" in 1916 in an article by J. Arthur Harris to speak of the abundance of living species in an environment "The Variable Desert" (Harris, 1916). Thomas Lovejoy popularized the term widely in 1980 in the preface to the book "Conservation Biology: An Evolutionary-Ecological Perspective" (Lovejoy, 1980). The concept then evolved to cover the variety of living organisms and the ecosystems in which they live and interact. The word "Biodiversity" was first printed in 1988 during the proceedings of the National Forum on Biological Diversity held in Washington, D.C. (September 21–25, 1986) jointly organized by the National Academy of Sciences and the Smithsonian Institution on the risk of extinction of living species. The word "Biological Diversity" was too long to be printed on the cover of the proceedings' book, so it was reduced to "Biodiversity" (Wilson, 1988).

The notion of GR appeared in the context of the emergence of economic and industrial interest in genes in the 1980s. The case of Sydney A. Diamond (U.S. Commissioner of Patents and Trademarks) and Ananda M. Chakrabarty (a geneticist working for General Electric) was a perfect illustration of the appeal of genetics and biotechnologies considered as an Eldorado (David, Wolfender, & Dias, 2015) capable of generating billions of dollars in annual revenues. For about 10 years Diamond opposed Chakrabarty by refusing any patent on living things. Chakrabarty/General Electric's patent application concerned a bacterium (*Pseudomonas putida*) genetically modified to metabolize oil spills. Finally Chakrabarty and General Electric won their case, with the U.S. Supreme Court ruling on June 16, 1980, with five votes to four in their favor. This decision had enormous consequences, as it contributed to the development of the genetics and biotechnology industries by authorizing patents on genes and genetically modified organisms (Kevles, 1994). This legal reversal regarding the intellectual property rights over living beings was based on the fact that GR genetically modified should be considered as patentable when the modification is new, inventive, and capable of industrial application (Pauchard, 2017).

The terms "Biodiversity" and "Genetic Resources" were then honored and popularized at the Earth Summit in Rio de Janeiro, Brazil in 1992.

Conscious of the environmental concern, the United Nations initiated a specific program for the environment at a First Earth Summit in Stockholm, June 5–16, 1972 (United Nations, 1972). This summit represented the first consideration of the global human impact on the environment and attempts to address the challenge of preserving it (Handl, 2012).

The second summit was held in Nairobi, Kenya in 1982 but no major decisions were taken.

Ten years later, concerned about the future of the Earth the international community, gathered at the bedside of our planet, adopted three international environmental conventions at the third Earth Summit in Rio de Janeiro, June 3–14, 1992: the CBD (United Nations, 1992), the Framework Convention on Climate Change and the Convention to Combat Desertification. During this international summit, biodiversity became the focus of worldwide scientific and political concerns with the awareness of the extent and irreversibility of its disappearance (Wilson, 1996).

The Convention on Biodiversity

The Convention on Biodiversity is an international agreement adopted in 1992, after several years of preparatory works. The first Ad Hoc Working Group of Experts on Biological Diversity met in November 1988 to think over an international CBD. In May 1992 at the Nairobi Conference the agreed text of the CBD was adopted in preparation of the upcoming Rio summit. The text was open for signatures on June 5, 1992, at the United Nations "Earth Summit" in Rio Brazil. The CBD entered into force on December 29, 1993, that is, 90 days after signature of the 30th ratification's instrument. The Convention now has 196 Parties: 195 nations + the European Union. The main absentee is the United States, which despite its signature on April 4, 1993, has never ratified the text, presumably to preserve its economic interests in the industrial and economic exploitation of functional units of heredity. The second absentee is the Holly See (The Vatican City) which is a tiny urban-only state.

The CBD managed to combine its three main objectives within a virtuous circle (United Nations, 1992) (Fig. 17.1). The conservation of biological diversity, the sustainable use of biodiversity and the fair and equitable sharing of benefits arising from the use of GR and traditional knowledge (TK) (United Nations, 1992). In this virtuous circle, the sustainable use of biodiversity is intended to generate economic means to preserve biodiversity for future

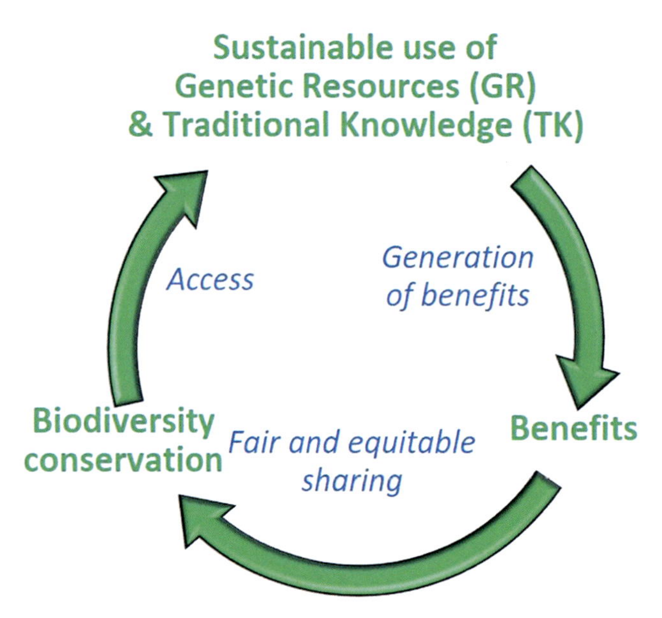

FIGURE 17.1 The virtuous circle of Access and Benefit Sharing (ABS).

generations. The provider nation and its local populations are therefore encouraged to preserve their GR capital, as it is a source of benefit. CBD article 1 specifies: "The objectives of this Convention, to be pursued in accordance with its relevant provisions, are the conservation of biological diversity, the sustainable use of its components and the fair and equitable sharing of the benefits arising out of the utilization of GR, including by appropriate access to GR and by appropriate transfer of relevant technologies, taking into account all rights over those resources and to technologies, and by appropriate funding" (United Nations, 1992).

It is important for understanding the historical background to note that in the very early 90s, following the fall of the Berlin Wall, the global success of finance and liberalism, it was key and trendy to give a monetary value to biodiversity and environmental services (David, 2018; Sullivan, 2013), goods with no financial value being considered as useless by definition. In this context it was decided to share the benefit with the source country of the GR which has gained the sovereignty over its biodiversity. This general trend to value biodiversity will lead several years later to the United Nations Millennium Ecosystem Assessment (Reid et al., 2005) and the Economics and Ecosystems of Biodiversity (TEEB, 2010) which calculated the economic value of each ecosystem service and developed natural capital accounting for policymakers.

At that time many thought that bioprospecting and valorization of natural resources will become a source of considerable economic wealth. Biodiversity-rich countries had the feeling of serving capitalist globalization which was controlling much of the extractive industries and international trade flows of natural resources. This perception was fueled by the belief in continuous "biopiracy" by industrialized countries and their large multinational firms. So after years of East-West political confrontations post World War II, we have witnessed a big turn with a focus on North-South relations after the fall of the Berlin Wall.

The notion of biodiversity resources at this time was restricted to GR. Indeed, during the negotiations, the economic potential of the nascent genetic revolution led to the intellectual fixation on "the functional units of heredity." The negotiators were all convinced that GR (genetic sequences) would generate immeasurable financial incomes. For this reason, the CBD only ruled on DNA and RNA. Plant extracts and natural molecules were completely ignored in the discussions.

And last but not least, an important advance of the CBD is the formal recognition of TK associated with GR through its article 8j. Thus TK can no longer be used or patented by companies without sharing it with the local communities that developed and preserved it. For centuries some biodiversity-rich countries such as Brazil and Peru have been opposed to the collection of seeds or plants from their endemic resources Rubber tree (*Hevea brasiliensis*) and Peruvian tree of fevers (*Cinchona* sp.), respectively, to prevent the establishment of production abroad. In Rio, the movements of the Amazonian tribes federated around the great Chief Kayapo Raoni Metuktire have drawn attention to the recognition of these traditional peoples, their ancestral knowledge, the fragility and destruction of their natural habitats (Conklin & Graham, 1995).

Since the CBD, several more specific international environmental conventions and agreements have been concluded, such as the Cartagena Protocol on Biosafety (May 2000) (United Nations, 2000); the International Treaty on Plant Genetic Resources for Food and Agriculture (FAO, 2001); the Bonn Guidelines, April 2002 (United Nations, 2002); Addis Ababa principles and guidelines for the sustainable use of biodiversity, May 2004 (United Nations, 2004); and the NP adopted on October 29, 2010 (United Nations, 2010).

It can be read between the lines of the CBD agreement an implicit recognition of the patent on GR as benefit sharing is expected by the source countries. Indeed, it is not conceivable to expect significative amounts of money from profit sharing without an essential patent protection of the activity of valorization of these GR in order to ensure a return on investment with protection against competitors.

Problems left unsolved by the CBD

The representatives gathered in Rio were focused and obsessed with genetic material, that is, "any material of plant, microbial, or other origin containing functional units of heredity" and the green and genetics revolution. These resources were supposed to be "green oil" and become a huge source of profit such as the petrol industry. Therefore, the megadiverse countries started to expect distribution of financial profits (benefit sharing) to solve development issues, very often with disregard for the biodiversity conservation which is the main goal of CBD as defined in its article 1 (Fig. 17.2).

Very soon too, everyone realized that the pharmaceutical, cosmetics, and perfumes industries do not use GR, but instead biological resources such as plant extracts and natural substances. From the users' perspective, academics and industries which valorize nongenetic resources, this situation was generating legal uncertainty (Dalton, 2004). In 2002 in Cancún (Mexico), several megadiverse countries set up the group of like-minded megadiverse countries to be a kind of megaphone for the developing countries in the international arena, to defend strong views on the benefits related issues and compliance measures in the GR user's

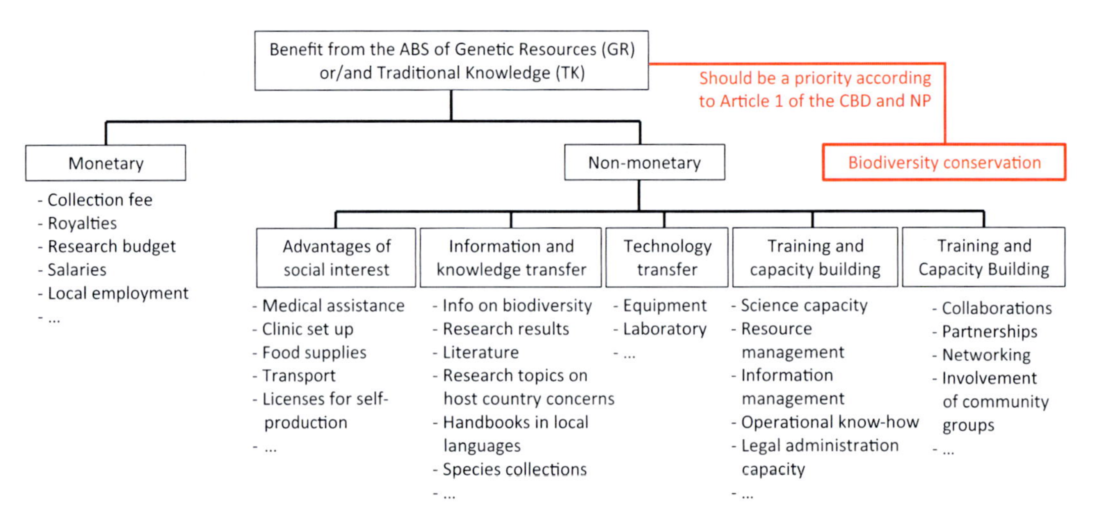

FIGURE 17.2 Monetary and nonmonetary benefits. Quickly the expectations exceeded the objectives of biodiversity conservation.

nations (Pauchard, 2017). Megadiverse countries were disappointed since the sought-after financial benefits expected for so long did not occur (Boyd, 2010). Since nations gained sovereignty over their natural resources upon CBD in Rio, some decided to control access of their nongenetic resources (vegetal material, plant extracts, and natural substances) according to the provisions of CBD article 11.1: "Recognizing the sovereign rights of States over their natural resources upon the CBD, the authority to determine access to GR rests with the national governments and is subject to national legislation." In this context, Andean nations (Bolivia, Columbia, Equator, and Peru) with Decision 391 (Comunidad Andina de Naciones, 1996) and other countries such as Costa Rica (1998), Brazil (2000), India (2002), South Africa (2004)... took legislative measures to control access to their nongenetic resources (Fig. 17.3).

In order to develop ABS, the CBD adopted in 2002 the Bonn guidelines on a voluntary basis (United Nations, 2002). But a clearer and more practical legislative instrument was needed to clarify the application of benefit sharing to research on the biological compounds (Oliva, 2011).

The Nagoya Protocol

Following more than 7 years of preparative discussions, exhausting late-night sessions, numerous parallel deliberations, intense down-to-the-wire negotiations, the Protocol was finally adopted on October 29, 2010, in Nagoya, Japan (United Nations, 2010). NP provided provisions about "access to GR and the fair and equitable sharing of benefits arising from their utilization" which is the third objective of the CBD.

The scope of the CBD was enlarged to nongenetic resources, providing the legal clarity awaited for so long by stakeholders through articles 2c, 2d, and 2e:

Article 2c: "Utilization of GR" means to conduct research and development on the genetic and/or biochemical composition of GR, including through the application of biotechnology as defined in Article 2 of the Convention;

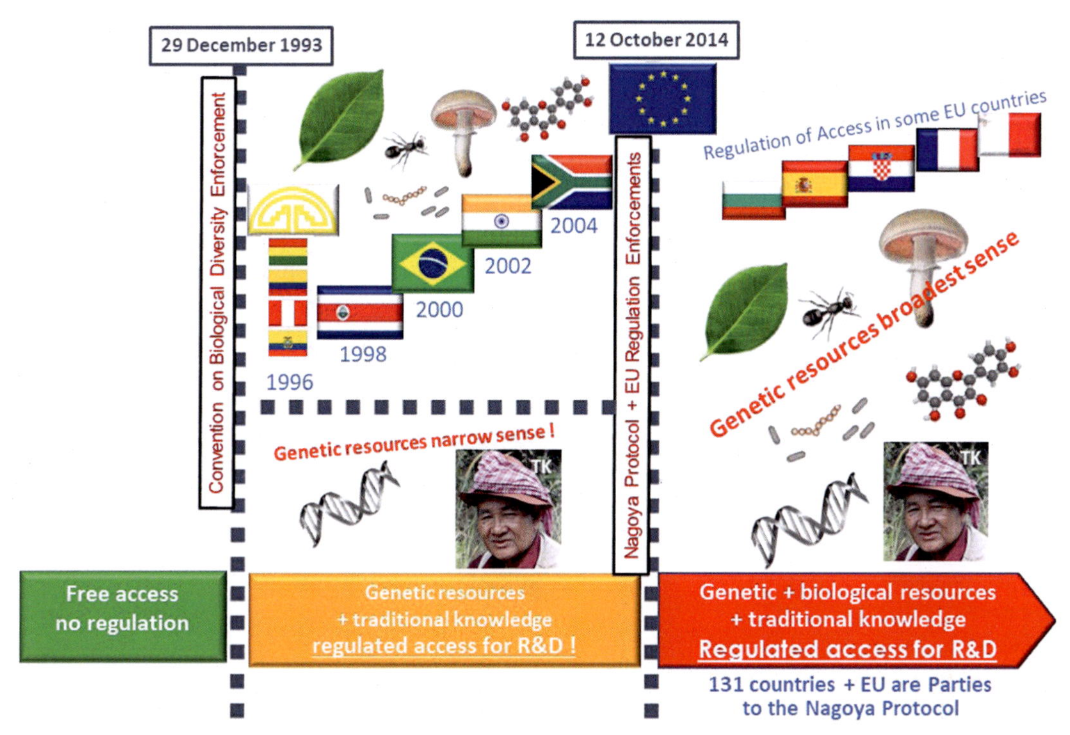

FIGURE 17.3 Evolution of access to GR and TK. *GR*, genetic resources; *TK*, traditional knowledge.

Article 2d: "Biotechnology" as defined in Article 2 of the Convention means any technological application that uses biological systems, living organisms, or derivatives thereof, to make or modify products or processes for specific use;

Article 2e: "Derivative" means a naturally occurring biochemical compound resulting from the genetic expression or metabolism of biological or GR, even if it does not contain functional units of heredity.

The final compromise text was characterized by the Earth Negotiation Bulletin as a "masterpiece in creative ambiguity" (Anonymous, 2010).

The NP sketched the outline of practical GR access regulations and possible benefit sharing. It also proposed several schemes and ABS promotion tools and created the ABS Clearing-House as a comprehensive information exchange center on these subjects (United Nations, 2014). While the Bonn guidelines in 2002 (United Nations, 2002) focused on access, that is "no sharing without access," the NP implements an approach focused on sharing. That is, "no access without sharing" in countries that control access to their GR and TK.

The Protocol entered into force on October 12, 2014, that is, 90 days after ratification by Uruguay, the 50th country to become a Party to PN. The countries that have ratified the Protocol commit to implement their own national regulations (United Nations, 2014). In December, 2021, 131 countries (66%) + the European Union are Parties to the Protocol. Only a minority of 67 countries (34%) on a total of 198 nations are not yet Parties.

The national biodiversity legislations

Neither the CBD nor the NP are juridical instruments that apply directly to users. These international agreements are binding, but only for the signatory countries that have committed themselves, by ratification to protecting biodiversity, facilitating access to their biodiversity, and implementing biodiversity legislations. Nations which are party to the NP are required according to article 15 of the NP to adopt enforcement legislation to ensure that the utilization of GR or TK within their jurisdictions are compliant with the provisions of the biodiversity legislations of the source countries. Therefore countries party to the NP have to control the use of GR within their territory and punish possible violations. But every country has leeway not to establish any access measures and not to request benefit sharing for access to its GR or TK. Nevertheless it must control (compliance) the legal access of foreign resources within its territory (Kamau, 2019).

The core instrument of the Nagoya regime is the establishment of bilateral ABS negotiations and permits between users and providers of GR/TK, rather than standard agreement between the GR/TK providing nations and the potential users (Pauchard, 2017).

Therefore, potential users first have to focus on the national Access regulations of the source country. Then the fact that a source country is Party to the NP only has an effect on due diligence and compliance measures in user's countries like EU member states. A country like Brazil which was, until very recently, not a party to the NP, had nevertheless access laws enforced since July 2000.

According to CBD and NP, access to GR/TK may remain "free" or difficult to get depending on the political strategy of the source country which is sovereign to implement control over access to biodiversity (CBD, Article 15). The CBD and the NP, therefore, have no direct impact on users, unlike national, supranational, and subnational laws that in practice regulate access to GR and ATK (Fig. 17.4).

Supranational laws are regulations elaborated and enforced by several states. The two most important examples of supranational laws are Decision No. 391 of the Andean Community of Nations and the European Regulation 511/2014. The former regulates access to biodiversity in Bolivia, Colombia, Ecuador and Peru and Venezuela since 1996, while the latter applies to all the 27 EU member states.

In contrast, infranational laws are only applicable to some subunits of one nation. For example, the three New Caledonian Provinces are French Overseas Territories but each of the three Provinces (South, North, and Loyalty Islands Provinces) have the legal competence to establish their own specific ABS laws.

The scope of ABS defined by the NP in international consensus is outlined in Fig. 17.5. But the general framework proposed by the NP can be adapted, each state being sovereign over its own GR and TK. For example, a State may decide to require benefit-sharing for conventional industrial commercial materials (commodities) used without any R&D valorization in the sense of article 2c of the NP (e.g., Brazil and India). Potential users, therefore, should be aware of the wide variety of legal situations depending on the source country (cf. "Non stabilized and heterogeneous regulations" section).

Each nation Party to the NP commits to control that the R&D on GR and/or TK conducted on its territory is compliant with the access legislations of the source countries.

International agreements applicable only to the signatory countries

Convention on Biological Diversity

Nagoya Protocol

Regulations applicable directly to users of GR or TK

Infranational regulations

National regulations

Supranational regulations

Documents not legally binding, indicative for users of GR or TK

European Commission Guidance Document C(2020) 8759
Professional associations' best practices guides, sectorial codes of conduct...

FIGURE 17.4 The different ABS legal layers. What is legally binding and applicable to whom. *ABS*, Access and Benefit Sharing.

Apply to

- Genetic resources (GR) subject to R&D = Plants, fungi, animals, microorganisms
- Wild or cultivated/farmed GR
- Traditional knowledge associated with GR
- Collections of GR
- GR/TK accessed where states exercise sovereign rights after national laws enforcement

Does not apply to

- GR or TK accessed before national laws enforcement
- GR from areas beyond national jurisdictions: International waters, space, Antarctic (to date)
- Commodities in the absence of R&D. Nevertheless benefit sharing is required by some national ABS laws even in absence of R&D on commodities (e.g. Brazil, India...) !
- *Ex-situ* collections. But some countries have claims on *ex situ* collections (e.g. Brazil) !
- Human genetic resources
- GR used as tool or reference
- Unintentional access (microorganisms)
- Digital Sequence Information (to date)

FIGURE 17.5 Scope of the ABS regulations according to NP. *ABS*, Access and Benefit Sharing; *NP*, Nagoya Protocol.

Moreover, users from member states of the European Union must declare via electronic portals, that is, exercise Due Diligence when receiving research funding to study GR or TK and/or before commercialization of GR-derived products (European Union, 2014; European Commission 2015, 2020).

Practical advice

Any person or organization planning to conduct R&D activities on GR (e.g., on wild plant samples) must imperatively answer the following questions:

- Is the species protected? At the international level, 28,000 plant species are concerned by the Convention on International Trade in Endangered Species of Wild Fauna and Flora (CITES, 1973). Protection may concern a country, a region, or even a very restricted administrative area.
- Has the owner of the land given his consent for a collection?
- What are the applicable phytosanitary, customs, and even narcotic substances legislations?
- What are the rules of access to GR in force in the source country to date?
- Are these ABS regulations applicable depending on the combination of source country + nature of the GR itself (wild, cultivated...) + intended use + date of access (Fig. 17.6).

All the information about the applicable legislation can be found on the ABSCH website (Access and Benefit Sharing Clearing House). This website (https://absch.cbd.int) is the reference tool for official information and to contact the National Competent Authority (NCA beside the national websites of provider countries). The main goal of the ABS

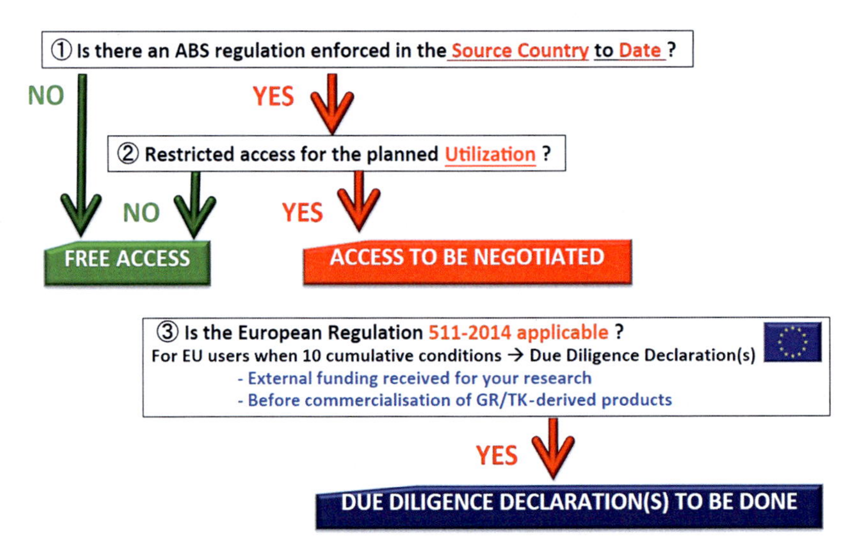

FIGURE 17.6 Key questions to be answered before accessing GR or TK. *GR*, genetic resource. *TK*, Traditional Knowledge.

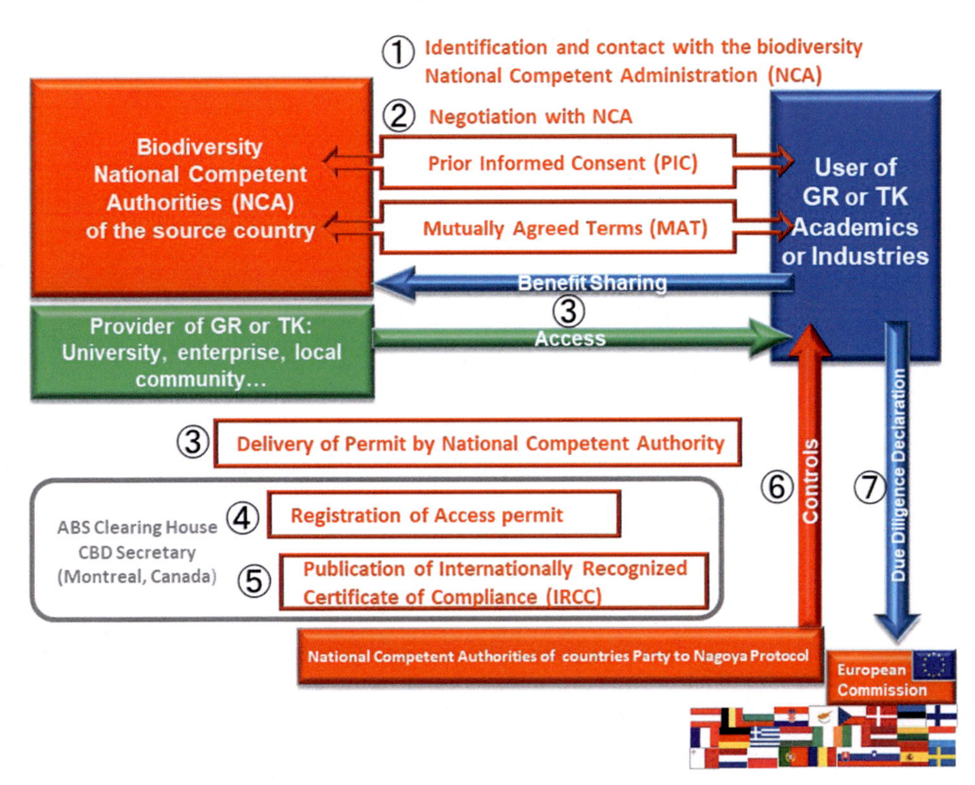

FIGURE 17.7 Administrative stages to be followed. All the steps from the identification of biodiversity National Competent Authorities of the source country to benefit sharing and potential controls.

Clearing House is to assist users in finding information on how to access GR and TK and source countries about the utilization of their GR (United Nations, 2014).

Negotiation for establishing the level of monetary benefits is an asymmetric discussion between the user and the NCA. When the expectations of the providing country are not realistic, as there is no arbitration system, the only possibility for the potential users is to refrain from accessing the GR. This issue has been a hurdle to successful ABS agreements in the past, often because of misunderstandings and too different expectations (Heinrich et al., 2020).

The NCA is the only official national representative entitled to negotiate and authorize access to the national GR on behalf of his/her nations. It is with this person that benefit sharing must be negotiated when required by legislation and after this ABS agreement the NCA will issue a permit. Benefit sharing can be monetary or nonmonetary (sharing of R&D results, training, knowledge transfer, etc.) (Fig. 17.2). The first step is to obtain the access permit, which may require a PIC and MAT depending on the country. The Biodiversity Convention Clearing-House, based in Montreal, Canada, will then register the source country's national permit and issue an IRCC, *i.e.* an "Internationally Recognized Certificate of Compliance". This document attests that the GR or TK has been legally acquired (Fig. 17.7).

An Internationally Recognized Certificate of Compliance (IRCC) presents several advantages: it is evidence of a legal access, a kind of international passport that testifies the user's compliance with ABS regulations. It facilitates the administrative process when the GR is transferred to subsequent users. The source countries can more easily track the use of their national GR. The IRCC enhances transparency and monitors the utilization of GR along the value chain (research, development, innovation, precommercialization, or commercialization). One key point is that some information may stay confidential to preserve the commercial or intellectual property of the users.

Discussion

Non stabilized and heterogeneous regulations

The NP applies to plants, animals, microorganisms, fungi used in R&D, whether academic or private. It should be emphasized that ABS is not about the value of the GR but about the added value generated by the R&D work. In exchange for access, biodiversity source countries receive benefits that may be monetary or nonmonetary (Fig. 17.2). Despite this general framework, some countries have chosen to regulate access beyond the NP's recommendations by requiring sharing for commercial uses (e.g.: commodities or dual sourcing). Commodities are basic raw materials that have been traded for many years, defined by their qualities and properties, and whose geographical origin is indifferent to industrial buyers.

By signing the Rio Convention in 1992, the countries rich in biodiversity accepted the principle of access to their resources and indirectly recognized the notion of patenting, which is indispensable for valuation and therefore for any benefit sharing. These source countries imagined that their biodiversity was going to generate billions of dollars each year via large companies like oil resources. Sharing the benefits of this "green gold" was supposed to bridge the gap between rich Western countries and developing countries of the South (Aubertin & Filloche, 2011). But in practice, the financial flow did not take place in the expected proportions. No comparison is possible between the rents from "black gold" and other subsoil resources or even the logging of tropical timber.

Source countries and users have radically different interests (Fig. 17.8). Everything separates them. GR/TK supplier countries need economic development and public or private research users are willing to share, but in realistic proportion to the activity generated and according to their means. The administration of the provider countries and the user scientists do not have the same expectations and interests, nor the same sense of time and priorities. Structurally, all the ingredients are present for tedious and unbalanced ABS negotiations and law interpretations.

Legislation on access to GR and TK is becoming better known to users, but in practice several points are still subject to divergent interpretations. For example, there is no consensual definition of R&D, the term is not specified in the CBD or the Protocol. The legal situations of isolated natural substances, preliminary screening, and the human microbiome are not the same in every country. In Europe these subjects are generally

Providers	Users
"Poor" countries of the South	"Rich" countries in the North
Biodiversity & TK (traditional knowledge) rich	Technologies, Industries rich
Technology & Industry poor	Biodiversity poor
Governments, local communities, NGOs	Academic or Industry researchers
Control over their GR & TK	Fair & Transparent access
Certainty of benefits sharing after access negotiation	Realistic timelines, clear and fair negotiation
Great expectation of funding	Limited funding opportunities
Interested in economic development, preservation	Interested in innovation
(Biodiversity) legislation often poorly developed	Need of legal security, clarity and transparency

FIGURE 17.8 Structural differences and conflicting interests between source countries and potential users.

considered to be outside the scope of ABS laws. It is important to note that the initial value of the resource is often confused with the value of the final good (the drug or cosmetic product). The value of the final products is in fact essentially immaterial due to years of R&D work by hundreds of specialists in different scientific and technical disciplines.

As NP implementation is being carried out at the national level, each country displays its unique national legislation, policies, or administrative processes (Heinrich et al., 2020). Therefore when we compare ABS regulations of one country with the legislations in neighboring nations, there is no uniformity. Each country presents its own unique mosaic of legal parameters. For some countries only native endemic GR are in the scope of their national ABS regulation. Cultivated/breeding GR are most of the time left out. In some countries like India, the law is not the same for Indian citizens living in India and foreigners or nonresident Indians. In Australia according to the land status, private or public owned, different laws apply...

The large array of regulations and competent authorities dealing with the NP and ABS matters at regional, national, and international levels leads to long and difficult ABS processes (Martins, Cruz, & Vasconcelos, 2020).

All these parameters regarding the type of GR, of users, of use... will have huge practical impacts on Access: free and fast or long, costly negotiations and demanding benefit sharing which may include compulsory collaboration with local R&D institutions and capacity building. All these aspects have a huge impact on the possibility of access to GR and ultimately on the feasibility of the research project (Fig. 17.9).

Ambiguities

Information needed by potential biodiversity users is not always clear, unambiguous even if parties to NP have commited themselves to make information available. It is even

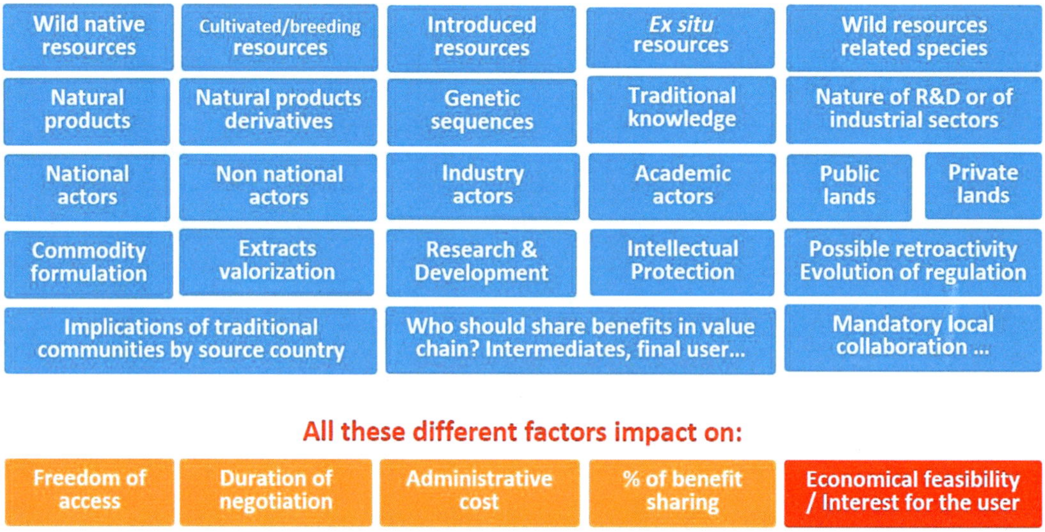

Wild native resources
Cultivated/breeding resources
Introduced resources
Ex situ resources
Wild resources related species
Natural products
Natural products derivatives
Genetic sequences
Traditional knowledge
Nature of R&D or of industrial sectors
National actors
Non national actors
Industry actors
Academic actors
Public lands
Private lands
Commodity formulation
Extracts valorization
Research & Development
Intellectual Protection
Possible retroactivity Evolution of regulation
Implications of traditional communities by source country
Who should share benefits in value chain? Intermediates, final user...
Mandatory local collaboration ...

All these different factors impact on:

Freedom of access
Duration of negotiation
Administrative cost
% of benefit sharing
Economical feasibility / Interest for the user

FIGURE 17.9 Complicated patchwork of administrative parameters specific of each source country.

sometimes unavailable on the ABSCH website or it is difficult to receive an answer from the NCA (United Nations, 2014). The lack of consensus between Parties upon the negotiation of NP has generated a lack of many definitions (e.g., R&D...) and the definitions provided are very broad and sometimes hardly operational (access, utilization...) (Delille, 2016).

The NP does not specify whether national regulations should apply to GR and TK accessed before its enforcement and utilized after. Many biodiversity-rich countries favored an inclusive approach as a huge amount of GR are available in *ex situ* collections. Users in developed countries are opposed to this option since the principle of nonretroactivity must prevail in international laws according to article 28 of the Vienna Convention on the Law of Treaties (United Nations, 1969). As no compromise could be found during the NP discussions, the Protocol left it up to each nation to clarify this ambiguity through their own legislation (Robinson & von Braun, 2019).

Before the ABS regulations access to biodiversity samples was much easier. With the implementation of national ABS regulations, researchers must engage more and more in potentially uncertain administrative procedures prior to any laboratory work. These procedures to obtain—or not—the necessary authorizations before starting research or development on GR can only add further delays to field collections and scientific studies.

The obligations are still not always very clear to users, since their practical implementation is relatively recent and not yet fully stabilized. For example, in France, access to cultivated plants is "free," but the practical definition of "cultivated plant" status is still subject to divergent interpretations. In addition, there are significant disparities between countries. In the European Union, some countries regulate access to GR (Spain, Croatia, France, Malta, and Bulgaria), unlike the majority of other member states where access is "free." Romania is contemplating regulating access to its GR.

Some paradoxical effects

These new constraints sometimes lead to paradoxical effects on conservation by discouraging purely academic researchers from conducting programs in ecology or botanical systematics because of complex bureaucracy (Gilbert, 2010).

Identification of species, systematics, and biological survey are essential prerequisites for biodiversity conservation, management, and sustainable utilization. In some biodiversity-rich countries like India or Brazil, national scientists have been disappointed by the paradoxical side effects of ABS laws. In India the Biological Diversity Act, 2002 was qualified as an imprudent and counterproductive legal measure which classes research as a criminal activity and stifle conservation studies (Pethiyagoda, 2004; Prathapan, Rajan, & Poorani, 2009; Prathapan et al., 2018).

In Brazil, enforced on November 16, 2015 (Brazil, 2015), the new ABS legislation (Law 13.123/2015) was feared to cause a bureaucratic collapse of environmental and biodiversity studies. Indeed this new regulation imposed severe administrative barriers to fundamental and applied research and could reduce international cooperation for Brazilian native species surveys (Alves et al., 2018). This was not new since van Rossmalen, a famous maverick Brazilian primatologist was jailed in 2007 for keeping monkeys in a rehabilitation facility in Manaus without appropriate permits (Check & Hayden, 2007; Jinnah & Jungcurt, 2009). Fortunately, the latest implementing decree of 25 October 2021 has simplified the administrative procedures for academic studies (Brazil, 2021).

Environmental scientists working in biodiversity-rich countries may face demanding and very restrictive administrations which crackdown on research fearing illegal activities.

It is not so easy for offices in charge of ABS to clearly identify commercial and noncommercial researchers (Gilbert, 2010). A balance has to be found between protecting a nation's intellectual property and collecting benefits but not impeding fundamental research. One historical reason is that the lab researchers initially were less involved in ABS exchanges than nongovernmental stakeholders, indigenous representatives, and trade associations (Jinnah & Jungcurt, 2009).

The objective of the ABS legislations should be to find a win-win-win balance between the control of GR and the facilitation of access for the sake of conservation (ten Kate, 2002).

Legal certainty

For academics and industry researchers, legal security is paramount but the nature of the R&D conducted is intrinsically more or less risky (Fig. 17.10). Research organizations need legal certainty in order to be able to invest in research programs that are long, expensive and already scientifically risky in terms of return on investment. Organizations do not wish to see their corporate image tarnished by accusations or bad publicity, press campaign, etc.

Biopiracy accusations: illegitimacy feeling or real legal infractions? For some organizations, biodiversity complexity is the last straw in dissuading researchers from practicing Drug Discovery with natural products, and bioprospection (David et al., 2015).

It is an offense to use GR or TK without having obtained a permit or legal authorization. In France, for example, it is punishable by imprisonment of 1 year and a fine of €150,000 to make use of GR or TK without having the mandatory documents; to fail to seek, keep or

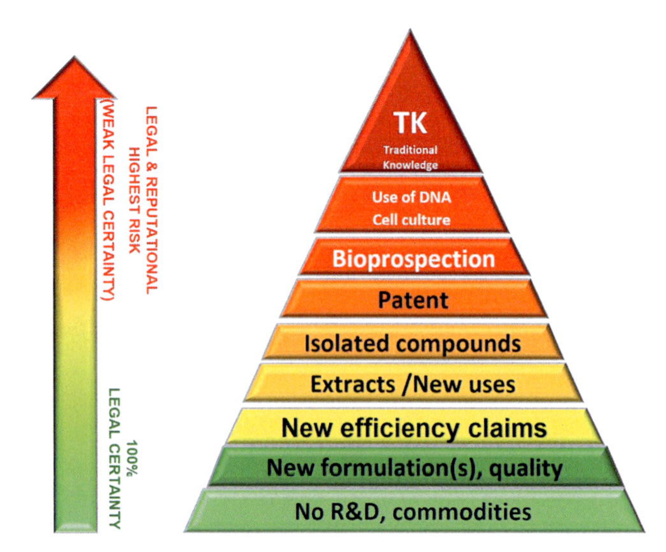

FIGURE 17.10 Level of legal risk according to the nature of activity. Source: *From David, B. (2018), New regulations for accessing plant biodiversity samples, what is ABS?.* Phytochemistry Reviews, 17, *1211–1223.*

transmit to subsequent users the relevant information on ABS for GR and the associated TK. The fine is increased to one million euros when the use of GR or TK has resulted in commercial use. The natural or legal persons guilty of the offenses are also liable, as an additional penalty, to a ban, for a period not exceeding 5 years, on requesting authorization to access GR and associated TK with a view to their commercial use (République Française, 2016). The offender will have to reimburse all external funds received for his research.

Countries party to the NP have to establish a regime of controls and sanctions according to the obligations under the NP. Germany, for example, where access to the national GR is free has established such a regime. The regulatory offenses may be sanctioned with an administrative fine of up to €50,000 (Federal Republic of Germany, 2015). Additionally, it is also an obligation under the transposition of the European Union Regulation No. 511/2014 in the European member states (European Union, 2014; European Commission, 2015, 2020).

Even the contract signed in 1991 between INBio, the Instituto Nacional de Biodiversidad of Costa Rica, and the American pharmaceutical laboratory MERCK, which was considered as a seminal example to follow during the CBD discussions in 1992 was accused of biopiracy. In fact, in such a tropical country, it is always possible to find an ethnic group or individuals who were not consulted for this agreement.

Countries that treat GR as an untapped gold mine with inherent, huge value or uncertain legislations might be overlooked in favor of those nations that seek an efficient, trusting, cooperative, and reasonable ABS agreements (Overmann & Scholz, 2017).

New trends and evolutions

Historically, the CBD and the NP only covered physical samples from territories where countries exercise sovereignty. But soon after the NP came into force, discussions expanded to include intangible resources, territories outside a nation's sovereignty such as the high seas and the Antarctic, and even undesired pathogens.

The curious case of pathogens

The NP has the laudable aim of cementing country sovereignty over its biological resources in order to preserve them. But obviously lethal pathogens as parasites, viruses, and bacteria cannot be regarded as GR to be preserved for human well-being. In fact pathogens were not directly included as GR in the NP text. On the contrary, NP mentions in its foreword the importance to "ensure access to human pathogens for public health preparedness and response purposes pathogens according to International Health Regulations" (WHO, 2005). Contrary to the spirit of the NP, some countries are requiring benefit sharing for access to pathogens samples. This position is explained by Indonesia's refusal in 2007 to share information on the lethal H_1N_1 virus, claiming that Indonesians would have to pay a high price to access the vaccine. Although this view appears understandable it could stifle the essential collaboration required to fight against human pathogens that we need to eradicate and not preserve (Cueni, 2021). In reaction to this situation, the WHO created in 2011, the Pandemic Influenza Preparedness framework, in order to promote international collaboration and access to vaccines (WHO, 2011). But some argued that this WHO framework applies only to influenza viruses and enshrined countries' sovereignty over their pathogens. No penalties are imposed to potential source countries for refusing to cooperate with the WHO (Kozlov, 2021).

Today, many countries which have implemented the NP in their national biodiversity legislation require negotiations and benefit sharing to access samples of pathogens. Fortunately, China shared the digital sequence of SARS-CoV-2 in almost real time and for free. This kickstarted the development of multiple vaccines in record time for the benefit of everyone. We cannot imagine the death toll if bilateral ABS negotiations with China had delayed vaccine research and development. Unfortunately, it is highly likely that a country could invoke NP to withhold parasites or pathogens samples or other lifesaving data until it receives monetary or nonmonetary benefits (Cueni, 2021).

Digital sequence information

Genetic digital information has become unavoidable nowadays for systematics and modern life sciences research. The historical and current model is "open access." An isolated individual sequence is not so useful since DSI requires a big data approach as open and as global as possible. Open access and freedom to use DSI create value for everyone, this being already aligned with the objectives of CBD and the NP.

Some source countries consider this openness as lost opportunities for benefit sharing.

Biology has become digital and DSI can be instantaneously and globally transferred. All the genomic information generated in research labs has the potential to be shared with anyone in real time. The difficult topic of digital biology had been discussed during Nagoya negotiations but was left out in order to reach an agreement (Phillips, Smyth, & de Beer, 2018).

Stifling the exchange and use of DSI would impact inventory, conservation and protection projects, and sustainable use of biodiversity which are the objectives of the CBD and the NP.

It is therefore essential that DSI, which are in the public domain, continues to remain in open access.

GR were initially defined as physical material and did not include immaterial information. The CBD and NP deal with regulation of the physical transfer of tangible genetic or biological material from a provider country to a user, pursuant to an ABS agreement. However, new genetics technologies fundamentally change that paradigm. Nowadays, the genome of a species can be easily sequenced anywhere and via databases valorized everywhere (Manheim, 2016).

The issue of DSI has been a controversial topic of discussion for years. Many options and scenarios are been discussed but no consensual agreement is to be found easily in the near future (Morgera, Switzer, & Geelhoed, 2020).

Inclusion of DSI in the scope of the NP and their definition, which is not yet consensual, will have implications on benefit-sharing options.

Even if the term DSI is widely used in CBD experts' discussions, it is not clearly defined as the distinction between "data" and "information" is still under discussion noting that the latter could imply more processing than the former. According to DSI experts there is no clear boundary between data and information (United Nations, 2020). The scope of DSI could range from nucleic acid or protein sequences to metabolites and macromolecules, and even all contextual information. Precise definition of DSI and the scope are expected to be clarified during the next CBD meeting in perspective of the 15th Conference of Parties in Kunming China postponed to 2022 due to Covid-19 pandemic.

Treating information as a resource derivative within the ABS transactions could generate unnecessary and inefficient burdens, a risk for valuing information and charge, taxes (Lawson, Humphries, & Rourke, 2019).

The future legislation on DSI will have significant impacts on how digital information is accessed, stored, shared, and used. Finally, it is the organization and structuring of research that will be affected by the extension of the scope of ABS (Laird et al., 2020).

Biodiversity beyond national jurisdiction

Biodiversity beyond national jurisdiction (BBNJ) is present in vast areas covering 65% of the world's oceans and also in polar zones. Until now these polar and marine environments are not in the scope of CBD and NP.

An international legally binding instrument on the conservation and sustainable use of marine BBNJ is under discussion since 2017 under the United Nations Convention on the Law of the Sea.

Since 2002 there have been international discussions on the introduction of a special ABS regime for the Antarctic (Katharina, 2020). Some believe that this issue of access to the Antarctic GR should be dealt with under the Antarctic Treaty (Conference on Antarctica, 1959) which is the most appropriate legal framework (Heinrich et al., 2020).

Conclusion

Biodiversity is a valuable common capital that provides ecosystem services essential to life and to our economies (Reid et al., 2005). Today more than ever, this biodiversity is under pressure and declining at unprecedented rates. This in turn is putting the livelihoods of billions of people around the world at risk (Potočnik, 2014). There is no doubt that we need to curb the unprecedented rate of biodiversity decline in order to preserve human well-being (Naeem, Chazdon, Duffy, Prager, & Worm, 2016). The recent emergence of the Covid-19 pandemic shows us that environmental and human health are closely linked (IUCN, 2020).

In the context of the alarming global decline in biodiversity (Butchart et al., 2010) benefit sharing can contribute to preserve natural areas, reduce artificialization of natural ecosystems, and develop awareness on these crucial environmental issues.

The new regulations on benefit sharing should therefore not be seen as a tax but as a voluntary commitment and an active contribution to the preservation of GR and GR-associated TK for future generations. This approach should harmonize the complementary strengths of every stakeholder in the process of economic development, knowledge production, at both local and global levels (Oguamanam, 2018).

Researchers and their structures working on bio-based projects (companies, public, and industry professional associations...) are in alignment with this. They have taken into account the ABS measures in their operations and activities and are compliant with national laws. For several years, even before enforcement of NP, they have proactively adopted corporate policies, codes of conduct, and guidelines (BIO, 2005; CETAF, 2018; EFPIA, 2013; ICC, 2013; IFPMA, 2013; Phillips, 2016).

The clarification and stabilization of access rules in each source country are creating on the ground opportunities for biodiversity that should finally achieve the virtuous circle of the CBD (Buck & Hamilton, 2011). ABS is now able to open wide-ranging opportunities for knowledge production on a sustainable scale across different industrial sectors (Oguamanam, 2018).

The implementation of the NP and respect of access laws will be fostered by efficient biodiversity administration in the source countries. Avilés-Polanco, Jefferson, Almendarez-Hernández, and Beltrán-Morales (2019) mentioned a larger number of IRCC associated with higher quality institutions. Less burdensome regulations and efficient legal frameworks favor commercial ABS agreements. Studies such as Sirakaya's analysis show that stakeholders are able to cooperate under the law when the right criteria are discussed (Sirakaya, De Brucker, & Vanagt, 2020).

Despite the objectives of biodiversity conservation clearly expressed in article 1 of CBD and of NP, conservation has faded from the ABS contracts and is nowadays a marginal concern at best (Laird et al., 2020). Unfortunately, after more than 25 years of developing ABS policies, tens of millions of dollars spent in meetings and discussions, the delivery in biodiversity preservation, support for technological development, and other benefit sharing is extremely low. Nevertheless, the main benefit is the environmental awareness and the scientific knowledge acquired through the ABS contracts (Richerzhagen & Holm-Mueller, 2005; Zheng, 2019). The huge financial means and time spent nursing the hope that indirect incentives from GR users through ABS will preserve biodiversity need to be revisited.

Benefits expected from the commercial use of GR have largely been exaggerated and not yet realized and cannot alone reach the goals of biodiversity preservation in addition to the global development agendas (Prathapan et al., 2018).

Moreover, there is a continuous decline of interest for research on GR, the so-called "green gold" of the 1980s (Chassagne, Cabanac, Hubert, David, & Marti, 2019; David et al., 2015; Firn, 2003; Pauchard, 2017).

ABS alone will not be enough to preserve biodiversity and to ensure economic development.

Unfortunately, this measure alone is not up to the challenge of biodiversity loss. The exploitation of underground natural resources (petrol, minerals, oils sands, and shale gas...) the development of industrial agriculture, the deforestation and logging of tropical timbers, the artificialization of the planet's soils, global warming... represent more significant direct levers than ABS. Even the economics of tourism could/should help to preserve the environment. It is true that promoting education and development can preserve biodiversity as providing productive opportunities is the best way to keep the poorest people from clearing fragile forests and ecosystems (Barrett & Lybbert, 2000). Contractual mechanism as payments for ecosystem services could be useful. In this scheme users of ecosystem services (such as the availability of drinking water) derived from environmentally sound activities reward the providers economically (Pavoni, 2013).

Legal certainty is a key issue. Researchers have become very reluctant to take ABS legal risk or be accused of biopiracy as sometimes even without having made any infringement, the users of GR can be accused of biopiracy by nongovernmental organizations, which tarnishes the image of the institutions and researchers involved (Bourdy, Aubertin, Jullian, & Deharo, 2017).

Biodiversity-rich countries could use their GR locally and develop quality phytomedicines, a green economy rather than expecting and relying on benefit sharing by multinationals to alleviate poverty, to raise living standards, and to develop public health (Hamilton et al., 2003). Academic research must be allowed to progress with minimal restraint in Southern nations, since economy and techniques are only emerging (Crouch, Douwes, Wolfson, Smith, & Edwards, 2008).

There are three key remaining issues to address to make the ABS regime more functional:

Operable legislations in both user countries and provider countries, fast line contractual mechanisms for ABS, and clarity on pending questions in grey zone arenas (Tvedt, 2014; Michiels et al., 2022).

It is now important to enforce the ABS legislations, to raise awareness among all concerned stakeholders: academic and industry people, GR collection managers, administration representatives, customs officers and to provide the long-needed legal clarity in win-win-win scenarios for all: source countries, researchers and users, people benefiting from valorization of GR, and of course the endangered biodiversity.

Acknowledgments

I would like to thank Ms Claire Mazars (Head of Nagoya/Access and Benefit Sharing, Pierre Fabre Green Mission) for fruitful exchanges.

References

Alves, R. J. V., Weksler, M., Oliveira, J. A., Buckup, P. A., Pombal, J. P., Santana, H. R. G., ... Grelle, C. E. V. (2018). Brazilian legislation on genetic heritage harms biodiversity convention goals and threatens basic biology research and education. *Anais Da Academia Brasileira de Ciências, 90*, 1279–1284. Available from https://doi.org/10.1590/0001-3765201820180460.

Anonymous. (2010). Summary of the tenth conference of the parties to the convention on biological diversity 18–29. *Earth Negotiations Bulletin, 9*, 1–30. Available from https://enb.iisd.org/download/pdf/enb09544e.pdf.

Aubertin, C., & Filloche, J. (2011). The Nagoya Protocol on the use of genetic resources: one embodiment of an endless discussion. *Sustentabilidade Em Debate, 2*, 51–63. Available from https://doi.org/10.18472/SustDeb.v2n1.2011.3906.

Avilés-Polanco, G., Jefferson, D. J., Almendarez-Hernández, M. A., & Beltrán-Morales, L. F. (2019). Factors that explain the utilization of the Nagoya Protocol framework for access and benefit sharing. *Sustainability (Switzerland), 11*(20), 5550. Available from https://doi.org/10.3390/su11205550.

Barrett, C. B., & Lybbert, T. J. (2000). Is bioprospecting a viable strategy for conserving tropical ecosystems? *Ecological Economics, 34*, 293–300. Available from https://doi.org/10.1016/S0921-8009(00)00188-9.

Bourdy, G., Aubertin, C., Jullian, V., & Deharo, E. (2017). Quassia "biopiracy" case and the Nagoya Protocol: A researcher's perspective. *Journal of Ethnopharmacology, 206*, 290–297. Available from https://doi.org/10.1016/j.jep.2017.05.030.

Boyd, R. (2010). Snaring the wealth: Can negotiators reach a uniform position on patenting the world's genetic resources. *Scientific American, 13*.

Brazil (2015). Law 13.123/2015. Available at http://www.planalto.gov.br/ccivil_03/_Ato2015-2018/2015/Lei/L13123.htm. (Accessed 13 December 2021).

Brazil (2021). Decree N° 10.844 from October 25, 2021. Available at https://www.in.gov.br/en/web/dou/-/decreto-n-10.844-de-25-de-outubro-de-2021-354622936. (Accessed 13 December 2021).

Buck, M., & Hamilton, C. (2011). The Nagoya Protocol on access to genetic resources and the fair and equitable sharing of benefits arising from their utilization to the convention on biological diversity. *Review of European Community and International Environmental Law, 20*, 47–61. Available from https://doi.org/10.1111/j.1467-9388.2011.00703.x.

Butchart, S. H. M., Walpole, M., Collen, B., Van Strien, A., Scharlemann, J. P. W., Almond, R. E. A., ... Watson, R. (2010). Global biodiversity: Indicators of recent declines. *Science (New York, N.Y.), 328*, 1164–1168. Available from https://doi.org/10.1126/science.1187512.

BIO. (2005). *Biotechnology innovation organization, guidelines for BIO members engaging in bioprospecting*. Retrieved from https://archive.bio.org/articles/guidelines-bio-members-engaging-bioprospecting. (Accessed 13 December 2021).

CETAF. (2018). *Consortium of European Taxonomic Facilities, Consortium of European Taxonomic Facilities: Best practice and code of conduct*. Retrieved from https://cetaf.org/sites/default/files/documents/cetaf_abs_code_of_conduct_all_annexes.pdf. (Accessed 13 December 2021).

CITES. (1973). *Convention on international trade in endangered species of wild fauna and flora*. Retrieved from https://cites.org/sites/default/files/eng/disc/CITES-Convention-EN.pdf.

Carson, R. (1962). *The silent spring (Fortieth Anniversary Edition)*. Boston: Mariner Book.

Chassagne, F., Cabanac, G., Hubert, G., David, B., & Marti, G. (2019). The landscape of natural product diversity and their pharmacological relevance from a focus on the Dictionary of Natural Products®. *Phytochemistry Reviews, 18*, 601–622. Available from https://doi.org/10.1007/s11101-019-09606-2.

Check, E., & Hayden, T. (2007). Strike threat over jailed primatologist. *Nature, 448*, 634. Available from https://doi.org/10.1038/448634a.

Collective. (1971). A Message to our 3.5 billion neighbours on planet earth, The Unesco Courier (4–5). https://en.unesco.org/courier/july-1971.

Comunidad Andina de Naciones. (1996). *Régimen común sobre acceso a los recursos genéticos*. Retrieved from https://www.ins.gov.co/Normatividad/Decisiones/DECISION 391 DE 1996 DE LA COMUNIDAD ANDINA DE NACIONES.pdf. (Accessed 13 December 2021).

Conference on Antartica. (1959). *The Antarctic treaty*. Retrieved from https://documents.ats.aq/ats/treaty_original.pdf. (Accessed 13 December 2021).

Conklin, B. A., & Graham, L. R. (1995). The shifting middle ground: Amazonian Indians and eco-politics. *American Anthropologist, 97*, 695–710. Available from https://doi.org/10.1525/aa.1995.97.4.02a00120.

Crouch, N. R., Douwes, E., Wolfson, M. M., Smith, G. F., & Edwards, T. J. (2008). South Africa's bioprospecting, access and benefit-sharing legislation: Current realities, future complications, and a proposed alternative. *South African Journal of Science*, *104*, 355–366.

Cueni, T. (2021). *This international agreement could lead to pandemics worse than COVID-19, International Business Times*. Retrieved from https://www.ibtimes.com/international-agreement-could-lead-pandemics-worse-covid-19-3178648. (Accessed 13 December 2021).

Dalton, R. (2004). Bioprospectors hunt for fair share of profits. *Nature*, *427*, 576. Available from https://doi.org/10.1038/427576a.

David, B. (2018). New regulations for accessing plant biodiversity samples, what is ABS? *Phytochemistry Reviews*, *17*, 1211–1223. Available from https://doi.org/10.1007/s11101-018-9573-1.

David, B., Wolfender, J. L., & Dias, D. A. (2015). The pharmaceutical industry and natural products: Historical status and new trends. *Phytochemistry Reviews*, *14*, 299–315. Available from https://doi.org/10.1007/s11101-014-9367-z.

Delille, T. (2016). Sharing benefits within a clear and sound legal framework. *Zeitschrift Für Stoffrecht*, *13*, 309–316. Available from https://stoffr.lexxion.eu/article/STOFFR/2016/6/4.

EFPIA. (2013). *Work on model contractual clauses, best practices etc. In view of the entry into force and implementation of the Nagoya Protocol*. Retrieved from https://www.cbd.int/abs/submissions/icnp-3/EU-IFPMA-EFIA-Letter.pdf. (Accessed 13 December 2021).

European Commission. (2015). *Commission implementing Regulation (EU) 2015/1866 of 13 October 2015 laying down detailed rules for the implementation of Regulation (EU) No 511/2014 of the European Parliament and of the Council as regards the register of collections, monitoring user compliance and best practices*. Retrieved from http://eur-lex.europa.eu/legal-content/EN/TXT/PDF/?uri = CELEX:32015R1866&from = EN. (Accessed 13 December 2021).

European Commission. (2020). *Guidance document on the scope of application and core obligations of Regulation (EU) No 511/2014 of the European Parliament and of the Council on the compliance measures for users from the Nagoya Protocol on Access to Genetic Resources and the Fair and Equitable Sharing of Benefits Arising from their Utilisation in the Union*. Retrieved from https://ec.europa.eu/transparency/regdoc/rep/3/2020/EN/C-2020-8759-F1-EN-MAIN-PART-1.PDF. (Accessed 13 December 2021).

European Union. (2014). *Regulation (EU) No. 511/2014 of the European Parliament and of the Council of 16 April 2014 on compliance measures for users from the Nagoya Protocol on Access to Genetic Resources and the Fair and Equitable Sharing of Benefits Arising from their Utilization in the Union*. Retrieved from http://eur-lex.europa.eu/legal-content/EN/TXT/PDF/?uri = CELEX:32014R0511&from = EN. (Accessed 13 December 2021).

FAO. (2001). *International treaty on plant genetic resources for food and agriculture*. Retrieved from http://www.fao.org/3/i0510e/i0510e.pdf.

Federal Republic of Germany. (2015). *Bundestag, Gesetz zur Umsetzung der Verpflichtungennach dem Nagoya-Protokoll, zur Durchführung der Verordnung (EU) Nr. 511/2014 und zur Änderung des Patentgesetzes sowie zur Änderung des Umweltauditgesetzes, Bundesgesetzblatt Jahrgang. 47*. Retrieved from https://absch.cbd.int/api/v2013/documents/CD747FBB-D813-EC08-9EF3-AE970436D156/attachments/DE_Umsetzungsgesetz.pdf.

Firn, R. D. (2003). Bioprospecting - Why is it so unrewarding? *Biodiversity and Conservation*, *12*, 207–216. Available from https://doi.org/10.1023/A:1021928209813.

Gilbert, N. (2010). Biodiversity law could stymie research. *Nature*, *463*, 598. Available from https://doi.org/10.1038/463598a.

Hamilton, A. C., Pei, S., Kessy, J., Khan, A. A., Lagos-Witte, S., & Shinwari, Z. K. (2003). *The purposes and teaching of applied ethnobotany*. Godalming, UK: WWF. Available from https://unesdoc.unesco.org/ark:/48223/pf0000145847.

Harris, J. A. (1916). The variable desert. *The Scientific Monthly*, *3*, 41–50. Available from http://www.jstor.org/stable/6182.

Heinrich, M., Scotti, F., Andrade-Cetto, A., Berger-Gonzalez, M., Echeverría, J., Friso, F., ... Spadafora, R. (2020). Access and benefit sharing under the Nagoya Protocol—Quo Vadis? Six Latin American case studies assessing opportunities and risk. *Frontiers in Pharmacology*, *11*, 765. Available from https://doi.org/10.3389/fphar.2020.00765.

Handl, G. (2012). *Declaration of the United Nations conference on the human environment (Stockholm declaration), 1972 and the Rio Declaration on environment and development, 1992. United Nations Audiovisual Library of International Law*. Retrieved from https://legal.un.org/avl/pdf/ha/dunche/dunche_e.pdf. (Accessed 13 December 2021).

ICC. (2013). *Views and information on the development, updating and use of sectoral and cross-sectoral model contractual clauses, voluntary codes of conduct, guidelines and best practices and/or standards (Article 19 and 20) in response to CBD request*. Retrieved from https://www.cbd.int/abs/submissions/icnp-3/ICC-letter.docx. (Accessed 13 December 2021).

IFPMA. (2013). *International Federation of Pharmaceutical Manufacturers & Associations, Guidelines for IFPMA Members on access to genetic resources and equitable sharing of benefits arising out of their utilization*. Retrieved from https://www.ifpma.org/wp-content/uploads/2016/05/IFPMA-Guidelines-on-ABS.pdf. (Accessed 13 December 2021).

IUCN. (2020). *International Union for Conservation of Nature and Natural Resour, International Union for Conservation of Nature annual report 2019*. Retrieved from https://portals.iucn.org/library/sites/library/files/documents/2020-025-En.pdf. (Accessed 13 December 2021).

IUCN. (n.d.). *International Union for Conservation of Nature, IUCN - A brief history*. Retrieved from https://www.iucn.org/about/iucn-a-brief-history. (Accessed 13 December 2021).

Jinnah, S., & Jungcurt, S. (2009). Global biological resources: Could access requirements stifle your research? *Science (New York, N.Y.), 323*, 464–465. Available from https://doi.org/10.1126/science.1167234.

Kamau, E. C. (2019). Implementation of the Nagoya Protocol - Fulfilling new obligations among emerging issues. *Bundesamt Für Naturschutz - Skripten, 564*. Available from https://www.bfn.de/fileadmin/BfN/service/Dokumente/skripten/Skript564.pdf.

Kate, K. T. (2002). Science and the convention on biological diversity. *Science (New York, N.Y.), 295*, 2371–2372. Available from https://doi.org/10.1126/science.1070725.

Katharina, H. (2020). Biological prospecting in Antarctica – A solution-based approach to regulating the collection and use of Antarctic Marine Biodiversity by taking the BBNJ process into account. *The Yearbook of Polar Law Online, 12*, 41–60. Available from https://doi.org/10.1163/22116427_012010005.

Kevles, D. J. (1994). Ananda Chakrabarty wins a patent: Biotechnology, law, and society, 1972-1980. *Historical Studies in the Physical and Biological Sciences, 25*, 111–135. Available from https://doi.org/10.2307/27757736.

Kozlov, M. (2021). *Science with borders: Researchers navigate red tape, The Scientist*. Retrieved from https://www.the-scientist.com/careers/science-with-borders-researchers-navigate-red-tape-68443. (Accessed 13 December 2021).

Laird, S., Wynberg, R., Rourke, M., Humphries, F., Muller, M. R., & Lawson, C. (2020). Rethink the expansion of access and benefit sharing. *Science (New York, N.Y.), 367*, 1200–1202. Available from https://doi.org/10.1126/science.aba9609.

Lawson, C., Humphries, F., & Rourke, M. (2019). The future of information under the CBD, Nagoya Protocol, Plant Treaty, and PIP Framework. *Journal of World Intellectual Property, 22*, 103–119. Available from https://doi.org/10.1111/jwip.12118.

Lovejoy, T. E. (1980). Foreword V-IX. In M. E. Soulé, & B. A. Wilcox (Eds.), *Conservation biology. An evolutionary-ecological perspective* (p. 395). Sunderland (MA, USA): Sinauer Associates Inc.

Manheim, B. S. (2016). Regulation of synthetic biology under the Nagoya Protocol. *Nature Biotechnology, 34*, 1104–1105. Available from https://doi.org/10.1038/nbt.3716.

Martins, J., Cruz, D., & Vasconcelos, V. (2020). The Nagoya Protocol and its implications on the EU atlantic area countries. *Journal of Marine Science and Engineering, 8*, 1–20. Available from https://doi.org/10.3390/jmse8020092.

Meadows, D. H., Meadows, D. L., Randers, J., & Behrens, W. W., III (1972). *The limits to growth*. Potomac Associates Book. Available from https://www.clubofrome.org/publication/the-limits-to-growth/.

Michiels, F., Feiter, U., Paquin-Jaloux, S., Jungmann D., Braun A., Sayoc, M. A. P., Armengol, R., Wyss, M. & David, B. (2022). Facing the harsh reality of Access and Benefit Sharing (ABS) legislation: An industry perspective. *Sustainability, 14*, 277. Available from https://doi.org/10.3390/su14010277.

Morgera, E., Switzer, S., & Geelhoed, M. (2020). *Study for the European Commission on 'Possible Ways to Address Digital Sequence Information - Legal and Policy Aspects, Strathclyde Centre for Environmental Law and Governance*. Retrieved from https://ec.europa.eu/environment/nature/biodiversity/international/abs/pdf/Final_study_legal_and_policy_aspects.pdf. (Accessed 13 December 2021).

Naeem, S., Chazdon, R., Duffy, J. E., Prager, C., & Worm, B. (2016). Biodiversity and human well-being: An essential link for sustainable development. *Proceedings of the Royal Society B, 283*, 20162091. Available from https://doi.org/10.1098/rspb.2016.2091.

Oguamanam, C. (Ed.), (2018). *Genetic resources, justice and reconciliation: Canada and global access and benefit sharing*. Cambridge: Cambridge University Press. Available from https://doi.org/10.1017/9781108557122.

Oliva, M. J. (2011). Sharing the benefits of biodiversity: A new international protocol and its implications for research and development. *Planta Medica, 77*, 1221–1227. Available from https://doi.org/10.1055/s-0031-1279978.

Overbye, D. (2018). *Apollo 8's earthrise: The shot seen round the world. The New York Times 18–12-2018*. Retrieved from https://www.nytimes.com/2018/12/21/science/earthrise-moon-apollo-nasa.html.

Overmann, J., & Scholz, A. H. (2017). Microbiological research under the Nagoya Protocol: Facts and fiction. *Trends in Microbiology, 25*, 85–88. Available from https://doi.org/10.1016/j.tim.2016.11.001.

Pauchard, N. (2017). Access and benefit sharing under the convention on biological diversity and its protocol: What can some numbers tell us about the effectiveness of the regulatory regime? *Resources*, *6*, 11. Available from https://doi.org/10.3390/resources6010011.

Pavoni, R. (2013). Channelling investment into biodiversity conservation: ABS and PES schemes. *Harnessing foreign investment to promote environmental protection incentives and safeguards* (pp. 206–227). Italy: Cambridge University Press. Available from https://doi.org/10.1017/CBO9781139344289.011.

Pethiyagoda, R. (2004). Biodiversity law has had some unintended effects. *Nature*, *429*, 129. Available from https://doi.org/10.1038/429129a.

Phillips, P. W. B., Smyth, S. J., & de Beer, J. (2018). Access and benefit-sharing in the age of digital biology. In C. Oguamanam (Ed.), *Genetic resources, justice and reconciliation: canada and global access and benefit sharing* (pp. 181–195). Cambridge: Cambridge University Press. Available from https://doi.org/10.1017/9781108557122.011.

Phillips, F.-K. (2016). Sustainable bio-based supply chains in light of the Nagoya Protocol. In W. Tate, et al. (Eds.), *Implementing triple bottom line sustainability into global supply chains*. London: Greenleaf Publishing. Available from https://papers.ssrn.com/sol3/papers.cfm?abstract_id = 3033039.

Potočnik, J. (2014). Foreword. In B. Coolsaet, et al. (Eds.), *Implementing the Nagoya Protocol: Comparing access and benefit-sharing regimes in Europe*. Leiden, Boston: Brill Nijhoff.

Prathapan, K. D., Rajan, P. D., & Poorani, J. (2009). Protectionism and natural history research in India. *Current Science*, *97*, 1411–1412. Available from http://www.ias.ac.in/currsci/nov252009/1411.pdf.

Prathapan, K. D., Pethiyagoda, R., Bawa, K. S., Raven, P. H., Rajan, P. D., Acosta, L. E., ... Hughes, L. E. (2018). When the cure kills—CBD limits biodiversity research. *Science (New York, N.Y.)*, *360*, 1405–1406. Available from https://doi.org/10.1126/science.aat9844.

Reid, W. V., Mooney, H. A., Cropper, A., Capistrano, D., Carpenter, S. R., Chopra, K., ... Zurek, M. B. (2005). *Ecosystems and human well-being: Synthesis, Millennium Ecosystem Assessment*. Washington: Island Press. Available from https://www.millenniumassessment.org/documents/document.356.aspx.pdf.

Richerzhagen, C., & Holm-Mueller, K. (2005). The effectiveness of access and benefit sharing in Costa Rica: Implications for national and international regimes. *Ecological Economics*, *53*, 445–460. Available from https://doi.org/10.1016/j.ecolecon.2004.06.031.

Robinson, D. F., & von Braun, J. (2019). In C. Correa, & X. Seuba (Eds.), *New challenges for the Nagoya Protocol: Diverging implementation regimes for access and benefit-sharing* (pp. 377–403). Singapore: Springer. Available from https://doi.org/10.1007/978-981-13-2856-5_16.

République Française. (2016). Loi no. 2016-1087 du 8 août 2016 pour la reconquête de la biodiversité, de la nature et des paysages. *Journal Officiel de La République Française*. Available from https://www.legifrance.gouv.fr/download/file/JO53V_65dij67lRvMQoMmHtIAj0JcaOEDqWIfclQeWk = /JOE_TEXTE.

Sirakaya, A., De Brucker, K., & Vanagt, T. (2020). Designing regulatory frameworks for access to genetic resources: A multi-stakeholder multi-criteria approach. *Frontiers in Genetics*, *11*, 549836. Available from https://doi.org/10.3389/fgene.2020.549836.

Sullivan, S. (2013). Banking nature? The spectacular financialisation of environmental conservation. *Antipode*, *45*, 198–217. Available from https://doi.org/10.1111/j.1467-8330.2012.00989.x.

Tvedt, M. W. (2014). Into ABS implementation: Challenges and opportunities for the Nagoya Protocol. *BIORES*, *8*, 8. Available from https://www1.sun.ac.za/awei/sites/default/files/FNI-2014-Tvedt-IntoABSImplementation.pdf.

TEEB. (2010). *The economics of ecosystems and biodiversity: Mainstreaming the economics of nature: A synthesis of the approach, conclusions and recommendations of TEEB*. Retrieved from http://www.teebweb.org/wp-content/uploads/Study and Reports/Reports/Synthesis report/TEEB Synthesis Report 2010.pdf. (Accessed 13 December 2021).

US Congress. (1864) *Yosemite Valley Grant Act, First Session of the 38th Congress, Congressional Globe*. Retrieved from https://memory.loc.gov/cgi-bin/query/r?ammem/AMALL:@field(NUMBER + @band(amrvl + vl499). (Accessed 13 December 2021).

United Nations. (1972). *Report of the United Nations conference on the human environment* (5–16). Retrieved from https://digitallibrary.un.org/record/523249/files/A_CONF.48_14_Rev.1-EN.pdf.

United Nations. (2002). *Bonn guidelines on access to genetic resources and fair and equitable sharing of the benefits arising out of their utilization*. Retrieved from https://www.cbd.int/doc/publications/cbd-bonn-gdls-en.pdf. (Accessed 13 December 2021).

United Nations. (2004). *Addis Ababa principles and guidelines for the sustainable use of biodiversity*. Retrieved from https://www.unep-aewa.org/sites/default/files/document/tc_inf7_5_addis_ababa_principles_gudelines_0.pdf. (Accessed 13 December 2021).

United Nations. (2010). *Nagoya Protocol on access to genetic resources and the fair and equitable sharing of benefits arising from their utilization to the convention on biological diversity*. Retrieved from https://www.cbd.int/abs/doc/protocol/nagoya-protocol-en.pdf. (Accessed 13 December 2021).

United Nations. (2014). *ABSCH - The access and benefit sharing Clearing-House*. Retrieved from https://absch.cbd.int/about. (Accessed 13 December 2021).

United Nations. (2020). *Report of the ad hoc technical expert group on digital sequence information on genetic resources. CBD/DSI/AHTEG/2020/1/7*. Retrieved from https://www.cbd.int/doc/c/ba60/7272/3260b5e396821d42bc21035a/dsi-ahteg-2020-01-07-en.pdf. (Accessed 13 December 2021).

United Nations. (2000). *Cartagena protocol on biosafety to the convention on biological diversity*. Retrieved from https://www.cbd.int/doc/legal/cartagena-protocol-en.pdf. (Accessed 13 December 2021).

United Nations. (1992). *Convention on biological diversity*. Retrieved from https://www.cbd.int/doc/legal/cbd-en.pdf. (Accessed 13 December 2021).

United Nations. (1969). *Vienna convention on the law of treaties*. Retrieved from https://legal.un.org/ilc/texts/instruments/english/conventions/1_1_1969.pdf. (Accessed 13 December 2021).

WHO. (2005). *International Health Regulations (Third Edition)*. Retrieved from https://www.who.int/publications/i/item/9789241580496. (Accessed 13 December 2021).

WHO. (2011). *World Health Organization, Pandemic influenza preparedness Framework for the sharing of influenza viruses and access to vaccines and other benefits*. Retrieved from http://whqlibdoc.who.int/publications/2011/9789241503082_eng.pdf.

Wilson, E. O. (1988). *Biodiversity*. Washington, DC: Smithsonian Institution & The National Academies Press. Available from https://www.nap.edu/catalog/989/biodiversity.

Wilson, E. O. (1996). Chapter 1: Introduction. In M. L. Reaka-Kudla, D. E. Wilson, & E. O. Wilson (Eds.), *Biodiversity II: Understanding and protecting our biological resources* (pp. 1–3). Washington: Joseph Henry Press.

Zheng, X. (2019). Key legal challenges and opportunities in the implementation of the nagoya protocol: The case of China. *Review of European, Comparative and International Environmental Law, 28*, 175–184. Available from https://doi.org/10.1111/reel.12282.

Index

Note: Page numbers followed by "*f*" and "*t*" refer to figures and tables, respectively.

A

Abies cilicica, 67–68
Abiotic stress, 385
Absorption, 513–514
Abyssinone V-4′ methyl ether, 169–170
Acacia sp., 421
Acalypha wilkesiana, 273
Access and Benefit Sharing (ABS), 529
  ambiguities, 541–542
  concept of biodiversity, 530–531
  convention on biodiversity, 531–533
  development of environmental awareness, 529–530
  legal certainty, 543–544
  national biodiversity legislations, 536–538
  new trends and evolutions, 544–546
    BBNJ, 546
    curious case of pathogens, 545
    digital sequence information, 545–546
  nonstabilized and heterogeneous regulations, 540–541
  practical advice, 538–540
  problems left unsolved by CBD, 533–534
  some paradoxical effects, 543
2β-acetoxy-5β-O-acetylhellebrigenin, 170–171
2β-acetoxymelianthusigenin, 170–171
3-acetylaleuritolic acid, 486
Achillea millefolium L. *See* Yarrow (*Achillea millefolium* L.)
Acinetobacter baumannii, 129, 388
Acmella oleracea, 485–486
Acnistus arborescens (Solanaceae), 46
Acquired immunodeficiency syndrome (AIDS), 183
Acremonium sp., 419
Acremonium zeae, 399
Active transport mechanisms, 514–515
Acyl-CoA, 400–401
Addis Ababa principles and guidelines, 533
Adequate clinical response (ACR), 295
Administrative subfile, 482
Aedes albopictus cell line (ATCC), 453
Aedes albopictus mosquito, 441–442
Aesculus indica Linn., 210
African rural communities, 479
Agathosma crenulata L., 73–74
Ageratum conyzoides L., 244
Agrimonia eupatoria L. *See* Agrimony (*Agrimonia eupatoria* L.)
Agrimony (*Agrimonia eupatoria* L.), 131–132
Aizoaceae family, 163
Alexander von Humboldt Biological Resources Research Institute, 4, 6–7
Algarrobo, 22–23
Alhabba-al-sawda family, 75–76
Alkaloids, 398–399, 486, 510
  antiplasmodial, 415–417
  bisindoles, 399
  piperine, 398–399
  pyrazin-2-one, 398
  pyrrocidines, 399
Alkylamides, 486
Allium cepa L., 62–63
Allium sativum L., 62–63, 104–126, 244, 250–251
Allocryptopine, 510
Alnus glutinosa L. *See* Black alder (*Alnus glutinosa* L.)
Aloe ferox. *See* Bitter aloe (*Aloe ferox*)
Aloe species, 166–167, 173
  A. barberae, 146–147
    traditional usage, 146–147, 147*f*
  A. excelsa, 147–161, 174
    traditional usage, 154–161
  A. ferox, 142–143, 161
    traditional usage, 161
  A. vera, 142–143
  compounds isolated from *Aloe* species, 167*f*
Aloe-emodin, 166–167, 194
Aloin, 194
Aloin A, 166–167
Aloysia citrodora, 69
Alpinia, 244
Alpinia galangal, 229
Alpumisoflavone, 169–170
Alstonia scholaris L., 244, 250
Altenusin, 423–424
Alternaria sp., 401
  A. alternata ZHJG5, 401–402
  A. brassicicola, 401

Alternaria sp. (*Continued*)
A. tenuis, 423–424
Althaea officinalis L. *See* Marshmallow (*Althaea officinalis* L.)
Amaryllidaceae, 62–63, 104–126
Allium cepa/*Allium sativum*, 62–63
selected endemic Lebanese medicinal plants, 62*f*
Allium sativum L., 104–126
Ambiguities, 541–542
American Gut Project, 514
Amoebic dysentery, 490
Amomum, 244
Amphotericin B, 411–412
Anacardiaceae, 63–64
Andrographis
A. alata, 374
A. echioides, 374
A. elongata, 374
A. glandulosa, 374
A. macrobotrys, 374
A. nallamalayana, 374
A. neesiana, 374
A. paniculata, 362, 368, 370, 374
A. serpyllifolia, 374
A. wightiana, 374
Andrographolide, 374
Angelica keiskei, 345
Angiosperm Phylogeny Group IV, 365
Animal-based studies, 242
Anise (*Pimpinella anisum* L.), 72–73
Annickia polycarpa, 273
Annona sp., 47
A. muricata, 46
A. purpurea, 46
Annotation/identification, 349–352
Anopheles, 267–268
Anthracenosides, 486
Anthraquinone compounds, 166–167, 173
Antibacterial activity, 44–47
Allium sativum, 63
Cistus salviifolius, 67
Conobea scoparioides, 45
Cucurbita moschata, 45
Cyclotrichium origanifolium, 69
Cymbopogon citratus, 45
Eucalyptus species, 73–74
Humulus lupulus, 66
lower Mekong basin, 242
Matricaria species, 65
Nigella sativa, 75–76
Otholobium mexicanum, 45
of plants used in South Africa, 148*t*
Rosmarinus officinalis, 45
Salvia species, 70
of significant medicinal herbs, 210–229
Thymbra spicata, 72–73
thymol/carvacrol rich species, 71
Za'atar plants, 71–72
Antibacterial compounds, 388–404, 389*t*
Antibacterial mechanism, 388, 403
Antibiotic resistance, 189–190
Antidengue
activity, 447
confirmation of top-ranked compounds, 450*t*
assay, 448–451
Antidiarrheal effect, diarrhea, 240
Antifungal activity
Cistus species, 67
Humulus lupulus, 66
Matricaria species, 65
Nigella sativa, 75–76
Salvia species, 70
thymol/carvacrol rich species, 71
Antifungal plants in Colombia, 6
Antigen-presenting cells (APCs), 188–189
Antiinfectious plants in Colombia, 6
Antiinfective agents, plants as sources of, 335–336
Antiinfective plants
annotation/identification, 349–352
bioassay-guided fractionation, 336
biochemometrics, 339–340
data analysis, 339
herbal products, commercialization, and quality issues of, 363
infectious diseases and, 362
metabolome coverage, 345–349
metabolomics, 336–338
metabolomics-driven antiinfective discovery from plants, 340–345
methods of detection, 338–339
plants as sources of antiinfective agents, 335–336
synergy, 352
Antileishmanial compounds, 405–413
in Colombia, 6
polyketide-alkaloids, 412–413
polyketides, 411–412
terpenoids, 413
Antimalarial drugs, 269
Antimalarial plants and compounds, 6, 295–298, 296*t*
Antimicrobial compounds from endophytic fungi, 388–425
antibacterial compounds, 388–404
alkaloids, 398–399
antivirulence compounds, 404–405
peptides, 399–401
polyketides, 401–404

terpenoids, 404
antiparasitic compounds, 405–425
antileishmanial compounds, 405–413
antiplasmodial, 414–423
antitrypanosomal/antiplasmodial/antileishmanial compounds, 423–425
Antimicrobial drugs, 384
Antimicrobial plants in Colombia, 6
Antimicrobial resistance, 383
Antiparasitic activity, 45
Annona muricata, 46
Annona purpurea, 46
Austroeupatorium inulifolium, 46
Campnosperma panamense, 46
Guatteria amplifolia, 46
Huberodendron patinoi, 46
Miconia theaezans, 45
Monochaetum myrtoideum and *Acnistus arborescens*, 46
Swinglea glutinosa, 47
Antiparasitic compounds, 405–425
antiparasitic compounds isolated from endophytic fungi, 406*t*
Antiplasmodial compounds, 414–423
alkaloids, 415–417
polyketide-alkaloid, 420
polyketides, 417–420
polypeptides, 421
terpenoids, 421–423
Antiretroviral HIV therapy, 513
Antitrypanosomal/antiplasmodial/antileishmanial compounds, 423–425
polyketides, 423–424
polypeptides, 425
Antiviral activity, 47
Annona sp. (Annonaceae), 47
Byrsonima verbascifolia L. (Malpighiaceae), 47
lower Mekong basin, 243
Vismia macrophylla Kunth. (Clusiaceae), 47
Antiviral plants in Colombia, 6
Antiviral therapy, 439–440
Antivirulence compounds, 404–405
APCs. *See* Antigen-presenting cells (APCs)
Apiaceae, 64, 127, 345
Petroselinum crispum (Mill.) Fuss, 127
Prangos asperula, 64
Apicidin B, 421
Arab-Muslim medicine, 461
Arabic medicine, 460–462
Arab sources of pharmacology, 462, 465*t*
list of sources, 460*t*
principles of, 461–462
Arabic pharmacology, 462
Arbovirus. *See* Arthropod-borne viral disease (arbovirus)
Argemone mexicana L., 295–298, 486–487, 506
Arisaema flavum, 210
Arisaema sinii K., 323–324
Aristolochia contorta, 373
Aristolochia debilis, 373
Aristolochic acid, 514
10(14)-aromadendrene-4β,15-diol, 171
Artemisia absinthium L. *See* Wormwood (*Artemisia absinthium* L.)
Artemisia afra, 268
Artemisia annua L., 210, 268–269, 298–299, 352, 362, 516–517
Artemisinin combination therapy (ACT), 509
Arthropod-borne viral disease (arbovirus), 439–440
Artichoke (*Cynara scolymus* L.), 7, 65
Ascaris suum, 345
Ascomycota spp., 401
Aspergillus, 66, 384, 398
A. fumigates, 64
A. fumigatus, 412
A. niger, 23, 62–63
A. sp. F1544, 412
A. sydowii, 412–413
A. terreus, 419
Asphodelaceae, 147–154, 192–193
Aspirin, 441
Asteraceae, 6–7, 68, 127–128, 244, 273, 343
Achillea millefolium L., 127
Artemisia absinthium L., 128
Calendula officinalis L., 128
compositae, 65
Matricaria species, 65
Matricaria chamomilla L., 128
Asterric acid, 419
ATCC. *See* Aedes albopictus cell line (ATCC)
Aurionite (Pb[OH]Cl), 472
Austroeupatorium inulaefolium, 23
Austroeupatorium inulifolium, 46
Autoinducer molecules, 311
Avicennia marina, 398, 416–417, 423
Ayurveda, 460–461, 504–505
Ayurvedic Pharmacopoeia, 365
Azaanthraquinone, 493
Azadirachta indica, 322, 362, 402
Azithromycin, 189–190

B

Bacillus sp., 400
B. cereus, 45, 62–63, 127–128, 130–132, 210–229, 231
B. macerans, 129
B. megaterium, 128, 130

Bacillus sp. (*Continued*)
B. psychrosaccharolyticus, 130
B. subtilis, 23, 70, 104, 128–132, 143–144, 166–167, 173, 210–229, 400–401, 403
B. subtilis ATCC 5230, 232
B. subtilis ATCC 6633, 230
B. subtilis CGMCC1. 1162, 401
Bacteria, 315, 362, 383, 388, 484, 530–531
Bacterial infections, 143–144
Bacterial quorum quenching, phytochemicals in, 326–328
Balkans, medicinal plants in, 103–104
medicinal plants with antimicrobial properties, 104–133, 105*t*
Amaryllidaceae, 104–126
Apiaceae, 127
Asteraceae, 127–128
Betulaceae, 128–129
Lamiaceae, 129–130
Malvaceae, 130–131
Pinaceae, 131
Rosaceae, 131–132
Urticaceae, 132–133
Bambusa vulgaris, 273
Basidiomycetes, 384
Basil (*Ocimum basilicum* L.), 130
Bauhinia, 244
Beauveria bassiana, 425
Beauvericin, 425
"Bedside-to-bench" workflow, 504–505
Benzofurans, 403
Benzophenones, 404
Berberidaceae, 65–66
Berberine (*Berberis*), 209–210, 510
Berberis species, 65–66
B. asiatica, 370
B. heterophylla, 66
B. libanotica, 60, 65–66
Besleria insolita, 399
β-sitosterol (BSS), 196
β-sitosterol glucoside (BSSG), 196
Betulaceae, 128–129
Bifidobacterium animalis subsp., 131–132
Binomial system, 361
Bioactive compounds, 343
"Bioactive molecular networking", 343, 345
Bioassay-guided fractionation (BGF), 335–336
Biochemometrics, 339–340
Biodiversity, 361
concept of, 530–531
convention on, 531–533
Biodiversity beyond national jurisdiction (BBNJ), 546
Biofilms, 315–317
formation process, 316*f*
and *Mycobacterium tuberculosis*, 317–319
Biofluid analyses, 512–513
Biological activity of plants in Colombia, 6
"Biological diversity", 530
Biopiracy, 532, 543
Biotic stress, 385
Biotransformation, 514
Bipolaris bicolor, 411–412
Bisindoles, 399
Bitter aloe (*Aloe ferox*), 161, 192–194
Black alder (*Alnus glutinosa* L.), 128–129
Black thyme (*T. spicata*), 72–73
Blackberry (*Rubus fruticosus* L.), 132
Blackthorn (*Prunus spinosa* L.), 132
Blastocladia pringsheimii, 77
Blastocystis hominis, 243
Bordetella bronchiseptica ATCC 4617, 230
Borreria verticillata, 405
Bottom-up approach, 511–512
Bronchial and respiratory diseases, 489
Brown alga, 343–344
Bufadienolides, 170–171
Bursaphelenchus xylophilus, 412
Byrsonima verbascifolia L., 47

C

C9 iridoid glycosides, 369–370
Caco-2 cell line model, 514–515
Caenorhabditis elegans, 345
Caesalpinia echinata, 425
Caesalpinia sappan L., 244, 252
Calamus, 244
Calcium-activated chloride channels (CaCCs), 241
Calendula (*Calendula officinalis* L.), 7
Calendula officinalis L., 128
Callicarpa acuminata, 412
Calligonum polygonoides L., 229
Calmogastryl, 491–492
Calophyllum sp., 421
Calotropis procera, 231
Cambodian medicine, 239
Campnosperma panamense, 46
Campylobacter, 237
Campylobacter jejuni/coli, 237
Cancer treatments, 513
Cancers, 513
Candida spp., 198, 493
C. albicans, 23, 63–64, 194, 399, 404
C. tropicalis, 63–64
Cannabaceae, 66
Cannabis sativa L., 229
Carboxylates, 472

Carcinoembryonic antigen-related cell adhesion molecule family (CEACAM), 187–188
Caribbean Sea, 3–4
Carica papaya, 273
Carissa opaca, 210
Cartagena Protocol on Biosafety, 533
Caseous necrosis, 320–321
Cassia abbreviata, 194
Catechin, 168–169, 451–452
Catecholates, 472
Cedrus libani, 67–68
Cefixime, 189
Ceftriaxone, 189
Celastrol, 451–452
Cell viability assay, 454
Cells, 453
CellTiterGlo Luminescent cell viability Assay kit, 454
Centella asiatica, 244, 251
Cerastium glomeratum, 232
Cercospora kikuchii, 424
Cercosporin, 424
Chaetomium sp., 424
  C. brasiliense, 419–420
  C. mollicellum, 419–420
Chaetoxanthone B, 419
Chaetoxanthone C, 424
Chagas disease, 384
CHAGO tumor cell line, 417–418
Chamomile tea (*Matricaria chamomilla* L.), 65, 128, 517
Chemical fingerprinting methods, 370
Chia (*Salvia hispanica* L.), 69–70
Chicory (*Cichorium intybus* L.), 345
Chinese Pharmacopoeia Commission, 363–364
Chitosan functionalized silver nanoparticles (CS-AgNPs), 231
Chlamydia, 183
Chlamydia trachomatis, 183
Chlamydomonas reinhardtii, 348–349
Chloridum sp., 402
2-chloro-5-methoxy-3-methylcyclohexa-2,5-diene-1,4-dione, 418
Chloroquine, 419
Chromobacterium violaceum, 321
Chromolaena odorata (Asteraceae), 249
Chromones, 401–402
Chronic wounds, 139
Chrysophanol, 166–167, 194
Cichorium intybus L. *See* Chicory (*Cichorium intybus* L.)
Cinnamomum camphora, 340–341
Cistaceae, 67
Cistus species, 67
  C. ladanifer, 67
  C. laurifolius L., 67
  C. monspeliensis L., 67
  C. populifolius, 67
  C. salviifolius L., 67
Citrobacter freundii, 132
Citrobacter rodentium, 248–249
Clary sage (*Salvia sclarea* L.), 69–70
Classical drug discovery, 509–510
Clematis cirrhosa L., 75
Clematis flammula L., 75
Clematis vitalba, 75
Clostridium perfringens ATCC 13124, 403
Coccoloba obtusifolia Jacq., 8–21
Cochliobolus sp., 411–412
Cochlioquinone A, 411–412
Cochlospermum vitifolium, 49
Codinaeopsis gonytrichoidesi, 420
Colombia
  antibacterial evaluation of 25 native plants of, 47–51
    Maclura tinctoria L. (Moraceae), 50–51
    Mammea americana L. (Calophyllaceae), 49–50
    plant extracts found in Colombian Caribbean, 48*t*
  biological evaluation as antimicrobials of plant extracts in, 24–47, 25*t*
    antibacterial activity, 44–47
    antiparasitic activity, 45
    antiviral activity, 47
  materials and methods, 6
    natural regions, 5*f*
  medicinal plants used as antimicrobials in, 3
  plants traditionally in Colombia as antimicrobials, 6–23
    Austroeupatorium inulaefolium (Asteraceae), 23
    Cymbopogon citratus (Poacea), 23
    Guazuma ulmifolia Lam. (Malvaceae), 23
    Hymenaea courbaril L. (Leguminosae), 22–23
    Jacaranda caucana Pittier (Bignoniaceae), 22
    plants for treatment of symptoms associated with infectious diseases, 8*t*
    Solanum nudum Dunal (Solanaceae), 22
    Xanthium strumarium L. (Asteraceae), 23
Colon microbiome, 516
Combretaceae, 270, 273
Combretum
  C. adenogonium, 195
  C. albiflorum, 321
  C. molle, 194–195
Complementary and alternative medicine (CAM), 363
Compositae, 65
Conifers, 67–68
Coniothyrium sp., 412
Conobea scoparioides (Scrophulariaceae), 45
Conocarpus erecta, 404
Convention on Biological Diversity (CBD), 529

Convention on International Trade in Endangered Species (CITES), 372
Conventional method, 504, 510
Conventional treatment, 441–442
Coriandrum sativum L., 230
Coscinium fenestratum, 244
Courbaril, 22–23
Covid-19 pandemic, 508
CpnT limits cytokine, 320–321
Crab louse (*Phiirius inguinalis*), 464
Crassocephalum crepidioides, 418–419
Cristacarpin, 169–170
Croton mutisianus, 8–21
Cryptococcus gattii, 22–23
Cryptococcus neoformans, 22–23
Cryptolepsis sanguinolenta, 273–295
Cucumis sativus L., 230
Cucurbita moschata Duchesne (Cucurbitaceae), 45
Culinary herbs, 68–69
Cultural belief system of Mekong area, 238–239
Cumin (*Cuminum cyminum* L.), 72–73
Cuminum cyminum L. *See* Cumin (*Cuminum cyminum* L.)
Cupressus sempervirens, 67–68
Curcuma aromatic, 244
Curcuma longa L., 229
Cutaneous infections and medications, 463–475
  characterization of plant–metal combinations, 474–475
  elementary metal particle, 470–472
  metal nanoparticles, 473–474
  organometallic molecule, 472–473
  plants and metals useful for skin diseases, 464–468
  renewed interest in metal–organic molecule combinations, 470
  specificity of skin and eye diseases in pharmacopeias, 463–464
  toxicity of metals, 468–469
Cyanella lutea, 163
Cyclotrichium origanifolium, 69
Cyclotrichium species, 69
Cymbopogon citratus, 23, 45
Cynara scolymus L. *See* Artichoke (*Cynara scolymus* L.)
Cystic fibrosis transmembrane conductance regulator (CFTR), 241
Cytochalasin D, 416–417
Cytochalasin H, 416–417
Cytochalasin J, 416–417
Cytochalasin O, 416–417
Cytochromes, 515–516
Cytokines, 187–188
Cytotoxic concentration (CC), 488

D

Danshen (*Salvia miltiorrhiza*), 69–70
Data analysis, 339
Data processing, 455
Dataset, 244–248
Datura innoxia, 229
De materia medica, 460–461
Debregeasia salicifolia, 210
Dengue virus (DENV), 439–440
  conventional treatment, 441–442
  dengue disease, 439–441
  materials and methods, 452–455
    cell viability assay, 454
    cells and virus, 453
    data processing, 455
    extracts preparation, 454
    identification of significant features, 455
    leaf extraction, 453
    plant collection, 452
    statistical analysis, 455
    UHPLC-HRMS profiling, 454
    virus infection, 454
  medicinal plants, 442
  metabolomics reveal antidengue compounds isolated from *Psidium guajava*, 442–455
  results, 445–451
    UHPLC-HRMS-based metabolomics approach, 445–446
  serotypes, 439–440
18-deoxy-19,20-epoxycytochalasin C, 415
Deoxyribonucleic acid (DNA), 363–364
12,13-deoxyroridin E, 423
Department of Traditional Medicine (DMT), 480
Dermal diseases, 464
Dermal wounds, bacteria associated with infections of, 143–146
  Bacillus subtilis, 144
  Pseudomonas aeruginosa, 145–146
  Staphylococcus aureus, 144–145
  Staphylococcus epidermidis, 145
Dermatological diseases, 463
Dermatophilus congolensis, 493
Dermatosis, ITM for management, 492–493
Diabetes, 513
Diaporthe miriciae, 417
Diarrhea, 243–244
  diarrheal diseases in Mekong Basin, 237–240
    role of traditional medicine, 237–238
  medicinal plants used for, 243–248
    dataset, 244–248
    literature search methodology, 243–244
  models assessing effect of plants on signs and symptoms of, 240–241, 240*f*

antidiarrheal effect, 240
spasmolytic activity, 240–241
pharmacological validation of plants used for, 239–240
Dicerandrol D, 419
Dichloromethane (DCM), 170–171, 194
Dickeya solani, 131
Dictionary of Natural Products (DNP), 447–448
2,3-didehydro-19a-hydroxy-14-epicochlioquinone, 402
Dietetics, 466–467
Digalactosyldiacylglycerol (DGDG), 343–344
Digital sequence information, 545–546
12,12a-dihydroantibiotic PI 016, 421
Dihydroartemisinin, 419
Dihydrokaempferol, 167–168
2,5-dihydroxy-1- (hydroxymethyl) pyridin-4-one, 416
2-(3,4-dihydroxyphenyl) ethanol, 167–168
2,5-diketopiperazine derivative, 399
Dillenia, 244
Diode array detector (DAD), 454
Dioncophylaceae, 273–295
Dioscorea, 244
Diospyros kaki, 229
Dipeptides, 399
Diterpenoids, 171
DNA barcoding
advancements in quality control methods, 363–365
herbal products, commercialization, and quality issues of antiinfective plants, 363
infectious diseases and antiinfective plants, 362
materials and methods, 365
results, 365–374
Andrographis paniculata, 374
Embelia ribes, 365–367
Paris polyphylla, 370–372
Picrorhiza kurroa, 369–370
Saussurea costus, 372–373
Swertia chirayita, 367–369
Syzygium aromaticum, 373–374
taxonomy and, 361–362
DNA metabarcoding method, 364–365, 374
DNA Sanger sequencing technique, 364–365
Dodonaea viscose, 229–230
Dog rose (*Rosa canina* L.), 132
Droplet-liquid microjunction-surface sampling probe, 349
Drug discovery process, 504–505, 543
Drug resistance, 388
Drug-resistant bacteria, 318
Dryopteridaceae family, 165
Dysentery, 237
dysenteral, 489–490
plant data, 489–490
presentation of ITM, 489
ITM for management, 489–490

E

Early Dengue infection and outcome study (EDEN study), 453
Early treatment failure (ETF), 295
"Earth Summit", 531
Ecopharmacognosy, 520
EDEN study. *See* Early Dengue infection and outcome study (EDEN study)
Edenia gomezpompae, 412
Elaeagnus angustifolia L., 69
Elaeodendron croceum, 195
Elaeodendron transvaalense, 195
Electronic databases, 243–244
Elementary metal particle, 470–472
Elephantorrhiza
E. burkhei, 162
E. elephantina, 162, 167–169, 168*f*, 173, 195
isolated compounds, 167–168
traditional usage, 162
E. goetzei, 162
E. suffruticosa, 162
Eleusine coracana, 411–412
Embelia ribes, 365–367
Embelin (2,5-dihydroxy-3-undecyl-1, 4-benzoquinone), 365
Enantia chlorantha, 416, 419
Enantia polycarpa, 273
"End TB" strategy, 309–312
Endophytes, 385
Endophytic fungi, 385–388
Endophytic microbes, 385
ent-3β-acetoxy-labda-8(17),12Z,14-trien-2α-ol, 171
ent-7β-hydroxy-15β,16β-epoxykauran-19-oic acid, 171
ent-7β-hydroxykaur-15-en-19-oic acid, 171
ent-12β-acetoxy-7β-hydroxykaur-16-en-19-oic acid, 171
ent-12β-acetoxy-15β-hydroxykaur-16-en-19-oate, 171
ent-12β-acetoxy-15β-hydroxykaur-16-en-19-oic acid, 171
ent-12β-acetoxy-17-oxokaur-15-en-19-oic acid, 171
ent-labda8(17),12Z,14-triene-2α,3β-dibenzoate, 171
ent-labda-8(17),12Z,14-triene-2α,3β-diol, 171
Entada abyssinica, 370
Entada africana, 490
Entamoeba histolytica, 243, 489–490
Enterobacter spp., 388
E. aerogenes, 72–73, 253–254
E. aerogenes ATCC 13048, 230
Enterococcus faecalis, 127–133, 231, 399, 403–404
Enterococcus faecium, 388, 399–400
Enteroinvasive *E. coli* (EIEC), 237
Enteropathogenic *E. coli* (EPEC), 237

Enteropooling assays, 241–242
Enterotoxigenic *E. coli* (ETEC), 237
Epidermal layer of skin, 60
Epigallocatechin-3-O-gallate (EGCG), 197
19,20-epoxycytochalasin C, 415
19,20-epoxycytochalasin D, 415
Epoxycytochalasin H, 417
Erwinia carotovora, 130–131
Erybraedin A, 169–170
Erythrina lysistemon, 162–163, 169–170, 169*f*, 173–174
  traditional usage, 163
Erythrophleum ivorense, 370
Eryzerin C, 169–170
Escherichia coli, 5, 23, 47–51, 62–63, 104, 128–130, 132–133, 186, 209–210, 229–231, 340–341, 402
  Escherichia coli ATCC 15224, 230
  Escherichia coli ATCC 8739, 232
  Escherichia coli CGMCC1.1571, 401
ESKAPE pathogens, 388
Essential oils (EO), 60, 64–77, 230, 340–341
Estrogen receptor-alpha (ERα), 402
Ethanolic plant, 196
Ethnobotanical surveys, 483
"Ethnomedicine", 209
Ethnopharmacology, 270, 322
Ethyl gallate, 167–169
Ethyl-1-O-β-D-galactopyranoside, 167–169
Etlingera elatior. *See* Torch ginger (*Etlingera elatior*)
Eucalyptus species, 73–74
  E. globulus, 229
Euglena gracilis, 348–349
Eukaryotic species, 361
Euphorbia, 244
  E. dendroides L., 343
  E. hirta, 451, 489
Euphorbiaceae, 6–7, 270, 273
European larch (*Larix decidua* Mill.), 131
European Medicines Agency, 363–364
Eurotium cristatum, 399
Evidence based medicine (EBM), 502
Ex vivo models, 240–241
Exacum tetragonum, 368
Exiguobacterium acetylicum, 130
Exserohilum rostratum, 417–418
Extensively drug-resistant (XDR), 190–192
Extracellular polymeric substance (EPS), 316

F

Fabaceae, 6–7, 162, 244, 273
Fagonia indica., 229
Ferula elaeochytris, 60
Ficus thonningii, 273
Filamentous fungi, 384
Flavivirus, 439–440
Flavonoids, 486
Food supplements, 502
Food-borne pathogens, 373
Formic acid (FA), 454
16β-formyloxymelianthugenin, 170–171
Fourier-transformed infrared spectroscopy, 338–339, 374
Fucus vesiculosus L., 343–344
Full width at half maximum (FWHM), 454
Fumiquinone B, 412
Functional foods, 502
Fungal endophyte, 413
  antimicrobial compounds from endophytic fungi, 388–425
  current situation of microbial infections, 383–384
  endophytic fungi, 385–388
  microbial natural products as sources of new drugs, 384
Fungal pathogen, 411–412
Fungi, 362, 383
Fusaripeptide A, 421
Fusarium
  F. oxysporum ZZP-R1, 404
  F. pallidoroseum, 421
  F. tricinctum, 400–401
Fusariumin C, 404

G

Galenia africana L., 163, 170, 170*f*, 173
  traditional usage, 163
Gallic acid, 167–169, 451–452
Ganoderma lucidum, 451–452
Ganodermanontriol, 451–452
Garcinia mangostana L., 349
Gardnerella vaginalis, 194
Gas chromatography-mass spectrometry (GC-MS), 340–341
Gastric ulcer, ITM for management of, 491–492
Gastrointestinal infections, 489
Gelatinase, 404
Genetic digital information, 545
Genetic resources (GR), 529
Germacrene D, 67
Giardia lamblia, 243
Ginger (*Zingiber officinale*), 63, 229–230
Ginseng polysaccharides, 516
"Global herbal supplement market", 502
Global Natural Product Social Molecular Networking system (GNPS system), 345–346
GMP. *See* Good manufacturing practices (GMP)
Gonorrhea, 183, 185–189
  Neisseria gonorrhoeae, 185–189

status of available treatments for, 189–192
first-line drugs, 190*f*
Good manufacturing practices (GMP), 482
Gram-negative bacteria, 64, 71, 209–210, 230
Gram-positive bacteria, 64, 209–210, 230
Grewia erythraea, 210–229
Grewia occidentalis L., 163–164
traditional usage, 164
Guápinol, 22–23
Guapira standleyana, 412
Guatteria amplifolia (Annonaceae), 46
Guava (*Psidium guajava*), 50–51, 248–249, 442
Guazuma ulmifolia Lam. (Malvaceae), 23

H

Haemonchus contortus, 341–343
Haemonchus placei, 341–343
Haemophilus influenza, 73–74
Halenia elliptica, 368–369
Hamelia patens, 21–22
Healthcare-associated infections, 143–144
Helianthus annuus L. *See* Sunflowers (*Helianthus annuus* L.)
Helicobacter pylori, 63–64, 470, 484
Heliotropium strigosum Willd., 229
Helminthosporium monoceras, 417–418
Helminths, 362
Hepatic biotransformations, 515–516
Hepatotoxic mechanism, 515–516
Herbal drugs, 363–364
Herbal hepatotoxicity, 515–516
Herbal medicine, 459–460
Herbal preparations, 502, 514–516
deciphering mode of action of, 516–517
examples of application of metabolomic studies in human, 517–519
perspectives on *Phaleria nisidai* decoction study, 519–520
models to study absorption and biotransformation of, 514–516
Herbal products, 363
Herpes Simplex Virus (HSV-1), 194
Hevea brasiliensis. *See* Rubber tree (*Hevea brasiliensis*)
Highresolution mass spectrometry (HRMS), 445
Himalaiella heteromalla, 232
Hippophae rhamnoides L., 322
Holarrhena pubescens, 253–254
Honey flower, 164
Host immune system
via nutrition immunity, 188
pathogenesis of, 187–188
Huberodendron patinoi Cuatrec. (Bombacaceae), 46
Human African trypanosomiasis, 384
Human cancer cell lines, 411–412
Human colon tumor cells (HCT-116), 400
Human immunodeficiency virus, acquired immunodeficiency syndrome (HIV/AIDS), 183
Human tumor cell lines BT474, 417–418
Humulus lupulus, 66
Hydroxamates, 472
5-hydroxy-1,4-naphthoquinone, 252
2β-hydroxy-3β,5β-di-O-acetylhellebrigenin, 170–171
7-hydroxy-3,4,5-trimethyl-6, 416
4-hydroxybenzoic acid, 167–169
16β-hydroxybersaldegenin 1,3,5-orthoacetate, 170–171
11-hydroxymonocerin, 417–418
Hymenaea courbaril L., 22–23
Hymenocrater sessilifolius, 210–229
Hypericum perforatum L., 348
Hypericum species, 348
Hypoxis hemerocallidea, 195–196
Hypoxis rooperi, 195–196
Hypoxoside, 196
Hypoxylon terricola, 416–417
Hypoxylon truncatum, 401

I

Improved traditional medicines (ITMs), 480–481, 509
categories of, 481
general information on, 480–481
in Mali
historical development of, 484
marketing authorization files for, 481–483
in management of infectious diseases, 484–493
for dermatosis management, 492–493
for dysentery management, 489–490
for malaria management, 485–488
for management of gastric ulcer associated with *Helicobacter pylori*, 491–492
for viral hepatitis management, 490–491
regulatory framework, 480–481
In murine models, 273
In vivo approach
clinical trials in humans for evaluation of antimalarial plants and compounds, 295–298
plant extracts and plant compounds validated by, 270–298
in vitro and in vivo evaluation of antimalarial compounds, 273–295
in vitro antimalarial evaluation of plant extracts, 272–273
in vivo antimalarial evaluation of plant extracts, 273
In vivo evaluation of antimalarial compounds, 273–295
Indian Medicinal Plant Database, 365
Indo-Persian therapeutic arsenal, 460–461

Infectious diseases, 362
Inhibitory concentration 50- (IC50), 6
Innovative analytical technologies, 363–364
INRSP. *See* National Institute for Research on Public Health (INRSP)
Interleukins (IL), 189
Internal diseases, 463–464
"International Health Regulations", 545
International Treaty on Plant Genetic Resources for Food and Agriculture, 533
International Union for Conservation of Nature (IUCN), 529–530
Intestinal membrane, 515
Inula helenium L., 373
Inula racemosa, 372–373
Isocochlioquinone A, 411–412
Iterative process, 506

J

Jacaranda caucana, 8, 22
Jacaranda mimosifolia, 22
Jacoumaric acid (JA), 447–448
Jatropha sp., 270
Javanicin, 402
Juglans regia L., 322

K

Kaempferol, 167–168, 173–174
Kandelia obovata, 423
Klebsiella aerogenes, 128–129
Klebsiella pneumonia, 5, 23, 47–51, 129–130, 132–133, 229, 231, 388

L

Lactobacillus, 243
  L. brevis, 127
  L. hilgardii, 127
  L. rhamnosus, 131–132
Lactococcus, 243
Lactoferrin, 188
Lactuca sativa L. *See* Lettuce (*Lactuca sativa* L.)
Lactuca triquetra, 60
Lagerstroemia, 244
Lagotis cashmeriana, 370
Lamiaceae, 6–7, 68–73, 129–130, 164–165
  Cyclotrichium species, 69
  Lamiaceae genera, 72
  Lavandula angustifolia Mill, 129
  Mentha longifolia (L.) L., 129
  Mentha x piperita L., 129–130
  Ocimum basilicum L., 130
  Phlomis species, 69
  Rosmarinus officinalis, 71
  Salvia species, 69–70
  Thymbra spicata, 72–73
  Thymol/carvacrol rich species, 71
  Za'atar plants: *Satureja thymbra; Origanum syriacum*, 71–72
Larix decidua, 131
Late treatment failure (LTF), 295
Lavandula, 68–69
Lavandula angustifolia. See Lavender (*Lavandula angustifolia*)
Lavandula stoechas L., 68
Lavender (*Lavandula angustifolia*), 68–69, 129, 230
Lawsonia inermis L., 230
LC–MS. *See* Liquid chromatography–mass spectrometry (LC–MS)
Leaf extraction, 453
Lebanese plants with antimicrobial activity, 62–77
  Amaryllidaceae, 62–63
  Anacardiaceae, 63–64
  Apiaceae, 64
  Asteraceae/Compositae, 65
  Berberidaceae, 65–66
  Cannabaceae, 66
  Cistaceae, 67
  Conifers, 67–68
  Lamiaceae, 68–73
  Myrtaceae, 73–74
  Portulacaceae, 74–75
  Ranunculaceae, 75–76
  Rosaceae, 77
  Rutaceae, 76
  traditional Lebanese medicinal plants, 78*t*
Lebanon, medicinal plants and traditional as antimicrobials in, 59–61
Leishmania sp., 22, 388
  L. amazonensis, 405–411, 423–424
  L. donovani, 405–411
  L. panamensis, 46
Lens culinaris Medik., 230
Leonotis leonurus, 324–325
Leprosy, 463–464
Leptadenia pyrotechnica, 210
Lethal dose 50 (LD50), 6, 488
Lettuce (*Lactuca sativa* L.), 65
Lewia infectoria, 399
Licarin A, 325–326
Licuala spinosa, 421–422
Lilium lancifolium Thunb. *See* Lily bulb (*Lilium lancifolium* Thunb)
Lily bulb (*Lilium lancifolium* Thunb), 518–519
Lingonberries (*Vaccinium vitis-idaea* L.), 517–518
Lipooligosaccharide (LOS), 186
Lippia chevalieri, 485

Liquid chromatography–mass spectrometry (LC–MS), 445, 511–512
Listeria innocua, 132
Listeria monocytogenes, 128–131
Litharge compounds, 472
Lobostemon fruticosus, 163
Lonicera nummulariifolia, 60
Lower Mekong basin, medicinal plants used as antidiarrheal agents in, 235
  medicinal plants used for diarrhea in, 243–248
  models assessing antiinfective properties, 242–243
    antibacterial activity, 242
    antiviral and antiparasitic activity, 243
    models, 243
  models assessing antimotility and antisecretory activities, 241–242
    antimotility activity, 241
    antisecretory activity, 241–242
  models assessing effect of plants on signs and symptoms of diarrhea, 240–241
  selected plant species, 248–254
  traditional medicine for diarrheal diseases, 237–240
    map, 237*f*
Ludwigia leptocarpa, 49
Lumnitzera racemosa, 416–417, 423
Lysiloma latisiliquum, 341–343
Lysisteisoflavone, 169–170

M

Maclura pomifera. *See* Osage orange (*Maclura pomifera*)
Maclura tinctoria, 49
Macrophages, 141–142, 320
Maesa indica, 365
Malaria (*Plasmodium* spp.), 383
Malaria, 268, 273, 373, 479–480
  ITM for management of, 485–488
  malarial 5, 485–486
    plant data, 485–486
    presentation of ITM, 485
  Sumafura Tiemoko Bengaly, 486–487
    plant data, 487
    presentation of ITM, 486–487
  Wolotisane, 487–488
    plant data, 488
    presentation of ITM, 487–488
Malva sylvestris L., 130–131
Malvaceae, 130–131, 244
  Althaea officinalis L., 130
  Malva sylvestris L., 130–131
Mammea americana, 49
Mangifera indica L. *See* Mango (*Mangifera indica* L.)
Mango (*Mangifera indica* L.), 232, 244, 252–253
Mangrove (*Rhizophora mangle* L.), 8–21
Marjoram (*Origanum*), 68–69
Marketing authorizations (MA), 480
  files, 481–483
Marrubium globosum, 60
Marshmallow (*Althaea officinalis* L.), 130
Mass spectrometry imaging (MSI), 347–348
Massamplification, 364–365
Mastic tree (*Pistacia lentiscus* L), 63–64
Matricaria chamomilla L. *See* Chamomile tea (*Matricaria chamomilla* L.)
Matricaria species, 65
Medicago falcata L., 210
Medicinal Flora of Colombia, 4
Medicinal herbs, 501
Medicinal plants, 5, 59–61, 104–133, 140, 185, 209–210, 311, 442
  antidengue medicinal plants, 443*t*
  in Colombia, 7
  for diarrhea, 239
  and mycobacterium quorum quenching, 322–326
  as quorum quenching agents, 321–322
  in West Africa, 269–270
Mediterranean diet, 74–75
Mekong. *See* Lower Mekong basin
Melaleuca alternifolia, 230
Melastoma, 244
Meliaceae, 270, 273
Melianthus comosus, 163–164, 170–171, 171*f*, 173
  traditional usage, 164
Melianthus major, 163
Melianthusigenin, 170–171
Menisporopsis theobromae, 421
Mentha. *See* Mint (*Mentha*)
Mentha aquatica L. *See* Watermint (*Mentha aquatica* L.)
Mentha spicata. *See* Spearmint (*Mentha spicata*)
Mentha x piperita L. *See* Peppermint (*Mentha x piperita* L.)
Metabolite imaging techniques, 347–348
Metabolome coverage, 345–349
Metabolomics, 336–338, 345, 445, 511–512
  metabolomics-driven antiinfective from plants, 340–345
Metabolomics approaches, 343
  in antiviral compound identification, 445
Metal, 468–469
  metal–organic molecule combinations, 470
  nanoparticles, 473–474
  for skin diseases, 464–468
  toxicity of, 468–469
Methanol (MeOH), 194
Methanolic extracts, 131, 229
Methicillin-resistant *Staphylococcus aureus* (MRSA), 73–74, 340–341

methyl ent-12β-acetoxy-7β-hydroxykaur-15-en-19-oate, 171
methyl ent-12β-acetoxy-17-oxokaur-15-en-19-oate, 171
methyl ent-12β-hydroxykaur-16-en-19-oate, 171
Methylene (CH_2), 350
Metric tons (MT), 367
MIC. *See* Minimum inhibitory concentration (MIC)
Miconia theaezans (Melastomataceae), 45
Microbial infections, 383–384
Microbial natural products, 384
Microbiome, 514
Micrococcus flavus, 131
Micrococcus luteus, 70, 132
Microextraction technique, 349
Microflora-mediated biotransformations, 514
Micromeria barbata, 73–74
Micromeria chamomilla, 69
Micromeria myrtifolia, 69
Microsomes, 515–516
Microsporum gypseum, 22–23
Minimum bactericidal concentration (MBC), 63
Minimum biofilm inhibitory concentration (MBIC), 325–326
Minimum inhibitory concentration (MIC), 6, 63, 72–73, 104, 316–317, 388–398, 492
Mint (*Mentha*), 68–69
Mitracarpus scaber, 492–493
Mitradermine, 492–493
  plant data, 492–493
  presentation of ITM, 492
Mobile phase A (MPA), 454
Molecular networking approach, 295, 343
Mollicellin E, 419–420
Mollicellin K, 419–420
Mollicellin L, 419–420
Mollicellin M, 419–420
Monocerin, 417–418
Monochaetia sp., 404
Monochaetum myrtoideum (Melastomataceae), 46
Monogalactosyldiacylglycerol (MGDG), 343–344
Multidimensional models, 339–340
Multidrug resistant (MDR), 190–192, 388
"Multiplex MS imaging" technique, 348–349
Multiplicity of infection (MOI), 454
Multivariate data analysis (MVA), 445, 455, 511–512
Mycobacterial biofilms, 317–319, 322
Mycobacterial quorum quenching, 322–326
  biofilms, 315–317
  biofilms and *Mycobacterium tuberculosis*, 317–319
  medicinal plants and, 322–326
    chemical structure, 323*f*
    licarin A, 325, 325*f*
    as quorum quenching agents, 321–322
    scanning electron microscopy images of Mtb, 324*f*
  phytochemicals in bacterial quorum quenching, 326–328
  quorum quenching research
    current state of, 312–314, 313*f*
    significance of, 311–312
  quorum sensing *vs.* quorum quenching, 315
  TB infection, 310*f*
  virulence factors, 319
  virulence factors and *Mycobacterium tuberculosis*, 319–321
Mycobacterium abscessus, 325
Mycobacterium fortuitum, 325
Mycobacterium massiliense, 325
Mycobacterium smegmatis, 318, 324–325
Mycobacterium spp., 398–399
Mycobacterium tuberculosis (Mtb), 72–73, 309–310, 317–321
Mycoleptodiscus indicus, 405
Myrsine africana L., 365
Myrtaceae (*Psidium guajava*), 73–74, 248–249

N

N-acyl-homoserine lactone, 311
n-hexane, 229
N-terminal exotoxin, 320–321
Nagoya Protocol, 529–540
  ambiguities, 541–542
  concept of biodiversity, 530–531
  convention on biodiversity, 531–533
  development of environmental awareness, 529–530
  legal certainty, 543–544
  national biodiversity legislations, 536–538
  nonstabilized and heterogeneous regulations, 540–541
  practical advice, 538–540
  problems left unsolved by CBD, 533–534
  some paradoxical effects, 543
  new trends and evolutions, 544–546
    BBNJ, 546
    curious case of pathogens, 545
    digital sequence information, 545–546
Naloxone, 241
Nanocarriers, 143
National biodiversity legislations, 536–538
National Competent Authority (NCA), 538–539
National Health Interview Survey, 363
National Institute for Communicable Diseases (NICD), 189–190
National Institute for Research on Public Health (INRSP), 480
National Institute of Health's National Metabolomics Data Repository, 345–346

National Institute of Standards and Technology Mass Spectral Library, 350
National Research Foundation, 140–141
Natural killer cells, 141–142
Natural products (NPs), 311, 322, 335, 511
  models to study absorption and biotransformation of, 514–516
Nauclea pobeguinii, 295–298
Negative ionization (NI), 445–446
Neisseria gonorrhoeae, 183, 185–189, 186*f*
  coinfections of, 188
  and evasion of host immune system, pathogenesis of, 187–188
    virulence factor of, 187*f*
  evasion of host immune system via nutrition immunity, 188
Nemania sp., 415
Neoaspergillic acid, 398
Nettle (*Urtica dioica* L.), 132–133
Network pharmacology, 513
Neutrophils, 141–142
Nigella sativa, 75–76
Nigrospora sp., 402
Noncommunicable diseases (NCDs), 508
Nonstabilized and heterogeneous regulations, 540–541
Nonstructural protein (NS protein), 441–442
14-norpseurotin A, 412–413
Norway spruce (*Picea abies*), 131
Nosocomial infections, 143–144, 321
Nuclear DNA sequences, 362
Nuclear factor kappa B (NF-κB), 187–188
Nuclear magnetic resonance, 511–512
Nuclear ribosomal internal transcribed spacer (nr-ITS), 368–369
Nucleic acid-based genotyping techniques, 363–364
Nutraceuticals, 502
Nutrients, 317–318

O

Ocimum basilicum L. *See* Basil (*Ocimum basilicum* L.)
Octaketides, 404
Oleanolic acid, 493
Oligarchic forests, 366
Omega-3 fatty acids, 74–75
Omics techniques, 511–512
Opacity-associated protein genes (*Opa* genes), 186
Oplopanax horridus, 340
Organelle DNA sequences, 362
Organometallic molecule, 472–473
Origanum species, 68–69, 72
  O. dictamnus L., 517–518
  O. ehrenbergii, 60, 72
  O. libanoticum, 60, 72
  O. minutiflorum, 72–73
  O. onites L., 68
  O. syriacum, 71–72
Orinoquia region, 8–21
Ornamentals, 68–69
Oroxylum indicum, 254
Orthogonal projection to latent structures (OPLS), 447
Osage orange (*Maclura pomifera*), 50–51
Otholobium mexicanum (Fabaceae), 45

P

P-glycoprotein (Pgp), 514–515
Pacific islands, 519–520
Paederia foetida L., 239
Pakistan, antibacterial properties of different medicinal plants from, 210–232
Palicourea marcgravii, 416–417
Palmarumycin CP2, 412
Palmarumycin CP17, 412
Palmarumycin CP18, 412
Palmitic acid, 194
Palmitoleic acid, 194
Panax ginseng, 243
Papaver somniferum L. *See* Sleeping poppy (*Papaver somniferum* L.)
Paradigm models, 511–512
  changes of, 511–513
Parallel artificial membrane permeability assay (PAMPA), 514–515
Parasites, 383, 484
Parasitic diseases, 464
Paris polyphylla, 370–372
Partial least square (PLS), 339, 445
Partial least squares regression discriminate analysis (PLSDA), 341–343
Pathogen-associated molecular patterns (PAMPs), 187–188
Pathogens, 141–143, 173–174
  curious case of, 545
Patulin, 404–405
Pectobacterium atrospecticum, 131
Peltophorum africanum, 196–197
Pelvic inflammatory disease (PID), 183–184
Penicillic acid, 404–405
Penicillium, 66, 384, 404–405
  P. coprobium, 404–405
  P. cyclopium, 62–63
  P. radicicola, 404–405
  P. restrictum G85ITS, 405
2′,3,4′,5,7-pentahydroxyflavone, 50–51
Peppermint (*Mentha x piperita* L.), 129–130
Peptides, 399–401
  dipeptides, 399

Peptides (*Continued*)
polypeptides, 400–401
Periconia sp., 398–399
Peristaltic index, 241
Persea americana, 8
Persicaria bistorta, 229–230
Pestalotiopsis sp., 404
Pesticides, 529–530
Petroselinum crispum (Mill.) Fuss, 127
Pgp. *See* P-glycoprotein (Pgp)
Phaleria nisidai, 509, 519–520
decoction study, 519–520
Pharmaceutical herbal products, 502
Pharmacognosy, 511
Pharmacokinetics (PK), 513–514, 516–517
synergies, 514
Pharmacological models, 243
Pharmacology, 460, 466–467, 511–512
Pharmacopeias, specificity of skin and eye diseases in, 463–464
Phaseollidin, 169–170
Phenolates, 472
Phiirius inguinalis. *See* Crab louse (*Phiirius inguinalis*)
Phlomis species, 69
Phlomis syriaca, 69
Phoma sp., 403
Phomoarcherin B, 422–423
Phomopsis sp., 419
P. archeri, 422–423
P. sojae, 417
Phomoxanthone A, 419
Phomoxanthone B, 419
Phosgenite ($Pb_2Cl_2CO_3$), 472
Phyllanthus emblica L., 229–230
Phyllanthus fraternus, 273
Phytochemicals in bacterial quorum quenching, 326–328, 327*t*
Phytochemistry, 322
Phytopharmaceuticals, 366
Phytosanitary laws, 362
Phytosterol, 196
Phytotherapy, 504
Phytovigilance system, 515–516
Picea abies. *See* Norway spruce (*Picea abies*)
Picea abies (L.), 131
Picrorhiza kurroa, 369–370
Picroside-I, 369–370
Picroside–II, 369–370
Pilin genes, 186
Pimpinella anisum L. *See* Anise (*Pimpinella anisum* L.)
Pinaceae, 131
Larix decidua Mill, 131
Picea abies (L.), 131
Piper longum L, 398–399
Piper methysticum, 519–520
Piper nigrum L., 244
Piperaceae, 6–7
Piperine, 398–399
Piptadenia adiantoides, 411–412
Pistacia lentiscus L. *See* Mastic tree (*Pistacia lentiscus* L)
Pistacia species, 63–64
P. chinensis, 229
P. chinensis subsp. *integerrima*, 229–230
P. integerrima, 210
P. lentiscus, 63–64
Planktonic cells, 316–317
Plant
as antiinfective agents, 335–336
characterization of plant–metal combinations, 474–475
collection, 452
extracts, 59–60
in vitro antimalarial evaluation of, 272–273
in vivo antimalarial evaluation of, 273
plant-based preparation, 270–272
plant-derived compounds, 268
for skin diseases, 464–468
as sources of antiinfective agents, 335–336
species, 244, 248–254
Allium sativum, 250–251
Alstonia scholaris, 250
antidiarrheal plant species, 248*f*
Caesalpinia sappan, 252
Centella asiatica, 251
Chromolaena odorata, 249
for diarrheal illnesses in lower Mekong basin, 245*t*
Holarrhena pubescens, 253–254
Mangifera indica, 252–253
Oroxylum indicum, 254
Psidium guajava, 248–249
Punica granatum, 251–252
of traditional in Colombia, 6
Plasmodium, 267–268, 273, 298–299
P. berghei, 273
P. berghei ANKA, 273
P. chabaudi, 273
P. chabaudi adami, 273
P. falciparum, 22, 46, 268, 343, 412, 414, 510
P. falciparum 3D7, 419, 423
P. falciparum FcB1, 488
P. falciparum K1, 417–418, 487
P. falciparum NF54, 419
P. falciparum W2, 488
P. vinckei, 273
P. yoelii, 273
P. yoelii 17XNL, 273

P. yoelii YM, 273
parasites, 388
Plasmodium chabaudi chabaudi (Pcc), 273, 488
Plasmodium spp. *See* Malaria (*Plasmodium* spp.)
Plectranthus fruticosus, 164–165, 171, 172*f*, 173
traditional usage, 165
PLS discriminant analysis (PLS-DA), 339–340
Polygonum paleaceum, 371–372
Polyherbal products, 364–365
Polyketide-alkaloids
antileishmanial compounds, 412–413
antiplasmodial, 420
Polyketides, 401–404
antileishmanial compounds, 411–412
antiplasmodial, 417–420
antitrypanosomal/antiplasmodial/antileishmanial compounds, 423–424
benzofurans, 403
benzophenones, 404
chromones, 401–402
octaketides, 404
quinones, 402
xanthones, 402–403
Polymerase chain reaction–restriction fragment length polymorphism method (PCR-RFLP method), 371–372
Polypeptides
antiplasmodial, 421
antitrypanosomal/antiplasmodial/antileishmanial compounds, 425
Polypeptides, 400–401
Polypharmacological mechanisms, 516–517
Polypharmacology, 512–513
Polysaccharides, 516
Polystichum pungens, 165, 173–174
traditional usage, 165
Pongamia pinnata, 402
Porcine tissues, 515
Portulaca oleracea L., 74–75
Portulacaceae, 74–75
Portulaca oleracea L., 74–75
Positive ionization (PI), 445–446
Postclinical studies, 509–510
Pot marigold (*Calendula officinalis* L.), 128
Potential antidengue medicinal plant, 442–445
Potential hepatotoxicity, 515–516
Potential urinary biomarkers, 518
Prangos asperula, 60
Prangos asperula, 64
Prangos pabularia, 64
Premenstrual symptoms, 508
Preussia sp., 419
Preussiafuran A Asterric acid, 419
Preussomerin EG1, 412
Principal component analysis (PCA), 446
Prodrugs, 512–513
Propionibacterium acnes, 77, 403
Propranolol, 241
Proteus mirabilis, 129, 132, 231
Proteus spp., 229
Protopine, 510
Protozoa, 362
Protozoal diseases, 383
Prunus spinosa L. *See* Blackthorn (*Prunus spinosa* L.)
Pseudeurotium ovalis, 412–413
Pseudomonas, 209–210, 402
P. aeruginosa, 5, 23, 47–51, 104, 128–130, 132–133, 143, 145–146, 173, 210–231, 321–322, 388, 402–404
P. aeruginosa ATCC 9027, 232
P. aeruginosa PA01 strain, 404–405
P. fluorescens, 62–63
P. fluorescens CGMCC1.1828, 401
P. pickettii ATCC 49129, 230
P. syringae, 65
Pseudomonas putida, 530–531
Pseurotin A, 412–413
Pseurotin D, 412–413
Psiadia arguta, 343
Psidium guajava, 248–249, 415, 442–455
metabolomic approach in antiviral compound identification, 445
potential antidengue medicinal plant, 442–445
Psidium guajava. *See* Guava (*Psidium guajava*); Myrtaceae (*Psidium guajava*)
Psidium guajava L. leaves (Pg leaves), 452
Psychotria horizontalis, 424
Pteleopsis suberosa, 491
Pubic louse, 464
Pullularin A, 421
Pullularin B, 421
Punica granatum L., 229, 244, 251–252
Purpureocillium lilacinum, 412
Putative antidengue compounds, 447–448, 449*t*
Pyrazin-2-one, 398
Pyrrocidines, 399
Pyrrolizidinic alkaloids, 515–516

Q

Quality control (QC), 446
Quantitative analyses, 510
Quassia amara, 273
Quercetin, 451–452
Quercetin-3-O-β-D-glucopyranoside, 167–169
Quinine, 209–210
Quinones, 402

Quorum quenching process, 311, 315
Quorum sensing process, 315, 326

R

Radix rehmanniae (*Rehmannia glutinosa* Libosch), 518–519
Random amplified polymorphic DNA (RAPD), 366–367
Randomized controlled studies, 504
Ranunculaceae, 75–76
  Clematis vitalba, 75
  Nigella sativa, 75–76
Rauvolfia macrophylla, 412
RDDP, 196
Reductionist approach, 511–513
Rehmannia glutinosa, 518–519
Rehmannia glutinosa Libosch. *See* Radix rehmanniae (*Rehmannia glutinosa* Libosch)
Respiratory diseases, 8
Retrospective treatment outcome study (RTO study), 505–508, 507*t*
Reverse pharmacology approach, 504–510
  clinical efficacy, 503–504
  clinical evaluations, 508–509
  defining preparations in medicinal plants, 503*t*
  laboratory stage, 509–510
  new and promising approaches for laboratory stage in, 510–520
    changes of paradigms, 511–513
    models to study absorption and biotransformation of natural products and pharmacokinetics, 513–514
  retrospective treatment outcome study, 505–508
Rhinacanthus nasutus, 374
Rhizophora mangle L. *See* Mangrove (*Rhizophora mangle* L.)
Rhizopus nigricans, 77
Rhodomyrtus tomentosa, 244
Ricinus sp., 270
Robustness, 513
"Roll back malaria", 267–268
Roridin E, 423
Rosa canina L. *See* Dog rose (*Rosa canina* L.)
Rosa damascene, 69, 77
Rosaceae, 6–7, 77, 131–132
  Agrimonia eupatoria L., 131–132
  Prunus spinosa L., 132
  Rosa canina L., 132
  Rosa damascene, 77
  Rubus fruticosus L., 132
Rosmarinus officinalis, 45, 68–69, 71
Rotaviruses, 243
Rubber tree (*Hevea brasiliensis*), 533
Rubiaceae, 6–7, 244, 270, 273
Rubus fruticosus L. *See* Blackberry (*Rubus fruticosus* L.)
Rumex hastatus D., 229
Rumex madaio, 404
Ruta chalepensis L., 76
Ruta graveolens L., 76
Ruta Montana L., 76
Ruta species, 76
Rutaceae, 76

S

Sage (*Salvia officinalis* L.), 68–70
Salmonella spp., 209–210, 237, 242
  S. enteritidis, 63–64
  S. paratyphi, 253–254
  S. setubal ATCC 19196, 230
  S. typhi, 128–130, 132–133, 210, 229, 231, 250–251
  S. typhi ATCC 14028, 232
  S. typhimurium, 66, 251–252, 340–341, 419–420
Salvia, 68–70
  S. africana-caerulea L., 70
  S. africana-lutea L., 70, 324–325
  S. chamelaegnea, 70
  S. fruticosa, 69–70
  S. miltiorrhiza, 516–517
  S. multicaulis, 70
  S. radula, 70
  S. runcinata, 70
  S. tomentosa, 69–70
  S. verbenaca L., 69–70
Salvia hispanica L. *See* Chia (*Salvia hispanica* L.)
Salvia lavandulifolia. *See* Spanish sage (*Salvia lavandulifolia*)
Salvia miltiorrhiza. *See* Danshen (*Salvia miltiorrhiza*)
Salvia officinalis L. *See* Sage (*Salvia officinalis* L.)
Salvia sclarea L. *See* Clary sage (*Salvia sclarea* L.)
Samanere, 490–491
Sandoricum koetjape, 418
Saponins, 486
Saponosides, 486
Sarcina lutea, 132
Sargassum thunbergii, 399
Satureja, 68–69, 72
  S. cuneifolia, 72–73
  S. thymbra L., 68, 71–72
Satureja hortensis L. *See* Summer savory (*Satureja hortensis* L.)
Saurauia scaberrima, 403
Saussurea costus, 372–373
Saxifraga, 368–369
Scabies, 373
Scopoletin, 486
Secondary metabolites, 371, 459–460

Selectivity index, 272
Senna occidentalis L., 485
Sequence-characterized amplified region (SCAR), 366–367
Serratia marcescens ATCC 13880, 232
Serratia rubidaea, 127
Sesamol, 323–324
Sexually transmitted diseases (STDs), 183
gonorrhea, 185–189
microorganisms associated with sexually transmitted diseases, 184*t*
selected South African plants used in traditional medicine, 192–198
status of available treatments for gonorrhea, 189–192
Shigella, 237
S. boydii, 242
S. flexneri, 128, 132–133, 229, 242
S. shiga, 128
S. sonnei, 128, 194, 242
Sideritis fruticosa, 68–69
Sideritis libanotica, 68
Sideroxylon mascatense, 229
Silver nanoparticles (AgNPs), 230
Silybum marianum, 405
Simian rotavirus (SA-11), 248–249
Siparuna sp., 405
Sleeping poppy (*Papaver somniferum* L.), 501, 516–517
Slevogtia orientalis, 368
Solanaceae, 6–7
Solanum nigrum L., 229
Solanum nudum Dunal (Solanaceae), 22
Solanum surattense Burm. f., 229
Solid-phase extraction (SPE), 445–446
South Africa
with activity against wound-associated bacteria, 146–166
antibacterial activity of plants used in South Africa, 148*t*
toxicity of South African medicinal plants used for wound healing, 155*t*
in vitro and in vivo wound-healing activity, 154*t*
bacteria associated with infections of dermal wounds, 143–146
compounds present in plants traditionally used for wound healing in, 166–172
currently available treatments and products, 142–143
topical creams, 142–143
transdermal drug delivery systems, 143
medicinal plants in, 139–141
pathophysiology of wound healing, 141–142
antibacterial activity of plants used in South Africa, 148*t*
plants used in traditional medicine, 191*t*, 192–198
Aloe ferox, 192–194
Cassia abbreviata, 194
Combretum molle, 194–195
Hypoxis hemerocallidea, 195–196
Peltophorum africanum, 196–197
Tabernaemontana elegans, 197–198
Terminalia sericea, 198
publications and citations relating of medicinal plants, 140*f*
wound infection, 142
South African healthcare systems, 139
Spanish sage (*Salvia lavandulifolia*), 69–70
Spearmint (*Mentha spicata*), 129–130
Spectroscopy, 338–339
Sphaeranthus indicus L., 229
Sphedamnocarpus pruriens, 324–325
Spilanthes oleracea L., 485
Spilanthol, 486
Sporothrix sp., 419
Staphylococcus, 209–210, 493
S. aureus, 5, 23, 45, 47–51, 63–64, 104, 128–133, 143–145, 173, 209–210, 229–232, 340–341, 345, 388, 398–399, 402, 404–405
S. epidermidis, 65, 128, 143–145, 173, 210, 402
S. pneumoniae, 400–401
State-of-the-art techniques, 516–517
Statistical analysis, 455
Stenotrophomonas maltophilia ATCC 13637, 232
"Stepchild of politics", 501–502
Sterol-type compounds, 45
Sterols, 486
Streptococcus spp., 230
S. agalactiae, 130–131
S. pneumonia, 73–74, 129–130, 229, 416
S. pyogenes, 73–74
Suaeda vermiculata, 229
Sumafura Tiemoko Bengaly, 486–487
Summer savory (*Satureja hortensis* L.), 72–73
Sunflowers (*Helianthus annuus* L.), 65
Superficial fungal diseases, 492–493
Sustainability, 520
Sustainable Development Goals (SDG), 309–310
Sutherlandia frutescens L., 165–166, 171–173, 172*f*
traditional usage, 166
Sutherlandioside A, B, C, and D, 171–172
Swertia
S. angustifolia, 368
S. bimaculata, 368
S. chirata, 229–230
S. chirayita, 367–369

Swertia (*Continued*)
S. ciliata, 368
S. cordata, 368
S. mussotii, 368
Swinglea glutinosa Merr (Rutaceae), 47
Synergy, 352, 512–513
Syphilis, 183
"Systems biology", 511–512
Syzygium aromaticum, 230–231, 362, 373–374

T
T cells, 187–188
Tabernaemontana elegans, 197–198
Tamsulosin, 241
Tandem mass spectrometry (MS/MS), 445
Tannins, 486
Tardioxopiperazine A, 399
Technology Innovation Agency, 140–141
Tecoma stans, 210
Tectona crispa, 419
Tectona grandis, 402–403
Tectona grandis, 273
Terminalia, 244, 447–448
T. macroptera, 487
T. paniculata, 198
T. sericea, 195, 198
T. sericea, 352
Terpenoids, 404, 413
antileishmanial compounds, 413
antiplasmodial, 421–423
2,3,4,6-tetrahydroisoquinoline-8-carboxylic acid, 416
Teucrium capitatum L., 67
Thin-layer chromatography (TLC), 166–167
Thymbra, 72
Thymbra spicata, 68–69, 72–73
Thymus, 68–69
T. cilicicus, 68
T. leucotrichus, 71
T. vulgaris L., 71, 229
Tinospora crispa, 416
TLC. *See* Thin-layer chromatography (TLC)
Toona ciliate, 210
Top-down approach, 511–512
Topical antibacterial treatments, 142–143
Topical creams, 142–143
Torch ginger (*Etlingera elatior*), 340–341
Torreya taxifolia, 415
Torulopsis glabrata, 63–64
Traditional and complementary medicines (T&CMs), 501–502
Traditional Chinese medicine (TCM), 452, 514
Traditional ecological knowledge (TEK), 103–104
Traditional Indian pharmacopeias, 474
Traditional medicine, 139–140, 165, 269, 363
in Colombia, 6
for diarrheal diseases, 237–240
cultural belief system of people living in Mekong area, 238–239
in management of diarrhea, role of, 237–238
pharmacological validation of plants used for diarrhea, 239–240
for treatment of STDs, 192–198
Traditional Medicine Strategy (WHO), 501–502
Traditional plant extracts, 502
Trans-isoferulic acid, 486
Transdermal drug delivery systems, 143
Transferrin (Tbps), 188
Transferulic acid, 486
Treponema pallidum, 183
Trichoderma, 384
T. asperellum, 400
T. viride, 62–63
Trichomonas vaginalis, 183
Trichomoniasis, 183
Trichophyton mentagrophytes, 22–23
Trichophyton rubrum, 22–23, 68–69
Trichophyton tonsurans, 22–23
Trigonelline, 518
Trillium govanianum, 371–372
Trillium tschonoskii, 371–372
Triterpenes, 486
Triumfetta cordifolia, 273
Trixis vauthieri DC, 423–424
Tropaeolum tuberosum, 8–21
Tropical forests, 384
Trypanosoma cruzi, 22, 423
Trypanosoma spp. *See* Trypanosomiasis (*Trypanosoma* spp.)
Trypanosomatidae parasites, 388
Trypanosomiasis (*Trypanosoma* spp.), 383
Tuberculosis (TB), 309–310, 373, 479–480, 513
Tuberculosis bacilli, 318
Tupistra spp, 371–372
Type 2 diabetes, 520

U
UK Medicines and Healthcare Products Regulatory Agency, 363–364
Ulcers, 463–464
Ultrahigh performance liquid chromatography-high-resolution mass spectrometry (UHPLC-HRMS), 445–446, 516
antidengue assay of pure authentic standards, 448–451
profiling, 454

UHPLC-HRMS-based metabolomics approach, 445–446
antidengue activity, 447
identification of putative antidengue compounds, 447–448
Ultraviolet-visible spectroscopy, 338–339
Uncaria stem (*U. rhynchophylla*), 516–517
Uncaria tomentosa, 8
Unicellular eukaryotic organisms, 383
United States Pharmacopoeia (USP), 363–364
Universal Natural Products Database (UNPD), 447–448
Untargeted metabolomic approaches, 349–350
Urinary tract infections (UTIs), 127
Ursolic acid, 70, 493
Urtica dioica L. *See* Nettle (*Urtica dioica* L.)
Urtica urens L., 166
traditional usage, 166
Urticaceae, 132–133
Urtica dioica L., 132–133
US Food and Drug Administration, 363–364
Usnic acid, 403
Ussing chamber system, 515

V

Vaccinium oxycoccos, 322
Vaccinium vitis-idaea L. *See* Lingonberries (*Vaccinium vitis-idaea* L.)
Valerian (*Valeriana officinalis* L.), 7
Valeriana jatamansi, 371–372
Vanilla albida, 422–423
Vanillic acid, 486
Variable importance in projection method (VIP method), 339–340
Vector control, 267–268
Vellozia gigantea, 417
Venereal diseases, 184
Verbascum leptostychum, 72
Vernonia amygdalina, 273, 295–298
Verticillium dahlia, 130–131
Vibrio cholerae, 237
Vincetoxicum stocksii, 210–229
Viola alba subsp. *dehnhardi*, 445
VIP method. *See* Variable importance in projection method (VIP method)
Viral hepatitis, ITM for management of, 490–491
Samanere, 490–491
plant data, 490–491
presentation of ITM, 490
Virulence factors, 319–321
Virus(es), 362, 383, 453, 484
infection, 454
Vismia macrophylla Kunth. (Clusiaceae), 47
Vladimiria berardioidea, 373
Vladimiria souliei, 373
Vochysia guatemalensis, 420

W

Water stem bark extracts of *C. abbreviata*, 194
Watermint (*Mentha aquatica* L.), 129–130
West Africa, medicinal plants from, 268
case of Artemisia, 298–299
plant extracts and plant compounds validated by in vivo approach, 270–298
antimalarial plants listed in West African, 271*f*
traditional use of medicinal plants, 269–270
medical doctors in West Africa, 270*f*
West African Health Organization (WAHO), 480
Western herbal market in medicinal plants, 104
White blood cells, 141–142
White oregano (*Origanum syriacum* L.), 71–72
Wild mint (*Mentha longifolia* (L.) L.), 129
Withania somnifera, 324–325, 372
Wolotisane, 487–488
World Health Organization (WHO), 185, 209, 237, 267, 439–440, 479
Wormwood (*Artemisia absinthium* L.), 128
Wound healing, 140, 142
activity, 146
pathophysiology of, 141–142
phases, 141*f*
in South Africa, 166–172
Aloe species, 166–167
Elephantorrhiza elephantine, 167–169
Erythrina lysistemon, 169–170
Galenia africana, 170
Melianthus comosus, 170–171
Plectranthus fruticosus, 171
Sutherlandia frutescens, 171–172
Wound infection, 142
Wound-associated bacteria
Aloe barberae Dyer, 146–147
Aloe excelsa Berger, 147–161
Aloe ferox Miller, 161
Elephantorrhiza elephantina, 162
Erythrina lysistemon Hutch, 162–163
Galenia africana L., 163
Grewia occidentalis L., 163–164
Melianthus comosus Vahl, 164
Plectranthus fruticosus L'Hér, 164–165
Polystichum pungens, 165
South African medicinal plant species with activity against, 146–166
Sutherlandia frutescens L., 165–166
Urtica urens L., 166

X

Xanthium strumarium L. (Asteraceae), 23
Xanthohumol, 66
Xanthomonas oryzae, 401–402
Xanthones, 402–403
Xanthorrhoeaceae, 161
Xylaria hypoxylon, 415–416
Xylaria sp., 405, 418, 421–422
Xylariaquinone, 418

Y

Yarrow (*Achillea millefolium* L.), 127
Yosemite Valley Grant Act, 529–530

Z

Zea mays L., 69
"Zimbabwe Aloe", 147–154
Zingiber officinale. See Ginger (*Zingiber officinale*)
Zingiberaceae, 244
Ziziphus, 244
Zygophyllum fabago L., 210–229

CPI Antony Rowe
Eastbourne, UK
April 05, 2022